Probability, Statistics, and Queueing Theory

With Computer Science Applications

Second Edition

This is a volume in
COMPUTER SCIENCE AND SCIENTIFIC COMPUTING

Werner Rheinboldt and Daniel Siewiorek, editors

Probability, Statistics, and Queueing Theory

With Computer Science Applications

Second Edition

Arnold O. Allen

Performance Technology Center
Hewlett-Packard
Roseville, California

ACADEMIC PRESS, INC.
Harcourt Brace Jovanovich, Publishers
Boston San Diego New York
London Sydney Tokyo Toronto

Copyright © 1990, 1977 by Academic Press, Inc.

MINITAB is a registered trademark of Minitab, Inc.

SAS and SAS/STAT are registered trademarks of SAS Institute, Inc.

Mathematica is a registered trademark of Wolfram Research

The poem by Samuel Hoffenstein on page 484 is from *Crown Treasury of Relevant Quotations,* **by Edward F. Murphy. Copyright © 1978 by Edward F. Murphy. Reprinted by permission of Crown Publishers Inc.**

ACADEMIC PRESS, INC.
1250 Sixth Avenue, San Diego, CA 92101

United Kingdom Edition published by
ACADEMIC PRESS LIMITED
24–28 Oval Road, London NW1 7DX

Library of Congress Cataloging-in-Publication Data

Allen, Arnold O.
    Probability, statistics, and queueing theory: with computer
science applications / Arnold O. Allen. — 2nd ed.
        p.   cm. — (Computer science and scientific computing)
    Includes bibliographical references and index.
    ISBN 0-12-051051-0 (acid-free paper)
    1. Probabilities   2. Queueing theory.   3. Mathematical statistics.
I. Title. II. Series.
QA273.P7955   1990
519.2—dc20
                                                                    90-732
                                                                    CIP

Printed in the United States of America

90   91   92   93   9 8 7 6 5 4 3 2 1

For my son, John,
and my colleagues
at the Hewlett-Packard
Performance Technology Center

Greek Alphabet

A α	alpha	N ν	nu
B β	beta	$\Xi\ \xi$	xi
$\Gamma\ \gamma$	gamma	O o	omicron
$\Delta\ \delta$	delta	$\Pi\ \pi$	pi
E ϵ	epsilon	P ρ	rho
Z ζ	zeta	$\Sigma\ \sigma$	sigma
H η	eta	T τ	tau
$\Theta\ \theta$	theta	$\Upsilon\ \upsilon$	upsilon
I ι	iota	$\Phi\ \phi$	phi
K κ	kappa	X χ	chi
$\Lambda\ \lambda$	lambda	$\Psi\ \psi$	psi
M μ	mu	$\Omega\ \omega$	omega

Table of Contents

Preface

Faith is belief without evidence in what is told by one who speaks without knowledge of things without parallel.

Ambrose Bierce

One must learn by doing the thing; for though you think you know it, you have no certainty until you try.

Sophocles

I am grateful to the many readers of the first edition of this book; the number of copies sold greatly exceeded my expectations. I am especially pleased by the number of readers who provided helpful comments, correction of typographical and other errors, and suggestions for the second edition.

The genesis of the first edition of this book is my experience in teaching the use of statistics and queueing theory for the design and analysis of computer communication systems at the Los Angeles IBM Information Systems Management Institute. After publication, the book was used for both technical and management courses in computer capacity planning at the Institute. Before attending the one-week technical course, students were asked to complete the IBM Independent Study Course *Capacity Planning: Basic Models* [1], which used my book as a textbook. Still later the book was used as one of the textbooks for the Self-Study Course *Introduction to Computer Modeling* [2]. The second edition evolved as a result of my experience in teaching courses at the IBM Information Systems Management Institute, the UCLA Extension Division, internal courses at the Hewlett-Packard company, as well as in writing the Independent Study Program [1] course for IBM and the Self-Study course [2] for Applied Computer Research.

The book is designed as a junior–senior level textbook on applied probability and statistics with computer science applications. While there are a number of examples of computer science applications, the book has been used successfully at a number of universities to teach probability and statistics classes with no emphasis on computer science. In addition, because of the prevalence of personal computers, most students of any discipline have no difficulty understanding examples with a computer science orientation. The book may also be used as a self-study book for the practicing computer science (data processing) professional. The assumed mathematical level of the reader who wants to read through all the proofs and do all the exercises is the traditional analytical geometry and calculus sequence. However, readers with only a college algebra background are able to follow much of the development and most of the examples; such readers should skip over most of the proofs.

I have attempted to state each theorem carefully so the reader will know when the theorem applies. I have omitted many of the proofs but have, in each such case, given a reference where the omitted proof can be found.[1] With a few exceptions I have provided the proof of a theorem only when the following conditions apply: (a) the proof is straightforward; (b) reading the proof will improve the reader's understanding; and (c) the proof is not long.

The emphasis in this book is on how the theorems and theory can be used to solve practical computer science problems. However, the book and a course based on the book should be useful for students who are not interested in computer science itself, but in using probability, statistics, and queueing theory to solve problems in other fields such as engineering, physics, operations research, and management science.

A great deal of computation is needed for many of the examples and exercises in this book because of the nature of the subject matter. The use of a computer is almost mandatory for the study of some of the queueing theory models. There are several queueing theory packages available for solving these models, such as the Best/1 series from BGS Systems, Inc., the MAP system from Performance Associates, and CMF MODEL from Boole & Babbage, but these packages are very expensive and may not be available to many readers. Another simpler (and an order of magnitude less expensive) queueing theory package is Myriad for the IBM PC or compatible from Pallas International of San Jose, California. To help readers

---

[1]Unless the proof is given as an exercise at the end of the chapter (with hints, of course).

who have no queueing theory package available I have included a number of APL programs in Appendix B as an aid in making queueing theory calculations. I wrote these programs in Kenneth Iverson's APL language because this language was the best available for such calculations when I started writing the second edition of this book. I have written the programs as directly as possible from the equations given in the text so that they can easily be converted to another language such as BASIC or PASCAL. Every APL program referred to in the text can be found in Appendix B. I received a copy of *Mathematica* in mid-1989 and wrote all new code after that time in *Mathematica*. For more traditional probability and statistical applications, a number of relatively low-cost packages are available for use on personal computers to ease the computational load for readers. We have demonstrated the use of some of these packages in this edition of the book and have some further comments about them below.

The excellent series of books by Donald Knuth [4–6] has influenced the writing of this book. I have adopted Knuth's technique of presenting complex procedures in an algorithmic way, that is, as a step by step process. His practice of rewarding the first finder of any error with $2 has also been adopted. I have followed his system of rating the exercises to encourage students to do at least the simpler ones. I believe the exercises are a valuable learning aid and have included more than twice as many in this edition as in the first edition. I believe Sophocles is right: you must do at least a few exercises to be sure that the material is understood.

Following Knuth, each exercise is given a rating number from 00 to 40. The rating numbers can be interpreted as follows: 00—a very easy problem that can be answered at a glance if the text has been read and understood; 10—a simple exercise, which can be done in a minute or so; 20—an exercise of moderate difficulty requiring 18 to 20 minutes of work to complete; 30—a problem of some difficulty requiring two or more hours of work; 40—a lengthy, difficult problem suitable for a term project. (All entries with ratings higher than 30 are "virtual.")

We precede the rating number by HM for "higher mathematics" if the problem is of some mathematical sophistication requiring an understanding of calculus, such as the evaluation of proper or improper integrals or summing an infinite series. The prefix C is used if the problem requires extensive computation, that would be laborious without computer aid such as a statistical package on a personal computer. T is used to indicate an exercise whose solution is basically tedious, even though the result may be important or exciting; that is, the required procedure is too complex to program for computer solution without more frustration than carrying it out manually with a pocket calculator.

The reader is assumed to have a basic knowledge of computer hardware and software. Computer illiteracy is now very rare. I recently was pleased to learn that a woman, who appeared to be in her eighties, kept the records of her women's club on her personal computer. She used an advanced data base management system to do it. She told me she bought an IBM PC when they first became available and had recently upgraded to a more powerful machine.

Statistical Computer Systems We Use in This Book

There are a number of valuable statistical computer systems available for assistance in making statistical calculations and for displaying data in various formats. These systems are especially useful for performing exploratory data analysis. I chose three of them to use in this book, not because everyone agrees they are the best, but because they were available to me and probably are available to most readers. My comments on the three systems are entirely subjective; your feelings about the systems may be different. My comments also apply only to the versions of the systems available at the time the book was written. Robin Raskin [9] tested the available statistical software for personal computers for *PC Magazine* in March 1989. Two of the three I chose were reviewed.

MINITAB [8], [10], [12]

The reference manual [8] for MINITAB is very readable and provides a good description of the MINITAB features. The book by Ryan, Joiner, and Ryan [10] was used by many as a reference manual before Release 7 and is still a valuable resource. MINITAB seems to be the statistical system most in use by textbook writers (at least according to the MINITAB advertisements). I use MINITAB for several examples because MINITAB *is* widely available, easy to learn, and easy to use. Instructors may ask their students to obtain the Student Edition described by Schaefer and Anderson [12]. There are some statistical procedures, such as the Kolmogorov–Smirnov goodness-of-fit test,[2] that are not directly available in MINITAB. However, by using macros, you can extend the capabilities of MINITAB to include this test as well as most others you may read about. Macros for the Kolmogorov–Smirnov test applied to the normal, Poisson, and continuous

---

[2]SAS/STAT doesn't provide a general procedure for this test either. Strangely enough, the newly announced Hewlett-Packard calculator, the HP-21S Stat/Math Calculator, does.

uniform distributions have been written by Joseph B. Van Matre of the University of Alabama in Birmingham. It is listed and described in the *Minitab Users' Group Newsletter* (MUG), Number 10 of September 1989. All macros published in MUG Newsletters are maintained on a macro library diskette, which is available, free, to members of the Minitab Users' Group. All the MINITAB examples in this book were run using MINITAB Release 7.2.

In Raskin [9] MINITAB was one of the two Editor's Choices in the Basic category. The Editor's Choice Citation for MINITAB is:

> We chose MINITAB Statistical Software as Editor's Choice because it makes some wise choices in terms of what to include. MINITAB offers consistency and simplicity, but its excellent command language and macro facilities make it possible to push the envelope. Because it runs on so many different machines it's an excellent choice for the "work wherever I can find a CPU" student.

Release 6 was the version of MINITAB reviewed by Raskin. Version 7.2 has some very nice additional features including high resolution graphics.

The EXPLORE Programs of Doane [3]

Doane's software is designed primarily for instruction. EXPLORE has a main menu that can take you to a help menu, a menu to choose one of the 24 EXPLORE programs, or to a file edit Menu. The file system makes it easy to enter and modify data. For a given statistical procedure, such as simple linear regression, it yields more information automatically than many expensive statistical systems and in a more pleasing format. It is indeed unfortunate that EXPLORE doesn't have more statistical routines. Professor Doane has assured me that new editions of his book with extensions to EXPLORE are in preparation. The EXPLORE examples in this book were done using the routines in the second edition. Since EXPLORE does not have much capability in calculating statistical distributions the Hewlett-Packard HP-21S provides a useful supplement to this package.

SAS/STAT for IBM PCs and Compatibles [11]

The SAS/STAT package is very powerful—it is true industrial strength. Unfortunately, it also is rather user-unfriendly and difficult to learn.[3] The

---

[3]It ran rather slowly on the 8 MHz IBM PC AT that I had in early 1989, too, and is noticeably slow on the 33 MHz IBM PC compatible with an Intel 80386 microprocessor that I had when I finished the book.

tutorial that is part of Release 6.03 is very good and the manuals have improved dramatically over Release 6.0. One can learn how to use the system for at least simple applications by going through the tutorial. In addition, the SAS Institute has published an introductory textbook by Schlotzhauer and Littell [13], which explains how to use SAS Procedures such as UNIVARIATE, MEANS, CHART, and TTEST to perform elementary statistical processes.[4] The Institute has also published a master index to the SAS documentation [7], which has gone a long way to alleviate the frustration I had with SAS/STAT 6.0—I never could guess which manual to consult. SAS/STAT, which, with Release 6.03, is identical to the mainframe version, contains a number of powerful statistical routines, each of which has several sophisticated options. It also offers flexible output and has a built-in programming language that makes it possible to construct any statistical procedure you can imagine.

SAS/STAT was reviewed by Frederick Barber in Raskin [9] as one of the advanced statistical systems available for the IBM PC. It was not the Editor's Choice although, Barber said, in part

> SAS computes a very wide range of descriptive and comparative statistics and performs ANOVA, MANOVA, factor and cluster analysis, plus least squares, GLM, and nonlinear regression, as well as many other procedures. An exhaustive set of options allows the user to customize the output and analyze statistical patterns in great depth. For the statistician who needs great depth and a wide range of statistical computing power, SAS is hard to beat.

SAS/STAT is more powerful than MINITAB. The mainframe version of SAS has been popular for years with performance analysts who work on large mainframe computers. If you are doing statistics on a daily basis, you may want to learn how to use SAS. If you do, be prepared for a steep learning curve. As this book goes to press (June 1990) the SAS Institute has announced version 6.04 of SAS/STAT for IBM PCs as well as a new mainframe version. The SAS Institute claims the new version is easier to use. I have not had an opportunity to try SAS/STAT 6.04.

All of my comments about statistical systems for the personal computer reflect my view of the situation in early 1990.

---

[4]Unfortunately, it fails to tell you what commands you need to give to make the procedures work. This is the truly difficult part about using SAS.

Mathematica, a System for Doing Mathematics by Computer [14]

Mathematica [14] is, strictly speaking, not a statistical system but rather a system for doing mathematics on the computer. It can be used as a super-calculator that is able to find the value of most well-known mathematical functions, real or complex. It also performs symbolic calculations such as finding integrals, derivatives, or infinite series. *Mathematica* makes it very easy to plot graphs of mathematical functions in two or three dimensions. It functions as a programming language and allows you to define new math-ematical functions in terms of those provided by *Mathematica* or those you have already defined. *Mathematica* became available to me in late July 1989 when the book was almost completed. It is a remarkable system and had a big influence on how I finished the book. I wrote no more APL code but wrote the *Mathematica* packages that appear in Appendix D.

Cited References

[1] Arnold O. Allen, *Capacity Planning: Basic Models*, Independent Study Program Course I0100, IBM, 1980.

[2] Arnold O. Allen, *Introduction to Computer Modeling*, Applied Computer Research, Phoenix, AZ, 1986.

[3] David P. Doane, *Exploring Statistics with the IBM PC, Second Edition*, Addison-Wesley, Reading, MA, 1987. Includes a diskette with EXPLORE statistical programs and a data diskette.

[4] Donald E. Knuth, *The Art of Computer Programming, Vol. 1, Fundamental Algorithms*, 2nd ed., Addison-Wesley, Reading, MA, 1973.

[5] Donald E. Knuth, *The Art of Computer Programming, Vol. 2, Seminumerical Algorithms*, 2nd ed., Addison-Wesley, Reading, MA, 1981.

[6] Donald E. Knuth, *The Art of Computer Programming, Vol. 3, Sorting and Searching*, Addison-Wesley, Reading, MA, 1973.

[7] *Master Index to SAS System Documentation for Personal Computers*, SAS Institute, Cary, NC, 1987.

[8] *MINITAB Reference Manual, Release 7*, Minitab, Inc., State College, PA, 1989.

[9] Robin Raskin, Statistical software for the PC: testing for significance, *PC Magazine*, **8(5)**, March 14, 1989.

[10] Barbara F. Ryan, Brian L. Joiner, and Thomas A. Ryan, Jr., *Minitab Handbook, Second Edition, Revised Printing*, Duxbury Press, Boston, 1985.

[11] *SAS Language Guide for Personal Computers, Release 6.03 Edition*, SAS Institute, Cary, NC, 1988.

[12] Robert L. Schaefer and Richard B. Anderson, *The Student Edition of MINITAB*, Addison-Wesley, Benjamin/Cummings, 1989.

[13] Sandra D. Schlotzhauer and Ramon C. Littell, *SAS System for Elementary Statistical Analysis*, SAS Institute, Cary, NC, 1987.

[14] Stephen Wolfram, *Mathematica: A System for Doing Mathematics by Computer*, Addison-Wesley, Redwood City, CA, 1988.

Acknowledgments

I want to thank Minitab, Inc. for providing me with a copy of MINITAB Release 7.2 for my personal computer to use for writing the book. Bob Graf of Minitab helped me make the best use of the system. Minitab's address is

Minitab, Inc.
3081 Enterprise Drive
State College, PA 16801
Telephone: (814) 238-3280
Telex: 881612

Thanks are due, too, to Wolfram Research, Inc. for providing me with the 386 version of *Mathematica*. Kevin McIsaac and others from Wolfram Research provided technical assistance in using *Mathematica*. The address of Wolfram Research is

Wolfram Research, Inc.
P. O. Box 6059
Champaign, Illinois 61821
Telephone: (217) 398-0700

I want to thank Pallas International for providing a copy of *Myriad*. The address of Pallas International is

Pallas International Corporation
1763 Valhalla Court
San Jose, CA 95132

I am grateful to The Institute of Statistical Mathematics in Tokyo for providing me with the Computer Science Monograph *Numerical Tables of The Queueing Systems I: $E_k/E_2/s$* by Hirotaka Sakasegawa.

I want to thank the Mathematica Association of America for permission to use the five mathematical clerihews by Karl David from Mathematics Magazine, April 1990.

The production staff of Academic Press, Boston, were supportive and helpful in the production of the book. Alice Peters, my publisher, encouraged me to write the book using LaTeX and made it possible to convert text that had already been written into LaTeX. She helped me in other ways, including assigning my production editor, Elizabeth Tustian, almost a year before the book was scheduled for production. Ms. Tustian helped me format the book. Carolyn Artin, the managing editor, also edited several chapters.

Elaine Barth typed several chapters in her usual professional manner. Beth Howard converted several of these chapters into LaTeX in an outstanding manner. I am grateful to her husband, Alan Howard, for suggesting that she learn LaTeX and do the typing for me. Thanks are due to James F. Dwinell, III, President of the Cambridge Trust Company of Cambridge, Massachusetts, for permission to use their advertisement on the title page for Part Two.

Several members of the Hewlett-Packard Performance Technology Center staff made contributions. Dan Sternadel recommended a more uniform method of setting up examples and did some proof reading. Tim Twietmeyer proofed several chapters, too. Gary Hynes made several excellent recommendations and discovered the solution to some key LaTeX programming problems. He also provided invaluable help with the *Mathematica* programs. David Ching, now with Borland International, helped me keep the software on my personal computer working properly through several changes of hardware and software. Jim Morris was my graphics consultant. Sharon Riddle helped me with some of the figures. Diane Karr, the librarian at the Hewlett-Packard Roseville Site, hunted down a number of references I needed. Wim Van Oudheusden of Hewlett-Packard, The Netherlands, suggested several quotations.

I am indebted to Ben Lutek for permission to use a number of his outstanding limericks.

Professor Darel J. Daley of the Australian National University provided advice on the queueing theory chapters. Professor Michael A. Stephens of Simon Fraser University advised me on goodness-of-fit tests. Shelly Weinberg of IBM assisted me with my APL programming. Professor Kishor Trivedi provided the solution to Example 4.4.3. Dr. Hanspeter Bieri of Universität Bern made helpful suggestions and obtained a key reference for me. Professor Richard E. Trueman of California State University, Northridge, suggested several exercises and provided me with his book *Quantitative Methods for Decision Making in Business* as well as the solutions manual.

Albert DiCanzio of ADASI Systems gave me a copy of *The Life and Works of A. K. Erlang* by Brockmeyer, Halstrøm, and Jensen when I lost mine.

I am most grateful for the invaluable services of Russell Ham. He read every line of every version of every chapter of the book. He found many errors of omission as well as commission. He offered valuable advice on every aspect of the book including typesetting, grammar, spelling, word choice, references, and quotations. I accepted most of his suggestions. His work made a difference!

Probability, Statistics, and Queueing Theory

With Computer Science Applications

Second Edition

Chapter 1

Introduction

"Is there any point to which you would wish to draw my attention?"
"To the curious incident of the dog in the night-time."
"The dog did nothing in the night-time."
"That was the curious incident," remarked Sherlock Holmes.

Come Watson, come! The game is afoot.
Sherlock Holmes

This chapter is a preview of the book. As the title of the book suggests, it is concerned with the application of probability, statistics, and queueing theory to computer science problems. The first edition was written primarily for the computer science (data processing) specialist or for one preparing for a career in this field. It was widely read by this audience but was also used for courses in applied probability as well as for introductory mathematical statistics courses. It was used, too, for queueing theory courses for operations research students. With the advent of the personal computer, the audience broadened to include many personal computer users with an interest in applied probability or statistics. This edition has the same emphasis as the first edition but makes more extensive use of available personal computer software such as MINITAB, SAS/STAT, EXPLORE, APL, and *Mathematica*. We have tried to make the book practical, interesting, and theoretically sound.

The book, like Julius Caesar's Gaul, is divided into three parts: Probability, Queueing Theory, and Statistical Inference.

There are three chapters in Part One. In Chapter 2 we discuss basic probability theory. Probability theory is important in computer science because most areas of computer science are concerned more with probabilistic rather than deterministic phenomena. The time it takes to write and check

1

out a computer program (especially those I write), the time it takes to run a program on any computer but the simplest personal computer, and the time it takes to retrieve information from a storage device are all examples of probabilistic or random variables. By this we mean that we cannot predict in advance exactly what these values will be. Such variables are called *random variables*. However, using basic probability theory, we can make probability estimates (that is, estimate the fraction of the time) that the values of a random variable will fall into certain ranges, exceed certain limits, etc. Thus, we may compute the 90th percentile value of response time, which is the value that is exceeded only one tenth of the time. In Chapter 2 we also discuss parameters of random variables, such as the mean or average value and the standard deviation. The standard deviation provides a measure of the spread of the values of the random variable about the mean. In the final part of Chapter 2 we discuss some powerful probability tools, including conditional expectation, the Law of Total Probability, transform methods, and inequalities. Transform methods are important for studying random variables. The transforms we define and illustrate include the moment generating function, the z-transform, and the Laplace–Stieltjes transform.

In Chapter 3 we study the probability distributions most commonly used in applied probability, particularly for computer science applications. We give examples of the use of all of the random variables except those used primarily in statistical inference, the subject of Part Three. A summary of the properties of the random variables studied in Chapter 3 is given in Tables 1 and 2 of Appendix A. In the last section of the chapter we provide further examples of the use of the transform methods that were introduced in Chapter 2.

In Chapter 4 the important concept of a stochastic process is defined, discussed, and illustrated with a number of examples. This chapter was written primarily as a support chapter for Part Two, Queueing Theory. We examine the Poisson process and the birth-and-death process because they are extremely important for queueing theory. We finish the chapter with a discussion of Markov processes and chains—subjects that are important not only for queueing theory but for much of computer science and operations research.

Part Two of this book is the subject area that is most likely to be unfamiliar to the reader. I didn't know queueing theory existed until I was tapped to teach it at the IBM System Science Institute. Queueing theory is a very useful branch of applied probability. However, some expressions, symbols, and words are used differently in queueing theory than they are in other areas of probability and statistics.

Figure 1.1. Elements of a queueing system.

Figure 1.1 shows the elements of a simple queueing system. There is a *customer* population where a customer may be an inquiry to be processed by an interactive computer system, a job to be processed by a batch computer system, a message or a packet to be transmitted over a communication link, a request for service by an input/output (*I/O*) channel, etc. Customers arrive in accordance with an *arrival process* of some type (a *Poisson arrival process* is one of the most common). Customers are provided service by a service facility, that has one or more *servers*, each capable of providing service to a customer. Thus, a server could be a program that processes an inquiry, a batch computer system, a communication link, an *I/O* channel, a central processing unit (CPU), etc. If all the servers in the service facility are busy when a customer arrives at the queueing system, that customer must queue for service. That is, the customer must join a queue (waiting line) until a server is available. In Chapter 5 we study the standard (one might say *canonical*) queueing systems and see how they can be applied to the study of computer systems. We have gathered most of the queueing theory formulas from Chapters 5 and 6 in Appendix C. You will find this appendix to be a useful reference section after you have mastered the material in the two queueing theory chapters. The APL programs for solving most of the models in Appendix C are displayed in Appendix B.

In Chapter 6, we discuss more sophisticated queueing theory models that have been developed to study computer and computer communication systems. A number of examples of how the models can be used are presented, too. Some *Mathematica* programs are displayed in Appendix D. Most of them are for the queueing network models of Chapter 6.

The subject matter of Part Three, statistical inference, is rather standard statistical fare but we have attempted to give it a computer science orientation. We also demonstrate how MINITAB, SAS, and EXPLORE can be used to remove much of the labor. Statistical inference could perhaps be defined as "the science of drawing conclusions about a population on the basis of a random sample from that population." For example, we may want to estimate the mean arrival rate of inquiries to an interactive inquiry system. We may want to deduce what type of arrival process is involved, as well. We can approach these tasks on the basis of a sample of the arrival times of inquiries during n randomly selected time periods. The first task is one of estimation. We want to estimate the mean arrival rate on the basis of the observed arrival rate during n time intervals. This is the subject of Chapter 7. In Chapter 7 we learn not only how to make estimates of parameters but also how to make probability judgments concerning the accuracy of these estimates. In Chapter 7 we also study exploratory data analysis and some of the tools that are used to study benchmarks.

Chapter 8 is about hypotheses testing. One of the important topics of this chapter is goodness-of-fit tests. We might want to test the hypothesis that the arrival pattern is a Poisson arrival pattern because Poisson arrival patterns have desirable mathematical properties. We discuss and illustrate the chi-square and Kolmogorov–Smirnov tests because they are popular and widely used. We will also discuss a class of EDF statistics called *quadratic statistics* because experts on goodness-of-fit tests tell us tests based on these statistics are more powerful than chi-square and Kolmogorov–Smirnov tests. In Chapter 8 we also study a number of classical statistical tests concerning means and variances. These tests can be used to study new paradigms or methodologies, such as those of software engineering or software design, to determine whether they are effective. We also provide an introduction to Analysis of Variance (ANOVA). Chapter 8 is a chapter in which the statistical functions of MINITAB, SAS, and EXPLORE are particularly useful.

Chapter 9 on regression is new to this edition of the book. Regression has many applications to computer science. It also provides many opportunities for making egregious errors. We attempt to show you how to make good use of regression without making errors.

This completes the summary of the book. We hope you will find the study of this book entertaining as well as educational. We have avoided being too solemn. We have chosen whimsical names for mythical companies and people in our examples. We made the examples as practical as possible within the constraints of a reasonably short description. We welcome your comments, suggestions, and observations. My address is: Dr. Arnold O. Allen, Hewlett-Packard, 8050 Foothills Blvd., Roseville, California 95678.

Part One:

Probability[1]

There once was a breathy baboon
Who always breathed down a bassoon,
  For he said, "It appears
  That in billions of years
I shall certainly hit on a tune."

Sir Arthur Eddington

---

[1] Figure provided by Mike Kury.

Preface to Part One: Probability

Probability is a mathematical discipline with aims akin to those, for example, of geometry or analytical mechanics. In each field we must carefully distinguish three aspects of the theory: (a) the formal logical content, (b) the intuitive background, (c) the applications. The character, and the charm, of the whole structure cannot be appreciated without considering all three aspects in their proper relation.

William Feller

The above quote is from William Feller's classic book.[2] Many mathematicians feel that Feller's is the finest mathematics book ever written; this author agrees. However, the revised printing of the third edition of this wonderful book was published in 1970 and there have been many advances in applied probability since then. Nevertheless, Feller's book is enjoyable and enlightening to peruse. It has also had a profound effect upon the attitude toward applied probability in the fields of mathematics, the physical sciences, and engineering. As Feller says in the preface to the third edition:

> When this book was first conceived (more than 25 years ago) few mathematicians outside the Soviet Union recognized probability as a legitimate branch of mathematics. Applications were limited in scope, and the treatment of individual problems often led to incredible complications. Under these circumstances the book could not be written for an existing audience, or to satisfy conscious needs. The hope was rather to attract attention to

---

[2]Reprinted by permission of the publisher from *An Introduction to Probability Theory and Its Applications*, Vol. I, 3rd ed., revised printing, John Wiley, New York, 1968 by William Feller.

7

little-known aspects of probability, to forge links between various parts, to develop unified methods, and to point to potential applications. Because of a growing interest in probability, the book found unexpectedly many users outside mathematical disciplines. Its widespread use was understandable as long as its point of view was new and its material was not otherwise available. But the popularity seems to persist even now, when the contents of most chapters are available in specialized works streamlined for particular needs. For this reason the character of the book remains unchanged in the new edition. I hope that it will continue to serve a variety of needs and, in particular, that it will continue to find readers who read it merely for enjoyment and enlightenment.

In Part One of this book we set up the concepts in probability and stochastic processes that we will need for the rest of the book. In Chapter 2 we consider the basics of probability and random variables and in Chapter 3 we consider a number of important probability distributions for applications. We also investigate the important concepts of inequalities, the Central Limit Theorem, and the application of transform techniques. In Chapter 4 we take up the important study of stochastic processes, which is very important for Part Two.

Chapter 2

Probability and Random Variables

Science is founded on uncertainty.
Lewis Thomas

Probability is the very guide of life.
Cicero

2.0   Introduction

One of the most noticeable aspects of many computer science related phenomena is the lack of certainty. When a job is submitted to a batch-oriented
computer system, the exact time the job will be completed is uncertain. The
number of jobs that will be submitted tomorrow is probably not known,
either. Similarly, the exact response time for an interactive inquiry system
cannot be predicted. If the terminals attached to a communication line
are polled until one is found that is ready to transmit, the required number of polls is not known in advance. Even the time it takes to retrieve
a record from a disk storage device cannot be predicted exactly. Each of
these phenomena has an underlying probabilistic mechanism. In order to
work constructively with such observed, uncertain processes, we need to put
them into a mathematical framework. That is the purpose of this chapter.

Experience enables you to recognize a mistake when you make it again.
Franklin P. Jones

9

2.1   Sample Spaces and Events

To apply probability theory to the process under study, we view it as a *random experiment*, that is, as an experiment whose outcome is not known in advance but for which the set of all possible individual outcomes is known. For example, if the 10 workstations on a communication line are polled in a specified order until either (a) all are polled or (b) one is found with a message ready for transmission, then the number of polls taken describes the outcome of the polling experiment and can only be an integer between 1 and 10. The *sample space* of a random experiment is the set of all possible simple outcomes of the experiment. These individual outcomes are also called *sample points* or *elementary events*. A sample space is a set and thus is defined by specifying what objects are in it. One way to do this, if the set is small, is to list them all, such as $\Omega = \{1,2,3\}$. When the set is large or infinite its elements are often specified by writing $\Omega = \{x : P(x)\}$, where $P(x)$ is a condition that x must satisfy to be an element of Ω. Thus, $\Omega = \{x : P(x)\}$ means, "Ω is the set of all x such that $P(x)$ is true." The set of all nonnegative integers could be specified by writing $\{n : n$ is an integer and $n \geq 0\}$. Some examples of sample spaces follow.

Example 2.1.1 If the random experiment consists of tossing a die, then $\Omega = \{1,2,3,4,5,6\}$ where the sample point n indicates that the die came to rest with n spots showing on the uppermost side. □

Example 2.1.2 If the random experiment consists of tossing two fair dice,[1] then one possible sample space $\Omega = \{(1,1),(1,2),\cdots,(6,6)\}$, where the outcome (i,j) means that the first die showed i spots uppermost and the second showed j. □

Example 2.1.3 If the random experiment consists of polling the terminals on a communication line in sequence until either (a) one of the seven terminals on the line is found to be ready to transmit or (b) all the terminals have been polled, the sample space could be represented by $\Omega = \{1,2,3,4,5,6,7,8\}$, where an 8 signifies that no terminal had a message ready, while an integer n between 1 and 7 means that the nth terminal polled was the first in sequence found in the ready state. □

Example 2.1.4 If the random experiment consists of tossing a fair coin again and again until the first head appears, the sample space can be represented by $\Omega = \{$ H, TH, TTH, TTTH, ...,$\}$, where the first sample point

---

[1]By a fair coin or a fair die we mean, of course, one for which each outcome is equally likely (whatever that means).

corresponds to a head on the first toss, the second sample point to a head on the second toss, etc. □

Example 2.1.5 The random experiment consists of measuring the elapsed time from the instant the last character of an inquiry is typed on an interactive terminal until the last character of the response from the computer system has been received and displayed at the terminal. This time is often called the "response time," although there are other useful definitions of response time. If it takes a minimum of one second for an inquiry to be transmitted to the central computer system, processed, a reply prepared, and the reply returned and displayed at the terminal, then $\Omega = \{$real $t : t \geq 1\}$. □

Thus, sample spaces can be *finite*, as in Examples 2.1.1–2.1.3, or *infinite*, as in Examples 2.1.4 and 2.1.5. Sample spaces are also classified as *discrete* if the number of sample points is finite or countably infinite (can be put into one-to-one correspondence with positive integers). The sample space of Example 2.1.4 is countably infinite since each sample point can be associated uniquely with the positive integer giving the number of tosses represented by the sample point. For example, the sample point TTTH represents four tosses. A sample space is *continuous* if its sample points consist of all the numbers on some finite or infinite interval of the real line. Thus, the sample space of Example 2.1.5 is continuous.

For discussing subsets of the real line, we use the notation (a, b) for the open interval $\{x : a < x < b\}$; $[a, b]$ for the closed interval $\{x : a \leq x \leq b\}$; $(a, b]$ for the half-open interval $\{x : a < x \leq b\}$; and $[a, b)$ for the half-open interval $\{x : a \leq x < b\}$, where all intervals are subsets of the real line. Note that a round bracket means the corresponding end point is *not* included and a square bracket means it is included.

An *event* is a subset of a sample space satisfying certain axioms (Axiom Set 2.2.1 described in Section 2.2). An event A is said to *occur* if the random experiment is performed and the observed outcome is in A.

Example 2.1.6 In Example 2.1.1, if $A = \{2, 3, 5\}$, then A is the event of rolling a prime number while the event $B = \{1, 3, 5\}$ is the event of rolling an odd number. □

Example 2.1.7 In Example 2.1.2, if $A = \{(1, 6), (2, 5), (3, 4), (4, 3), (5, 2), (6, 1)\}$, then A is the event of rolling a seven. The event $B = \{(5, 6), (6, 5)\}$ corresponds to rolling an 11. □

Example 2.1.8 In Example 2.1.3, if $A = \{1, 2, 3, 4, 5\}$, then A is the event of requiring five polls or less, while $B = \{6, 7, 8\}$ is the event of requiring more than five polls. □

Example 2.1.9 In Example 2.1.4, $A = \{\text{TTH, TTTH}\}$ is the event that three or four tosses are required; $B = \{\text{H, TH, TTH}\}$ is the event that not more than three tosses are needed. □

Example 2.1.10 In Example 2.1.5, $A = \{t : 20 \leq t \leq 30\}$ is the event that the response time is between 20 and 30 seconds. □

Since a sample space Ω is a set and an event A is a subset of Ω, we can form new events by using the usual operations of set theory. For some of these operations a somewhat different terminology is used in probability theory than in set theory—a terminology more indicative of the intuitive meaning of the operations in probability theory. Some of the set operations and corresponding probability statements are shown in Table 2.1.1. We will use probability statements and set theory statements interchangeably in this book.

Table 2.1.1. Set Operations and Probability Statements

Set Operation	Probability Statement
$A \cup B$	At least one of A or B occurs
$A \cap B$	Both A and B occur
$\bar{A}$	A does not occur
$\emptyset$	The impossible event
$A \cap B = \emptyset$	A and B are mutually exclusive
$A \cap \bar{B}$	A occurs and B does not occur
$A \subset B$	If A occurs so does B

We indicate that the outcome ω is a sample point of event A by writing $\omega \in A$. We write $A = \emptyset$ to indicate that the event A contains no sample points. Here $\emptyset$ is the empty set, called the *impossible event* in probability theory. The impossible event, $\emptyset$, is considered to be an event just as Ω itself is. The reader should note that $\emptyset$ is not the Greek letter phi but rather a Danish letter pronounced "ugh," the sound one makes upon receiving an unexpected blow to the solar plexus. It has been rumored that the prevalence of $\emptyset$ in Danish words has been a leading cause of emphysema in Denmark. Professor Richard Arens of UCLA has recommended a new symbol for the empty set constructed by adding a second slash mark to $\emptyset$ and pronounced "uh uh," of course.[2]

---

[2]I do not know how to construct this symbol with LaTeX. The typesetter of the first edition of this book knew how to construct it.

To every event A there corresponds the event $\bar{A}$, called the *complement of A*, consisting of all sample points of Ω that are not in A. Thus, $\bar{A}$ is defined by the condition "*A* does not occur." As particular cases of the complement, $\bar{\Omega} = \emptyset$ and $\bar{\emptyset} = \Omega$. The concept of complement is illustrated by the *Venn diagram* of Figure 2.1.1. In each Venn diagram that follows, the large rectangle will represent the sample space Ω, and simple geometric figures will be used to represent other events. A point thus represents an elementary event or outcome and the inside of a figure represents a collection of them (an event).

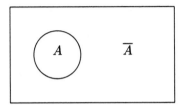

Figure 2.1.1. A and $\bar{A}$.

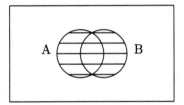

Figure 2.1.2. The event $A \cup B$.

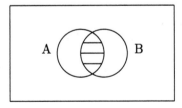

Figure 2.1.3. The event $A \cap B$.

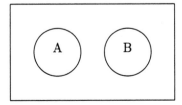

Figure 2.1.4. Mutually exclusive events.

With each two events A and B are associated two new events, that correspond to the intuitive ideas "either A or B occurs" and "both A and B occur." The first of these events, $A \cup B$ (read: "A or B") is the ordinary set union consisting of all sample points that are either in A or in B or (possibly) in both A and B. The second event $A \cap B$ (read: "A and B") is the ordinary set intersection, that is, all sample points which belong both to A and to B. If A and B have no sample points in common, that is, if $A \cap B = \emptyset$, we say that A and B are *mutually exclusive events*. Clearly, if A and B are mutually exclusive, then the occurrence of one of them precludes the occurrence of the other. These concepts are illustrated in the Venn diagrams of Figures 2.1.1–2.1.4. In Figure 2.1.2, $A \cup B$ is represented by the shaded area. $A \cap B$ is shaded in Figure 2.1.3. The events A and B of Figure 2.1.4 are mutually exclusive.

The concepts of union and intersection can be extended in a similar way to any finite collection of events such as $A \cup B \cup C$ or $A \cap B \cap C \cap D$. For a countable collection of events $A_1, A_2, A_3, \ldots$, the union $\bigcup_{n=1}^{\infty} A_n$ of the events is defined to be the event consisting of all sample points that belong to at least one of the sets $A_n, n = 1, 2, \ldots$; the intersection $\bigcap_{n=1}^{\infty} A_n$ of the events is the event consisting of all sample points that belong to each of the events $A_n, n = 1, 2, \ldots$.

If every sample point of event A is also a sample point of event B, so that A is a subset of B, we write $A \subset B$ and say that "if event A occurs, so does event B." In this case, $B - A$ is defined to be the set of all sample points in B that are not in A. Thus, $\bar{A} = \Omega - A$ for every event A.

Example 2.1.11 Consider the sample space of Example 2.1.3. Let A be the event that at least five polls are required and B the event that not more than four polls are required ($A = \{5, 6, 7, 8\}, B = \{1, 2, 3, 4\}$). Then $A \cup B = \Omega$ and $A \cap B = \emptyset$, so A and B are mutually exclusive. They are also complements ($\bar{A} = B$ and $\bar{B} = A$), although mutually exclusive events are not necessarily complementary. $\square$

> *The heights by great men reached and kept*
> *Were not attained by sudden flight,*
> *But they, while their companions slept,*
> *Were toiling upward in the night.*

                                           Henry Wadsworth Longfellow

2.2 Probability Measures

1st Die	Sum of Spots					
6	7	8	9	10	11	12
5	6	7	8	9	10	11
4	5	6	7	8	9	10
3	4	5	6	7	8	9
2	3	4	5	6	7	8
1	2	3	4	5	6	7
	1	2	3	4	5	6
	2nd Die					

Figure 2.2.1. Two-dice experiment.

In the early or classical days of probability, there was much concern with games of chance. Early workers in the field, such as Cardano and Pascal, were occupied with questions about the likelihood of winning in various games and in how to divide the purse if the game were discontinued before completion (called the "division problem" or the "problem of points"). See Ore [17] for a discussion of Pascal's role in the invention of probability theory and Ore [16] for a description of Cardano's work. Ore's book also contains a translation of Cardano's book *The Book of Games of Chance*. See Snell [23, pages 2–6] for a further discussion of the history of probability theory. The sample spaces for gambling problems were constructed in such a way that each elementary event or outcome was equally likely. In Example 2.1.2, if the two dice are perfectly formed, each of the 36 elementary outcomes is equally likely to occur on any given trial of the experiment, so a probability of 1/36 is assigned to each sample point. Thus, for finite sample spaces with n equiprobable sample points, each event A was assigned the *probability* $P[A] = n_A/n$, where n_A is the number of sample points in A.

Example 2.2.1 Consider the two-dice experiment of Example 2.1.2. We can construct the table of Figure 2.2.1 to help in calculating probabilities. Thus, if A is the event of rolling 11, we can see from Figure 2.2.1 that A consists of the sample points (5, 6) and (6, 5), so $P[A] = n_A/36 = 2/36 = 1/18$. Likewise, if B is the event of rolling 7 or 11, $P[B] = n_B/36 = (6+2)/36 = 2/9$. Other probabilities for this experiment can be calculated in a similar manner. □

The classical definition of probability worked well for the kind of problem for which it was designed. However, the classical theory would not suffice to assign probabilities to the events of Example 2.1.3 because the elementary events are not equiprobable. Likewise, it would not help in Examples 2.1.4 or 2.1.5, because these sample spaces are infinite. The classical definition has been generalized into a set of axioms that every probability measure should satisfy in assigning probabilities to events. Some additional conditions, however, must be imposed on the collection of events of a sample space before we can assign probabilities to them.

The family $\mathcal{F}$ of events of a sample space Ω is assumed to satisfy the following axioms (and thus form a σ-algebra):

Axiom Set 2.2.1 (*Axioms of a σ-Algebra*)[3]

A1 $\emptyset$ and Ω are elements of $\mathcal{F}$.

A2 If $A \in \mathcal{F}$, then $\overline{A} \in \mathcal{F}$.

A3 If $A_1, A_2, A_3, \ldots$ are elements of $\mathcal{F}$, so is $\bigcup_{n=1}^{\infty} A_n$.

It can be shown that these axioms also imply that, if each of the events $A_1, A_2, \ldots$ belongs to $\mathcal{F}$, then $\bigcap_{n=1}^{\infty} A_n$ is an element of $\mathcal{F}$, and similarly for finite intersections (see Exercise 7). Also, $A_1, \cup \cdots \cup A_n$ is in $\mathcal{F}$ if each A_i is. Likewise, if A, B are in $\mathcal{F}$, then $B - A = B \cap \overline{A}$ and thus is in $\mathcal{F}$.

A probability measure $P[\cdot]$, regarded as a function on the family $\mathcal{F}$ of events of a sample space Ω, is assumed to satisfy the following axioms:

Axiom Set 2.2.2 (*Axioms of a Probability Measure*)

P1 $0 \leq P[A]$ for every event A.

P2 $P[\Omega] = 1$.

P3 $P[A \cup B] = P[A] + P[B]$ if the events A and B are mutually exclusive.

P4 If the events $A_1, A_2, A_3, \ldots$ are mutually exclusive (that is, $A_i \cap A_j = \emptyset$ if $i \neq j$), then

$$P\left[\bigcup_{n=1}^{\infty} A_n\right] = \sum_{n=1}^{\infty} P[A_n].$$

It is immediate from **P3** by mathematical induction that for any finite collection $A_1, A_2, \ldots, A_n$ of mutually exclusive events

$$P[A_1 \cup A_2 \cup \cdots \cup A_n] = P[A_1] + \cdots + P[A_n]. \quad \square$$

---

[3]Here we use the symbol $\in$ in the usual set theoretic sense; that is, it means "is an element of" or "belongs to."

Although there is not general agreement among statisticians and philosophers as to exactly what probability is, there *is* general agreement that a probability measure $P[\cdot]$ should satisfy the above axioms. The axioms are satisfied for the classical theory defined above. These axioms lead immediately to some consequences that are useful in computing probabilities. Some of them are listed in the following theorem.

Theorem 2.2.1 *Let $P[\cdot]$ be a probability measure defined on the family $\mathcal{F}$ of events of a sample space Ω. Then*

(a) $P[\emptyset] = 0$;

(b) $P[A] = 1 - P[\overline{A}]$ *for every event A;*

(c) $P[A \cup B] = P[A] + P[B] - P[A \cap B]$ *for any events A, B;*

(d) $A \subset B$ *implies $P[A] \leq P[B]$ for any events A, B.*

Proof (a) $A \cup \emptyset = A$. A and $\emptyset$ are mutually exclusive ($A \cap \emptyset = \emptyset$) so, by Axiom **P3**, $P[A] = P[A \cup \emptyset] = P[A] + P[\emptyset]$. Hence, $P[\emptyset] = 0$.

(b) A and $\overline{A}$ are mutually exclusive by the definition of $\overline{A}$. Hence, by Axioms **P2** and **P3**,

$$1 = P[\Omega] = P[A \cup \overline{A}] = P[A] + P[\overline{A}].$$

Hence,

$$P[A] = 1 - P[\overline{A}].$$

(c) $A \cup B$ is the union of the mutually exclusive events $A \cap B$, $\overline{A} \cap B$, and $A \cap \overline{B}$; that is,

$$A \cup B = (A \cap B) \cup (\overline{A} \cap B) \cup (A \cap \overline{B}).$$

Therefore,

$$P[A \cup B] = P[A \cap B] + P[\overline{A} \cap B] + P[A \cap \overline{B}]. \tag{2.1}$$

In addition, $A \cap B$ and $A \cap \overline{B}$ are disjoint events whose union is A. Hence,

$$P[A] = P[A \cap B] + P[A \cap \overline{B}]. \tag{2.2}$$

Similarly,

$$P[B] = P[A \cap B] + P[\overline{A} \cap B]. \tag{2.3}$$

Adding (2.2) to (2.3) yields

$$P[A] + P[B] = 2P[A \cap B] + P[A \cap \overline{B}] + P[\overline{A} \cap B]. \tag{2.4}$$

Substituting (2.1) into (2.4) yields

$$P[A] + P[B] = P[A \cup B] + P[A \cap B] \tag{2.5}$$

or

$$P[A \cup B] = P[A] + P[B] - P[A \cap B]. \tag{2.6}$$

(d) Since $A \subset B$, B is the union of the disjoint events A and $B - A$. Thus,

$$P[B] = P[A] + P[B - A]. \tag{2.7}$$

Since $P[B - A] \geq 0$, this means that $P[A] \leq P[B]$. ∎

Example 2.2.2 A collection of 100 computer programs was examined for various types of errors (bugs). It was found that 20 of them had syntax errors, 10 had input/output *(I/O)* errors that were not syntactical, five had other types of errors, six programs had both syntax errors and *I/O* errors, three had both syntax errors and other errors, two had both *I/O* and other errors, while one had all three types of error. A program is selected at random from this collection, that is, it is selected in such a way that each program is equally likely to be chosen. Let S be the event that the selected program has errors in syntax, I be the event it has *I/O* errors, and O the event that it has other errors. Table 2.2.1 gives the probabilities associated with some of the events.

The probability that the program will have a syntax error or an *I/O* error or both is

$$
\begin{aligned}
P[S \cup I] &= P[S] + P[I] - P[S \cap I] \\
&= \frac{20}{100} + \frac{10}{100} - \frac{6}{100} = \frac{24}{100} = \frac{6}{25}, \quad \text{by Theorem 2.2.1(c).}
\end{aligned}
$$

The probability that it will have some type of error is given by

$$
\begin{aligned}
P[S \cup I \cup O] &= P[S] + P[I] + P[O] - P[S \cap I] \\
&\quad - P[S \cap O] - P[I \cap O] + P[S \cap I \cap O] \\
&= \frac{20}{100} + \frac{10}{100} + \frac{5}{100} - \frac{6}{100} - \frac{3}{100} - \frac{2}{100} + \frac{1}{100} \\
&= \frac{25}{100} = \frac{1}{4}.
\end{aligned}
$$

In making the last calculation we used the formula

$$P[A \cup B \cup C] = P[A] + P[B] + P[C] - P[A \cap B]$$

$$-P[A \cap C] - P[B \cap C] + P[A \cap B \cap C].$$

This formula follows from Theorem 2.2.1(c) and the *distributive law* $(A \cup B) \cap C = (A \cap C) \cup (B \cap C)$ (see Exercises 3 and 4). □

Table 2.2.1. Probabilities for Example 2.2.2

Event	S	I	O	$S \cap I$	$S \cap O$	$I \cap O$	$S \cap I \cap O$
Prob.	20/100	10/100	5/100	6/100	3/100	2/100	1/100

Sometimes probabilities are stated in terms of *odds*, especially by professional gamblers or bookmakers. If an event has probability $P[A]$ of occurring, the *odds for* A are defined by the following ratio:

$$\text{odds for } A = \frac{P[A]}{1 - P[A]}, \tag{2.8}$$

and the *odds against* A by

$$\text{odds against } A = \frac{1 - P[A]}{P[A]}. \tag{2.9}$$

The odds are expressed, whenever possible, by the ratio of whole numbers. For example, if $P[A] = \frac{1}{3}$, then

$$\text{odds for } A = \frac{1/3}{2/3} = \frac{1}{2},$$

and we use the notation 1:2 for these odds (read as "one to two"). In this case the odds against A are 2:1. It is easy to show that, if the odds for the event A are $a : b$, then

$$P[A] = \frac{a}{a + b}. \tag{2.10}$$

Thus, for example, if event A has odds 7:3 *against*, it has odds 3:7 for and $P[A] = 3/(3 + 7) = 0.3$.

Bookmakers make this sound even more complicated by telling the bettor the odds against an event rather than the odds for it. The numerator in the odds statement then reflects the profit on a successful bet in the amount of the denominator. Thus, a $2 wager at 7:2 odds (against) will, if successful, result in the return of the $2 stake plus a $7 profit. *The Odds On Virtually Everything*, (Verstappen [24]), does do much of what the title suggests. For example, it claims that the odds are 423:1 against becoming a millionaire in the United States. (However, there are 5,262 citizens for every millionaire in Wyoming compared to approximately one millionaire for every 37 citizens in Idaho.) (See also Siskin and Staller [22] for other estimates of odds on everyday happenings.)

For probability calculations involving finite sample spaces, we need some results from combinatorial analysis.

> *How do I love thee?*
> *Let me count the ways.*
> Elizabeth Barrett Browning

> *The best way to be brief is to leave things out.*
> A. S. C. Ehrenberg

2.3   Combinatorial Analysis

Combinatorial analysis is the science of counting—the number of elements in prescribed sets, the number of ways a particular selection can be made, etc.

One activity that is frequently employed in probability and statistics is drawing a few elements or items (a sample) from a collection or source (a population). Such a selection can be made with or without replacement. For example, if two cards are to be drawn from a 52-card bridge deck[4] without replacement, one card is removed and then another without putting the first card back. Drawing with replacement requires that a card be drawn, recorded, and returned to the deck before the second card is drawn, so that the two cards drawn may be identical. We assume in all drawing, with or without replacement, that the collection Ω from which a drawing is made consists of n distinct objects $O_1, O_2, \ldots, O_n$. A *permutation of order* k is an ordered selection of k elements from Ω, where $0 \leq k \leq n$. A *combination of order* k is an unordered selection of k elements from Ω, that is, a subset of k elements. The selections for both permutations and combinations can be made with or without replacement but are assumed to be made without replacement, unless otherwise stated.

---

[4]For the definition of a "bridge deck" see the discussion just after Exercise 13 at the end of this chapter.

Example 2.3.1 Suppose $\Omega = \{x, y, z\}$ and we draw two letters from Ω ($k = 2$). There are nine permutations of order 2 with replacement:

$$xx, \ xy, \ xz, \ yx, \ yy, \ yz, \ zx, \ zy, \ zz. \qquad (2.11)$$

There are six permutations made without replacement: xy, xz, yx, yz, zx, zy. There are six combinations made with replacement: xx, xy, xz, yy, yz, zz (the permutations xy and yx, for example, are not distinguished, because combinations are unordered). There are three combinations without replacement: xy, xz, yz. $\square$

One of the fundamental tools in combinatorics is the *multiplication principle*, which we state formally as a theorem.

Theorem 2.3.1 (*Multiplication Principle*) *If a task A can be done in m different ways and, after it is completed in any of these ways, task B can be completed in n different ways, then A and B, together, can be performed in m × n ways.*

Corollary *Suppose k tasks $A_1, A_2, \ldots, A_k$ are to be done and that A_1 can be completed in n_1 ways, A_2 in n_2 ways after A_1 is completed, A_3 in n_3 ways after A_1 and A_2 are completed, $\ldots, A_k$ in n_k ways after $A_1, A_2, \ldots, A_{k-1}$ are completed. Then the total task, $A_1, A_2, \ldots, A_k$ in succession, can be performed in $n_1 \times n_2 \times \cdots \times n_k$ ways.*

Proof The corollary follows immediately from the theorem by mathematical induction. The theorem itself follows from simple enumeration; that is, it is completely obvious. ∎

Hereafter we will refer to the multiplication principle, even when, strictly speaking, we use the corollary to it.

We define $n!$ (pronounced "n factorial") for each nonnegative integer n by $0! = 1, n! = n(n-1)!$ for $n > 0$. Thus $1! = 1, 2! = 2, 3! = 6, 4! = 24$, etc., and we can write

$$n! = n \times (n-1) \times (n-2) \times \cdots \times 2 \times 1.$$

Theorem 2.3.2 *The number of permutations of n elements, taken k at a time, without replacement, is*

$$P(n, k) = \frac{n!}{(n-k)!} = n(n-1)(n-2) \cdots (n-k+1).$$

With replacement allowed, the number of permutations is n^k.

Proof The first element in the permutation can be selected in n different ways from the n elements in Ω. After the first selection is made, there are $n - 1$ elements left in Ω from which to make the second selection if replacement is not allowed. After the second selection, there are $n - 2$ elements left in Ω from which to make the third selection, etc. Hence, by the multiplication principle,

$$P(n, k) = n(n - 1)(n - 2) \cdots (n - k + 1) = \frac{n!}{(n - k)!}.$$

If replacement is allowed there are n choices for each selection, so

$$P(n, k) = n^k. \qquad \blacksquare$$

Both the symbols $C(n, k)$ and $\binom{n}{k}$ are used to designate the number of combinations of k objects selected from a set of n elements.

Theorem 2.3.3 *There are*

$$C(n, k) = \binom{n}{k} = \frac{n!}{k!(n - k)!}$$

combinations of n objects, taken k at a time without replacement.

Corollary $\binom{n}{k}$ *is the coefficient of $x^k y^{n-k}$ in the expansion of $(x + y)^n$, that is,*

$$(x + y)^n = \sum_{k=0}^{n} \binom{n}{k} x^k y^{n-k}.$$

(This is why $\binom{n}{k}$ is often called a binomial coefficient.)

Proof of Theorem If replacement is not allowed, each combination of k elements forms $k!$ permutations of order k. Hence,

$$(k!)\binom{n}{k} = P(n, k) \quad \text{or} \quad \binom{n}{k} = \frac{P(n, k)}{k!} = \frac{n!}{k!(n - k)!}. \qquad \blacksquare$$

Proof of Corollary $(x + y)^n$ can be written as

$$(x + y)(x + y) \cdots (x + y) \quad (n \text{ factors}),$$

and the coefficient of $x^k y^{n-k}$ in the expansion is the number of ways we can choose x from k of these factors and y from the remaining $n - k$ factors. This is precisely $\binom{n}{k}$. $\blacksquare$

Example 2.3.2 Suppose 5 terminals are connected to an on-line computer system by attachment to one communication line. When the line is polled to find a terminal ready to transmit, there may be 0, 1, 2, 3, 4, or 5 terminals in the ready state. One possible sample space to describe the system state is $\Omega = \{(x_1, x_2, x_3, x_4, x_5)$: each x_i is either 0 or 1$\}$, where $x_i = 1$ means "terminal i is ready" and $x_i = 0$ means "terminal i is not ready." The sample point $(0, 1, 1, 0, 0)$ corresponds to terminals 2 and 3 ready to transmit but terminals 1, 4, and 5 not ready. By the multiplication principle, the number of sample points is $2^5 = 32$, since each x_i of $(x_1, x_2, x_3, x_4, x_5)$ can be selected in 2 ways. However, if we assume that exactly 3 terminals are in the ready state, then

$$\Omega = \{(x_1, x_2, x_3, x_4, x_5) : \text{exactly 3 of the } x_i\text{'s are 1 and 2 are 0}\}.$$

In this case the number of sample points of Ω is the number of ways that the three terminals that are ready can be chosen from the five available, that is,

$$\binom{5}{3} = \frac{5!}{3!(5-3)!} = \frac{5!}{3!2!} = \frac{5 \times 4}{2} = 10.$$

If the terminals are polled sequentially until a ready terminal is found, the number of polls required can be 1, 2, or 3. Let A_1, A_2, A_3 be the events that the required number of polls is 1, 2, 3, respectively. A_1 can occur only if $x_1 = 1$ and the other two 1's occur in the remaining 4 positions. Hence, the number of sample points favorable to A_1, n_1, is calculated as

$$n_1 = \binom{4}{2} = \frac{4!}{2!2!} = \frac{4 \times 3}{2} = 6 \quad \text{and} \quad P[A_1] = \frac{n_1}{n} = \frac{6}{10}.$$

A_2 can occur only if $x_1 = 0, x_2 = 1$, and the remaining two 1's are distributed in positions 3 through 5. Hence,

$$P[A_2] = \frac{\binom{3}{2}}{10} = \frac{3}{10}.$$

Similarly,

$$P[A_3] = \frac{\binom{2}{2}}{10} = \frac{1}{10}.$$

We have assumed, of course, that each terminal is equally likely to be in the ready condition. $\square$

Since many have trouble remembering the difference between a permutation and a combination, as well as the symbols used, we provide Table 2.3.1 as a quick reference guide.

Table 2.3.1. Permutations and Combinations

Permutations	Combinations
Number of ways of selecting k items out of n items	
Repetitions are not allowed	
Order is important	Order is not important
Arrangements of n items taken k at a time	Subsets of n items taken k at a time
$P(n, k) = \dfrac{n!}{(n - k)!}$	$C(n, k) = \dbinom{n}{k} = \dfrac{n!}{k!(n - k)!}$

Is life so dear or peace so sweet, as to be purchased at the price of chains and slavery? Forbid it, Almighty God! I know not what course others may take, but as for me, give me liberty or give me death!

Patrick Henry

2.4   Conditional Probability

It is often useful to calculate the probability that an event A occurs when it is known that an event B has occurred, where B has positive probability. The symbol for this probability is $P[A|B]$ and reads "the conditional probability of A, given B."

Example 2.4.1 If, in Example 2.2.2, it is known that the program that was drawn has an error in syntax, what is the probability that it has an I/O error, also?

Solution Twenty programs have errors in syntax, and six of these also have I/O errors. Hence, the required probability is $6/20 = 3/10$. The knowledge that the selected program has a syntactical error effectively reduced the size of the sample space from 100 to 20. □

In general, to calculate the probability that A occurs, given that B has occurred, means reevaluating the probability of A in the light of the information that B has occurred. Thus, B becomes our new sample space and we are interested only in that part of A that occurs with B, that is, $A \cap B$. Thus, we must have the formula

$$P[A|B] = \frac{P[A \cap B]}{P[B]}, \tag{2.12}$$

if $P[B] > 0$. The conditional probability of A given B is not defined if $P[B] = 0$. In (2.12), $P[A \cap B]$ was divided by $P[B]$ so that $P[B|B] = 1$, making $P[\cdot|B]$ a probability measure. The event B in (2.12) is often called the *conditioning event*.

Equation 2.12 can be used to make the probability calculation in Example 2.4.1, with B the event that the program has at least one error in syntax and A the event that the program has at least one *I/O* error. Then,

$$P[A|B] = \frac{P[A \cap B]}{P[B]} = \frac{6/100}{20/100} = \frac{6}{20} = \frac{3}{10},$$

as before.

Equation 2.12 can be rewritten in a form called the *multiplication rule*.

Theorem 2.4.1 (*Multiplication Rule*) *For events A and B*

$$P[A \cap B] = P[A]P[B|A], \tag{2.13}$$

if $P[A] \neq 0$, and

$$P[A \cap B] = P[B]P[A|B], \tag{2.14}$$

if $P[B] \neq 0$. (If either $P[A] = 0$ or $P[B] = 0$ then $P[A \cap B] = 0$ by Theorem 2.2.1(d).)

Corollary (*General Multiplication Rule*) *For events $A_1, A_2, \ldots, A_n$,*

$$\begin{aligned}
P[A_1 \cap A_2 \cap \cdots \cap A_n] &= P[A_1]P[A_2|A_1]P[A_3|A_1 \cap A_2] \cdots \\
&\times P[A_n|A_1 \cap \cdots \cap A_{n-1}]
\end{aligned} \tag{2.15}$$

provided all the probabilities on the right are defined. A sufficient condition for this is that $P[A_1 \cap A_2 \cap \cdots \cap A_{n-1}] > 0$, since $P[A_1] \geq P[A_1 \cap A_2] \geq \cdots \geq P[A_1 \cap A_2 \cap \cdots \cap A_{n-1}]$.

Proof Equations 2.13 and 2.14 are true by the definition of conditional probability, (2.12). The corollary follows by mathematical induction on n: For $n = 2$ the result is the theorem and Thus, is true. Now suppose the corollary is true for $n = k \geq 2$ and $A_1, A_2, \ldots, A_k, A_{k+1}$ are events. Let $A = A_1 \cap A_2 \cap \ldots \cap A_k$ and $B = A_{k+1}$.

Then, by (2.13),

$$
\begin{aligned}
P[A_1 \cap A_2 \cap \cdots \cap A_k \cap A_{k+1}] &= P[A \cap B] = P[A]P[B|A] \\
&= P[A_1]P[A_2|A_1]P[A_3|A_1 \cap A_2] \cdots \\
&\quad \times P[A_k|A_1 \cap \cdots \cap A_{k-1}] \\
&\quad \times P[A_{k+1}|A_1 \cap \cdots \cap A_k],
\end{aligned}
$$

where the last equality follows from the inductive assumption that

$$
\begin{aligned}
P[A_1 \cap A_2 \cap \cdots \cap A_k] &= P[A_1]P[A_2|A_1]P[A_3|A_1 \cap A_2] \cdots \\
&\quad \times P[A_k|A_1 \cap A_2 \cap \cdots \cap A_{k-1}].
\end{aligned}
$$

This completes the proof. ∎

Example 2.4.2 Suppose a survey of 100 computer installations in a certain city shows that 75 of them have at least one brand X computer. If three of these installations are chosen at random without replacement, what is the probability that each of them has at least one brand X machine?

Solution Let A_1, A_2, A_3 be the event that the first, second, third, selection, respectively, has a brand X computer. The required probability is

$$
P[A_1 \cap A_2 \cap A_3] = P[A_1]P[A_2|A_1]P[A_3|A_1 \cap A_2]
$$

by the general multiplication rule.

This value is

$$
\frac{75}{100} \times \frac{74}{99} \times \frac{73}{98} = 0.418,
$$

which is somewhat lower than intuition might lead one to believe. □

One of the main uses of conditional probability is to assist in the calculation of unconditional probability by the use of the following theorem.

Theorem 2.4.2 (*Law of Total Probability*) *Let* $A_1, A_2, \ldots, A_n$ *be events such that*

(*a*) $A_i \cap A_j = \emptyset$ *if* $i \neq j$ (*mutually exclusive events*),

(*b*) $P[A_i] > 0, i = 1, 2, \ldots, n,$

(*c*) $A_1 \cup A_2 \cup \cdots \cup A_n = \Omega.$

(*A family of events satisfying* (*a*)–(*c*) *is called a* partition *of* Ω.)
   Then, for any event A,

$$
P[A] = P[A_1]P[A|A_1] + P[A_2]P[A|A_2] + \cdots + P[A_n]P[A|A_n]. \quad (2.16)
$$

Proof Let $B_i = A \cap A_i, i = 1, 2, 3, \ldots, n$. Then $B_i \cap B_j = \emptyset$ if $i \neq j$ because the events $A_1, A_2, \ldots, A_n$ are mutually exclusive, and

$$A = B_1 \cup B_2 \cup \cdots \cup B_n, \tag{2.17}$$

because each element of A is in exactly one B_j.

Hence,

$$P[A] = P[B_1] + P[B_2] + \cdots + P[B_n]. \tag{2.18}$$

But

$$P[B_i] = P[A \cap A_i] = P[A_i]P[A|A_i], \quad i = 1, 2, \ldots, n. \tag{2.19}$$

Substituting (2.19) into (2.18) yields (2.16) and completes the proof. ∎

Example 2.4.3 Inquiries to an on-line computer system arrive on five communication lines. The percentage of messages received from lines 1, 2, 3, 4, 5, are 20, 30, 10, 15, and 25, respectively. The corresponding probabilities that the length of an inquiry will exceed 100 characters are 0.4, 0.6, 0.2, 0.8, and 0.9. What is the probability that a randomly selected inquiry will be longer than 100 characters?

Solution Let A be the event that the selected message has more than 100 characters and A_i the event that it was received on line i ($i = 1, 2, 3, 4, 5$). Then, by the law of total probability,

$$\begin{aligned} P[A] &= P[A_i]P[A|A_i] + \cdots + P[A_5]P[A|A_5] \\ &= 0.2 \times 0.4 + 0.3 \times 0.6 + 0.1 \times 0.2 \\ &\quad + 0.15 \times 0.8 + 0.25 \times 0.9 = 0.625. \ \square \end{aligned} \tag{2.20}$$

Two events A and B are said to be *independent* if $P[A \cap B] = P[A]P[B]$. This implies the usual meaning of independence; namely, that neither event influences the occurrence of the other. For, if A and B are independent (and both have nonzero probability), then

$$P[A|B] = \frac{P[A \cap B]}{P[B]} = \frac{P[A]P[B]}{P[B]} = P[A], \tag{2.21}$$

and

$$P[B|A] = \frac{P[A \cap B]}{P[A]} = \frac{P[A]P[B]}{P[A]} = P[B]. \tag{2.22}$$

The concept of two events A and B being independent should not be confused with the concept of their being mutually exclusive. In fact, if A and B are mutually exclusive,

$$0 = P[\emptyset] = P[A \cap B],$$

and thus, $P[A \cap B]$ can be equal to $P[A]P[B]$ only if at least one of them has probability zero. Hence, mutually exclusive events are *not* independent except in the trivial case that at least one of them has zero probability.[5] The next example ties together some of the concepts discussed so far.

Example 2.4.4 Suppose that a certain department has 3 unbuffered terminals, that can be connected to a computer via 2 communication lines. Terminal 1 has its own leased line while terminals 2 and 3 share a leased line so that at most one of the two can be in use at any particular time. During the working day terminal 1 is in use 30 minutes of each hour, terminal 2 is used 10 minutes of each hour, and terminal 3 is used 5 minutes of each hour—all times being average times. Assuming the communication lines operate independently, what is the probability that at least one terminal is in operation at a random time during the working day? If the operation of the two lines is not independent with the conditional probability that terminal 2 is in use given that terminal 1 is in operation equal to 1/3, and the corresponding conditional probability that terminal 3 is in use equal to 1/12, what is the probability that at least one line is in use?

Solution Case 1: The lines operate independently. Let A, B, C be the events that terminals 1, 2, 3, respectively, are in use. The event that the first line is in use is A and the event that the second line is in use is $B \cup C$, and these events are independent. The event U, that at least one terminal is in use, is $A \cup (B \cup C)$.

By Theorem 2.2.1(c),

$$P[U] = P[A \cup (B \cup C)] = P[A] + P[B \cup C] - P[A \cap (B \cup C)]. \quad (2.23)$$

By the independence of A and $B \cup C$,

$$P[A \cap (B \cup C)] = P[A]P[B \cup C]. \quad (2.24)$$

Substituting (2.24) into (2.23) yields

$$P[U] = P[A] + P[B \cup C] - P[A]P[B \cup C]. \quad (2.25)$$

Since B and C are mutually exclusive, Axiom **P3** yields

$$P[B \cup C] = P[B] + P[C] = \frac{10}{60} + \frac{5}{60} = \frac{1}{4}. \quad (2.26)$$

Since $P[A] = 0.5$, substitution into (2.25) gives

$$P[U] = 0.5 + 0.25 - 0.5 \times 0.25 = 0.625. \quad (2.27)$$

---

[5]Since the occurrence of either of these events precludes the occurrence of the other, we would not expect them to be independent on purely intuitive grounds.

Case 2: The communication lines are not independent. In this case (2.23) still applies but $P[A \cap (B \cup C)]$ has the formula,

$$P[A \cap (B \cup C)] = P[(A \cap B) \cup (A \cap C)], \qquad (2.28)$$

by the distributive law for events (see Exercise 3).

The events $A \cap B$ and $A \cap C$ are mutually exclusive, since B and C are. Hence,

$$
\begin{aligned}
P[A \cap (B \cup C)] &= P[(A \cap B) \cup (A \cap C)] = P[A \cap B] + P[A \cap C] \\
&= P[A]P[B|A] + P[A]P[C|A] = \frac{1}{2} \times \frac{1}{3} + \frac{1}{2} \times \frac{1}{12} \\
&= \frac{5}{24}, \qquad (2.29)
\end{aligned}
$$

by the multiplication rule. Substituting (2.29) into (2.23) gives

$$P[U] = \frac{1}{2} + \frac{1}{4} - \frac{5}{24} = \frac{13}{24} = 0.542. \ \square \qquad (2.30)$$

Theorem 2.4.3 (*Bayes' Theorem*) *Suppose the events* $A_1, A_2, \ldots, A_n$ *form a partition of* Ω *(for the definition of a partition, see Theorem 2.4.2). Then, for any event A with $P[A] > 0$,*

$$P[A_i|A] = \frac{P[A_i]P[A|A_i]}{P[A_1]P[A|A_1] + P[A_2]P[A|A_2] + \cdots + P[A_n]P[A|A_n]},$$
$$i = 1, 2, \ldots, n. \qquad (2.31)$$

Proof For each i,

$$P[A_i|A] = \frac{P[A_i \cap A]}{P[A]} = \frac{P[A_i]P[A|A_i]}{P[A]}. \qquad (2.32)$$

Equation 2.31 now follows from (2.32) by applying the law of total probability, Theorem 2.4.2, to calculate $P[A]$. ∎

The $P[A_i], i = 1, 2, \ldots, n$, are called *prior* (or *a priori*) probabilities and the $P[A_i|A], i = 1, 2, \ldots, n$, are called *posterior* (or *a posteriori*) probabilities. To calculate the posterior probabilities using Bayes' theorem, we must know both the prior probabilities $P[A_1], P[A_2], \ldots, P[A_n]$ and the conditional probabilities

$$P[A|A_1], \ldots, P[A|A_n].$$

Bayes' theorem[6] is often called Bayes' rule or Bayes' formula. We give some examples of its use below.

One important application of Bayes' theorem is in screening tests. We use the terminology of the excellent paper by Gastwirth [9]. A screening test is used to determine whether a person belongs to the class D of those who have a specific disease such as cancer or AIDS. The test result that indicates that the person is a member of class D is denoted by S; a result indicating nonmembership is denoted by $\overline{S}$. The accuracy of a test is specified by two probabilities. The first is called the *sensitivity of the test*, defined to be the probability that a person with the disease is correctly diagnosed, or

$$\eta = P[S|D].$$

The second is called the *specificity*, defined to be the probability that a person who does not have the disease is correctly diagnosed, or

$$\theta = P[\overline{S}|\overline{D}].$$

For good tests both η and θ should be very close to one. Another item of interest is $\pi = P[D]$, that is, the probability that a randomly selected person in the population has the disease. The most critical problem in screening tests, as we shall see in the example below, is that when π is small there can be a large number of false positives, that is healthy people whose test results indicate they have the disease (are members of D). The probability of this happening to a healthy individual is

$$P[S|\overline{D}] = 1 - P[\overline{S}|\overline{D}] = 1 - \theta.$$

The probability of most interest to a person who takes the test and gets a positive reading (the result S) is the probability that such a person actually has the disease. By Bayes' theorem this is

$$
\begin{aligned}
P[D|S] &= \frac{P[D]P[S|D]}{P[D]P[S|D] + P[\overline{D}]P[S|\overline{D}]} \\
&= \frac{\pi\eta}{\pi\eta + (1 - \pi)(1 - \theta)}.
\end{aligned}
\tag{2.33}
$$

$P[D|S]$ is called the "predictive value of a positive test" and abbreviated PVP. Gastwirth [9] cites a study of the ELISA test for AIDS used to screen donated blood for the AIDS antibody in which the estimated value for η was 0.977 and for θ was 0.926.

---

[6]Named for the Reverend Thomas Bayes; it was published (posthumously) in 1763.

Example 2.4.5 Suppose the ELISA test mentioned above is used to screen donated blood from a population in which the probability of an individual having the AIDS antibody is 0.0001. Suppose $\eta = 0.977$ and $\theta = 0.926$. Then, by (2.33)

$$P[D|S] = \frac{0.0001 \times 0.977}{0.0001 \times 0.977 + 0.9999 \times 0.074} = 0.001319.$$

If this test were performed on 100,000 blood samples, there would be about $100,000 \times 0.0001 = 10$ with the AIDS antibody of which 9.77 would be diagnosed correctly, on the average. However, of the 99,990 samples with no antibody, $99,990 \times 0.074 = 7,399.26$ would be incorrectly diagnosed as having the antibody. Note that

$$\frac{9.77}{9.77 + 7,399.26} = 0.001319.$$

Fortunately, there is another (more expensive) test that can be used to confirm or reject positive ELISA tests. □

Example 2.4.6 Suppose a pair of fair dice are tossed. Let A be the event that "the first die turns up odd," B be the event that "the second die turns up odd," while C is the event "the total number of spots showing uppermost is odd." Then clearly A and B are independent with $P[A] = P[B] = \frac{1}{2}$. From Figure 2.2.1 it is clear that $P[C] = \frac{1}{2}$. Given that A has occurred, C can occur only if the second die turns up even. Hence,

$$P[C|A] = \frac{1}{2}$$

and, similarly,

$$P[C|B] = \frac{1}{2}.$$

Hence,

$$P[C|A] = P[C],$$

and

$$P[C|B] = P[C],$$

so the events A and C are independent, as are B and C. (We showed earlier that A and B are independent.) Thus, the three events A, B, C are pairwise independent. Since C cannot occur if A and B both do,

$$P[A \cap B \cap C] = 0.$$

However,

$$P[A]P[B]P[C] = \frac{1}{2} \times \frac{1}{2} \times \frac{1}{2} = \frac{1}{8},$$

so that

$$P[A \cap B \cap C] \neq P[A]P[B]P[C]. \quad \square$$

Definition 2.4.1 The events $A_1, A_2, \ldots, A_n$ are *mutually independent* if

$$P[A_i \cap A_j] = P[A_i]P[A_j],$$
$$P[A_i \cap A_j \cap A_k] = P[A_i P[A_j]P[A_k],$$
$$\cdots$$
$$P[A_1 \cap A_2 \cap \cdots \cap A_n] = P[A_1]P[A_2] \cdots P[A_n]$$

for all combinations of indices such that

$$1 = i < j < \cdots < k \leq n.$$

This definition can be extended to an infinite sequence of events as follow:

Definition 2.4.2 Given an infinite sequence of events $A_1, A_2, \ldots$, such that the events $A_1, A_2, \ldots, A_n$ are mutually independent for every n, then $A_1, A_2, \ldots$ is said to be a *sequence of independent events*.

We note that the events of Example 2.4.6 are pairwise independent but not mutually independent.

Example 2.4.7 (*The Birthday Problem*) This problem is one of the most famous in probability theory and illustrates the fact that our intuition can sometimes lead us astray. Suppose there are n people in a room, that no one was born on February 29th, and that this is not a leap year. Everyone can see that, if $n > 365$, then at least two people have the same birthday; that is, have their birthdays on the same month and same day of the month. (The year is not considered.) Now suppose n is at least 2 but less than 365 ($2 \leq n < 365$). Then:

(a) What is the probability that at least two people have the same birthday?

(b) What is the smallest n such that this probability exceeds 0.5?

Solution We assume there are 365 equally likely days for each person's birthday.[7] Thus, by the multiplication principle (the corollary to Theorem

---

[7]Berresford [3] has shown that adjusting birth date frequencies to match actual observed frequencies does not significantly change the results of this exercise. In particular, it does not change the value of n in (b).

2.3.1), there are 365^n possible choices for the birthdays of n people. Let the sample space Ω consist of the 365^n n-tuples $(k_1, k_2, \ldots, k_n)$ where each k_i is an integer between 1 and 365. (We assume the days of the year are numbered from 1 to 365 with 1 representing January 1 and 365 representing December 31.) Then k_1 represents the birthday of the first person, k_2 represents the birthday of the second person, $\ldots$, k_n represents the birthday of the nth person. We assume each of the sample points has probability $1/365^n$. Let E be the event that no two of the n people have the same birthday, that is, if $(k_1, k_2, \ldots, k_n)$ is a sample point of E, then

$$k_i \neq k_j \quad \text{for } i \neq j.$$

Hence, if $k = (k_1, k_2, \ldots, k_n) \in E$, k_1 can assume 365 possible values, k_2 can assume 364 values because it must be different from k_1, k_3 can assume 363 possible values because it must be different from both k_1 and k_2, $\ldots$, k_n can assume $365 - (n-1) = 365 - n + 1$ possible values. Therefore, by the multiplication principle, the number of sample points in E is

$$365 \times 364 \times 363 \times \cdots \times (365 - n + 1).$$

If we let q_n be the probability that no two of the n people have the same birthday, then

$$
\begin{aligned}
q_n = P[E] \quad &= \quad \frac{365 \times 364 \times 363 \times \cdots \times (365 - n + 1)}{365^n} \\
&= \quad 1 \times \left(1 - \frac{1}{365}\right) \times \left(1 - \frac{2}{365}\right) \times \cdots \times \left(1 - \frac{(n-1)}{365}\right).
\end{aligned}
$$

The above formula for q_n could also be derived using the general multiplication rule. Let p_n be the probability that at least two people have the same birthday. Then

$$p_n = 1 - q_n.$$

This is the solution to (a). Some values of p_n are shown in Table 2.4.1, below. It is rather surprising to most people to find that the answer to (b), above, is only 23. We have verified this fact, experimentally, in a number of classes taught at the Los Angeles IBM Information Systems Management Institute. We found that in classes of 30 or more students multiple birthdays were very common. With 40 or more students we almost always found two or more students with the same birthday. Moser [15] discusses more surprises concerning the birthday problem. $\square$

Table 2.4.1. Birthday p_n

n	p_n	n	p_n
2	0.00274	23	0.50730
3	0.00820	25	0.56870
5	0.02714	30	0.70632
10	0.11695	40	0.89123
15	0.25290	50	0.97037
20	0.41144	75	0.99972
22	0.47570	100	1.00000

If a man does not keep pace with his companions, perhaps it is because he hears a different drummer. Let him step to the music that he hears, however measured or far away.

Henry David Thoreau

2.5   Random Variables

In many random experiments we are interested in some number associated with the experiment rather than the actual outcome. Thus, in Example 2.1.2, we may be interested in the sum of the numbers shown on the dice. In Example 2.3.2 we may be interested in the number of polls taken to find the first ready terminal. We are thus interested in a function that associates a number with the outcome of an experiment—such a function is called a *random variable*. Formally, a *random variable* X is a real-valued function defined on a sample space Ω. Some examples of random variables of interest to computer science follow.

Example 2.5.1 Let X be the number of jobs processed by a computer center in one day. The sample space Ω might consist of collections of job numbers—an outcome is the set of job numbers of jobs run during the day. (We assume each job number is unique.) Thus, if $\omega = \{ x_1, x_2, \ldots, x_n \}$ is a sample point consisting of the set of job numbers of jobs run during the day, then $X(\omega) = n$. $\square$

Example 2.5.2 Let X be the number of communication lines in operation in an on-line computer system of n lines. The sample space Ω could be the collection of n-tuples $(x_1, x_2, \ldots, x_n)$ where each x_i is 1 if line i is in operation and otherwise x_i is 0. $\square$

To avoid cumbersome notation, we will use abbreviations to denote some special events. If X is a random variable and x is a real number, we write

$$X = x$$

for the event

$$\{\omega : \omega \in \Omega \text{ and } X(\omega) = x\}.$$

Similarly, we write

$$X \leq x$$

for the event

$$\{\omega : \omega \in \Omega \text{ and } X(\omega) \leq x\},$$

and

$$y < X \leq x$$

for the event

$$\{\omega : \omega \in \Omega \text{ and } y < X(\omega) \leq x\}.$$

Another property required of a random variable is that the set $X \leq x$ be an event for each real x, that is, $X \leq x$ be an element of $\mathcal{F}$ for each real x.[8] This is necessary so that probability calculations can be made. A function having this property is said to be a *measurable function* or *measurable in the Borel sense* (see Cramér [5, page 37]).

For each random variable X we define its *distribution function F* for each real x by

$$F(x) = P[X \leq x].$$

Some intuitively clear properties of a distribution function are stated in the following proposition.

Proposition 2.5.1 (*Properties of a Distribution Function*)

D1 *F is a nondecreasing function; that is, $x < y$ implies $F(x) \leq F(y)$.*

D2 $\lim_{x \to +\infty} F(x) = 1$.

D3 $\lim_{x \to -\infty} F(x) = 0$.

Proof For the proof of Proposition 2.5.1, see Apostol [1]. ∎

---

[8]Recall, by Axiom Set 2.2.1, that $\mathcal{F}$ is the family of events in Ω.

The distribution function can be used to make certain probability calculations. For example, if $x < y$, then

$$P[x < X \leq y] = F(y) - F(x).$$

This is true because the events $x < X \leq y$ and $X \leq x$ are disjoint; their union is $X \leq y$; and thus,

$$F(y) = P[x < X \leq y] + P[X \leq x] = P[x < X \leq y] + F(x). \qquad (2.34)$$

With each random variable X, we associate another function $p(\cdot)$, called the *probability mass function* of X (abbreviated as pmf), defined for all real x by

$$P(x) = P[X = x].$$

Thus, if x is a value that X cannot assume, then $p(x) = 0$. The set T of all x with $p(x) > 0$ is either finite or countably infinite. For a proof see Cramér [5, page 52]. (A set is countably infinite or denumerable if it can be put into one-to-one correspondence with the positive integers and thus enumerated $x_1, x_2, x_3, \ldots$.) The random variable X is said to be *discrete* if

$$\sum_{x \in T} p(x) = 1,$$

where $T = \{x : p(x) > 0\}$. Thus, X is discrete if T consists of either (a) a finite set, say, $x_1, x_2, \ldots, x_n$, or (b) an infinite set, say, $x_1, x_2, x_3, \ldots$, and, in addition,

$$\sum_{x_i} p(x_i) = 1.$$

Thus, a real-valued function $p(\cdot)$ defined on the whole real line is the probability mass function of a discrete random variable if and only if the following three conditions hold:

(i) $p(x) \geq 0$ for all real x.

(ii) $T = \{x | p(x) > 0\}$ is finite or countably infinite; that is, $T = \{x_1, x_2, \ldots\}$.

(iii) $\displaystyle\sum_{x_i \in T} p(x_i) = 1.$

If X is a discrete random variable, the elements of T are called the *mass points of X,* and we say, "X assumes the values $x_1, x_2, x_3, \ldots$."

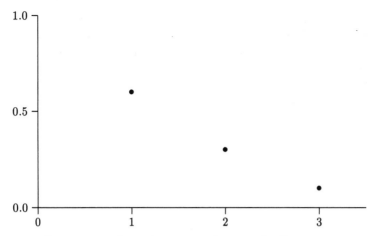

Figure 2.5.1. Probability mass function for Example 2.5.3.

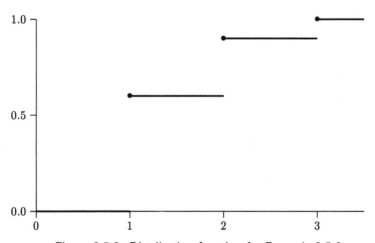

Figure 2.5.2. Distribution function for Example 2.5.3.

Example 2.5.3 In Example 2.3.2 we implicitly define a random variable X, which counts the number of polls until a ready terminal is found. X is a discrete random variable that assumes only the values 1, 2, 3. The probability mass function is defined by $p(1) = 0.6$, $p(2) = 0.3$, and $p(3) = 0.1$. The pmf $p(\cdot)$ of X is shown graphically in Figure 2.5.1; the distribution function F is shown in Figure 2.5.2. Thus, the probability that two or fewer polls are required is $F(2) = P(1) + p(2) = 0.9$, which can be read from Figure 2.5.2 or calculated from the probability mass function. □

A random variable X is *continuous* if $p(x) = 0$ for all real x. The reason for the terminology is that the distribution function for a continuous random variable is a continuous function in the usual sense. By contrast, the distribution function for a discrete random variable has a discontinuity at each point of positive probability (mass point). We will be concerned only with those continuous random variables X that have a *density function* f with the following properties:

(a) $f(x) \geq 0$ for all real x.

(b) f is integrable and $P[a \leq X \leq b] = \int_a^b f(x)dx$ if $a < b$.[9]

(c) $\displaystyle\int_{-\infty}^{\infty} f(x)dx = 1$.

(d) $F(x) = \int_{-\infty}^{x} f(t)dt$ for each real x.

By the fundamental theorem of calculus, at each point x where f is continuous,
$$\frac{dF}{dx} = f(x).$$

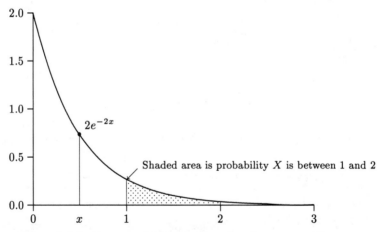

Figure 2.5.3. Exponential density function (parameter 2).

Example 2.5.4 Let $\alpha > 0$. The random variable X is said to be an *exponential random variable with parameter* α or to have an *exponential*

---

[9]This means that the required probability is the area under the curve $y = f(x)$ between a and b.

distribution with parameter α if it has the distribution function

$$F(x) = \begin{cases} 1 - e^{-\alpha x} & \text{for } x > 0 \\ 0 & \text{for } x \leq 0. \end{cases} \qquad (2.35)$$

The density function $f = dF/dx$ is given by

$$f(x) = \begin{cases} \alpha e^{-\alpha x} & \text{for } x > 0 \\ 0 & \text{for } x \leq 0. \end{cases} \qquad (2.36)$$

Suppose, for example, that $\alpha = 2$ and we wish to calculate the probability that X assumes a value between 1 and 2. This probability is the area under the curve $y = 2e^{-2x}$ (shown in Figure 2.5.3) between $x = 1$ and $x = 2$. The probability may also be computed using the distribution function $F(x) = 1 - e^{-2x}$ shown in Figure 2.5.4. We calculate

$$\begin{aligned} P[1 \leq X \leq 2] &= F(2) - F(1) = (1 - e^{-4}) - (1 - e^{-2}) = e^{-2} - e^{-4} \\ &= 0.135335283 - 0.018315639 = 0.117019644. \quad \square \end{aligned}$$

In making this calculation we have used the fact that, for a continuous random variable X with $a < b$,

$$P[a \leq X \leq b] = P[a < X \leq b] = P[a \leq X < b] = P[a < X < b], \quad (2.37)$$

since

$$P[X = a] = P[X = b] = 0.$$

(This is true because, by definition, a continuous random variable X has the property that $P[X = x] = 0$ for all real x.)

Most random variables are either discrete or continuous but, occasionally, we shall encounter a random variable of mixed type: a random variable that is continuous for some range of values and discrete for others. Usually a discrete random variable comes about when something is counted, such as number of jobs, inquiries, messages, etc.[10] A continuous random variable often occurs when something is measured, such as the time between the arrival of two consecutive inquiries, the response time at a terminal, the time it takes to process a job, etc. Unless otherwise noted, we will assume that all random variables under consideration are either discrete or continuous and not of mixed type.

The distribution function F of a random variable X describes how the probability mass of X is distributed along the real line. From this point

---

[10]It should be noted that, between any two distinct real numbers that a discrete random variable assumes, there exist other values it doesn't assume.

of view a random variable X determines how the probability mass of one unit is apportioned or spread out over the real numbers. A discrete random variable allocates the mass in nuggets or mass points, while a continuous random variable diffuses the probability mass out in a continuous manner.

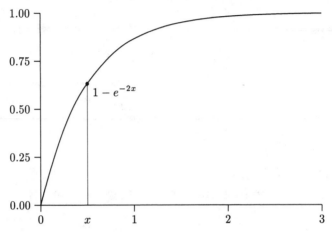

Figure 2.5.4. Exponential Distribution Function (Parameter 2)

> *"I know what you're thinking about," said Tweedledum;*
> *"but it isn't so, nohow." "Contrariwise," continued Tweedledee,*
> *"if it was so, it might be; and if it were so, it would be;*
> *but as it isn't, it ain't. That's logic."*
> Lewis Carroll
> *Through the Looking Glass*

2.6   Parameters of Random Variables

All possible probability calculations involving a random variable X can be made from its pmf $p(\cdot)$, if it is discrete; from its density function f, if it is continuous; or from its distribution function F in either case. However, there are some parameters of a random variable that are important in summarizing its properties in a way that is easy to comprehend and to use for making probability estimates.

Let X be a discrete random variable with pmf $p(\cdot)$. We define the *mean* or *expected value* of X, $\mu = E[X]$ by the formula

$$\mu = E[X] = \sum_{x_i} x_i p(x_i) = x_1 p(x_1) + x_2 p(x_2) + \cdots, \qquad (2.38)$$

provided

$$\sum_{x_i} |x_i| p(x_i) < \infty.$$

(We require absolute convergence of that the sum in (2.38) to guarantee the sum does not change if the x_i's are reordered.) If X is a continuous random variable with density function $f(\cdot)$, we define $\mu = E[X]$ by

$$\mu = E[X] = \int_{-\infty}^{\infty} x f(x) ds, \qquad (2.39)$$

provided

$$\int_{-\infty}^{\infty} |x| f(x) dx < \infty.$$

If h is a real-valued function of a real variable, and X is a random variable, then $h(X)$ is a new random variable defined for all $\omega \in \Omega$ by $h(X)(\omega) = h(X(\omega))$. Thus, if $h(x) = e^x$, then $h(X)(\omega) = e^{X(\omega)}$. The following lemma shows how to calculate the expected value of $h(X)$.

Lemma 2.6.1 (*Law of the Unconscious Statistician*).[11] *Suppose h is a real-valued function of a real variable.*

(a) If X is discrete,

$$E[h(X)] = \sum_{x_i} h(x_i) p(x_i), \qquad (2.40)$$

provided

$$\sum_{x_i} |h(x_i)| p(x_i) < \infty.$$

(b) If X is continuous,

$$E[h(X)] = \int_{-\infty}^{\infty} h(x) f(x) dx, \qquad (2.41)$$

provided

$$\int_{-\infty}^{\infty} |h(x)| f(x) dx < \infty.$$

Proof See Ross [21] or Grimmett and Stirzaker [13]. ■

---

[11]This law received its name because of "unconscious" statisticians who have used it as if it were the definition.

The two most important parameters used to describe or summarize the properties of a random variable X are the mean $\mu = E[X]$ defined by (2.38) or (2.39) and *standard deviation* σ, where σ^2 is the *variance of* X (we also use Var$[X]$ for σ^2) defined by

$$\sigma^2 = \text{Var}[X] = E[(X - \mu)^2]. \tag{2.42}$$

Thus, for a discrete random variable X, Var$[X]$ is given by

$$\sigma^2 = \text{Var}[X] = \sum_i (x_i - \mu)^2 p(x_i),$$

while, if X is continuous,

$$\sigma^2 = \text{Var}[X] = \int_{-\infty}^{\infty} (x - \mu)^2 f(x) dx. \tag{2.43}$$

The reason that the mean or expected value is important is intuitively clear to almost everyone. For example, if X is a discrete random variable that describes how much one is to win for each outcome in a sample space Ω, then

$$\mu = E[X] = \sum_i x_i p(x_i)$$

is the weighted average of what one is to win. Thus, calling μ the mean or expected value seems proper. The mean is a summary of what we expect of the random variable. If we have only one number to use to describe a random variable, the mean would seem to be the proper one. However, the standard deviation, σ, does not have such an intuitive meaning to many of us, at first. However, as we will see in Chebyshev's inequality (Theorem 2.10.2) and the one-sided inequality (Theorem 2.10.3), the standard deviation is the natural unit to measure the deviation of a random variable from its mean. Thus, if you were told that "Rockefeller and I, together, had an average income last year of $1,010,000 with a standard deviation of $990,000," then it would be clear that our individual incomes were significantly different.[12]

It is best not to worry as to exactly what σ is. We will show that it is a most useful quantity for many applied probability applications. We must ask the reader to "have faith that truth will be revealed."

---

[12]See Exercise 60.

Example 2.6.1 Referring to Example 2.5.2, we see that

$$\mu = E[X] = \sum_{i=1}^{3} x_i p(x_i) = 1 \times 0.6 + 2 \times 0.3 + 3 \times 0.1 = 1.5, \qquad (2.44)$$

$$\sigma^2 = \text{Var}[X] = \sum_{i=1}^{3} (x_i - 1.5)^2 p(x_i)$$

$$= (1 - 1.5)^2 \times 0.6 + (2 - 1.5)^2 \times 0.3 + (3 - 1.5)^2 \times 0.1 = 0.45, \qquad (2.45)$$

and thus,

$$\sigma = (\text{Var}[X])^{1/2} = (0.45)^{1/2} = 0.6708. \quad \square \qquad (2.46)$$

This example can be generalized. Consider a communication line with m terminals attached, of which n are ready to transmit.[13] Let X be the number of polls required to find the first terminal that is ready. Then X can assume only the values $i = 1, 2, \ldots, m - n + 1$ with

$$p(i) = P[X = i] = \frac{\binom{m-i}{n-1}}{\binom{m}{n}}. \qquad (2.47)$$

See Exercise 35 for the proof of (2.47). $E[X]$ and $\text{Var}[X]$ can then be calculated by the formulas

$$\mu = E[X] = \sum_{i=1}^{m-n+1} i p(i) \qquad (2.48)$$

and

$$\sigma^2 = \text{Var}[X] = \sum_{i=1}^{m-n+1} (i - \mu)^2 p(i). \qquad (2.49)$$

In Exercise 37 you are asked to derive a simple, closed-form formula for $E[X]$ and for $E[X^2]$.

The APL function POLL (shown in Appendix B) with parameters m and n can be used in the general case to compute the probabilities that X assumes the values $1, 2, 3, \ldots, m - n + 1$. It also computes the expected value and standard deviation of X.

The APL function PARAM can be used to calculate the mean and standard deviation of a discrete random variable from the set of its possible values and the corresponding set of probabilities.

---

[13]We assume that n is at least one.

Example 2.6.2 Suppose X has an exponential distribution with parameter $\alpha > 0$. (See Example 2.5.3.) Then

$$\mu = E[X] = \int_0^\infty x\alpha e^{-\alpha x} dx. \qquad (2.50)$$

Using integration by parts (the formula $\int u\,dv = uv - \int v\,du$) with $u = x$ and $dv = \alpha e^{-\alpha x} dx$, together with one application of the formula

$$\lim_{x \to \infty} x e^{-\alpha x} = 0, \qquad (2.51)$$

brings (2.50) to the form

$$\mu = E[X] = \int_0^\infty e^{-\alpha x} dx = \left. \frac{-e^{-\alpha x}}{\alpha} \right|_0^\infty = -\lim_{x \to \infty} \frac{e^{-\alpha x}}{\alpha} + \frac{1}{\alpha} = \frac{1}{\alpha}. \quad (2.52)$$

Similarly,

$$\sigma^2 = \int_0^\infty \left(x - \frac{1}{\alpha} \right)^2 \alpha e^{-\alpha x} dx = \frac{1}{\alpha^2}, \qquad (2.53)$$

so that $\sigma = 1/\alpha = \mu$. Thus, for example, the exponential random variable with $\alpha = 2$ has $\mu = 0.5$ and $\sigma = 0.5$. $\square$

The sequence of *moments* of X defined by $E[X^k], k = 1, 2, 3, \ldots$, is sometimes of interest. The first moment coincides with the mean or expected value. From the definition, we see that if X is discrete, then for each $k = 1, 2, 3, \ldots$,

$$E[X^k] = \sum_i x_i^k p(x_i) = x_1^k p(x_1) + x_2^k p(x_2) + \cdots, \qquad (2.54)$$

while, if X is continuous,

$$E[X^k] = \int_{-\infty}^\infty x^k f(x) dx, \quad k = 1, 2, 3, \ldots. \qquad (2.55)$$

It can be shown that under very general conditions (see Feller [8, pages 227–228]), if all the moments for X exist, they uniquely determine the distribution of X; that is, if X and Y have the same sequence of moments, then $F_X = F_Y$. The following examples show that a random variable may not have *any* moments—not even a mean value.

Example 2.6.3 Let X assume the values $2, 2^2, 2^3, \ldots, 2^k, \ldots$ with the probability mass function $p(\cdot)$, defined by

$$p(x_k) = p(2^k) = \frac{1}{2^k}, \quad k = 1, 2, 3, \ldots. \qquad (2.56)$$

Then $p(\cdot)$ is a true probability mass function, since

$$\sum_{k=1}^{\infty} p(x_k) = \sum_{k=1}^{\infty} \frac{1}{2^k} = \frac{1}{2} \sum_{k=0}^{\infty} \frac{1}{2^k} = \frac{1}{2} \times \frac{1}{1-\frac{1}{2}} = \frac{1}{2} \times 2 = 1. \qquad (2.57)$$

However,

$$\sum_{i=1}^{\infty} x_i p(x_i) = 1 + 1 + 1 + \cdots \qquad (2.58)$$

diverges so that even the first moment fails to exist. $\square$

Example 2.6.4 Let X be the continuous random variable with density function f defined by

$$f(x) = \begin{cases} 0 & \text{for } x < 1 \\ \frac{1}{x^2} & \text{for } x \geq 1. \end{cases} \qquad (2.59)$$

Then

$$\int_{-\infty}^{\infty} f(x)dx = \int_{1}^{\infty} \frac{dx}{x^2} = -\frac{1}{x}\Big|_{1}^{\infty} = -\lim_{x \to \infty} \frac{1}{x} + 1 = 1, \qquad (2.60)$$

so f is a density function. However,

$$\int_{-\infty}^{\infty} xf(x)dx = \int_{1}^{\infty} \frac{dx}{x} = \ln x\Big|_{1}^{\infty} = +\infty, \qquad (2.61)$$

so that the first moment of X fails to exist. Clearly no higher order moments exist, either. $\square$

Although Examples 2.6.3 and 2.6.4 show that not all random variables have moments, most useful theoretical and empirically derived random variables do have moments. The *squared coefficient of variation* is a parameter widely (some even claim *wildly*) used by computer system modelers and queueing theory aficionados. It is defined by

$$C_X^2 = \frac{\text{Var}[X]}{E[X]^2}. \qquad (2.62)$$

It is used for measuring the degree of irregularity of a positive random variable X (positive means $P[X < 0] = 0$) compared to the exponential random variable for which $C_X^2 = 1$. The hypoexponential Erlang-k distribution described in Section 3.2.6 has a squared coefficient of variation $C_X^2 = 1/k$, while the hyperexponential distribution H_2 of Section 3.2.10 has

$C_X^2 \geq 1$. A gamma distribution, described in Section 3.2.5, can have C_X^2 values of any positive value; that is, we can have $0 \leq C_X^2 < 1$ or $1 \leq C_X^2$. This makes the gamma distribution very useful for modeling studies.

One must use care in reading papers in which the C^2 concept is used because some authors use the *coefficient of variation* rather than the *squared coefficient of variation*, although they do not make this clear. We prefer the squared form because $C_X^2 = 1/k$ for the Erlang-k distribution.

> *"Data! data! data!" he cried impatiently.*
> *"I can't make bricks without clay."*
> Sherlock Holmes

2.7   Jointly Distributed Random Variables

Sometimes it is of interest to investigate two or more random variables simultaneously. Thus, if X and Y are two random variables defined on the same sample space Ω, we define the *joint distribution function F of X and Y* for all real x and y by

$$F(x,y) = P[X \leq x, Y \leq y] = P[(X \leq x) \cap (Y \leq y)]. \qquad (2.63)$$

Sometimes we write $F_{X,Y}$ for the joint distribution function of X and Y to emphasize that it is a joint distribution. Given $F_{X,Y}$, the individual distribution functions F_X and F_Y can be computed as follows:

$$F_X(x) = \lim_{y \to \infty} F_{X,Y}(x,y) \quad \text{for each real } x, \qquad (2.64)$$

and

$$F_y(y) = \lim_{x \to \infty} F_{X,Y}(x,y) \quad \text{for each real } y. \qquad (2.65)$$

F_X and F_Y are called the *marginal distribution functions of X and Y*, respectively, corresponding to the joint distribution $F_{X,Y}$. Then the *joint probability mass function* $p(\cdot,\cdot)$ *of X and Y* is defined by

$$p(x,y) = P[X = x, Y = y]. \qquad (2.66)$$

Let $T = \{(x,y) : p(x,y) > 0\}$. Then, if $\sum_{(x,y) \in T} p(x,y) = 1$, we say that X and Y are *jointly discrete*. (It can be shown that T is either finite or countable.) If X and Y are jointly discrete, the probability mass functions p_X and p_Y of X and Y can be calculated as

$$p_X(x) = \sum_{\substack{y \text{ such that} \\ P(x,y) > 0}} p(x,y), \qquad (2.67)$$

$$p_Y(y) = \sum_{\substack{x \text{ such that} \\ P(x,y) > 0}} p(x,y). \qquad (2.68)$$

In this case p_X and p_Y are called *marginal probability mass functions*.

Example 2.7.1 Suppose a communication line is to be polled when it is known that two of the four terminals on the line are ready to transmit; using the notation of Example 2.3.2, Ω consists of six sample points $(1,1,0,0)$, $(1,0,1,0)$, $(1,0,0,1)$, $(0,1,1,0)$, $(0,1,0,1)$, $(0,0,1,1)$. Let X be the number of polls until the first ready terminal is found and Y the number until the second ready terminal is found. The joint probability mass function and the marginal probability mass functions are shown in Table 2.7.1. It shows that $p(1,2) = P[X = 1, Y = 2] = 1/6, p(2,2) = p(3,2) = p(3,3) = 0$, etc. Also $p_X(1) = p(1,2)+p(1,3)+p(1,4) = 1/2, p_X(2) = p(2,3)+p(2,4) = 1/3$, etc. (Obviously Y assumes only the values 2, 3, 4.)

Table 2.7.1. Joint Probability
Mass Function

X	Y			
	2	3	4	p_X
1	1/6	1/6	1/6	1/2
2	0	1/6	1/6	1/3
3	0	0	1/6	1/6
p_Y	1/6	1/3	1/2	

It is also true, for two jointly discrete random variables, that

$$F(x,y) = \sum_{x_i \leq x} \sum_{y_j \leq y} p(x_i, y_j). \qquad (2.69)$$

Thus, from Table 2.7.1, we see that

$$F(2,3) = p(1,2) + p(1,3) + p(2,2) + p(2,3) = 0.5, \qquad (2.70)$$

and

$$F(2,4) = F(2,3) + p(1,4) + p(2,4) = \frac{5}{6}. \ \square \qquad (2.71)$$

X and Y are *jointly continuous* if their joint distribution function F is continuous on the whole plane. We are interested only in jointly continuous random variables with a *joint density function* f such that, if A is a set of real numbers as is B, then

$$P[X \in A, Y \in B] = \int_B \int_A f(x,y) dx \, dy. \qquad (2.72)$$

In this case, for any real u and v,

$$F(u, v) = \int_{-\infty}^{v} \int_{-\infty}^{u} f(x, y) dx\, dy. \qquad (2.73)$$

If the random variables X and Y are jointly continuous, then each is a continuous random variable and for each real x and y,

$$f_X(x) = \int_{-\infty}^{\infty} f(x, y) dy. \qquad (2.74)$$

$$F_Y(y) = \int_{-\infty}^{\infty} f(x, y) dx. \qquad (2.75)$$

The equations (2.74) and (2.75) follow from (2.64), (2.65), and (2.73). For example,

$$
\begin{aligned}
F_X(u) &= \lim_{v \to \infty} F(u, v) = \lim_{v \to \infty} \int_{-\infty}^{v} \int_{-\infty}^{u} f(x, y) dx\, dy \\
&= \lim_{v \to \infty} \int_{-\infty}^{u} dx \int_{-\infty}^{v} f(x, y) dy = \int_{-\infty}^{u} dx \int_{-\infty}^{\infty} f(x, y) dy.
\end{aligned}
$$

Hence, $f_X(x) = \int_{-\infty}^{\infty} f(x, y) dy$, since

$$f_X(x) = \frac{dF_X}{dx}(x).$$

Example 2.7.2 Suppose the random variables X and Y have the joint density function

$$F(x, y) = \begin{cases} xy \exp[-\frac{1}{2}(x^2 + y^2)] & \text{for } x > 0 \text{ and } y > 0 \\ 0 & \text{otherwise.} \end{cases} \qquad (2.76)$$

Find $F_X(x)$, $f_Y(y)$, and $F(1, 1) = P[X \leq 1, Y \leq 1]$. This example is due to Parzen [page 291, 19].

Solution By (2.74),

$$
\begin{aligned}
f_X(x) &= \int_0^{\infty} f(x, y) dy = \int_0^{\infty} xy \exp[-\frac{1}{2}(x^2 + y^2)] dy \\
&= x \exp[-\frac{1}{2}x^2] \int_0^{\infty} y \exp[-\frac{1}{2}y^2] dy \\
&= x \exp[-\frac{1}{2}x^2] \left(-\exp[-\frac{1}{2}y^2] |_0^{\infty} \right) = x \exp[-\frac{1}{2}x^2].
\end{aligned}
$$

Similarly,

$$f_Y(y) = \int_0^\infty xy \exp[-\frac{1}{2}(x^2 + y^2)]dx = y \exp[-\frac{1}{2}y^2].$$

By (2.73),

$$
\begin{aligned}
F(1,1) &= \int_0^1 \int_0^1 f(x,y)dx\,dy = \int_0^1 x \exp[-\frac{1}{2}x^2]dx \int_0^1 y \exp[-\frac{1}{2}y^2]dy \\
&= (-\exp[-\frac{1}{2}x^2]|_0^1)(-\exp[-\frac{1}{2}y^2]|_0^1) \\
&= (1 - \exp[-0.5])(1 - \exp[-0.5]) \\
&= (1 - 0.606531)(1 - 0.606531) = 0.154818. \quad \square
\end{aligned}
$$

Suppose g is a function of two variables. Then, if X and Y are jointly distributed random variables, the mathematical expectation of $g(X,Y)$, $E[g(X,Y)]$, is defined as

$$E[g(X,Y)] = \sum_{\substack{\text{all } (x,y) \\ \text{such that} \\ p(x,y) > 0}} g(x,y)p(x,y), \tag{2.77}$$

if X and Y are jointly discrete and $p(\cdot,\cdot)$ is the joint probability mass function, or as

$$E[g(X,Y)] = \int_{-\infty}^\infty \int_{-\infty}^\infty g(x,y)f(x,y)dx\,dy, \tag{2.78}$$

if X and Y are jointly continuous with joint density function f.

Two random variables X and Y are said to be *independent* if any of the following relations hold:

(a) Their joint distribution function can be expressed as a product:

$$F(x,y) = F_X(x)F_Y(y) \quad \text{for all real } x \text{ and } y. \tag{2.79}$$

(b) They are jointly discrete and their joint probability mass function can be written,

$$p(x,y) = p_X(x)p_Y(y) \quad \text{for all real } x \text{ and } y. \tag{2.80}$$

(c) They are jointly continuous and their joint density function can be written,

$$f(x, y) = f_X(x)f_Y(y) \quad \text{for all real } x \text{ and } y. \qquad (2.81)$$

Two random variables X and Y that are not independent by one of the above criteria are said to be *dependent*.

The next theorem gives some of the properties of the expected value operator $E[\cdot]$.

Theorem 2.7.1 *Suppose X and Y are random variables; c is a constant; g and h are arbitrary measurable functions.*[14] *Then*

(a) $E[c] = c$. *(The expected value of a constant random variable is the constant.)*

(b) $E[cX] = cE[X]$.

(c) $E[X + Y] = E[X] + E[Y]$. *(X and Y need not be independent.)*

(d) $E[g(X)h(Y)] = E[g(X)]E[h(Y)]$ *if X and Y are independent and the expectations on the right exist.*

Proof (a) Suppose X is the constant random variable c; that is, $P[X = c] = 1$. X is discrete so, by definition, $E[X] = E[c] = cP[X = c] = c$.

(b) If X is discrete with pmf $p(\cdot)$,

$$E[cX] = \sum_{x_i} cX_i p(x_i) = c \sum_{x_i} x_i p(x_i) = cE[X].$$

If X is continuous,

$$E[cX] = \int_{-\infty}^{\infty} cxf(x)dx = c \int_{-\infty}^{\infty} xf(x)dx = cE[X].$$

(c) We give the proof for the case that X and Y are jointly continuous. If they are jointly discrete the proof is similar.

$$
\begin{aligned}
E[X + Y] &= \int_{-\infty}^{\infty} \int_{-\infty}^{\infty} (x + y)f(x, y)dx\,dy \\
&= \int_{-\infty}^{\infty} x\,dx \int_{-\infty}^{\infty} f(x, y)dy + \int_{-\infty}^{\infty} y\,dy \int_{-\infty}^{\infty} f(x, y)dx \\
&= \int_{-\infty}^{\infty} xf_X(x)dx + \int_{-\infty}^{\infty} yf_Y(y)dy = E[X] + E[Y].
\end{aligned}
$$

---

[14]See the definition of measurable function two paragraphs before Proposition 2.5.1.

The next to last equality follows from (2.74) and (2.75).

(d) We prove the theorem for the case that X and Y are jointly continuous. The jointly discrete case is similar.

$$
\begin{aligned}
E[g(X)h(Y)] &= \int_{-\infty}^{\infty} \int_{-\infty}^{\infty} g(x)h(y)f(x,y)dx\,dy \\
&= \int_{-\infty}^{\infty} \int_{-\infty}^{\infty} g(x)h(y)f_X(x)f_Y(y)dy\,dx \\
&= \int_{-\infty}^{\infty} g(x)f_X(x)dx \int_{-\infty}^{\infty} h(y)f_Y(y)dy \\
&= E[g(X)]E[h(Y)]. \quad\blacksquare
\end{aligned}
$$

The *covariance* of X and Y, written $\mathrm{Cov}[X,Y]$, is defined by

$$
\begin{aligned}
\mathrm{Cov}[X,Y] &= E[(X-E[X])(Y-E[Y])] \\
&= E[XY - XE[Y] - YE[X] + E[X]E[Y]] \\
&= E[XY] - E[X]E[Y] - E[Y]E[X] + E[X]E[Y] \\
&= E[XY] - E[X] - E[X]E[Y]. \quad (2.82)
\end{aligned}
$$

The *correlation (coefficient)* of X and Y, written $\rho(X,Y)$ is defined by

$$
\rho(X,Y) = \frac{\mathrm{Cov}(X,Y)}{(\mathrm{Var}[X]\mathrm{Var}[Y])^{1/2}}, \quad (2.83)
$$

provided both variances are nonzero. Rice [20, pages 125–127] shows that $|\rho(X,Y)| \leq 1$ with equality if and only if $P[Y = aX + b] = 1$ for some a and b.[15]

If $\mathrm{Cov}[X,Y] = 0$, then X and Y are said to be *uncorrelated*. Theorem 2.7.1(d) implies that any two independent random variables X and Y are uncorrelated. However, not all uncorrelated random variables are independent.[16]

The next theorem gives some useful properties of the variance operator $\mathrm{Var}[\cdot]$.

Theorem 2.7.2 *Suppose X and Y are random variables; c is a constant; and all the variances in the formulas below exist. Then*

(a) $\mathrm{Var}[c] = 0$.

(b) $\mathrm{Var}[cX] = c^2\mathrm{Var}[X]$.

---

[15]See, for example, Grimmett and Stirzaker [13, page 42].
[16]See Exercise 43.

(c) $\text{Var}[X + Y] = \text{Var}[X] + \text{Var}[Y] + 2\text{Cov}[X, Y]$.

(d) $\text{Var}[X] = E[X^2] - (E[X])^2$.

Proof

(a) If c is a constant random variable, then by Theorem 2.7.1 (a) $E[c] = c$. Hence, $\text{Var}[c] = E[(c - c)^2] = 0$.

(b) $\text{Var}[cX] = E[cX])^2] = E[(c(X - E[X]))^2] = c^2 E[(X - E[X])^2]$

$\qquad = c^2 \text{Var}[X]$.

(c) $\begin{aligned}
\text{Var}[X + Y] &= E[\{(X + Y) - (E[X] + E[Y])\}^2] \\
&= E[\{(X - E[X]) + (Y - E[Y])\}^2] \\
&= E[(X - E[X])^2 + (Y - E[Y])^2 + 2(X - E[X]) \\
&\quad \times (Y - E[Y]) \\
&= E[(X - E[X])^2] + E[(Y - E[Y])^2 + 2(X - E[X]) \\
&\quad \times (Y - E[Y]) \\
&= \text{Var}[X] + \text{Var}[Y] + 2\text{Cov}[X, Y].
\end{aligned}$

(d) $\begin{aligned}
\text{Var}[X] &= E[(x - E[X])^2] = E[X^2 - 2xE[X] + (E[X])^2] \\
&= E[X^2] - 2E[X]E[X] + (E[X])^2 = E[X^2] - (E[X])^2.
\end{aligned}$

We have used Theorem 2.7.1 freely in the above equalities. ∎

It should be noted that if X and Y are independent and thus uncorrelated, then

$$\text{Var}[X + Y] = \text{Var}[X] + \text{Var}[Y]. \qquad (2.84)$$

An application of mathematical induction shows that

$$\text{Var}[X_1 + X_2 + \cdots + X_n] = \text{Var}[X_1] + \text{Var}[X_2] + \cdots + \text{Var}[X_n], \qquad (2.85)$$

for any finite collection of mutually independent random variables.

Although we defined the joint distribution function F for only two random variables in (2.63), the concept can be extended, in a natural way, to any finite number of random variables. Thus, if $X_1, X_2, X_3, \ldots, X_n$ are random variables, their *joint distribution function* F (or $F_{X_1, X_2, \ldots, X_n}$ if it is desired to make names of the random variables explicit) is defined by

$$\begin{aligned}
F(x_1, x_2, \ldots, x_n) &= P[X_1 \le x_1, X_2, \le x_2, \cdots, X_n \le x_n], \\
&\quad \text{for all real } x_1, x_2, \ldots, x_n.
\end{aligned} \qquad (2.86)$$

All the other concepts we have discussed are then defined just as they were for two random variables. Thus, for example, one condition that means the random variables $X_1, X_2, \ldots, X_n$ are independent is that

$$F(x_1, x_2, \ldots, x_n) = F_{X_1}(x_1) F_{X_2}(x_2) \cdots F_{X_n}(x_n)$$
$$\text{for all real } x_1, x_2, \ldots, x_n. \tag{2.87}$$

Some particular functions of several random variables are important for applications. The properties of one such function are given in the next theorem.

Theorem 2.7.3 *Let $X_1, X_2, \ldots, X_n$ be n independent random variables with distribution functions $F_{X_1}, F_{X_2}, \cdots, F_{X_n}$. Let $Y = g(X_1, \cdots, X_n)$ be the random variable defined by*

$$Y(\omega) = \max\{X_1(\omega), X_2(\omega), \cdots, X_n(\omega)\} \quad \text{for each } \omega \in \Omega. \tag{2.88}$$

Then the distribution function F_Y is given by

$$F_Y(y) = F_{X_1}(y) F_{X_2}(y) \cdots F_{X_n}(y) \quad \text{for each real } y. \tag{2.89}$$

Proof From the definition of Y, we know that $Y \leq y$ if and only if $X_1 \leq y$, $X_2 \leq y$, ..., $X_n \leq y$. Hence,

$$
\begin{aligned}
F_Y(y) &= P[Y \leq y] = P[X_1 \leq y, \ldots, X_n \leq y] \\
&= P[X_1 \leq y] P[X_2 \leq y] \cdots P[X_n \leq y] \\
&= F_{X_1}(y) F_{X_2}(y) \cdots F_{X_n}(y). \tag{2.90}
\end{aligned}
$$

In the next-to-last equality in (2.90) we used the independence of the random variables. ■

Example 2.7.3 An on-line airline reservation system uses two identical duplexed computer systems, each of which has an exponential time to failure with a mean of 2000 hours. Each computer system has built-in redundancy so failures are rare. The system fails only if both computers fail. What is the probability that the system will not fail during one week (168 hours) of continuous operation? 30 days?

Solution Theorem 2.7.3 applies. The distribution function of the time to failure X in one system is

$$F(t) = P[X \leq t] = 1 - e^{-t/2000}, \quad t \text{ in hours.} \tag{2.91}$$

$F(t)$ in (2.91) is the probability that a failure will occur *before* time t in one of the systems. Hence, the probability of a system failure (both computer systems down) within a week is, by Theorem 2.7.3,

$$(1 - e^{-168/2000})^2 = (0.080569)^2 = 0.006491.$$

Thus, the probability of no system failure for at least a week is

$$1 - 0.006491 = 0.993509.$$

The corresponding probability for 30 days is

$$1 - (1 - e^{-720/2000})^2 = 1 - (0.302324)^2 = 1 - 0.0914 = 0.9086.$$

If the system was not duplexed, that is, consisted of only one computer system, the probability of no failure within 30 days is 0.69768. Thus, if it is desired that the probability of failure-free operation for at least a week is to exceed 0.95, a duplex system is required. □

Let us consider the general case of two identical computer systems each of which has an exponential time to failure with a mean of $1/\lambda$ time units. By Theorem 2.7.3, the distribution function for Y, the joint time to failure, is given by

$$F(t) = (1 - e^{-\lambda t})^2.$$

Hence, the density function of Y is

$$2\lambda(1 - e^{-\lambda t})e^{-\lambda t}.$$

Therefore, the mean time to failure, often abbreviated MTTF, is

$$E[Y] = \int_0^\infty t\, f(t)\, dt = 2\lambda \int_0^\infty (1 - e^{-\lambda t})e^{-\lambda t} t\, dt.$$

If we make the substitution $u = \lambda t$, or $t = u/\lambda$, in this integral so that $dt = du/\lambda$, we obtain

$$
\begin{aligned}
E[Y] &= \frac{2}{\lambda} \int_0^\infty (1 - e^{-u})e^{-u} u\, du \\
&= \left(\frac{2}{\lambda}\right)\left(\int_0^\infty u e^{-u}\, du - \int_0^\infty u e^{-2u}\, du\right) = \frac{3}{2\lambda},
\end{aligned}
$$

by Formula 3.351.3 of Gradshteyn and Ryzhik [10], which claims that

$$\int_0^\infty x^n e^{-\mu x}\, dx = n!\, \mu^{-n-1}.$$

In this example

$$\frac{1}{\lambda} = 2000 \text{ hours},$$

so

$$E[Y] = 3000 \text{ hours}.$$

Example 2.7.3 is not realistic in the sense that we did not account for the fact that a failed computer normally would be repaired. This enhanced realism is considered in Example 4.3.3.

The next theorem is similar to Theorem 2.7.3 but describes the distribution of the minimum of several random variables.

Theorem 2.7.4 *Let $X_1, X_2, \ldots, X_n$ be independent random variables. Let $Y = g(X_1, \ldots X_n)$ be the random variable defined by*

$$Y(\omega) = \min\{X_1(\omega), X_2(\omega), \ldots, X_n(\omega)\} \quad \text{for each } \omega \in \Omega. \tag{2.92}$$

Then the distribution function F_Y is given by

$$
\begin{aligned}
F_Y(y) &= 1 - (1 - F_{X_1}(y))(1 - F_{x_2}(y)) \cdots (1 - F_{X_n}(y)) \\
&\text{for each real } y.
\end{aligned} \tag{2.93}
$$

Proof For each real y, $Y > y$ if and only if $X_1 > y$, $X_2 > y$, $\ldots$, $X_n > y$. Hence,

$$
\begin{aligned}
P[Y > y] &= P[X_1 > y, X_2 > y, \ldots, X_n > y] \\
&= P[X_1 > y]P[X_2 > y] \cdots P[X_n > y] \\
&= (1 - F_{X_1}(y))(1 - F_{X_2}(y)) \cdots (1 - F_{X_n}(y)).
\end{aligned}
$$

Therefore,

$$F_Y(y) = 1 - P[Y > y] = 1 - (1 - F_{X_1}(y))(1 - F_{X_2}(y)) \cdots (1 - F_{X_n}(y)).$$

∎

Example 2.7.4 A computer system consists of n subsystems, each of which has the same exponential distribution of time to failure. Each subsystem is independent but the whole computer system fails if any of the subsystems do. Find the distribution function F for system time to failure. If the mean time to failure of each subsystem is 2000 hours, and there are four subsystems, find the mean time to system failure and the probability that the time to failure exceeds 100 hours.

Solution By Theorem 2.7.4, if there were n subsystems, then

$$F(t) = 1 - (e^{-\mu t})^n = 1 - e^{-n\mu t}, \tag{2.94}$$

where $1/\mu$ is the average time to failure, since the distribution function for time to failure is

$$F_X(t) = 1 - e^{-\mu t} \tag{2.95}$$

for each subsystem. Thus, the system time to failure has an exponential distribution with mean value $1/(n\mu) = (1/\mu)/n$. If $1/\mu = 2000$ and $n = 4$, then the system mean time to failure is $2000/4 = 500$ hours. Hence, the distribution function for system time to failure is

$$F(t) = 1 - e^{-t/500}, \tag{2.96}$$

and thus the probability that it exceeds 100 hours is $e^{-100/500} = e^{-0.2} = 0.8187.$ □

The individual subsystems need not have the same mean time to failure in order for the overall system time to failure to have an exponential distribution. If each subsystem has an exponentially distributed time to failure with mean values $1/\mu_i, i = 1, 2, \ldots, n$, then the distribution function F for the system time to failure is given by

$$F(t) = 1 - \exp\{-t(\mu_1 + \mu_2 + \cdots + \mu_n)]; \tag{2.97}$$

that is, the time to failure has an exponential distribution with mean value

$$\frac{1}{\mu_1 + \mu_2 + \cdots + \mu_n}.$$

In the above example with $n = 4$, if the mean time to failure has been 1000 hours, 2000 hours, 3000 hours, 4000 hours, respectively, then the average time to failure would be

$$\frac{1}{\frac{1}{1000} + \frac{1}{2000} + \frac{1}{3000} + \frac{1}{4000}} = \frac{1}{\frac{(12+6+4+3)}{12000}}$$

$$= \frac{12000}{25} = 480 \text{ hours}.$$

By (2.97) the probability that the system time to failure exceeds 100 hours is

$$e^{-100/480} = 0.81194.$$

In the next theorem the convolution method is given for calculating the pmf $p(\cdot)$ or the density function $f(\cdot)$ for the sum of two independent random variables. Later we will see that transform methods make it easier to find $p(\cdot)$ or $f(\cdot)$ than the method of Theorem 2.7.5.

Theorem 2.7.5 (*Convolution Theorem*) *Let X and Y be jointly distributed random variables with $Z = X + Y$. Then the following hold.*

(a) *If X and Y are independent discrete random variables, each taking on the values $0, 1, 2, 3, 4, \ldots$, then Z takes on the values $k = i + j$ $(i, j = 0, 1, 2, 3, \ldots)$ and*

$$P[Z = k] = \sum_{i+j=k} p_X(i)p_Y(j) = \sum_{i=0}^{k} p_X(i)p_Y(k - i). \qquad (2.98)$$

(b) *If X and Y are independent continuous random variables,*

$$P[Z \le z] = \int_{-\infty}^{\infty} f_X(x)F_Y(z - x)dx = \int_{-\infty}^{z} f_Z(x)dx \qquad (2.99)$$

where the density function of Z is given by

$$f_Z(z) = \int_{-\infty}^{\infty} f_X(x)f_Y(z - x)dx = \int_{-\infty}^{\infty} f_X(z - y)f_Y(y)dy. \quad (2.100)$$

Proof (a) The event $Z = k$ can be represented as

$$
\begin{aligned}
[Z = k] &= [(X = 0) \cap (Y = k)] \cup [(X = 1) \cap (Y = k - 1)] \\
&\quad \cup [(X = 2) \cap (Y = k - 2)] \cup \cdots \cup [(X = k) \cap (Y = 0)].
\end{aligned}
$$

Hence,

$$P[Z = k] = p(0, k) + p(1, k - 1) + \cdots + p(k, 0) = \sum_{i=0}^{k} p(i, k - i), \quad (2.101)$$

where $p(\cdot, \cdot)$ is the joint density function of X and Y. But X and Y are independent so (2.101) yields (2.98), because $p(i, j) = p_X(i)p_Y(j)$ for each i and j.

(b) The equation

$$
\begin{aligned}
P[Z \le z] &= \iint_{x+y \le z} f_{X,Y}(x, y)dx\, dy \\
&= \int_{-\infty}^{\infty} dx \int_{-\infty}^{z-x} f_{X,Y}(x, y)dy \qquad (2.102)
\end{aligned}
$$

is valid for all jointly continuous random variables, independent or not. If we assume X and Y are independent, so that $f_{X,Y}(x, y) = f_X(x)f_Y(y)$, then (2.102) yields

$$
\begin{aligned}
P[Z \le z] &= \int_{-\infty}^{\infty} dx \int_{-\infty}^{z-x} f_X(x) f_Y(y) dy \\
&= \int_{-\infty}^{\infty} f_X(x) dx \int_{-\infty}^{z-x} f_Y(y) dy \\
&= \int_{-\infty}^{\infty} f_X(x) F_Y(z - x) dx. \qquad (2.103)
\end{aligned}
$$

Differentiation of (2.103) gives (2.100). ■

The sum $\sum_{i=0}^{k} p_X(i) p_Y(k - i)$ is called the *convolution* of p_X and p_Y and designated $p_X * p_Y(k)$. Thus, Theorem 2.7.5 asserts that the pmf of the sum of two independent discrete random variables, each of which assumes only nonnegative integer values, is the convolution of the individual pmf's. Similarly, the integral $\int_{-\infty}^{\infty} f_X(x) f_Y(z-x) dx$ is called the *convolution* of f_X and f_Y. By symmetry it can also be calculated as $\int_{-\infty}^{\infty} f_X(z - y) f_Y(y) dy$. Theorem 2.7.5 shows that the density of the sum of two independent continuous random variables is the convolution of the individual densities.

Example 2.7.5 Let X and Y be independent random variables, each having an exponential distribution with parameter α. Find the density function of $Z = X + Y$.

Solution The density function is the same for each random variable:

$$
F_X(x) = f_Y(x) = \begin{cases} \alpha e^{-\alpha x} & \text{for } x > 0. \\ 0 & \text{for } x \le 0. \end{cases}
$$

Thus, $f_{X+Y}(z) = 0$ for $z \le 0$ and by (2.100), for $z > 0$,

$$
f_{X+Y}(z) = \int_{0}^{z} f_X(x) f_Y(z - x) dx.
$$

(We have used the fact that $f_Y(z) = 0$ for $z < 0$.) Hence,

$$
f_{X+Y}(z) = \int_{0}^{z} \alpha e^{-\alpha x} \alpha e^{-\alpha(z-x)} dx = \alpha^2 e^{-\alpha z} \int_{0}^{z} dx = \alpha^2 z e^{-\alpha z}.
$$

Thus, $Z = X + Y$ has what is called an Erlang-2 distribution with parameter $\alpha/2$. We will study the Erlang family of random variables in Chapter 3. □

Example 2.7.6 If $\alpha > 0$, the discrete random variable that assumes the values $0, 1, 2, 3, \ldots$ and that has the pmf $p(\cdot)$ defined by

$$p(k) = e^{-\alpha} \frac{\alpha^k}{k!}, \quad k = 0, 1, 2, \ldots, \tag{2.104}$$

is called a *Poisson random variable with parameter* α or is said to have a *Poisson distribution*. We will investigate Poisson random variables in more detail in Chapter 3. Assuming that (2.104) does define a pmf, let us calculate the pmf $p(\cdot)$ for the sum of two independent Poisson distributed random variables, one with parameter α and one with parameter β. If X is the first random variable and Y the second, then by (2.98) of and (2.104),

$$
\begin{aligned}
p(k) &= \sum_{i=0}^{k} p_X(i) p_Y(k-i) \\
&= \sum_{i=0}^{k} e^{-\alpha} \frac{\alpha^i}{i!} e^{-\beta} \frac{\beta^{k-i}}{(k-i)!} \\
&= \frac{e^{-(\alpha+\beta)}}{k!} \sum_{i=0}^{k} \frac{k!}{i!(k-i)!} \alpha^i \beta^{k-i} \\
&= \frac{e^{-(\alpha+\beta)}}{k!} \sum_{i=0}^{k} \binom{k}{i} \alpha^i \beta^{k-i} \\
&= \frac{e^{-(\alpha+\beta)}}{k!} (\alpha+\beta)^k.
\end{aligned}
$$

The last equality is true by the corollary to Theorem 2.3.3. Thus, the sum of two independent Poisson random variables is another Poisson random variable whose parameter is the sum of the original parameters; that is, $X + Y$ is Poisson with parameter $\alpha + \beta$. For this reason Poisson random variables are said to have the *reproductive property*. Exponential random variables do *not* have this property, as we saw in Example 2.7.5. Consider Theorem 2.7.3. Gravey [12] considered the special case of Theorem 2.7.3 in which all the random variables have the same distribution. In several special cases he found exact values for $E[Y]$. We state his results for these special cases in the following proposition. $\square$

Proposition 2.7.1 *Suppose $X_1, X_1, \cdots, X_n$ are independent identically distributed random variables. We consider two special cases.*

(a) Suppose each X_i has a geometric distribution with probability p of success on each trial (and thus probability $q = 1 - p$ of failure). Then,

if Y is defined by

$$Y = \max\{X_1(\omega), X_2(\omega), \cdots, X_n(\omega)\} \quad \text{for each } \omega \in \Omega, \qquad (2.105)$$

we have

$$E[Y] = \sum_{k=1}^{n} \binom{n}{k} (-1)^{k+1} \frac{q^k}{1 - q^k}. \qquad (2.106)$$

(b) Suppose each X_i has an exponential distribution with mean $E[X] = 1/\alpha$. Then, if Y is defined by (2.105), we have

$$P[Y > x] = \sum_{k=1}^{n} \binom{n}{k} (-1)^{k+1} \exp(-\alpha k x), \qquad (2.107)$$

for each $x \geq 0$ and

$$E[Y] = \frac{1}{\alpha}(1 + \frac{1}{2} + \cdots + \frac{1}{n}). \qquad (2.108)$$

Proof See Gravey [12]. ∎

> *"What's one and one and one and one and one and one and one and one and one and one?"*
> *"I don't know," said Alice. "I lost count."*
> *"She can't do Addition," the Red Queen interrupted.*
> Lewis Carroll
> *Through the Looking Glass*

2.8   Conditional Expectation

In Example 2.6.1 we assumed that a communication line had five terminals attached, three of which were ready to transmit, and we calculated the mean and standard deviation of X, the number of polls required to find the first ready terminal.[17] X depends upon the number of terminals in the ready state. Thus, if we let Y be the random variable giving the number of ready terminals, we are interested in the average value of X, given that Y assumes one of the values 0, 1, 2, 3, 4, or 5. This is the *conditional expectation of X given Y*, which we now define formally.

Suppose X and Y are discrete random variables assuming the values $x_1, x_2, \ldots$, and $y_1, y_2 \ldots$, respectively. Then for each y_j such that $p_Y(y_j) >$

---

[17]See also Example 2.3.2.

0, we define the *conditional probability mass function of X given that Y = y_j* by

$$p_{X|Y}(x_i|y_j) = \frac{p(x_i, y_j)}{p_Y(y_j)}, \quad i = 1, 2, 3, \ldots, \quad (2.109)$$

where $p(\cdot, \cdot)$ is the joint probability mass function of X and Y. We then define the *conditional expectation of X given that Y = y_j*, for all y_j such that $p_Y(y_j) > 0$, by

$$E[X|Y = y_j] = \sum_{x_i} x_i p_{X|Y}(x_i|y_j). \quad (2.110)$$

Similarly, we define the *conditional kth moment of X given that Y = y_j* by

$$E[X^k|Y = y_j] = \sum_{x_i} x_i^k p_{X|Y}(x_i|y_j), \quad k = 1, 2, \ldots. \quad (2.111)$$

The motivation for (2.109) is that

$$P[X = x_i|Y = y_j] = \frac{P[X = x_i, Y = y_j]}{P[Y = y_j]} = \frac{p(x_i, y_j)}{p_Y(y_j)}.$$

In reality the pmf $p(\cdot)$ we defined in Example 2.6.1 is $p_{X|Y}(\cdot|3)$, and we calculated the expected value of X given that $Y = 3$. We provide the other conditional expectation values for Example 2.6.1 in Table 2.8.1. We use (2.47) which is a general formula for the case of m terminals on a line with n ready to transmit where $1 \leq n \leq m$. Equation (2.47) is *not* valid when $n = 0$, that is, when no terminals are ready to transmit; we assume that it takes m polls to discover this fact, so that

$$E[X|Y = 0] = m.$$

The values shown in Table 2.8.1 for $y \geq 1$ can easily be calculated using the APL function POLL in Appendix B. A formula of Exercise 36, discovered by Russell Ham, can also be used to make the calculations.

Table 2.8.1. Conditional expectation of number of polls for Example 2.6.1.

y:	0	1	2	3	4	5
$E[X|Y = y]$:	5.0	3.0	2.0	1.5	1.2	1.0

Suppose X and Y are jointly continuous with the joint density function f. Then the *conditional probability density function of X given that Y = y*, is defined for all values of y such that $f_Y(y) > 0$, by

$$f_{X|Y}(x|y) = \frac{f(x, y)}{f_Y(y)}. \quad (2.112)$$

The *conditional expectation of X, given that $Y = y$*, is defined for all values of y such that $f_Y(y) > 0$, by

$$E[X|Y = y] = \int_{-\infty}^{\infty} x f_{X|Y}(x|y)dx. \qquad (2.113)$$

The *conditional kth moment of X, given that $Y = y$*, is defined for all values of y such that $f_Y(y) > 0$, by

$$E[X^k|Y = y] = \int_{-\infty}^{\infty} x^k f_{X|Y}(x|y)dx, \quad k = 1, 2, 3, \ldots . \qquad (2.114)$$

Thus, the first conditional moment is the conditional expectation.

Example 2.8.1 The jointly continuous random variables X and Y of Example 2.7.2 have the joint density function

$$f(x, y) = \begin{cases} xy \exp[-\frac{1}{2}(x^2 + y^2)] & \text{for } x > 0 \text{ and } y > 0 \\ 0 & \text{otherwise,} \end{cases} \qquad (2.115)$$

and

$$f_Y(y) = \begin{cases} y \exp[-\frac{1}{2}y^2] & \text{for } y > 0 \\ 0 & \text{otherwise.} \end{cases} \qquad (2.116)$$

Hence, if $y > 0$,

$$f_{X|Y}(x|y) = \frac{f(x, y)}{f_Y(y)} = x \exp[-\frac{1}{2}x^2] = f_X(x) \quad \text{for all } y > 0. \qquad (2.117)$$

Thus, $f_{X|Y}(x, y)$ is independent of the particular value of y. This is to be expected since X and Y are independent. (They *are* independent, since $f(x, y) = f_X(x)f_Y(y)$.) Hence, for each $y > 0$,

$$E[X|Y = y] = E[X] = \int_0^{\infty} x^2 \exp[-\frac{1}{2}x^2]dx.$$

This integral is difficult to evaluate but its value is $\sqrt{2\pi}/2$. □

Given jointly distributed random variables X and Y, $E[X|Y = y]$ is a function of the random variable Y, let us say $h(Y)$. Thus, $h(Y)$ is a random variable having an expected value $E[h(Y)]$. If Y is discrete and assumes the values $y_1, y_2, \ldots$, then

$$E[h(Y)] = E[E[X|Y = y]] = \sum_{y_j} E[X|Y = y_j]P[Y = y_j], \qquad (2.118)$$

while if Y is continuous with density function f_Y, then

$$E[h(Y)] = E[E[X|Y = y]] = \int_{-\infty}^{\infty} E[X|Y = y]f_Y(y)dy. \qquad (2.119)$$

Equations (2.118) and (2.119) can be formally represented by the equation

$$E[h(Y)] = E[E[X|Y = y]] = \int_{-\infty}^{\infty} E[X|Y = y]dF_Y(y), \qquad (2.120)$$

where the integral in question is a *Stieltjes* integral, which of course is calculated by (2.118), when Y is discrete, and by (2.119) when Y is continuous. The Stieltjes integral can also be used to evaluate $E[h(Y)]$ when Y is neither discrete nor continuous, but this is beyond the scope of the book. The interested reader can consult Parzen [19, pages 233–235] or Apostol [2, chapter 7].

The next theorem shows how to evaluate $E[X], E[X^k]$ in terms of $E[X|Y = y]$ and $E[X^k|Y = y]$.

Theorem 2.8.1 *Let X and Y be jointly distributed random variables. Then*

$$E[E[X|Y = y]] = \int_{-\infty}^{\infty} E[X|Y = y]dF_Y(y) = E[X] \qquad (2.121)$$

and

$$E[E[X^k|y = y]] = \int_{-\infty}^{\infty} E[X^k|Y = y]dF_Y(y) = E[X^k], \ k = 1, 2, 3, \ldots. \qquad (2.122)$$

(Equation (2.121) is known as the *law of total expectation* and (2.122) as the *law of total moments*.)

Proof We prove (2.122) for the cases

(a) X and Y are discrete, and

(b) X and Y are continuous.

Equation (2.121) is a special case of (2.122). We omit the proof for the cases when one of X or Y is continuous and the other is discrete.

Case (a) Suppose that X and Y are discrete. Then

$$
\begin{aligned}
E[E[X^k|Y=y]] &= \sum_{y_j} E[X^k|Y=y_j]p_Y(y_j) \\
&= \sum_{y_j}\sum_{x_i} x_i^k P[X=x_i|Y=y_j]p_Y(y_j) \\
&= \sum_{y_j}\sum_{x_i} x_i^k \frac{p(x_i,y_j)}{p_Y(y_j)}p_Y(y_j) \\
&= \sum_{y_j}\sum_{x_i} x_i^k p(x_i,y_j) \\
&= \sum_{x_i} x_i^k \sum_{y_j} p(x_i,y_j) \\
&= \sum_{x_i} x_i^k p_X(x_i) \\
&= E[X^k].
\end{aligned}
$$

This proves that $E[X^k] = E[E[X^k|Y=y]]$ when X and Y are both discrete.

Case (b) Suppose that X and Y are continuous. Then

$$
\begin{aligned}
E[E[X^k|Y=y]] &= \int_{-\infty}^{\infty} E[X^k|Y=y]f_Y(y)dy \\
&= \int_{-\infty}^{\infty}\left[\int_{-\infty}^{\infty} x^k f_{X|Y}(x|y)dx\right]f_Y(y)dy \\
&= \int_{-\infty}^{\infty}\int_{-\infty}^{\infty} x^k \frac{f(x,y)}{f_Y(y)}f_Y(y)dx\,dy \\
&= \int_{-\infty}^{\infty}\int_{-\infty}^{\infty} x^k f(x,y)dx\,dy \\
&= \int_{-\infty}^{\infty} x^k \int_{-\infty}^{\infty} f(x,y)dy\,dx \\
&= \int_{-\infty}^{\infty} x^k f_X(x)dx \\
&= E[X^k]. \quad \blacksquare
\end{aligned}
$$

Example 2.8.2 Consider Example 2.6.1, in which five terminals were connected to one communication line. Let the value of X be the number of polls until the first ready terminal is found (when there is a ready terminal) and 5, otherwise; let Y be the number of terminals ready. We can find $E[X]$ by Theorem 2.8.1 if we know the pmf for Y. Let us assume that $p_Y(\cdot)$ is as shown in Table 2.8.2. This pmf was calculated assuming that each terminal was independent of the others and had probability 0.5 of being ready to transmit. Thus, Y has a *binomial distribution,* which is discussed in Example 2.9.5 and in Chapter 3. Using the data in the table, we see that

$$E[X] = \frac{(5 \times 1 + 3 \times 5 + 2 \times 10 + 1.5 \times 10 + 1.2 \times 5 + 1)}{32} = 1.9375$$

and

$$E[X^2] = \frac{(25 + 11 \times 5 + 5 \times 10 + 2.7 \times 10 + 1.6 \times 5 + 1)}{32} = 5.1875.$$

The APL function MPOLL calculates the expected number of polls and second moment of number of polls, as well as $E[X|Y = n]$ and $E[X^2|Y = n]$ for $n = 0, 1, \ldots, m$ given the pmf of Y, using the APL functions POLLM and POLL2M. (The pmf of X is given by (2.47) which the reader is asked to prove in Exercise 36.) The formulas of Exercise 36 can be used to calculate the last two columns of Table 2.8.2. $\square$

Table 2.8.2. Data for Example 2.8.2

n	$p_Y(n)$	$E[X|Y = n]$	$E[X^2|Y = n]$
0	1/32	5.0	25
1	5/32	3.0	11
2	10/32	2.0	5.0
3	10/32	1.5	2.7
4	5/32	1.2	1.6
5	1/32	1.0	1.0

Example 2.8.3 An on-line computer system receives inquiry messages of n different types. The message length distribution for each type i is a random variable X_i, $i = 1, 2, \ldots, n$. The message type of the current message is given by the discrete random variable Y, which assumes the values 1 through n. The message length X of the current message is determined by X_i if and only if $Y = i$. Thus, $E[X|Y = i] = E[X_i]$, and $E[X^2|Y = i] = E[X_i^2]$. Therefore, by Theorem 2.8.1,

$$E[X] = \sum_{i=1}^{n} E[X|Y = i]p_Y(i) = \sum_{i=1}^{n} E[X_i]p_Y(i), \tag{2.123}$$

and

$$E[X^2] = \sum_{i=1}^{n} E[X^2|Y = i]p_Y(i) = \sum_{i=1}^{n} E[X_i^2]p_Y(i). \qquad (2.124)$$

From (2.123) and (2.124) we can calculate

$$\text{Var}[X] = E[X^2] - (E[X])^2.$$

As an example, suppose 10 different types of message arrive at the central computer system. The fraction of each type, as well as the mean and standard deviation of the message length (in characters) of each type, are shown in Table 2.8.3. Find the mean and the standard deviation of the message length for all messages that arrive at the central computer system.

Table 2.8.3. Message length data for Example 2.8.3.

Message type	$p_Y(i)$ (Fraction of type)	$E[X_i]$ (Mean length)	σ_{X_i} (Standard deviation)
1	0.100	100	10
2	0.050	120	12
3	0.200	200	20
4	0.050	75	5
5	0.025	300	25
6	0.075	160	40
7	0.150	360	36
8	0.050	50	4
9	0.150	60	3
10	0.150	130	10

Solution Theorem 2.8.1 applies as outlined above. The expected value $E[X]$ is calculated by (2.123) to yield 164.25 characters. To apply (2.124), we calculate $E[X_i^2]$ by the formula $E[X_i^2] = E[X_i]^2 + Var[X_i]$ for $i = 1, 2, \ldots, 10$ to obtain the respective values 10,100; 14,544; 40,400; 5,650; 90,625; 27,200; 130,896; 2,516; 3,609; and 17,000.

Then by (2.124), we calculate

$$E[X^2] = \sum_{i=1}^{10} E[X_i^2]p_Y(i) = 37,256.875.$$

Thus,

$$\text{Var}[X] = E[X^2] - E[X]^2 = 37,256.875 - 26,978.0625 = 10,278.8125,$$

so that $\sigma_X = 101.38$ characters. The reader might be tempted to conclude that

$$\text{Var}[X] = \text{Var}[X_1]p_Y(1) + \text{Var}[X_2]p_Y(2) + \cdots + \text{Var}[X_{10}]p_Y(10),$$

but this formula is incorrect. In the present case it could yield $\text{Var}[X] = 445.625$ or $\sigma_X = 21.11$, although the correct value of σ_X is 101.38. The APL function CONDEXPECT can be used to make calculations such as those in this example. $\square$

> *I've been rich and I've been poor; rich is better.*
> Sophie Tucker

summation convention *n. A mathematicians' shindig held each year in the Kronecker Delta.*
Stan Kelly-Bootle
The Devil's DP Dictionary

2.9 Transform Methods

Calculating the mean, the variance, and the moments of a random variable can be a tedious process. In the case of a discrete random variable, it is often true that complicated sums must be evaluated; for a continuous random variable, the integrals involved may be difficult to evaluate. These difficulties can often be overcome by transform methods.

One of the first transform methods used successfully was the logarithm. Using the identity $\log(A \times B) = \log(A) + \log(B)$ converted the problem of multiplying two large numbers A and B into the simpler problem of adding the two numbers $a = \log(A)$ and $b = \log(B)$. To complete the operation, of course, it was necessary to find the *inverse transform or antilogarithm* to obtain the value of $A \times B$. In a similar way, by using some transformations to be described below, we can transform a random variable into a transformed function with a different domain in which it is easier to perform such operations as taking the convolution of two or more random variables (see Theorem 2.7.5) or finding moments. Of course, after the operation is complete, we must be able to make the inverse transform. In many cases this can be done by inspection. In other cases it may be more complex. We will discuss this further in Section 3.4 of Chapter 3.

The *moment generating function* $\psi[\cdot]$ (or $\psi_X[\cdot]$) of a random variable X is defined by $\psi[\theta] = E[e^{\theta X}]$ for all real θ such that $E[e^{\theta X}]$ is finite. Thus,

$$\psi[\theta] = \begin{cases} \displaystyle\sum_{x_i} e^{\theta x_i} p(x_i) & \text{if } X \text{ is discrete} \\[2em] \displaystyle\int_{-\infty}^{\infty} e^{\theta X} f(x) dx & \text{if } X \text{ is continuous.} \end{cases} \qquad (2.125)$$

We note that the moment generating function is always defined for $\theta = 0$ and that $\psi[0] = 1$.

We say that X *has a moment generating function* if there exists a $\delta > 0$ such that $\psi[\theta]$ is finite for all $|\theta| \leq \delta$. There are random variables without moment generating functions, such as the random variable of Example 2.6.3 and that of Example 2.6.4. However, most random variables of concern to us have moment generating functions. A random variable X has a moment generating function if and only if all the moments of X exist (are finite). In defining the moment generating function $\psi_X[\cdot]$, we have *transformed* the random variable X, which is defined on a sample space, into the function $\psi_X[\cdot]$ defined for some set of real numbers. The next theorem gives some important properties of the moment generating function.

Theorem 2.9.1 (*Properties of the Moment Generating Function*) *Let X and Y be random variables for which the moment generating functions $\psi_X[\cdot]$ and $\psi_Y[\cdot]$ exist. Then the following hold:*

(a) $F_X = F_Y$ *if and only if* $\psi_X[\cdot] = \psi_Y[\cdot]$ *(uniqueness).*

(b) $E[X^n]$ *exists for* $n = 1, 2, 3, \ldots$. *The coefficient of* θ^n *in the power series for* $\psi_X[\theta]$,

$$\psi_X[\theta] = \sum_{n=0}^{\infty} \psi_X^{(n)}[0] \frac{\theta^n}{n!}, \qquad (2.126)$$

is $E[X^n]/n!$, *so that*

$$E[X^n] = \left. \frac{d^n \psi_X[\theta]}{d\theta^n} \right|_{\theta=0}. \qquad (2.127)$$

Hence,

$$E[X] = \psi_X'[0], \quad E[X^2] = \psi_X''[0], \qquad (2.128)$$

and

$$\sigma_X^2 = \psi_X''[0] - \left(\psi_X'[0] \right)^2. \qquad (2.129)$$

(c) $\psi_{X+Y}[\theta] = \psi_X[\theta]\psi_Y[\theta]$ *for all θ, if X and Y are independent.*

Proof

(a) The proof is beyond the scope of this book but may be found in Feller [8]. The proof is carried out by showing that an inverse transform exists, which maps $\psi_X[\cdot]$ to $F_X[\cdot]$.

(b) There exists a $\delta > 0$ such that the power series

$$e^{\theta x} = 1 + x\theta + x^2 \frac{\theta^2}{2!} + \cdots + x^n \frac{\theta^n}{n!} + \cdots \qquad (2.130)$$

converges uniformly in x for $|\theta| \leq \delta_1 < \delta$. Hence, we calculate the expectation term by term to get

$$\psi_X[\theta] = E[e^{\theta x}] = 1 + E[X]\theta + \cdots + E[X^n]\frac{\theta^n}{n!} + \cdots . \qquad (2.131)$$

Since the infinite series representation of a function is unique, a comparison of the coefficients of θ^n in (2.131) with those of (2.126) shows that (2.127) is true.

(c) If X and Y are independent,

$$\psi_{X+Y}[\theta] = E[e^{\theta(X+Y)}] = E[e^{\theta X} e^{\theta Y}] = E[e^{\theta x}]E[e^{\theta Y}] = \psi_X[\theta]\psi_Y[\theta]. \qquad (2.132)$$

The third equality is true by Theorem 2.7.1. ∎

It is immediate by mathematical induction that if $X_1, X_2, \ldots, X_n$ are independent random variables, then

$$\psi_{X_1+X_2+\cdots+X_n}[\theta] = \psi_{X_1}[\theta]\psi_{X_2}[\theta] \cdots \psi_{X_n}[\theta],$$

for all θ such that

$$\psi_{X_i}[\theta] \text{ is defined for } i = 1, 2, \ldots, n.$$

We now give some examples of how Theorem 2.9.1 can be applied.

Example 2.9.1 Let X be an exponential random variable with parameter α (see Examples 2.5.3 and 2.6.2). Then

$$\psi[\theta] = \int_0^\infty \alpha e^{\theta x} e^{-\alpha x} dx = \alpha \int_0^\infty e^{-x(\alpha-\theta)} dx.$$

If $\theta < \alpha$, then

$$\psi[\theta] = \frac{-\alpha}{\alpha - \theta} e^{-x(\alpha-\theta)} \Big|_0^\infty = \frac{\alpha}{\alpha - \theta}. \qquad (2.133)$$

Hence,

$$\frac{d\psi}{d\theta} = \frac{\alpha}{(\alpha - \theta)^2},$$

so that by Theorem 2.9.1(b),

$$E[X] = \left.\frac{d\psi}{d\theta}\right|_{\theta=0} = \frac{1}{\alpha}.$$

Also,

$$\frac{d^2\psi}{d\theta^2} = \frac{2\alpha}{(\alpha - \theta)^3},$$

so that again by Theorem 2.9.1(b),

$$E[X^2] = \left.\frac{d^2\psi}{d\theta^2}\right|_{\theta=0} = \frac{2}{\alpha^2}.$$

Thus,

$$\text{Var}[X] = E[X^2] - (E[X])^2 = \frac{2}{\alpha^2} - \frac{1}{\alpha^2} = \frac{1}{\alpha^2}.$$

We can use (2.133) to generate a simple formula for all the moments of X. If $\theta < \alpha$, then

$$\psi[\theta] = \frac{\alpha}{\alpha - \theta} = \frac{1}{1 - \left(\dfrac{\theta}{\alpha}\right)} = 1 + \frac{\theta}{\alpha} + \left(\frac{\theta}{\alpha}\right)^2 + \cdots + \left(\frac{\theta}{\alpha}\right)^n + \cdots. \quad (2.134)$$

Here, we have used the fact that if $|x| < 1$,

$$\frac{1}{1 - x} = 1 + x + x^2 + \cdots + x^n + \cdots \text{(the geometric series)}.$$

Equating the coefficients of θ^n in (2.131) and (2.134) yields

$$\frac{E[X^n]}{n!} = \frac{1}{\alpha^n} \quad \text{or} \quad E[X^n] = \frac{n!}{\alpha^n} = n!E[X]^n, \quad n = 1, 2, 3, \ldots.$$

Thus, we have found all the moments of the exponential distribution with very little effort. $\square$

Example 2.9.2 Let X be a Poisson random variable with parameter α (see Example 2.7.6). Then

$$\psi[\theta] = \sum_{k=0}^{\infty} e^{\theta k} p(k) = \sum_{k=0}^{\infty} e^{k\theta} e^{-\alpha} \frac{\alpha^k}{k!} = e^{-\alpha} \sum_{k=0}^{\infty} \frac{(\alpha e^{\theta})^k}{k!} = e^{-\alpha} e^{\alpha e^{\theta}} = e^{\alpha(e^{\theta} - 1)}.$$

Thus,

$$E[X] = \frac{d\psi}{d\theta}\Big|_{\theta=0} = \alpha e^\theta e^{\alpha(e^\theta-1)}\Big|_{\theta=0} = \alpha,$$

and

$$E[X^2] = \frac{d^2\psi}{d\theta^2}\Big|_{\theta=0} = \alpha e^\theta e^{\alpha(e^\theta-1)}(1+\alpha e^\theta)\Big|_{\theta=0} = \alpha(1+\alpha) = \alpha + \alpha^2.$$

Hence,

$$\text{Var}[X] = E[X^2] - (E[X])^2 = \alpha^2 + \alpha - \alpha^2 = \alpha.$$

Thus, both the mean and variance of a Poisson random variable with parameter α are equal to α. □

Example 2.9.3 Let X and Y be independent Poisson random variables with parameters α and β, respectively. Using moment generating functions show that the random variable $X + Y$ is also a Poisson random variable with parameter $\alpha + \beta$.

Solution (We have already proven the result in Example 2.7.6 using the method of convolutions.) By Theorem 2.9.1 and Example 2.9.2,

$$\psi_{X+Y}[\theta] = \psi_X[\theta]\psi_Y[\theta] = e^{\alpha(e^\theta-1)}e^{\beta(e^\theta-1)} = e^{(\alpha+\beta)(e^\theta-1)}. \qquad (2.135)$$

Since (2.135) is the moment generating function of a Poisson random variable with parameter $\alpha + \beta$, we conclude, by the uniqueness of the moment generating function, that $X + Y$ has a Poisson distribution with parameter $\alpha + \beta$. That is,

$$P[X + Y = k] = e^{-(\alpha+\beta)}\frac{(\alpha+\beta)^k}{k!}, \quad k = 0,1,2,\ldots. \qquad (2.136)$$

The uniqueness of the moment generating function guarantees that no random variable which does *not* have a Poisson distribution can have the *same* moment generating function as a Poisson random variable. □

Let X be a discrete random variable assuming only nonnegative integer values and let $p(j) = P[X = j] = p_j$, $j = 0,1,2,\ldots$. Then the function $g[z] = g_X[z]$ defined by

$$g[z] = E[z^x] = \sum_{j=0}^{\infty} p_j z^j = p_0 + p_1 z + p_2 z^2 + \cdots \qquad (2.137)$$

is called the *generating function* of X or the *z-transform of X.*[18] Since $g[1] = p_0 + p_1 + p_2 + \cdots = 1$, $g[z]$ converges for $|z| \leq 1$. The next theorem states some of the useful properties of the generating function.

Theorem 2.9.2 (*Properties of the Generating Function or z-transform*) *Let X and Y be discrete random variables assuming only nonnegative integer values. Then the following hold.*

(a) X and Y have the same distribution if and only if $g_X[\cdot] = g_Y[\cdot]$ (*uniqueness*).

(b) $p_n = \dfrac{1}{n!} \dfrac{d^n g_X[z]}{dz^n}\bigg|_{z=0}$, $n = 0, 1, 2, \ldots$.

(c) $E[X] = g'_X[1]$ and $\operatorname{Var}[X] = g''_X[1] + g'_X[1] - (g'_X[1])^2$.

(d) $g_{X+Y}[z] = g_X[z]g_Y[z]$, if X and Y are independent.

Proof

(a) Since
$$g_X[z] = p_0 + p_1^z + p_2 z^2 + \cdots$$

is a convergent power series for $|z| \leq 1$, $g_X[z]$ is unique by the uniqueness of power series.

(b) Also true by uniqueness of power series.

(c) $g'_X[z] = \displaystyle\sum_{j=1}^{\infty} j p_j z^{j-1} = p_1 + 2P_2 z + 3p_3 z^2 + \cdots$.
Hence,
$$g'_X[1] = \sum_{j=1}^{\infty} j p_j = E[X].$$

$$g''_X[z] = 2P_2 + 3 \times 2P_3 z + 4 \times 3p_4 z^2 + 5 \times 4p_5 z^3 + \cdots$$
$$= \sum_{j=2}^{\infty} j(j-1)p_j z^{j-2}.$$

---

[18]Some authors use z-transform for $g[-z] = p_0 + p_1 z^{-1} + p_2 z^{-2} + \cdots$. If $\pi = (\pi_0, \pi_1, \pi_2, \cdots)$ is a probability distribution (that is, $\sum_{n=0}^{\infty} \pi_n = 1$) it is also common to write its z-transform as $\pi[z] = \sum_{n=0}^{\infty} \pi_n z^n$.

Thus,

$$g_X''[1] = \sum_{j=1}^{\infty} j(j-1)p_j = E[X(X-1)] = E[X^2] - E[X].$$

Therefore,

$$\text{Var}[X] = E[X^2] - (E[X])^2 = g_X''[1] + g_X'[1] - (g_X'[1])^2.$$

(d) Let $a_k = P[X = k]$, $b_k = P[Y = k]$, $c_k = P[X + Y = k]$, $k = 0, 1, 2, \ldots$. Then we know by Theorem 2.7.5 that the sequence $\{c_k\}$ is the convolution of the sequences $\{a_k\}$ and $\{b_k\}$; that is,

$$c_k = \sum_{i=0}^{k} a_i b_{k-i}, \quad k = 0, 1, 2, \ldots. \tag{2.138}$$

Moreover, if we formally multiply together the power series for $g_X[z]$ and $g_Y[z]$, we get

$$\begin{aligned}
g_X[z]g_Y[z] &= \left(\sum_{k=0}^{\infty} a_k z^k\right)\left(\sum_{k=0}^{\infty} b_k z^k\right) \\
&= \sum_{k=0}^{\infty}\left(\sum_{i=0}^{k} a_i b_{k-i}\right) z^k = g_{X+Y}[z]. \blacksquare
\end{aligned}$$

The reader should note that a discrete random variable that takes on only nonnegative integer values has the moment generating function

$$\psi[\theta] = \sum_{k=0}^{\infty} e^{k\theta} p_k = \sum_{k=0}^{\infty} (e^\theta)^k p_k = g[e^\theta]. \tag{2.139}$$

That is, the moment generating function is obtained from the generating function (z-transform) by a simple change of variable. Similarly, $g[z] = \psi[\ln[z]]$, if $z > 0$.

Example 2.9.4 A random variable X is called a *Bernoulli random variable* (has a Bernoulli distribution) if it can assume only two values, usually taken to be 1 and 0, the first with probability p and the second with probability $q = 1 - p$. Find the mean and variance of a Bernoulli random variable X, using its generating function.

Solution The generating function of X is

$$g[z] = q + pz. \tag{2.140}$$

Hence, $g'[z] = p$ and $g''[z] = 0$. Therefore, by Theorem 2.9.2,

$$E[X] = g'[1] = p$$

and

$$\text{Var}[X] = g''[1] + g'[1] - (g'[1])^2 = 0 + p - p^2 = p(1 - p) = pq. \quad \square$$

Example 2.9.5 The Bernoulli random variable X, discussed in Example 2.9.4, is often used to describe a random experiment with but two outcomes, success or failure. We define X to be 1 for a success and to be 0 for a failure. Such an experiment is called a *Bernoulli trial*. A sequence of n such trials is called a *Bernoulli sequence of trials* if the probability of success does not change from trial to trial. An example is tossing a coin repeatedly, with a head considered a success. Let Y be the random variable that counts the number of successes in a Bernoulli sequence of n trials, where $n \geq 1$. Then we can write

$$Y = X_1 + X_2 + \cdots + X_n, \tag{2.141}$$

where $X_1, X_2, \ldots, X_n$ is a collection of identical Bernoulli random variables. Hence, by Theorem 2.9.2(d),

$$g_Y[z] = (q + pz)^n = \sum_{k=0}^{\infty} P[Y = k] z^k. \tag{2.142}$$

But, by the binomial theorem (the corollary to Theorem 2.3.3),

$$(pz + q)^n = \sum_{k=0}^{n} \binom{n}{k} (pz)^k q^{n-k}. \tag{2.143}$$

Equating coefficients of z^k in (2.143) and (2.142) gives

$$p(k) = P[Y = k] = \binom{n}{k} p^k q^{n-k}, \quad k = 0, 1, 2, \ldots, n. \tag{2.144}$$

Y is called a *binomial random variable* or said to have a *binomial distribution*. (The binomial distribution will be discussed more completely in Chapter 3.)

Since $g_Y[z] = (q+pz)^n$, then $g_Y'[z] = n(q+pz)^{n-1}p$ and $E[Y] = g_Y'[1] = np$. Also, $g_Y''[z] = n(n-1)(q+pz)^{n-2}p^2$. Hence, by Theorem 2.9.2(c),

$$
\begin{aligned}
\text{Var}[Y] &= g_Y''[1] + g_Y'[1] - (g_Y'[1])^2 = n(n-1)p^2 + np - n^2p^2 \\
&= np(1-p) = npq. \quad \Box
\end{aligned}
$$

The transform most widely used by engineers and applied mathematicians is the *Laplace–Stieltjes* transformation defined below.

Let X be a random variable such that $P[X < 0] = 0$. then the Laplace–Stieltjes transform of X is defined for $\theta \geq 0$ by

$$
X^*[\theta] = E[e^{-\theta x}] = \begin{cases} \displaystyle\int_0^\infty e^{-\theta x} f(x)dx & \text{if } X \text{ is continuous} \\[2ex] \displaystyle\sum_{x_i} e^{-\theta x} p(x_i) & \text{if } X \text{ is discrete.} \end{cases}
\tag{2.145}
$$

Sometimes $X^*[\theta]$ is called the Laplace–Stieltjes transform of F. The integral $\int_0^\infty e^{-\theta x} f(x)dx$ is called the *Laplace transform of* f. Many authors write

$$
X^*[\theta] = \int_0^\infty e^{-\theta x} dF(x),
\tag{2.146}
$$

where the integral is called a *Stieltjes integral*. However, the integral is always evaluated as we have shown in (2.145), that is, as $\int_0^\infty e^{-\theta x} f(x)dx$ if X is continuous and as $\sum_{x_i} e^{-\theta x_i} p(x_i)$ if X is discrete.

Theorem 2.9.3 (*Properties of the Laplace–Stieltjes Transform*) *Let X and Y be random variables with Laplace–Stieltjes transforms $X^*[\cdot]$ and $Y^*[\cdot]$. Then the following hold.*

(a) $F_X = F_Y$ *if and only if* $X^*[\cdot] = Y^*[\cdot]$ (*uniqueness*).

(b) *For* $\theta > 0$, $X^*[\theta]$ *has derivatives of all orders given by*

$$
\frac{d^n X^*}{d\theta^n} = \begin{cases} \displaystyle (-1)^n \int_0^\infty e^{-\theta x} x^n f(x)dx & \text{if } X \text{ is continuous} \\[2ex] \displaystyle (-1)^n \sum_{x_i} e^{-\theta x_i} x_i^n p(x_i) & \text{if } X \text{ is discrete.} \end{cases}
$$

$$
\tag{2.147}
$$

(c) *If* $E[X^n]$ *exists, then*

$$
E[X^n] = (-1)^n \left. \frac{d^n X^*[\theta]}{d\theta^n} \right|_{\theta=0}
\tag{2.148}
$$

In particular, if $E[X]$ and $E[X^2]$ exist, then

$$E[X] = -\frac{dX^*}{d\theta}[0], \quad E[X^2] = \frac{d^2X^*}{d\theta^2}[0]. \qquad (2.149)$$

(d) $(X+Y)^*[\theta] = X^*[\theta]Y^*[\theta]$, if X and Y are independent.

The proof of Theorem 2.9.3 is beyond the scope of this book but may be found in Feller [8]. For more information on transform methods see Section 3.4 of Chapter 3.

Example 2.9.6 Let X be an exponential random variable with parameter α, that is,

$$f(x) = \begin{cases} \alpha e^{-\alpha x} & \text{if } 0 < x. \\ 0 & \text{otherwise.} \end{cases} \qquad (2.150)$$

Then, if $\theta < \alpha$,

$$X^*[\theta] = \int_0^\infty \alpha e^{-\theta x} e^{-\alpha x} dx = \alpha \int_0^\infty e^{-(\theta+\alpha)x} dx = \frac{\alpha}{\alpha+\theta}. \quad \square \qquad (2.151)$$

> *A mathematician in Reno,*
> *Overcome by the heat and the vino,*
>   *Became quite unroulli*
>   *Expounding Bernoulli,*
> *And was killed by the crowd playing Keno.*

> Stan Kelly-Bootle
> *The Devil's DP Dictionary*

2.10   Inequalities

In this section we consider some inequalities and their uses. One important application is the derivation of the law of large numbers.

Theorem 2.10.1 (*Markov's Inequality*) *Let X be a random variable with expected value $E[X]$ and such that $P[X < 0] = 0$. Then, for each $t > 0$,*

$$P[X \geq t] \leq \frac{E[X]}{t}. \qquad (2.152)$$

Proof We give the proof for discrete X. The proof when X is continuous is similar.

$$
\begin{aligned}
E[X] &= \sum_{x_i} x_i p(x_i) \\
&= \sum_{x_i < t} x_i p(x_i) + \sum_{t \le x_i} x_i p(x_i) \\
&\ge \sum_{t \le x_i} x_i p(x_i) \\
&\ge \sum_{t \le x_i} t p(x_i) \quad = \quad t P[X \ge t].
\end{aligned}
$$

Hence,

$$
P[X \ge t] \le \frac{E[X]}{t},
$$

and the proof is complete. ∎

Example 2.10.1 Suppose an interactive computer system is proposed for which it is estimated that the mean response time $E[T]$ is 0.5 seconds. Use Markov's inequality to estimate the probability that the response time T will be 2 seconds or more.

Solution By Markov's inequality,

$$
P[T \ge 2] \le \frac{E[T]}{2} = \frac{0.5}{2} = \frac{1}{4}. \quad \square
$$

It should be noted that Markov's inequality implies that

$$
P[X \ge kE[X]] \le \frac{1}{k}, \quad k > 0.
$$

This inequality usually gives rather crude estimates because only the value of $E[X]$ is assumed to be known. Chebyshev's inequality, in which the standard deviation is also assumed to be known, gives better probability estimates. $\square$

Theorem 2.10.2 (*Chebyshev's Inequality*)[19] *Let X be a random variable with finite mean $E[X]$ and standard deviation $\sigma > 0$. Then for every $t > 0$,*

$$
P[|X - E[X]| \ge t] \le \frac{\sigma^2}{t^2}, \tag{2.153}
$$

---

[19]P. L. Chebyshev (1821–1894), whose name is also spelled Tchebychev, Tchebycheff, and several other ways, was one of Russia's finest mathematicians. For a fascinating account of why he thinks Chebyshev should be spelled Tschebyscheff, see Davis [6].

or

$$P[|X - E[X]| \geq t\sigma] \leq \frac{1}{t^2}. \qquad (2.154)$$

Proof Applying Markov's inequality to $(X - E[X])^2$ with t^2 in place of t yields

$$P[(X - E[X])^2 \geq t^2] \leq \frac{E[(X - E[X])^2]}{t^2} = \frac{\sigma^2}{t^2}. \qquad (2.155)$$

However, $(X - E[X])^2 \geq t^2$ if and only if $|X - E[X]| \geq t$. Substituting this relation into (2.155) yields (2.153). Equation (2.154) follows from (2.153) by using $t\sigma$ in place of t in (2.153). ■

Example 2.10.2 Suppose that, for the proposed interactive computer system of Example 2.10.1, it is estimated that the standard deviation of response time is 0.1 seconds. Use Chebyshev's inequality to estimate the probability that the response time will be between 0.25 and 0.75 seconds.

Solution

$$P[(T \leq 0.25) \cup (T \geq 0.75)] = P[|T - 0.5| \geq 0.25] \leq \frac{0.1^2}{0.25^2} = \left(\frac{4}{10}\right)^2 = 0.16.$$

Hence,

$$P[0.25 < T < 0.75] = 1 - P[|T - 0.5| \geq 0.25] =\geq 1 - 0.16 = 0.84. \quad \square$$

Chebyshev's inequality often gives poor probability estimates. For example, if X has an exponential distribution with mean $E[X] = 2$, then

$$\begin{aligned} P[|X - E[X]| \geq 4] &= P[|X - 2| \geq 4] = 1 - P[X \leq 6] \\ &= 1 - (1 - e^{-6/2}) = e^{-3} = 0.0498, \end{aligned}$$

although the Chebyshev inequality shows only that this probability does not exceed 0.25. However, the next example shows that the Chebyshev inequality cannot be improved without strengthening the hypotheses.

Example 2.10.3 Suppose a discrete random variable X can assume only the values $-2, 0, 2$ with $p(-2) = p(2) = \frac{1}{8}$ and $p(0) = \frac{3}{4}$. Then

$$E[X] = -2 \times \frac{1}{8} + 0 \times \frac{3}{4} + 2 \times \frac{1}{8} = 0,$$

$$E[X^2] = 4 \times \frac{1}{8} + 0 + 4 \times \frac{1}{8} = 1,$$

$$\text{Var}[X] = E[X^2] - (E[X])^2 = 1 - 0 = 1,$$

and $\sigma = 1$. Then, by Chebyshev's inequality, $P[|X - E[X]| \geq 2] \leq \frac{1}{4}$. However,

$$P[|X - E[X]| \geq 2] = P[(X = 2) \cup (X = -2)] = \frac{1}{8} + \frac{1}{8} = \frac{1}{4}.$$

Hence, the value estimated by Chebyshev's inequality is the exact value. $\square$

In many computer science applications we are more interested in calculating one tail of a probability distribution than in calculating both tails; that is, we want an estimate of the size of $P[X - E[X] > t]$ or $P[X - E[X] < t]$ rather than the estimate of $P[|X - E[X]| \geq t]$ provided by Chebyshev's inequality. The one-sided inequality gives us this estimate.

Theorem 2.10.3 (*One-Sided Inequality*) *Let X be a random variable with finite mean $E[X]$ and variance σ^2. Then,*

$$P[X \leq t] \leq \frac{\sigma^2}{\sigma^2 + (t - E[X])^2} \quad \text{if } t < E[X], \tag{2.156}$$

and

$$P[X > t] \leq \frac{\sigma^2}{\sigma^2 + (t - E[X])^2} \quad \text{if } t > E[X]. \tag{2.157}$$

Proof Cramér [5], using advanced methods, shows that if X is a random variable with mean 0 and standard deviation σ, then

$$P[X \leq t] \leq \frac{\sigma^2}{\sigma^2 + t^2} \quad \text{for } t < 0, \tag{2.158}$$

and

$$P[X > t] \leq \frac{\sigma^2}{\sigma^2 + t^2} \quad \text{for } t > 0. \tag{2.159}$$

Now, if X is an arbitrary random variable with a finite mean and variance, then $E[X - E[X]] = E[X] - E[X] = 0$, and

$$\text{Var}[X - E[X]] = E[(X - E[X])^2] = \text{Var}[X].$$

Hence, if $t < E[X]$, then $t - E[X] < 0$ and by (2.158),

$$P[X \leq t] = P[X - E[X] \leq t - E[X]] \leq \frac{\sigma^2}{\sigma^2 + (t - E[X])^2}.$$

If $t > E[X]$, then $t - E[X] > 0$ and (2.159) gives

$$P[X > t] = P[X - E[X] > t - E[X]] \leq \frac{\sigma^2}{\sigma^2 + (t - E[X])^2}. \quad \blacksquare$$

Example 2.10.4 A mathematical model of a proposed interactive computer system gives a mean time to retrieve a record from a direct access storage device as 400 milliseconds with a standard deviation of 116 milliseconds. One design criterion requires that 90% of all retrieval times must not exceed 750 milliseconds. Use the one-sided inequality to test the design criterion.

Solution Let T be the retrieval time. The design criterion is that $P[T \leq 750$ milliseconds$] \geq 0.90$.

By the one-sided inequality,

$$P[T > 750] \leq \frac{116^2}{116^2 + (750 - 400)^2} = \frac{1}{1 + (350/116)^2} = 0.09897.$$

Hence,

$$P[T \leq 750] \geq 1 - 0.09897 = 0.90103,$$

and the design criterion is met. The best estimate we could make with Chebyshev's inequality is

$$
\begin{aligned}
P[T \geq 750] &= P[T - 400 \geq 350] \leq P[|T - 400| \geq 350] \leq (116/350)^2 \\
&= 0.1098.
\end{aligned}
$$

This does not indicate that the design criterion has been met. $\square$

Example 2.10.5 Professor Frank N. Stein has a favorite random variable. He uses an "unfair" coin, which comes up heads with probability 0.9 and tails with probability 0.1 (it balances on edge with probability zero). Each time you toss the coin he pays \$9.00 for a tail but charges you \$1.00 for a head. If X is the amount you receive per toss, then

$$E[X] = (-1) \times 0.9 + 9 \times 0.1 = 0,$$

and

$$
\begin{aligned}
E[X^2] &= \text{Var}[X] \quad (\text{since } E[X] = 0) \\
&= (-1)^2 \times 0.9 + 9^2 \times 0.1 = 9,
\end{aligned}
$$

so

$$\sigma = 3.$$

Thus, 10 percent of the probability mass is at least three standard deviations from the mean. By contrast, for the normal distribution (which we consider in Chapter 3), only 0.27 percent of the probability mass is three standard deviations or more from the mean.

The one-sided inequality gives

$$P[X \leq -1] \leq \frac{3^2}{3^2 + (-1)^2} = 0.9,$$

which is exact.

It also gives

$$P[X > 8.99999] \leq \frac{9}{9 + (8.99999)^2} = 0.1000002,$$

but

$$P[X > 9] \leq 0.1,$$

although the true probability that X exceeds 9 is zero. $\square$

We have discussed some useful inequalities for estimating the probability that a single event occurs, such as $T \geq 20$ seconds. Sometimes we are interested in the simultaneous occurrence of several events. For example, suppose $Z = X + Y$; and we know that (a) $P[X \leq 2] = 0.9$ and (b) $P[Y \leq 4] = 0.9$. We would like to be able to say something about $P[Z \leq 6]$. (No, $P[Z \leq 6] \neq 0.9$.) Some inequalities, called Bonferroni's inequalities, enable us to do that. Before we state them, we need a result that generalizes the formula of Theorem 2.2.1(c) and the result of Exercise 4.

Theorem 2.10.4 (*Poincare's Formula*) *Suppose $A_1, A_2, \ldots, A_n$ are events in a sample space Ω. Then*

$$
\begin{aligned}
P\left[\bigcup_{j=1}^{n} A_j\right] &= P[A_1 \cup \cdots \cup A_n] \\
&= \sum_j P[A_j] - \sum_{j,k} P[A_j \cap A_k] \\
&\quad + \sum_{j,k,l} P[A_j \cap A_k \cap A_l] - \sum_{j,k,l,m} P[A_j \cap A_k \cap A_l \cap A_m] \\
&\quad + \cdots \\
&\quad + (-1)^{n-1} P[A_1 \cap A_2 \cap \cdots \cap A_n], \qquad (2.160)
\end{aligned}
$$

where the indices in each sum are distinct and range from 1 to n.

Proof See Chung [4, pages 162–163] or Feller [7, pages 99–100]. ∎

In some textbooks (2.160) is written as

$$P\left[\bigcup_{j=1}^{n} A_j\right] = \sum_{r=1}^{n} (-1)^{r-1} S_r = S_1 - S_2 + \cdots + (-1)^{n-1} S_n, \qquad (2.161)$$

where

$$S_r = \sum_{1 \le i_1 < \cdots < i_r \le n} P[A_{i_1} \cap \cdots \cap A_{i_r}]. \qquad (2.162)$$

Thus,

$$S_1 = P[A_1] + P[A_2] + \cdots + P[A_n], \qquad (2.163)$$

$$S_2 = \sum_{i=1}^{n-1} \sum_{j=i+1}^{n} P[A_i \cap A_j], \qquad (2.164)$$

and

$$S_n = P[A_1 \cap A_2 \cap \cdots \cap A_n]. \qquad (2.165)$$

It is easy to see that the sum A_1 has n terms, the sum S_2 has $\binom{n}{2}$ terms, and the general sum S_r has $\binom{n}{r}$ terms, for the rth sum is the sum of the numbers $P[A_{i_1} \cap A_{i_2} \cap \cdots \cap A_{i_r}]$ over all the indices $i_1, i_2, \ldots, i_r$ such that $i_1 < i_2 < \cdots < i_r$. Since the indices are chosen from the numbers from 1 to n, there are exactly $\binom{n}{r}$ choices.

Theorem 2.10.4 yields the following set of inequalities.

Theorem 2.10.5 (*Bonferroni's Inequalities*) *Suppose* $A_1, A_2, A_3, \ldots, A_n$ *are events in a sample space* Ω *and that* $A = \bigcup_{j=1}^{n} A_j$. *Then, in the notation of* (2.161)–(2.165),

$$P[A] \le S_1. \qquad (2.166)$$

$$S_1 - S_2 \le P[A]. \qquad (2.167)$$

$$P[A] \le S_1 - S_2 + S_3. \qquad (2.168)$$

$$\cdots$$

$$P[A] \le S_1 - S_2 + \cdots + S_n, \qquad (2.169)$$

when n *is odd, and*

$$S_1 - S_2 + \cdots - S_n \le P[A], \qquad (2.170)$$

when n *is even.*

Proof See Feller [7]. ■

Corollary to Theorem 2.10.5 (*Bonferroni's Inequality*) *Suppose the events* $A_1, A_2, \cdots, A_n$ *and* A *are as in Theorem 2.10.5. Let* $p_i = P[A_i]$ *and* $q_i = 1 - p_i$ *for* $i = 1, 2, \cdots, n$. *Then*

$$P[\text{all } n \text{ events occur}] \ge 1 - (q_1 + q_2 + \cdots + q_n). \qquad (2.171)$$

Proof If we call the occurrence of one of the A_i's a success, then the event A is "at least one success." Hence, (2.166) yields

$$P[\text{at least one success}] \le p_1 + p_2 + \cdots + p_n, \qquad (2.172)$$

and, taking complements,

$$P[\text{no success}] \ge 1 - (p_1 + p_2 + \cdots + p_n). \qquad (2.173)$$

If we now take the complements of the A_i's, and replace each p_i by q_i, then (2.173) becomes

$$P[\text{no } \bar{A_i} \text{ occurs}] = P[\text{all } n \text{ events occur}] \ge 1 - (q_1 + q_2 + \cdots + q_n). \quad (2.174)$$

∎

Note that the $A_1, A_2, \ldots, A_n$ need not be independent. We consider an example.

Example 2.10.6 Suppose $Z = X + Y$, $P[X \le 2] = 0.9$, and $P[Y \le 4] = 0.9$. Let $A_1 = \{X \le 2\}$ and $A_2 = \{Y \le 4\}$. Then, by (2.171),

$$P[Z \le 6] \ge 1 - (0.1 + 0.1) = 0.8. \qquad \square$$

Suppose, in Example 2.10.6, we seek a value of z such that $P[Z \le z] \ge 0.9$. We could do this by finding a value x and a value y such that $P[X \le x] = 0.95$ and $P[Y \le y] = 0.95$, for then

$$P[Z \le x + y] \ge 1 - (0.05 + 0.05) = 0.9.$$

In the more general case of (2.174), if we want $P[\text{all events occur}] \ge 1 - \alpha$, and we can control the probabilities p_i that individual events occur, we take $p_i = 1 - \alpha/n$ so that $q_i = \alpha/n$.

Example 2.10.7 (a) What choice of p will guarantee a probability of at least 0.9 that each of four equiprobable events occur simultaneously? (b) With this value of p, what is the probability that all four events occur simultaneously, if the events are independent?

Solution

(a) $p = 1 - \frac{0.1}{4} = 0.975$.

(b) $p^4 = 0.903688$. $\square$

We will see other applications of Bonferroni's inequality.

Most of us have an intuitive feel for what the probability of an event A, such as rolling a 7 with a pair of dice, "really" is, which is close to the "relative frequency" school of thought about probability. We have the feeling that, if we perform the random experiment n times and let S_n be the number of times that event A occurs, then S_n/n is approximately $P[A]$, at least in the sense that $\lim_{n\to\infty} S_n/n = P[A]$. The *law of large numbers* makes this intuitive notion more precise (and shows that it is true).

Let A be an event, that has probability $P[A]$, and suppose we perform a Bernoulli sequence of n trials as described in Example 2.9.5, where a success corresponds to the occurrence of event A. Let S_n be the number of successes in the n trials. As we saw in Example 2.9.5, S_n has a binomial distribution with $E[S_n] = nP[A]$ and $\mathrm{Var}[S_n] = nP[A](1 - P[A])$. We are interested in the ratio S_n/n. We calculate

$$E\left[\frac{S_n}{n}\right] = \frac{1}{n}E[S_n] = P[A]$$

and

$$\mathrm{Var}\left[\frac{S_n}{n}\right] = \frac{1}{n^2}\mathrm{Var}[S_n] = \frac{P[A](1 - P[A])}{n}.$$

Let $\epsilon > 0$ be arbitrary. Then by Chebyshev's inequality,

$$P\left[\left|\frac{S_n}{n} - P[A]\right| \geq \epsilon\right] \leq \frac{P[A](1 - P[A])}{n\epsilon^2}. \tag{2.175}$$

The expression on the right of (2.175) can be made as small as desired, for fixed values of ϵ and A, by choosing n sufficiently large. This proves the following theorem.

Theorem 2.10.6 (*Weak Law of Large Numbers*) *Let A be an event and S_n the number of times that A occurs in a Bernoulli sequence of n trials. Then for each $\epsilon > 0$,*

$$\lim_{n\to\infty} P\left[\left|\frac{S_n}{n} - P[A]\right| \geq \epsilon\right] = 0. \tag{2.176}$$

■

There is a stronger form of Theorem 2.10.6 called the *strong law of large numbers*, which uses a more restrictive definition of the intuitive idea that $\lim_{n\to\infty} S_n/n = P[A]$ (see Feller [7, pages 202–204]). However, a more useful form of the law is immediate from the *central limit theorem*, which is discussed in Chapter 3.

The weak law of large numbers shows that $p = P[A]$ can be estimated by S_n/n and that this estimate converges to p. However, it does not give any information as to how large n should be to guarantee that the error is less than a given value for a certain probability level. The Chebyshev inequality does provide crude estimates, for if $\delta > 0$ then, by Chebyshev's inequality,

$$P\left[\left|\frac{S_n}{n} - p\right| \geq \delta\right] \leq \frac{\text{Var}\left[\frac{S_n}{n}\right]}{\delta^2} = \frac{p(1-p)}{n\delta^2}. \qquad (2.177)$$

It is easy to show that $p(1 - p)$ has its maximum value at $p = 1/2$ (see Exercise 2). Hence, no matter what value p actually has, we have

$$P\left[\left|\frac{S_n}{n} - p\right| \geq \delta\right] \leq \frac{1}{4n\delta^2}. \qquad (2.178)$$

Suppose now that δ and $\epsilon > 0$ are given and we want to find how many trials of the experiment we need to be sure that

$$P\left[\left|\frac{S_n}{n} - p\right| \geq \delta\right] \leq \epsilon. \qquad (2.179)$$

If we know approximately what the value of p is, we see that (2.179) will be satisfied if $p(1-p)/n\delta^2 \leq \epsilon$ or $n \geq p(1-p)/\epsilon\delta^2$. If we have no idea what the value of p is, we can use (2.178) to conclude that $n \geq 1/(4\epsilon\delta^2)$ trials will suffice. Since Chebyshev's inequality usually yields poor estimates, we would expect either of these estimates to yield conservative estimates for n. In Chapter 3 we show that the central limit theorem can be applied to give a better estimate.

Example 2.10.8 Assuming that each terminal in an interactive system has the same probability p of being in use during the peak period of the day (the load is evenly distributed over the terminals), we want to know how many observations n need be made so that

$$P\left[\left|\frac{S_n}{n} - p\right| \geq 0.1\right] \leq 0.05.$$

If the first 100 observations indicate that p is approximately 0.2, how many more trials are needed?

Solution The estimate, based on (2.178), is $n = 1/(4 \times 0.05 \times 0.01) = 500$. If p is approximately 0.2, then we can use (2.177) to conclude that we need a total of $n = 0.2 \times 0.8/(0.05 \times 0.01) = 320$ observations or an additional 220. In Example 3.3.3 we show that this estimate can be improved by using the central limit theorem. $\square$

Student Sayings

Our observation of Nancy's distribution has given us many fine moments.

> *An exterminator made this contribution*
> *On rats arriving in random profusion*
> *   "I know nothing of math,*
> *   Probability or stats,*
> *But I handle 'em with Poisson distribution."*

A. Student

2.11   Exercises

1. [C20] The interactive order entry system of the WEWE Diaper Company can receive order messages from Los Angeles, San Diego, Bakersfield, and San Francisco. Ordering activity in each city is independent of that from the other cities. The probability that the system receives one or more orders during any one minute time interval (during the peak period of the day) from Los Angeles, San Diego, Bakersfield, or San Francisco, respectively, is 0.8, 0.3, 0.05, 0.5.

   (a) What is the probability that ordering activity occurs from exactly one of the cities during any one minute period?

   (b) Exactly two cities?

   (c) Not more than two cities?

   (d) No city?

2. [HM05] In discussing the weak law of large numbers we claimed that the function $pq = p(1-p)$ has a unique maximum value of $\frac{1}{4}$ at $p = \frac{1}{2}$. Prove this claim.

3. [20] Suppose A, B, and C are events in some sample space Ω, and thus are subsets of Ω. Prove the distributive law

$$(A \cup B) \cap C = (A \cap C) \cup (B \cap C).$$

4. [18] Prove that, if A, B, and C are events, then

$$\begin{aligned}
P[A \cup B \cup C] &= P[A] + P[B] + P[C] - P[A \cap B] - P[A \cap C] \\
&\quad - P[B \cap C] + P[A \cap B \cap C].
\end{aligned}$$

[*Hint:* Use Theorem 2.2.1(c) and the result of Exercise 3.]

5. [15] Assume that a single depth charge has a probability of $\frac{1}{3}$ of sinking a submarine, $\frac{1}{2}$ of damage, and $\frac{1}{6}$ of missing. Assume also that two damaging explosions sink the sub. If four depth charges are dropped on a submarine, what is the probability that the sub sinks?

6. [18] Assume $A_1, A_2, A_3, \ldots$ are subsets of some set Ω. Prove De Morgan's formulas:

   (a) $\overline{A_1 \cup A_2 \cup \cdots \cup A_N} = \overline{A_1} \cap \overline{A_2} \cap \cdots \cap \overline{A_N}$.

   (b) $\overline{A_1 \cap A_2 \cap \cdots \cap A_N} = \overline{A_1} \cup \overline{A_2} \cup \cdots \cup \overline{A_N}$.

   (c) $\overline{\bigcup_{n=1}^{\infty} A_n} = \bigcap_{n=1}^{\infty} \overline{A_n}$.

   (d) $\overline{\bigcap_{n=1}^{\infty} A_n} = \bigcup_{n=1}^{\infty} \overline{A_n}$.

7. [15] Let $A_1, A_2, \ldots$ be events in some sample space Ω. Use Axiom Set 2.2.1 and the results of Exercise 6 to prove that

   (a) $A_1 \cap A_2 \cap \cdots \cap A_N$ is an event for each positive integer N.

   (b) $\bigcap_{n=1}^{\infty} A_n$ is an event.

8. [10] An on-line computer system has four incoming communication lines with the properties described in the table below. What is the probability that a randomly chosen message has been received without error?

Line	Fraction of traffic	Fraction of messages without error
1	0.4	0.9998
2	0.3	0.9999
3	0.1	0.9997
4	0.2	0.9996

9. [15] Twas Brillig has a drawer containing a mixture of 15 black and 20 blue socks. Twas is sick in bed when his friend Slithy Toves comes to visit.

   (a) Twas asks Slithy to get him a pair of matched socks from the drawer (either a black pair or a blue pair). It is too dark for Slithy to distinguish the colors. How many socks must Slithy remove from the drawer to be sure of getting a matched pair?

(b) Suppose now there are an equal number of black and blue socks in the drawer. Suppose the minimum number of socks Slithy must draw to be sure of getting a pair is the same as the minimum number he must draw to be sure of getting at least one black sock and one blue sock. How many socks are in the drawer?

10. [15] Big Bored Securities has two brands of personal computers in the Information Center to use for demonstrations, brand y and brand z. If two personal computers are selected at random, the probability that both are brand y is $1/2$. What is the smallest number of personal computers that could be in the Information Center?

11. [20] Suppose the random variable X has finite mean, μ, and finite standard deviation, σ. Suppose also that

$$P[|X - \mu| > K] = 0.$$

Prove that $\sigma \leq K$.

12. [18] Calculate

(a) the probability of getting at least one ace by rolling four dice and

(b) the probability of rolling at least one double ace (popularly known as "snake eyes") in 24 throws of two dice. The fact that the first number is larger than the second is known as de Méré's paradox. See Feller [7, page 56] and Chung [4, pages 138–139].

13. [5] A box contains 50 washers of which 3 are defective. If 2 are randomly chosen what is the probability they will both be good?

It is traditional in any discussion of probability to include some examples from card games. For the benefit of sheltered readers we include a definition of bridge and poker.

Definition of Bridge and Poker

A pack or deck of bridge cards contains 52 cards arranged in four suits of thirteen each. The four suits are known as spades, clubs, hearts and diamonds. The first two are black, the last two are red. There are thirteen face values (2, 3, ..., 10, jack, queen, king, ace) in each suit. Cards having the same face value are said to be of the same kind. Playing bridge, by definition, means dealing (distributing) the cards to four players known as North, South, East, and West (or N, S, E, W, for short) so that each player receives thirteen cards. The deck is assumed to be well shuffled before the cards are dealt. Playing poker, by convention, means choosing five cards randomly from a bridge deck.

We show a picture of an unshuffled deck of cards.

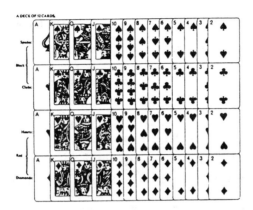

An unshuffled deck of bridge cards.

We illustrate card calculations with two examples.

Example 2.11.1 What is the probability that in a bridge game North and South, between them, have an equal number of black and red cards?

Solution We need to calculate the probability that 13 of the 26 cards will be red and 13 black. The number of ways the two hands of 26 cards can be drawn from the deck is $\binom{52}{26}$. A collection of 13 red cards can be drawn from the 26 red cards in the deck in $\binom{26}{13}$ ways. Since for each drawing of 13 given red cards, 13 black ones can be drawn in $\binom{26}{13}$ different ways, the total number of ways to draw 13 black and 13 red cards is $\binom{26}{13} \times \binom{26}{13}$. Therefore, the desired probability is

$$\frac{\binom{26}{13} \times \binom{26}{13}}{\binom{52}{26}} = \frac{(26!)^4}{(52!) \times (13!)^4} = 0.2181255. \quad \square$$

Example 2.11.2 Find the probability of drawing a royal flush in poker. (A royal flush consists of 10, jack, queen, king, and ace, all in a single suit.)

Solution There are $\binom{52}{5}$ possible poker hands. There are only 4 possible ways of drawing a royal flush, since there is just one in each suit. Hence the required probability is

$$\frac{4}{\binom{52}{5}} = \frac{4 \times 5! \times 47!}{52!} = 0.000001539077. \ \square$$

14. [28] Find the probability of getting each of the following poker hands:

   (a) A straight flush (five cards in sequence in a single suit, but not a royal flush. Since an ace can also be thought of as a one, the sequence ace, 2, 3, 4, 5 in one suit is a straight flush).

   (b) Four of a kind (four cards with the same face value).

   (c) Full house (one pair and one triple of the same face value, such as ace, ace, king, king, king).

   (d) Flush (five cards in one suit but not a straight or royal flush).

   (e) Straight (five cards in sequence, not all of the same suit).

15. [22] Find the probability of *not* drawing a pair in poker. (Of course you still could have a straight or a flush, etc., but not three or four of a kind.)

16. [18] Find the probability of getting a real "bust" hand in poker. A "bust" hand has no pair and is neither a straight, a flush, a straight flush nor a royal flush. [The ranking of poker hands from high to low is royal flush, straight flush, four of a kind, full house, flush, straight, three of a kind, two pairs, one pair, and, in the case of a bust hand, the highest ranked single card. Since single cards are ranked ace, king, queen, jack, 10, 9, ..., 2 without regard to suit, the best bust hand is an "ace high."]

17. [25] For poker calculate the probability of drawing

   (a) exactly one pair.

   (b) two pairs.

   (c) three of a kind.

18. [15] Calculate the probability that a bridge hand

   (a) will be all spades.

   (b) will contain no spades.

   (c) will consist entirely of one suit.

   [See Exercises 20, 22, 51–54 for more card game problems.]

19. [13] Suppose A and B are independent events. Prove that

    (a) A and $\overline{B}$ are independent and

    (b) $\overline{A}$ and $\overline{B}$ are independent.

20. [28] Suppose a pack of eight cards is formed from the kings and queens of a bridge deck. Two cards are drawn from it. Show that no two of the following events are independent. A: At least one of the cards is black. B: One of the cards is the queen of spades. C: Both cards are kings. D: Both cards are queens.

21. [10] Prove that $P[A|B] = 1$ if and only if $P[B] \neq 0$ and $P[\overline{A} \cap B] = 0$.

22. [15] Suppose two cards are drawn from the deck considered in Exercise 20. Calculate

    (a) the probability that both cards are queens, given that one of the cards is a queen.

    (b) the probability that both cards are queens, given that one of them is a red queen.

    (c) the probability that both are queens, given that one of them is the queen of hearts.

23. [12] Fred Poisson, the chief statistician at Disneyland, has found that 72% of the visitors go on the Jungle Cruise, 56% ride the Monorail, 60% take the Matterhorn ride, 50% go on the Jungle Cruise and ride the Monorail, 45% go on the Jungle Cruise and on the Matterhorn ride, 40% ride the Monorail and take the Matterhorn ride, and 30% take all three rides. Assuming Poisson's figures are correct, calculate the probability that a visitor to Disneyland will

    (a) go on at least one of the three rides.

    (b) ride the Monorail given that the Jungle Cruise was taken.

    (c) take the Matterhorn ride given that both the Jungle Cruise and Monorail rides were taken.

24. [12] All the families in Dogpatch have exactly two children. For these families we can represent the children by bb, bg, gb, gg. In each pair b stands for boy and g for girl; the first letter in each pair represents the older child. We assume boys and girls are equally likely so that probability of each sample point is $1/4$.

   (a) Given that a family has a boy (event B), what is the probability that both children are boys (event A)?

   (b) Given that the older child is a boy (event C), what is the probability that both children are boys (event A)?

   (c) Let A be the event that "the family has children of both sexes," and B the event "there is at most one girl." Are A and B independent?

25. [15] The families of workers at Tiny Timber have at most 3 children each. The probability distribution for the number of children per family is given by

Number of children:	0	1	2	3
Probability:	0.20	0.50	0.25	0.05

The probability that a child is a boy is the same as the probability a child is a girl.

   (a) Calculate the probability that a family has exactly one boy. (There may be girls too.)

   (b) Calculate the probability there are two children in a family given that the family has exactly one boy.

26. [10] The employees parking lot at the Buss Stout Fence Company has 50 percent U.S. cars, of which 15 percent are compact; 30 percent of the cars are European, of which 40 percent are compact; and 20 percent of the cars are Japanese, of which 60 percent are compact. If a car is randomly selected from the lot, calculate

   (a) The probability it is a compact.

   (b) Given that the car is a compact, the probability that it is European.

27. [8] Belchfire Motors automobiles are equally likely to be manufactured on Monday, Tuesday, Wednesday, Thursday, or Friday; no cars are constructed on weekends. Ralph Wader, the company statistician, has determined that 4 percent of the cars produced on Monday are "lemons"; 1 percent of the cars made on Tuesday, Wednesday, or Thursday are lemons; and 2 percent of cars manufactured on Friday are lemons. You find that your Belchfire car is truly a lemon. What is the probability it was manufactured on Monday?

In the next five exercises we will consider the matching problem, sometimes called *Montmort's problem*. In one version of the problem a jar contains n balls numbered from 1 to n. The balls are mixed thoroughly and drawn out one at a time. What is the probability that no ball is drawn in the order of its label? That is, on no draw, say the ith draw, is it true that ball number i is drawn. In another version n letters are typed and n envelopes addressed to n different people. The letters are randomly stuffed in envelopes. The question then becomes the probability that none of the addressees receive the correct letter. In the "mixed-up hats" version, n men check their hats. When they reclaim them each man is given a hat randomly selected from those remaining. All versions of the matching problem can be modeled as n urns and n numbered balls with one ball inserted into each urn. A match occurs if ball k is put into urn k. We let A_k be the event that there is a match in the kth urn. Then $P[A_k]$ is the probability that the match occurs; for the hat problem it is the probability that the kth man gets his own hat.

28. [8] For the sample space of inserting n balls into n urns let each sample point be an n-tuple $(x_1, x_2, \ldots, x_n)$, where x_j represents the number of the ball put into the jth urn (sometimes, unromantically, called a pot). Thus, each component is a number from 1 to n and no two components are equal. Then the event $A_k = \{(x_1, x_2, \ldots, x_n) \in \Omega : x_k = k\}$. Prove that $P[A_k] = 1/n$. Thus, the probability of a man getting his own hat does not depend on whether he gets to make the first, second, or even last choice.

29. [10] Noah Peale and Mail Chauvinist are part of a group of 6 people who have put their hats on a table. Everyone then selects a hat randomly from those on the table. Calculate the probability that

    (a) Noah gets his own hat.
    (b) both Noah and Mail get their own hats.
    (c) at least one, either Noah or Mail, will get his own hat.

30. [25] Consider the matching problem of n urns and n numbered balls. Prove that the probability that there is at least one match is given by

$$1 - \frac{1}{2!} + \frac{1}{3!} - \cdots + \frac{(-1)^{n-1}}{n!} \approx 1 - e^{-1} = 0.632120559.$$

31. [12] Consider Exercise 30.

    (a) Calculate the probability of at least one match for $n = 2, 3, 4$, and compare it to $1 - e^{-1}$.

(b) Show that

$$\left| 1 - e^{-1} - \left(1 - \frac{1}{2!} + \frac{1}{3!} - \cdots + \frac{(-1)^{n-1}}{n!} \right) \right| \leq \frac{1}{(n+1)!}.$$

Conclude that, for $n \geq 4$, the probability of at least one match differs from $1 - e^{-1} \approx 0.63$ by less than 0.01; that is, the probability of no match is about 0.63 for all $n \geq 4$.

32. [10] Bigg Fakir claims that by clairvoyance he can tell the numbers of four cards numbered one to four, that are laid face down on a table. If he has no special powers and guesses at random, calculate the following:

   (a) the probability that Bigg gets at least one right.

   (b) the probability he gets two right.

   (c) the probability Bigg gets them all right.

33. [3]

   (a) Suppose $\binom{n}{11} = \binom{n}{7}$. What is n?

   (b) Suppose $\binom{18}{r} = \binom{18}{r-2}$. what is r?

34. [22] Following Knuth [14, page 51], we define $\binom{r}{k}$ for all real r and all integers k by

$$\binom{r}{k} = \frac{r(r-1)\cdots(r-k+1)}{k(k-1)\cdots(1)} = \prod_{1 \leq j \leq k} \left(\frac{r+1-j}{j} \right),$$

when k is a nonnegative integer and

$$\binom{r}{k} = 0$$

when k is negative.[20] Thus,

$$\binom{-7.2}{2} = \frac{(-7.2)(-8.2)}{2} = 29.52,$$

and $\binom{r}{0} = 1$ for all r, by the convention that an empty product in the definition of $\binom{r}{k}$ is one. Prove

---

[20]For even more about binomial coefficients see Graham *et al.* [11, chapter 5].

(a) $\binom{r}{k} = \frac{r}{k}\binom{r-1}{k-1}$ if k is a nonzero integer.

(b) $\binom{r}{k} = \frac{r}{r-k}\binom{r-k}{k}$, when k is an integer and $k \neq r$.

(c) $\binom{r}{k} = \binom{r-1}{k} + \binom{r-1}{k-1}$, when k is any integer.

(d) $\binom{-r}{k} = (-1)^k\binom{r+k-1}{k}$, when k is any integer.

35. [C25] Seven terminals of an interactive system at Crocker Ship are attached by a communication line to the central computer. Exactly four of the seven terminals are ready to transmit a message. Assume that each terminal is equally likely to be in the ready state. Let X be the random variable whose value is the number of terminals polled until the first ready terminal is located.

   (a) What values may X assume?

   (b) What is the probability that X will assume each of these values? Assume that terminals are polled in a fixed sequence without repetition.

   (c) Suppose the communication line has m terminals attached, of which n are ready to transmit where $n \geq 1$. Show that X can assume only the values $i = 1, 2, \ldots, m-n+1$ with $p(i) = P[X = i] = \binom{m-i}{n-1}/\binom{m}{n}$.

36. [35] Assume, as in Exercise 35(c), that m terminals at Transend Realty are attached to a communication line linked to a computer. Suppose also that Y terminals are ready to transmit, where $Y \geq 1$. Let X be the number of polls required to find the first terminal in the ready state. Prove the following results (due to Russell Ham):

   (a) $E[X|Y = n] = \left(\frac{m+1}{n+1}\right)$.

   (b) $E[X^2|Y = n] = \left[1 + 2\left(\frac{m-n}{n+1}\right)\right]\left(\frac{m+1}{n+1}\right)$.

37. [C20] (This rating is [T30] if nothing more powerful than a nonprogrammable calculator is available and [15] if APL is available so that you can copy and use the APL functions BINOMIAL and MPOLL.) Suppose seven terminals are connected to a communication line of an interactive computer system. Each terminal operates independently and has probability 0.2 of being ready to transmit. Thus, if Y is the random variable that counts the number of terminals ready to transmit, Y has a binomial distribution with parameters $n = 7$ and $p = 0.2$. Find the mean and standard deviation of the number of polls necessary to find the first ready terminal. Assume that 7 polls are required to discover that no terminal is ready.

38. [20] Swann Dive, a systems programmer at Poly Unsaturated, offered his friend Charlie Tuna, an application programmer, the following proposition. On each roll of three dice Swann would pay Charlie one dollar for each ace that showed; if no aces were turned Charlie would pay Swann one dollar. Charlie reasons that the probability of rolling an ace on the first die is $\frac{1}{6}$; similarly for the second and third die. Hence the probability $3 \times \frac{1}{6} = \frac{1}{2}$ of getting at least one ace and he might get two or even three of them.

   (a) Is Charlie right—that is, is it a good proposition for him?

   (b) What is the probability that Charlie will roll one, two, or three aces, respectively?

   (c) What is the average amount of money Charlie can expect to win each time the dice are rolled? (Swann didn't tell him, but this game is known as chuck-a-luck at carnivals.)

39. [35] A single disk storage device has N concentric tracks and one access arm. It has been loaded with data in such a way that successive movements of the access arm (called track seeks) are independent of one another. The probability that a randomly chosen seek will take the arm to track i is p_i. Let X represent the number of tracks passed between consecutive seeks, assuming that no physical repositioning of the access arm takes place between successive seek operations. Show the following:

   (a) X assumes the values $0, 1, \ldots, N-1$ and has the pmf $p(\cdot)$ defined by

   $$p(j) = P[X = j] = \begin{cases} \displaystyle\sum_{i=1}^{N} p_i^2, & j = 0 \\[2em] \displaystyle 2\sum_{i=1}^{N-j} p_i p_{i+j}, & j = 1, 2, \ldots, N-1. \end{cases}$$

   (b) For the case that $p_i = 1/N$ for all i, it is true that

   $$E[X] = \frac{(N^2 - 1)}{3N} \approx \frac{N}{3},$$

   $$E[X^2] = \frac{(N^2 - 1)}{6} \approx \frac{N^2}{6},$$

and

$$\text{Var}[X] \approx \frac{N^2}{18}.$$

(c) Suppose T, the seek time, is a linear function of X; that is,

$$T = AX + B,$$

where A and B are constants. (A is then given by

$$A = \frac{\text{maximum seek time} - \text{minimum seek time}}{N - 1},$$

and B is minimum seek time.)

Then it is true that

$$E[T] = \frac{A(N^2 - 1)}{3N} + B \approx \frac{AN}{3} + B,$$

and

$$\text{Var}[T] = A^2 \text{Var}[X] \approx \frac{A^2 N^2}{18}.$$

40. [5] Refer to Example 2.4.6. Calculate

$$P[A_2|A], \ P[A_3|A], \text{ and } P[A_4|A].$$

41. [20] Dusty Page, a librarian at Hard Core Computer (makers of solid state memory), tripped over the discrete random variables X and Y when he stepped from his office. These random variables have the joint probability mass function shown in the table below. Thus, X assumes the values 0 and 1, and Y assumes the values 0, 1, and 2.

X \ Y	0	1	2
0	1/8	1/4	1/8
1	0	1/8	3/8

Help Dusty out by doing or answering the following:

(a) Find the marginal probability mass functions p_X and p_Y.

(b) Find the conditional probability mass function of X, given that $Y = 2$.

(c) Are X and Y independent random variables? Why?

(d) Calculate $E[X], E[Y], \text{Var}[X],$ and $\text{Var}[Y]$.

(e) Find the probability mass function for $Z = X + Y$.

42. [HM30] Suppose X and Y are independent random variables, each with the density function f given by

$$f(x) = \begin{cases} 1 & \text{for } 0 < x < 1 \\ 0 & \text{otherwise.} \end{cases}$$

Show that the density function of $Z = X + Y$ is given by

$$f_Z(z) = \begin{cases} z & \text{for } 0 \leq z \leq 1 \\ 2 - z & \text{for } 1 < z \leq 2 \\ 0 & \text{otherwise.} \end{cases}$$

Thus, Z has a triangular distribution. (As we shall see in Chapter 3, X and Y are said to be *uniformly distributed*.) *Hint:* This Exercise can be solved by using convolution (Theorem 2.7.5) or by using the Laplace–Stieltjes transform (Theorem 2.9.3(a) and (d)).

43. [20] Suppose X and Y have the joint discrete distribution shown in the table. Show that X and Y are uncorrelated but not independent.

X \ Y	-1	0	1
-1	0	$\frac{1}{4}$	0
0	$\frac{1}{4}$	0	$\frac{1}{4}$
1	0	$\frac{1}{4}$	0

44. [12] Suppose X is an arbitrary random variable such that the mean $E[X]$ and standard deviation σ are defined (finite). For any p such that $0 < p < 0.5$, find $x_p > E[X]$ so that $P[X > x_p] \leq p$.

45. [10] Suppose X is a random variable with finite mean and variance. For $50 \leq r < 1$, we define the rth percentile value $\pi_X(r)$ by

$$P[X \leq \pi_X(r)] = \frac{r}{100}.$$

Thus, the 90th percentile value $\pi_X(90)$ is defined by

$$P[X \leq \pi_X(90)] = 0.90.$$

Show that

$$\pi_X(90) \leq E[X] + 3\sigma,$$

and

$$\pi_X(95) \leq E[X] + \sigma\sqrt{19}.$$

46. [5] A discrete random variable X is called a *truncated Poisson* random variable if its mass points are $0, 1, 2, \ldots, N$ and its probability mass function $p(\cdot)$ is given by $p(k) = Ce^{-\alpha}\alpha^k/k!$, $k = 0, 1, 2, \ldots, N$. What is the value of the constant C?

47. [5] The average length of messages received at a message switching center is 50 characters with a standard deviation of 10 characters. How many bytes (characters) of storage should be provided for each message buffer to ensure that 95% of all messages fit into one buffer?

48. [15] A certain access method, called method A, has been found to give a mean record retrieval time of 36 milliseconds with a standard deviation of 7 milliseconds, while method B has a mean retrieval time of 42 milliseconds with a standard deviation of 4 milliseconds.

    (a) If a major design objective is to have 90% of all individual retrievals completed in 55 milliseconds or less, which method should be selected?

    (b) Does the chosen method meet the objective?

49. [15] Inquiries to an interactive computer system at Rhode Block Security are of four types and make reference to different data bases as follows:

Inquiry type	Percent of type	Mean reference time (msec)	Standard deviation of reference time(msec)
A	40	100	80
B	30	120	30
C	20	80	40
D	10	40	20

For the entire collection of inquiries, what is

(a) the mean reference time?

(b) the variance of reference time?

50. [HM15] Use the Laplace–Stieltjes transform to derive the results of Example 2.9.1, that is, to prove that, for an exponential random variable X, we have

    (a) $\text{Var}[X] = E[X]^2$.
    (b) $E[X^n] = n!E[X]^n$, $n = 1, 2, \ldots$.

51. [30] Suppose a sequence of bridge hands is dealt. Let A be the event that each player is dealt one ace on a particular deal.

   (a) Show that A has a probability of about one-tenth. (Actually 0.1054981993.)

   (b) What is the probability that one particular player gets no ace for three consecutive deals?

   (c) Show that the probability that event A occurs at least once in seven deals is about one-half. (Actually, 0.54178581 or 0.5217031 if 0.1 is used as the probability of A.) *Hint:* By the general multiplication rule (Corollary to Theorem 2.4.1), the number of ways of dealing one bridge hand is

$$\binom{52}{13}\binom{39}{13}\binom{26}{13} = \frac{52!}{(13!)^4}.$$

52. [30] Recall that $P(n, k)$ is the symbol for the number of permutations of n objects taken k at a time and $C(n, k) = \binom{n}{k}$ is the symbol for the number of combinations of n objects taken k at a time. Using this notation we see that the number of different bridge hands is $C(52, 13) = 52!/(13!)(39!) = 6.350135 \times 10^{11}$. We can compute the probability of a given distribution of cards by suit (such as 12 in one suit and one in another) in a randomly chosen hand by dividing the number of possible hands with such a distribution by $C(52, 13)$. Consider a 5–4–3–1 distribution. If the suits are given (say the five-card suit is hearts, the four-card suit diamonds, the three-card suit clubs, and the remaining card is a spade), there are

$$
\begin{aligned}
C(13, 5)C(13, 4)C(13, 3)C(13, 1) &= 1,287 \times 715 \times 286 \times 13 \\
&= 3,421,322,190
\end{aligned}
$$

such hands. But there are $P(4, 4) = 24$ ways of permuting the 4 different sized suits in a 5–4–3–1 distribution so we have

$$
\begin{aligned}
P[\text{5–4–3–1 distribution}] &= \frac{P(4, 4)C(13, 5)C(13, 4)C(13, 3)C(13, 1)}{C(52, 13)} \\
&= 0.129307054.
\end{aligned}
$$

   (a) Show that the probability of a 4–4–3–2 distribution is

$$
\begin{aligned}
P[\text{4–4–3–2 distribution}] &= \frac{P(4, 2)C(13, 4)^2 C(13, 3)C(13, 2)}{C(52, 13)} \\
&= 0.215511757.
\end{aligned}
$$

(b) Show that the probability of a 4–3–3–3 distribution is

$$P[\text{4-3-3-3 distribution}] = \frac{P(4,1)C(13,4)C(13,3)^3}{P(52,13)}$$
$$= 0.105361303.$$

(c) Show that for any specific x–y–z–w distribution (where $x + y + z + w = 13$), we have

$$P[\text{x-y-z-w distribution}] = \frac{nC(13,x)C(13,y)C(13,z)C(13,w)}{C(52,13)}$$

$$n = \begin{cases} P(4,4) = 24 \text{ if all suits are of different size} \\ P(4,2) = 12 \text{ if exactly 2 suits are of the same size} \\ P(4,1) = 4 \text{ if 3 suits are of the same size.} \end{cases}$$

Of course, n is the number of different suit arrangements for a given x–y–z–w distribution.

53. [25] You are West in a bridge game and have no ace.

(a) What is the probability that your partner, East, has no ace?

(b) What is the probability that East has two or more aces?

54. [25] In a bridge game North and South have 10 spades between them.

(a) What is the probability that the three remaining spades are all in one hand (that is, that either East or West has no spades)?

(b) If the king of spades is one of the three spades, what is the probability that one player has the king and the other has the remaining two spades?

55. [25] What is the probability that in a hand of bridge each player has all cards in one suit; that is, one player has all spades, one all hearts, one all clubs, and one all diamonds?

56. [25] During the winter season at the Fearsome Peaks Ski Resort, each of the two roads from Area A to Area B has probability p of being blocked by snow. The same can be said of the two roads that lead from Area B to Area C; that is, all roads, independently, have probability p of being blocked by snow.

(a) What is the probability that there is an open path from Area A to Area C?

(b) Having calculated the probability in part (a) when $p = 1/2$, the owners of FPSR decide to build a direct road from Area A to Area C, which, independently of the other roads, is blocked with probability p. What is the new probability there is an open road from Area A to Area C?

(c) If $p = 0.25$, calculate the probabilities of part (a) and part (b).

57. [20] Consider the following.

(a) Suppose a coin that has probability p of turning up heads is tossed once. If X is the number of heads, and Y the number of tails show that X and Y are not independent.

(b) Let the coin of part (a) be tossed a random number of times N, where N is a Poisson random variable with parameter α, (see Example 2.7.6 for the definition). Let X and Y be the resulting numbers of heads and tails, respectively. Show that X and Y *are* independent.

58. [10] Kollossal Airways and Teeny Weeny Airlines compete for passengers from Pointaye to Pointbee. It is known that each passenger who makes reservations fails to show up with probability 1/10 independently of other passengers so Kollossal always books 20 passengers for their 18 seat airplane and Teeny books 10 for their nine-seat airliner. What is the probability that each is overbooked on a randomly chosen flight?

59. [M22] Prove that

$$\sum_{i=0}^{n} \binom{n}{i}^2 = \binom{2n}{n}.$$

[*Hint:* Use the identity $(1+z)^n(1+z^{-1})^n = z^{-n}(1+z)^{2n}$.]

60. [02] Show that, if you had an income of $20,000 last year and Rockefeller had an income of $2,000,000, then your joint average income would be $1,010,000 with a standard deviation of $990,000.

61. [HM12] Suppose the joint density function of the continuous random variables X and Y is given by

$$f(x,y) = \begin{cases} 2 - x - y, & \text{if } 0 < x < 1 \text{ and } 0 < y < 1, \\ 0 & \text{otherwise.} \end{cases}$$

    (a) Find the marginal density functions $f_X(\cdot)$ and $f_Y(\cdot)$ of X and Y, respectively. Are X and Y independent?

    (b) Find the conditional density functions $f_{X|Y}(x|y)$ and $f_{Y|X}(y|x)$.

    (c) Calculate $E[X|Y = y]$ and $E[Y|X = x]$.

62. [HM12] Suppose the joint density function of the continuous random variables X and Y is given by

$$f(x,y) = \begin{cases} e^{-x-y}, & \text{if } x > 0 \text{ and } y > 0, \\ 0, & \text{otherwise.} \end{cases}$$

Answer (a), (b), and (c) of Exercise 61 for the above X and Y.

63. [5] Swann Dive (see Exercise 38) offers his friend, Charlie Tuna, a new proposition. Charlie will deal himself 2 cards from a well-shuffled deck of bridge cards. If Charlie has one or more hearts, Swann will give him a dollar; otherwise Charlie must pay Swann a dollar. What is the probability that Charlie will win on one play?

64. [6] Swann (see Exercise 63) shuffles 5 black cards and 5 red cards and lets Charlie randomly choose 2 of the cards. If they are both red or both black, Swann gives Charlie a dollar; otherwise he takes a dollar from him. What is the probability that Charlie will win a dollar on one play?

65. [6] Swan asks Charlie to toss an honest coin three times. Charlie must call heads or tails before each toss. If he is right at least two out of three times he wins a dollar; otherwise he loses a dollar. What is the probability that Charlie wins?

66. [20] Swann Dive has three cards, which he shuffles in a hat. One of his cards is red on both sides, one red on one side and black on the other; the third is black on both sides. Swann randomly selects a card and places it face down on the table. The side that shows is black. Swann offers to pay his friend, Charlie, a dollar if the other side is black; otherwise he takes a dollar from Charlie. What is the probability that Charlie wins?[21]

---

[21]Weaver [25, page 126] makes the following statement about this problem: "Forty years ago, when graduate students had to work for their living, the author used to teach this particular problem, at reasonable rates and using the experimental method, to his college friends."

67. [10] Charlie Tuna puts two decks of well shuffled playing cards side by side in front of you. He begins by simultaneously turning over a card in each deck. He does this, over and over, until all cards have been turned over in pairs. If, on any turn, Charlie hits the same card in both decks you win a dollar. If he has no matches you lose a dollar. (You win only one dollar if he has multiple matches.) What is the probability you will win? [*Hint:* See Exercise 30.]

68. [25] John and Mark found 16 dollars in a paper bag. Rather than splitting the cash they decided to flip a coin for it. They decided that the one who first wins 10 tosses gets all the money. After 15 tosses of the coin John has won eight times and Mark seven times. On the 16th flip the coin rolled away and was lost, so they decided to divide the 16 dollars based on their respective chances of winning if they started up again. Clearly, John should get more than Mark, but exactly how much should each receive? Note: This is a special case of a general problem called the "problem of points" first solved successfully by Pascal. [*Hint:* What is the maximum number of tosses that remain to be made to settle the winner?]

69. [10] You decide to offer a gambling game with cards to your friend, Amos. You mark each card with a number from 1 to 52; that is, you write 1 on the first card in the deck, 2 on the second card, etc., to 52 on the last card. You shuffle the cards. If the top four cards are in ascending order you pay Amos $20; otherwise he pays you a dollar. (By ascending order we mean, for example, the top card is 7, the second card 12, the third card is 40, and the fourth card is 47.) What is the probability that Amos will win? What is your average winning per play? [*Hint:* In how many orders can the top 4 cards be arranged?]

70. [15] You allow your friend, Sally, to shuffle a three-card deck consisting of an ace, a king, and a queen. Sally chooses two of the cards at random and discards the third. She shows you a queen when you ask for a picture card. What is the probability that she also has the king?

71. [5] The weather forecaster on TV reported that the probability of rain tomorrow is 1/4. Find

    (a) the odds in favor of rain tomorrow, and

    (b) the odds against rain tomorrow.

72. [5] Suppose the odds in favor of Barry Blunt marrying Sally Sharp are 3:5 (3 to 5). Find the probability that they will get married.

73. [5] Consider Example 2.10.2.

    (a) Use Chebyshev's inequality (Theorem 2.10.2) to show that the probability the response time is one second or more is 0.04.

    (b) Use the one-sided inequality (Theorem 2.10.3) to show that the probability that the response time will exceed one second is 4/104.

74. [10] Moon Systems, a manufacturer of scientific workstations, produces its Model 13 System at sites A, B, and C; 20% at A, 35% at B, and the remaining 45% at C. The probability that a Model 13 System will be found defective upon receipt by a customer is 0.01 if shipped from site A, 0.06 if from site B, and 0.03 from site C.

    (a) What is the probability that a Model 13 System selected at random at a customer location will be found defective?

    (b) Suppose a Moon Model 13 System selected at random is found to be defective upon arrival at a customer location. What is the probability that it was manufactured at site B?

75. [8] Suppose a bookmaker tells you the odds against the Washington Redskins beating the Dallas Cowboys next week is 3:2. Assuming the odds are correct, (a) what is the probability that the Redskins will win and (b) if the Redskins win and you have bet $10 that they will win, how much will you win?

With the jargon of Math, do you find
That some lessons can lessen your mind?
  Turn over a leaf:
  Study Laughing and Grief...
See your Reeling and Writhing unwind!
(with apologies to Lewis Carroll)

Ben W. Lutek

In the long run, we are all dead.
John Maynard Keynes

There were so many kinsmen Bernoulli
That keeping them straight would unduly
Have tired and worn to a frazzle
The record-keepers of Basel.

Karl David
Wells College

References

[1] Tom M. Apostol, *Calculus*, Vol. II, 2nd ed., Ginn (Blaisdell), Boston, 1969.

[2] Tom M. Apostol, *Mathematical Analysis*, 2nd ed., Addison-Wesley, Reading, MA, 1974.

[3] G. C. Berresford, The uniformity assumption in the birthday problem, *Math. Mag.*, **53**(5), (Nov. 1980).

[4] Kai Lai Chung, *Elementary Probability Theory with Stochastic Processes*, 3rd ed., Springer-Verlag, New York, 1979.

[5] Harald Cramér, *Mathematical Methods of Statistics*, Princeton University Press, Princeton, 1946.

[6] Phillip J. Davis, *The Thread: A Mathematical Yarn*, 2nd ed., Harcourt Brace Jovanovich, Inc., Cambridge, MA, 1989.

[7] William Feller, *An Introduction to Probability Theory and Its Applications*, Vol. I, 3rd ed., John Wiley, New York, 1968.

[8] William Feller, *An Introduction to Probability Theory and Its Applications*, Vol. II, 2nd ed., John Wiley, New York, 1971.

[9] Joseph L. Gastwirth, The statistical precision of medical screening procedures: application to polygraph and AIDS antibodies test data, *Statistical Science*, **2(3)**, August 1987.

[10] I. S. Gradshteyn and I. M. Ryzhik, *Table of Integrals, Series, and Products*, Academic Press, New York, 1980.

[11] Ronald L. Graham, Donald E. Knuth, and Oren Patashnik, *Concrete Mathematics: A Foundation for Computer Science*, Addison-Wesley, Reading, MA, 1989.

[12] A. Gravey, A simple construction of an upper bound for the mean of the maximum of n identically distributed random variables, *Journal of Applied Prob.*, **22**(4), (Dec. 1985), 844–851.

[13] G. R. Grimmett and D. R. Stirzaker, *Probability and Random Processes*, Oxford Univ. Press, Oxford, 1982.

[14] Donald E. Knuth, *The Art of Computer Programming, Vol. 1, Fundamental Algorithms*, 2nd ed., Addison-Wesley, Reading, MA, 1973.

[15] W. Moser, It's not a coincidence, but it *is* a surprise, *CRUX MATHEMATICORUM*, **10(1)**, (Jan. 1984), 210-213.

[16] Oystein Ore, *Cardano, The Gambling Scholar*, Princeton Univ. Press, Princeton, 1953. Also published by Dover Publications, New York.

[17] Oystein Ore, Pascal and the invention of probability theory, *Amer. Math. Month.*, **67**, (1960), 409–419.

[18] Edward Packel, *The Mathematics of Games and Gambling*, The Mathematical Association of America, 1981.

[19] E. Parzen, *Modern Probability Theory and Its Applications*, John Wiley, New York, 1960.

[20] John A. Rice, *Mathematical Statistics and Data Analysis*, Wadsworth & Brooks/Cole, Belmont, CA, 1988.

[21] Sheldon M. Ross, *Introduction to Probability Models*, 4th ed., Academic Press, Orlando, 1989.

[22] Bernard Siskin and Jerome Staller, *What Are The Chances?*, Crown Publishers, New York, 1989.

[23] J. Laurie Snell, *Introduction to Probability*, Random House/Birkhäuser, New York, 1988.

[24] Peter J. Verstappen (commentaries), *The Odds On Virtually Everything*, G. P. Putnam's Sons, New York, 1980.

[25] Warren Weaver, *Lady Luck: The Theory of Probability*, Doubleday, New York, 1963. Republished by Dover Publications, New York, 1982.

Chapter 3

Probability Distributions

All business proceeds on beliefs,
or judgments of probabilities,
and not on certainties.
Charles Eliot

3.0 Introduction

In Chapter 2 we defined a random variable X to be a real-valued function defined on a sample space. Thus, to each outcome ω of a random experiment, the random variable X assigns the value $X(\omega)$. The "randomness" in the name *random variable* comes about because of the uncertainty of the outcome of the experiment *before* the experiment is performed; once the outcome of the experiment has been determined, so has the value of the random variable. Thus, if the random variable X counts the number of spots turned up when two dice are tossed, as soon as the dice are tossed, the value of X is known. The usefulness of the random variable concept depends upon the ability to determine the probability that the *values* of the random variable occur in a given set of real numbers. That is, the *probability distribution* of a random variable is its most important property. For this reason the two statements

1. "X is a Poisson random variable," and

2. "X has a Poisson distribution,"

are used interchangeably. The same is true for any other type of random variable, of course. If the probability distribution of a random variable is known, the actual underlying sample space is not important. Thus, if we know the distribution function $F(\cdot)$ of X, defined for all real x by $F(x) = P[X \leq x]$, we can calculate the probability $P[X \in A]$ where A is a set of real numbers satisfying very weak restrictions. In most practical cases we are interested in probabilities such as $P[a < X \leq b] = F(b) - F(a)$; the distribution function $F(\cdot)$ enables us to make this type of calculation easily. If X is a discrete random variable, its distribution function can be calculated from its probability mass function $p(\cdot)$ defined for all real x by $p(x) = P[X = x]$. If X is continuous, its distribution function can be calculated from its density function $f(\cdot)$, which is characterized by the properties:

(i) $f(x) \geq 0$ for all real x,

(ii) $\int_{-\infty}^{\infty} f(x)\,dx = 1$, and

(iii) $P[a \leq X \leq b] = \int_{a}^{b} f(x)\,dx$ for all real a, b with $a < b$.

Then

$$F(x) = \int_{-\infty}^{x} f(t)\,dt. \tag{3.1}$$

In this chapter we will study some common random variables, which are especially useful for computer science applications.

> *Make things as simple as possible but no simpler.*
> Albert Einstein

3.1   Discrete Random Variables

A random variable X is discrete if

$$\sum_{x \in T} p(x) = \sum_{x \in T} P[X = x] = 1, \tag{3.2}$$

where $T = \{\text{real } x : p(x) > 0\}$. The set T is either finite or countably infinite. (For a proof see Apostol [3, page 511].) Each point of T is called a *mass point of X*. We sometimes indicate the mass points of X by writing "X assumes the values $x_1, x_2, x_3, \cdots$." Just as the distribution function of X can be calculated from the pmf $p(\cdot)$ by the formula

$$F(x) = \sum_{x_i \leq x} p(x_i), \tag{3.3}$$

the pmf $p(\cdot)$ can be calculated from $F(\cdot)$ at all mass points by

$$p(x_i) = \lim_{x \to x_i^-} \{F(x_i) - F(x)\}. \tag{3.4}$$

That is, the graph of $F(\cdot)$ is a step function with a jump at each mass point x_i, the jump having magnitude $p(x_i)$.

We summarize the properties of some useful discrete random variables in Table 1 of Appendix A.

3.1.1 Bernoulli Random Variables

Several important discrete random variables are derived from the concept of a Bernoulli sequence of trials.

A *Bernoulli trial* is a random experiment in which there are only two possible outcomes, usually called success or failure, with respective probabilities p and q, where $p + q = 1$. We assume $0 < p < 1$, for otherwise the results are trivial. A sequence of such trials is a *Bernoulli sequence* if the trials are independent and the probability of success (or failure) is constant from trial to trial. A *Bernoulli random variable* describes a Bernoulli trial and thus assumes only two values: 1 (for success) with probability p and 0 (for failure) with probability $q = 1 - p$.

An example of a Bernoulli trial can be constructed from any sample space Ω that has an event A such that $0 < P[A] < 1$, by identifying the occurrence of A with success and $\overline{A}$ with failure. The corresponding Bernoulli random variable X is defined to be 1 for every point of A and 0 for every point of $\overline{A}$.

A Bernoulli random variable is completely determined by the value of p and therefore is said to have one parameter, namely, p. As we saw in Example 2.9.4, such a random variable has the z-transform or generating function

$$g(z) = q + pz. \tag{3.5}$$

We also showed that $E[X] = p$ and $\sigma^2 = pq$. The Bernoulli random variable is the basis for other important random variables, including the binomial and geometric random variables.

3.1.2 Binomial Random Variables

Consider a Bernoulli sequence of n trials where the probability of success on each trial is p. The random variable X, that counts the number of successes in the n trials is called a *binomial random variable* with parameters n and p. Thus, X can assume only the values $0, 1, 2, \ldots, n$. A Bernoulli sequence of n trials can be represented as a string $a_1, a_2, \cdots, a_n$ where each a_i is either s for a success or f for a failure. Thus, a sequence of 5 trials in which 2 successes are followed by a failure, a success, and a failure, would be represented as ssfsf. If now the binomial random variable X has parameters

n and p, and k is an integer between 0 and n (inclusive), then any string $a_1, a_2, \ldots, a_n$ representing k successes and $n - k$ failures has probability $p^k q^{n-k}$, since each trial is independent. (The probability can be calculated by replacing each s in $a_1, a_2, \ldots, a_n$ by p, each f by q and multiplying the resulting numbers.) The number of strings $a_1, a_2, \ldots, a_n$ representing k successes and $n - k$ failures is just the number of ways the k indices representing success can be chosen from the n indices, that is, $\binom{n}{k}$. Hence, the pmf $b(\cdot; n, p)$ of a binomial random variable with parameters n and p is defined by

$$b(k; n, p) = \binom{n}{k} p^k q^{n-k}, \quad k = 0, 1, \cdots, n, \tag{3.6}$$

where $q = 1 - p$.

X can be represented as

$$X = X_1 + X_2 + \cdots + X_n, \tag{3.7}$$

where $X_1, X_2, \ldots, X_n$ are independent identically distributed Bernoulli random variables. By Theorem 2.7.1 we have

$$E[X] = E[X_1] + E[X_2] + \cdots + E[X_n] = np, \tag{3.8}$$

since $E[X_i] = p$ for each i.

By Theorem 2.7.2,

$$\text{Var}[X] = \text{Var}[X_1] + \text{Var}[X_2] + \cdots + \text{Var}[X_n] = npq, \tag{3.9}$$

since $\text{Var}[X_i] = pq$ for each i. (In Example 2.9.5 we calculated the mean and variance of X using generating functions.)

The APL function BINOMIAL can be used to calculate binomial probabilities and BINSUM can be used to sum binomial probabilities.

Example 3.1.1 A master file of 120,000 records is stored as a sequential file on a direct-access storage device in blocks of six records. Each day the transaction file is run against the master file and approximately 5 percent of the records are updated. The records to be updated are assumed to be distributed uniformly over the master file. An entire block of records must be updated if one or more records in the block need updating. What is the mean and standard deviation of the number of blocks that must be updated? Use Chebyshev's inequality to estimate the probability that 5,200 to 5,400 blocks must be updated.

Solution Let X be the random variable that counts the number of records in a block that must be updated. It is reasonable to assume that X is a binomial random variable with parameters $n = 6$ and $p = 0.05$. (A Bernoulli trial consists of checking a record to see if it must be updated, that is, whether or not it is listed in the transaction file.) A given block must be updated if $X \geq 1$, that is, with probability $P[X \geq 1] = 1 - P[X = 0]$. Hence, the probability that any given block must be updated is

$$1 - b(0; 6, 0.05) = 1 - (0.95)^6 = 1 - 0.735092 = 0.264908. \tag{3.10}$$

Let Y be the number of blocks that need to be updated. Y is a binomial random variable with parameters $n = 20,000$ and $p = 0.264908$. Therefore, the average number of blocks to be updated is $E[Y] = 20,000 \times 0.264908 = 5,298.16 \approx 5,300$ with standard deviation

$$\sigma = (20,000 \times 0.264908 \times 0.735092)^{1/2} \approx 62.41 \quad \text{blocks.} \tag{3.11}$$

By Chebyshev's inequality the probability that Y is between 5,200 and 5,400 blocks is

$$P[|Y - 5,300| \leq 100] \geq 1 - \left(\frac{62.41}{100}\right)^2 = 0.6105. \tag{3.12}$$

Using the APL function BINSUM, the author has calculated the correct probability that Y lies between 5,200 and 5,400 to be 0.890677863. This value was confirmed by the SAS/STAT [28] function PROBBNML. Thus, the above estimate of the probability is not very good.[1] In Exercise 44 we ask you to use the normal approximation. □

Example 3.1.2 The interactive computer system at Gnu Glue has 20 communication lines to the central computer system. The lines operate independently and the probability that any particular line is in use is 0.6. What is the probability that 10 or more lines are in use?

Solution The number of lines in operation X has a binomial distribution with parameters $n = 20$ and $p = 0.6$. The required probability is thus

$$P[X \geq 10] = \sum_{k=10}^{20} \binom{20}{k} (0.6)^k (0.4)^{20-k} = 0.872479. \tag{3.13}$$

Equation (3.13) is a tedious calculation to carry out with a pocket calculator (unless, of course, it is programmed to calculate binomial probabilities) but it can be done easily using SAS/STAT [28], MINITAB [20], an HP-21S calculator with its binomial library program, or the EXPLORE program BINOM of Doane [9]. The probability can also be approximated by the *normal distribution*, as we will show later in the chapter. The APL function BINSUM provides another way to make the above calculation. □

---

[1] Even the HP-21S calculator could not handle this calculation.

3.1.3 Geometric Random Variables

Suppose a sequence of Bernoulli trials is continued until the first success occurs. Let X be the random variable that counts the number of trials *before* the trial at which the first success occurs. Then X can assume the values $0, 1, 2, 3, \ldots$. X assumes the value zero if and only if the first trial yields a success; hence, with probability p. X assumes the value 1 if and only if the first trial yields a failure and the second a success; hence the probability qp, where $q = 1 - p$. Continuing in this way we see that the pmf of X is given by

$$p(k) = q^k p, \quad k = 0, 1, 2, \ldots. \tag{3.14}$$

The probability-generating function of X is thus

$$g(z) = \sum_{k=0}^{\infty} q^k p z^k = p \sum_{k=0}^{\infty} (qz)^k = \frac{p}{1 - qz}. \tag{3.15}$$

In order for (3.15) to hold, we must have $qz < 1$ or $z < 1/q$. Then

$$g'(z) = \frac{pq}{(1 - qz)^2} \quad \text{and} \quad g''(z) = \frac{2pq^2}{(1 - qz)^3}.$$

Hence, by Theorem 2.9.2,

$$E[X] = g'(1) = \frac{pq}{(1 - q)^2} = \frac{q}{p}, \tag{3.16}$$

and

$$
\begin{aligned}
\text{Var}[X] &= g''(1) + g'(1) - (g'(1))^2 \\
&= 2\frac{q^2}{p^2} + \frac{q}{p} - \frac{q^2}{p^2} \\
&= \frac{q^2}{p^2} + \frac{q}{p} = \frac{q(q + p)}{p^2} \\
&= \frac{q}{p^2}.
\end{aligned} \tag{3.17}
$$

The geometric random variable is important in queueing theory and other areas of applied probability.

Example 3.1.3 Consider Example 3.1.1. Let X be the number of blocks of the master file that are read before the first block that must be updated is found. Then X is a geometric random variable with parameter $p = 0.265$. The expected value of X is $q/p = 0.735/0.265 = 2.774$ blocks with standard deviation $\sqrt{q}/p = 3.235$. $\square$

Suppose now that during each time interval of a fixed length, say h, that an event of some kind, called an *arrival*, may or may not occur. Suppose further that the occurrence or nonoccurrence of the arrival in each interval is determined by a Bernoulli random variable with a fixed probability p of success from one interval to the next, that is, by a Bernoulli sequence of trials. Then the *interarrival time* T is defined to be the number of trials (time intervals of length h) before the first success (arrival). Thus, T has a geometric distribution with parameter p. Now suppose we are given that there were no arrivals during the first m intervals of length h and we wish to calculate the probability that there will be k more time intervals with no arrivals before the next arrival, that is, $P[T = k + m | T \geq m]$ for $k = 0, 1, 2, \ldots$. By the definition of conditional probability (see Section 2.4), we have

$$P[T = k + m | T \geq m] = \frac{P[(T = k + m) \cap (T \geq m)]}{P[T \geq m]}. \qquad (3.18)$$

But

$$(T = k + m) \cap (T \geq m) = (T = k + m), \qquad (3.19)$$

and

$$P[T \geq m] = pq^m(1 + q + q^2 + \cdots) = \frac{pq^m}{(1 - q)} = q^m. \qquad (3.20)$$

Hence,

$$P[T = k + m | T \geq m] = \frac{pq^{m+k}}{q^m} = pq^k = P[T = k]. \qquad (3.21)$$

Equation (3.21) shows the Markov or memoryless property of the geometric distribution, that is, the presence or absence of an arrival at any point in time has no effect on the interarrival time to the next arrival. The system simply does not "remember" when the last arrival (success) occurred. Thus, in Example 3.1.3, the average interarrival time between any two successive blocks that must be updated is 2.774; also at any arbitrary point in time, it is the average number of blocks to be read before the next block is found that must be updated.

3.1.4 Poisson Random Variables

We say that a random variable X is a Poisson random variable with parameter $\alpha > 0$ if X has the mass points $0, 1, 2, 3, \ldots$, and if its probability mass function $p(\cdot; \alpha)$ is given by

$$p(k; \alpha) = P[X = k] = e^{-\alpha}\frac{\alpha^k}{k!}, \quad k = 0, 1, 2, \ldots. \qquad (3.22)$$

This probability distribution was discovered by the French mathematician Siméon D. Poisson (1781–1840).

Equation (3.22) does define a pmf because $p(k; \alpha)$ is clearly nonnegative for all nonnegative integers k, and

$$\sum_{k=0}^{\infty} p(k; \alpha) = e^{-\alpha} \sum_{k=0}^{\infty} \frac{\alpha^k}{k!} = e^{-\alpha} e^{\alpha} = e^0 = 1.$$

In Example 2.9.2 we showed that a Poisson random variable with parameter α has the moment generating function

$$\psi(\theta) = \exp[\alpha(e^{\theta} - 1)], \tag{3.23}$$

and furthermore, that

$$E[X] = \alpha \quad \text{and} \quad \text{Var}[X] = \alpha.$$

In Example 2.9.3 we showed that, if $Z = X + Y$, where X is Poisson distributed with parameter α and Y is Poisson distributed with parameter β where X and Y are independent, then Z is Poisson distributed with parameter $\alpha + \beta$. Thus, independent Poisson random variables have the reproductive property.

The Poisson random variable is one of the four or five most important random variables for applied probability and statistics. One reason for this importance is that a great many natural and man-made phenomena are described by Poisson random variables.

The following phenomena have Poisson distributions.

(a) The number of alpha particles emitted from a radioactive substance per unit time (see Bateman [5], Rutherford and Geiger [26], and Lippman [18, pages 76–77]). (Geiger is the inventor of the celebrated Geiger counter, that counts not geigers but rather radiation levels.)

(b) The number of flying-bomb hits in the south of London during World War II (see Clarke [7], and Feller [10, pages 160–161]).

(c) The number of vacancies per year in the United States Supreme Court (see Wallis [37] and Parzen [22, pages 256–257]).

Other examples include misprints per page of a book, raisins per cubic inch of raisin bread, deaths caused by horse kicks per corps-year in the Prussian cavalry (see Bortkiewicz [6]), and the number of chromosome interchanges in organic cells caused by X-ray radiation.

A number of random variables of interest to computer science have been found to have Poisson distributions. We shall discuss some of these in this book.

Another reason for the importance of the Poisson distribution is that its pmf, given by (3.22), is easy to calculate. Furthermore, a binomial random variable can often be approximated by a Poisson random variable—in fact, this is the way Poisson originally conceived the probability distribution that bears his name.

Theorem 3.1.1 *Suppose X has a binomial distribution with parameters n and p. Then, if n is large and p is small with $\alpha = np$, $b(k; n, p)$ is approximately $p(k; \alpha)$ in the sense that*

$$\lim_{n \to \infty} b\left(k; n, \frac{\alpha}{n}\right) = p(k; \alpha), \quad k = 0, 1, 2, \ldots.$$

Proof Fix k with $0 \le k \le n$. Then

$$
\begin{aligned}
b(k; n, p) &= \binom{n}{k} \left(\frac{\alpha}{n}\right)^k \left(1 - \frac{\alpha}{n}\right)^{n-k} \\
&= \frac{n! \, \alpha^k}{k!(n-k)! \, n^k} \frac{\left(1 - \frac{\alpha}{n}\right)^n}{\left(1 - \frac{\alpha}{n}\right)^k} \\
&= \frac{\alpha^k}{k!} \left(1 - \frac{\alpha}{n}\right)^n \left[\frac{n!}{n^k (n-k)!}\right] \left(1 - \frac{\alpha}{n}\right)^{-k} \quad (3.24)
\end{aligned}
$$

Consider the term in square brackets in (3.24). It can be written as

$$
\begin{aligned}
\frac{n!}{n^k (n-k)!} &= \frac{n(n-1)(n-2) \cdots (n-k+1)}{n^k} \\
&= \left(1 - \frac{1}{n}\right)\left(1 - \frac{2}{n}\right) \cdots \left(1 - \frac{k-1}{n}\right). \quad (3.25)
\end{aligned}
$$

Hence,

$$\lim_{n \to \infty} \frac{n!}{n^k (n-k)!} = 1. \quad (3.26)$$

Also, since k is fixed,

$$\lim_{n \to \infty} \left(1 - \frac{\alpha}{n}\right)^{-k} = \lim_{n \to \infty} \frac{1}{\left(1 - \frac{\alpha}{n}\right)^k} = \frac{1}{\lim_{n \to \infty} \left(1 - \frac{\alpha}{n}\right)^k} = \frac{1}{1} = 1 \quad (3.27)$$

By a well-known property of the exponential function

$$\lim_{n\to\infty} \left(1 - \frac{\alpha}{n}\right)^n = e^{-\alpha} \tag{3.28}$$

Combining (3.24)-(3.28), we see that

$$\lim_{n\to\infty} b\left(k; n, \frac{\alpha}{n}\right) = e^{-\alpha}\frac{\alpha^k}{k!}, \quad k = 0, 1, 2, \ldots. \ \blacksquare \tag{3.29}$$

The import of Theorem 3.1.1 is that, if n is large and p is small so that np is not close to either p or n, then the binomial random variable with parameters n and p can be approximated by a Poisson random variable with the parameter $\alpha = np$. (Zehna [39] claims that electronics has ended the usefulness of the Poisson approximation to the binomial. Nevertheless, it is sometimes useful, and the Poisson distribution is very important in its own right.)

The APL functions POISSON and POISSONΔDIST can be used to make Poisson probability calculations. $\square$

Table 3.1.1. Example 3.1.4

k	$P[X = k]$	Poisson approximation $e^{-2}2^k/k!$
0	0.13262	0.13534
1	0.27065	0.27067
2	0.27341	0.27067
3	0.18228	0.18045
Total probability	0.85896	0.85713

Example 3.1.4 Suppose the Wildgoose Errcraft computer installation has a library of 100 subroutines and that each week, on the average, bugs are found (and corrected) in two of the subroutines. Assuming that the number of subroutines per week with newly discovered and corrected bugs has a binomial distribution, use the Poisson approximation to calculate the probability that errors will be found in not more than three subroutines next week.

Solution Using the APL functions POISSON and BINOMIAL, and rounding to five decimal places, we compute the values in Table 3.1.1. The true value of the required probability is 0.85896; the value given by the Poisson approximation is 0.85713. These values are close, although some individual probabilities are a little off; for example, for $k = 0$ and 2, the error of the approximation about 0.0027. $\square$

Sheu [32] has proven the following theorem, which gives us a bound on the possible error made by the Poisson approximation to the binomial.

Theorem 3.1.2 *Let* $b(k; n, p) = \binom{n}{k} p^k (1 - p)^{n-k}$, $1 \leq n$, $0 \leq p \leq 1$; $b(k; n, p) = 0$, *for* $k > n$. *Let* $p(k, \alpha) = e^{-\alpha} \frac{\alpha^k}{k!}$, *for* $k = 0, 1, \ldots$. *Then we have*

$$\sum_{k=0}^{\infty} |b(k; n, p) - p(k, np)| \leq \min\{2np^2, 3p\}.$$

Proof See Sheu [32]. ∎

For Example 3.1.4, Theorem 3.1.2 guarantees that the sum of absolute values of all the errors in the Poisson approximation will not exceed 0.06; individual errors must, of course, also be less than 0.06. The sum of the absolute values of the observed errors is 0.00731.

The value of the Poisson distribution as a means of approximating the binomial distribution is minor compared to its value in describing random variables that occur in computer science and other sciences.

Example 3.1.5 Suppose it has been determined that the number of inquiries that arrive per second at the central computer installation of the Varoom Broom on-line computer system can be described by a Poisson random variable with an average rate of 10 messages per second. What is the probability that no inquiries arrive in a one second period? What is the probability that 15 or fewer inquiries arrive in a one-second period?

Solution By hypothesis

$$P[X = k] = e^{-10} \frac{10^k}{k!}, \quad k = 0, 1, 2, \ldots.$$

Hence, the probability that no inquiry arrives in a one-second period is $e^{-10} = 4.54 \times 10^{-5}$. The answer to the second question is

$$e^{-10} \sum_{k=0}^{15} \frac{10^k}{k!} = 0.95126.$$

This is a laborious calculation to make without a computer but can be made easily with the APL function POISSONΔDIST. $\square$

In Chapter 4 we give the conditions that characterize a random phenomenon that is described by a Poisson random variable. It will be evident that these conditions are characteristic of many real-life situations.

It is interesting to note that the value of this distribution was not recognized for many years after it was discovered—even by Poisson himself. Stigler [33], in discussing the major book by Poisson on probability, (Poisson [23]), mentions the distribution and goes on to say

> ...in a section of the book concerned with the form of the binomial distribution for large numbers of trials, Poisson does in fact derive this distribution in its cumulative form, as a limit to the binomial distribution when the chance of a success is very small (Stigler, 1982a). The distribution appears on only one page in all of Poisson's work (Poisson, 1837, p. 206).

Ladislaus von Bortkiewicz in his famous monograph Bortkiewicz [6] showed that the Poisson distribution is valuable for modeling many real-world phenomena. Bortkiewicz is considered by most statisticians to be the first to demonstrate the value of the Poisson distribution. Now this distribution is one of the most widely used in applied probability and statistics. The example in Bortkiewicz's book that is most cited (probably because it brings such vivid pictures to mind) is the model of deaths by horse kick in the Prussian cavalry. We consider the Prussian cavalry data in Example 8.4.2.

3.1.5   Discrete Uniform Random Variables

A random variable X that assumes a finite number of values $x_1, x_2, \ldots, x_n$ each with the same probability, $1/n$, is called a *discrete uniform* random variable. Often the values are taken to be multiples of some value L, such as $L, 2L, 3L, \ldots, nL$. The expected value is given by

$$E[X] = \frac{1}{n} \sum_{i=1}^{n} x^i. \tag{3.30}$$

The second moment $E[X^2]$ is given by

$$E[X^2] = \frac{1}{n} \sum_{i=1}^{n} x_i^2, \tag{3.31}$$

and the variance can be calculated by the formula

$$\mathrm{Var}[X] = E[X^2] - (E[X])^2. \tag{3.32}$$

God may be subtle, but he is not malicious.
Albert Einstein
LaTeX, *on the other hand is* both *subtle and malicious.*
Russell Ham

3.2 Continuous Random Variables

A continuous random variable X is characterized by the property that $P[X = x] = 0$ for all real x; that is, its probability mass function assumes only the value zero. In this book each continuous random variable we consider is described by a density function $f(\cdot)$, with properties defined in Section 2.5. It is *not* true, in general, that $f(x) = P[X = x]$; it *is* true that, for each real x and for small Δx, the probability that the value of X lies between x and $x + \Delta x$ is about $f(x)\Delta x$. Some of the properties of the continuous random variables we discuss is this section are summarized in Table 2 of Appendix A. (It is the author's belief that continuous random variables should be known as *indiscreet random variables*, since they clearly are not discrete; this would also add a little spice to a subject with a reputation for dullness.)

3.2.1 Continuous Uniform Random Variables

A continuous random variable X is said to be a *uniform random variable on the interval a to b* or to be *uniformly distributed on the interval a to b*, if its density function is given by

$$f(x) = \begin{cases} \dfrac{1}{b-a} & \text{for } a < x < b \\ 0 & \text{otherwise.} \end{cases} \tag{3.33}$$

The corresponding distribution function is easily calculated by integration to give

$$F(x) = \begin{cases} 0 & \text{for } x \leq a \\ \dfrac{x-a}{b-a} & \text{for } a < x < b \\ 1 & \text{for } x \geq b. \end{cases} \tag{3.34}$$

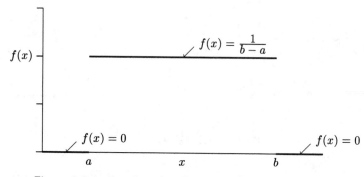

Figure 3.2.1. Density of uniform distribution on a to b.

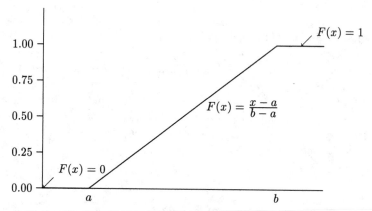

Figure 3.2.2. Distribution function of uniform distribution.

Figure 3.2.1 is a graph of the density function of a random variable that is uniformly distributed on the interval a to b, and Figure 3.2.2 is the corresponding distribution function. Thus, the probability that the values of X will lie in any subinterval of the interval from a to b is merely the ratio of the length of the subinterval to the length of the whole interval, that is, the probability that X will lie in any subinterval of length δ is $\delta/(b-a)$.

It is an easy exercise to show (see Exercise 41) that

$$E[X] = \frac{a+b}{2}, \quad \sigma^2 = \frac{(b-a)^2}{12}. \tag{3.35}$$

Example 3.2.1 Suppose the disks in a disk memory device rotate once every 25 milliseconds. When a read/write head is positioned over a track to read a particular record from the track, the record can be anywhere on the track. Hence, the rotational delay, T, before the required record is in position to be read is uniformly distributed on the interval from 0 to 25 milliseconds. Thus, $E[T] = 12.5$ milliseconds,

$$\sigma_T^2 = \frac{25^2}{12} = 52.0833,$$

and

$$\sigma_T = \sqrt{52.0833} = 7.2169 \text{ milliseconds.}$$

The probability that the rotational delay is between 5 and 15 milliseconds is $10.25 = 0.4$, the same as the probability that it is between 15 and 25 milliseconds. $\square$

3.2.2 Exponential Random Variables

A continuous random variable X has an *exponential distribution with parameter* $\alpha > 0$ if its density function f is defined by

$$f(x) = \begin{cases} \alpha e^{-\alpha x}, & x > 0 \\ 0, & x \leq 0. \end{cases} \tag{3.36}$$

The distribution function F is then given by

$$F(x) = \begin{cases} 1 - e^{-\alpha x} = 1 - \exp(-x/E[X]), & x > 0 \\ 0, & x \leq 0. \end{cases} \tag{3.37}$$

(Students who have learned the importance of the exponential distribution have been known to shout "Eureka!" upon seeing

Equation (3.37) appear. Therefore, I call this formula the "Eureka formula." It should not be confused with the motto of the state of California.)

Figure 2.5.3 shows the density function for an exponential random variable with $\alpha = 2$ and Fig. 2.5.4 is the graph of the corresponding distribution function. As shown in Example 2.6.2, an exponential random variable with parameter α has mean $E[X] = 1/\alpha$ and $\text{Var}[X] = 1/\alpha^2 = E[X]^2$.

In Example 2.9.1 we proved that the moments are given by

$$E[X^k] = \frac{k!}{\alpha^k} = k! \, E[X]^k, \quad k = 1, 2, 3, \ldots. \tag{3.38}$$

One reason for the importance of the exponential distribution to queueing theory and elsewhere is the *Markov property*, sometimes called the *memoryless* property, given by

$$P[X > t + h \mid X > t] = P[X > h], \quad t > 0, \, h > 0. \tag{3.39}$$

One interpretation of (3.39) is that, if X is the waiting time until a particular event occurs and t units of time have produced no event, then the distribution of further waiting time is the same as it would be if no waiting time had passed—that is, the system does not remember that t time units have produced no "arrival." To prove (3.39) we note that, by (3.37), $P[X > x] = e^{-\alpha x}$ for all real positive x. Hence,

$$
\begin{aligned}
P[X > t + h | X > t] &= \frac{P[(X > t + h) \cap (X > t)]}{P[X > t]} \\
&= \frac{P[X > t + h]}{P[X > t]} \\
&= \frac{e^{-\alpha(t+h)}}{e^{-\alpha t}} \\
&= \frac{e^{-\alpha t} e^{-\alpha h}}{e^{-\alpha t}} \\
&= e^{-\alpha h} = P[X > h].
\end{aligned}
$$

Figure 2.5.3 shows that the density function for the exponential distribution is not symmetrical about the mean but is highly skewed. In fact, for an exponential random variable X it is true that

$$P[X \le E[X]] = 1 - e^{-E[X]/E[X]} = 1 - e^{-1} = 0.63212. \tag{3.40}$$

For any random variable, such as a uniform random variable, which is symmetrical about the mean, $P[X \le E[X]] = 0.5$. Thus, for an exponential random variable, values of X between 0 and $E[X]$ are more likely to occur than values between $E[X]$ and $2E[X]$, although each interval is one standard deviation long ($\text{Var}[X] = E[X]^2$, so that $\sigma_x = E[X]$).

For any random variable X, its rth *percentile value*, $\pi[r]$, is defined by $P[X \le \pi[r]] = r/100$. Thus, the 90th percentile value of an exponential distributed random variable is defined by

$$P[X \le \pi[90]] = 0.9 \quad \text{or} \quad 1 - e^{-\alpha \pi[90]} = 0.9.$$

Hence,

$$e^{-\alpha \pi[90]} = 0.1. \tag{3.41}$$

By taking the natural logarithm of both sides of (3.41) and solving for $\pi[90]$, we get

$$\pi[90] = -\frac{\ln(0.1)}{\alpha} = E[X] \ln(10) = 2.30259\, E[X] \approx 2.3\, E[X]. \tag{3.42}$$

(Here $\ln(x)$ means the logarithm of x to the base e.)

Similarly,

$$\pi[95] = E[X] \ln(20) = 2.99573\, E[X] \approx 3\, E[X], \qquad (3.43)$$

and for $r > 0$,

$$\pi[r] = E[X] \ln\left(\frac{100}{100 - r}\right). \qquad (3.44)$$

In this book we will use the approximations

$$\pi[90] \approx E[X] + 1.3\sigma_x = 2.3E[X], \qquad (3.45)$$

and

$$\pi[95] \approx E[X] + 2\sigma_x = 3E[X] \qquad (3.46)$$

for the exponential distribution.

Example 3.2.2 Personnel of the Farout Engineering company use an on-line terminal to make routine engineering calculations. If the time each engineer spends in a session at a terminal has an exponential distribution with an average value of 36 minutes, find

(a) The probability that an engineer will spend 30 minutes or less at the terminal,

(b) The probability that an engineer will use it for more than an hour.

(c) If an engineer has already been at the terminal for 30 minutes, what is the probability that he or she will spend more than another hour at the terminal?

(d) Ninety percent of the sessions end in less than R minutes. What is R?

Solution Let T be the time an engineer spends using the terminal. By (3.37) the probability that T does not exceed t minutes is

$$1 - e^{-t/36}.$$

Hence, the probability that T is not more than 30 minutes is

$$1 - e^{-30/36} = 1 - e^{-5/6} = 1 - 0.43460 = 0.5654.$$

By taking complements, (3.37) yields

$$P[T > t] = e^{-t/36}, \quad t \text{ in minutes}.$$

Hence, the probability that over an hour is spent at the terminal in one session is

$$e^{-60/36} = e^{-5/3} = 0.1889,$$

or slightly less than 20% of the time. By the Markov property, the fact that an engineer has already been using the terminal for 30 minutes has no effect on the probability that he or she will use it for at least another hour. Hence, this probability is 0.1889. R is $\pi[90]$ or about $2.3E[X] = 2.3 \times 36 = 82.8$ minutes. $\square$

We summarize the properties of the exponential distribution in the following theorem.

Theorem 3.2.1 (*Properties of the Exponential Distribution*) *Let X be an exponential random variable with parameter $\alpha > 0$. Then the following hold.*

(a) If $\theta < \alpha$, then the moment generating function $\psi(\cdot)$ is given by

$$\psi(\theta) = \frac{\alpha}{(\alpha - \theta)}, \tag{3.47}$$

and the Laplace–Stieltjes transform, $X^*[\theta]$, by

$$X^*[\theta] = \frac{\alpha}{\alpha + \theta}. \tag{3.48}$$

(b) $E[X^k] = \dfrac{k!}{\alpha^k} = k!\, E[X]^k, \; k = 1, 2, 3, \ldots.$

(c) $E[X] = \dfrac{1}{\alpha}, \; \text{Var}[X] = \dfrac{1}{\alpha^2} = E[X]^2.$

(d) X has the Markov property

$$P[X > t + h \mid X > t] = P[X > h], \quad t > 0, \; h > 0.$$

(e) The rth percentile value $\pi[r]$ defined by $P[X \le \pi[r]] = r/100$ is given by

$$\pi[r] = E[X] \ln\left(\frac{100}{100 - r}\right).$$

(f) Suppose the number of arrivals, Y, of some entity per unit of time is described by a Poisson random variable with parameter λ. Then the time T between any two successive arrivals (the interarrival time) is independent of the interarrival time of any other successive arrivals and has an exponential distribution with parameter λ. Thus, $E[T] = 1/\lambda$, and $p[T \le t] = 1 - e^{-\lambda t}$ for $t \ge 0$.

(g) *Suppose interarrival times of customers to a queueing system are independent, identically distributed, exponential random variables, each with mean $1/\lambda$. Then the number of arrivals, Y_t, in any interval of length $t > 0$, has a Poisson distribution with parameter λt; that is,*

$$P[Y_t = k] = e^{-\lambda t} \frac{(\lambda t)^k}{k!},$$

for $k = 0, 1, 2, \cdots$.

(h) *Suppose $X_1, X_2, \ldots, X_n$ are independent exponential random variables with parameters $\alpha_1, \alpha_2, \alpha_3, \cdots, \alpha_n$, respectively, and $Y = \min\{X_1, X_2, \cdots, X_n\}$. Then Y has an exponential distribution with parameter $\alpha = \alpha_1, +\alpha_2, + \cdots + \alpha_n$. In particular, if each $\alpha_i = \alpha$, then Y is exponential with parameter $n\alpha$.*

Proof Items (a)–(e) have been proven, above, except for (3.48), which was calculated in Example 2.9.6. The proof of (f) is given in Chapter 4 (Theorem 4.2.2). Item (g) follows from Theorems 4.2.3 and 4.2.1. Item (h) follows from Theorem 2.7.4. ■

3.2.3 Shifted Exponential Random Variables

We occasionally find use for a random variable that is *almost exponential*. Perhaps a service facility provides service that must be at least D but has an exponential distribution for service time $s > D$. Another case is response time at a terminal of an on-line system, which may be approximately exponential but has a minimum positive value D. The minimum, D, represents the minimum transmission time to send a request to the central processor, plus the minimum service time at the processor, added to the minimum time to return the response to a terminal. These random variables could be described by a *shifted exponential distribution* with density function given by

$$f(s) = \alpha e^{-\alpha(s - D)}, \quad \text{for } s \geq D. \tag{3.49}$$

The properties of a shifted exponential random variable are given in the following theorem.

Theorem 3.2.2 *Let Y be a shifted exponential random variable with the density function given in (3.49) with $D \geq 0$. Then*

$$Y^*[\theta] = e^{-\theta D} \times \frac{\alpha}{\alpha + \theta}, \tag{3.50}$$

$$E[Y] = D + \frac{1}{\alpha}, \tag{3.51}$$

$$\text{Var}[Y] = \frac{1}{\alpha^2}, \tag{3.52}$$

$$F(x) = P[Y \le x] = 1 - e^{-\alpha(x-D)}, \quad for\ x \ge D \tag{3.53}$$

and

$$0 \le C_Y^2 \le 1. \tag{3.54}$$

Furthermore, if X is a positive random variable with $C_X^2 \le 1$, then a shifted exponential random variable Y can be constructed with the same two first moments as X by setting

$$\alpha = \frac{1}{C_X E[X]}, \tag{3.55}$$

and

$$D = E[X] - \frac{1}{\alpha}. \tag{3.56}$$

Proof It is easy to show (see Exercise 70) that

$$Y^*[\theta] = e^{-\theta D} \times \frac{\alpha}{\alpha + \theta}. \tag{3.57}$$

Equations (3.51) and (3.52) follow from Theorem 2.9.3 and some easy differentiation. Equation (3.53) follows from

$$P[Y \le x] = \int_D^x \alpha e^{-\alpha(s-D)}\, ds = 1 - e^{-\alpha(x-D)}.$$

The formula (3.54) for C_Y^2 follows from the formula

$$C_Y^2 = \frac{\dfrac{1}{\alpha^2}}{D^2 + \dfrac{2D}{\alpha^1} + \dfrac{1}{\alpha^2}},$$

since $D \ge 0$ and $\alpha > 0$. The claim that the shifted exponential random variable constructed using the parameters defined by (3.55) and (3.56) has the same first two moments as X can be verified by simple substitution into (3.51) and (3.52) and the formula

$$E[Y^2] = \text{Var}[Y] + E[Y]^2. \quad \blacksquare$$

Example 3.2.3 The performance analysts at Alcapones Loansharking Services estimate the mean response time at their computer center for customers with the highest priority is 0.75 seconds and the minimum response time is 0.25 seconds. They decide to model this random variable as a shifted exponential. Thus, $D = 0.25$ seconds and $1/\alpha = 1/2$, or $\alpha = 2$. The analysts want to estimate the probability that a response time will not exceed one second and to calculate the 90th percentile value of the response time. We know the distribution function is given by

$$F(x) = P[Y \leq x] = 1 - \exp(-2(x - 0.25)), \quad \text{for } x \geq 0.25 \text{ seconds. } (3.58)$$

Hence, the probability that Y does not exceed one second is

$$F(1) = P[Y \leq 1] = 1 - e^{-1.5} = 0.77687.$$

Let x be the 90th percentile response time. Then, by (3.58),

$$0.9 = 1 - \exp(-2(x - 0.25)). \tag{3.59}$$

Solving (3.59) for x yields

$$x = \frac{\ln(10)}{2} + 0.25 = 1.40 \quad \text{seconds.} \quad \square$$

Example 3.2.4 Suppose the analysts in the above example had chosen a shifted exponential with mean 0.75 seconds and $C_X^2 = 0.25$. Then, by Theorem 3.2.2,

$$\alpha = \frac{1}{0.5 \times 0.75} = \frac{8}{3},$$

and

$$D = \frac{3}{4} - \frac{3}{8} = \frac{3}{8}.$$

Hence,

$$P[Y \leq 1] = 1 - \exp\left(-\frac{8}{3}\left(1 - \frac{3}{8}\right)\right) = 0.811.$$

We also calculate the 90th percentile response time to be

$$x = \frac{3}{8}\ln(10) + \frac{3}{8} = 1.24 \quad \text{seconds.} \quad \square$$

3.2.4 Normal Random Variables

A continuous random variable X is said to be a *normal* random variable with parameters μ and $\sigma > 0$ if it has the density function

$$f(x) = \frac{1}{\sigma\sqrt{2\pi}} \exp[-\frac{1}{2}\left(\frac{x - \mu}{\sigma}\right)^2], \quad x \text{ real.} \tag{3.60}$$

We indicate this fact by writing "X is $N(\mu, \sigma^2)$." A *standard normal* random variable is one with parameters $\mu = 0$ and $\sigma = 1$. Thus, the standard normal density $\varphi(\cdot)$ is defined by

$$\varphi(x) = \frac{1}{\sqrt{2\pi}} \exp[-\frac{1}{2}x^2], \quad x \text{ real.} \tag{3.61}$$

The corresponding *standard normal distribution function* $\Phi(\cdot)$ is therefore defined by

$$\Phi(x) = \int_{-\infty}^{x} \varphi(t)dt = \int_{-\infty}^{x} \frac{\exp[-\frac{1}{2}t^2]}{\sqrt{2\pi}} dt. \tag{3.62}$$

The standard normal distribution is important because every normal distribution can be calculated in terms of it. Thus, if X is normally distributed with parameters μ and σ (X is $N(\mu, \sigma^2)$), then

$$
\begin{aligned}
F_X(x) &= \frac{1}{\sigma\sqrt{2\pi}} \int_{-\infty}^{x} \exp[-\frac{1}{2}((t-\mu)/\sigma)^2] dt \\
&= \frac{1}{\sqrt{2\pi}} \int_{-\infty}^{(x-\mu)/\sigma} \exp[-\frac{1}{2}y^2] dy \\
&= \Phi\left(\frac{x-\mu}{\sigma}\right).
\end{aligned}
\tag{3.63}
$$

The second integral in (3.63) is the result of the change of variable $y = (t-\mu)/\sigma$. Unfortunately, $\Phi(\cdot)$ cannot be calculated in closed form but must be approximated using numerical methods. The APL function NDIST calculates values of the standard normal distribution using formula (26.2.17) of Abramowitz and Stegun [1]. It was used to create Table 3 of Appendix A, a table of values of the standard normal distribution function $\Phi(\cdot)$. Of course $\Phi(x)$ can also be calculated using MINITAB, SAS/STAT, *Mathematica*, and the HP-21S. In order to prove a number of useful properties of the normal distribution, we first need to prove that

$$\int_{-\infty}^{\infty} \varphi(x)dx = \int_{-\infty}^{\infty} \frac{\exp[-z^2/2]}{\sqrt{2\pi}} dz = 1. \tag{3.64}$$

This, in particular, will show that (3.60) defines a density function, since $f(x) > 0$ for all real x, because the exponential function assumes only positive values, and

$$\int_{-\infty}^{\infty} \frac{\exp[-\frac{1}{2}((x-\mu)/\sigma)^2]}{\sigma\sqrt{2\pi}} dx = \int_{-\infty}^{\infty} \frac{\exp[-z^2/2]}{\sqrt{2\pi}} dz, \tag{3.65}$$

under the change of variable $z = (x - \mu)/\sigma$.

To prove (3.64), we write

$$\left(\int_{-\infty}^{\infty} \varphi(x)dx\right)^2 = \left(\frac{1}{\sqrt{2\pi}} \int_{-\infty}^{\infty} \exp[-x^2/2]dx\right)$$

$$\times \left(\frac{1}{\sqrt{2\pi}} \int_{-\infty}^{\infty} \exp[-y^2/2]dy\right)$$

$$= \frac{1}{2\pi} \int_{-\infty}^{\infty} \int_{-\infty}^{\infty} \exp[-(x^2+y^2)/2]dx\,dy. \quad (3.66)$$

Now we can transform to polar coordinates. Thus, $r^2 = x^2 + y^2$ and $dx\,dy = r\,dr\,d\theta$. (See Apostol [3] for a discussion of how to convert from Cartesian to polar coordinates.) Making the polar coordinates substitution in (3.66) gives

$$\left(\int_{-\infty}^{\infty} \varphi(x)dx\right)^2 = \frac{1}{2\pi} \int_0^{2\pi} \int_0^{\infty} \exp[-r^2/2]r\,dr\,d\theta$$

$$= \frac{1}{2\pi} \int_0^{2\pi} [-\exp[-r^2/2]]\big]_{r=0}^{\infty} \, d\theta$$

$$= \frac{1}{2\pi} \int_0^{2\pi} d\theta = 1. \quad (3.67)$$

This proves (3.64).

If X is $N(\mu, \sigma^2)$, then the moment generating function of X (see Section 2.9) is given by

$$\psi(\theta) = E[\exp(\theta X)] = \frac{1}{\sigma\sqrt{2\pi}} \int_{-\infty}^{\infty} \exp(\theta x)\exp[-\frac{1}{2}((x-\mu)/\sigma)^2]dx. \quad (3.68)$$

Let $z = (x - \mu)/\sigma$. Then (3.68) yields

$$\psi(\theta) = \frac{\exp(\mu\theta)}{\sqrt{2\pi}} \int_{-\infty}^{\infty} \exp\left(\sigma\theta z - \frac{z^2}{2}\right) dz$$

$$= \frac{\exp(\mu\theta)}{\sqrt{2\pi}} \int_{-\infty}^{\infty} \exp[-\frac{1}{2}(z^2 - 2\sigma\theta z + \sigma^2\theta^2 - \sigma^2\theta^2)]dz$$

$$= \frac{\exp(\mu\theta)}{\sqrt{2\pi}} \exp[(\sigma\theta)^2/2] \int_{-\infty}^{\infty} \exp[-\frac{1}{2}(z - \sigma\theta)^2]dz$$

$$= \exp(\mu\theta)\exp[(\sigma\theta)^2/2] \int_{-\infty}^{\infty} \frac{\exp[-w^2/2]}{\sqrt{2\pi}}dw$$

$$= \exp[\mu\theta + (\sigma^2\theta^2)/2]. \quad (3.69)$$

In the next-to-last integral we substituted $w = z - \sigma\theta$, and used (3.64). Thus, the moment generating function of a normal random variable X with parameters μ and σ is $\exp[\mu\theta + (\sigma^2\theta^2)/2)]$. Hence,

$$\frac{d\psi}{d\theta} = (\mu + \sigma^2\theta)\exp[\mu\theta + (\sigma^2\theta^2)/2],$$

and

$$\frac{d^2\psi}{d\theta^2} = \sigma^2\exp[\mu\theta + (\sigma^2\theta^2)/2] + (\mu + \sigma^2\theta)^2\exp[\mu\theta + (\sigma^2\theta^2)/2].$$

Therefore, by Theorem 2.9.1,

$$E[X] = \frac{d\psi}{d\theta}\bigg|_{\theta=0} = \mu, \quad E[X^2] = \frac{d^2\psi}{d\theta^2}\bigg|_{\theta=0} = \sigma^2 + \mu^2,$$

and

$$\mathrm{Var}[X] = E[X^2] - (E[X])^2 = \sigma^2 + \mu^2 - \mu^2 = \sigma^2.$$

Thus, the parameters μ and σ are, respectively, the mean and standard deviation of X.

We summarize what we have just shown in the following theorem.

Theorem 3.2.3 (*Properties of a Normal Random Variable*) *Suppose X is a normal random variable with parameters μ and σ (X is $N(\mu, \sigma^2)$). Then*

$$E[X] = \mu, \quad \mathrm{Var}[X] = \sigma^2, \tag{3.70}$$

and

$$\psi(\theta) = \exp[\mu\,\theta + (\sigma^2\theta^2)/2]. \ \blacksquare \tag{3.71}$$

If $X_1, X_2, \ldots, X_n$ are n independent random variables having normal $N(\mu_1, \sigma_1^2)$, $N(\mu_2, \sigma_2^2), \ldots, N(\mu_n, \sigma_n^2)$ distributions, respectively, the moment generating function of

$$Y = X_1 + X_2 + \cdots X_n,$$

is, by Theorem 2.9.1,

$$\begin{aligned}
\psi_Y(\theta) &= \psi_{X_1}(\theta)\psi_{X_2}(\theta)\cdots\psi_{X_n}(\theta) \\
&= \exp\left[\theta\sum_{i=1}^{n}\mu_i + \frac{\theta^2}{2}\sum_{i=1}^{n}\sigma_i^2\right],
\end{aligned} \tag{3.72}$$

which is the moment generating function of an

$$N\left(\sum_{i=1}^{n}\mu_i, \sum_{i=1}^{n}\sigma_i^2\right)$$

random variable. We have proven the following theorem.

Theorem 3.2.4 *Suppose* $X_1, X_2, \ldots, X_n$ *are* n *independent random variables such that* X_1 *is* $N(\mu_1, \sigma_1^2)$, X_2 *is* $N(\mu_2, \sigma_2^2)$, $\ldots$, X_n *is* $N(\mu_n, \sigma_n^2)$. *Then* $Y = X_1 + X_2 + \cdots + X_n$ *is normally distributed with mean* $\mu_1 + \mu_2 + \cdots + \mu_n$ *and variance* $\sigma_1^2 + \sigma_2^2, + \cdots + \sigma_n^2$. ∎

The symmetry of the normal densities about the mean follows from (3.60); that is, $f(\mu + x) = f(\mu - x)$ for all real x. As we saw from Equation (3.63), probability calculations for any normal distribution can be made from the standard normal distribution. If X is $N(\mu, \sigma^2)$, the change of variable

$$Z = \frac{(X - \mu)}{\sigma} \tag{3.73}$$

yields a normalized random variable that is $N(0, 1)$. The numerator in (3.73) is a shift of origin transformation which transforms the mean value to 0. Division by σ converts the value of $X - \mu$ into units of standard deviation, σ. The fact that Z is normally distributed follows from the uniqueness of the moment generating function as follows:

$$\psi(\theta) = E[e^{\theta((X-\mu)/\sigma)}] = e^{-\mu\theta/\sigma} E[e^{\theta X/\sigma}], \tag{3.74}$$

by Theorem 2.7.1(b). Now X/σ is a normal random variable with mean μ/σ and variance 1 by the properties of mean and variance (Theorems 2.7.1 and 2.7.2) and by the fact that dividing a random variable by a constant does not change the nature of the random variable, but only the scale. Hence, the moment generating function of X/σ is

$$E[\exp(\theta x/\sigma)] = \exp[(\mu\theta/\sigma) + \theta^2/2], \tag{3.75}$$

by (3.71). Substituting (3.75) into (3.74) yields

$$\psi(\theta) = \exp(-\mu\theta/\sigma) \exp[(\mu\theta/\sigma) + \theta^2/2] = \exp(\theta^2/2). \tag{3.76}$$

Since (3.76) is the moment generating function of a standard normal random variable, Z is $N(0, 1)$.

As illustrated in Figure 3.2.3, the probability that a randomly selected value x of a normal random variable X will fall within one standard deviation of the mean μ is 0.68268 or 68.268 percent. This can be calculated, using Table 3 or Appendix A, by noting that the probability that z is greater than 1 is $1 - 0.84134 = 0.15866$, so that, by the symmetry of the standard normal density,

$$P[\mu - \sigma \le X \le \mu + \sigma] = P[-1 \le Z \le 1] = 1 - 2(0.15866) = 0.68268.$$

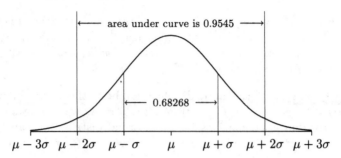

Figure 3.2.3. Area under normal density curve.

Similarly,

$$P[\mu - 1.96\sigma \leq X \leq \mu + 1.96\sigma] = 0.95.$$

Shah [31] has developed an easy way to compute an approximation for the area under the standard normal density curve from 0 to z. He claims the maximum absolute error is only 0.0052. This is certainly accurate enough for many purposes. Shah's approximation is

$$P[0 \leq Z \leq z] = \begin{cases} z(4.4 - z)/10, & \text{for } 0 \leq z \leq 2.2, \\ 0.49, & \text{for } 2.2 < z < 2.6, \\ 0.50, & \text{for } 2.6 \leq z. \end{cases}$$

We ask you to test the accuracy of the above approximation in Exercise 58.

Example 3.2.5 Suppose the number of message buffers in use in the Levy Stress interactive inquiry system, X, has a normal distribution with a mean of 100 and a standard deviation of 10. Calculate the probability that the number of buffers in use does not exceed 120; lies between 80 and 120; exceeds 130; respectively.

Solution Because 120 is 2 standard deviations above the mean, the first probability requested is the probability that z does not exceed 2, which, by Table 3 of Appendix A, is 0.97725. If x is between 80 and 120, it is not more than 2 standard deviations from the mean; hence, the second probability is $0.9545 = P[-2 \leq Z \leq 2]$. Since 130 is 3 standard deviations above the mean,

$$P[X > 130] = P[Z > 3] = 1 - 0.99865 = 0.00135. \quad \square$$

The normal distribution is the most important distribution in applied probability and statistics because many useful random variables have normal or nearly normal distributions and (more important) because of the central limit theorem, which is discussed in Section 5 of this chapter.

As W. J. Youden[2] said

THE

NORMAL

LAW OF ERROR

STANDS OUT IN THE

EXPERIENCE OF MANKIND

AS ONE OF THE BROADEST

GENERALIZATIONS OF NATURAL

PHILOSOPHY ◇ IT SERVES AS THE

GUIDING INSTRUMENT IN RESEARCHES

IN THE PHYSICAL AND SOCIAL SCIENCES AND

IN MEDICINE AGRICULTURE AND ENGINEERING ◇

IT IS AN INDISPENSABLE TOOL FOR THE ANALYSIS AND THE

INTERPRETATION OF DATA OBTAINED BY OBSERVATION AND EXPERIMENT

3.2.4.1 Bivariate Normal Random Variables

The jointly distributed random variable (X, Y) has a *bivariate normal distribution* if it has the joint density function

$$f(x,y) = \frac{1}{2\pi\sigma_X\,\sigma_Y\sqrt{1-\rho^2}} \exp\left\{ -\frac{1}{2(1-\rho^2)} \left[\left(\frac{x-\mu_X}{\sigma_X}\right)^2 - \frac{2\rho(x-\mu_X)(y-\mu_Y)}{\sigma_X\sigma_Y} + \left(\frac{y-\mu_Y}{\sigma_Y}\right)^2 \right] \right\}, \tag{3.77}$$

where μ_X, μ_Y, σ_X, σ_Y, and ρ are constants with $0 < \rho < 1$. The parameter ρ is called the *correlation coefficient* of X and Y.

The bivariate normal distribution is an important distribution for regression analysis, the subject of Chapter 9 of this book. We list the properties of the bivariate normal distribution in the following theorem.

Theorem 3.2.5 (*Properties of the bivariate normal distribution*) *Suppose* (X, Y) *is a jointly distributed random variable with joint density function given by (3.77). Then the following are true.*

(a) The marginal distribution of X is normal with mean μ_X and standard deviation σ_X; that is, X is $N(\mu_X, \sigma_X^2)$.

(b) The marginal distribution of Y is normal with mean μ_Y and standard deviation σ_Y; that is, Y is $N(\mu_Y, \sigma_Y^2)$.

---

[2]William John Youden (1900–1971) was a statistician and chemist.

(c) *The conditional density function of Y, given that $X = x$, is given by*

$$f_{Y|X}(y|x) = \frac{\exp\left[-\frac{1}{2}\left\{\frac{y - \mu_Y - \rho\frac{\sigma_Y}{\sigma_X}(x - \mu_X)}{\sigma_Y\sqrt{1 - \rho^2}}\right\}^2\right]}{\sqrt{2\pi}\sigma_Y\sqrt{1 - \rho^2}}. \quad (3.78)$$

(*Thus, for a fixed x, $Y = Y_x$ is a normal random variable with mean $\mu_Y + \rho(\sigma_Y/\sigma_X)(x - \mu_X)$ and standard deviation $\sigma_Y\sqrt{1 - \rho^2}$.*)

(d) *The conditional density function of X, given that $Y = y$, is given by*

$$f_{X|Y}(x|y) = \frac{\exp\left[-\frac{1}{2}\left\{\frac{x - \mu_X - \rho\frac{\sigma_X}{\sigma_Y}(y - \mu_Y)}{\sigma_X\sqrt{1 - \rho^2}}\right\}^2\right]}{\sqrt{2\pi}\sigma_X\sqrt{1 - \rho^2}} \quad (3.79)$$

(*Thus, for a fixed y, $X = X_y$ is a normal random variable with mean $\mu_X + \rho(\sigma_x/\sigma_Y)(y - \mu_Y)$ and standard deviation $\sigma_X\sqrt{1 - \rho^2}$.*)

Proof See Exercises 54, 55. ■

If X and Y are jointly distributed random variables, we define the *curve of regression of Y on X* by

$$E[Y|X = x] = \int_{-\infty}^{\infty} y f_{Y|X}(y|x)dy. \quad (3.80)$$

The result of Theorem 3.2.5(c) is that, if X and Y have a bivariate normal distribution, the curve of regression of Y on X is the straight line

$$E[Y|X = x] = \mu_{y|x} = \mu_Y + \frac{\rho\sigma_Y}{\sigma_X}(x - \mu_X), \quad (3.81)$$

and, for each x, $Y = Y_x$ is a normal random variable with mean

$$E[Y_x] = \mu_Y + \frac{\rho\sigma_Y(x - \mu_X)}{\sigma_X}, \quad (3.82)$$

and standard deviation

$$\sigma_{Y_x} = \sigma_Y\sqrt{1 - \rho^2}. \quad (3.83)$$

Similar remarks apply to the curve of regression X on Y.

3.2.5 Gamma Random Variables

A continuous random variable X is said to have a *gamma distribution with parameters* $\beta > 0$ and $\alpha > 0$ if its density function f is given by

$$f(x) = \begin{cases} \dfrac{\alpha(\alpha x)^{\beta-1}e^{-\alpha x}}{\Gamma(\beta)}, & x > 0 \\ 0, & x \leq 0. \end{cases} \qquad (3.84)$$

Here β is the *shape parameter* and α is the *scale parameter*. Varying β changes the shape of the density function, while changing α corresponds to changing the units of measurement (say, from minutes to hours) and does not affect the shape of the density. $\Gamma(\cdot)$ is the celebrated gamma function defined by

$$\Gamma(t) = \int_0^\infty x^{t-1}e^{-x}dx, \quad t > 0. \qquad (3.85)$$

It can be shown (see Exercise 50) that

$$\Gamma(n+1) = n!, \quad n = 0, 1, 2, \dots, \qquad (3.86)$$

and that

$$\Gamma(t+1) = t\Gamma(t) \quad \text{for all } t > 0. \qquad (3.87)$$

For an excellent discussion of the gamma function, see Parzen [22].

If X has a gamma distribution with parameters β and α, then its moment generating function $\psi(\cdot)$ is given by

$$\begin{aligned} \psi(\theta) &= E[e^{\theta X}] = \int_0^\infty \frac{e^{\theta x}\alpha^\beta x^{\beta-1}e^{-\alpha x}}{\Gamma(\beta)}dx \\ &= \frac{\alpha^\beta}{\Gamma(\beta)}\int_0^\infty x^{\beta-1}e^{-(\alpha-\theta)x}dx. \end{aligned} \qquad (3.88)$$

This integral converges if $\theta < \alpha$. Making the substitution $y = (\alpha - \theta)x$ in (3.88), yields

$$\begin{aligned} \psi(\theta) &= \frac{\alpha^\beta}{\Gamma(\beta)(\alpha-\theta)^\beta}\int_0^\infty y^{\beta-1}e^{-y}dy \\ &= \frac{\alpha^\beta}{\Gamma(\beta)(\alpha-\theta)^\beta}\Gamma(\beta) = \frac{\alpha^\beta}{(\alpha-\theta)^\beta}, \quad \theta < \alpha. \end{aligned} \qquad (3.89)$$

Similarly, the Laplace–Stieltjes transform, $X^*[\theta]$, is given by

$$X^*[\theta] = \left(\frac{\alpha}{\alpha+\theta}\right)^\beta, \quad \theta < \alpha. \qquad (3.90)$$

We summarize some of the important properties of gamma random variables in the next theorem.

Theorem 3.2.6 (*Properties of Gamma Random Variables*) *Suppose X is a gamma random variable with parameters β and α; that is, its density function f is given by* (3.84). *Then the following are true.*

(a) *The moment generating function $\psi(\cdot)$ is defined for all $\theta < \alpha$ by*

$$\psi(\theta) = \frac{\alpha^\beta}{(\alpha - \theta)^\beta},\qquad(3.91)$$

with Laplace–Stieltjes transform

$$X^*[\theta] = \left(\frac{\alpha}{\alpha + \theta}\right)^\beta.\qquad(3.92)$$

(b) $E[X] = \dfrac{\beta}{\alpha},\quad \text{Var}[X] = \dfrac{\beta}{\alpha^2},\quad E[X^3] = \dfrac{\beta(\beta + 1)(\beta + 2)}{\alpha^3}.$

(c) *If Y is independent of X and has a gamma distribution with parameters γ and α, then $Z = X + Y$ has a gamma distribution with parameters $\beta + \gamma$ and α. (Gamma random variables are reproductive with respect to β.)*

(d) *If $X_1, X_2, \ldots, X_n$ are mutually independent random variables, each with an exponential distribution with parameter α, then their sum Y has a gamma distribution with parameters n and α. Furthermore, the distribution function of Y is given by*

$$F_Y(x) \;=\; G_n(x) = 1 - e^{\alpha x}\left\{1 + \alpha x + \frac{(\alpha x)^2}{2!} + \cdots\right.$$
$$\left. + \frac{(\alpha x)^{n-1}}{(n-1)!}\right\},\quad x \geq 0.\qquad(3.93)$$

Proof Item (a) was proven by (3.89) and (3.90). The proofs of (b) and (c) are simple exercises in the use of Theorem 2.9.1 (see Exercise 36). For the proof of (d), see Feller [11, page 11]. ∎

Example 3.2.6 Suppose the time, X, between inquiries in the Cutrate Construction interactive system has an exponential distribution with an average value of one second. Let t be an arbitrary point in time and T the elapsed time until the fifth inquiry arrives (after time t). Find the expected value and variance of T. What is the probability that T does not exceed 6 seconds? That it exceeds 9 seconds?

Solution By Theorem 3.2.6(d), since $T = X_1 + X_2 + X_3 + X_4 + X_5$, where X_1, X_2, X_3, X_4, X_5 are independent identically distributed exponential random variables, each with an average value of one second, T is a gamma random variable with parameters $\beta = 5$ and $\alpha = 1$. Hence,

$$E[T] = \frac{\beta}{\alpha} = 5 \text{ seconds} \quad \text{and} \quad \text{Var}[T] = \frac{\beta}{\alpha^2} = 5 \text{ seconds}^2.$$

By (3.93),

$$P[T \leq 6] = G_5(6) = 1 - e^{-6} \left(1 + 6 + \frac{6^2}{2!} + \frac{6^3}{3!} + \frac{6^4}{4!} \right) = 0.7149.$$

By the same formula,

$$P[T > 9] = 1 - P[T \leq 9] = e^{-9} \left(1 + 9 + \frac{9^2}{2!} + \frac{9^3}{3!} + \frac{9^4}{4!} \right) = 0.055. \quad \square$$

If X is a gamma random variable with parameter $\beta = n$, where n is a small positive integer, then the values of the distribution function of X, G_n, can be calculated fairly easily using (3.93). However, if the parameters β and α are arbitrary positive numbers, probability calculations are more difficult. The APL function GΔDIST calculates the distribution function of a gamma random variable using some approximate formulas from Abramowitz and Stegun [1] as implemented by Anscombe [2]. Both SAS/STAT and MINITAB provide the distribution function for the gamma distribution. However, they both use different parameters to describe the gamma distribution than we do in this book. To calculate the value $P[X \leq t]$ using SAS/STAT, we execute the SAS/STAT statement

$$P = \text{PROBGAM}(\alpha t, \beta); \tag{3.94}$$

and then print out the value of P. (The semicolon in 3.94) is part of the command.) To calculate this value with our *Mathematica* function *gammadist*, type gammadist$[\alpha, \beta, t]$ or, using the *Mathematica* commands directly type GammaRegularized$[\beta, 0, \alpha t]$. For example, the *Mathematica* to calculate the value of $G_5(6)$ is

```
In[10]:= N[GammaRegularized[5, 0, 6]]
```

```
Out[10]= 0.714943
```

This agrees with the value of $G_5(6)$ calculated in Example 3.2.6. The MINITAB procedure for calculating $P[X \leq t]$ where X is gamma with parameters β and α, is to issue the MINITAB command "**CD** t;" followed by the subcommand "**GAMMA** β $\frac{1}{\alpha}$." where β and $\frac{1}{\alpha}$ are the decimal number values of β and $\frac{1}{\alpha}$, respectively. We demonstrate the MINITAB procedure, below. MINITAB[3] responds by typing the two numbers t and $P[X \leq t]$.

Example 3.2.7 The response time at the terminal of the Hopdup Autos interactive system has a gamma distribution with an average value of 0.5 seconds and a variance of 0.1 seconds[2]. What is the probability that the response time of a randomly selected inquiry will not exceed 0.72 seconds? 1.0 second?

Solution Let X be the response time. Since X has a gamma distribution, we must have $\beta/\alpha = 0.5$ and $\beta/\alpha^2 = 0.1$. Solving for β and α yields $\beta = 2.5$ and $\alpha = 5.0$. We calculate $P[X \leq 0.72]$, using MINITAB, as follows

```
MTB > CDF 0.72;
SUBC> GAMMA 2.5 0.2.
   0.7200 0.7938
```

Similarly, MINITAB found that $P[X \leq 1.0] = 0.9248$. The value of $P[X \leq 0.72]$ can be calculated by the SAS/STAT statement

$$P = \text{PROBGAM}(3.6, 2.5);$$

yielding 0.7938140803. Similarly, SAS/STAT found that $P[X \leq 1.0] = 0.924764754$.[4] □

The gamma random variable is useful for approximating other positive random variables. For example, it is easy to construct a gamma random variable X with a given positive mean $K > 0$ and squared coefficient of variation $C^2 > 0$. The following algorithm shows how.

Algorithm 3.2.1 (*Algorithm G*) *Given* $C^2 > 0$ *and* $K > 0$, *this algorithm will produce a gamma random variable* X *with squared coefficient of variation* $C_X^2 = C^2$ *and mean* $E[X] = K$.

Step 1 [*Calculate the parameter* β] Set

$$\beta = \frac{1}{C^2}.$$

---

[3]The Student Edition of MINITAB does not support the GAMMA subcommand.
[4]The APL function GΔDIST and *Mathematica* provide the same values for $P[X \leq 0.72]$ and $P[X \leq 1.0]$ as SAS/STAT, of course.

Step 2 [*Calculate* α] *Set*

$$\alpha = \frac{1}{K \times C^2}.$$

Step 3 [*Produce F*] *The distribution function F with the density function given by (3.81) with parameters calculated in Step 1 and Step 2, is the distribution function of a gamma random variable having the required properties.*

Proof The proof follows immediately from the formulas

$$E[X] = \frac{\beta}{\alpha},$$

and

$$C_X^2 = \frac{1}{\beta}. \quad \blacksquare$$

3.2.6 Erlang-k Random Variables

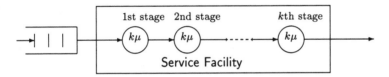

Figure 3.2.4. Erlang's model of his distribution.

The Danish mathematician A. K. Erlang used a special class of gamma random variables, now often called Erlang-k random variables, in his study of delays in telephone traffic. A random variable, T, is said to be an *Erlang-k random variable with parameter μ* or to have an *Erlang distribution with parameters k and μ* if T is a gamma random variable with the density function f given by

$$f(t) = \begin{cases} \dfrac{\mu k (\mu k t)^{k-1}}{(k-1)!} e^{-\mu k t} & \text{for } t > 0 \\ 0 & \text{for } t \le 0. \end{cases} \qquad (3.95)$$

The physical model that Erlang had in mind was a service facility consisting of k identical independent stages, each with an exponential distribution of service time as shown in Figure 3.2.4. He wanted this special facility to have the same average service time as a single facility whose service time was exponential with parameter μ. Thus, the service time, T, for the facility with k stages could be written as the sum of k exponential random variables, each with parameter μk. Hence, by Theorem 3.2.6(d), T has a gamma distribution with parameters k and μk. Thus,

$$E[T] = \frac{1}{\mu}, \quad \text{Var}[T] = \frac{1}{k\mu^2} = \frac{E[T]^2}{k},$$

and

$$F(t) = P[T \le t] = 1 - e^{-yt}\left[1 + \frac{yt}{1!} + \frac{(yt)^2}{2!} + \cdots + \frac{(yt)^{k-1}}{(k-1)!}\right],$$

where $y = \mu k$.

It can also be shown (see Exercise 39) that

$$
\begin{aligned}
E[T^n] &= \frac{k(k+1)\cdots(k+n-1)}{(k\mu)^n} \\
&= \left(1 + \frac{1}{k}\right)\left(1 + \frac{2}{k}\right)\cdots\left(1 + \frac{n-1}{k}\right)E[T]^n, \qquad (3.96)
\end{aligned}
$$

and thus

$$E[T^2] = \left(1 + \frac{1}{k}\right)(E[T])^2 \quad \text{and} \quad E[T^3] = \left(1 + \frac{1}{k}\right)\left(1 + \frac{2}{k}\right)(E[T])^3.$$
$$\tag{3.97}$$

It should be noted that, for a fixed average value $E[T]$, the variance of T decreases as k increases and, in the limit, goes to 0. Thus, an Erlang-k distribution can be used to approximate any nonnegative random variable whose variance does not exceed the square of its mean. The random variable T in Example 3.2.6 has an Erlang-5 distribution with parameter $\mu = 1/5 = 0.2$ (mean value 5).

Example 3.2.8 There are five independent stages in the repair of a certain piece of computer equipment. The repair time for each stage is exponentially distributed with an average value of 10 minutes. What is the probability that a customer engineer can repair the equipment in an hour or less? Not more than 90 minutes?

Solution The repair time, T, has an Erlang-5 distribution with average value 50 minutes (parameter $\mu = 0.02$) and a variance of $E[T]^2/k = 2500/5 = 500$ minutes^2. (T has the same distribution as the random variable T in Example 3.2.6, if the unit of time there is taken as seconds and here as tens of minutes. The change of scale leads to the appearance of the term $\frac{t}{10}$ in (3.98).) Thus, the distribution function F is given by

$$F(t) = 1 - e^{-t/10} \left[1 + \frac{t}{10} + \frac{1}{2} \left(\frac{t}{10} \right)^2 + \frac{1}{6} \left(\frac{t}{10} \right)^3 + \frac{1}{24} \left(\frac{t}{10} \right)^4 \right], \quad (3.98)$$

where t is in minutes. Hence, the probability that the repair time will not exceed 60 minutes is $F(60) = 0.7149$. (The details of this calculation are shown in Example 3.2.6.) The probability that the repair time will not exceed 90 minutes is

$$F(90) = 1 - e^{-9} \left[1 + 9 + \frac{9^2}{2} + \frac{9^3}{6} + \frac{9^4}{24} \right] = 0.945. \quad \Box$$

Random variables which are useful in computer science can often be approximated by Erlang-k random variables, thereby simplifying calculations. In addition, some useful mathematical models, particularly in queueing theory, assume Erlang-k probability distributions. Thus, if an empirically determined random variable can be approximated by an Erlang-k distributed random variable, well-known mathematical models can be applied to make useful predictions.

The usual procedure for selecting an Erlang-k random variable Y to approximate a given random variable X is as follows:

(1) Let $\mu = 1/E[X]$.

(2) Let k be the largest integer less than or equal to $E[X]^2/\text{Var}[X]$ (the "floor" of this quantity). Then Y is the Erlang-k random variable with parameters k and μ.

Table 3.2.1. Message Length Distribution

Message length in characters:	25	50	70	100	140
Fraction with this length:	0.4	0.3	0.1	0.15	0.05

Example 3.2.9 Message lengths for the Euphoria State on-line system have the distribution shown in Table 3.2.1. Approximate this message length distribution by an Erlang-k random variable Y.

Solution Let X be the message length. Then

$$
\begin{aligned}
E[X] &= 25 \times 0.4 + 50 \times 0.3 + 70 \times 0.1 + 100 \times 0.15 \\
&\quad + 140 \times 0.05 = 54 \text{ characters,}
\end{aligned}
$$

and

$$\begin{aligned}
\text{Var}[X] \;=\; & (25-54)^2 \times 0.4 + (50-54)^2 \times 0.3 + (70-54)^2 \times 0.1 \\
& +(100-54)^2 \times 0.15 + (140-54)^2 \times 0.05 = 1054.
\end{aligned}$$

Since $E[X]^2/\text{Var}[X] = 2.77$, we let Y be the Erlang-2 random variable with average value 54 ($\mu = 1/54$). Thus, Y has the same mean as X but its variance is slightly different; it is 1458 rather than the correct 1054. $\square$

The reader should verify that the gamma random variable, Y, with $\beta = 54^2/1024$ and $\alpha = 54/1024$ has exactly the same mean and variance as the message length distribution of Example 3.2.9. This is the random variable generated by Algorithm 3.2.1. The APL function GΔDIST can be used to calculate values of the distribution function of Y. So can MINITAB, SAS/STAT, *Mathematica*, and the HP-21S.

3.2.7 Chi-Square Random Variables

A random variable Y is said to have a *chi-square distribution with n degrees of freedom* if it can be represented as

$$Y = X_1^2 + X_2^2 + \cdots + X_n^2, \tag{3.99}$$

where $X_1, X_2, \ldots, X_n$ are independent standard normal random variables. Thus, it is evident that Y can assume only nonnegative values. To discuss the properties of a chi-square random variable, we need the following theorem.

Theorem 3.2.7 *Let X be a continuous random variable with density function f and distribution function F. Then the density function g of the random variable $Y = X^2$ is defined by*

$$g(y) = \begin{cases} \frac{1}{2\sqrt{y}} \left(f\left(\sqrt{y}\right) + f\left(-\sqrt{y}\right) \right), & \text{for } y > 0 \\ 0 & \text{for } y \leq 0, \end{cases} \tag{3.100}$$

and the distribution function G is given by

$$G(y) = \begin{cases} F\left(\sqrt{y}\right) - F\left(-\sqrt{y}\right) & \text{for } y > 0 \\ 0 & \text{for } y \leq 0. \end{cases} \tag{3.101}$$

Proof Since Y cannot assume negative values, $G(y) = 0$ for $y \leq 0$. For $y > 0$, $Y = X^2 \leq y$ is equivalent to $-\sqrt{y} \leq X \leq \sqrt{y}$. Hence,

$$\begin{aligned}
G(y) \;&=\; P[Y \leq y] = P[-\sqrt{y} \leq X \leq \sqrt{y}] \\
&=\; F\left(\sqrt{y}\right) - F\left(-\sqrt{y}\right).
\end{aligned}$$

This proves (3.101). By differentiation we calculate

$$g(y) = G'(y) = \frac{1}{2\sqrt{y}}\left(F'\left(\sqrt{y}\right) + F'\left(-\sqrt{y}\right)\right) = \frac{1}{2\sqrt{y}}\left(f\left(\sqrt{y}\right) + f\left(-\sqrt{y}\right)\right),$$

which completes the proof. ∎

Suppose now that Y has a chi-square distribution with one degree of freedom, that is, that $Y = X^2$ where X is a standard normal random variable. Then, by Theorem 3.2.5, the density function g for Y is given by

$$
\begin{aligned}
g(y) &= \frac{1}{2\sqrt{y}}\left(\frac{e^{-y/2}}{\sqrt{2\pi}} + \frac{e^{-y/2}}{\sqrt{2\pi}}\right) = \frac{e^{-y/2}}{\sqrt{2}\sqrt{\pi}\sqrt{y}} \\
&= \frac{(\frac{1}{2})(y/2)^{(1/2)-1}e^{-y/2}}{\Gamma(1/2)} \quad \text{for } y > 0, \qquad (3.102)
\end{aligned}
$$

since $\Gamma(1/2) = \sqrt{\pi}$. (For a proof that $\Gamma(1/2) = \sqrt{\pi}$, see Exercise 50.) A comparison of (3.102) with (3.81) shows that Y is a gamma random variable with parameters $\beta = 1/2$ and $\alpha = 1/2$. Hence, the moment generating function of Y is given by

$$\psi(\theta) = (1 - 2\theta)^{-1/2}. \qquad (3.103)$$

If now Y is a chi-square random variable with n degrees of freedom, we can apply Theorem 2.9.1(c) to conclude that the moment generating function of Y is

$$\psi(\theta) = ((1 - 2\theta)^{-1/2})^n = (1 - 2\theta)^{-n/2}. \qquad (3.104)$$

The moment generating function given by (3.104) is that of a gamma random variable with parameters $\beta = n/2$ and $\alpha = 1/2$. We summarize the properties of chi-square random variables in the following theorem.

Theorem 3.2.8 (*Properties of Chi-Square Random Variables*) *Let X be a chi-square random variable with n degrees of freedom (and thus, a gamma random variable with parameters $\beta = n/2$ and $\alpha = 1/2$). Then the following statements are true.*

(a): $\psi(\theta) = (1 - 2\theta)^{-n/2}, E[X] = n, \mathrm{Var}[X] = 2n$ *and the density function of X is given by*

$$f(x) = \begin{cases} \dfrac{x^{(n/2)-1}e^{-x/2}}{2^{n/2}\Gamma(n/2)} & \text{for } x > 0 \\ 0 & \text{for } x \leq 0. \end{cases} \qquad (3.105)$$

(b): If Y is an independent chi-square random variable with m degrees of freedom, then X+Y is a chi-square random variable with n+m degrees of freedom (the chi-square distribution has the reproductive property).

(c): As n increases, X approaches a normal distribution; that is, for large n, X is approximately $N(n, 2n)$.

Proof Item (a) follows immediately from the fact that X has a gamma distribution with parameters $\beta = n/2$ and $\alpha = 1/2$, as we showed above. Now

$$\psi_{X+Y}(\theta) = (1 - 2\theta)^{n/2}(1 - 2\theta)^{m/2} = (1 - 2\theta)^{(n+m)/2}$$

and therefore, (b) holds by the uniqueness of the moment generating function (Theorem 2.9.1); and (c) is a consequence of the central limit theorem, which is discussed in Section 3.3. ■

The chi-square distribution is best known for its use in "chi-square tests," which are used to test various statistical hypotheses about observed random variables. Some of these tests are discussed in Chapter 8.

Table 4 of Appendix A gives critical values, χ_α^2, of the chi-square distributed random variable χ^2, defined by $P[\chi^2 > \chi_\alpha^2] = \alpha$. For example, the table shows that, if X has a chi-square distribution with 25 degrees of freedom, then the probability that X assumes a value greater than 37.653 is 0.05. Critical values, χ_α^2, can also be calculated using MINITAB, SAS/STAT, *Mathematica*, and the HP-21S. The same can be said for the critical values, t_α, and $f_\alpha(n, m)$, to be discussed later in this chapter.

3.2.8   Student's t Distribution

A continuous random variable X is said to have a *Student's t distribution with n degrees of freedom* if its density function is given by

$$f_n(x) = \frac{1}{\sqrt{n\pi}} \frac{\Gamma((n+1)/2)}{\Gamma(n/2)} \left(1 + \frac{x^2}{n}\right)^{-(n+1)/2} \qquad \text{for all real } x. \quad (3.106)$$

This distribution was discovered in 1908 by William S. Gosset, who used the pen name "A. Student" [35]. (Gosset worked for the Guiness brewery in Dublin, which at that time did not allow its employees to publish research papers under their own names.) It is evident from (3.106) that f_n is symmetric about $x = 0$, and it is easy to show that it assumes a maximum value there (see Exercise 49).

Theorem 3.2.9 (*Properties of a Student's t Random Variable*) *Let X be a Student's t random variable with n degrees of freedom as defined above. Then the following statements are true.*

(a): *For $n = 1$, X has no expected value; for $n > 1$, $E[X] = 0$.*

(b): *For $n = 1, 2$, the second moment does not exist; for $n > 2$,*

$$\text{Var}[X] = \frac{n}{n-2}.$$

(c): *For large values of n, X can be approximated by a standard normal random variable, that is,*

$$\lim_{n \to \infty} X = Y,$$

where Y is $N(0, 1)$.

Proof The proof can be found in Stuart and Ord [34]. ∎

The Student's t distribution is used primarily in dealing with small samples from a normal population. This is discussed in Chapters 7 and 8.

Table 5 of Appendix A gives critical values, t_α, of a Student's t distributed random variable X, defined by $P[X > t_\alpha] = \alpha$.

3.2.9 F-Distributed Random Variables

A continuous random variable X has an *F distribution with (n, m) degrees of freedom* if it has the density function f_{nm} given by

$$f_{nm}(x) = \begin{cases} \dfrac{(n/m)^{n/2}\Gamma((n+m)/2)x^{(n/2)-1}}{\Gamma(n/2)\Gamma(m/2)(1 + (n/m)x)^{(n+m)/2}} & \text{for} \quad x > 0 \\ 0 & \text{for} \quad x \leq 0. \end{cases} \qquad (3.107)$$

Sometimes the F-distribution is called *Snedecor-F*.

Theorem 3.2.10 *(Properties of the F Distribution) Suppose U has a chi-square distribution with n degrees of freedom and V a chi-square distribution with m degrees of freedom, with U and V independent. Then $Y = (U/n)/(V/m)$ has an F distribution with (n, m) degrees of freedom and thus has the density (3.107). If we define $f_\alpha(n, m)$ to be the unique number such that $P[Y > f_\alpha(n, m)] = \alpha$, then*

$$f_{1-\alpha}(n, m) = \frac{1}{f_\alpha(m, n)}.$$

If $m > 2$, then $E[Y] = m/(m - 2)$. If $m > 4$, then

$$\text{Var}[Y] = \frac{m^2(2n + 2m - 4)}{n(m-2)^2(m-4)}.$$

Proof The proof of this theorem can be found in Stuart and Ord [34]. ∎

Selected values of $f_\alpha(n, m)$ are given in Table 6 of Appendix A.

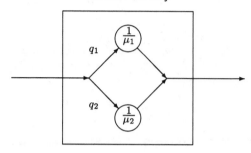

Figure 3.2.5. Hyperexponential model.

3.2.10 Hyperexponential Random Variables

If the service time of a queueing system has a large standard deviation relative to the mean value, it can often be approximated by a hyperexponential distribution. Hyperexponential, in this case, means *super exponential*. It would seem entirely proper to call a distribution for which the standard deviation is less than the mean *hypoexponential*. Thus, the constant distribution is the most hypoexponential of all! A hyperexponential random variable may, of course, represent many other interesting phenomena besides the service time of a queueing system; however, this provides an intuitively appealing way of describing a hyperexponential distribution, so we use it. The model representing the simplest hyperexponential distribution is shown in Figure 3.2.5. This model has two parallel stages in the facility; the top one providing exponential service with parameter μ_1 and the bottom stage providing exponential service with parameter μ_2. A customer entering the service facility chooses the top stage with probability q_1 or the bottom stage with probability q_2, where $q_1 + q_2 = 1$. After receiving service at the chosen stage, the service time being exponentially distributed with average service rate μ_i (average service time $1/\mu_i$), the customer leaves the service facility. A new customer is not allowed to enter the facility until the original customer has completed service. Thus, the density function for the service time is given by

$$f_s(t) = q_1\mu_1 e^{-\mu_1 t} + q_2\mu_2 e^{-\mu_2 t}, \quad t \geq 0. \tag{3.108}$$

Therefore, by integration,

$$W_s = \frac{q_1}{\mu_1} + \frac{q_2}{\mu_2}, \tag{3.109}$$

$$E[s^2] = 2\frac{q_1}{\mu_1^2} + 2\frac{q_2}{\mu_2^2}, \qquad (3.110)$$

and

$$E[s^3] = 6\frac{q_1}{\mu_1^3} + 6\frac{q_2}{\mu_2^3}. \qquad (3.111)$$

Hence, we calculate

$$\mathrm{Var}[s] = E[s^2] - E[s]^2 = \frac{2q_1}{\mu_1^2} + \frac{2q_2}{\mu_2^2} - \left(\frac{q_1}{\mu_1} + \frac{q_2}{\mu_2}\right)^2. \qquad (3.112)$$

The distribution function of a two-stage hyperexponential service time can be calculated by integrating (3.108), yielding

$$W_s[t] = 1 - q_1 e^{-\mu_1 t} - q_2 e^{-\mu_2 t}. \qquad (3.113)$$

Similarly, the moment generating function of the service time, $\psi_s(\cdot)$, is

$$\psi_s(\theta) = \frac{q_1 \mu_1}{\mu_1 - \theta} + \frac{q_2 \mu_2}{\mu_2 - \theta}, \quad \text{if} \quad \theta < \mu_1 \quad \text{and} \quad \theta < \mu_2. \qquad (3.114)$$

Likewise, the Laplace–Stieltjes transform, $W_s^*[\theta]$, is

$$W_s^*[\theta] = \frac{q_1 \mu_1}{\mu_1 + \theta} + \frac{q_2 \mu_2}{\mu_2 + \theta} \quad \text{if} \quad \theta < \mu_1 \quad \text{and} \quad \theta < \mu_2. \qquad (3.115)$$

The following algorithm creates an H_2 random variable X with a given mean $E[X] = 1/\mu$ and $C_X^2 = C^2 \geq 1$. It is said to have *balanced means*, since

$$\frac{q_1}{\mu_1} = \frac{q_2}{\mu_2}.$$

Algorithm 3.2.2 (*Algorithm H*) *Given* $C^2 \geq 1$ *and* $\mu > 0$, *this algorithm will produce a two-stage hyperexponential random variable* X *with squared coefficient of variation* $C_X^2 = C^2$ *and mean* $E[X] = \frac{1}{\mu}$, *such that*

$$\frac{q_1}{\mu_1} = \frac{q_2}{\mu_2}. \qquad (3.116)$$

The distribution function of X *is given by*

$$F(x) = 1 - q_1 e^{-\mu_1 x} - q_2 e^{-\mu_2 x}. \qquad (3.117)$$

Step 1 [*Calculate* q_1 *and* q_2] *Set*

$$q_1 = \frac{1}{2}\left(1 - \left[\frac{(C^2-1)}{(C^2+1)}\right]^{\frac{1}{2}}\right) \qquad (3.118)$$

and

$$q_2 = 1 - q_1. \qquad (3.119)$$

Step 2 [*Calculate μ_1 and μ_2*] *Set*

$$\mu_1 = 2q_1\mu \tag{3.120}$$

and

$$\mu_2 = 2q_2\mu. \tag{3.121}$$

Step 3 [*Produce F*] *The distribution function F defined by (3.117), with parameters calculated in Step 1 and Step 2, is the distribution function of a two-stage hyperexponential random variable having the required properties.*

Proof The proof is a simple exercise using (3.109) and (3.110) and some algebra (See Exercise 56(a)). ∎

The APL function BH2 implements Algorithm H; that is, it produces the parameters for an H_2 distribution with balanced means. The APL function H2ΔDIST is the distribution function for an H_2 random variable.

The following algorithm produces the H_2 *distribution with the gamma normalization*, since it produces an H_2 random variable with a given mean $E[X] = \frac{1}{\mu}$ and squared coefficient of variation $C_X^2 = C^2$ that also has the same third moment as the gamma distribution with this C_X^2 and $E[X]$.

Algorithm 3.2.3 (*Algorithm HG*) *Given $C^2 \geq 1$ and $\mu > 0$, this algorithm will produce a two-stage hyperexponential random variable X with squared coefficient of variation $C_X^2 = C^2$ and mean $E[X] = \frac{1}{\mu}$. Furthermore, X has the same third moment as the gamma random variable Y, such that*

$$E[Y] = \frac{1}{\mu},$$

and $C_Y^2 = C^2$. The distribution function of X is given by

$$F(x) = 1 - q_1 e^{-\mu_1 x} - q_2 e^{-\mu_2 x}. \tag{3.122}$$

Step 1 [*Calculate μ_1 and μ_2*] *Set*

$$\mu_1 = \frac{2}{E[X]} \times \left(1 + \sqrt{\frac{(C^2 - \frac{1}{2})}{(C^2 + 1)}}\right), \tag{3.123}$$

and

$$\mu_2 = \frac{4}{E[X]} - \mu_1. \tag{3.124}$$

Step 2 [*Calculate q_1 and q_2*] *Set*

$$q_1 = \frac{\mu_1(\mu_2 \times E[X] - 1)}{\mu_2 - \mu_1},$$ (3.125)

and

$$q_2 = 1 - q_1.$$ (3.126)

Step 3 [*Produce F*] *The distribution function F defined by (3.122), with parameters calculated in Step 1 and Step 2, is the distribution function of a two-stage hyperexponential random variable having the required properties.*

Proof We ask the reader to show that the above algorithm does what we claim it does in Exercise 56(b). ∎

The APL function GH2 implements Algorithm HG; that is, it calculates the parameters of an H_2 random variable with the gamma normalization.

You may be thinking, "The H_2 probability distribution has 3 independent parameters (q_1 or q_2, μ_1, and μ_2). Couldn't we use the three parameters to construct an H_2 distribution X such that $E[X] = k_1$, $E[X^2] = k_2$, and $E[X^3] = k_3$ for any three positive numbers k_1, k_2, k_3?" A little thought should convince you that *some* restrictions must apply, other than that the numbers are positive. (After all, we must have $C_X^2 \geq 1$.) Whitt [38] has worked out what the restrictions are and what the algorithm is to construct the two-stage hyperexponential distribution.

Algorithm 3.2.4 (*Algorithm HW, Whitt's Algorithm*) *Given $k_1 > 0$, $k_2 > 0$, and $k_3 > 0$, satisfying the conditions*

$$C^2 = \frac{k_2}{k_1^2} - 1 \geq 1,$$ (3.127)

and

$$k_3 k_1 \geq 1.5 k_2^2,$$ (3.128)

this algorithm will produce a two-stage hyperexponential random variable X (an H_2 random variable) with squared coefficient of variation $C_X^2 = C^2$ and the following moments:

$$E[X] = k_1,$$ (3.129)

$$E[X^2] = k_2,$$ (3.130)

$$E[X^3] = k_3.$$ (3.131)

The distribution function of X is given by

$$F(x) = 1 - q_1 e^{-\mu_1 x} - q_2 e^{-\mu_2 x}.$$ (3.132)

Step 1 [*Calculate μ_1 and μ_2*] *Calculate*

$$x = k_1 k_3 - 1.5k_2^2, \tag{3.133}$$

$$y = k_2 - 2k_1^2, \tag{3.134}$$

and

$$v = \left(x + 1.5y^2 - 3k_1^2 y\right)^2 + 18k_1^2 y^3. \tag{3.135}$$

Then set

$$\frac{1}{\mu_1} = \frac{x + 1.5y^2 + 3k_1^2 y + \sqrt{v}}{6k_1 y}, \tag{3.136}$$

and

$$\frac{1}{\mu_2} = \frac{x + 1.5y^2 + 3k_1^2 y - \sqrt{v}}{6k_1 y}. \tag{3.137}$$

Step 2 [*Calculate q_1 and q_2*] *Set*

$$q_1 = \frac{k_1 - \dfrac{1}{\mu_2}}{\dfrac{1}{\mu_1} - \dfrac{1}{\mu_2}} \tag{3.138}$$

and

$$q_2 = 1 - q_1. \tag{3.139}$$

Step 3 [*Produce F*] *The distribution function F defined by (3.132), with parameters calculated in Step 1 and Step 2, is the distribution function of a two-stage hyperexponential random variable having the required properties.*

Proof We ask the reader to show that the above algorithm does what we claim it does in Exercise 56(c). ■

The APL function WH can be used to make the calculations of this algorithm. The APL function MOMENTS can be used to check the answer. That is, MOMENTS will calculate the first three moments of an H_2 distribution.

Example 3.2.10 Helga Tooterfluz, the lead computer performance analyst at Sanitary Sewer Sweepers, decides to do the following:

(a) Construct a gamma distributed random variable X with $E[X] = 5$ and $C_X^2 = 2$. She also wants to calculate

$$P[X \le 7].$$

(b) Construct a two-stage hyperexponential random variable Y with balanced means, which has the same mean and squared coefficient of variation as X of part (a). Helga also wants to calculate

$$P[Y \leq 7].$$

(c) Construct a two-stage hyperexponential random variable U with the gamma normalization that has the same mean and squared coefficient of variation as X of part (a). Ms. Tooterfluz also wants to calculate

$$P[U \leq 7].$$

(d) Compute $E[X^3]$, $E[Y^3]$, and $E[U^3]$.

Solution

(a) Helga applied the formulas of Algorithm G (Algorithm 3.2.1) to compute the parameters of the gamma random variable X, obtaining

$$\beta = \frac{1}{C_X^2} = 0.5,$$

and

$$\alpha = \frac{\beta}{E[X]} = 0.1.$$

Then, using MINITAB, SAS/STAT, *Mathematica*, or GΔDIST, she computes
$$P[X \leq 7] = 0.7632764294.$$

(b) Using Algorithm H, Ms. Tooterfluz calculates $q_1 = 0.2113248654$, $q_2 = 0.7886751346$, $\mu_1 = 0.08452994616$, and $\mu_2 = 0.3154700538$. Then she uses $(3.117)^5$ to calculate

$$P[Y \leq 7] = 0.796390747.$$

(c) Helga uses the formulas of Algorithm HG to calculate $\mu_1 = 0.6828427125$, $\mu_2 = 0.1171572875$, $q_1 = 0.5$, and $q_2 = 0.5$. Formula (3.122) then yields
$$P[U \leq 7] = 0.7756079927.$$

---

[5]Actually, she uses the APL function BH2 to calculate the parameters and the APL function H2ΔDIST to calculate the probability $P[Y \leq 7]$. For (c), she used GH2 and H2ΔDIST.

(d) Since X has a gamma distribution,

$$E[X^3] = \frac{\beta(\beta + 1)(\beta + 2)}{\alpha^3} = 1,875.$$

By (3.111),

$$E[Y^3] = 6 \left[\frac{q_1}{\mu_1^3} + \frac{q_2}{\mu_2^3} \right] = 2,250.$$

Again, by (3.111),

$$E[U^3] = 6 \left[\frac{q_1}{\mu_1^3} + \frac{q_2}{\mu_2^3} \right] = 1,875. \quad \square$$

Note that

$$E[X^3] = E[U^3].$$

3.2.11   Coxian Random Variables

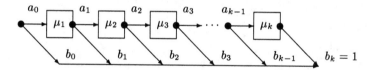

Figure 3.2.6. Cox's method of stages for a service facility.

The Coxian distribution is a generalization of both the Erlang-k and the hyperexponential distributions. Cox [8] proposed that a service center be represented as a network of stages or nodes, as shown in Figure 3.2.6.

Only one customer at a time is allowed in the facility. A customer initially entering the service facility will receive service at stage 1 with probability a_0 or leave the facility without service with probability $b_0 = 1 - a_0$. (For modeling queueing theory service centers, we are interested only in cases where $a_0 = 1$ and $b_0 = 0$.) After receiving service at stage i (distributed exponentially with mean $1/\mu_i$), a customer leaves the facility with probability b_i or proceeds to stage $i + 1$ with probability a_i, $i = 1, 2, \cdots, k - 1$. Naturally, $a_i + b_i = 1$ for all i. A customer completing service at stage k leaves the service facility. The probability that a customer reaches stage i is $A_i = a_0 a_1 \ldots a_{i-1}$ $(i = 1, 2, \cdots, k)$ and the probability that a customer visits stages $1, 2, \ldots, i$ and leaves the facility is $A_i b_i$. Hence, the time s a customer spends in the service facility is, with probability $A_i b_i$, the sum of i independent, exponentially distributed random variables. Hence, by Theorem 2.8.1 and Example 2.9.1, the expected value of s is given by

$$E[s] = \sum_{i=1}^{k} A_i b_i E[\sum_{j=1}^{i} X_j] = \sum_{i=1}^{k} A_i b_i \sum_{j=1}^{i} \frac{1}{\mu_j}. \tag{3.140}$$

Similarly, $E[s^2]$ is given by

$$\sum_{i=1}^{k} A_i b_i E[Y_i^2], \tag{3.141}$$

where

$$Y_i = X_1 + X_2 + \cdots + X_i. \tag{3.142}$$

But

$$
\begin{aligned}
E[Y_i^2] &= E[X_1^2 + X_2^2 + \cdots + X_i^2 + 2\sum_{l \neq m} X_l X_m] \\
&= \frac{2}{\mu_1^2} + \frac{2}{\mu_2^2} + \cdots + \frac{2}{\mu_i} \\
&\quad + \frac{2}{\mu_1 \mu_2} + \frac{2}{\mu_1 \mu_3} + \cdots + \frac{2}{\mu_1 \mu_i} \\
&\quad + \frac{2}{\mu_2 \mu_3} + \cdots + \frac{2}{\mu_{i-1} \mu_i}.
\end{aligned} \tag{3.143}
$$

It is not difficult to show that the Laplace–Stieltjes transform of a Coxian distribution is given by

$$X^*[\theta] = b_0 + \sum_{i=1}^{k} a_0 \ldots a_{i-1} b_i \prod_{j=1}^{i} \frac{\mu_j}{\theta + \mu_j}. \tag{3.144}$$

Cox [8] shows that his method of stages representation is the most general way of constructing a random variable from independent exponential stages. Thus, an Erlang-k, a hyperexponential, or any nonnegative random variable having a rational Laplace–Stieltjes transform is a Coxian distribution. The latter claim is proven in his paper. The method of stages representation of a service facility makes it easy to handle, mathematically, in a queueing network model. Thus, it is the method used by Baskett, Chandy, Muntz, and Palacios [4] in their classic paper. Likewise, Khomonenko and Bubnov [15] were able to use Cox's distribution to solve a queueing theory model. Sometimes it is somewhat difficult to get a given probability distribution into the Cox method of stages format, that is, to match the given distribution by a Coxian distribution. It is *not* difficult for the Erlang-k distribution. For this distribution $b_0 = 0$, $a_0 = a_1 = \cdots a_{k-1} = 1$, $b_k = 1$, and $\mu_i = k\mu$ for $i = 1, 2, \cdots, k$. The Erlang-k distribution *is* a special case of the Coxian distribution. The two-stage hyperexponential is *not* a two-stage Coxian distribution[6] but can be represented by one. We will demonstrate one way to do this in the following example. It is not a trivial exercise.

Example 3.2.11 Consider the H_2 distribution Y generated by Algorithm 3.2.2 with $C_Y^2 = 5$ and $E[Y] = 1$. Using the APL function BH2, we obtain $q_1 = 0.09175170954$, $q_2 = 0.9082482905$, $\mu_1 = 0.1835034191$, and $\mu_2 = 1.816496581$. Using the APL function MOMENTS, we calculate $E[Y^2] = 6$ and $E[Y^3] = 90$. Let us approximate Y with a Coxian distribution with two stages. Let us choose the Coxian distribution X with the same first three moments as Y. We will let $a_0 = 1$ and $b_2 = 1$. By the use of (3.123), Theorem 2.9.3, and some tedious algebra, we obtained the following set of equations:

$$a_1 + b_1 = 1.$$

$$E[X] = \frac{1}{\mu_1} + \frac{a_1}{\mu_2} = 1.$$

$$E[X^2] = \frac{2}{\mu_1^2} + \frac{2a_1}{\mu_1\mu_2} + \frac{2a_1}{\mu_2^2} = 6.$$

$$E[x^3] = \frac{6}{\mu_1^3} + \frac{6a_1}{\mu_1^2\mu_2} + \frac{6a_1}{\mu_1\mu_2^2} + \frac{6a_1}{\mu_2^3} = 90.$$

The solution to this set of equations is :

$$a_1 = 0.0824829046378,$$

$$\mu_1 = 1.81649658092,$$

and

$$\mu_2 = 0.183503419072.$$

The Coxian distribution uses the same two exponential distributions as were used by the H_2 distribution! However, for the Coxian distribution, the customer *always* receives service from the faster server followed, occasionally, (with probability 0.0824829) by service from the slower server. By

---

[6]All textbooks, including this one, say that the hyperexponential is Coxian but this should not be interpreted in the literal sense. A Coxian distribution can represent a hyperexponential distribution but the flow of customers through such a service facility is not exactly as it would be for a hyperexponential facility. However, the Laplace–Stieltjes transforms, and therefore the probability distributions, are identical.

(3.124), the Coxian distribution has the Laplace–Stieltjes transform

$$X^*[\theta] = \frac{1.66666667}{\theta + 1.816496581} + \frac{0.0274943}{(\theta + 1.816496581)(\theta + 0.183503419072)},$$
(3.145)

while the Laplace–Stieltjes Transform of the original H_2 distribution is

$$Y^*[\theta] = \frac{1.6498299}{\theta + 1.816496581} + \frac{0.0168367524}{(\theta + 0.183503419072)},$$
(3.146)

by (3.115). The two distributions certainly look different! However, if they are both rearranged into the rational format, we find that:

$$Y^*[\theta] = X^*[\theta] = \frac{1.666667\theta + 0.33333333}{(\theta + 1.816496581)(\theta + 0.183503419072)}.$$
(3.147)

The probability distributions are the same, although the two method of stages models that produced them are different! □

Marie [19] provides an easier way to construct a two stage Coxian distribution with a given mean $1/\mu$ and squared coefficient of variation C^2. Of course, it will not have the same third moment as the random variable we constructed in the above example.

Algorithm 3.2.5 (*Algorithm M, Marie's Algorithm*) *Given $\mu > 0$ and $C^2 \geq 0.5$, this algorithm will generate the parameters a_1, μ_1, and μ_2 for a two-stage Coxian random variable X, such that $E[X] = 1/\mu$ and $C_X^2 = C^2$.*

Step 1 *Let $\mu_1 = 2\mu$.*

Step 2 *Let $\mu_2 = \mu/C^2$.*

Step 3 *Let $a_1 = 1/(2C^2)$.*

Proof See Marie [19]. ■

Example 3.2.12 Consider Example 3.2.11. Let us use Marie's algorithm to construct a two-stage Coxian random variable X with $E[X] = 1$ and $C_X^2 = 5$. We choose $\mu_1 = 2$, $\mu_2 = 0.2$, and $a_1 = 0.1$. Then, by the equations we developed in the above example,

$$\begin{aligned} E[X] &= 0.5 + 0.5 = 1, \\ E[X^2] &= 0.5 + 0.5 + 5 = 6, \end{aligned}$$

so that

$$\text{Var}[X] = E[X^2] - E[X]^2 = 6 - 1 = 5,$$

and

$$C_X^2 = \frac{\text{Var}[X]}{E[X]^2} = 5.$$

We also have

$$E[X^3] = \frac{6}{8} + \frac{0.6}{0.8} + \frac{0.6}{2(0.2)^2} + \frac{0.6}{0.2^3} = 84.$$

The random variable is not very different from the one we constructed in Example 3.2.11. □

The two-stage hyperexponential distribution can be generalized to k stages for any positive integer ≥ 2. The extension to k stages is straightforward but rarely of practical importance, so we will not discuss it here.

When you can measure what you are speaking about and express it in numbers you know something about it; but when you cannot express it in numbers, your knowledge is of a meager and unsatisfactory kind.
Lord Kelvin

3.3 Central Limit Theorem

In Chapter 2 we discussed the weak law of large numbers, which indicates, roughly speaking, that the probability $P[A]$ of an event A can be estimated by S_n/n, where S_n is the number of times the event A occurs in n independent trials of the basic experiment. Unfortunately, the law of large numbers does not provide a method for estimating how close we are to the true probability, although we saw in Section 2.10 that by using Chebyshev's inequality, we could make a crude estimate of how large n need be so that

$$P\left[\left|\frac{S_n}{n} - p\right| \geq \delta\right] \leq \epsilon,$$

for given positive δ and ϵ.

The central limit theorem allows us to improve this estimate. It also allows us to make probability judgments about other types of estimates. This theorem is one of the most important in applied probability theory.

Theorem 3.3.1 *(Central Limit Theorem) Let $X_1, X_2, \ldots$ be independent, identically distributed random variables, each having mean μ and standard deviation $\sigma > 0$. Let $S_n = X_1 + \cdots + X_n$. Then for each $x < y$,*

$$\lim_{n \to \infty} P\left[x \leq \frac{S_n - n\mu}{\sigma\sqrt{n}} \leq y\right] = \Phi(y) - \Phi(x), \tag{3.148}$$

where Φ is the standard normal distribution function.

Proof The proof of this theorem may be found in Parzen [22]. ∎

This theorem is truly remarkable in that no special assumptions need be made about the character of X_1. It can be discrete, continuous, or of mixed type. No matter what the form of X_1, the sum S_n approaches a normal distribution with mean $n\sigma^2$ (S_n is approximately $N(n\mu, n\sigma^2)$). Of course, the rate of convergence of X_n to a normal distribution depends on X_1. For example, if X_1 is normally distributed, then by Theorem 3.2.3, S_n is normally distributed for all n—no approximation is involved. However, if X_1 is a discrete uniform distribution, then n must be somewhat large before S_n can be reasonably approximated by a normal random variable. The result of the central limit theorem is true, under rather general conditions, even if each X_k has a different distribution with mean μ_k and standard deviation σ_k, if $\sum_{k=1}^{n} \mu_k$ is substituted for $n\mu$ and $(\sum_{k=1}^{n} \sigma_k^2)^{1/2}$ is substituted for $\sigma\sqrt{n}$ in (3.149); that is, (3.149) becomes

$$\lim_{n \to \infty} P\left[x \leq \frac{S_n - E[S_n]}{(\text{Var}[S_n])^{1/2}} \leq y\right] = \Phi(y) - \Phi(x). \tag{3.149}$$

This version of the central limit theorem is the basis for an explanation of the observed fact that many random variables such as the height and weight of humans, the yields of crops, the temperature at a certain geographical location for a given day of the year, etc., tend to be normally distributed. Each of these random variables can be represented as the sum of a large number of independent random variables.

The central limit theorem has a special case now called the *DeMoivre-Laplace limit theorem*. It was originally proved by Abraham DeMoivre (1667–1754) in his *Doctrine of Chances*, which was published in 1714. Pierre Simon Laplace (1749–1827) extended DeMoivre's result in his famous treatise *Théorie Analytique des Probabilités*, published in 1812.

Let $X_1, X_2, \ldots, X_n$ be independent Bernoulli random variables, each with probability p of success. Then $S_n = X_1, X_2, \ldots, X_n$ is a binomial random variable and the following theorem follows from Theorem 3.3.1.

Theorem 3.3.2 (*DeMoivre–Laplace Limit Theorem*) *Let S_n be a binomial random variable with parameters n and p. Then for any nonnegative integers a and b, with $a \leq b$, as $n \to \infty$,*

$$P[a \leq S_n \leq b] \to \Phi\left(\frac{b - np}{\sqrt{npq}}\right) - \Phi\left(\frac{a - np}{\sqrt{npq}}\right). \tag{3.150}$$

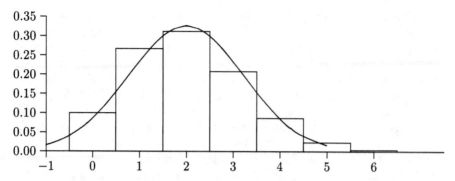

Figure 3.3.1. Normal approximation of binomial.

The following corollary tells us how to approximate the binomial distribution with the normal distribution.

Corollary (The normal approximation of the binomial distribution) Suppose a and b are nonnegative integers with $a \leq b$. Then it is approximately true that

$$\sum_{k=a}^{b} \binom{n}{k} p^k q^{n-k} = \Phi\left(\frac{b - np + \frac{1}{2}}{\sqrt{npq}}\right) - \Phi\left(\frac{a - np - \frac{1}{2}}{\sqrt{npq}}\right). \quad \blacksquare \quad (3.151)$$

Note that the right hand sides of (3.151) and (3.152) differ by the $\frac{1}{2}$ terms, which are called the *continuity corrections*. The reason for the continuity corrections is that, if we use the normal distribution to approximate the discrete binomial distribution, we are, in effect, fitting a continuous distribution to a discrete distribution, as suggested by Figure 3.3.1. In this figure, the step function gives the probabilities of k successes in eight Bernoulli trials with $p = 0.25$. That is, for each k, the area under the binomial graph between $k - \frac{1}{2}$ and $k + \frac{1}{2}$ is the probability of k successes. The density function for the approximating normal random variable has mean $np = 2$ and standard deviation $\sqrt{npq} = \sqrt{1.5} = 1.225$. The true probability that S_n is between 1 and 3 inclusive is $\sum_{k=1}^{3} \binom{8}{k}(0.25)^k(0.75)^{8-k} = 0.7861$. If we use (3.152) with the continuity correction, we approximate this probability by the area under the normal density curve from 0.5 to 3.5, which is

$$\Phi\left(\frac{3.5 - 2}{1.225}\right) - \Phi\left(\frac{0.5 - 2}{1.225}\right) = \Phi(1.224) - \Phi(-1.224) = 2\Phi(1.224) - 1$$

$$= 2 \times 0.88952 - 1 = 0.77904.$$

This is a fairly good approximation because, as Figure 3.3.1 shows, this binomial distribution is not very symmetrical. For example,

$$P[X = 1] = 0.2671 \neq 0.2076 = P[X = 3].$$

The accuracy of the approximation formula (3.152) improves with the size of n, and in the limit the error goes to zero. The accuracy also improves with the degree of symmetry of S_n. A number of rules have been devised to ensure that the approximation is reasonably good. Freund and Walpole [12] require both np and nq to exceed 5. Lippman [18] asks that $npq \geq 10$. Hoel [14] says the normal approximation is fairly good as long as $np > 5$ when $p \leq \frac{1}{2}$ and $nq > 5$ when $p > \frac{1}{2}$. Stuart and Ord [34] claim that if $np^{3/2} > 1.07$, then the error in the normal approximation for any $b(k; n, p)$ is less than 0.05. Ostle and Mensing [21] assert that, if $npq > 25$, the error in the normal approximation is less than $0.15/\sqrt{npq}$. They point out, however, that for values of p very close to 0 or 1, the approximation will be less reliable in the tail than the center of the distribution, and in these cases one should either use the Poisson approximation or calculate exact probabilities.

Feller [10, chapter VII] has some excellent examples of the use of the normal approximation. In Table 2 he compares some exact probabilities together with the normal approximations for the binomial distribution with parameters $n = 100$ and $p = 0.3$ (and thus with mean 30 and variance 21). This distribution satisfies the rules of thumb of Freund and Walpole, Lippman, and Hoel. However, although the approximation error is zero for $P[21 \leq S_n \leq 21] = P[32 \leq S_n \leq 39]$, there is a 400 percent error in the approximation for $P[9 \leq S_n \leq 11]$ and a 100 percent error in the approximation for $P[12 \leq S_n \leq 14]$. Ostle and Mensing's warning about the tails is certainly correct! It should be pointed out that the Poisson approximation is not so good for $P[9 \leq S_n \leq 11]$, either. It yields a value of 0.000061831 versus the correct value of 0.000005575 for a relative error of 1,009 percent. The calculation of $P[9 \leq S_n \leq 11]$ is not as formidable as it may first seem. $P[9 \leq S_n \leq 11] = P[S_n = 9] + P[S_n = 10] + P[S_n = 11]$, where, for example, we can calculate

$$P[S_n = 9] = \binom{100}{9} \times 0.3^9 \times 0.7^{91}, \qquad (3.152)$$

where $\binom{100}{9}$ can be calculated by the expression

$$\frac{100}{9} \times \frac{99}{8} \times \frac{98}{7} \times \frac{97}{6} \times \frac{96}{5} \times \frac{95}{4} \times \frac{94}{3} \times \frac{93}{2} \times \frac{92}{1}. \qquad (3.153)$$

The calculation of (3.154) is routine on most pocket calculators (especially those with RPN logic) and most calculators have a y^x key, that enables one to calculate the last two factors of (3.153) with ease. In fact, some pocket calculators, such as the Hewlett-Packard HP-32S, allow one to calculate $\binom{100}{9}$ directly.

Example 3.3.1 In Example 3.1.2, we considered a binomial random variable with parameters 20 and 0.6, that described the number of communication lines in use. The probability that 10 or more lines are in operation was found to be 0.872479. The normal approximation for this probability is

$$\Phi\left(\frac{20 - 12 + 0.5}{2.19}\right) - \Phi\left(\frac{10 - 12 - .05}{2.19}\right) = \Phi(3.881) + \Phi(1.142) - 1$$

$$= 0.99995 + 0.87327 - 1 = 0.87322,$$

a fairly good approximation. This random variable satisfies most of the rules of thumb we gave above. Of the ones involving np, nq, or npq, it fails only Lippman's requirement and that of Ostle and Mensing. However, the error is only -0.00074, which in absolute value is less than the Ostle and Mensing error bound of $0.15/\sqrt{npq} = 0.0685$. $\square$

The normal distribution can also be used to approximate a Poisson distribution. This follows from the fact that both the Poisson and normal distributions may be used to approximate the binomial distribution.

Example 3.3.2 In Example 3.1.5, X is a Poisson random variable with $\alpha = 10$. We calculated $P[X \leq 15]$ to be 0.95126 by using the APL function POISSONΔDIST. Approximate the answer using the normal distribution.

Solution We use the normal distribution with $\mu = 10$ and $\sigma = \sqrt{10}$. Then

$$P[X \leq 15] \approx P\left[z \leq \frac{15.5 - 10}{\sqrt{10}}\right] = P[z \leq 1.739] = 0.95898. \ \square$$

Suppose we want to estimate $p = P[A]$ for some event A, where we know that $0 < P[A] < 1$. We can let X be the Bernoulli random variable, that is 1 when the event A occurs on a particular trial of the experiment and 0 otherwise. Successive independent trials of the experiment yield the sequence of independent Bernoulli random variables $X_1, X_2, X_3, \ldots$. If we let $S_n = X_1 + X_2 + \cdots X_n$, then S_n counts the number of times that the event A occurred in n trials of the experiment. The weak law of large numbers, Theorem 2.10.4, indicated that the ratio S_n/n converges to

$p = P[A]$. In Example 2.10.6 we saw that Chebyshev's inequality enabled us to make a crude estimate of the required value of n, such that

$$P\left[\left|\frac{S_n}{n} - p\right| \geq \delta\right] \leq \epsilon, \tag{3.154}$$

where δ and ϵ are given positive numbers. We will now show how the central limit theorem allows us to improve that estimate. Let $p = P[A]$ and $q = 1 - p$. S_n is a binomial random variable with parameters n and p, so $E[S_n] = npq$. Suppose δ and ϵ are given positive numbers. Then

$$P\left[\left|\frac{S_n}{n} - p\right| \geq \delta\right] = P\left[\frac{S_n}{n} - p \leq -\delta\right] + P\left[\frac{S_n}{n} - p \geq \delta\right]. \tag{3.155}$$

Some algebraic manipulation of (3.156) yields

$$P\left[\left|\frac{S_n}{n} - p\right| \geq \delta\right] = P\left[\frac{S_n - np}{\sqrt{npq}} \leq -\frac{\delta\sqrt{n}}{\sqrt{pq}}\right] + P\left[\frac{S_n - np}{\sqrt{npq}} \geq \frac{\delta\sqrt{n}}{\sqrt{pq}}\right]. \tag{3.156}$$

The right side of (3.157) is now in a form for which we can apply the central limit theorem (the mean of X_1 is p and the standard deviation is $\sqrt{pq}$). Hence, we conclude that

$$P\left[\left|\frac{S_n}{n} - p\right| \geq \delta\right] \approx \Phi\left(\frac{-\delta\sqrt{n}}{\sqrt{pq}}\right) + 1 - \Phi\left(\frac{\delta\sqrt{n}}{\sqrt{pq}}\right) = 2\left(1 - \Phi\left(\frac{\delta\sqrt{n}}{\sqrt{pq}}\right)\right). \tag{3.157}$$

To find n such that (3.155) is valid, we set the right side of (3.158) to ϵ and $r = \delta\sqrt{n}/\sqrt{pq}$ to arrive at the equation

$$2(1 - \Phi(r)) = \epsilon, \tag{3.158}$$

or

$$\Phi(r) = (2 - \epsilon)/2. \tag{3.159}$$

The value of r that makes (3.160) true can be found from Table 3 of Appendix A. The definition of r then yields the following estimate for n:

$$n = r^2 pq/\delta^2 \leq r^2/4\delta^2, \tag{3.160}$$

since $pq = p(1-p)$ has a maximum value of $1/4$, achieved when $p = q = 1/2$ (see Exercise 2 of Chapter 2).

Example 3.3.3 In Example 2.10.8 we wanted to estimate the probability, p, that a randomly selected terminal chosen during the peak period was busy. The estimation method was to choose a terminal randomly n times during this period, count the number of times S_n that a selected terminal was in use, and use the ratio S_n/n as the estimate of p. Chebyshev's inequality was used to find the smallest n such that (3.155) would be true with $\delta = 0.1$ and $\epsilon = 0.05$. The value of n was 500 if no knowledge of p was assumed, and 320 if it were known that p was approximately 0.2. If we apply (3.160) and (3.161) for the first case, we get $r = 1.96$ and $n = 96$. If we assumed that p was about 0.2, then (3.161) yields $n = 62$. Thus, for the given requirements, 100 samples should suffice. However, an error of 0.1 in a quantity with a magnitude of only 0.2 is a large relative error! Let us turn the question about and ask, "If we make 500 observations to estimate p and let $\epsilon = 0.05$ in (3.155), what is the value of δ? That is, what is the maximum error in the estimate at the 5 percent level of uncertainty?" As before, (3.160) yields $r = 1.96$ and

$$\delta = \frac{r\sqrt{pq}}{\sqrt{n}} = \frac{1.96}{\sqrt{500}}\sqrt{pq},$$

which has the value 0.0438 or 0.0351, depending on our assumption about the value of p. For 100 observations, these values of δ are 0.098 and 0.0784, respectively. □

Nothing so needs reforming as other people's habits.
Mark Twain

3.4   Applied Transforms

In Section 2.9 we defined some useful transforms, including the moment generating function, the generating function (z-transform), the Laplace–Stieltjes transform, and the Laplace transform. We also explained the primary properties of these transforms and demonstrated their usefulness. In this section we will show further applications of transform methods in applied probability.

For any nonnegative random variable, X, we defined the Laplace–Stieltjes transform of X, X^*, by

$$X^*[\theta] = \begin{cases} \int_0^\infty e^{-\theta x} f(x)\, dx & \text{if } X \text{ is continuous} \\ \sum_{x_i} e^{-\theta x_i} p(x_i) & \text{if } X \text{ is discrete,} \end{cases} \qquad (3.161)$$

sometimes written in the more symbolic form

$$X^*[\theta] = \int_0^\infty e^{-\theta x}\, dF(x). \qquad (3.162)$$

In doing so we *transformed* the random variable X into the real (or complex) variable X^*. The reason for doing this is that operations on the variable X^* that correspond to operations on X are often much simpler to perform. For example, we saw in Theorem 2.9.3 that we can calculate moments of X by the formula

$$E[X^n] = (-1)^n \frac{d^n X^*[\theta]}{d\theta^n}\bigg|_{\theta = 0}, \qquad n = 1, 2, \cdots . \tag{3.163}$$

This is usually much easier than it is to calculate $E[X^n]$ directly from the definition. Theorem 2.7.5 showed that to get the probability mass function or the density function of the sum of the two independent random variables X and Y, we calculate the convolution of X and Y. Convolution is a rather difficult operation to perform. However, the Laplace–Stieltjes transform of the sum is merely the product of the transforms. That is,

$$(X + Y)^*[\theta] = X^*[\theta]Y^*[\theta], \tag{3.164}$$

if X and Y are independent, by Theorem 2.9.3(d). We can now get the density function (or the probability mass function) of $X + Y$ by inverting the transform $X^*[\theta]Y^*[\theta]$.

If f is any real-valued function, the *Laplace transform of* f is defined by

$$f^*[\theta] = \int_0^\infty e^{-\theta x} f(x) \, dx, \tag{3.165}$$

provided the integral in (3.166) exists. Thus, if $f(x) = \frac{dF}{dx}(x)$ is the density function of a continuous random variable X, then the Laplace–Stieltjes transform of X is the Laplace transform of f; that is,

$$X^*[\theta] = \int_0^\infty e^{-\theta x} \, dF(x) = \int_0^\infty e^{-\theta x} f(x) \, dx = f^*[\theta]. \tag{3.166}$$

The Laplace transform of the density function, f, of a nonnegative random variable, X, always exists for any $\theta \geq 0$ because

$$|f^*(\theta)| \leq \int_0^\infty |e^{-\theta t} f(t)| \, dt \leq \int_0^\infty f(t) \, dt = 1.$$

Since f^* exists, by (3.167), the Laplace–Stieltjes transform of X must exist, also. We use the convention that the lower limit of each of the integrals in (3.167) is to be evaluated so as to include the jump at the origin if $F(0) > 0$; that is, we define $\int_0^\infty e^{-\theta x} \, dF(x)$ to mean $\int_{0-}^\infty e^{-\theta x} \, dF(x)$ and

$\int_0^\infty e^{-\theta x} f(x)\, dx$ to mean $\int_{0-}^\infty e^{-\theta x} f(x)\, dx$, where by convention for any integral,

$$\int_{0-}^b = \lim_{\epsilon \to 0} \int_{-\epsilon}^b \qquad \text{for values of } \epsilon > 0.$$

$X^*[\theta]$ is also defined for complex θ when the real part of θ is positive. (Recall that any complex number z can be written as $z = a + bi$, where $i = \sqrt{-1}$; a is a real number called the *real part of z*, and b is a real number called *the imaginary part of z*.)

In Table 10 of Appendix A, we have indicated some useful properties and identities for the Laplace transform. The proofs can be found in Giffin [13]. Table 11 of Appendix A gives some transform pairs.

It is often useful to invert Laplace–Stieltjes transforms, that is, to find the probability distribution that has a given transform. From a table of probability distributions and their transforms, we can often find the inverse transformation by inspection since, by Theorem 2.9.3(a), a probability distribution is determined by its Laplace–Stieltjes transform. The Laplace–Stieltjes transforms for most of the continuous random variables of interest in this book are shown in Table 2 of Appendix A. It is important to note, that, by (3.167), inverting the Laplace–Stieltjes transform of a continuous random variable X means inverting the Laplace transform of its density function, and thus yields the density function.

One of the most useful properties of the mapping $f \to f^*$, which carries a real-valued function into its Laplace transform is linearity; that is, if a and b are constants and f and g are real-valued functions having Laplace transforms, then $af + bg \to af^* + bg^*$. This property makes it much easier to find the inverse of a transform, as the following example shows.

Example 3.4.1 Suppose X and Y are independent, exponential random variables with parameters α and β, respectively, where $\alpha \neq \beta$. Then, by Theorem 2.9.3(d) and Example 2.9.6, we can write

$$(X + Y)^*[\theta] = X^*[\theta]Y^*[\theta] = \left(\frac{\alpha}{\theta + \alpha}\right)\left(\frac{\beta}{\theta + \beta}\right). \tag{3.167}$$

It is not clear from (3.168) what the inverse transform is. We can write (3.168) in a simpler form by the use of partial fractions, which will make the inverse transform clear. We can write

$$\begin{aligned}
(X + Y)^*[\theta] &= \left(\frac{\alpha}{\theta + \alpha}\right)\left(\frac{\beta}{\theta + \beta}\right) = \frac{c_1}{\theta + \alpha} + \frac{c_2}{\theta + \beta} \\
&= \frac{(c_1 + c_2)\theta + c_1\beta + c_2\alpha}{(\theta + \alpha)(\theta + \beta)}.
\end{aligned} \tag{3.168}$$

This yields the equations

$$c_1 + c_2 = 0,$$

and

$$c_1\beta + c_2\alpha = \alpha\beta.$$

The solution of these equations is

$$c_1 = \frac{\alpha\beta}{\beta - \alpha},$$

$$c_2 = -c_1.$$

In Table 2 of Appendix A, we see that the Laplace–Stieltjes transform of the exponential random variable with parameter α is $\alpha/(\theta + \alpha)$. This means that the inverse transform of $\alpha/(\theta + \alpha)$ is $\alpha e^{-\alpha t}$. We conclude that the inverse transform of $(X + Y)^*$ is the density function f_{X+Y} given by

$$f_{X+Y}(t) = \frac{\alpha\beta}{\alpha - \beta}(e^{-\beta t} - e^{-\alpha t}). \quad \Box$$

Another useful property of the Laplace transform is the formula for the transform of the derivative given by

$$\frac{df}{dt}^*[\theta] = \theta f^*[\theta] - f(0). \tag{3.169}$$

(See Kleinrock [16] or Giffin [13] for a proof of (3.170).) We will illustrate how (3.170) can be used in the following example.

Example 3.4.2 Consider Example 4.3.3 where we have the following set of differential equations:

$$\frac{dP_2}{dt}(t) = -2\lambda P_2(t) + \mu P_1(t),$$

$$\frac{dP_1}{dt}(t) = 2\lambda P_2(t) - (\lambda + \mu)P_1(t), \tag{3.170}$$

$$\frac{dP_0}{dt}(t) = \lambda P_1(t).$$

The initial conditions are

$$P_2(0) = 1,$$

$$P_1(0) = 0, \tag{3.171}$$

$$P_0(0) = 0.$$

If we take the Laplace transform of each of the differential equations of
(3.171), applying (3.170) to each left side, we obtain

$$
\begin{aligned}
\theta P_2^*[\theta] - 1 &= -2\lambda P_2^*[\theta] + \mu P_1^*[\theta], \\
\theta P_1^*[\theta] &= 2\lambda P_2^*[\theta] - (\lambda + \mu)P_1^*[\theta], \\
\theta P_0^*[\theta] &= \lambda P_1^*[\theta].
\end{aligned}
\tag{3.172}
$$

Solving (3.173) for $P_0^*[\theta]$, we obtain

$$
P_0^*[\theta] = \frac{2\lambda^2}{\theta[\theta^2 + (3\lambda + \mu)\theta + 2\lambda^2]}.
\tag{3.173}
$$

Let Y be the random variable that gives the time of failure of the system.
Clearly, the density function of Y, f_Y, is given by

$$
f_Y(t) = \frac{dP_0}{dt}(t),
$$

so that by (3.170),

$$
f_Y^*[\theta] = \theta P_0^*[\theta] - P_0(0) = \frac{2\lambda^2}{\theta^2 + (3\lambda + \mu)\theta + 2\lambda^2}.
\tag{3.174}
$$

We can factor the denominator of (3.175) so that it can be written as

$$
f_Y^*[\theta] = \frac{2\lambda^2}{\alpha_1 - \alpha_2}\left(\frac{1}{\theta + \alpha_2} - \frac{1}{\theta + \alpha_1}\right),
\tag{3.175}
$$

where

$$
\begin{aligned}
\alpha_1 &= \frac{3\lambda + \mu + \alpha_3}{2}, \\
\alpha_2 &= \frac{3\lambda + \mu - \alpha_3}{2}, \\
\alpha_3 &= \sqrt{\lambda^2 + 6\lambda\mu + \mu^2}.
\end{aligned}
$$

Table 11 of Appendix A shows that the inverse transform of $1/(\theta - a)$ is
e^{at}. Using this fact and the linearity of the transform, we conclude that

$$
f_Y(t) = \frac{2\lambda^2}{\alpha_1 - \alpha_2}\left(e^{-\alpha_2 t} - e^{-\alpha_1 t}\right). \quad \square
\tag{3.176}
$$

There is a special function called the *Dirac delta function* or the *unit
impulse function*, which is of great utility in working with Laplace–Stieltjes
or Laplace transforms. This function is not a function in the usual sense,

because it has "magical properties" not possessed by any ordinary function. It has been made mathematically legitimate by Schwartz's theory of distributions (Schwartz [30]) and can be thought·of as being defined by the properties

$$\delta(t) = 0 \quad \text{for } t \neq 0, \tag{3.177}$$

and

$$\int_{a-\epsilon}^{a+\epsilon} \delta(t-a)f(t)\,dt = f(a), \tag{3.178}$$

for any constant a, any function f continuous at a, and any $\epsilon > 0$.

What makes the Dirac delta function especially useful is that its Laplace transform is the constant 1. Because of this, the inverse Laplace transform of any constant c is $c\delta(t)$. Similarly, the inverse Laplace transform of $e^{-a\theta}$ is $\delta(t-a)$.

We will illustrate the use of the Dirac delta function in the following example from Chapter 5.

Example 3.4.3 In Section 5.3 we show that the Laplace–Stieltjes transform of the queueing time (time spent waiting for service to begin) for the M/G/1 queueing system is given by

$$W_q^*[\theta] = \frac{(1-\rho)\theta}{\theta + \lambda[W_s^*[\theta] - 1]}, \tag{3.179}$$

where ρ is the server utilization (fraction of time the server is busy), λ is the average arrival rate of customers to the system, and $W_s^*[\theta]$ is the Laplace–Stieltjes transform of the service time. If the service time is exponential (so that M/G/1 becomes M/M/1), then

$$W_s^*[\theta] = \frac{\mu}{\theta + \mu}, \tag{3.180}$$

and by (3.180),

$$
\begin{aligned}
W_q^*[\theta] &= \frac{(1-\rho)\theta}{\theta + \lambda\left(\frac{\mu}{\theta+\mu} - 1\right)} \\
&= \frac{(1-\rho)(\theta+\mu)}{\theta + (\mu - \lambda)}. \tag{3.181}
\end{aligned}
$$

This expression for $W_q^*[\theta]$ is not in the proper form for using Table 11 of Appendix A, since the numerator and denominator are of the same degree.

We perform division to obtain

$$
\begin{aligned}
W_q * [\theta] &= (1 - \rho) + \frac{\lambda(1 - \rho)}{\theta + \mu - \lambda} \\
&= (1 - \rho) + \frac{\lambda(1 - \rho)}{\theta + \mu(1 - \rho)}.
\end{aligned}
\tag{3.182}
$$

We have used the formula

$$
\rho = \frac{\lambda}{\mu}
$$

in simplifying (3.183). We can now invert (3.183) by inverting the two terms individually and adding the results together, by the linearity of the Laplace transform. The constant term, $(1 - \rho)$, has the inverse transform $(1 - \rho)\delta(t)$ by the above discussion of the Dirac delta function (see, also, Entry 8 in Table 11 of Appendix A). The second term in (3.183) has the inverse transform

$$
\lambda(1 - \rho)e^{-\mu(1-\rho)t},
$$

by Entry 5 of Table 11 (remember that $cf \rightarrow cf^*$). The inverse transform can also be obtained for the second term in (3.183) by the *Mathematica* function InverseLaplace in the package Calculus/InverseL.m as follows

```
In[3]:= lambda (1-rho)/(theta + mu (1 - rho))

         lambda (1 - rho)
Out[3]= --------------------
         mu (1 - rho) + theta

In[4]:= InverseLaplace[%, theta, t]

         lambda (1 - rho)
Out[4]= ----------------
          mu (1 - rho) t
         E
```

Hence, the density function of q, $f_q(t)$, is given by

$$
f_q(t) = (1 - \rho)\delta(t) + \lambda(1 - \rho)e^{-\mu(1-\rho)t}, \quad t \geq 0.
\tag{3.183}
$$

We integrate (3.184) to obtain

$$
\begin{aligned}
W_q[t] &= P[q \leq t] \\
&= \int_0^t f_q(x)\, dx \\
&= (1 - \rho) + \lambda(1 - \rho) \left[\frac{-e^{-\mu(1-\rho)x}}{\mu(1 - \rho)} \right]\Bigg|_0^t
\end{aligned}
\tag{3.184}
$$

$$= (1 - \rho) + \rho(1 - e^{-\mu(1-\rho)t})$$
$$= 1 - \rho e^{-\mu(1-\rho)t}, \quad t \geq 0.$$

This agrees with the result obtained by other means in Section 5.2.1. $\square$

We have considered sums of random variables where the number of variables in the sum is fixed. There are cases in which it is advantageous to consider random sums in which the number of random variables in the sum is itself random. That is, suppose we have a sequence $X_1, X_2, \ldots$ of independent and identically distributed random variables. Let N be a discrete random variable, independent of $X_1, X_2, \ldots$ having the probability mass function $p_N(n) = P[N = n]$ for $n = 0, 1, \ldots$. We define the random sum S_N by

$$S_N = X_1 + X_2 + \cdots + X_N, \tag{3.185}$$

where S_N is assumed to be zero when $N = 0$. Some examples of the use of (3.186) follow.

Examples of Random Sums

Queueing Theory

Suppose N is the number of customers arriving at a service facility in some specified period of time, and X_i is the service time required by the ith customer. Then $S_N = X_1 + X_2 + \cdots + X_N$ is the total service time demand, often called *virtual service.*

Accidents

Suppose X_i denotes the number of persons injured in the ith traffic accident on a day in Los Angeles and N is the random variable describing the number of accidents per day. Then $S_N = X_1 + X_2 + \cdots + X_N$ is the total number of persons injured in traffic accidents on a day in Los Angeles.

Insurance Risk

Let N be the number of claims that arrive at an insurance company per working day. Suppose X_i is the amount of the ith claim. Then $S_N = X_1 + X_2 + \cdots + X_N$ is the total liability of the insurance company.

Banking

Let N be the number of requests for cash made at the ATMs of a certain bank for the city of San Francisco in a day. Let X_i be the amount of cash requested by the ith customer. Then $S_N = X_1 + X_2 + \cdots + X_N$ is the total amount of cash requested.

In the next theorem we will look at the case of a random sum in which the random variables in the sum are discrete.

Theorem 3.4.1 *If $X_1, X_2, \ldots$ is a sequence of independent identically distributed discrete random variables with common generating function g_X, and N is a discrete random variable with generating function g_N, then*

$$S_N = X_1 + X_2 + \cdots + X_N$$

has a generating function given by

$$g_{S_N}(z) = g_N(g_X(z)). \tag{3.186}$$

Furthermore,

$$E[S_N] = E[X]E[N], \tag{3.187}$$

and

$$\text{Var}[S_N] = E[N]\text{Var}[X] + E[X]^2\text{Var}[N]. \tag{3.188}$$

Proof We use conditional expectation and the law of total expectation (see Theorem 2.8.1) to obtain

$$
\begin{aligned}
g_{S_N}(z) &= E[z^{S_N}] = E[E[z^{S_N}|N]] \\
&= \sum_n E[z^{S_N}|N = n]\,P[N = n] \\
&= \sum_n E\left[z^{X_1 + X_2 + \cdots + X_n}\right]\,P[N = n] \\
&= \sum_n E[z^{X_1}] \cdots E[z^{X_n}]\,P[N = n] \quad \text{by independence} \\
&= \sum_n (g_X(z))^n\,P[N = n] \\
&= g_N(g_X(z)).
\end{aligned}
$$

This proves (3.187). The proof of (3.188) and (3.189) is given by Taylor and Karlin [36]. ∎

Example 3.4.4 Suppose Y has a binomial distribution with parameters p and N, where N has a binomial distribution with parameters $q = 1 - p$ and M. What is the marginal distribution of Y?

Solution Y can be written as the random sum

$$Y = X_1 + X_2 + \cdots + X_N,$$

where the X_i are independent Bernoulli random variables, each with parameter p. We can write N as the (nonrandom) sum

$$N = Z_1 + Z_2 + \cdots + Z_M,$$

of identical independent Bernoulli random variables, each with parameter $q = 1 - p$. Hence, by Table 2 of Appendix A, we can write the generating functions

$$g_X(z) = q + pz, \tag{3.189}$$

and

$$g_N(z) = (p + qz)^M. \tag{3.190}$$

Therefore, by Theorem 3.4.1,

$$
\begin{aligned}
g_Y(z) &= g_N(g_X(z)) \\
&= [p + q(q + pz)]^M \\
&= (p + q^2 + pqz)^M \\
&= [(1 - pq) + pqz]^M,
\end{aligned}
\tag{3.191}
$$

since

$$p + q^2 = p + q(1 - p) = p + q - pq = 1 - pq.$$

Note that (3.192) is the generating function of a binomial random variable with parameters pq and M. Hence, by the uniqueness of the generating function, Y has a marginal distribution that is binomial with these parameters. $\square$

In the next theorem we look at random sums in which $X_1, X_2, \ldots$ are continuous nonnegative random variables. We show how to calculate the Laplace–Stieltjes transform of the random sum. We also show that the formulas for calculating the mean and the variance of the random sum are the same as for the discrete case. In fact, these formulas are true even if the underlying continuous random variables are not necessarily nonnegative.

Theorem 3.4.2 *Suppose $X_1, X_2, \ldots$ is a sequence of independent identically distributed continuous random variables with common Laplace–Stieltjes transform $X^*[\theta]$, and N is a discrete random variable with generating function g_N. Then, if*

$$S_N = X_1 + X_2 + \cdots + X_N, \tag{3.192}$$

we have

$$E[S_N] = E[X]E[N], \tag{3.193}$$

and

$$\mathrm{Var}[S_N] = E[N]\,\mathrm{Var}[X] + E[X]^2\,\mathrm{Var}[N]. \tag{3.194}$$

Furthermore, the Laplace–Stieltjes transform of S_N is given by

$$S_N^*[\theta] = g_N(X^*[\theta]). \tag{3.195}$$

Proof For a fixed value of $N = n$ we calculate the conditional expectation

$$E[S_N|N = n] = \sum_1^n E[X_i] = n\,E[X]. \tag{3.196}$$

Then, by the law of total expectation, (see Theorem 2.8.1)

$$\begin{aligned}
E[S_N] &= \sum_1^N n\,E[X]\,p_N(n) \\
&= E[X]\sum_1^N n\,p_N(n) \\
&= E[X]\,E[N],
\end{aligned} \tag{3.197}$$

which proves (3.194). By Theorem 2.7.2(d),

$$E[S_N^2|N = n] = \mathrm{Var}[S_N|N = n] + E[S_N|N = n]^2, \tag{3.198}$$

and, because of the independence of the X_i,

$$\mathrm{Var}[S_N|N = n] = \sum_1^n \mathrm{Var}[X_i] = n \times \mathrm{Var}[X]. \tag{3.199}$$

Substituting (3.197) and (3.200) into (3.199) yields

$$E[S_N^2|N = n] = n\,\mathrm{Var}[N] + n^2\,E[X]^2.$$

Hence, by Theorem 2.8.1,

$$\begin{aligned}
E[S_N^2] &= \sum_1^\infty [n\,\mathrm{Var}[X] + n^2\,E[X]^2]\,p_N(n) \\
&= \mathrm{Var}[X]\,E[N] + E[N^2]\,E[X]^2 \\
&= \mathrm{Var}[X]\,E[N] + (\mathrm{Var}[N] + E[N]^2)\,E[X]^2.
\end{aligned}$$

Finally, we calculate

$$\begin{aligned}
\mathrm{Var}[S_N] &= E[S_N^2] - E[S_N]^2 \\
&= E[N]\,\mathrm{Var}[X] + E[X]^2\,\mathrm{Var}[N],
\end{aligned}$$

which proves (3.195). Let us write $S_{N|N}^*[\theta|n]$ for the conditional Laplace–Stieltjes transform of S_N given that $N = n$. Then, by Theorem 2.8.1,

$$S_N^*[\theta] = \sum_1^\infty S_{N|N}^*[\theta|n]\,p_N(n)$$

$$= \sum_{1}^{\infty} (X^*[\theta])^n \, p_N(n)$$

$$= g_N(X^*[\theta]).$$

This completes the proof. ∎

Example 3.4.5 Suppose Y can be written as the random sum

$$Y = X_1 + X_2 + \cdots + X_N,$$

where each X_i has an exponential distribution with parameter α and N has the geometric distribution described in Exercise 4. That is,

$$P[N = k] = p\,q^{k-1}, \quad k = 1, 2, \ldots.$$

This means that

$$g_N(z) = pz + pqz^2 + pq^2z^3 + \cdots$$

$$= pz \sum_{0}^{\infty} (qz)^n$$

$$= \frac{pz}{1 - qz}.$$

Therefore, by Theorem 3.4.2,

$$Y^*[\theta] = \frac{p\left(\dfrac{\alpha}{\alpha + \theta}\right)}{1 - q\left(\dfrac{\alpha}{\alpha + \theta}\right)} = \frac{p\alpha}{p\alpha + \theta}.$$

Therefore, by the uniqueness of the Laplace–Stieltjes transform, Y is exponential with parameter $p\alpha$. □

Example 3.4.5 can be interpreted as follows. Suppose a stream of customers arrives at a fork or switching point. A customer is sent along the left path (path 1) with probability p or along the right path (path 2) with probability $q = 1 - p$. We assume the arrival process is an *exponential renewal process*; that is, the interarrival times for successive customers at the fork are described by independent identically distributed exponential random variables, each with parameter α. Let the random variable Y describe the time between the arrival of a customer who is sent along path 1 and the arrival of the next customer who is sent along this same path. Then Y can be represented by the random sum

$$Y = X_1 + X_2 + \cdots + X_N,$$

where N has the geometric distribution described in Example 3.4.5. Furthermore, Example 3.4.5 shows that the interarrival time along the first path has an exponential distribution with parameter $p\alpha$. The same argument, of course, shows that the interarrival time along path 2 has an exponential distribution with parameter $q\alpha$. In Exercise 53 we ask you to generalize this result to an r-way junction.

In Example 2.7.5, we saw that if Y was the *fixed* sum of two exponential random variables, which corresponds to sending every other customer along path 1, then Y would have the Erlang-2 distribution, which is more regular (more nearly constant) than the exponential distribution. In Exercise 52 we ask you to generalize this result to show that if a stream of arriving customers having an exponential interarrival time distribution is split deterministically into k streams, then the interarrival times along each new stream have an Erlang-k distribution.

In Chapter 5 we study the M/G/1 queueing system and show that the formula for the generating function or z-transform of the steady state number of customers in the system, $g_N(z)$, is given by the Pollaczek–Khintchine transform equation,

$$g_N(z) = \frac{(1-\rho)(1-z)W_s^*[\lambda(1-z)]}{W_s^*[\lambda(1-z)] - z}, \tag{3.200}$$

where $W_s^*[\theta]$ is the Laplace–Stieltjes transform of the service time. As a special case, let us consider the M/D/1 queueing system. Then $W_s^*[\theta] = e^{-\theta W_s}$. Substituting this formula into (3.201) yields

$$\begin{aligned} g_N(z) &= \frac{(1-\rho)(1-z)e^{-\rho(1-z)}}{e^{-\rho(1-z)} - z} \\ &= \frac{(1-\rho)(1-z)}{1 - ze^{\rho(1-z)}}. \end{aligned} \tag{3.201}$$

If we assume

$$|ze^{\rho(1-z)}| < 1,$$

we can expand (3.202) in the geometric series

$$g_N(z) = (1-\rho)(1-z)\sum_{j=0}^{\infty}\left[ze^{\rho(1-z)}\right]^j. \tag{3.202}$$

Kobayashi [17, pages 196-198] proved that, by comparing the coefficients of z^n in (3.203) and in the definition

$$g_N(z) = \sum_{n=0}^{\infty} p_n z^n, \tag{3.203}$$

it can be shown that

$$p_0 = 1 - \rho,' \qquad (3.204)$$

$$p_1 = (1 - \rho)(e^\rho - 1), \qquad (3.205)$$

and

$$p_n = (1 - \rho) \sum_{j=1}^{n} \frac{(-1)^{n-j}(j\rho)^{n-j-1}(j\rho + n - j)e^{j\rho}}{(n-j)!} \qquad n = 2, 3, \cdots. \qquad (3.206)$$

Example 3.4.6 Suppose the Wringing Wet Wardrobe Company (a manufacturer of swimming suits–the type worn by swimmers rather than the type that swims) discovers that one of their computer I/O subsystems can be modeled as an M/D/1 queueing system with $\rho = 0.9$. Then, by (3.205), $p_0 = 1 - \rho = 0.1$. By (3.206), $p_1 = (1 - \rho)(e^\rho - 1) = 0.14596$. By (3.207)

$$p_2 = (1 - \rho) \sum_{j=1}^{2} \frac{(-1)^{2-j}(j\rho)^{2-j-1}(j\rho + 2 - j)e^{j\rho}}{(2-j)!} = 0.13764.$$

We show more values of p_n in the table below. These values were calculated with the aid of the APL function PN. In Exercise 52 of Chapter 5, we show how to use

$$p_n, \quad n = 0, 1, 2, \cdots,$$

to calculate the distribution function of the time in the system. □

Table 3.4.1.

n	p_n	n	p_n
0	0.10000	8	0.04096
1	0.14596	9	0.03330
2	0.13764	10	0.02707
3	0.11505	11	0.02200
4	0.09380	12	0.01789
5	0.07625	13	0.01454
6	0.06198	14	0.01182
7	0.05039	15	0.00961

3.5 Summary

The name of this chapter is *Probability Distributions* which is meant to suggest that the most important property of a random variable X is how it distributes probability. By this we mean how the probability associated with the values of X is distributed over these values. For a discrete random variable, the most convenient way to describe this distribution, usually, is by an analytical formula for the probability mass function. That is, given a mass point x_n of X, the pmf $p(\cdot)$ describes how to calculate the associated probability. For example, if X has a Poisson distribution with parameter α, and k is a nonnegative integer, then $p(k) = P[X = k] = e^{-\alpha}\alpha^k/k!$. For a continuous random variable, the most convenient method of describing the probability distribution is by means of the distribution function F, defined for all real x by $F(x) = P[X \leq x]$. Thus, if X is an exponential random variable with parameter α, then $F(x) = 1 - e^{-\alpha x} = 1 - \exp(-x/E[X])$ for $x \geq 0$.

In the first two sections of this chapter we considered some discrete and continuous random variables that have been found to be especially useful in applied probability theory, and of special importance to computer science applications. Each of these random variables is determined by either one or two parameters; that is, given the parameter or parameters, the entire probability distribution is known. This makes it relatively easy to fit one of these distributions to an empirical distribution. In Chapter 7 we discuss the problem of how to estimate the parameters necessary to fit a well-known distribution to an empirically derived one, and in Chapter 8 we address the problem of judging how good the fit is.

A summary of the properties of the random variables discussed in this chapter is given in Tables 1 and 2 of Appendix A. Examples are given in the text of the use of most of these random variables.

In the third section of this chapter we discussed the central limit theorem and some of its applications. The basic idea of the theorem is that the sum of independent random variables tends toward a normal random variable under very weak restrictions. This explains the special importance of the normal distribution. Several examples were given of the use of the central limit theorem.

In the last section we gave a number of examples of how some of the transforms we defined in Chapter 2 can be applied to solve fairly difficult problems with ease.

Student Sayings

Socrates took Poisson.
Monique is exponentially distributed!
No μs is good μs.
Keep your hyperexponential away from me!

3.6 Exercises

1. [10] One-fourth of the source programs submitted by Jumpin Jack compiled successfully. What is the probability that exactly one of Jumpin's next five programs will compile? That three out of five will?

2. [10] Six programmers from Alfa Romalfa decide to toss coins on an "odd person out" basis to determine who will buy the coffee. Thus, there will be a loser if exactly one of the coins falls heads or exactly one falls tails.[7] What is the probability that the outcome will be decided on the first toss? What is the probability that exactly four trials will be required? Not more than four?

3. [15] Sayure Praers, a small-plane, short-haul airline, has found that approximately 5% of all persons holding reservations on a certain flight do not show up. If the plane holds 50 passengers and Sayure takes reservations for 53 (this is called *overbooking*), what is the probability that every passenger who arrives on time for the flight will have a seat? (Assume there are no walk-ins.)

4. [HM22] Some authors modify our definition of a geometric random variable X so that it counts the number of trials *including* the trial at which the first success occurs. Thus, X can assume the values $1, 2, 3, \ldots$. For this modified geometric random variable, find the pmf $p(\cdot)$, the expected value, and the variance in terms of the probability of success on each trial p and of $q = 1 - p$.

5. [15] Jumpin Jill finds that, when she is developing a program module, syntax errors are discovered by the compiler on 60% of the runs she makes. Furthermore, this percentage is independent of the number of runs made on the same module. How many runs does she need to make of one module, on the average, to get a run with no syntax errors? What is the probability that more than 4 runs will be required. [*Hint*: Use the result of Exercise 4.]

6. [C18] Ms. Nancy Nevermiss can put 10 shots in succession through the center of a target (the bull's-eye) one-fifth of the time. This is called a *possible*. Suppose Nancy independently fires 10 sequences of 10 shots each, each sequence at a fresh target.

   (a) What is the probability that she will get at least two possibles?

---

[7]We assume that no coin will stand on an edge.

(b) What is the conditional probability that Nancy will get at least two possibles, given that she gets at least one?

7. [18] About one percent of all teller transactions at Chaste National Bank have a certain type of error. How large a random sample (with replacement) must be taken if the probability of its containing at least one transaction with an error is to be not less than 0.95? [*Hint:* Use the Poisson distribution.]

Multinomial Distribution There is an important generalization of the binomial distribution called the *multinomial distribution*. Suppose each of n independent repeated trials can have one of several outcomes, which we label $E_1, E_2, \cdots, E_r$. Let the probability that E_i will occur on any trial be p_i for $i = 1, 2, \cdots, r$. For $r = 2$ we have the binomial case. We assume that

$$p_1 + p_2 \cdots + p_r = 1,$$

where $p_i \geq 0$ for all i. The result of n trials is a sequence of n events such as $E_2 E_1 E_2 \cdots E_1$. The probability that in n trials E_1 occurs k_1 times, E_2 occurs k_2 times, etc. is

$$\frac{n!}{k_1! k_2! \cdots k_r!} p_1^{k_1} p_2^{k_2} \cdots p_r^{k_r}, \qquad (3.207)$$

where the k_i are nonnegative numbers satisfying

$$k_1 + k_2 + \cdots + k_r = n.$$

We will ask you to prove (3.208) in Exercise 8. We now give an example of the use of the multinomial distribution.

Example 3.6.1 A card is drawn with replacement five times from a well-shuffled bridge deck. What is the probability of obtaining 2 clubs and 1 diamond?

Solution We let E_1 be the event of drawing a club, E_2 the event of drawing a diamond, and E_3 the event of drawing a spade or a heart. Since $p_1 = 1/4$, $p_2 = 1/4$, and $p_3 = 1/2$, the required probability is

$$\frac{5!}{2! 1! 2!} \left(\frac{1}{4}\right)^2 \left(\frac{1}{4}\right)^1 \left(\frac{1}{2}\right)^2 = 30 \times 0.25^3 \times 0.5^2$$

$$= 0.1171875.$$

8. [12] Prove (3.208) for the multinomial distribution using the following theorem from Feller [10, page 37]. Let $k_1, k_2, \cdots, k_r$ be integers such that

$$k_1 + k_2 + \cdots + k_r = n, \quad k_i \geq 0.$$

The number of ways a population of n elements can be divided into r subpopulations of which the first contains k_1 elements, the second k_2 elements, etc., is

$$\frac{n!}{k_1! k_2! \cdots k_r!}.$$

9. [10] The interactive system at Banker's Tryst can process 5 kinds of inquiries; the respective probabilities are 0.1, 0.15, 0.4, 0.25, and 0.1. What is the probability the next 10 inquiries will include 1 of the first type, 2 of the second, 3 of the third, 3 of the fourth, and one of the fifth?

10. [5] In Kleen City on Thursday night, half of the TV audience watches Channel 6, 40 percent watches Channel 12, and the remaining 10 percent watch Channel 13 (a channel for the lucky!). Find the probability that of 10 Thursday night TV viewers, 5 will be watching Channel 6, 4 will be looking at Channel 12, and one will be viewing Channel 13.

The Hypergeometric Distribution Suppose a collection of N elements contains r elements that are red and $N - r$ elements that are black. Suppose we choose n elements from this set, without replacement. If k is an integer such that $k \le n$ and $k \le r$, then there are $\binom{r}{k}$ ways of choosing k red elements and $\binom{N-r}{n-k}$ ways of choosing $n-k$ of the black elements or $\binom{r}{k}\binom{N-r}{n-k}$ ways of choosing k red elements and $n - k$ black elements. If we let X be the number of red elements in the sample of size n and assume the sample is chosen at random (without replacement), then $p_k = P[X = k]$ is given by

$$p_k = \frac{\binom{r}{k}\binom{N - r}{n - k}}{\binom{N}{n}}, \tag{3.208}$$

provided $k \le r$ and $n - k \le N - r$. The random variable X is said to be a *hypergeometric random variable* with parameters n, N, and r. You can test your skills with this distribution by doing Exercises 11 through 16.

11. [25] Prove that for a hypergeometric random variable X with parameters n, N, and r,

$$E[X] = \frac{nr}{N}$$

and

$$\text{Var}[X] = \frac{nr(N - r)(N - n)}{N^2(N - 1)}.$$

12. [10] An inspector at Keypon Trucking checks the exhaust fumes of 5 of the company's 30 trucks. If 3 of the 30 trucks have truly dirty exhausts, what is the probability that none of them will be tested, that is, that none of them will appear in the inspector's sample?

13. [8] To avoid being caught by customs inspectors, Able Smuggler puts 6 narcotic tablets in a a bottle containing 9 vitamin pills of similar appearance. If a customs inspector chooses three of the tablets at random for analysis, what is the probability that Mr. Smuggler will be arrested?

14. [C5] Big Byte ships identical computer components in boxes of 50. Before shipment a random sample of 5 components is tested; the box is shipped if no more than 1 component is found to be defective. If a box contains 20% defectives, what is the probability it will be shipped?

15. [6] Digitizing Dingleberry Doodlers randomly chooses a committee of 3 people from 4 analysts and 2 systems programmers.

   (a) Write the pmf of the random variable X, that counts the number of analysts on the committee.

   (b) Find $P[2 \leq X \leq 3]$.

16. [5] In the description of a hypergeometric distribution X, given just before Exercise 11, we see that the sample of size n is taken without replacement. If the sample were taken with replacement, X would be binomial with parameters n and $p = r/N$. If n is small relative to N, there is not much difference between the two methods of sampling, so a hypergeometric random variable can be approximated by a binomial random variable with parameters n and $p = r/N$. (The usual rule of thumb is that n should not exceed 5 percent of N.) Let X be a hypergeometric random variable with parameters $n = 5$, $N = 500$, and $r = 20$. Calculate the probability that $X = 2$ and the binomial approximation to this probability.

17. [C10] In the experiment of Example 7.1.8, suppose 200 animals are tagged, 20 are captured or recaptured, and 4 of the 20 are discovered to be tagged so that the maximum likelihood estimate of the population size is

$$\widehat{N} = \left[\frac{200 \times 20}{4} \right] = 1,000 \text{ animals}.$$

(a) If the actual value of N is 503, calculate the probability that 4 or fewer tagged animals are found in a sample of 20 captured or recaptured animals. Calculate the binomial approximation as well.

(b) If $N = 2,790$, what is the probability that 4 or more tagged animals are found in a sample of 20 recaptured animals? What is the binomial approximation of this value?

The Multivariate Hypergeometric Distribution Suppose a set of N elements contains r_1 elements of the first kind, r_2 elements of the second kind, ..., and r_l elements of the lth kind, so that $\sum_{i=1}^{l} r_i = N$. We are interested in the probability of getting k_1 elements of the first kind, k_2 elements of the second kind,..., and k_l elements of the lth kind from a random sample of size n chosen without replacement, from the original N elements. If $X_1, X_2, \cdots, X_l$ are random variables that count the number of elements in the sample of type $1, 2, \cdots, l$, respectively, then it is easy to see that

$$P[X_1 = k_1, X_2 = k_2, \cdots, X_l = r_l] = \frac{\binom{r_1}{k_1}\binom{r_2}{k_2}\cdots\binom{r_l}{k_l}}{\binom{N}{n}}, \qquad (3.209)$$

where each k_i satisfies $0 \le k_i \le n$, and $k_i \le r_i$ for each i, and where

$$\sum_{i=1}^{l} k_i = n, \quad \text{and} \quad \sum_{i=1}^{l} r_i = N.$$

The random variables $X_1, X_2, \cdots, X_l$ are said to have a *multivariate hypergeometric distribution* if and only if their joint probability distribution is given by (3.210).

18. [C5] Find the probability that a bridge hand of 13 cards consists of four spades, five hearts, one diamond, and three clubs.

19. [11] Ecstasy Products sells a certain product to pharmacies in boxes of 100 with a guarantee that at most 10 items in a box are defective. Debilitating Drugs has a buyer who accepts a box of the product only if a random sample of 10 items chosen without replacement from the box contains no defective items. What is the probability that a box will be rejected although it contains exactly 10 defective items and thus meets the conditions of the guarantee?

The Pascal (Negative Binomial) Distribution Consider a sequence of Bernoulli trials with probability of success p on each trial (and thus, probability $q = 1 - p$ of failure). Let r be a fixed positive integer. Let $p(k; r, p)$ denote the probability that the rth success occurs at trial number $r + k$ (where $k = 0, 1, \ldots$). This is the probability that k failures occur before the rth success; thus, there must be k failures among the $r + k - 1$ trials before the $(r + k)$th trial results in the rth success. The probability of the former is $\binom{r+k-1}{k}p^{r-1}q^k$ and of the latter is p, so we must have

$$p(k; r, p) = \binom{r + k - 1}{k}p^r q^k, \quad k = 0, 1, \cdots. \tag{3.210}$$

A discrete random variable X with the pmf given by (3.211) is called a *Pascal* or *negative binomial* random variable with parameters r and p. The geometric random variable of Section 3.1.3 is a special case with $r = 1$.

20. [HM15] Find $E[X]$ and $\text{Var}[X]$ for a Pascal random variable X with parameters r and p. *Hint*: Use Theorem 2.9.2(c) and the fact that, by the result of Exercise 35(d) of Chapter 2, Formula (3.211) can be written as

$$p(k; r, p) = \binom{-r}{k}p^r(-q)^r \quad k = 0, 1, \ldots.$$

21. [12] Big Blast, Inc. is responsible for launching some special top-secret satellites. Five of the satellites have been constructed. It is desired that three of them be placed in orbit. If the probability of successfully launching an individual satellite is 0.95, what is the probability that Big Blast can carry out its mission without more satellites?

22. [8] If X has a Poisson distribution and $P[X = 0] = P[X = 1]$, find $E[X]$.

23. [10] The average number of traffic accidents per week at Coroner's Corner is 14. What is the probability that there will be 3 or more accidents at this curve on any given day?

24. [10] Suppose X is a Poisson random variable with $E[X] = \alpha$.

    (a) Prove that, if $P[X = k] = P[X = k + 1]$, then $\alpha = k + 1$.

    (b) Prove that, if $\alpha = k + 1$, then $P[X = k] = P[X = k + 1]$.

25. [12] Suppose X is a Poisson random variable with $E[X] = \alpha$.

(a) Show that $P[X = k] \geq P[X = k+1]$ implies that $k + 1 \geq \alpha$ and conversely.

(b) Show that $P[X = k - 1] \leq P[X = k]$ implies that $k \leq \alpha$ and conversely.

(c) Use (a) and (b) to show that the pmf of X, $P[X = k]$ first increases monotonically, then decreases monotonically, reaching its greatest value when $\alpha - 1 \leq k \leq \alpha$. For example, if $\alpha = 4$, then the maximum values of the pmf occur at $k = 3, 4$. The values are

$$e^{-4}\frac{4^3}{3!} = e^{-4}\frac{4^4}{4!} = 0.19537.$$

26. [18] Consider Example 3.1.5.

(a) For what values of k does the probability mass function of X assume its maximum value? Calculate the value.

(b) Using the APL function POISSONΔDIST, it was shown that $P[X \leq 15] = 0.95126$. Estimate this value using the one-sided inequality (Theorem 2.10.3).

(c) The APL function POISSONΔDIST shows that $P[4 \leq X \leq 16] = 0.96262$. Estimate this value using

   (i) the Chebyshev inequality, and

   (ii) the normal approximation.

27. [15] As discussed by Clarke [7], the number of V_2 flying bomb hits in London during World War II had a Poisson distribution. Assume that in the area affected the average time between bomb hits was 2.5 hours.

(a) Using the Poisson distribution, calculate the probability of no hits during a six hour period.

(b) Make the calculation of part (a) using the exponential distribution.

28. [15] Let X be a Poisson random variable with parameter α. Prove that

(a) $P[X \leq \alpha/2] \leq 4/(\alpha + 4) < 4/\alpha$ and

(b) $P[X \geq 2\alpha] \leq 1/(1 + \alpha) < 1/\alpha$.

[*Hint*: Use the one-sided inequality, Theorem 2.10.3.]

29. [12] There are 125 misprints in a 250 page user manual for the EZASPI System. What is the probability that there are at least two misprints on a given page?

30. [15] The arrival pattern of order messages of the interactive order entry system of the Dizzy Disc record company has a Poisson distribution with an average of 25 arrivals per minute during the peak period. What is the probability that more than 30 orders will arrive in one minute of the peak period? Use the normal approximation if you don't have facilities such as APL, MINITAB, or EXPLORE to make the exact calculation.

31. [15] Recall that $b(k; n, p)$ is the notation for the probability that a binomial random variable with parameters n and p assumes the value k; that is,

$$b(k; n, p) = \binom{n}{k} p^k (1 - p)^{n-k}.$$

Consider the sequence $b(0; n, p), b(1; n, p), \ldots, b(n; n, p)$.

   (a) Show that the term $b(k; n, p)$ is greater than $b(k - 1; n, p)$ for $1 \le k < (n + 1)p$ and is smaller for $k > (n + 1)p$.

   (b) Show also that, if $(n + 1)p = m$ is an integer, then $b(m; n, p) = b(m - 1; n, p)$.

Note: Since there is exactly one integer m such that $(n + 1)p - 1 < m \le (n + 1)p$, we see by (a) that the maximum value in the sequence is $b(m; n, p)$. It is called the *central term* or the "the most probable number of successes". For example, if $n = 19$ and $p = 0.4$, then $m = 8$ and the maximum value of $b(k; 19, 0.4)$ is

$$
\begin{aligned}
b(8; 19, 0.4) &= \binom{19}{8} 0.4^8 0.6^{11} \\
&= 0.179705788 \\
&= \binom{19}{7} 0.4^7 0.6^{12} \\
&= b(7; 19, 0.4).
\end{aligned}
$$

32. [15] About 1% of the population of a certain country is left handed. What is the probability that at least four out of 200 people at Kysquare Testing (located in this country) are left handed?

33. [18] An interactive system at Flybynight Airlines has 200 workstations each connected by a local area network (LAN) to the local computer center. Each workstation independently has probability 0.05 of being signed on to the computer center. What is the probability that 20 or more of the workstations are signed on? Use the normal approximation if you can't make the exact calculation.

34. [C20, if you have only a pocket calculator; 10 if you have APL or MINITAB available.]

    (a) Write a formula for p_k, the probability that in a group of 500 people, exactly k will have birthdays on Valentine's Day. Assume the 500 people are chosen at random with each of them having probability $p = 1/365$ of being born on Valentine's Day.

    (b) Calculate p_k for $k = 0, 1, 2, 3, 4, 5$.

    (c) Make the calculation of part (b) using the Poisson approximation.

35. [15] Inquiries of the Poisson Portal interactive query system arrive at the central computer in a Poisson pattern at an average rate of 12 inquiries per minute.

    (a) What is the probability that the time interval between the next two inquiries will be less than 7.5 seconds?

    (b) More than 10 seconds?

    (c) What is the 90th percentile value for interarrival time?

36. [HM18] Prove Theorem 3.2.6(b) and 3.2.6(c) using Theorem 2.9.1.

37. [20] Suppose entries to an order-entry system of the Shootemup Arms Company arrive at the central processor with a Poisson pattern at an average rate of 30 per minute.

    (a) Given that an order entry transaction has just arrived, what is the average time until the fourth succeeding transaction arrives?

    (b) What is the probability that it will take longer than 10 seconds for this entry to arrive? Less than 5 seconds?

    (c) Will the answers to the above questions change if the point in time at which measurement begins is 1 second after a transaction arrives?

38. [18] Cookie Crumbles wants to put enough raisins in its raisin cookie dough so that not more than one cookie in a hundred will have no raisins. How many raisins should an average cookie contain, assuming a random distribution of raisins in the dough?

39. [HM25] Suppose X has a gamma distribution. Prove that its moments are given by

$$E[X^n] = \left[\prod_{k=1}^{n-1}(1 + kC_X^2)\right] E[X]^n \quad n = 2, 3, \cdots.$$

Since the squared coefficient of variation is given by

$$C_X^2 = \frac{1}{\beta},$$

this means that we can write the above formula as

$$E[X^n] = \frac{\beta(\beta+1)(\beta+2)\cdots(\beta+n-1)}{\alpha^n}, \quad n = 1, 2, 3, \cdots.$$

This result also implies that for the Erlang-k random variable, we have

$$E[X^n] = \left(1 + \frac{1}{k}\right)\left(1 + \frac{2}{k}\right)\cdots\left(1 + \frac{n-1}{k}\right)E[X]^n, \quad n = 1, 2, 3, \cdots.$$

[*Hint*: Use Theorem 2.9.1(b).]

40. [5] Let X be a discrete uniform random variable assuming only the value c (X is thus a constant random variable). Show that for each positive integer n,

$$E[X^n] = E[X]^n = c^n.$$

41. [HM15] Suppose X is uniformly distributed on the interval a to b. Show that

$$E[X] = \frac{a+b}{2} \quad \text{and} \quad \sigma^2 = \frac{(b-a)^2}{12}.$$

42. [15] Suppose a discrete uniform random variable X assumes only the values $C + L, C + 2L, \cdots, C + nL$, where C, n, and L are constants. Show that

$$E[X] = C + \frac{(n+1)}{2}L, \quad E[X^2] = C^2 + (n+1)LC + \frac{(n+1)(2n+1)}{6}L^2,$$

and

$$\text{Var}[X] = \frac{n^2 - 1}{12}L^2.$$

[*Hint*:

$$\sum_{i=1}^{n} i = \frac{n(n+1)}{2} \quad \text{and} \quad \sum_{i=1}^{n} i^2 = \frac{n(n+1)(2n+1)}{6}.]$$

43. [12] The simulation model of a proposed computer system for Students Gosset uses a discrete approximation of a continuous uniform distribution on the interval 10 to 30. Find the mean and variance of the continuous uniform distribution and compare these values to those for the discrete approximation if

   (a) the eleven values $10, 12, \ldots, 30$ are used for the discrete distribution,

   (b) the 101 values $10, 10.2, \ldots, 30$ are used. [Hint: See the previous exercise.]

44. [10] Consider Example 3.1.1. Use the normal approximation to estimate the probability that between 5,200 and 5,400 blocks must be updated. For the estimate assume that the mean number of blocks to be updated is 5,300.

45. [15] The message length distribution for the incoming messages of an interactive system for the Sockituem Finance Company has a mean of 90 characters and a variance of 1500. Fit an Erlang distribution to this message length distribution.

46. [HM15] Show that the density function for a chi-square random variable has a unique maximum at $x = n$, if $n > 2$.

47. [18] Every fifth customer arriving at Pourboy Finance is given a prize.

   (a) If the number of customers who arrive in a one minute period has a Poisson distribution with mean λ, describe the interarrival time distribution for the customers who receive gifts.

   (b) If $\lambda = 5$ customers per minute, what is the probability that the time between two successive winners exceeds 1 minute?

48. [15] A simulation model of a proposed new computer system for the Hunkydory Boat Company has been constructed. The model provides an estimate of the utilization, ρ, of the central processing unit (CPU) by testing one hundred times every millisecond to determine whether or not it is busy and using the formula

$$\rho = P[\text{CPU is busy}] = \frac{S_n}{n},$$

where n is the number of samples and S_n the number of times the CPU is busy. How many samples should be made if $\delta = 0.005$ and $\epsilon = 0.001$ in the formula

$$P\left[\left|\frac{S_n}{n}\right| \geq \delta\right] \leq \epsilon?$$

Assume ρ is near 0.5.

49. [HM18] Show that the density function f_n for a Student's t distribution with n degrees of freedom assumes a unique maximum when $x = 0$.

50. [HM30] Consider the gamma function defined by

$$\Gamma(t) = \int_0^\infty x^{t-1} e^{-x} dx, \quad t > 0.$$

Prove the following:

(a) $\Gamma(t+1) = t\Gamma(t)$ for all $t > 0$. (Since $\Gamma(1) = 1$, this implies that $\Gamma(n+1) = n!$, $n = 1, 2, \cdots$.)

(b) Show that $\Gamma(t)$ can be written as

$$\Gamma(t) = 2^{1-t} \int_0^\infty z^{2t-1} e^{-\frac{1}{2}z^2} dz,$$

for all $t > 0$.

(c) Using (b), we can write

$$\Gamma\left(\frac{1}{2}\right) = \sqrt{2} \int_0^\infty e^{-\frac{1}{2}z^2} dz,$$

and thus,

$$\left[\Gamma\left(\frac{1}{2}\right)\right]^2 = 2\left\{\int_0^\infty e^{-\frac{1}{2}x^2} dx\right\}\left\{\int_0^\infty e^{-\frac{1}{2}y^2} dy\right\}$$
$$= 2\int_0^\infty \int_0^\infty e^{-\frac{1}{2}(x^2+y^2)} dx\, dy.$$

Now use polar coordinates to evaluate the double integral and thereby show that $\Gamma\left(\frac{1}{2}\right) = \sqrt{\pi}$.

51. [15] Get High Airlines wants to estimate the fraction of smokers p among their passenger population. The airline plans to use sampling with replacement to determine their estimate $\hat{p}$ of p. They set up the requirement that

$$P[|\hat{p} - p| \geq 0.005] \leq 0.05.$$

(a) How large a sample should Get High take if nothing is assumed about the size of p?

(b) How large a sample is required if Get High knows that p is very close to 0.4?

52. [18] The result of Example 2.7.5 can be interpreted as follows. A stream of entities arrive at a junction in such a way that the time between successive arrivals (interarrival time) has an exponential distribution with mean $1/\lambda$. If the entities (customers) are alternately routed along two separate paths such that the first customer takes the first path, the second the second path, the third the first path, the fourth the second path etc., so that the odd-numbered arrivals take the first path and the even-numbered ones take the second path, then the interarrival time on each path has an Erlang-2 distribution with average value $2/\lambda$. Generalize the above result to show that if a stream of customers having an exponential interarrival time is split deterministically into k streams, then the interarrival times along each new stream have an Erlang-k distribution with mean k/λ.

53. [M30] Suppose, as in Exercise 52, that a stream of customers arrives at an r-way junction, so that the time between successive arrivals has an exponential distribution with mean $1/\lambda$. Then, by Theorem 3.2.1(g), the number of arrivals per unit of time has a Poisson distribution with mean λ. Suppose the branch selected by each arrival is chosen independently with the probability that an arrival takes path i equal to p_i for $i = 1, 2, \cdots, r$. We can imagine a random number generator that chooses 1 with probability p_1, 2 with probability p_2, ..., r with probability p_r, where $\sum_{i=1}^{r} p_r = 1$. Each customer then takes the path chosen by the random number generator and the generator makes a new choice for each arriving customer. Prove that the ith output stream has a Poisson pattern with mean rate $p_i \lambda$. [*Hint*: Let $N(t)$ be the number of customer arrivals to the junction in t time units. (We assume the observations begin when $t = 0$.) Let $N_i(t)$ be the number of these arrivals that take the ith path. Then the conditional joint distribution of $N_i(t)$ $(i = 1, 2, \cdots, r)$ given that $N(t) = n$,

$$P[N_1(t) = k_1, N_2(t) = k_2, \ldots, N_r(t) = k_r | N(t) = n],$$

has a multinomial distribution (see Exercise 8 where event E_i is the event that a customer takes path i). Multiplying this probability by the probability that $N(t) = n$, which has a Poisson distribution with mean λt by Theorem 3.2.1(g), we obtain the joint probability distribution $P(k_1, k_2, \ldots, k_r)$. $P(k_1, k_2, \ldots, k_r)$ expresses the probability that k_1 customers take the first path, k_2 take the second path, etc. Show that the joint probability factors into the product of r Poisson probabilities.]

54. [HM20] Prove (a) (and thus, by symmetry, (b), also) of Theorem 3.2.5.
    Hint: Recall that

$$f_X(x) = \int_{-\infty}^{\infty} f(x,y)dy, \qquad (3.211)$$

where $f(x,y)$ is given by

$$\begin{aligned}
f(x,y) &= \frac{1}{2\pi\sigma_X\,\sigma_Y\sqrt{1-\rho^2}} \exp\left\{-\frac{1}{2(1-\rho^2)}\left[\left(\frac{x-\mu_X}{\sigma_X}\right)^2\right.\right. \\
&\quad \left.\left. -\frac{2\rho(x-\mu_X)(y-\mu_Y)}{\sigma_X\sigma_Y} + \left(\frac{y-\mu_Y}{\sigma_Y}\right)^2\right]\right\},
\end{aligned} \qquad (3.212)$$

To simplify the integration of (3.212) let $u = (x-\mu_X)/\sigma_X$ and $v = (y-\mu_Y)/\sigma_Y$. Then, since $dy = \sigma_Y\,dv$, (3.212) reduces to

$$f_X(x) = \frac{\int_{-\infty}^{\infty} \exp\left[-\left(u^2 - 2\rho\,u\,v + v^2\right)/2(1-\rho^2)\right]\,dv}{2\pi\,\sigma_X\sqrt{1-\rho^2}}. \qquad (3.213)$$

Adding and subtracting $\rho^2 u^2$ gives

$$\begin{aligned}
u^2 - 2\rho uv + v^2 &= v^2 - 2\rho uv + \rho^2 u^2 - \rho^2 u^2 + u^2 \\
&= (v - \rho u)^2 + u^2(1-\rho^2)
\end{aligned}$$

and thus, (3.214) becomes

$$f_X(x) = \frac{e^{-\frac{u^2}{2}}}{2\pi\sigma_X\sqrt{1-\rho^2}} \int_{-\infty}^{\infty} \exp\left\{\frac{-(v-\sigma u)^2}{2(1-\rho^2)}\right\}\,dv. \qquad (3.214)$$

Now let $z = (v - \rho u)/\sqrt{1-\rho^2}$, and using the fact that

$$\int_{-\infty}^{\infty} \exp\left(\frac{-z^2}{2}\right)\,dz = \sqrt{2\pi},$$

show that

$$f_X(x) = \frac{\exp\left[-\frac{1}{2}\left(\frac{x-\mu_X}{\sigma_X}\right)^2\right]}{\sqrt{2\pi}\sigma_X}. \qquad (3.215)$$

55. [HM20] Prove that (c) (and thus, by symmetry, (d)) holds in Theorem 3.2.5.

56. [18] Prove the following:

(a) A random variable created by Algorithm 3.2.2 is a two-stage hyperexponential random variable having the properties claimed.

(b) A random variable created by Algorithm 3.2.3 is a two-stage hyperexponential random variable having the properties claimed.

(c) A random variable created by Algorithm 3.2.4 is a two-stage hyperexponential random variable having the properties claimed.

57. [C10] Professor Stanley Pennypacker, who collected the azalea data of Exercise 12–19 of Ryan *et al.* [27], discovered the distance he walks between discoveries of ozone damaged azaleas is Erlang-4 with a mean of 20 feet. Find the following probabilities:

(a) the probability that Professor Pennypacker walks not farther than 30 feet to the next damaged azalea. ($P[X \leq 30]$.)

(b) the probability he must walk more than 10 feet to find the next damaged azalea. ($P[X > 10]$.)

58. [7] Consider the approximation due to Arvind K. Shah that appears in Section 3.2.4. Calculate the following probabilities, below

(a) Using Table 3 of Appendix A (or the APL function NDIST), and

(b) using Shah's approximation.
   (i) $P[-2 \leq X \leq 1.5]$.
   (ii) $P[-1 \leq X \leq 1.28]$.
   (iii) $P[Z \leq 1.28]$.
   (iv) $P[Z \leq 1.64]$.

59. [C15]

(a) Use Algorithm 3.2.1 (Algorithm G) to construct a gamma random variable X with mean 10 and $C_X^2 = 4$. Calculate $E[X^3]$ and $P[X \leq 15]$.

(b) Use Algorithm 3.2.2 (Algorithm H) to construct a random variable X with mean 10 and $C_X^2 = 4$. Calculate $E[X^3]$ and $P[X \leq 15]$.

(c) Use Algorithm 3.2.3 (Algorithm HG) to construct a random variable X with mean 10 and $C_X^2 = 4$. Calculate $E[X^3]$ and $P[X \leq 15]$.

60. [C15 If you use the APL function WH, it is 5.] Consider Algorithm 3.2.4. Use this algorithm to find the H_2 distribution X for which $E[X] = 1$, $E[X^2] = 5$, and $E[X^3] = 40$.

61. [TM25] Consider Example 3.2.11. Find a Coxian distribution to match the given H_2 distribution by first finding the Laplace–Stieltjes transform of the given distribution. Then find the two-stage Coxian distribution with the same Laplace–Stieltjes transform.

62. [HM18] Use convolution to show that if X and Y are independent exponential random variables with parameters α and β, respectively, where $\alpha \neq \beta$, then the density function f of their sum $X + Y$ is given by

$$f_{X+Y}(t) = \frac{\alpha\beta}{\alpha - \beta}(e^{-\beta t} - e^{-\alpha t}), \quad t \geq 0.$$

[We showed this, using the Laplace–Stieltjes transform, in Example 3.4.1.]

63. [10] Let X be an Erlang-k random variable with parameter μ. Show that

$$X^*[\theta] = \left(\frac{k\mu}{k\mu + \theta}\right)^k.$$

Hint: Use the fact that the Laplace–Stieltjes transform of an exponential random variable with parameter α is

$$\frac{\alpha}{\theta + \alpha},$$

as well as the fact that X can be represented as shown in Figure 3.2.5.

64. [HM15] Suppose X is a gamma random variable with parameters β and α. Show directly from the definition that

$$X^*[\theta] = \left(\frac{\alpha}{\theta + \alpha}\right)^\beta.$$

65. [M15] Assume that Y has a binomial distribution with parameters p and N, where N has a Poisson distribution with parameter λ. Find the marginal distribution for Y.

66. [10] Assume that a hen lays N eggs, where N has a Poisson distribution with parameter λ. Suppose each egg hatches with probability p independently of the other eggs. Let Y be the number of chicks that hatch. We can write Y as the random sum

$$Y = X_1 + \cdots + X_N,$$

where $X_1, X_2, \ldots$ are independent Bernoulli random variables with parameter p.

(a) What is the distribution of Y?

(b) If $\lambda = 4$ and $p = 0.25$, what is the average and variance of the number of chicks hatched? That is, what is $E[Y]$ and $\mathrm{Var}[Y]$?

67. [HM20] The number of automobile accidents in Los Angeles per week, N, has a Poisson distribution with mean 100. The number of persons injured in each such accident has a binomial distribution with parameters $p = 0.2$ and N.

(a) What is the distribution of Y, the total number of persons injured in automobile accidents per week?

(b) What is the mean and variance of Y?

68. [HM15] Suppose a nonnegative random variable X has the Laplace–Stieltjes transform $K/(\theta + 2)$. Find K, the density function f of X, and $E[X^3]$.

69. [HM8] Suppose X has the generating function $(1 + z^2)/2$. Find $E[X]$, $P[X = E[X]]$, and $\mathrm{Var}[X]$.

70. [HM10] Let Y be a shifted exponential random variable with density function given by

$$f(s) = \alpha e^{-\alpha(s-D)}, \quad \text{for } s \geq D.$$

Prove the following:

$$Y^*[\theta] = e^{-\theta D} \times \frac{\alpha}{\alpha + \theta},$$

$$E[Y] = D + \frac{1}{\alpha},$$

$$\mathrm{Var}[Y] = \frac{1}{\alpha^2},$$

and

$$F(x) = P[Y \leq x] = 1 - e^{-\alpha(x-D)}, \quad \text{for } x \geq D.$$

One machine can do the work of 50 ordinary men. No machine can do the work of one extraordinary man.
Elbert Hubbard, 1913

Imagination is more important than knowledge.
Albert Einstein

Art is a lie that makes us realize truth.
Picasso

Take a chance! All life is a chance. The man who goes furthest is generally the one who is willing to do and dare. The "sure thing" boat never gets far from the shore.
Dale Carnegie

References

[1] Milton Abramowitz and Irene Stegun, *Handbook of Mathematical Functions*, National Bureau of Standard, Washington DC, 1964. Also published by Dover Publishing Company of New York.

[2] Francis John Anscombe, *Computing In Statistical Science through APL*, Springer-Verlag, New York, 1981.

[3] Tom M. Apostol, *Calculus*, Vol. II, 2nd ed., Ginn (Blaisdell), Boston, 1969.

[4] F. Baskett, K. M. Chandy, R. R. Muntz, and F. G. Palacios, Open, closed, and mixed networks of queues with different classes of customers, *JACM*, **22(2)**, (April 1975), 248–260.

[5] H. Bateman, On the probability distribution of α particles, *Philos. Mag. Ser. 6*, **20**, (1910), 704–707.

[6] Ladislaus von Bortkiewicz, *Das Gesetz der Kleinen Zahlen*, Teubner, Leipzig, 1898.

[7] R. D. Clarke, An application of the Poisson distribution, *J. Inst. of Actuaries*, **72**, (1946), 48.

[8] D. R. Cox, A use of complex probabilities in the theory of stochastic processes, *Proc. Cambridge Phil. Soc.*, **51**, (1955), 313-319.

[9] David P. Doane, *Exploring Statistics with the IBM P C*, 2nd ed., Addison-Wesley, Reading, MA, 1987. Includes a diskette with EXPLORE statistical programs and a data diskette.

[10] William Feller, *An Introduction to Probability Theory and Its Applications*, Vol. I, 3rd ed. John Wiley, New York, 1968.

[11] William Feller, *An Introduction to Probability Theory and Its Applications*, Vol. II, 2nd ed. John Wiley, New York, 1971.

[12] John E. Freund and Ronald E. Walpole, *Mathematical Statistics*, 3rd ed., Prentice-Hall, Englewood Cliffs, NJ, 1980.

[13] William C. Giffin, *Transform Techniques for Probability Modeling*, Academic Press, New York, 1975.

[14] Paul G. Hoel, *Introduction to Mathematical Statistics*, 4th ed., John Wiley, New York, 1971.

[15] A. D. Khomonenko, and V. P. Bubnov, A use of Coxian distribution law for iterative solution of $M/G/n/R \leq \infty$, *Problems of Control and Information Theory*, **14(2)**, (1985), 143–153.

[16] Leonard Kleinrock, *Queueing Systems, Volume 1: Theory*, John Wiley and Sons, New York, 1975.

[17] Hisashi Kobayashi, *Modeling and Analysis: An Introduction to System Performance Evaluation Methodology*, Addison-Wesley, Reading, MA, 1978.

[18] S. A. Lippman, *Elements of Probability and Statistics*, Holt, New York, 1971.

[19] Raymond Marie, Méthodes iteratives de resolution de modèles mathematiques de systèmes informatiques, *R A I R O Informatique/Comput. Sci.*, **12**, (1978), 107–122.

[20] *MINITAB Reference Manual Release 7*, Minitab Inc., 1989.

[21] Bernard Ostle and Richard W. Mensing, *Statistics in Research*, 3rd ed., The Iowa State University Press, Ames, 1975.

[22] E. Parzen, *Modern Probability Theory and Its Applications*, John Wiley, New York, 1960.

[23] Siméon Denis Poisson, *Recherches sur la probabilité des jugements en matière criminelle et en matière civile, précédés des règles générales du calcul des probabilités*, Bachelier, Paris, 1837.

[24] William H. Press, Brian P. Flannery, Saul A. Teukolsky, and William T. Vettering, *Numerical Recipes: The Art of Scientific Computing*, Cambridge University Press, 1986.

[25] William H. Press, Brian P. Flannery, Saul A. Teukolsky, and William T. Vettering, *Numerical Recipes in C: The Art of Scientific Computing*, Cambridge University Press, 1988.

[26] E. Rutherford and H. Geiger, The probability variations in the distribution of α particles, *Philos. Mag. Ser. 6*, **20**, (1910), 700–704.

[27] Barbara F. Ryan, Brian L. Joiner, and Thomas A. Ryan, Jr., *Minitab Handbook*, 2nd ed., Revised Printing, Duxbury Press, Boston, 1989.

[28] *SAS Language Guide For Personal Computers, Release 6.03 Edition*, SAS Institute, Cary, NC, 1988.

[29] Robert L. Schaefer and Richard B. Anderson, *The Student Edition of MINITAB*, Addison-Wesley, Benjamin/Cummings, 1989.

[30] L. Schwartz, *Théorie des Distributions, Volume 2*. Actualités scientifiques et industrielles No. 1122, Hermann et Cie, Paris, 1959.

[31] A. K. Shah, A simpler approximation for areas under the standard normal curve, *The American Statistician*, **39(1)**, (Feb. 1985), 80.

[32] Shey Shiung Sheu, The Poisson approximation to the binomial distribution, *The American Statistician*, **38(3)**, (Aug. 1984), 206–207.

[33] Stephen M. Stigler, *The History of Statistics: The Measurement of Uncertainty Before 1900*, Harvard University Press, 1986.

[34] Alan Stuart and J. Keith Ord, *Kendall's Advanced Theory of Statistics Vol. I Distribution Theory*, 5th ed., Oxford University Press, New York, 1987.

[35] "A. Student" (William S. Gosset), The probable error of a mean, *Biometrika*, **6**, 1908.

[36] Howard M. Taylor and Samuel Karlin, *An Introduction To Stochastic Modeling*, Academic Press, Orlando, FL, 1984.

[37] W. A. Wallis, The Poisson distribution and the Supreme Court, *J. Amer. Stat. Assoc.* **31** (1936), 376–380.

[38] Ward Whitt, Approximating a point process by a renewal process, I: two basic methods, *Opns. Res.*, **30(1)**, (Jan.–Feb. 1982), 125–147.

[39] Peter W. Zehna, How electronics ended the Poisson approximation to the binomial distribution, *The Amer. Math. Monthly*, **94(10)**, (Dec. 1987), 984–987.

Chapter 4

Stochastic Processes

Fate laughs at probabilities.
E. G. Bulwer-Lytton

Statistics are no substitute for judgment.
Henry Clay

4.0   Introduction

In Chapter 3 we considered some common random variables that are useful in investigating probabilistic computer science phenomena, such as the number of jobs waiting to be processed, the response time for an interactive inquiry system, the time between messages in an order entry system, etc. When we considered a random variable, such as the number of jobs, N, waiting to be processed, we did not allow for the fact that the probability distribution of N may change with time. That is, if we let N_1 be the number of jobs in the job queue at 8 A.M. and N_2 the corresponding number at 11 A.M., then N_1 and N_2 probably have different probability distributions. (To investigate the nature of the distribution of N_1, we could note the number of jobs at 8 A.M. each day for a number of days; for N_2 we could do the same thing at 11 A.M.) Thus, we have a family of random variables $\{N(t), t \in T\}$, where T is the set of all times during the day that the computer center is in operation. Such a family of random variables is called a *stochastic process.*

> *We are the music-makers,*
>   *And we are the dreamers of dreams,*
> *Wandering by lone sea-breakers,*
>   *And sitting by desolate streams;*
> *World-losers and world-forsakers,*
>   *On whom the pale moon gleams:*
> *Yet we are the movers and shakers*
>   *Of the world for ever, it seems.*

Arthur O'Shaughnessy

4.1   Definitions

A family of random variables $\{X(t), t \in T\}$ is called a *stochastic process*. Thus, for each $t \in T$, where T is the *index set* of the process, $X(t)$ is a random variable. An element of T is usually referred to as a time parameter and we often refer to t as time, although this is not part of the definition. The *state space* of the process is the set of all possible values that the random variables $X(t)$ can assume. Each of these values is called a *state* of the process.

Stochastic processes are classified in a number of ways, such as by the index set and by the state space. If $T = \{0, 1, 2, \cdots\}$ or $T = \{0, \pm 1, \pm 2, \cdots\}$, the stochastic process is said to be a *discrete parameter process* and we will usually indicate the process by $\{X_n\}$. If $T = \{t : -\infty < t < \infty\}$ or $T = \{t : t \geq 0\}$, the stochastic process is said to be a *continuous parameter process* and will be indicated by $\{X(t), -\infty < t < \infty\}$ or $\{X(t), t \geq 0\}$. The state space is classified as *discrete* if it is finite or countable; it is *continuous* if it consists of an interval (finite or infinite) of the real line. For a stochastic process $\{X(t)\}$, for each t, $X(t)$ is a random variable and thus a function from th‚e underlying sample space, Ω, into the state space. For any $\omega \in \Omega$, there is a corresponding collection $\{X(t)(\omega), t \in T\}$ called a *realization* or *sample path* of X at ω (usually the ω is elided.)

Example 4.1.1 The waiting time of an arriving inquiry message until processing is begun, is $\{W(t), t \geq 0\}$. The arrival time, t, of the message is the continuous parameter. The state space is also continuous. $\square$

Example 4.1.2 The number of messages that arrive in the time period from 0 to t, is $\{N(t), t \geq 0\}$. This is a continuous parameter, discrete state space process. $\square$

Example 4.1.3 Let $\{X_n\}, n = 1, 2, 3, 4, 5, 6, 7\}$ denote the average time to run a batch job at the computer center on the nth day of the week. Thus,

X_1 is the average job execution time on Sunday, X_2 on Monday, etc. Then $\{X_n\}$ is a discrete parameter, continuous state space process. □

Example 4.1.4 Let $\{X_n, n = 1, 2, \ldots, 365(366)\}$ denote the number of batch jobs run at a computer center on the nth day of the year. This is a discrete parameter, discrete state space process. □

Consider random (unpredictable) events such as

(a) the arrival of an inquiry at the central processing system of an inter-active computer system,

(b) a telephone call to an airline reservation center,

(c) and end-of-file interrupt, or

(d) the occurrence of a hardware or software failure in a computer system.

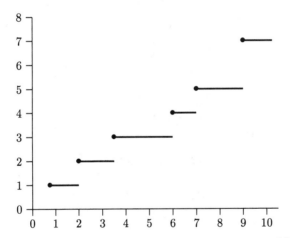

Figure 4.1.1. Realization of counting process $N(t)$.

Such events can be described by a *counting process* $\{N(t), t \geq 0\}$, where $N(t)$ is the number of events that have occurred after time 0 but not later than time t. (The realization of a typical counting process is shown in Figure 4.1.1.)

The idea of a counting process is formalized in the following definition.

Definition 4.1.1 $\{N(t), t \geq 0\}$ constitutes a *counting process* provided that

1. $N(0) = 0$,

2. $N(t)$ assumes only nonnegative integer values,

3. $s < t$ implies that $N(s) \le N(t)$, and

4. $N(t) - N(s)$ is the number of events that have occurred after s but not later than t, that is, in the interval $(s, t]$.

The next definition formalizes the idea that "one quantity is small relative to another quantity" and makes it possible to indicate this fact without specifying the exact relationship between the two quantities.

Definition 4.1.2 The function f is $o(h)$ (read "f is little-oh of h" and written "$f = o(h)$") if

$$\lim_{h \to 0} \frac{f(h)}{h} = 0,$$

that is, if given $\epsilon > 0$, there exists $\delta > 0$ such that $0 < |h| < \delta$ implies

$$\left| \frac{f(h)}{h} \right| < \epsilon.$$

Example 4.1.5   (a) The function $f(x) = x$ is not $o(h)$, since

$$\lim_{h \to 0} \frac{f(h)}{h} = \lim_{h \to 0} \frac{h}{h} = 1 \ne 0.$$

(b) The function $f(x) = x^2$ is $o(h)$, since

$$\lim_{h \to 0} \frac{f(h)}{h} = \lim_{h \to 0} \frac{h^2}{h} = \lim_{h \to 0} h = 0.$$

(c) The function $f(x) = x^r$ where $r > 1$ is $o(h)$, since

$$\lim_{h \to 0} \frac{f(h)}{h} = \lim_{h \to 0} h^{r-1} = 0.$$

This generalizes (b).

(d) If f is $o(h)$ and g is $o(h)$, then $f + g$ is $o(h)$, since

$$\lim_{h \to 0} \frac{f(h) + g(h)}{h} = \lim_{h \to 0} \frac{f(h)}{h} + \lim_{h \to 0} \frac{g(h)}{h} = 0 + 0 = 0.$$

(e) If f is $o(h)$ and c is a constant, then cf is $o(h)$, since

$$\lim_{h \to 0} \frac{cf(h)}{h} = c \lim_{h \to 0} \frac{f(h)}{h} = c \times 0 = 0.$$

(f) It follows from (d) and (e), by mathematical induction, that any finite linear combination of functions, each of which is $o(h)$, is also $o(h)$. That is, if $c_1, c_2, \ldots, c_n$ are n constants and $f_1, f_2, \ldots, f_n$ are n functions, each of which is $o(h)$, then

$$\sum_{i=1}^{n} c_i f_i$$

is $o(h)$.

(g) Suppose X is an exponential distribution with parameter α and $h > 0$. Then

$$P[X \le t + h | X > t] = P[X \le h],$$

by the Markov property of the exponential distribution (Theorem 3.3.1(d)). But

$$
\begin{aligned}
P[X \le h] &= 1 - e^{-\alpha h} \\
&= 1 - \left[1 - \alpha h + \sum_{n=2}^{\infty} \frac{(-\alpha h)^n}{n!} \right] \\
&= \alpha h - (\alpha h)^2 \times \sum_{n=2}^{\infty} \frac{(-\alpha h)^{n-2}}{n!} \\
&= \alpha h + o(h),
\end{aligned}
$$

so that

$$P[X \le t + h | X > t] = \alpha h + o(h). \quad \square$$

A continuous parameter stochastic process $\{X(t), t \ge 0\}$ has *independent increments* if events occurring in nonoverlapping time intervals are *independent; that is, if* $(a_1, b_1), \ldots, (a_n, b_n)$ are n nonoverlapping intervals, then the n random variables

$$X(b_1) - X(a_1), X(b_2) - X(a_2), \ldots, X(b_n) - X(a_n)$$

are independent. The process has *stationary increments* if $X(t+h) - X(s+h)$ has the same distribution as $X(t) - X(s)$ for each choice of indices s and t (with $s < t$) and for every $h > 0$; that is, the distribution $X(t) - X(s)$ depends only on the length of the interval from s to t and not on the particular value of s.

Example 4.1.6 Suppose $X_1, X_2, \ldots$ are independent identically distributed Bernoulli random variables, each with probability p of success (that is, of assuming the value 1). Let $S_n = X_1 + \cdots + X_n$, that is, the number of successes in n Bernoulli trials. Then $\{S_n, n = 1, 2, 3, \ldots\}$ is called a *Bernoulli* process. It has the state space $\{0, 1, 2, 3, \ldots\}$ so it is a discrete parameter discrete state space process. For each n, S_n has a binomial distribution with pmf $p(\cdot)$ defined by $p(k) = P[S_n = k] = \binom{n}{k} p^k q^{n-k}, k = 0, 1, \ldots, n$, where $q = 1 - p$. As we saw in Section 3.1.3, starting at any particular Bernoulli trial, the number of succeeding trials, Y, before the next success has geometric distribution; that is,

$$P[Y = k] = q^k p, \quad k = 0, 1, 2, \ldots. \ \square$$

I returned, and saw under the sun, that the race is not to the swift, nor the battle to the strong, neither yet bread to the wise, nor yet riches to men of understanding, nor yet favour to men of skill; but time and chance happeneth to them all.
Ecclesiastes 9:11

4.2 The Poisson Process

Definition 4.2.1 A counting process $\{N(t), t \geq 0\}$ (see Definition 4.1.1) is a *Poisson process with rate* $\lambda > 0$, if (a)–(d), below are true.

(a) The process has independent increments. (Events occurring in nonoverlapping intervals of time are independent of each other.)

(b) The increments of the process are stationary. (The distribution of the number of events in any interval of time depends only on the length of the interval and not on when the interval begins.)

(c) The probability that exactly one event occurs in any time interval of length h is $\lambda h + o(h)$, that is,

$$P[N(h) = 1] = \lambda h + o(h).$$

(d) The probability that more than one event occurs in any time interval of length h is $o(h)$, that is,

$$P[N(h) \geq 2] = o(h).$$

Note that (c) and (d) together imply that

$$P[N(h) = 0] = 1 - \lambda h + o(h),$$

since

$$
\begin{aligned}
P[N(h) = 0] &= 1 - P[N(h) = 1] - P[N(h) \geq 2] \\
&= 1 - \lambda h - o(h) - o(h) = 1 - \lambda h + o(h). \quad (4.1)
\end{aligned}
$$

The last equality follows from Example 4.1.5(d) and (e).

The following theorem shows that Definition 4.2.1 of a Poisson process is descriptive.

Theorem 4.2.1 *Let $\{N(t), t \geq 0\}$ be a Poisson process with rate $\lambda > 0$. Then the random variable Y describing the number of events in any time interval of length $t > 0$ has a Poisson distribution with parameter λt. That is,*

$$
P[Y = k] = e^{-\lambda t} \frac{(\lambda t)^k}{k!}, \quad k = 0, 1, 2, \ldots. \quad (4.2)
$$

Thus, the average number of events occurring in any time interval of length t is λt.

Proof Let $t > 0$. By the definition of a Poisson process the number of events occurring in any time interval is independent of those in any nonoverlapping interval and depends only upon the length of the given interval. Therefore, we can assume without loss of generality, that the interval of interest extends from 0 to t. We define

$$
P_n(t) = P[N(t) = n] \quad \text{for each nonnegative integer } n. \quad (4.3)
$$

It is true that no events occur by time $t + h$ only if no events occur by time t and no events occur in the interval from t to $t + h$. Hence,

$$
\begin{aligned}
P_0(t + h) &= P_0(t)P[N(t + h) - N(t) = 0] \\
&= P_0(t)P[N(h) = 0] = P_0(t)(1 - \lambda h + o(h)). \quad (4.4)
\end{aligned}
$$

The first equality in (4.4) follows from the fact that the process has independent increments; the second equality follows from the stationarity of the increments. The last equality follows from (4.1).

Equation (4.4) and Example 4.1.5(e), together, yield

$$
\frac{P_0(t + h) - P_0(t)}{h} = -\lambda P_0(t) + \frac{o(h)}{h}. \quad (4.5)
$$

Letting $h \to 0$ in (4.5), we arrive at the differential equation

$$
\frac{dP_0(t)}{dt} = -\lambda P_0(t). \quad (4.6)
$$

The solution of (4.6) with the initial condition $P_0(0) = P[N(0) = 0] = 1$ is given by

$$P_0(t) = e^{-\lambda t},$$

as can be verified by direct substitution in (4.6).

Now suppose $n > 0$. Then n events can occur by time $t+h(N(t+h) = n)$ in three mutually exclusive ways.

(a) n events occur by time t and no event occurs in the interval from t to $t + h$.

(b) $n-1$ events occur by time t and exactly one event occurs in the interval between t and $t + h$.

(c) $n - k$ events occur by time t for some k from the set $\{2, 3, \ldots, n\}$ and exactly k events occur in the interval from t to $t + h$.

Hence, summing up the probabilities associated with (a), (b), and (c) yields

$$P_n(t + h) = P_n(t)(1 - \lambda h + o(h)) + \lambda h P_{n-1}(t) + o(h). \qquad (4.7)$$

Therefore,

$$\frac{P_n(t + h) - P_n(t)}{h} = -\lambda P_n(t) + \lambda P_{n-1}(t) + \frac{o(h)}{h}, \qquad (4.8)$$

and taking the limit as $h \to 0$ gives

$$\frac{dP_n(t)}{dt} = -\lambda P_n(t) + \lambda P_{n-1}(t). \qquad (4.9)$$

The solution to (4.9) subject to the initial condition $P_n(0) = 0$ is given by

$$P_n(t) = \frac{e^{-\lambda t}(\lambda t)^n}{n!}, \qquad (4.10)$$

as can be checked by direct substitution.

Since $P[Y = k] = P_k(t) = e^{-\lambda t}(\lambda t)^k / k!$, $k = 0, 1, 2, \ldots$, this completes the proof. ∎

It is important to note that according to Theorem 4.2.1, the number of events occurring in *each* time interval of length t has a Poisson distribution with an average value of λt. Hence, the average number of events occurring per unit time is $\lambda t/t = \lambda$. The next theorem gives another important attribute of a Poisson process. It was stated before as Theorem 3.2.1(f).

Theorem 4.2.2 *Let $\{N(t), t \geq 0\}$ be a Poisson process with rate λ. Let $0 < t_1 < t_2 < t_3 < \cdots$ be the successive occurrence times of events, and let the interarrival times $\{\tau_n\}$ be defined by $\tau_1 = t_1, \tau_2 = t_2 - t_1, \ldots, \tau_k = t_k - t_{k-1}, \ldots$. Then the interarrival times $\{\tau_n\}$ are mutually independent, identically distributed, exponential random variables, each with mean $1/\lambda$.*

Proof Since a Poisson process has independent increments ((a) of Definition 4.2.1), events occurring after t_n are independent of those occurring before $t_n, n = 1, 2, \ldots$. This proves that $\tau_1, \tau_2, \ldots$ are independent random variables. For any $s \geq 0$ and any $n \geq 1$, the events $\{\tau_n > s\}$ and $\{N(t_{n-1} + s) - N(t_{n-1}) = 0\}$ are equivalent (we define t_0 to be zero so that t_{n-1} is defined when $n = 1$). The events in brackets are equivalent, since $\{\tau_n > s\}$ is true, if and only if the nth event has not yet occurred s time units after the occurrence of the $(n - 1)$th event; but this is the same as the requirement that $\{N(t_{n-1} + s) - N(t_{n-1}) = 0\}$ is true. Thus, it is true that

$$P[\tau_n > s] = P[N(t_{n-1} + s) - N(t_{n-1}) = 0] = P[N(s) = 0] = e^{-\lambda s}, \quad (4.11)$$

by Theorem 4.2.1 and the fact that the process has stationary increments. Therefore,

$$P[\tau_n \leq s] = 1 - e^{-\lambda s}, \quad s \geq 0. \quad \blacksquare \quad (4.12)$$

The next theorem shows that the converse of Theorem 4.2.2 also is true.

Theorem 4.2.3 *Let $\{N(t), t \geq 0\}$ be a counting process such that the interarrival times of events, $\{\tau_n\}$, are independent, identically distributed, exponential random variables, each with the average value $1/\lambda$. Then $\{N(t), t \geq 0\}$ is a Poisson process with rate λ..*

Proof We omit the proof of this theorem. The proof can be found in Chung [1, pages 200–202]. $\blacksquare$

When the occurrence of some event, such as the arrival of an inquiry to an inquiry system, the arrival of customers at a bank, the arrival of messages to a message switching center, etc., is described by a Poisson process, we often hear the events described as "random" with some sort of selection process in which either (a) each of a finite number of elements has the same probability of selection or (b) a time is chosen in some interval of time so that each subinterval of the same length has the same probability of containing the selected point.

The next theorem shows us that the word *random is* appropriate for describing a Poisson process.

Theorem 4.2.4 *Suppose* $\{N(t), t \geq 0\}$ *is a Poisson process and one event has taken place in the interval from 0 to t. Then Y, the random variable describing the time of occurrence of this Poisson event, has a continuous uniform distribution on the interval from 0 to t; that is, if* $0 < \delta < t$*, any subinterval of* $(0, t]$ *of length* δ *has probability of* δ/t *of containing the time of occurrence of the event.*

Proof Let $0 < x < t$. By the definition of Y,

$$P[Y \leq x] = P[\tau_1 \leq x | N(t) = 1]. \qquad (4.13)$$

But, by the definition of conditional probability,

$$
\begin{aligned}
P[\tau_1 \leq x | N(t) = 1] &= \frac{P[(N(x) = 1) \text{ and } (N(t) - N(x) = 0)]}{P[N(t) = 1]} \\
&= \frac{P[N(x) = 1]P[N(t-x) = 0]}{P[N(t) = 1]} \\
&= \frac{\lambda x e^{-\lambda x} e^{-\lambda(t-x)}}{\lambda t e^{-\lambda t}} = \frac{x}{t},
\end{aligned} \qquad (4.14)
$$

where the next to last equality in (4.14) follows from Theorem 4.2.1. ∎

The Poisson process is a special case of a general type of stochastic process that is important to queueing theory. In the next section we study this type of process called a *birth-and-death process*.

My mother groan'd, my father wept;
Into the dangerous world I leapt,
Helpless, naked, piping loud,
Like a fiend hid in a cloud.

                                                        William Blake

First our pleasures die—and then
Our hope, and then our fears—and when
These are dead, the debt is due,
Dust claims dust—and we die too.

                                                    Percy Bysshe Shelley

4.3   Birth-and-Death Process

In the last section, we studied a Poisson process $\{N(t), t \geq 0\}$ that counted
the number of occurrences of some type of event, that could also be interpreted as an arrival of some entity at an average rate λ. We think of such
an arrival as a birth. For a Poisson process, the probability of one birth
in a short time interval h is $\lambda h e^{-\lambda h} = \lambda h + o(h)$, and this probability is
independent of how many births have occurred. We can think of λ as the
birth rate. For some systems, such as a biological species or a queueing
system, it might be reasonable to suppose that the birth rate depends on
the number of the population present, that is, that the probability of a
birth in a short time interval h must be $\lambda_n h + o(h)$, where n is the size
of the population, and the birth rate λ_n depends upon this number n. It
is also reasonable to allow deaths or decreases in the population with the
probability of a death in an interval of length h equal to $\mu_n h + o(h)$. Thus,
the intuitive idea behind a birth-and-death process is that of some type of
a population, that is simultaneously gaining new members through births
and losing old members through deaths—such as the human population of
the earth. The population we have in mind for most applications of birth-
and-death processes to computer science is that of customers in a queueing
system. Of course customer is a generic word here, and could correspond
to a computer job to be processed, an I/O request, a message arrival to a
communication system, etc. Customer arrivals correspond to births, and
customer departures (after receiving service) correspond to deaths.

Definition 4.3.1 Consider a continuous parameter stochastic process $\{X(t),$
$t \geq 0\}$ with the discrete state space $0, 1, 2, \ldots$. Suppose this process describes a system that is in state $E_n, n = 0, 1, 2, \ldots$ at time t, if and only if
$X(t) = n$ (the system has a population of n elements or customers at time
t). Then the system is said to be described by a *birth-and-death process*
if there exist nonnegative birth rates $\{\lambda_n, n = 0, 1, 2, \ldots\}$ and nonnegative
death rates $\{\mu_n, n = 1, 2, \ldots\}$ such that the following postulates (sometimes
called the *nearest-neighbor* assumptions) are satisfied.

1. State changes are only allowed from state E_n to state E_{n+1} or from
   state E_n to E_{n-1} if $n \geq 1$, but from state E_0 to state E_1 only.

2. If at time t the system is in state E_n, the probability that between
   time t and time $t+h$ the transition from state E_n to state E_{n+1} occurs
   (indicated by $E_n \rightarrow E_{n+1}$) equals $\lambda_n h + o(h)$, and the probability that
   the transition $E_n \rightarrow E_{n-1}$ occurs (if $n \geq 1$) equals $\mu_n h + o(h)$.

3. The probability that, in the time interval from t to $t + h$, more than
   one transition occurs is $o(h)$.

Postulate (1) allows only one birth or death to occur at a time and states that no death can occur if the system is empty. Postulate (2) gives the transition probabilities, that is, the probability of a birth or death in a small time interval when the system population is n. The last postulate states that the probability of more than one birth or death in a short time interval is negligible.

When we describe a queueing system as a birth-and-death process, we think of state E_n as corresponding to n customers in the system, either waiting for or receiving service.

We will now derive the differential-difference equations for $P_n(t) = P[X(t) = n]$, the probability that the system is in state E_n at time t. The procedure is very similar to the method we used in the proof of Theorem 4.2.1. In fact, the differential-difference equations we derived there are a special case of the equations we derive here.

If $n \geq 1$ the probability $P_n(t+h)$ that at time $t+h$ the system will be in state E_n has four components.

1. The probability that it was in state n at time t and no transitions occurred, either births or deaths. This probability is the product of (a) $P_n(t)$, (b) the probability that the transition $E_n \to E_{n+1}$ did *not* occur, or $1 - \lambda_n h + o(h)$, and (c) the probability that the transition $E_n \to E_{n-1}$ did *not* occur, or $1 - \mu_n h + o(h)$. Hence, the required probability is

$$P_n(t)(1 - \lambda_n h + o(h))(1 - \mu_n h + o(h))$$
$$= P_n(t)[1 - \mu_n h + o(h) - \lambda_n h + \lambda_n \mu_n h^2 - \lambda_n h o(h) + o(h)]$$
$$= P_n(t)[1 - \mu_n h - \lambda_n h + o(h)] = P_n(t)(1 - \lambda_n h - \mu_n h) + o(h),$$

since, by Example 4.2.5,

$$o(h)(1 - \mu_n h + o(h)) = o(h),$$
$$\lambda_n \mu_n h^2 - \lambda_n h o(h) + o(h) = o(h), \quad \text{and} \quad P_n(t)o(h) = o(h).$$

2. The probability $P_{n-1}(t)$ that the system was in state E_{n-1} at time t, times the probability that the transition $E_{n-1} \to E_n$ occurred in the interval from t to $t+h$. This latter probability equals $\lambda_{n-1}h + o(h)$, so the total contribution is

$$P_{n-1}(t)(\lambda_{n-1}h + o(h)) = P_{n-1}(t)\lambda_{n-1}h + o(h). \tag{4.15}$$

3. The probability $P_{n+1}(t)$ that the system was in state E_{n+1} at time t, multiplied by the probability that the transition $E_{n+1} \to E_n$ occurred during the interval from t to $t+h$. The contribution is thus

$$P_{n+1}(t)\mu_{n+1}h + o(h). \tag{4.16}$$

4. The probability that two or more transitions occur between times t and $t + h$, that leave the system in state E_n. (For example, the two transitions $E_{n+2} \to E_{n+1}$ and $E_{n+1} \to E_n$.) By hypothesis this probability is $o(h)$.

Since the events leading to the four components are mutually exclusive, the result is that

$$
\begin{aligned}
P_n(t + h) &= [1 - \lambda_n h - \mu_n h]P_n(t) + \lambda_{n-1}hP_{n-1}(t) \\
&\quad + \mu_{n+1}hP_{n+1}(t) + o(h).
\end{aligned} \tag{4.17}
$$

Transposing the term $P_n(t)$ and dividing by h, we get

$$
\begin{aligned}
\frac{P_n(t + h) - P_n(t)}{h} &= -(\lambda_n + \mu_n)P_n(t) \\
&\quad + \lambda_{n-1}P_{n-1}(t) + \mu_{n+1}P_{n+1}(t) + \frac{o(h)}{h}.
\end{aligned}
$$

Taking the limit as $h \to 0$ gives us the equation

$$
\frac{d_n(t)}{dt} = -(\lambda_n + \mu_n)P_n(t) + \lambda_{n-1}P_{n-1}(t) + \mu_{n+1}P_{n+1}(t). \tag{4.18}
$$

This equation is valid for $n \geq 1$. For $n = 0$, by similar reasoning, we get

$$
\frac{dP_0(t)}{dt} = -\lambda_0 P_0(t) + \mu_1 P_1(t). \tag{4.19}
$$

If the initial state is E_i, then the initial conditions are given by

$$
P_i(0) = 1 \quad \text{and} \quad P_j(0) = 0 \quad \text{for} \quad j \neq i. \tag{4.20}
$$

The birth-and-death process depends on the infinite set of differential-difference equations (4.18) and (4.19) with initial conditions (4.20). It can be shown that this set of equations has a solution $P_n(t)$ for all n and t under very general conditions. However, the solutions are very difficult to obtain analytically, except for some very special cases.

One such special case is the pure-birth process with $\lambda_n = \lambda > 0$, $\mu_n = 0$ for all n, and the initial conditions $P_0(0) = 1, P_j(0) = 0$ for $j \neq 0$. (Any process for which all the μ_n are zero is called a *pure-birth process*; any for which all the λ_n are zero is called a *pure-death process*.) This leads to the set of equations

$$
\begin{aligned}
\frac{dP_N(t)}{dt} &= -\lambda p_n(t) + \lambda p_{n-1}(t), \quad n \geq 1 \\
\frac{dP_0(t)}{dt} &= -\lambda p_0(t).
\end{aligned} \tag{4.21}
$$

As we saw in the proof of Theorem 4.2.1, the solution of (4.21) satisfying
the given initial conditions is given by

$$P_n(t) = \frac{e^{-\lambda t}(\lambda t)^n}{n!}, \quad n \geq 0, \ t \geq 0. \tag{4.22}$$

Thus, the process is a Poisson process and we have a new characterization
of a Poisson process. It is a pure-birth process with a constant birth rate.

In general, finding the time-dependent solutions of a birth-and-death
process is very difficult. However, if $P_n(t)$ approaches a constant value p_n
as $t \to \infty$ for each n, then we say that the system is in *statistical equilibrium*.
Under very general conditions these limits exist and are independent of the
initial conditions. When a system is in statistical equilibrium, we sometimes
say the system is in the *steady state* or that the system is *stationary*, because the state of the system does not depend on time.[1] If we could obtain
the time-dependent solutions $\{P_n(t)\}$, we could solve for the steady state
solutions $\{p_n\}$ by the equations $\lim_{t\to\infty} P_n(t) = p_n, n = 0, 1, 2, \ldots$. Since
we cannot, in general, find the time-dependent or transient solutions to the
birth-and-death differential-difference equations (4.18) and (4.19), analytically, we will take limits as $t \to \infty$ on both sides of these equations and,
using the fact that $\lim_{t\to\infty} dP_n(t)/dt = 0$ for all n and $lim_{t\to\infty} P_n(t) = p_n$
(we assume that the steady state solutions do exist), we obtain the set of
difference equations

$$0 = \lambda_{n-1}p_{n-1} + \mu_{n+1}p_{n+1} - (\lambda_n + \mu_n)p_n, \quad n \geq 1 \tag{4.23}$$

$$0 = \mu_1 p_1 - \lambda_0 p_0, \quad n = 0. \tag{4.24}$$

The last equation yields

$$p_1 = (\lambda_0/\mu_1)p_0. \tag{4.25}$$

Equation (4.23) can be written as

$$\mu_{n+1}p_{n+1} - \lambda_n p_n = \mu_n p_n - \lambda_{n-1}p_{n-1}, \quad n \geq 1. \tag{4.26}$$

If we define $g_n = \mu_n p_n - \lambda_{n-1}p_{n-1}$ for $n = 1, 2, 3, \ldots$ we see that (4.26) can
be written as

$$g_{n+1} = g_n, \quad n \geq 1. \tag{4.27}$$

Hence, $g_n = $ constant and, by (4.24), $g_1 = $ constant $= 0$. Hence, $g_n = 0$ for
all n or (assuming $\mu_n > 0$ for all n)

$$p_{n+1} = \frac{\lambda_n}{\mu_{n+1}}p_n, \quad n \geq 0. \tag{4.28}$$

---

[1] We discuss stationary processes in more detail in Section 4.4.

Thus, we compute successively

$$p_1 = \frac{\lambda_0}{\mu_1}p_0, \quad p_2 = \frac{\lambda_1}{\mu_2}p_1 = \frac{\lambda_0\lambda_1}{\mu_1\mu_2}p_0, \quad p_3 = \frac{\lambda_2}{\mu_3}p_2 = \frac{\lambda_0\lambda_1\lambda_2}{\mu_1\mu_2\mu_3}p_0.$$

Continuing for $n = 4, 5, 6, \ldots$, we see by induction that

$$p_n = \frac{\lambda_0\lambda_1 \cdots \lambda_{n-1}}{\mu_1\mu_2 \cdots \mu_n}p_0, \quad n \geq 1. \tag{4.29}$$

This gives the solutions in terms of p_0, the probability that the system is in state E_0 (the system is empty), that is determined by the condition

$$\sum_{n=1}^{\infty} p_n = p_0 + p_1 + p_2 + \cdots = 1. \tag{4.30}$$

If we substitute (4.29) into (4.30) we obtain

$$p_0 \left(1 + \frac{\lambda_0}{\mu_1} + \frac{\lambda_0\lambda_1}{\mu_1\mu_2} + \cdots + \frac{\lambda_0\lambda_1 \cdots \lambda_{n-1}}{\mu_1\mu_2 \cdots \mu_n} + \cdots \right) = 1. \tag{4.31}$$

Hence, the steady state probabilities (4.29) exist if the series

$$S = 1 + \frac{\lambda_0}{\mu_1} + \frac{\lambda_0\lambda_1}{\mu_1\mu_2} + \cdots + \frac{\lambda_0\lambda_1 \cdots \lambda_{n-1}}{\mu_1\mu_2 \cdots \mu_n} + \cdots < \infty. \tag{4.32}$$

(We assume the λ_n and μ_n are nonnegative.) When this is true, $p_0 = 1/S > 0$, or the probability that the system is empty is positive. In the case of a queueing system, this means that the service facility sometimes "catches up" or gets all the customers processed. On the other hand, if the series for S diverges, this is an indication that the queueing system is unstable because arrivals are occurring faster, on the average, than departures. For actual real-life queueing systems described by birth-and-death processes, we may safely assume that the steady state probabilities $\{p_n\}$ exist if and only if the series for S converges and then they are given by (4.29) with $p_0 = 1/S$.

Before we study the steady state solutions to some important birth-and-death processes, we consider a simple queueing system for which we *can* calculate the time dependent functions $P_n(t)$ for all n.

Example 4.3.1 Consider a queueing system with one server and no waiting line. We assume a Poisson arrival process with parameter λ and an exponential service time distribution with parameter μ. The former means, by definition, that the probability of an arrival in the interval $(0, h]$ is $\lambda h + o(h)$.

If the server is busy at time t, then the probability that service for the customer will be completed by time $t + h$ is $1 - e^{-\mu h} = \mu h + o(h)$. (Here we have used the lack of memory or Markov property of the exponential distribution.) Thus, our birth-and-death model has only states E_0 and E_1, with E_0 corresponding to the server being idle and E_1 to the server being busy. An arrival that occurs when the server is busy is turned away and thus has no effect on the system. Therefore, an arrival will cause a state transition (from E_0 to E_1) only if the arrival occurs when the server is idle. Hence, $\lambda_0 = \lambda, \lambda_n = 0$ for $n \neq 0, \mu_1 = \mu$ and $\mu_n = 0$ for $n \neq 1$. The birth-and-death differential-difference equations (4.18) and (4.19) become

$$\frac{dP_1(t)}{dt} = \lambda P_0(t) - \mu P_1(t), \quad \frac{dP_0(t)}{dt} = -\lambda P_0(t) + \mu P_1(t). \qquad (4.33)$$

We can set the initial conditions to be $P_0(0) + P_1(0) = 1$. Since

$$\frac{d(P_0(t) + P_1(t))}{dt} = 0$$

by (4.33), we have

$$P_0(t) + P_1(t) = 1, \quad t \geq 0. \qquad (4.34)$$

If we substitute (4.34) into the second equation of (4.33), we obtain

$$\frac{dP_0(t)}{dt} + (\lambda + \mu)P_0(t) = \mu. \qquad (4.35)$$

By elementary differential equation theory (see, for example, Coddington [2, page 41]), we have

$$P_0(t) = \frac{\mu}{\lambda + \mu} + \left(P_0(0) - \frac{\mu}{\lambda + \mu}\right) e^{-(\lambda+\mu)t}. \qquad (4.36)$$

By symmetry we also have

$$P_1(t) = \frac{\lambda}{\lambda + \mu} + \left(P_1(0) - \frac{\lambda}{\lambda + \mu}\right) e^{-(\lambda+\mu)t}. \qquad (4.37)$$

Now, if we take the limit as $t \to \infty$ in (4.36) and (4.37), we obtain the steady state probabilities p_0 and p_1 as

$$p_0 = \lim_{t \to \infty} P_0(t) = \frac{\mu}{\lambda + \mu}, \qquad (4.38)$$

and

$$p_1 = \lim_{t \to \infty} P_1(t) = \frac{\lambda}{\lambda + \mu}. \qquad (4.39)$$

On the other hand, if we set the time derivatives to 0 in (4.33) and replace $P_0(t)$ by p_0, $P_1(t)$ by p_1, we obtain

$$0 = \lambda p_0 - \mu p_1, \quad 0 = -\lambda p_0 + \mu p_1. \tag{4.40}$$

Either equation of (4.40) gives

$$p_1 = (\lambda/\mu)p_0 \tag{4.41}$$

(the equations of (4.40) are equivalent). Substituting (4.41) into the equation $p_0 + p_1 = 1$ gives $p_0(1 + \lambda/\mu) = 1$, or

$$p_0 = \mu/(\lambda + \mu). \tag{4.42}$$

Substituting (4.42) into (4.41) gives

$$p_1 = \lambda/(\lambda + \mu). \ \Box \tag{4.43}$$

The latter method of obtaining the solution was obviously more straightforward than the former.

A useful, intuitively appealing technique has been devised to derive the steady state difference equations (4.23) and (4.24). It involves the use of a state-transition rate diagram, that graphically illustrates the postulates for a birth-and-death system. Figure 4.3.1 is a general *state-transition rate diagram* for a birth and death system. In this diagram a state E_n is represented by a circle or oval labeled with the number n. The arrows in this diagram show the only state transitions allowed and are labeled with the mean transition rates (either birth or death). Since we assume the system is in the steady state we assume the following principle: *Flow Rate In = Flow Rate Out Principle.* If a birth-and-death system has reached the steady state (equilibrium) condition, then for every state of the system n ($n = 0, 1, 2, \ldots$), the mean flow rate of the population into the state must equal the mean flow rate out. The equations expressing this condition are called the *balance equations.* Consider first a state E_n with $n \geq 1$. Then by Figure 4.3.1, we see that the mean flow rate of the population into state n is $\lambda_{n-1}p_{n-1} + \mu_{n+1}p_{n+1}$, while the mean flow rate of the population out of the state is $\mu_n p_n + \lambda_n p_n = (\mu_n + \lambda_n)p_n$. Therefore, by the Flow Rate In = Flow Rate Out Principle, we have the balance equation

$$\lambda_{n-1}p_{n-1} + \mu_{n+1}p_{n+1} = (\lambda_n + \mu_n)p_n, \quad n \geq 1. \tag{4.44}$$

Equation (4.44) is equivalent to (4.23). For state E_0 the above Principle immediately gives

$$\mu_1 p_1 = \lambda_0 p_0, \tag{4.45}$$

that is the same as (4.24).

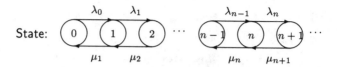

Figure 4.3.1. State-transition rate diagram for birth-and-death process.

Figure 4.3.2. State-transition rate diagram for Example 4.3.2.

Example 4.3.2 The queueing system of Example 4.3.1 has the state transition rate diagram of Figure 4.3.2. We can write the following balance equation by inspection:

$$\mu p_1 = \lambda p_0. \tag{4.46}$$

Hence,

$$p_1 = (\lambda/\mu)p_0. \tag{4.47}$$

But,

$$1 = p_0 + p_1 = p_0\left(1 + \frac{\lambda}{\mu}\right), \tag{4.48}$$

so

$$p_0 = \mu/(\lambda + \mu), \tag{4.49}$$

and

$$p_1 = (\lambda/\mu)p_0 = \lambda/(\lambda + \mu), \tag{4.50}$$

as we got in Example 4.3.1. □

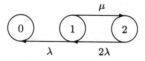

Figure 4.3.3. State-transition rate diagram for Example 4.3.3.

Example 4.3.3 In Example 2.7.3 we did not include a computer repair facility. One would expect that a computer that failed would be repaired as soon as possible, usually before the other computer went out of commission. Suppose each failed computer of Example 2.7.3 has an exponentially distributed repair time, R, with a mean of 10 hours. This is the *mean time to repair*, written MTTR. Thus, $E[R] = \text{MTTR} = 10$ hours.

We can model the duplexed computer system with a repair facility as a birth-and-death process model with states E_0, E_1, and E_2, where being in state E_j means that j of the computers are in operation. State E_0 corresponds to a system failure; such a failure requires that an emergency back-up procedure be put into operation. The state-transition rate diagram is shown in Figure 4.3.3. We assume the failure rate per machine (computer) is λ and the repair rate per machine is μ. Since the whole system has failed if both computers are simultaneously out of service, we assume that $\lambda_0 = 0$. State 0 is an *absorbing state*, since once the system enters the state, it never leaves it. In our notation

$$\lambda_0 \;=\; 0, \quad \lambda_1 = \mu, \quad \lambda_n = 0, \quad n \neq 1.$$
$$\mu_1 \;=\; \lambda, \quad \mu_2 = 2\lambda, \quad \mu_n = 0, \quad n \neq 1, 2.$$

Equations (4.18) and (4.19) then become

$$\frac{dP_2}{dt}(t) \;=\; -2\lambda P_2(t) + \mu P_1(t),$$
$$\frac{dP_1}{dt}(t) \;=\; 2\lambda P_2(t) - (\lambda + \mu)P_1(t), \qquad (4.51)$$
$$\frac{dP_0}{dt}(t) \;=\; \lambda P_1(t).$$

The initial conditions are

$$P_2(0) \;=\; 1,$$
$$P_1(0) \;=\; 0, \qquad (4.52)$$
$$P_0(0) \;=\; 0.$$

If we let Y be the random variable that describes the time to system failure, then

$$f_Y(t) = \frac{dP_0}{dt}(t).$$

As we show in Example 3.4.2, Y has the density function

$$f_Y(t) = \frac{2\lambda^2}{\alpha_1 - \alpha_2}\left(e^{-\alpha_2 t} - e^{-\alpha_1 t}\right), \qquad (4.53)$$

where

$$\alpha_1 = \frac{3\lambda + \mu + \alpha_3}{2},$$

$$\alpha_2 = \frac{3\lambda + \mu - \alpha_3}{2},$$

$$\alpha_3 = \sqrt{\lambda^2 + 6\lambda\mu + \mu^2}.$$

Thus, the distribution function for Y is given by

$$F(t) = \int_0^t f_Y(x)\,dx$$

$$= \frac{2\lambda^2}{\alpha_1 - \alpha_2} \int_0^t \left(e^{-\alpha_2 x} - e^{-\alpha_1 x}\right) dx$$

$$= \frac{2\lambda^2}{\alpha_1 - \alpha_2} \left[-\frac{1}{\alpha_2}\left(e^{-\alpha_2 t} - 1\right) + \frac{1}{\alpha_1}\left(e^{-\alpha_1 t} - 1\right) \right] \quad (4.54)$$

$$= \frac{2\lambda^2}{\alpha_1 - \alpha_2} \left[\frac{1}{\alpha_2} - \frac{1}{\alpha_1} + e^{-\alpha_1 t} - e^{-\alpha_2 t} \right], \quad t \ge 0$$

The MTTF $= E[Y]$ is given by

$$E[Y] = \int_0^\infty y f_Y(y)\,dy = \frac{2\lambda^2}{\alpha_1 - \alpha_2} \left[\int_0^\infty y e^{-\alpha_2 y}\,dy - \int_0^\infty y e^{-\alpha_1 y}\,dy \right]. \tag{4.55}$$

By Formula 3.351.3 of Gradshteyn [5][2]

$$\int_0^\infty y e^{-\alpha y}\,dy = \frac{1}{\alpha^2}.$$

Hence, (4.55) becomes

$$E[Y] = \frac{2\lambda^2}{\alpha_1 - \alpha_2} \left[\frac{1}{\alpha_2^2} - \frac{1}{\alpha_1^2} \right] = \frac{2\lambda^2(\alpha_1 + \alpha_2)}{\alpha_1^2 \alpha_2^2} = \frac{3}{2\lambda} + \frac{\mu}{2\lambda^2}. \tag{4.56}$$

Recall from Example 2.7.3 that $3/2\lambda$ was the mean time to failure (MTTF) if no repair facility is provided, so the term $\mu/2\lambda^2$ represents the improvement in MTTF due to the repair capability. In our example,

$$\frac{1}{\lambda} = 2,000 \text{ hours,}$$

---

[2]The actual formula is

$$\int_0^\infty x^n e^{-\mu x}\,dx = n!\mu^{-n-1}.$$

while

$$\frac{1}{\mu} = 10 \text{ hours},$$

so that

$$\frac{3}{2\lambda} = 3,000 \text{ hours},$$

and

$$\frac{\mu}{2\lambda^2} = 200,000 \text{ hours},$$

for a total of 203,000 hours. The repair capability has dramatically improved the MTTF.

The probability of no system failure by time t is $1 - F(t)$, where $F(t)$ is given by (4.54). For this example, the probability of no failure within a week is 0.99922 and within 30 days is 0.99651. This is a tremendous improvement over the figures for Example 2.7.3! □

I've known what it is to be hungry, but I always went right to a restaurant.
Ring Lardner

4.4 Markov Chains

A stochastic process $\{X(t), t \in T\}$ is a *Markov process* if for any set of $n + 1$ values $t_1 < t_2 < \cdots < t_n < t_{n+1}$ in the index set and any set $\{x_1, x_2, \ldots, x_{n+1}\}$ of $n + 1$ states, we have

$$P[X(t_{n+1}) = x_{n+1}|X(t_1) = x_1, X(t_2) = x_2, \ldots, X(t_n) = x_n]$$
$$= P[X(t_{n+1}) = x_{n+1}|X(t_n) = x_n]. \tag{4.57}$$

Intuitively, (4.57) indicates that the future of the process depends only on the present state and not upon the history of the process. That is, the entire history of the process is summarized in the present state.

A Markov process is called a *Markov chain* if its state space is discrete. Markov processes can thus be classified as in Table 4.4.1, where we assume the index set is time.

Table 4.4.1. Classification of Markov Processes

Type of parameter	State space	
	Discrete	Continuous
Discrete time	Discrete time Markov chain	Discrete time Markov process
Continuous time	Continuous time Markov chain	Continuous time Markov process

Since a Markov chain by definition has a discrete state space, we could label the states $\{E_0, E_1, E_2, \ldots\}$. However, for notational convenience, we will usually assume the states are the nonnegative integers $\{0, 1, 2, \ldots\}$. For the general case it is easy to associate each integer i with the corresponding state E_i.

For a discrete time Markov chain, it is fruitful to think of the process as making *state transitions* at times $t_n, n = 1, 2, 3, \ldots$ (possibly into the same state). Thus, the discrete time Markov chain $\{X_n\}$ (we write X_n for $X(t_n)$) starts in an initial state, say i, when $t = t_0$ ($X_0 = i$), and makes a state transition at the next step (time in the sequence); that is, when $t = t_1$, so that $X_1 = j$, etc. The one-step transition probabilities defined by

$$P[X_{n+1} = j | X_n = i], \quad n, i, j = 0, 1, 2, \ldots.$$

generally depend upon the index n. However, we are interested primarily in Markov chains for which the one-step transition probabilities are independent of n and thus can be denoted by P_{ij}. Such a Markov chain is said to have *stationary transition probabilities* or to be *homogeneous in time*. All discrete Markov chains in the sequel are assumed to be time homogeneous, unless the contrary is stated. The transition probabilities P_{ij} can be exhibited as a square matrix

$$P = \begin{bmatrix} P_{00} & P_{01} & P_{02} & P_{03} & \cdots \\ P_{10} & P_{11} & P_{12} & P_{13} & \cdots \\ P_{20} & P_{21} & P_{22} & P_{23} & \cdots \\ \vdots & \vdots & \vdots & \vdots & \\ P_{i0} & P_{i1} & P_{i2} & P_{i3} & \cdots \\ \vdots & \vdots & \vdots & \vdots & \end{bmatrix},$$

called the *transition probability matrix* of the chain. If the number of states is finite, say n, then there will be n rows and n columns in the matrix P;

otherwise the matrix will be infinite. We must have

$$P_{ij} \geq 0, \quad i, j = 0, 1, 2, \ldots, \tag{4.58}$$

and

$$\sum_{j=0}^{\infty} P_{ij} = 1. \quad i = 0, 1, 2, \ldots. \tag{4.59}$$

Equation (4.58) is true because each P_{ij} is a transition probability. The $(i+1)$st row of P represents the probabilities that the process will make the transition from state i to state j, $0, 1, 2, \ldots$, at the next step. That is, this row gives the probability distribution of X_{n+1} given that $X_n = i$. Thus, (4.59) must hold, since some transition occurs (possibly back to state i).

Example 4.4.1 Consider a sequence of Bernoulli trials in which the probability of success on each trial is p and of failure is q, where $p + q = 1$ and $0 < p < 1$. Let the state of the process at trial n be the number of uninterrupted successes that have been completed at this point (a sequence of such successes is called a success run). Thus, if the first 5 outcomes were SFSSF (a success followed by a failure, two successes, and another failure), we would have $X_0 = 1$, $X_1 = 0$, $X_2 = 1$, $X_3 = 2$, and $X_4 = 0$. The transition probability matrix is given by

$$P = [P_{ij}] = \begin{bmatrix} q & p & 0 & 0 & 0 & \cdots \\ q & 0 & p & 0 & 0 & \cdots \\ q & 0 & 0 & p & 0 & \cdots \\ q & 0 & 0 & 0 & p & \cdots \\ \vdots & & & & & \end{bmatrix}.$$

The state 0 can be reached in one transition from any state while the state $i+1$ can only be reached (in one transition) from state i, $i = 0, 1, 2, \ldots$. This Markov chain is clearly homogeneous in time. $\square$

We now define the n-step transition probabilities $P_{ij}^{(n)}$ by

$$P_{ij}^{(n)} = P[X_n = j | X_0 = i]. \tag{4.60}$$

Since the Markov chain $\{X_n\}$ has stationary transition probabilities, we have

$$P_{ij}^{(n)} = P[X_{m+n} = j | X_m = i] \quad \text{for all } m \geq 0 \text{ and } n > 0. \tag{4.61}$$

It is true, of course, that $P_{ij}^{(1)} = P_{ij}$.

For convenience, we define

$$P_{ij}^{(0)} = \delta_{ij} = \begin{cases} 1 & \text{if } i = j \\ 0 & \text{if } i \neq j. \end{cases}$$

(δ_{ij} is called the Kronecker delta function.) The n-step transition probabilities can be computed using the *Chapman–Kolmogorov equations*, that are

$$P_{ij}^{(n+m)} = \sum_{k=0}^{\infty} P_{ik}^{(n)} P_{kj}^{(m)} \quad \text{for all } n, m, i, j \geq 0. \tag{4.62}$$

In particular, when $m = 0$,

$$P_{ij}^{(n)} = \sum_{k=0}^{\infty} P_{ik}^{(n-1)} P_{kj}, \quad n = 2, 3, \ldots \quad \text{and all } i, j \geq 0. \tag{4.63}$$

To see that the Chapman–Kolmogorov equations (4.62) are true we reason as follows: $P_{ik}^{(n)} P_{kj}^{(m)}$ is the probability that, starting in state i, the process will go to state j in $n + m$ transitions through a path that takes it into state k at the nth transition. Hence, summing over all the intermediate states k provides the probability that the process will be in state j after $n + m$ transitions. That is, we have

$$
\begin{aligned}
P_{ij}^{(n+m)} &= P[X_{n+m} = j | X_0 = i] \\
&= \sum_{k=0}^{\infty} P[X_{n+m} = j, X_n = k | X_0 = i] \\
&= \sum_{k=0}^{\infty} P[X_{n+m} = j | X_n = k, X_0 = i] P[X_n = k | X_0 = i] \\
&= \sum_{k=0}^{\infty} P_{ik}^{(n)} P_{kj}^{(m)}. \tag{4.64}
\end{aligned}
$$

If we denote the matrix of n-step transition probabilities, $P_{ij}^{(n)}$, by $P^{(n)}$, then (4.63) shows that $P^{(n)} = P^{(n-1)} P = P^{(n)}$, that is, the matrix product of $P^{(n-1)}$ by P. Hence, $P^{(n)}$ can be calculated as the nth power of the matrix P.

Example 4.4.2 Consider a communication system that transmits the digits 0 and 1 through several stages. At each stage the probability that the same digit will be received by the next stage, as transmitted, is 0.75. What is the probability that a 0 that is entered at the first stage is received as a 0 by the fifth stage?

Solution We want to find $P_{00}^{(5)}$. The transition probability matrix P is given by

$$P = \begin{bmatrix} 0.75 & 0.25 \\ 0.25 & 0.75 \end{bmatrix}.$$

Hence,

$$P^{(2)} = \begin{bmatrix} 0.625 & 0.375 \\ 0.375 & 0.625 \end{bmatrix}, \text{ and } P^{(4)} = P^{(2)}P^{(2)} = \begin{bmatrix} 0.53125 & 0.46875 \\ 0.46875 & 0.53125 \end{bmatrix}.$$

Thus,

$$P^{(5)} = P^{(1)}P^{(4)} = \begin{bmatrix} 0.515625 & 0.484375 \\ 0.484375 & 0.515625 \end{bmatrix}.$$

Therefore, the probability that a zero will be transmitted through five stages as a zero is

$$P_{00}^{(5)} = 0.515625. \quad \Box$$

We have been dealing only with conditional probabilities up to now. For example, P_{ij}^n is the probability that the state at time n is j, given that the state at time 0 is i. To calculate the unconditional distribution of the state at time n, we must specify the initial probability distribution of the states $0, 1, 2, \ldots$. That is, we must provide

$$P[X_0 = i] = \pi_i, \quad \text{for } i \geq 0,$$

where, of course,

$$\sum_{i=0}^{\infty} \pi_i = 1.$$

Then we can calculate, for any j,

$$\begin{aligned} P[X_n = j] &= \sum_{i=0}^{\infty} P[X_n = j | X_0 = i] P[X_0 = i] \\ &= \sum_{i=0}^{\infty} P_{ij}^{(n)} \pi_i. \end{aligned}$$

The last sum can be obtained by multiplying the row vector $\pi = (\pi_0, \pi_1, \ldots)$ by the jth column of P using matrix multiplication. Therefore, the probability distribution of X_n is given by

$$\beta = (\beta_0, \beta_1, \ldots) = \pi P,$$

where

$$\beta_i = P[X_n = i] \quad \text{for } i \geq 0.$$

For example, if the initial distribution of 0's and 1's in Example 4.4.2 is given by $\pi_0 = P[X_0 = 0] = 0.75$ and $\pi_1 = P[X_0 = 1] = 0.25$, then $\beta_0 = P[X_5 = 0] = 0.5078125$ and $\beta_1 = P[X_5 = 1] = 0.4921875$.

State j of a Markov chain $\{X_n\}$ is said to be *reachable* from state i if it is possible for the chain to proceed from state i to state j in a finite number of transitions, that is, if $P_{ij}^{(n)} > 0$ for some $n \geq 0$. If every state is reachable from every other state, the chain is said to be *irreducible*. We shall be concerned primarily with irreducible Markov chains.

Example 4.4.3 The Markov chain of Example 4.4.1 is irreducible because there is a one-step transition from any state i to state 0, while if $j > 0$ there is clearly a j-step transition from state 0 to state j. $\square$

Example 4.4.4 Suppose a Markov chain with three states has the probability transition matrix

$$P = \begin{bmatrix} 0 & 1 & 0 \\ 0.5 & 0 & 0.5 \\ 1 & 0 & 0 \end{bmatrix}.$$

Determine whether or not this Markov chain is irreducible.[3]

Solution The answer is not obvious from looking at P, so let us take some powers of P. We see that

$$P^{(2)} = \begin{bmatrix} 0.5 & 0 & 0.5 \\ 0.5 & 0.5 & 0 \\ 0 & 1 & 0 \end{bmatrix}, \quad P^{(3)} = \begin{bmatrix} 0.5 & 0.5 & 0 \\ 0.25 & 0.5 & 0.25 \\ 0.5 & 0 & 0.5 \end{bmatrix},$$

while

$$P^{(4)} = \begin{bmatrix} 0.25 & 0.5 & 0.25 \\ 0.5 & 0.25 & 0.25 \\ 0.5 & 0.5 & 0 \end{bmatrix}, \quad \text{and} \quad P^{(5)} = \begin{bmatrix} 0.5 & 0.25 & 0.25 \\ 0.375 & 0.5 & 0.125 \\ 0.25 & 0.5 & 0.25 \end{bmatrix}.$$

$P^{(5)}$ is a *positive matrix*; that is, each of its elements is positive. Therefore, a transition can be made between any two states in five steps, so the Markov chain is irreducible. $\square$

The *period* of state i is the greatest common divisor of the set of all positive integers n such that $P_{ii}^{(n)} > 0$. [If $P_{ii}^{(n)} = 0$ for all $n \geq 1$ we define $d(i) = 0$.] If $d(i) > 1$ state i is said to be *periodic*, while if $d = 1$, then i is an *aperiodic state*. A state i for which $P_{ii} > 0$ has period 1, since the system can remain in state i indefinitely. Hence, it is an aperiodic state. A Markov chain is aperiodic if every state has period 1 (is aperiodic).

---

[3]This example was provided by Peter F. Brown.

Example 4.4.5 Consider the Markov chain whose transition probability matrix is

$$P = \begin{bmatrix} 0 & 1 \\ 1 & 0 \end{bmatrix}.$$

Then

$$P^{(2)} = \begin{bmatrix} 1 & 0 \\ 0 & 1 \end{bmatrix}.$$

In fact, every even power of P is the identity matrix and every odd power is P itself. Thus, every state is periodic with period 2. □

Two states i and j are said to *communicate* if i is reachable from j and j is reachable from i; that is, there exist integers m and n such that $P_{ji}^{(m)} > 0$ and $P_{ij}^{(n)} > 0$. We indicate that states i and j communicate by writing $i \leftrightarrow j$. An important theorem about the states of a Markov chain follows.

Theorem 4.4.1 *Let $\{X_n\}$ be a Markov chain and $i \leftrightarrow j$ mean that states i and j communicate. Then the relation $i \leftrightarrow j$ is an equivalence relation. That is, (a) $i \leftrightarrow i$ for each state i, (b) if $i \leftrightarrow j$, then $j \leftrightarrow i$, and (c) if $i \leftrightarrow j$ and $j \leftrightarrow k$, then $i \leftrightarrow k$. Thus, the states of a Markov chain can be partitioned into equivalence classes of states such that two states i and j are in the same class if and only if $i \leftrightarrow j$. A Markov chain is irreducible if and only if there is exactly one equivalence class. Furthermore, if $i \leftrightarrow j$, then i and j have the same period.*

Proof See Parzen [7, pages 208, 209, 262]. ∎

Taylor and Karlin [9, page 147] make the following claims for the properties of the period of a state (in addition to the last claim of Theorem 4.4.1):

1. If state i has period $d(i)$, then there exists an integer N depending on i such that for all integers $n \geq N$,

$$P_{ii}^{(nd(i))} > 0.$$

That is, a return to state i can occur at all sufficiently large multiples of the period.

2. If

$$P_{ji}^{(m)} > 0,$$

then

$$P_{ji}^{(m+nd(i))} > 0$$

for all sufficiently large n (a positive integer) .

For each state i of a Markov chain we define $f_i^{(n)}$ to be the probability that the first return to state i occurs n steps or transitions after leaving i. That is, $f_i^{(n)} = P[X_n = i, X_k \neq i \text{ for } k = 1, 2, \ldots, n-1 | X_0 = i]$. (We define $f^{(0)} = 1$ for all i.) The probability of ever returning to state i is given by $f_i = \sum_{n=1}^{\infty} f_i^{(n)}$. If $f_i < 1$, then i is said to be a *transient state*. If $f_i = 1$, then state i is said to be *recurrent*. If state i is recurrent, we define the *mean recurrence time of i*, that is, the average time to return to state i by

$$m_i = \sum_{n=1}^{\infty} n f_i^{(n)}. \qquad (4.65)$$

If $m_i = \infty$, then state i is said to be *recurrent null*, while, if $m_i < \infty$, then state i is said to be *positive recurrent* or *recurrent nonnull*. Thus, each recurrent state i is either positive recurrent or recurrent null.

The next theorem provides a criterion for the recurrence of a state i.

Theorem 4.4.2 *A state i is recurrent if and only if*

$$\sum_{n=1}^{\infty} P_{ii}^{(n)} = \infty.$$

Equivalently, state i is transient if and only if

$$\sum_{n=1}^{\infty} P_{ii}^{(n)} < \infty.$$

Corollary If $i \leftrightarrow j$ and i is recurrent, then so is j.

Proof See the proofs of Theorem 4.2 and Corollary 4.1 in Taylor and Karlin [9]. ∎

Ross [8, page 141] shows that, with probability 1, a transient state can be visited only a finite number of times, as the name suggests. He shows that a consequence of this fact is that a finite state Markov chain cannot have all transient states—at least one must be recurrent.

Example 4.4.6 (*A One-Dimensional Random Walk*) Consider a Markov chain with state space the set of all integers, positive, negative, and zero. Suppose on each transition the state moves one state to the right with probability p or one state to the left with probability $q = 1 - p$, where $0 < p < 1$. One interpretation is that the Markov chain represents the walk of a drunken man as he staggers backward and forward along a straight line. Another is that it represents the winnings of a gambler who on each play of a game wins or loses a dollar.

It is clear that all states communicate, so, by Theorem 4.4.2, either all states are recurrent or all are transient. Let us consider state 0 and test to see whether $\sum_{n=1}^{\infty} P_{00}^{(n)}$ is finite or infinite.

By the gambling interpretation it is impossible to be even after an odd number of plays, so

$$P_{00}^{(2n+1)} = 0, \quad \text{for } n = 0, 1, 2, \ldots.$$

We would be even after $2n$ trials if and only if we won n and lost n. Hence, we can use the binomial probability to calculate

$$P_{00}^{(2n)} = \binom{2n}{n} p^n q^n = \frac{(2n)!}{n!n!} p^n q^n.$$

We now appeal to Stirling's formula

$$n! \sim n^{n+1/2} e^{-n} \sqrt{2\pi},$$

where

$$a_n \sim b_n$$

means that

$$\lim_{n \to \infty} \frac{a_n}{b_n} = 1.$$

Applying Stirling's formula yields

$$P_{00}^{(2n)} \sim \frac{(4pq)^n}{\sqrt{\pi n}}.$$

It is easy to show (see Exercise 2 of Chapter 2) that $p(1-p) = pq \leq \frac{1}{4}$ with equality holding if and only if $p = q = \frac{1}{2}$. Hence, $\sum_{n=0}^{\infty} P_{00}^{(n)} = \infty$ if and only if $p = \frac{1}{2}$. Therefore, by Theorem 4.4.2, the one-dimensional random walk is recurrent if and only if $p = q = \frac{1}{2}$. Intuitively, we can see that if $p \neq q$, there is a positive probability that if the initial state is the origin, the state of the system will drift to $+\infty$ or $-\infty$ without returning to the origin. $\square$

Theorem 4.4.3 *Let $\{X_n\}$ be an irreducible Markov chain. Then exactly one of the following holds:*

1. *all states are positive recurrent,*

2. *all states are recurrent null,*

3. *all states are transient.*

Proof See Feller [4, page 391]. ∎

Let us define the probability that the discrete Markov chain $\{X_n\}$ is in the state j at the nth step by $\pi_j^{(n)}$. That is,

$$\pi_j^{(n)} = P[X_n = j].$$

Thus, our *initial distribution* of the states $0, 1, 2, \ldots$ is given by

$$\pi_j^{(0)} = P[X_0 = j], \quad j = 0, 1, \ldots.$$

A discrete Markov chain is said to have a *stationary probability distribution* $\pi = (\pi_0, \pi_1, \ldots)$ if the matrix equation $\pi = \pi P$ is satisfied, where each $\pi_i \geq 0$ and $\sum_i \pi_i = 1$. The matrix equation $\pi = \pi P$ can be written as the set of equations

$$\pi_j = \sum_i \pi_i P_{ij}, \quad j = 0, 1, 2, \ldots. \tag{4.66}$$

The reason such a probability distribution π is called a stationary distribution is that if $\pi_j^{(0)} = \pi_j$, $j = 0, 1, \ldots$, for such a distribution, then $\pi_j^{(n)} = \pi_j$ for all n and j; the probabilities $\pi_j^{(n)}$ do not change with time, but are stationary.

A Markov chain is said to have a *long-run* or *limiting* probability distribution $\pi = (\pi_0, \pi_1, \ldots)$ if

$$\lim_{n \to \infty} \pi_j^{(n)} = \lim_{n \to \infty} P[X_n = j] = \pi_j, \quad j = 0, 1, \ldots.$$

A discrete Markov chain that is irreducible, aperiodic, and for which all states are positive recurrent is said to be *ergodic*. The next theorem is important for the application of Markov chains to queueing theory and is stated without proof. (The proof can be found in Feller [4, Section XV.7].)

Theorem 4.4.4 *If $\{X_n\}$ is an irreducible, aperiodic, time homogeneous Markov chain, then the limiting probabilities*

$$\pi_j = \lim_{n \to \infty} \pi_j^{(n)}, \quad j = 0, 1, \ldots$$

always exist and are independent of the initial state probability distribution $\pi^{(0)} = (\pi_0^{(0)}, \pi_1^{(0)}, \ldots)$. If all the states are not positive recurrent (and thus either all states are recurrent null or all are transient), then $\pi_j = 0$ for all j and no stationary probability distribution exists. However, if all the states of $\{X_n\}$ are positive recurrent so that the chain is ergodic, then $\pi_j > 0$ for all j and $\pi = (\pi_0, \pi_1, \ldots)$ forms a stationary probability distribution where

$$\pi_j = \frac{1}{m_j}, \quad j = 0, 1, \ldots. \tag{4.67}$$

In this latter case, the limiting distribution is the unique solution of the set of equations

$$\sum_i \pi_i = 1 \qquad (4.68)$$

$$\pi_j = \sum_i \pi_i P_{ij}, \quad j = 0, 1, 2, \ldots. \ \blacksquare \qquad (4.69)$$

Note that part of the conclusion is that for an ergodic Markov chain the stationary probability distribution and the long-run (limiting) probability distribution are the same. Such probability distributions are also called *equilibrium* or *steady state* distributions. In Markov chain applications to queueing theory, these are the distributions of most interest. Thus, it is important to know under what conditions a Markov chain is ergodic and thus has a steady state distribution. The next two theorems provide some answers.

Theorem 4.4.5 *The irreducible, aperiodic Markov chain $\{X_n\}$ is positive recurrent (and thus ergodic) if there exists a nonnegative solution of the system*

$$\sum_j P_{ij} x_j \leq x_i - 1 \quad (i \neq 0), \qquad (4.70)$$

such that

$$\sum_j P_{0j} x_j < \infty. \qquad (4.71)$$

Theorem 4.4.6 *The irreducible, aperiodic Markov chain $\{X_n\}$ is positive recurrent (and thus ergodic) if and only if there exists a nonnull solution of the equations*

$$\sum_j x_j P_{ji} = x_i, \quad i = 0, 1, 2, \ldots, \qquad (4.72)$$

such that

$$\sum_j |x_j| < \infty. \qquad (4.73)$$

Proof The proofs of the last two theorems are given in Parzen [7, Chapter 6]. ■

It is important to note that a limiting (steady state) distribution is also a stationary distribution but the converse is not true. There may be a stationary distribution but not a limiting distribution. For example, consider the Markov chain of Example 4.5.5 with transition probability matrix

$$P = \begin{bmatrix} 0 & 1 \\ 1 & 0 \end{bmatrix}.$$

Then $\pi = (\frac{1}{2}, \frac{1}{2})$ is a stationary distribution, since $\pi P = \pi$, but there is no limiting distribution.

Parzen [7, Chapter 6] shows that if $\pi = (\pi_0, \pi_1, \ldots)$ is the limiting and thus stationary distribution of an irreducible, positive recurrent Markov chain then for each j, the limiting probability that the process will be in state j at time n also equals the long-run proportion of time that the process will be in state j.

The next theorem shows that the behavior of finite-state Markov chains is somewhat simpler than that of those with infinitely many states.

Theorem 4.4.7 *A finite-state Markov chain $\{X_n\}$ that is irreducible and aperiodic is ergodic.*

Proof The theorem follows immediately from Theorem 4 of Feller [4, page 392]. ∎

It follows from Theorem 4.4.4 that such a Markov chain has a steady state (limiting) probability distribution.

Example 4.4.7 Consider again the Markov chain of Example 4.4.2 with state-transition matrix

$$P = \begin{bmatrix} 0.75 & 0.25 \\ 0.25 & 0.75 \end{bmatrix}.$$

It is clear that this Markov chain is irreducible and aperiodic. Hence, by Theorem 4.4.7 it is also ergodic and thus has a limiting probability distribution, that is also a stationary distribution. We can apply Theorem 4.4.4 to calculate the equilibrium probability distribution $\pi = (\pi_0, \pi_1)$.

We have the equations

$$\pi_0 + \pi_1 = 1, \quad \pi_0 = 0.75\pi_0 + 0.25\pi_1, \quad \pi_1 = 0.25\pi_0 + 0.75\pi_1.$$

The unique solution of these equations is $\pi_0 = 0.5$, $\pi_1 = 0.5$. This means that if data are passed through a large number of stages, the output is independent of the original input and each digit received is equally likely to be a 0 or a 1. This also means that

$$\lim_{n \to \infty} P^n = \begin{bmatrix} 0.5 & 0.5 \\ 0.5 & 0.5 \end{bmatrix}.$$

(We saw, in Example 4.4.2, that

$$P^4 = \begin{bmatrix} 0.53125 & 0.46875 \\ 0.46875 & 0.53125 \end{bmatrix},$$

so that

$$P^8 = \begin{bmatrix} 0.501953125 & 0.498046875 \\ 0.498046875 & 0.501953125 \end{bmatrix},$$

and the convergence is rapid.)

Note also that

$$\pi P = (0.5, 0.5) = \pi,$$

so π is a stationary distribution as claimed. $\square$

For completeness we list another theorem on recurrence, although it is not as useful as the preceding three theorems.

Theorem 4.4.8 *The irreducible Markov chain* $\{X_n\}$ *is recurrent if there exists a sequence* $\{y_i\}$ *such that*

$$\sum_j P_{ij} y_j \le y_i \quad \text{for } i \ne 0, \tag{4.74}$$

with

$$\lim_{i \to \infty} y_i = \infty.$$

Proof See Parzen [7, Chapter 6]. ■

Note that the recurrent chain could be recurrent null.

Sometimes we are interested in showing that a particular Markov chain is *not* recurrent but transient. The next theorem gives a necessary and sufficient condition for this.

Theorem 4.4.9 *An irreducible Markov chain is transient if and only if there exists a bounded nonconstant solution of the equations*

$$\sum_j P_{ij} y_j = y_i, \quad i \ne 0. \tag{4.75}$$

Proof See Parzen [7, Chapter 6]. ■

We now consider several examples of Markov chains that are part of the folklore. (Example 4.4.6 was one of these, too.)

Example 4.4.8 (*Gambler's Ruin*) Suppose a gambler has probability p of winning one unit on each play of a game and probability $q = 1 - p$ of losing one unit. We assume that successive plays of the game are independent. The problem is to determine the probability that if the gambler starts with i units, her fortune will reach N before it reaches 0.

If we let X_n denote the player's fortune at time n, the process $\{X_n, n = 0, 1, 2, \ldots\}$ is a Markov chain with transition probabilities

$$P_{00} = P_{NN} = 1,$$

and

$$P_{i,i+1} = p = 1 - P_{i,i-1}, \quad i = 1, 2, \ldots, N - 1.$$

Therefore, the transition probability matrix is

$$P = \begin{bmatrix} 1 & 0 & 0 & 0 & \cdots & 0 & 0 \\ q & 0 & p & 0 & \cdots & 0 & 0 \\ 0 & q & 0 & p & \cdots & 0 & 0 \\ \vdots & \vdots & \vdots & \vdots & \vdots & \vdots & \vdots \\ 0 & 0 & 0 & 0 & \cdots & 0 & p \\ 0 & 0 & 0 & 0 & \cdots & 0 & 1 \end{bmatrix}.$$

We have the three classes $\{0\}$, $\{1, 2, \ldots, N - 1\}$, and $\{n\}$. It is possible to reach the first or third class from the second class, but it is not possible to return. Reaching the first class corresponds to losing and reaching the third class means winning. The first and third classes are recurrent and the second class is transient. Each transient state $1, 2, \ldots, N - 1$ has period 2. As we noted after Theorem 4.4.2, a transient state of a Markov process is visited only finitely many times (with probability 1). Hence, after a finite amount of time, the gambler witl either reach her goal of N units or lose all her money; it is impossible for the game to continue indefinitely.[4]

Let $P_i, i = 0, 1, \ldots, N$ denote the probability that, starting with i units, the gambler's fortune will grow to N before she loses all her resources. Ross [8, Example 5a, page 155] shows that

$$P_i = \begin{cases} \dfrac{1 - (q/p)^i}{1 - (q/p)^N}, & \text{if } p \neq \dfrac{1}{2} \\[3mm] \dfrac{i}{N}, & \text{if } p = \dfrac{1}{2}. \end{cases}$$

---

[4] Actually, this is technically possible but the probability of it happening is 0.

If we let $N \to \infty$ in the above formulas, we obtain

$$
P_i \to
\begin{cases}
1 - \left(\dfrac{q}{p}\right)^i, & \text{if } p > \dfrac{1}{2} \\[3mm]
0, & \text{if } p \le \dfrac{1}{2}.
\end{cases}
$$

This result agrees with intuition. It says that if $p > \frac{1}{2}$, then it is possible that the gambler's fortune can grow without limit. However, if $p \le \frac{1}{2}$, then the gambler is sure to lose; that is, she will lose all her money with probability 1 against an infinitely wealthy opponent.

Note that this game can be viewed as a game between two players, say player A and Player B. Then we assume that the total wealth of the two players is N units. Initially player A has i units and player B $N - i$ units. The game is over when one of the players has all the money. Player A has it all if state N is reached while player B wins it all if state 0 is reached. $\square$

Example 4.4.9 (*The Ehrenfest Urn Model*) There are several equivalent (or nearly so) definitions of this model. It originated with a mathematical model developed by the physicists P. and T. Ehrenfest [3] to describe diffusion through a membrane. We imagine two urns containing a total of $2N$ balls (molecules). Suppose the first container, which we label A, holds i balls and the second container (urn B) holds the remaining $2N - i$ balls. A ball is selected at random (each ball has the probability $1/2N$ of being selected), removed from its urn, and moved to the other one. (This corresponds to a molecule diffusing at random through the membrane.) Each selection generates a transition of the process. Let $\{X_n, n = 0, 1, \ldots, 2N\}$ be the Markov chain describing the number of balls in urn A at the nth stage. The transition probabilities are given by

$$P_{00} = P_{NN} = 0,$$

$$P_{01} = 1 = P_{N,N-1},$$

$$P_{i,i+1} = \frac{2N - i}{2N} = 1 - P_{i,i-1}, \quad i = 1, 2, \ldots, 2N - 1,$$

$$Pij = 0 \quad \text{otherwise.}$$

Ross [8, pages 167–168] shows there is a steady state distribution π of the balls with

$$\pi_i = \binom{2N}{i}\left(\frac{1}{2}\right)^{2N}, \quad i = 0, 1, 2, \ldots, 2N.$$

By symmetry the number of balls in urn B has the same distribution. $\square$

4.5   Renewal Theory

A Poisson process can be characterized as a counting process for which the times between successive events are independent, identically distributed, exponential random variables. A renewal process is a generalization of the Poisson process.

Definition 4.5.1 Let $\{N(t), t \geq 0\}$ be a counting process and X_1 the time of occurrence of the first event. Let X_n be the random variable that measures the time between the $(n-1)$th and the nth event of this process for $n \geq 2$. If the sequence of nonnegative random variables $\{X_n, n \geq 1\}$ is independent and identically distributed, then $\{N(t), t \geq 0\}$ is a *renewal process*.

The common distribution function is

$$F(x) = P[X_n \leq x], \quad n = 1, 2, 3, \ldots.$$

We will use μ for the common mean and σ^2 for the common variance of the sequence $\{X_n, n \geq 1\}$.

Since we have specified that $\{X_n, n \geq 1\}$ is a collection of positive random variables, $F(0) = 0$.

When an event counted by $N(t)$ occurs, we say that a *renewal* has taken place. The sum

$$W_0 = 0, \quad W_n = X_1 + X_2 + \cdots + X_n, \quad n \geq 1$$

is called the *waiting time* until the nth renewal.

The counting process $\{N(t), t \geq 0\}$ and the process $\{W_n, n \geq 0\}$ are both called the *renewal process*.

The following example is almost universally used as a prototype to describe a renewal process.

Example 4.5.1 (*The Light Bulb Example*) A light bulb is installed at time $W_0 = 0$. When it burns out, at time $W_1 = X_1$, it is replaced by a new bulb, that burns out at time $W_2 = X_1 + X_2$. This process continues indefinitely; as each bulb burns out it is replaced with a fresh bulb. It is assumed that successive bulb lifetimes $\{X_n, n \geq 1\}$ are independent and have the same distribution. That is,

$$P[X_n \leq x] = F(x), \quad n = 1, 2, 3, \ldots.$$

$N(t)$ is the number of light bulb replacements that occur no later than time t and $\{N(t), t \geq 0\}$ is a renewal process. $\square$

Definition 4.5.2 Let $\{N(t), t \geq 0\}$ be a renewal process. Then the function $M(t)$, defined for all $t > 0$ by

$$M(t) = E[N(t)],$$

is the *renewal function* of the renewal process.

Proposition 4.5.1 *Let $\{N(t), t \geq 0\}$ be a renewal process. Then*

$$N(t) \geq k \quad \text{if and only if} \quad W_k \leq t. \tag{4.76}$$

Proof Suppose $N(t) \geq k$. Then there have been at least k renewals by time t so that $W_k \leq t$. Conversely, suppose $W_k \leq t$. This means that at least k renewals have occurred by time t, or $N(t) \geq k$. ∎

Since $W_n = X_1 + X_2 + \cdots + X_n$, it follows from Theorem 2.7.5 that

$$P[W_n \leq x] = F_n(x), \quad n = 1, 2, \ldots,$$

where F_n is the n-fold convolution of F with itself. Therefore, it follows from (4.76) that

$$
\begin{aligned}
P[N(t) \geq n] &= P[W_n \leq t] \\
&= F_n(t) \quad t \geq 0, \quad n = 1, 2, \ldots.
\end{aligned}
\tag{4.77}
$$

From (4.77) we see that

$$
\begin{aligned}
P[N(t) = n] &= P[N(t) \geq n] - P[N(t) \geq n + 1] \tag{4.78} \\
&= F_n(t) - F_{n+1}(t), \quad t \geq 0, \quad n = 1, 2, \ldots. \tag{4.79}
\end{aligned}
$$

Proposition 4.5.2 *Let X be a nonnegative integer valued random variable. Then*

$$E[X] = \sum_{n=0}^{\infty} P[X > k] = \sum_{n=1}^{\infty} P[X \geq n]. \tag{4.80}$$

Proof Let us write p_n in place of $P[X = n]$. Then we can write

$$
\begin{aligned}
E[X] &= 0P_0 + 1P_1 + 2p_2 + 3p_3 + \cdots \\
&= p_1 + p_2 + p_3 + p_4 + \cdots \\
&\quad + p_2 + p_3 + p_4 + \cdots \\
&\quad\quad + p_3 + p_4 + \cdots
\end{aligned}
$$

$$+p_4 + \cdots$$

$$\vdots$$

$$= \ P[X \geq 1] + P[X \geq 2] + P[X \geq 3] + \cdots$$

$$= \ \sum_{n=1}^{\infty} P[X \geq n].$$

∎

We can now use Proposition 4.5.2 to calculate

$$M(t) = E[N(t)] \ = \ \sum_{n=1}^{\infty} P[N(t) \geq n]$$

(4.81)

$$= \ \sum_{n=1}^{\infty} P[W_n \leq t] = \sum_{n=1}^{\infty} F_n(t).$$

Theorem 4.5.1 *Let* $\{N(t), t \geq 0\}$ *be a renewal process with renewal function* $M(t)$. *Then* $M(t)$ *and the* $F(\cdot)$ *can be determined, uniquely, one from the other.*

Proof Let F_n be the n-fold convolution of F with itself. Taking the Laplace–Stieltjes transform of each side of (4.81) yields

$$M^*[\theta] \ = \ \sum_{n=1}^{\infty} F_n^*[\theta] = \sum_{n=1}^{\infty} (F^*[\theta])^n$$

$$= \ F^*[\theta] \sum_{n=0}^{\infty} (F^*[\theta])^n = \frac{F^*[\theta]}{1 - F^*[\theta]}.$$

(4.82)

Solving (4.82) for $F^*[\theta]$ produces

$$F^*[\theta] = \frac{M^*[\theta]}{1 + M^*[\theta]}.$$

(4.83)

Since the Laplace–Stieltjes transform of a distribution is unique, this completes the proof. ∎

Example 4.5.2 (*The Poisson Renewal Process*) Suppose the renewal process $\{N(t), t \geq 0\}$ is Poisson with parameter λ.[5] Then, the process *is a*

---
[5]See Definition 4.2.1.

renewal process by Theorem 4.2.2. By Theorem 4.2.1,

$$P[N(t) = k] = e^{-\lambda} \frac{(\lambda t)^k}{k!}, \quad k = 0, 1, 2, \ldots, \tag{4.84}$$

so that

$$M(t) = \lambda t. \tag{4.85}$$

This result can also be obtained by inverting the transform (4.82). □

By the uniqueness of the renewal function, the Poisson process is the *only* renewal process with a linear renewal function.

Definition 4.5.3 Suppose $\{N(t), t \geq 0\}$ is a renewal process. We define the random variables γ_t (*excess life* or *excess random variable*), δ_t (*current life* or *age random variable*), and β_t (*total life*), respectively, by the formulas

$$\begin{aligned} \gamma_t &= W_{N(t)+1} - t \quad (\textit{excess life}) \\ \delta_t &= t - W_{N(t)} \quad (\textit{current life}) \\ \beta_t &= \gamma_t + \delta_t \quad (\textit{total life}). \end{aligned}$$

Figure 4.5.1 indicates the relationships between the random variables of Definition 4.5.3. If we think of these definitions in terms of Example 4.5.1 and think of t as the current time, then excess life, η_t, is the length of time from time t (now) until the current bulb must be replaced. The current life, δ_t, is the length of time the current bulb has been in use. The total life, β_t, is the total length of time the current bulb will have been in service when it finally burns out.

Proposition 4.5.3 *Let $\{N(t), t \geq 0\}$ be a Poisson process with parameter λ. Then*

$$P[\gamma_t \leq x] = 1 - e^{-\lambda x}, \quad \textit{for } x \geq 0,, \tag{4.86}$$

$$P[\delta_t \leq x] = \begin{cases} 1 - e^{-\lambda x} & \textit{for } 0 \leq x < t \\ 1 & \textit{for } t \leq x, \end{cases} \tag{4.87}$$

and

$$E[\beta_t] = \frac{1}{\lambda} + \frac{1}{\lambda}(1 - e^{-\lambda x}). \tag{4.88}$$

Proof See Taylor and Karlin [9, pages 283–284]. ■

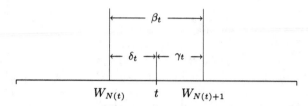

Figure 4.5.1. Definition 4.5.3.

Note that the expected value of the total life, $E[\delta_t]$, is larger than the mean life $E[X_n] = 1/\lambda$ of any renewal interval. In fact, when t becomes large, the mean total life is almost twice as long.

Proposition 4.5.4 *Let $\{N(t), t \geq 0\}$ be a renewal process with $E[X_n] = \mu$ for all n. Then with probability 1,*

$$\frac{N(t)}{t} \rightarrow \frac{1}{\mu} \quad as \quad t \rightarrow \infty. \tag{4.89}$$

Proof See Ross [8, pages 273-274]. ∎

As Ross notes, the above proposition is true when $\mu = \infty$. In this case the $N(t)/t$ approaches the limit 0. The number $1/\mu$ is called the *rate* of the process. The next theorem describes the renewal function for large values of t.

Theorem 4.5.2 (*Elementary Renewal Theorem*) *Let $\{N(t), t \geq 0\}$ be a renewal process with $E[X_n] = \mu$ for all n. Then*

$$\frac{M(t)}{t} \rightarrow \frac{1}{\mu} \quad as \quad t \rightarrow \infty. \tag{4.90}$$

Proof See Grimmett and Stirzaker [6, pages 284–287]. ∎

4.6   Exercises

1. [00] Is a constant function, say $f(x) = c \neq 0$, $o(h)$?

2. [C12] Suppose that in New Zealand, home of the Gala apple, years for these wonderful apples can be described as great, average, or poor. Suppose that following a great year the probabilities of great, average, or poor years are 0.5, 0.3, and 0.2, respectively. Suppose, also, that following an average year the probabilities of great, average, or poor years are 0.2, 0.5, and 0.3, respectively. Finally, suppose that following a poor year the probabilities for great, good, or poor years are 0.2, 0.2, and 0.6, respectively. Assume we can describe the situation from year to year by a Markov chain with the states 0, 1, and 2 corresponding to great, average, and poor years, respectively. Please do the following:

   (a) Set up the transition probability matrix P of this Markov chain.

   (b) Suppose the initial probability for a great year is 0.2, for an average year is 0.5, and for a poor year is 0.3. Calculate the probability distribution after one year and after 5 years.

3. [15] Consider the Markov chain with states 0, 1, 2, 3 with transition probability matrix

$$P = \begin{bmatrix} 0 & 0 & \frac{1}{2} & \frac{1}{2} \\ 1 & 0 & 0 & 0 \\ 0 & 1 & 0 & 0 \\ 0 & 1 & 0 & 0 \end{bmatrix}.$$

Determine which states are transient and which are recurrent.

4. [15] Consider the Markov chain with states 0, 1, 2, 3, 4 with transition probability matrix

$$P = \begin{bmatrix} \frac{3}{4} & \frac{1}{4} & 0 & 0 & 0 \\ \frac{3}{4} & \frac{1}{4} & 0 & 0 & 0 \\ 0 & 0 & \frac{3}{4} & \frac{1}{4} & 0 \\ 0 & 0 & \frac{3}{4} & \frac{1}{4} & 0 \\ \frac{1}{4} & \frac{1}{4} & 0 & 0 & \frac{1}{2} \end{bmatrix}.$$

Determine the classes of this chain and whether each is transient or recurrent.

5. 15] Lucky Lily and Winning William decide to play the following game based on an urn containing nine white balls and eleven black ones.

The play proceeds as follows. A ball is drawn and replaced. If it is white, Lili wins a dollar from Winning. If the ball is black, Winning wins a dollar from Lili. Lili starts with 20 dollars and Winning with 10 dollars. The game continues until one player wins all of the other player's money. What is the probability that Winning wins?

6. [15] Two groups, Group Able and Group Baker, are competing for the same responsibility at Consolidated Craven. Group Able has a head count of 50, that is, is authorized to have 50 people. Group Baker has a head count of 20. Each year one person is taken from one group and given to the other. If the probability that the shift is from Able to Baker is 0.52, show that one group will disappear. Calculate the probability that Group Able will survive.

7. [HM25] Consider Example 4.4.2. Suppose now that the probability that a 0 is received as a 1 is α and the probability a 1 is received as a 0 is β, so that the transition probability matrix P is given by

$$P = \left[\begin{array}{cc} 1 - \alpha & \alpha \\ \beta & 1 - \beta \end{array} \right].$$

(a) Show that $(\pi_0, \pi_1) = (\beta/(\alpha + \beta), \alpha/(\alpha + \beta)$ is a stationary distribution.

(b) Show that $f_0^{(1)} = (1 - \alpha)$ and $f_0^{(n)} = \alpha\beta(1 - \beta)^{n-2}$ for $n = 2, 3, \ldots$.

(c) Calculate the mean recurrence time $m_0 = \sum_{n=1}^{\infty} n f_0^{(n)}$ and verify that $\pi_0 = 1/m_0$.

8. [15] Suppose $\{N(t), t \geq 0\}$ is a renewal process with renewal function $M(t) = 5t$. What is the probability distribution of the number of renewals by time 15?

References

[1] Kai Lai Chung, *Elementary Probability Theory with Stochastic Processes*, 3rd ed., Springer-Verlag, New York, 1979.

[2] Earl A. Coddington, *An Introduction to Ordinary Differential Equations*, Prentice-Hall, Englewood Cliffs, NJ, 1961.

[3] P. and T. Ehrenfest, Über zwei bekannte Einwände gegen das Boltzmannsche H-Theorem, *Physikalische Zeitschrift*, Vol. 8 (1907).

[4] William Feller, *An Introduction to Probability Theory and Its Applications*, Vol. I, 3rd ed. John Wiley, New York, 1968.

[5] I. S. Gradshteyn and I. M. Ryzhik, *Table of Integrals, Series, and Products*, Academic Press, New York, 1980.

[6] Geoffrey R. Grimmett and David R. Stirzaker, *Probability and Random Processes*, Oxford University Press, Oxford, 1983.

[7] Emanuel Parzen, *Stochastic Processes*, Holden-Day, San Francisco, 1962.

[8] Sheldon M. Ross, *Stochastic Processes*, John Wiley, New York, 1983.

[9] Howard M. Taylor and Samuel Karlin, *An Introduction To Stochastic Modeling*, Academic Press, Orlando, 1984.

Carl Friedrich Gauss
Brought down Euclid's house,
Which would have pleased nary
Lambert nor Saccheri.

Karl David
Wells College

Computers are useless. They can only give you answers.
Pablo Picasso

A private railroad car is not an acquired taste. One takes to it
immediately.
Eleanor R. Belmont

Part Two:

Queueing Models[6]

"They also serve who only stand and wait."
– Milton

We would like to relieve our customers from having to "serve" and therefore want to shorten our teller lines. If you, or if you know someone who, might like to start a career in banking as a teller, see our receptionist.

Our present President once served his time as a Cambridge Trust Company teller.

Cambridge Trust Company

---

[6]The advertisement by the Cambridge (Massachusetts) Trust Company for new tellers is used by permission.

Preface to Part Two: Queueing Models

The title of Barrer's article "Queuing with Impatient Customers and Indifferent Clerks,"[7] captures the essence of the feelings of most people toward systems in which queueing for service is necessary. (Who could guess from the title that Barrer's article is concerned with applying antiaircraft fire and guided missiles to defend against attacking aircraft?)

Some students are intimidated by queueing theory because it is often presented as a purely abstract mathematical discipline rather than as a useful tool for computer science, applied mathematics, and engineering. To help students understand the material, we have provided many examples of how the material can be applied to real problems of our everyday world. We have also supplied a great many exercises. We discuss the simple models consisting (mostly) of a single service center with one or more servers (providers of service) in Chapter 5. Many of the models in this chapter can be solved easily with a pocket calculator. Some of the models may require a programmable calculator. For some, a personal computer is very desirable. The formulas for each of the models of Chapter 5 have been collected in Appendix C. APL programs for most of the models are given in Appendix B. *Mathematica* programs are provided in Appendix D for some of the models. Readers with an IBM PC or compatible or an IBM PS/2 may be interested in the *Myriad* modeling software package available from PALLAS International Corporation, San Jose, California.

In Chapter 6 we shall consider networks of queues, that is, queueing systems like those of Chapter 5 that have been connected together to model model computer systems.

---

[7] "Queuing with impatient customers and indifferent clerks," by D. Y. Barrer, *Opns. Res.*, **5**, (1957), 644–649.

Chapter 5

Queueing Theory

5.0   Introduction

One of the most fruitful areas of applied probability theory is that of queueing theory or the study of waiting line phenomena (a queue is a waiting line). Waiting in line (queueing) for service is one of the most unpleasant experiences of life on this planet. Barrer [2] says it all in the title of his paper, "Queueing[1] with Impatient Customers and Indifferent Clerks." Barrer says,

---

[1] I cannot bring myself to spell queueing "queuing" as Barrer did.

In certain queueing processes a potential customer is considered
"lost" if the system is busy at the time service is demanded. The
telephone subscriber hangs up when he gets a busy signal. A
man trying to get a haircut during his lunch hour does not wait
unless a chair is immediately available. Another form of this
general situation is that in which customers wait for service,
but wait for a limited time only. If not served during this time,
the customer leaves the system and is considered lost. Such
situations occur in the processing or merchandising of perish-
able goods. Many types of military engagements are similarly
characterized. An attacking airplane engaged by antiaircraft or
guided missiles is available for "service," i. e., is within range,
for only a limited time.

In spite of the catchy title, which is descriptive of the common feeling
about queues, Barrer's paper is an innovative application of queueing theory
to the destruction of attacking warplanes, not to general queueing theory.

We must join a queue when we want to get cash from an automatic
teller machine (ATM), buy stamps, pay for our groceries, purchase a movie
ticket, obtain a table in a crowded restaurant, etc. Larson [38] discusses
some of the psychological implications of queues. He says,

> Queues involve waiting, to be sure, but one's attitudes to-
> ward queues may be influenced more strongly by other factors.
> For instance, customers may become infuriated if they experi-
> ence *social injustice*, defined as violation of first in, first out.
> *Queueing environment* and *feedback regarding the likely mag-
> nitude of the delay* can also influence customer attitudes and
> ultimately, in many instances, a firm's market share. Even if
> we focus on the wait itself, the "outcome" of the queueing ex-
> perience may vary nonlinearly with the delay, thus reducing the
> importance of average time in queue, the traditional measure
> of queueing performance. This speculative paper uses personal
> experiences, published and unpublished cases, and occasionally
> "the literature" to begin to organize our thoughts on the im-
> portant attributes of queueing.

Larson discusses some techniques that help to make queues more bearable
for humans.

Queues are also common in computer systems. Thus, there are queues
of inquiries waiting to be processed by an interactive computer system,
queues of data base requests, queues of I/O (input/output) requests, etc.

Figure 5.0.1. Elements of a queueing system.

Figure 5.0.1 represents the elements of a basic queueing system, pictorially. We consider a queueing system to be basic if it has only one service facility, although there may be more than one server in the facility. (The reader may note that queueing is spelled "queuing" in some publications (the last e is elided) but I prefer "queueing" because (1) that is the way most queueing theory authorities spell it, and (2) it is a delightful and rare word having five consecutive vowels.)

Customers from a *population* or *source* enter a queueing system to receive some type of service. The word *customer* is used in the generic sense and thus may be an inquiry message requiring transmission and processing, a program requiring *I/O* service, a program in a multiprogramming computer system requiring CPU service, etc. The *service facility* of the queueing system has one or more *servers* (sometimes called *channels*). A server is an entity capable of performing the required service for a customer. If all servers in the service center are busy when a customer enters the queueing system, the customer must join the queue until a server is free.

In any system that can be modeled as a queueing system, there are trade-offs to be considered. If the service facility of the system has such a large capacity that queues rarely form, then the service facility is likely to be idle a large fraction of the time so that unused capacity exists. Conversely, if almost all customers must join a queue (wait for service) and the servers are rarely idle, there may be customer dissatisfaction and possibly lost customers as Barrer [2] noted.

In Table 5.0.1 we list some typical computer queueing systems.

Table 5.0.1. Typical Queueing Systems

Queueing System	Customer	Server(s)
Airline reservation system	Traveler wanting information and/or reservations	Agent plus terminal to a computer reservation system
Interactive inquiry system	Inquiry from terminal	Communication line plus a computer
Interactive order entry system	Order	Communication line plus a computer
DASD (direct access storage device) queueing system	Request for records from DASD	Channel plus control unit and DASD
Message buffering system	Message (incoming or outgoing)	Message buffer(s) (all of them together form the service facility)

Queueing theory, in many cases, enables a designer to ensure that the proper level of service is provided in terms of response time requirements (response time is the sum of customer queueing time and service time) while avoiding excessive cost. The designer can do this by considering several alternative systems and evaluating them by analytic queueing theory models. The future performance of an existing system can be predicted so that upgrading of the system can be done on a timely basis. For example, an analytical model of an interactive system may indicate that the expected load a year in the future will swamp the present system; the model may make it possible to evaluate different alternatives for increased capacity, such as adding more main memory, getting a faster CPU, providing more auxiliary storage, replacing some disk drives by drums, etc. We shall give a number of practical examples of how queueing theory can help one explore the alternatives available in an informed way. For very large computer systems, commercial analytical queueing theory modeling packages exist. Most of these are described in Howard and Butler [23].

In this chapter we discuss the elements of queueing theory and study some basic queueing models that are of great utility in the study of computer systems. These basic models can be used to study subsystems of large computer systems, such as the *I/O* subsystem. In the next chapter we show how some of these basic queueing models can be combined to study more complex systems in which the output of one service center may be the input to another (queues in tandem and networks of queues). In such systems there is a slight "abuse of notation" in which we refer to a "queue" to describe a service center plus the associated queue or queues. The modeling packages we mentioned above model networks of queues.

5.1 Describing a Queueing System

Figure 5.1.1 illustrates the primary random variables in a queueing system.

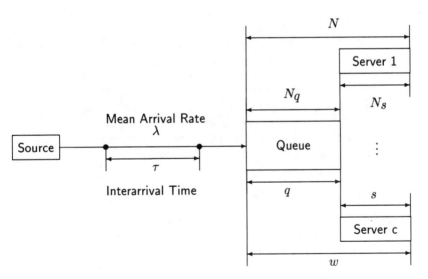

Figure 5.1.1. Queueing theory random variables.

The basic queueing theory definitions and notation are listed in Table 5.1.1. A more complete set of definitions and notation is given in Table 1 of Appendix C.

Table 5.1.1. Basic Queueing Theory Notation and Definitions

c	Number of identical servers.
L	Expected steady state number of customers in the queueing system, $E[N]$.
L_q	Expected steady state number of customers in the queue, $E[N_q]$. Does not include those receiving service, $E[N_s]$.
L_s	Expected steady state number of customers receiving service, $E[N_s]$.
λ	Mean (average) arrival rate of customers to the system.
μ	Mean (average) service rate per server, that is, the mean rate of service completions while the server is busy.
$N[t]$	Random variable describing the number of customers in the system at time t.
N	Random variable describing the steady state number of customers in the system.
$N_q[t]$	Random variable describing the number of customers in the queue at time t.
N_q	Random variable describing the steady state number of customers in the queue.
$N_s[t]$	Random variable describing the number of customers receiving service at time t.
N_s	Random variable describing the steady state number of customers in the service facility.
$p_n[t]$	Probability there are n customers in the system at time t.
p_n	Steady state probability there are n customers in the system.
q	Random variable describing the time a customer spends in the queue (waiting line) before service begins.
ρ	Server utilization $= \dfrac{\lambda}{c\mu} = \dfrac{E[N_s]}{c}$.
s	Random variable describing the service time. $E[s] = \frac{1}{\mu}$.
τ	Random variable describing interarrival time. $E[\tau] = \frac{1}{\lambda}$.
w	Random variable describing the total time a customer spends in the queueing system. $w = q + s$.
W	Expected (mean or average) steady state time a customer spends in the system. $W = E[w] = W_q + W_s$.
W_q	Expected (mean or average) steady state time a customer spends in the queue. $W_q = E[q] = W - W_s$.
W_s	Expected (mean or average) customer service time, $E[s]$.

There are some obvious relations between some of the random variables shown in Figure 5.0.1. With respect to the number of customers in the queueing system, we must have

$$N[t] = N_q[t] + N_s[t], \qquad (5.1)$$

$$N = N_q + N_s. \qquad (5.2)$$

In (5.2) we assume the queueing system has reached the *steady state*, that we now describe. When a queueing system is first put into operation and for some time afterwards, the number of customers in the queue and in service depends strongly on both the initial conditions (such as the number of customers queued up waiting for the facility to begin operation) and on how long the system has been in operation (the time parameter t). After the system has been in operation for some time, however, the influences of the initial conditions have "damped out" and state of the system is independent of time—the system is in equilibrium or the steady state. However, N, N_q, and N_s *are* random variables; that is, they are not constant but have probability distributions.

Equation (5.2) of course implies that

$$E[N] = E[N_q] + E[N_s], \qquad (5.3)$$

that is often written

$$L = L_q + L_s. \qquad (5.4)$$

There are some obvious relationships between the random variables describing time, also; clearly the total time in the queueing system for a customer is the sum of the queueing time (time spent in line waiting for service) and service time; that is,

$$w = q + s, \qquad (5.5)$$

and

$$E[w] = E[q] + E[s]. \qquad (5.6)$$

Equation (5.6) is often written

$$W = W_q + W_s. \qquad (5.7)$$

Some common English words have a special meaning in queueing theory. A customer who refuses to enter a queueing system because the queue is too long is said to be *balking* while one who leaves the queue without receiving service because of excessive queueing time is said to have *reneged*.

Customers may *jockey* from one system to another with a shorter queue for service.

In order to describe a queueing system, analytically, a number of elements of the system must be known. We consider the most important of these below.

Population or Source

The primary characterization of the population or source of potential customers is whether it is finite or infinite. An infinite source system is easier to describe mathematically than one with a finite source. The reason for this is that, in a finite source system, the number of customers in the system affects the arrival rate; indeed, if every potential customer is already in the system, the arrival rate drops to zero. For infinite population systems the number of customers in the system has no effect on the arrival pattern. If the customer population is finite but large, we sometimes assume an infinite source to simplify the mathematics.

Arrival Pattern

The ability of a queueing system to provide service for an arriving stream of customers depends not only on the mean arrival rate, λ, but also on the pattern in which they arrive. Thus, if customer arrivals are evenly spaced in time, say every h time units, the service facility can provide better service than if customers arrive in clusters. (The extreme case of clustering in which a number of customers arrive simultaneously is called *bulk arrivals*.) We assume customer arrivals at times

$$0 \leq t_0 < t_1 < t_2 < \cdots < t_n < \cdots \tag{5.8}$$

(We always assume that observation of a queueing system begins at time $t = 0$.) The random variables $\tau_k = t_k - t_{k-1}$, $(k = 1, 2, 3, \ldots)$, are called *interarrival times*. We usually assume that $\tau_1, \tau_2, \ldots$ is a sequence of independent identically distributed random variables and use the symbol τ for an arbitrary interarrival time. The usual method of specifying the arrival pattern is to give the distribution function, $A[\cdot]$, of the interarrival time. The arrival pattern most commonly assumed for applied queueing models (because of its pleasant mathematical properties) is the exponential pattern: $A[t] = 1 - e^{-\lambda t}$, where λ is the average arrival rate. (If you are rusty on the exponential distribution, you should review Section 3.2.2. The exponential distribution is of paramount importance in queueing theory.)

Because of the properties of the exponential distribution, summarized in Theorem 3.2.1, if the interarrival time of customers to a queueing system has an exponential distribution, the arrival pattern is called a *Poisson arrival pattern* (or *process*) or a *random arrival pattern* (or sometimes just said to be *random*). Other commonly assumed arrival patterns are constant, Erlang-k, and hyperexponential.

The symbol λ is reserved (except for finite queue systems and loss systems) for the mean or average rate into the system; therefore, the average interarrival time, $E[\tau]$, equals $1/\lambda$. (For finite queue systems (Section 5.2.2) and loss systems (examined in Section 5.2.4), λ_a is used for average arrival rate into the system.)

Service Time Distribution

The exponential distribution is often used to describe the service time of a server because of the Markov or "memoryless" property of this distribution (Theorem 3.2.1(d)). Thus, if the service time is exponential, the expected time remaining to complete a customer service is independent of the service already provided. Suppose now that the queueing system has several identical servers, each with an exponential service time with parameter μ, and that n of the servers are now busy. Let T_i be the remaining service time for server i $(i = 1, 2, \ldots, n)$. By the Markov property, each T_i has an exponential distribution with parameter μ. T, the time until the next service completion, is the minimum of $\{T_1, T_2, \ldots, T_n\}$. Hence, by Theorem 3.2.1(h), T has an exponential distribution with parameter $n\mu$; the service facility is performing like a single exponential server with mean service rate $n\mu$. In queueing theory exponentially distributed service time is called *random service* and the distribution function, $W_S[\cdot]$, is given by

$$W_S[t] = P[s \le t] = 1 - e^{-\mu t} = 1 - e^{-t/W_S}. \tag{5.9}$$

Here μ is called the *average service rate*. The average (mean) service rate, $\mu = 1/W_s$, is the average rate at which a server processes customers when the server is busy. This definition is valid for all service time distributions. Other common service time distributions are Erlang-k, constant, and hyperexponential. The hyperexponential distribution is useful to describe a service time distribution with a large variance relative to the mean (see Section 3.2.9).

The *squared coefficient of variation*, C_X^2, defined for a random variable X with $E[X] \neq 0$ by

$$C_X^2 = \frac{\text{Var}[X]}{E[X]^2}, \tag{5.10}$$

is a useful parameter to measure the character of probability distributions used to represent service time or interarrival time.[2] If X is constant,

---

[2]We have discussed C_X^2 for some special distributions in Chapter 2. It is often used by queueing theory specialists to compare distributions to the exponential distribution.

$C_X^2 = 0$; if X has an exponential distribution, then $C_X^2 = 1$; if X has an Erlang-k distribution, then $C_X^2 = 1/k$; and, if X has a k-stage hyperexponential distribution, then $C_X^2 \geq 1$. (For example, if X has a two-stage hyperexponential distribution with $q_1 = 0.4$, $q_2 = 0.6$, $\mu_1 = 0.5$, and $\mu_2 = 0.01$, then $E[X] = 60.8$, $E[X^2] = 12,003.2$, $\text{Var}[X] = 8,306.56$, and $C_X^2 = 2.25$.) We conclude that, for C_s^2 close to zero, the service time is almost constant; if C_s^2 is close to one, the service time is approximately exponential; if C_s^2 is close to $1/k$ for some positive integer k, then s can be approximated by an Erlang-k distribution; and, finally, if $C_s^2 > 1$ then s has a great deal of variability and can be approximated by a two-stage hyperexponential distribution. Similarly if C_τ^2 is close to zero, the arrival process has a regular pattern; if C_τ^2 is close to one, the arrival pattern is nearly random; if C_τ^2 is close to $1/k$ for some positive integer k, then τ can be approximated by an Erlang-k distribution; while if $C_\tau^2 > 1$, then the arrivals tend to cluster.

Maximum Queueing System Capacity

In some systems the queue capacity is assumed to be infinite; that is, every arriving customer is allowed to wait until service can be provided. Other queueing systems, called "loss systems," have zero queue capacity; thus, if a customer arrives when the service facility is fully utilized (all the servers are busy), the customer is turned away. For example, some dial-up telephone systems are loss systems. Still other queueing systems, such as a message buffering system, have a positive but not infinite capacity queue. We use K to represent the maximum number of customers allowed in such a system.

Number of Servers

The simplest queueing system, in this sense, is the *single-server system* that can serve only one customer at a time. A *multiserver system* has c (usually) identical servers and can provide service to as many as c customers simultaneously. In an *infinite server system* each arriving customer is immediately provided with a server. Although there cannot actually be infinitely many servers in any system, there are queueing systems that have sufficient servers that they *appear* to have infinitely many.

Queue Discipline (Service Discipline)

This is the rule for selecting the next customer to receive service. The most common queue discipline is "first-come, first-served," abbreviated as FCFS or sometime called "first-in, first-out" and abbreviated FIFO. Other common queue disciplines include "last-come, first- served," LCFS (or "last-in, first-out," LIFO); "random-selection-for-service," RSS (or "service-in-random order," SIRO), that means that each customer in the queue has the same probability of being selected for service; or "priority service," PRI. Priority service means that some customers get preferential treatment, just as in George Orwell's *Animal Farm* some animals (the pigs) were "more equal" than others. In a priority queueing system, customers are divided into priority classes with preferential treatment afforded by class. We study priority queueing systems in Section 5.4.

A special notation, called the Kendall notation, after David Kendall [26], its originator, has been developed to describe queueing systems. The notation has the form $A/B/c/K/m/Z$; where A describes the interarrival time distribution, B the service time distribution, c the number of servers, K the system capacity (maximum number of customers allowed in the system), m the number in the population or source, and Z the queue discipline. Usually the shorter notation $A/B/c$ is used, and it is assumed that there is no limit to the length of the queue, the customer source is infinite, and the queue discipline is FCFS. The symbols chosen by Kendall and traditionally used for A and B are

GI	general independent interarrival time
G	general service time
H_k	k-stage hyperexponential interarrival or service time distribution
E_k	Erlang-k interarrival or service time distribution
M	exponential interarrival or service time distribution
D	deterministic (constant) interarrival or service time distribution[3]
U	uniform interarrival or service time distribution

When we say a queueing model, such as M/G/1, has a general *service time distribution*, we mean the equations of the model are valid for general service time distributions (make few assumptions about the service time distribution) and thus, in particular, the equations are valid for the M/M/1 system. However, equations developed specifically to describe an M/M/1 queueing system would give more information than the general equations developed for the M/G/1 model and applied, as a special case, to M/M/1. Similar remarks apply to the phrase *general independent interarrival time distribution*.

An example of the full Kendall notation is $M/E_4/3/20/\infty/SIRO$. For this system the interarrival time is exponential, the service time is Erlang-4 for each of the three servers, the maximum system capacity is 20 (3 in service and 17 in the queue), the source is infinite, and the queue discipline is *service in random order.*

---

[3]Kendall used the word *deterministic* to describe D but *degenerate* is more descriptive.

As the Kendall notation suggests, certain properties of a queueing system are assumed known; it is desired to calculate measures of performance of the queueing system from these known parameters. It is usually assumed that the average arrival rate λ (or, equivalently, the average interarrival time, $E[\tau]$) and the average service rate per server μ (or the average service time per server, W_s) are known. It is also assumed that the arrival and service time distributions are known. (For the M/G/1 model, only some of the moments need to be known to compute useful performance information.) One fundamental measure of queueing system performance is the *traffic intensity*[4] $a = W_s/E[\tau]$, also known as the *offered load*. It should be noted that W_s is the average service time per server, while $E[\tau]$ is the average interarrival time for all customers entering the queueing system and not just for the customers who are serviced by a particular server (unless, of course, there is but one server). Since $\lambda = 1/E[\tau]$ and $\mu = 1/W_s$, the traffic intensity can also be written as λW_s or λ/μ. The quantity $\rho = a/c = \lambda/(c\mu)$ is called the *server utilization* because it represents the average fraction of the time that each server is busy (assuming the traffic is evenly distributed to the servers); that is, it is the probability that a given server is busy (as observed by an outside observer).

Example 5.1.1 Consider a D/D/1 queueing system with a constant interarrival time of 20 seconds and a constant service time of 10 seconds. Then the server is busy half of the time, since $\rho = a = 10/20 = 0.5$. If the server is replaced by one that requires exactly 15 seconds to service a customer, then $\rho = 15/20 = 0.75$ and this server is busy three-fourths of the time. Replacing this server by one requiring exactly 30 seconds to service a customer may save some money but the traffic intensity $a = 30/20 = 1.5$. In order to keep up the server must provide 30 seconds of service every 20 seconds! This is impossible. Two servers must be provided. Thus, the traffic intensity a is a measure of the required number of servers and ρ (when it is less than one) is a measure of congestion. In general we can argue that if customers are arriving at the rate λ and the c servers serve them at the rate $c\mu$, then we must have $\lambda < c\mu$ if the servers are to keep up. But this means that $a = \lambda/\mu < c$. $\square$

Although server utilization, ρ, is a measure of congestion, there are some other useful measures of queueing system performance including the following steady state values:

---

[4]Although traffic intensity is dimensionless, it is often referred to in units of erlangs in honor of the queueing theory pioneer Agner Krarup Erlang.

W	average customer time in the system (queueing for and in service)
W_q	average customer queueing time
$\pi_w[90]$	90th percentile value of w
$\pi_q[90]$	90th percentile value of q
L	average number of customers in system
L_q	average number of customers in the queue
p_n	probability there are n customers in the queueing system

Little's Law

One of the foundations of queueing theory is the formula

$$L = \lambda W. \tag{5.11}$$

The formula (5.11) applies to any system in equilibrium in which customers arrive, spend a certain amount of time, and then depart. In (5.11) we assume that λ is the average arrival rate, W is the average time a customer spends in the system, and L is the average number of customers in the system. The formula goes by a number of names, including "Little's law", "Little's formula", and "Little's theorem." It was first proven by John D. C. Little [42] in the context of a steady state queueing system in which L, λ, and W have the queueing theory definitions. However, (5.11) holds in more general situations that need not have anything to do with queues. Although Little's law is easy to state and intuitively reasonable, the proof is difficult. Little [42] provided the first known proof. However, it is a rather formal, nonintuitive proof using the mathematical concept of metric transitivity. Stidham [56] has published a simpler proof that is quite general and more intuitive than Little's proof. We state Stidham's version of Little's theorem without proof.

Theorem 5.1.1 (*Little's Theorem According to Stidham*) *Let $L(x)$ be the number of customers present at time x. Define L by*

$$L = \lim_{t \to \infty} \frac{1}{t} \int_0^t L(x)dx. \tag{5.12}$$

Define λ by

$$\lambda = \lim_{t \to \infty} \frac{N(t)}{t}, \tag{5.13}$$

where $N(t)$ is the number of customers who arrive in the interval $[0, t]$. Let W_i be the time in the system for the ith customer and define the mean time *in the system* W *by*

$$W = \lim_{n \to \infty} \frac{1}{n} \sum_{i=1}^{n} W_i. \tag{5.14}$$

If λ and W exist and are finite, then so does L, and

$$L = \lambda W. \tag{5.15}$$

Proof See Stidham [56]. ∎

In the following example we show some simple applications of Little's law.[5]

Example 5.1.2 Little's law can be applied to the queue, itself, to prove that

$$L_q = \lambda W_q, \tag{5.16}$$

and to the service center, alone, to prove that

$$L_s = \lambda W_s = a, \tag{5.17}$$

for any number of servers. If λ and W_s are known for any steady state queueing system, then Little's law allows us to calculate all of the primary performance measures L, L_q, W, and W_q, if any one of them is known. For example, if W is known then

$$L = \lambda W, \qquad W_q = W - W_s, \qquad \text{and} \qquad L_q = \lambda W_q. \quad \square$$

Many phenomena encountered in queueing theory are not intuitive. The following example is one such case.

Example 5.1.3   (*A Queueing Theory Paradox*)   Taxis pass a certain corner with an average interarrival time of 20 seconds. What is the average time that one would expect to wait for a taxi? (Assume that you are in New York City so you can't telephone for a taxi.)

Solution Intuitively, it would seem that a taxi is just as likely to arrive at one point in time between arrivals as any other; that is, by symmetry, the distribution of arrival time should be uniform on the interval from 0 to 20 seconds. Thus, the average waiting time should be 10 seconds. This is true, however, only when the taxis arrive exactly 20 seconds apart. In fact, as is shown by Takács [59, page 10], if w is the length of time until the next

---

[5]I asked John D. C. Little which of the appellations Little's law, Little's formula, or Little's theorem he preferred. Little said that he had no preference; he just hoped his name would be spelled correctly.

arrival of a taxi measured from the time of arrival of the person seeking a cab, then

$$E[w] = \frac{E[\tau]}{2} \left\{ 1 + C_\tau^2 \right\}, \tag{5.18}$$

where τ is the taxi interarrival time. Thus, if the interarrival time is exponential one would expect to wait 20 seconds, on the average, for a taxi. It follows from Proposition 4.5.3 that, if τ is exponential, so is w with the same parameters as τ. If the interarrival time is hyperexponential with $C_\tau^2 = 3$. one would expect to wait 40 seconds, on the average! This well-known "waiting time paradox" is discussed by Feller [16, pages 11, 23] (he uses a bus in his example rather than a taxi). Snell [55] also calls it the "bus paradox" and provides a BASIC program to simulate the waiting time for exponential τ and to draw a graph of the result. □

The above example highlights the fact that in queueing theory, intuition is often misleading. One *can* get an appropriate intuitive picture of what is happening in this example by thinking of the taxi arrivals as being appropriately scattered along the time axis and realizing that a randomly chosen point on this axis is more likely to fall in a long interval between two arrivals than in a short one. Also, the larger C_τ^2 is, the more clustered the arrivals are. If arrivals are clustered, then there must be some very large gaps between some of the arrivals to make up for the short interarrival times in the clusters.

5.2   Birth-and-Death Process Models

A number of important queueing theory models fit the birth-and-death process description of Section 4.3. A queueing system based on this process is in state E_n at time t if the number of customers in the system is n, that is, if $N[t] = n$. A birth is a customer arrival and a death occurs when a customer leaves the system after completing service. We consider only steady state solutions to the queueing model. Thus, given the birth rates $\{\lambda_n\}$ and death rates $\{\mu_n\}$, and assuming that

$$S = 1 + C_1 + C_2 + \cdots < \infty, \tag{5.19}$$

where

$$C_n = \frac{\lambda_0 \lambda_1 \cdots \lambda_{n-1}}{\mu_1 \mu_2 \cdots \mu_n}, \quad n = 1, 2, \ldots, \tag{5.20}$$

we calculate

$$p_0 = \frac{1}{S}, \tag{5.21}$$

and

$$p_n = P[N = n] = C_n\, p_0, \quad n = 1, 2, 3, \ldots. \tag{5.22}$$

From the probabilities calculated by (5.22), we can generate measures of queueing system performance.

5.2.1   The M/M/1 Queueing System

This model assumes a random (Poisson) arrival pattern and a random (exponential) service time distribution. The arrival rate does not depend on the number of customers in the system and, by Theorem 4.2.1, the probability of an arrival in a time interval of length $h > 0$ is given by

$$
\begin{aligned}
e^{-\lambda h}(\lambda h) &= \lambda h \left(1 - \lambda h + \frac{(\lambda h)^2}{2!} - \cdots \right) \\
&= \lambda h - (\lambda h)^2 + \frac{(\lambda h)^3}{2!} - \cdots + (-1)^{n+1}\frac{(\lambda h)^n}{(n-1)!} + \cdots \\
&= \lambda h + o(h).
\end{aligned}
\tag{5.23}
$$

Thus, we have

$$\lambda_n = \lambda, \quad n = 0, 1, 2, \ldots. \tag{5.24}$$

By hypothesis, the service time distribution is given by

$$W_s[t] = P[s \le t] = 1 - e^{-\mu t}, \quad t \ge 0. \tag{5.25}$$

Hence, if a customer is receiving service, the probability of a service completion (death) in a short time interval, h, is given by

$$1 - e^{-\mu h} = 1 - \left(1 - \mu h + \frac{(\mu h)^2}{2!} - \cdots \right) = \mu h + o(h). \tag{5.26}$$

(Here we have used the memoryless property of the exponential distribution in neglecting the service already completed.)

Thus,

$$\mu_n = \mu, \quad n = 1, 2, 3, \ldots. \tag{5.27}$$

Therefore, the state-transition diagram for the M/M/1 queueing system is given by Figure 5.2.1 and, since $\lambda/\mu = \rho$ and each C_n is equal to ρ^n, we have

$$S = 1 + \rho + \rho^2 + \cdots + \rho^n + \cdots = \frac{1}{(1-\rho)}. \tag{5.28}$$

(We assumed that $\rho < 1$, so that a steady state solution does exist.)

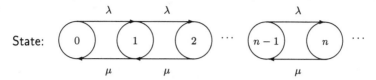

Figure 5.2.1. State-transition rate diagram for M/M/1 system.

To sum the series for S we have used the well-known summation formula for the geometric series, namely,

$$\sum_{n=0}^{\infty} x^n = 1 + x + x^2 + \cdots = \frac{1}{(1-x)} \quad \text{if } |x| < 1. \tag{5.29}$$

Hence, by (5.21) and (5.22).

$$p_n = P[N = n] = (1 - \rho)\rho^n, \quad n = 0, 1, 2, \ldots. \tag{5.30}$$

But (5.30) is the pmf for a geometric random variable; that is, N has a geometric distribution with $p = 1 - \rho$ and $q = \rho$. Hence, by Table 1 of Appendix A,

$$L = E[N] = \frac{q}{p} = \frac{\rho}{(1 - \rho)}, \tag{5.31}$$

and

$$\sigma_N^2 = \frac{\rho}{(1 - \rho)^2}. \tag{5.32}$$

By Little's law,

$$W = E[w] = \frac{L}{\lambda} = \frac{W_s}{(1 - \rho)}, \tag{5.33}$$

since

$$\rho = \lambda W_s. \tag{5.34}$$

Now,

$$W_q = E[q] = W - W_s = \frac{\rho W_s}{(1 - \rho)}. \tag{5.35}$$

Applying Little's law, again, gives,

$$L_q = E[N_q] = \lambda W_q = \frac{\rho^2}{(1 - \rho)}. \tag{5.36}$$

By (5.30), we calculate

$$P[\text{server busy}] = 1 - P[N = 0] = 1 - (1 - \rho) = \rho. \tag{5.37}$$

By the law of large numbers this probability can be interpreted as the fraction of time that the server is busy; it *is* appropriate to call ρ the "server utilization."

We now have the four parameters most commonly used to measure the performance of a queueing system W, W_q, L, and L_q, as well as the pmf, p_n, of the number in the system, N. For the M/M/1 system, we can also derive the exact distribution of w and q.

If an arriving customer finds no customers in the system, $(N = 0)$, then there is no queueing for service, so $W_q[0] = P[q = 0] = P[N = 0] = 1 - \rho$. Thus, $W_q[\cdot]$ has a probability concentration or mass at $t = 0$. However, if a customer arrives when n customers are already in the system, then this arrival has to wait through n exponential service times; that is, the conditional waiting time in the queue, given that n customers are in the system, is given by

$$q = s_1 + s_2 + \cdots + s_n, \qquad (5.38)$$

where $s_1, s_2, \ldots, s_n$ are n independent identically distributed exponential random variables, each with expected value $1/\mu$. (By the memoryless property of the exponential distribution, there is no need to account for the service time already expended on the customer receiving service.) By Theorem 3.2.6(d), q has a gamma distribution with parameters n and μ, that means that the conditional density function of q is given by

$$f_{q,n}(t) = \mu e^{-\mu t} \frac{(\mu t)^{n-1}}{(n-1)!}, \qquad t \geq 0, \ n \geq 1. \qquad (5.39)$$

Thus, if $n > 0$,

$$P[q \leq t | N = n] = \int_0^t \frac{\mu^n x^{n-1} e^{-\mu x}}{(n-1)!} \, dx. \qquad (5.40)$$

Therefore, by the law of total probability, Theorem 2.4.2,

$$P[0 < q \leq t] = \sum_{n=1}^{\infty} P[q \leq t | N = n] P[N = n]. \qquad (5.41)$$

Substituting (5.40) into (5.41) and using the fact that

$$p_n = P[N = n] = (1 - \rho)\rho^n = \left(1 - \frac{\lambda}{\mu}\right)\left(\frac{\lambda}{\mu}\right)^n, \qquad (5.42)$$

yields

$$P[0 < q \leq t] = \sum_{n=1}^{\infty} \int_0^t \frac{\mu^n x^{n-1} e^{-\mu x}}{(n-1)!} \left(1 - \frac{\lambda}{\mu}\right)\left(\frac{\lambda}{\mu}\right)^n \, dx$$

$$
\begin{aligned}
&= \int_0^t \lambda e^{-\mu x} \left(1 - \frac{\lambda}{\mu}\right) \sum_{n=1}^{\infty} \frac{(\lambda x)^{n-1}}{(n-1)!} \, dx \\
&= \int_0^t \lambda e^{-\mu x} \left(1 - \frac{\lambda}{\mu}\right) e^{\lambda x} \, dx \\
&= \int_0^t \lambda \left(1 - \frac{\lambda}{\mu}\right) e^{-x(\mu - \lambda)} \, dx \\
&= \frac{\lambda}{\mu} \int_0^t (\mu - \lambda) e^{-x(\mu - \lambda)} \, dx \\
&= \rho[1 - e^{-\mu(1-\rho)t}] \\
&= \rho[1 - e^{-t/W}].
\end{aligned}
\tag{5.43}
$$

We have already shown that $W_q[0] = 1 - \rho$. If $t > 0$, we can use (5.43) to calculate

$$
\begin{aligned}
W_q[t] &= P[q = 0] + P[0 < q \leq t] \\
&= 1 - \rho + \rho[1 - e^{-t/W}] \\
&= 1 - \rho e^{-t/W}
\end{aligned}
\tag{5.44}
$$

Note that, although q is discrete at the origin and continuous for $t > 0$, (5.44) is valid for all values of $t \geq 0$.

The average waiting time in queue for those who must queue is of particular interest for queueing systems in which people are the customers. As Kolesar [34] noted, if people are kept waiting too long in the queue for an automated teller machine (ATM), they will renege; that is, leave without receiving service. People will not wait indefinitely for any type of service. By Theorem 2.8.1,

$$
\begin{aligned}
W_q &= P[q = 0] \times E[q|q = 0] + P[q > 0] \times E[q|q > 0] \\
&= (1 - \rho) \times 0 + \rho \times E[q|q > 0].
\end{aligned}
\tag{5.45}
$$

Hence,

$$
E[q|q > 0] = \frac{W_q}{\rho} = \frac{W_s}{(1 - \rho)} = W.
\tag{5.46}
$$

Since $W = W_q + W_s$, this means that, on the average, customers who must wait will wait in the queue one average service time longer than the average customer waits.

The derivation of the distribution function $W[\cdot]$ is similar to that for $W_q[\cdot]$. If a customer arrives when there are already n customers in the system, then the total time this customer spends in the system is the sum of $n + 1$ independent exponential random variables, each with mean $1/\mu$. Hence, the density function, $f_{w,n}(\cdot)$, is the gamma density

$$f_{w,n}(t) = \frac{\mu^{n+1} t^n e^{-\mu t}}{n!}, \quad t \geq 0. \tag{5.47}$$

By the law of total probability, we have

$$
\begin{aligned}
W[t] &= \sum_{n=0}^{\infty} P[w \leq t | n = n] P[N = n] \\
&= \sum_{n=0}^{\infty} \int_0^t \frac{\mu^{n+1} x^n e^{-\mu x}}{n!} \left(1 - \frac{\lambda}{\mu}\right) \left(\frac{\lambda}{\mu}\right)^n dx \\
&= \int_0^t \mu e^{-\mu x} \left(1 - \frac{\lambda}{\mu}\right) \sum_{n=0}^{\infty} \frac{(\lambda x)^n}{n!} dx \\
&= \int_0^t \mu e^{-\mu x} \left(1 - \frac{\lambda}{\mu}\right) e^{\lambda x} dx \\
&= \int_0^t \mu \left(1 - \frac{\lambda}{\mu}\right) e^{-\mu(1-\lambda/\mu)x} dx \\
&= 1 - e^{-\mu(1-\rho)t} \\
&= 1 - e^{-t/W}. \tag{5.48}
\end{aligned}
$$

This shows that w has an exponential distribution and that

$$W = \frac{1}{\mu(1-\rho)} = \frac{W_s}{1-\rho}, \tag{5.49}$$

as we calculated before.

Because w has an exponential distribution, we know that

$$\sigma_w = W^2 = \left(\frac{W_s}{1-\rho}\right)^2. \tag{5.50}$$

We also know, by Theorem 3.2.1, that the rth percentile value of w, $\pi_w[r]$, is given by

$$\pi_w[r] = W \ln\left(\frac{100}{100-r}\right), \tag{5.51}$$

so that, in particular,

$$\pi_w[90] = W \ln 10 \approx 2.3W, \tag{5.52}$$

and

$$\pi_w[95] = W \ln 20 \approx 3W. \tag{5.53}$$

Similarly, using the distribution function of q, we calculate

$$\pi q[r] = \frac{W_s}{1 - \rho} \ln\left(\frac{100\rho}{100 - r}\right) = W \ln\left(\frac{100\rho}{100 - r}\right). \tag{5.54}$$

A number of formulas for an M/M/1 queueing system are shown in Table 3 of Appendix C and can be evaluated by the APL function MΔMΔ1 of Appendix B.

Example 5.2.1 Traffic to a message switching center for Extraterrestrial Communications Corporation arrives in a random pattern (remember that "random pattern" means exponential interarrival time) at an average rate of 240 messages per minute. The line has a transmission rate of 800 characters per second. The message length distribution (including control characters) is approximately exponential with an average length of 176 characters. Calculate the principal statistical measures of system performance assuming that a very large number of message buffers is provided. What is the probability that 10 or more messages are waiting to be transmitted? What would be the average response time, W, if the traffic rate into the center increased by 10%?

Solution The average service time is the average time required to transmit a message or

$$W_s = \frac{\text{average message length}}{\text{line speed}} = \frac{176 \text{ char.}}{800\dfrac{\text{char.}}{\text{sec.}}} = 0.22 \text{ seconds.} \tag{5.55}$$

Hence, since the average arrival rate

$$\lambda = \frac{240 \text{ messages}}{\text{minute}} = 4 \frac{\text{messages}}{\text{second}}, \tag{5.56}$$

the server utilization

$$\rho = \lambda W_s = 4 \times 0.22 = 0.88; \tag{5.57}$$

that is, the communication line is transmitting outgoing messages 88% of the time. Using the M/M/1 formulas of Table 3, Appendix C, we calculate the following.[6]

---

[6]The APL function MΔMΔ1 could be used to make the computations.

$L = \rho/(1 - \rho) = 7.33$ messages.   Average number of
                                                messages in the system.

$L_q = \rho^2/(1 - \rho) = 6.45$ messages.   Average number of
                                                messages in queue.

$W = W_s/(1 - \rho) = 1.83$ seconds.   Average response time.

$W_q = \rho \times W = 1.61$ seconds.   Average queueing time.

$\pi_w[90] = W \ln 10 = 4.21$ seconds.   90th percentile time
                                                in the system.

$\pi_q[90] = W \ln(10\rho) = 3.98$ seconds.   90th percentile queueing time.

Since 10 or more messages are queueing if and only if 11 or more messages
are in the system, the required probability is $\rho^{11} = 0.245$. If λ is increased
by 10%, so is ρ, which becomes 0.968, so that

$$W = \frac{W_s}{1 - \rho} = 6.875 \text{ seconds.} \tag{5.58}$$

This is a considerable increase from 1.83 seconds! $\square$

Graph of $\dfrac{W}{W_s} = \dfrac{1}{1 - \rho}$ versus ρ.

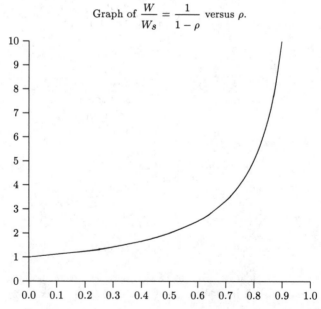

Figure 5.2.2. $\dfrac{W}{W_s}$ versus ρ for M/M/1 queueing system.

The reason for the large increase in response time as the utilization increased by 10% from 0.88 to 0.968 is shown graphically in Figure 5.2.2. The curve of W/W_s rises sharply as ρ approaches 1. That is, the slope of the curve increases rapidly as ρ grows beyond about 0.8. Since

$$\frac{dW}{d\rho} = W_s \times (1 - \lambda W_s)^{-2}, \tag{5.59}$$

a small change in ρ (due to a small change in λ, assuming W_s is fixed) causes a change in W given approximately by

$$\left(\frac{dW}{d\rho}\right) \Delta\rho = \left(\frac{dW}{d\rho}\right) W_s \, \Delta\lambda = W_s^2 (1 - \lambda W_s)^{-2} \Delta\lambda. \tag{5.60}$$

Thus, if $\rho = 0.5$, a change $\Delta\lambda$ in λ will cause a change in W of about $4 \times W_s^2 \times \Delta\lambda$, while, if $\rho = 0.88$, the change in W will be about $69.44 \times W_s^2 \times \Delta\lambda$, or 17.36 times the size of the change that occurred when $\rho = 0.5$!

We have discussed the M/M/1 queueing system in more detail than we will for most of the queueing models. It is a simple but important model to study. It has some pleasant properties that many of the queueing models we will consider lack. For example, we have been able to find the complete probability distributions for the random variables N, w, and q; for many queueing models we have difficulty in computing the average values L, L_q, W, and W_q. A number of systems in the world around us can be at least approximated by the M/M/1 system.

5.2.2 The M/M/1/K Queueing System

Example 5.2.1 was somewhat unrealistic in the sense that no message switching system can have an unlimited number of buffers. The M/M/1/K system is a more accurate model of this type of system in which at most K customers are allowed in the system: 1 in service and $K - 1$ in the queue. When there are K customers in the system, arriving customers are turned away. Figure 5.2.3 is the state-transition diagram for this model. Thus, as a birth-and-death process, the coefficients are

$$\lambda_n = \begin{cases} \lambda & \text{for } n = 0, 1, 2, \ldots, K - 1 \\ 0 & \text{for } n \geq K, \end{cases} \tag{5.61}$$

$$\mu_n = \begin{cases} \mu & \text{for } n = 0, 1, 2, \ldots, K \\ 0 & \text{for } n > K. \end{cases} \tag{5.62}$$

This gives the steady state probabilities

$$p_n = p_0 \left(\frac{\lambda}{\mu}\right)^n = p_0 \, a^n \quad \text{for } n = 0, 1, 2, \ldots, K, \tag{5.63}$$

where

$$a = \lambda W_s = \frac{\lambda}{\mu}. \tag{5.64}$$

Since

$$1 = p_0 + p_1 + \cdots + p_K = p_0 \sum_{n=0}^{K} a^n = \left(\frac{1 - a^{K+1}}{1 - a}\right) p_0, \tag{5.65}$$

if $\lambda \neq \mu$, we have,

$$p_0 = \frac{1 - a}{1 - a^{K+1}}. \tag{5.66}$$

Since there are never more than K customers in the system, the system reaches a steady state for all values of λ and μ. That is, we need not assume that $\lambda < \mu$ for the system to achieve the steady state. If $\lambda = \mu$, then $a = 1$ and

$$p_0 = \frac{1}{K+1} = p_n \quad \text{for } n = 1, 2, \ldots, K. \tag{5.67}$$

Thus, the steady state probabilities are

$$p_n = \begin{cases} \dfrac{(1-a)\, a^n}{1 - a^{K+1}} & \text{for } \lambda \neq \mu, \quad n = 0, 1, \ldots, K \\[2mm] \dfrac{1}{K+1} & \text{for } \lambda = \mu, \quad n = 0, 1, \ldots, K. \end{cases} \tag{5.68}$$

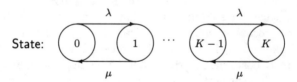

Figure 5.2.3. State-transition rate diagram for M/M/1/K system.

It should be noted that, if $\lambda < \mu$, as $K \to \infty$, each p_n in (5.68) approaches the correct value for the M/M/1 system.

If $\lambda \neq \mu$, then

$$L = \sum_{n=0}^{K} n\, p_n$$

$$= \left(\frac{1-a}{1-a^{K+1}}\right) \sum_{n=1}^{K} (na^n)$$

$$= \left(\frac{1-a}{1-a^{K+1}}\right) a \sum_{n=1}^{K} na^{n-1}$$

$$= \left(\frac{1-a}{1-a^{K+1}}\right) a \sum_{n=1}^{K} \frac{d\,a^n}{d\,a}$$

$$= \left(\frac{1-a}{1-a^{K+1}}\right) a \frac{d}{d\,a} \sum_{n=0}^{K} a^n$$

$$= \left(\frac{1-a}{1-a^{K+1}}\right) a \frac{d}{d\,a} \left(\frac{1-a^{K+1}}{1-a}\right)$$

$$= \frac{a}{1-a} - \frac{(K+1)a^{K+1}}{1-a^{K+1}}. \tag{5.69}$$

Thus, if $\lambda < \mu$, the expected number in the system, L, is always less than for the unlimited queue length case (where $L = a/(1-a)$). If $\lambda = \mu$, then $a = 1$ and

$$L = \sum_{n=0}^{K} n\, p_n = \frac{1+2+\cdots+K}{K+1} = \frac{K}{2}. \tag{5.70}$$

Thus, (5.69) and (5.70) can be summarized by

$$L = \begin{cases} \dfrac{a}{1-a} - \dfrac{(K+1)a^{K+1}}{1-a^{K+1}} & \text{if } \lambda \neq \mu \\[2ex] \dfrac{K}{2} & \text{if } \lambda = \mu. \end{cases} \tag{5.71}$$

In either case,

$$L_q = L - (1 - p_0) \tag{5.72}$$

because

$$\begin{aligned} E[N_S] &= P[N = 0] \times E[N_S|N = 0] + P[N > 0] \times E[N_S|N > 0] \\ &= p_0 \times 0 + (1 - p_0) \times 1 \\ &= 1 - p_0. \end{aligned} \tag{5.73}$$

All the traffic reaching the system does not enter the system because customers are not allowed admission when there are K customers in the system, that is, with probability p_K. Thus, if λ_a is the average rate of customers *into* the system,

$$\lambda_a = \lambda(1 - p_K). \tag{5.74}$$

We can then apply Little's law to obtain

$$W = \frac{L}{\lambda_a}, \tag{5.75}$$

and

$$W_q = \frac{L_q}{\lambda_a}. \tag{5.76}$$

The true server utilization, ρ, which is the probability that the server is busy, is given by

$$\rho = \lambda_a W_s = \lambda(1 - p_K) W_s = (1 - p_K)a. \tag{5.77}$$

The derivation of the distribution functions of q and w is more complex for the M/M/1/K model than it was for the M/M/1 model. For $n = 0, 1, \ldots, K - 1$ let q_n be the probability that an arriving customer who enters the system finds n customers already in the system. It can be shown by Bayes' theorem (Theorem 2.4.3) (see Exercise 12) that

$$q_n = \frac{p_n}{1 - p_K}, \quad n = 0, 1, \ldots, K - 1. \tag{5.78}$$

Proceeding as we did in deriving $W[\cdot]$ for the M/M/1 model, we calculate (see Exercise 13)

$$
\begin{aligned}
W[t] &= \sum_{n=0}^{K-1} P[w \le t | N_a = n] \times P[N_a = n] \\
&= \sum_{n=0}^{K-1} \left\{ \int_0^t \frac{\mu(\mu x)^n}{n!} e^{-\mu x} \, dx \right\} q_n \\
&= 1 - \sum_{n=0}^{K-1} q_n Q[n; \mu t],
\end{aligned}
\tag{5.79}
$$

where

$$Q[n; \mu t] = e^{-\mu t} \sum_{k=0}^{n} \frac{(\mu t)^k}{k!} \tag{5.80}$$

is the Poisson distribution function and N_a is the random variable that counts the number of customers in an M/M/1/K queueing system just before a customer arrives to enter the system. (Thus, N_a assumes the values $0, 1, \ldots, K - 1$ and $P[N_a = n] = q_n$.) $W[t]$ can be calculated with the aid of tables of the Poisson distribution function or by using an APL function such as POISSONΔDIST. The APL function DMΔMΔ1ΔK computes the values of $W[t]$ and $W_q[t]$ for the M/M/1/K model. The same reasoning that led to $W_q[\cdot]$ for the M/M/1 queueing system shows that $W_q[t]$ is given by

$$
\begin{aligned}
W_q[t] &= W_q[0] + \sum_{n=1}^{K-1} P[q \le t | N_q = n] \times q_n \\
&= q_0 + \sum_{n=1}^{K-1} q_n \int_0^t \frac{\mu(\mu x)^{n-1}}{(n-1)!} e^{-\mu x} \, dx \\
&= 1 - \sum_{n=0}^{K-2} q_{n+1} \times Q[n; \mu t].
\end{aligned}
\tag{5.81}
$$

($Q[n; \mu t]$ was defined above.)

Example 5.2.2 Consider Example 5.2.1. Suppose Extraterrestrial Communications has the same arrival pattern, message length distribution, and line speed as described in the example. Suppose, however, that it is desired to provide only the minimum number of message buffers required to guarantee that

$$
p_K < 0.005. \tag{5.82}
$$

How many buffers should be provided? For this number of buffers calculate L, L_q, W, and W_q. What is the probability that the time an arriving message spends in the system does not exceed 2.5 seconds? What is the probability that the queueing time of a message before transmission is begun does not exceed 2.5 seconds?

Solution The M/M/1/K model fits this system with $a = \lambda W_s = 0.88$ erlangs.

The probability that all the buffers are filled, given $K - 1$ buffers are provided, is

$$
p_K = \frac{(1 - a)a^K}{1 - a^{K+1}}, \tag{5.83}
$$

where $a = 0.88$. Equation (5.83) can be solved for $p_K = 0.005$ and $a = 0.88$ to obtain $K = 25.142607$. Thus, we need to have 25 buffers ($K = 26$) to make $p_K < 0.005$. Using the APL function MΔMΔ1ΔK, that makes the calculations for the M/M/1/K system using the formulas of Table 4 of Appendix C, we obtain the following.

$L = 6.449$ messages.   Average number of messages in the system.

$L_q = 5.573$ messages.   Average number of messages queued for the line.

$W = 1.62$ seconds.   Average time a message spends in the system (queueing for the line and in transmission).

$W_q = 1.40$ seconds.   Average time a message queues for the line.

All of these performance numbers are smaller than the corresponding values for the M/M/1 model with the same λ and W_s. Using the APL function DMΔMΔ1ΔK, we calculate $W[2.5] = P[w \leq 2.5] = 0.77208$, while $W_q[2.5] = P[q \leq 2.5] = 0.8039$. All the performance metrics for the $M/M/1/26$ system are superior to those for the M/M/1 system of Example 5.2.1. The penalty for this improved performance is that $100 \times p_K = 0.4464\%$ of the messages are refused and must be sent again at a later time. $\square$

5.2.3   The M/M/c Queueing System

For this model we assume random (exponential) interarrival and service times with c identical servers. This system can be modeled as a birth-and-death process with the coefficients

$$\lambda_n = \lambda \quad n = 0, 1, 2, \ldots, \tag{5.84}$$

and

$$\mu_n = \begin{cases} n\mu, & n = 1, 2 \ldots, c, \\ c\mu, & n \geq c. \end{cases} \tag{5.85}$$

The state-transition diagram is shown in Figure 5.2.4. Thus, by (5.20), with $a = \lambda/\mu$ and $\rho = a/c$,

$$C_n = \begin{cases} \dfrac{a^n}{n!}, & n = 1, 2, \ldots, c, \\ \rho^{n-c} \dfrac{a^c}{c!}, & n = c+1, c+2, \ldots. \end{cases} \tag{5.86}$$

Hence, if $\rho < 1$ so that the steady state exists, then

$$\begin{aligned} S &= \frac{1}{p_0} \\ &= 1 + a + \frac{a^2}{2!} + \cdots + \frac{a^{c-1}}{(c-1)!} + \frac{a^c}{c!}\left(1 + \rho + \rho^2 + \cdots\right) \end{aligned}$$

$$= \sum_{n=0}^{c-1} \frac{a^n}{n!} + \frac{a^c}{c!} \sum_{n=0}^{\infty} \rho^n$$

$$= \sum_{n=0}^{c-1} \frac{a^n}{n!} + \frac{a^c}{c!\,(1-\rho)}. \tag{5.87}$$

Hence,

$$p_0 = \left\{ \sum_{n=0}^{c-1} \frac{a^n}{n!} + \frac{a^c}{c!\,(1-\rho)} \right\}^{-1}, \tag{5.88}$$

and

$$p_n = \begin{cases} \dfrac{a^n}{n!}\, p_0 & \text{if } n = 0, 1, \ldots, c-1, \\[2mm] \dfrac{a^n}{c!\, c^{n-c}}\, p_0 & \text{if } n \geq c. \end{cases} \tag{5.89}$$

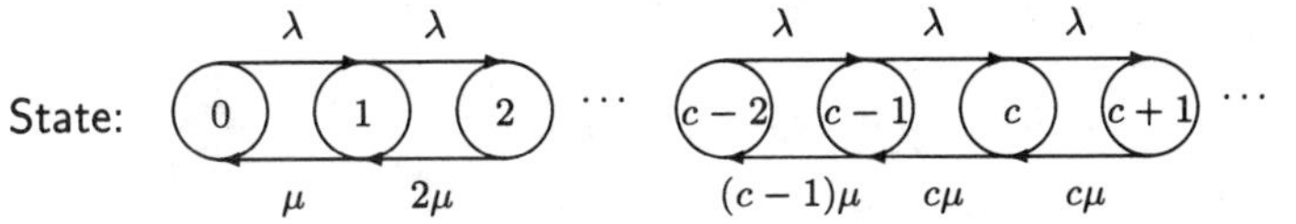

Figure 5.2.4. State-transition rate diagram for M/M/c system.

We shall now derive the formula for one of the key parameters of the M/M/c queueing system. It is the probability that an arriving customer must queue for service. The formula for this number is known as *Erlang's C formula* or *Erlang's delay formula* and written $C[c, a]$. Since a customer must queue for service if and only if there are c or more customers already in the system, we know that $C[c, a] = P[N \geq c]$.[7] Hence, we calculate

$$C[c, a] = \sum_{n=c}^{\infty} p_n$$

$$= 1 - \sum_{n=0}^{c-1} p_n$$

$$= 1 - p_0 \sum_{n=0}^{c-1} \frac{a^n}{n!}$$

[7]It follows from Wolff's paper, Wolff [66], that p_c is the probability that an arriving customer will find exactly c customers in the system *before* the arriving customer enters. Hence, the arriving customer must queue for service. This explains why the summation in (5.90) starts at $n = c$ and not at $n = c + 1$.

$$
\begin{aligned}
&= 1 - \frac{\displaystyle\sum_{n=0}^{c-1} \frac{a^n}{n!}}{\displaystyle\sum_{n=0}^{c-1} \frac{a^n}{n!} + \frac{a^c}{c!\,(1-\rho)}} \\[2em]
&= \frac{\dfrac{a^c}{c!}}{(1-\rho)\displaystyle\sum_{n=0}^{c-1} \frac{a^n}{n!} + \frac{a^c}{c!}}
\end{aligned}
\tag{5.90}
$$

Some excellent algorithms have been developed for evaluating $C[c, a]$, that makes it useful for computing other performance parameters. For example, it is easy to show that

$$
p_0 = \frac{c!\,(1-\rho)C[c,a]}{a^c}.
\tag{5.91}
$$

Also, by the definition of $C[c, a]$, we know that

$$
W_q[0] = P[q = 0] = 1 - C[c, a].
\tag{5.92}
$$

We can now derive the formula for L_q:

$$
\begin{aligned}
L_q &= \sum_{n=c}^{\infty} (n - c)p_n \\
&= \sum_{k=0}^{\infty} k p_{c+k} \\
&= p_0 \frac{a^c}{c!} \sum_{k=0}^{\infty} k\rho^k \\
&= p_0 \frac{a^c}{c!} \{0 + 1\rho + 2\rho^2 + 3\rho^3 + \cdots\} \\
&= p_0 \frac{a^c}{c!} \rho \frac{d}{d\rho} \{1 + \rho + \rho^2 + \cdots\} \\
&= p_0 \frac{a^c}{c!} \rho \frac{d}{d\rho} \left(\frac{1}{1-\rho}\right) \\
&= p_0 \frac{a^c}{c!(1-\rho)^2} \rho \\
&= \frac{\rho C[c, a]}{1 - \rho}.
\end{aligned}
\tag{5.93}
$$

We used (5.91) in the last step of (5.93). By Little's law,

$$W_q = \frac{L_q}{\lambda} = \frac{C[c, a]W_s}{c(1 - \rho)} = \frac{P[N \geq c]W_s}{c(1 - \rho)}, \qquad (5.94)$$

since

$$\rho = \frac{\lambda W_s}{c}. \qquad (5.95)$$

Then we also have

$$W = W_q + W_s, \qquad (5.96)$$

and

$$L = \lambda W. \qquad (5.97)$$

We have developed the formula for $W_q[0]$. Now we derive the formula for $W_q[t]$ when $t > 0$. An arriving customer must queue for service only if $N = n \geq c$ when the customer arrives at the system. All c servers are busy, so, as we explained earlier (in the discussion of "Service Time Distribution" in Section 5.1), the time between service completions in the service center has an exponential distribution with average value $1/(c\mu)$. There are c customers receiving service and $n - c$ customers waiting in the queue. Therefore, the new arrival must wait for $n-c+1$ service completions before receiving service. (If $n = c$, so that no customers are queueing for service, the new arrival must wait for one service completion. If $n = c + 1$, two service completions are required, etc.) Hence, the waiting time in queue is the sum of $n-c+1$ independent exponential random variables, each with mean $1/(c\mu)$; that is, it is a gamma distribution with parameters $n - c + 1$ and $c\mu$. Hence, if $t > 0$, we can write, since $\Gamma(n - c + 1) = (n - c)!$, that

$$
\begin{aligned}
W_q[t] &= W_q[0] + \sum_{n=c}^{\infty} P[q \leq t | N = n] \times p_n \\
&= W_q[0] + \sum_{n=c}^{\infty} \frac{p_0 a^n}{c! \, c^{n-c}} \int_0^t \frac{c\mu(c\mu x)^{n-c}}{(n-c)!} e^{-c\mu x} \, dx \\
&= W_q[0] + \frac{p_0 \, a^c}{(c-1)!} \int_0^t \mu e^{-c\mu x} \left(\sum_{n=c}^{\infty} \frac{(a\mu x)^{n-c}}{(n-c)!} \right) dx \\
&= W_q[0] + \frac{p_0 \, a^c}{(c-1)!} \int_0^t \mu e^{-c\mu x} e^{a\mu x} \, dx \\
&= W_q[0] + \frac{p_0 \, a^c}{(c-1)!} \int_0^t \mu e^{-\mu x(c-a)} \, dx \\
&= W_q[0] + \frac{p_0 \, a^c}{(c-a)(c-1)!} (1 - e^{-\mu t(c-a)})
\end{aligned}
$$

$$
\begin{aligned}
&= \quad 1 - C[c,a] + C[c,a](1 - e^{-\mu t(c-a)}) \\
&= \quad 1 - C[c,a]e^{-\mu t(c-a)}.
\end{aligned}
\tag{5.98}
$$

We have used (5.91) and (5.92) in the last two steps of (5.98). Although q has a probability mass at the origin and is continuous for $t > 0$, formula (5.98) is valid for all $t \geq 0$.

The distribution function $W[\cdot]$ for waiting time in the system is given by

$$
W[t] = \begin{cases}
1 - \dfrac{(a - c + W_q[0])}{(a + 1 - c)} e^{-\mu t} - \dfrac{C[c,a]}{(a + 1 - c)} e^{-c\mu t(1-\rho)}, & \text{if } a \neq c - 1, \\[3mm]
1 - (1 + C[c,a]\mu t)e^{-\mu t} & \text{if } a = c - 1.
\end{cases}
\tag{5.99}
$$

(The formulas from which these formulas follow are derived in Section 5.3.2. See Exercise 26.)

Formula (5.98) can be used to calculate the rth percentile value of q, that is

$$
\pi_q[r] = \frac{W_s}{c(1 - \rho)} \ln\left(\frac{100\, C[c,a]}{100 - r} \right).
\tag{5.100}
$$

The formulas for the M/M/c queueing system are given in Table 5 of Appendix C. Table 6 gives the formulas for the special case that $c = 2$. Figure 1 of Appendix C is a graph of $C[c,a]$ versus a for selected values of c.

Example 5.2.3 KAMAKAZY AIRLINES is planning a new telephone reservation center. Each agent will have a reservations terminal and can serve a typical caller in 5 minutes, the service time being exponentially distributed. Calls arrive randomly and the system has a large message buffering system to hold calls that arrive when no agent is free. An average of 36 calls per hour is expected during the peak period of the day. The three design criteria for the new facility are

1. The probability a caller will find all agents busy must not exceed 0.1 (10%).

2. The average waiting time for those who must wait is not to exceed one minute.

3. Less than 5% of all callers must wait more than one minute for an agent.

How many agents (and terminals) should be provided? How will this system perform if the number of callers per hour is 10% higher than anticipated?

Solution The expected peak period average arrival rate, λ, is 36 calls per hour or 0.6 calls per minute. Therefore, the traffic intensity $a = \lambda W_s = 3$ erlangs. Thus, 4 agents are needed just to keep up with the callers.

Translated into technical queueing theory notation the requirements for the system are

1. $P[N \geq 3] = C[c, 3] \leq 0.1$.

2. $E[q|q > 0] \leq 1$ minute.

3. $P[q > 1] \leq 0.05$.

We seek the minimum c such that $C[c, 3] \leq 0.1$ and that the other two requirements are met, as well. Using Figure 1 of Appendix C,[8] it appears that $C[6, 3] \approx 0.1$. Direct calculation of $C[6, 3]$, using Erlang's C formula shows that it is 0.0991. Since $C[5, 3] = 0.236$, at least 6 agents are required to satisfy the first design criterion. The APL function MΔMΔC indicates that for 6 agents $E[q|q > 0] = 1.67$ minutes. We find that eight agents are required, since for seven agents $E[q|q > 0] = 1.25$ minutes, while, for eight agents, $E[q|q > 0]$ is exactly one minute. Eight agents satisfies all the criteria, since for eight agents,

$$P[q > 1] = C[8, 3] e^{-0.2(8-3)} = 0.00476. \qquad (5.101)$$

If the peak traffic is 10% higher than 36 calls per hour, then the probability that all eight agents are busy is 0.022, the average queueing time for callers delayed is 1.06 minutes, and the fraction of callers who must wait more than 1 minute is 0.0085. Thus, the proposed system looks good, even if the traffic is slightly higher than anticipated. Since $\rho = a/8 = 3/8$, each agent is busy only three-eighths of the time during the peak period—such is the price of good service! As shown in Table 5.2.1, with six agents only one of the design criteria is met and with four agents the performance is deplorable. Eight agents is a good choice. We have shown that eight agents should be on duty during the peak period. More than this number may be needed to provide for coffee breaks, I/O breaks, etc., so that eight agents are available for duty. $\square$

If the reader has calculated the performance figures in the last example, he or she probably found that the most difficult part was calculating $C[c, a]$.[9] Russell Ham, a colleague of mine, developed an algorithm that

---

[8]Using graphs like Figure 1 induces vertigo in some people. I have this problem, too.

[9]Some readers report that calculating Erlang's C formula using formula (5.90) seems to be an unnatural act.

Table 5.2.1. Example 5.2.3

c	$C[c,a]$	$E[q\mid q>0]$	$P[q>1]$
8	0.0129	1.00	0.0048
7	0.0376	1.25	0.0169
6	0.0991	1.67	0.0544
5	0.2362	2.50	0.1583
4	0.5094	5.00	0.4171

makes it easy to calculate $C[c,a]$ with a programmable calculator. Recently, it has come to my attention that the algorithm was discovered in the 1920s but is still not widely known. It is based on Erlang's B formula, that I will discuss in the next section. Erlang's B formula, $B[c,a]$, is given by

$$B[c,a] = \frac{\dfrac{a^c}{c!}}{1 + a + \dfrac{a^2}{2!} + \cdots + \dfrac{a^c}{c!}}. \tag{5.102}$$

Ham's algorithm depends on the following facts that we ask you to prove in Exercise 25.

(a)

$$\frac{1}{C[c,a]} = \rho + \frac{(1-\rho)}{B[c,a]} \tag{5.103}$$

(b)

$$\frac{1}{B[1,a]} = 1 + \frac{1}{a} \tag{5.104}$$

(c)

$$\frac{1}{B[n,a]} = 1 + \frac{n}{a} \times \frac{1}{B[n-1,a]} \quad \text{for } n = 2,3,\cdots,c. \tag{5.105}$$

Ham's original algorithm follows.

Algorithm 5.2.1 *HO (Ham's original algorithm) Given the traffic intensity $a = \lambda W_s$ and the number of servers c, this algorithm will generate $B[c,a]$ and $C[c,a]$.*

Step 1. *Set*

$$\frac{1}{B[c,a]} = 1 + \frac{1}{a}. \tag{5.106}$$

Step 2. *Calculate*

$$\frac{1}{B[n,a]} = 1 + \frac{n}{a} \times \frac{1}{B[n-1,a]} \quad \textit{for } n = 2, 3, \cdots, c. \tag{5.107}$$

Step 3. *Set*

$$B[c,a] = \left(\frac{1}{B[c,a]}\right)^{-1}. \tag{5.108}$$

Step 4. *Calculate*

$$C[c,a] = \frac{B[c,a]}{\rho B[c,a] + 1 - \rho}. \tag{5.109}$$

Ham's original algorithm is efficient and works well on a pocket calculator or a computer as long as $B[c,a]$ is large enough so that none of the terms $1/B[n,a]$ causes an overflow condition. Tom Warner of IBM Canada wrote the following one-line APL program that will compute $B[c,a]$ very quickly.

```
[0]  Z ← C  WBCA  A
[1]    Z ← ÷ + / × \1, Φ(ιC) ÷ A
```

If the algorithm is used with most calculators and for most computer implementations to calculate $B[40, 0.01]$, it will fail because $B[40, 0.01]$ is approximately

$$e^{-0.01} \times \frac{(0.01)^{40}}{40!} = \frac{e^{-0.01}}{40!} \times 10^{-80} = 1.234 \times 10^{-128}. \tag{5.110}$$

However, Warner's one-liner will handle this calculation easily! The modified form of the algorithm does not have overflow problems. However, I have not been able to write a one-line APL program to implement the modified algorithm.

Algorithm 5.2.2 *HM (Ham's modified algorithm) Given the traffic intensity $a = \lambda W_s$ and the number of servers c, this algorithm will generate $B[c,a]$ and $C[c,a]$.*

Step 1. *Set*

$$B[1,a] = \frac{a}{1+a}. \tag{5.111}$$

Step 2. *Calculate*

$$B[n,a] = \frac{a\,B[n-1,a]}{n + a\,B[n-1,a]} \quad \textit{for } n = 2, 3, \cdots, c. \tag{5.112}$$

Step 3. *Calculate*

$$C[c,a] = \frac{B[c,a]}{\rho B[c,a] + 1 - \rho}. \tag{5.113}$$

5.2.4   The M/M/c/c Queueing System (M/M/c Loss System)

The M/M/c/c queueing system is often called the M/M/c loss system because customers who arrive when all the servers are busy are not allowed to wait for service and thus are lost to the system.

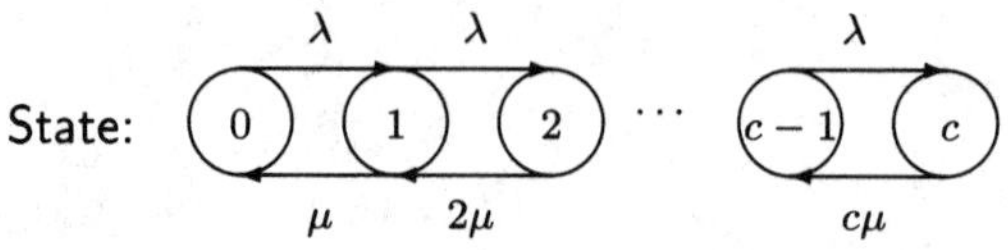

Figure 5.2.5. State-transition rate diagram for M/M/c/c system.

The state-transition diagram is given in Figure 5.2.5. From the diagram we see that

$$C_n = \frac{a^n}{n!} \quad \text{for } n = 1, 2, 3, \cdots, c, \tag{5.114}$$

where $a = \lambda W_s = \lambda/\mu$. We calculate

$$\begin{aligned} S &= \frac{1}{p_0} \\ &= 1 + a + \frac{a^2}{2!} + \frac{a^3}{3!} + \cdots + \frac{a^c}{c!}. \end{aligned} \tag{5.115}$$

Hence,

$$\begin{aligned} p_n &= C_n \times p_0 \\ &= \frac{\dfrac{a^c}{c!}}{1 + a + \dfrac{a^2}{2!} + \cdots + \dfrac{a^c}{c!}} \quad \text{for } n = 0, 1, \cdots, c. \end{aligned} \tag{5.116}$$

The distribution given by (5.116) is called the "truncated Poisson distribution." In particular, the probability that all c servers are busy, so that an arriving customer is lost, is given by *Erlang's B formula* or *Erlang's loss formula*:

$$B[c, a] = \frac{\dfrac{a^c}{c!}}{1 + a + \dfrac{a^2}{2!} + \cdots + \dfrac{a^c}{c!}}. \tag{5.117}$$

$B[c, a]$ was, of course, discovered by the great queueing theory pioneer, A. K. Erlang. Just as with the M/M/1/K queueing model, the actual average arrival rate into the system, λ_a, is less than λ because some arrivals are turned away. We see that

$$\lambda_a = \lambda(1 - B[c, a]). \tag{5.118}$$

Since no customers are allowed to wait for service, W_q and L_q are zero. We calculate

$$L = E[N] = \sum_{n=0}^{c} np_n = p_0 \sum_{n=1}^{c} n\frac{a^n}{n!} = ap_0 \sum_{n=0}^{c-1} \frac{a^n}{n!} = a(1 - B[c, a]). \tag{5.119}$$

By Little's law,

$$W = \frac{L}{\lambda_a} = W_s. \tag{5.120}$$

Of course (5.120) is obvious because there is no queueing for service. Therefore, w has the same distribution as s and we can write

$$W[t] = 1 - e^{\mu t} = 1 - e^{t/W_s}. \tag{5.121}$$

It was conjectured by Erlang and later proven by others[10] that all the formulas we have given for the M/M/c/c queueing system (except (5.121), of course, that becomes $W[t] = P[s \leq t]$) are also true for the M/G/c/c queueing system. That is, only the average value of service time is important. Such queueing systems are called "robust systems." For a proof see Gross and Harris [18].

Example 5.2.4 The Sad Sack Clothing Company has decided to install a tie-line telephone system between its east coast and west coast facilities. A caller receives a busy signal if the call is dialed when all the lines are in use. An average of 105 calls per hour, with an average length of 4 minutes, is expected. Enough lines are to be provided to ensure that the probability of getting a busy signal will not exceed 0.005. How many lines should be provided? How many lines are required if the probability of a busy signal is not to exceed 0.01? What would the performance be with 10 lines?

Solution The traffic intensity $a = (105/60) \times 4 = 7$ erlangs. The APL function BCA shows that $B[15, 7] = 0.00332$ while $B[14, 7] = 0.00713$, so 15 lines are required. The smallest c such that $B[c, 7] \leq 0.01$ is 14, so we save only one line if we double the allowed probability of a busy signal. If only 10 tie-lines are provided, the probability of a busy signal is $B[10, 7] = 0.07874$. $\square$

---

[10]According to Takács [59, page 186], B. A. Sevastyanov gave the first correct proof of Erlang's conjecture in 1957.

The formulas for the $M/M/c/K$ queueing system can be derived much as they were for the M/M/c/c system and are given in Table 8 of Appendix C.

5.2.5   M/M/∞ Queueing System

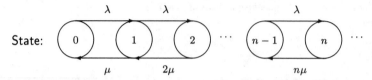

Figure 5.2.6. State-transition rate diagram for M/M/∞ system.

No real life queueing system can have an infinite number of servers; what is meant, here, is that a server is immediately provided for each arriving customer. The state-transition diagram for this model is shown in Figure 5.2.6. We can read from the figure that

$$C_n = \frac{a^n}{n!}, \quad n = 1, 2, 3, \cdots,$$

so that

$$S = \frac{1}{p_0} = \sum_{n=0}^{\infty} \frac{a^n}{n!} = e^a.$$

Hence,

$$p_n = e^{-a} \frac{a^n}{n!}, \quad n = 0, 1, 2; \cdots, \tag{5.122}$$

that is, N has a Poisson distribution! It can be shown (see Gross and Harris [18]) that (5.122) is also true for the $M/G/\infty$ queueing system. The fact that N has a Poisson distribution tells us that $L = a$ is the average number of busy servers, with $\sigma_N^2 = a$. The M/M/∞ queueing model can be used to estimate the number of lines in use in a large communication network or as a gross estimate of values in an M/M/c or M/M/c/c queueing system for large values of c. In Example 5.2.4, a was 7 erlangs, which is close to the average number of servers in use for the M/M/15/15 queueing system. (The exact value is 6.97676.) For the M/M/15/15 system,

$$B[15, 7] = 0.0033186 \approx p_{15} = e^{-7} \times \frac{7^{15}}{15!} = 0.0033106. \tag{5.123}$$

Example 5.2.5 Calls in a telephone system arrive randomly at an exchange at the rate of 140 per hour. If there is a very large number of lines available to handle the calls, that last an average of 3 minutes, what is the average number of lines in use? Estimate the 90th and 95th percentile of number of lines in use.

Solution The $M/M/\infty$ queueing model can be used for the estimates. For this example $a = \lambda W_s = (140/60) \times 3 = 7$ erlangs. Hence, the average number of lines in use is 7. We can use the normal approximation as an estimate of percentile values. The 90th percentile value of the normal distribution is the mean plus 1.28 standard deviations; the 95th percentile value is the mean plus 1.645 standard deviations. Thus, the 90th percentile value of number of lines is $7 + 1.28 \times \sqrt{7} = 10.38$ or 10 lines; the 95th percentile value is $7 + 1.645 \times \sqrt{7} = 11.35$ or 11 lines. The APL function POISSONΔDIST shows that $P[N \leq 10] = 0.901479$ and $P[N \leq 11] = 0.946650$. $\square$

5.2.6 The M/M/1/K/K Queueing System (Machine Repair with One Repairman)

(We are guilty of "abuse of notation" here. The arrival pattern to the repairman is not random but quasi-random; that is, the repair interarrival time is what is called quasi-random, as described at the end of this section.) This model, a limited source model in which there are only K customers, is variously called the machine repair model, the machine interference model, or even the cyclic queue model. It is one of the most useful of all queueing theory models. One way to view this model is shown in Figure 5.2.7. The population of potential customers for this queueing system consists of K identical devices, each of which has an operating time of O time units between breakdowns, O having an exponential distribution with average value $E[O] = 1/\alpha$. ($E[O]$ is sometimes called the "mean time to failure" and abbreviated MTTF.) The repairman repairs the machines at an exponential rate with an average repair time of $1/\mu$ time units. The operating machines are outside the queueing system (outlined by dashed lines) and enter the system only when they break down and thus require repair. The queueing system always reaches the steady state because there can be no more than K customers in the system (one machine being repaired and $K - 1$ waiting for repairs). When n of the machines are down (not operating) then $K - n$ of them are operating and the time until the next machine breaks down is the minimum of $K - n$ identical exponential distributions and thus, by Theorem 3.2.1(h), is exponential with parameter $(K - n)\alpha$. Hence, the

state-transition diagram to describe the system is given by Figure 5.2.8.
From the figure we see that

$$p_n = \frac{K!}{(K-n)!} \times \left(\frac{W_s}{E[O]}\right)^n \times p_0, \quad n = 0, 1, \ldots, K \qquad (5.124)$$

where

$$p_0 = \left\{\sum_{k=0}^{K} \frac{K!}{(K-k)!} \times \left(\frac{W_s}{E[O]}\right)^k\right\}^{-1}. \qquad (5.125)$$

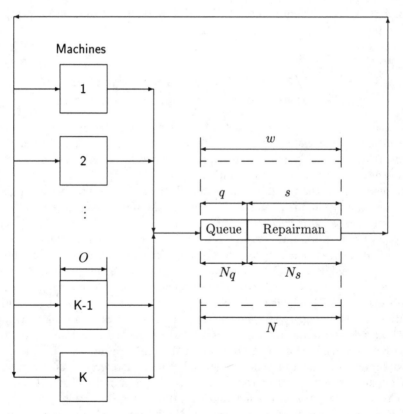

Figure 5.2.7. Machine repair queueing system.

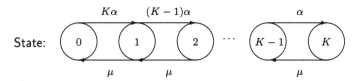

Figure 5.2.8. State-transition rate diagram for M/M/1/K/K system.

If we define $z = E[O]/W_s$, then (5.125) can be written as

$$p_0 = \frac{\dfrac{z^K}{K!}}{\displaystyle\sum_{n=0}^{K} \dfrac{z^n}{n!}} = B[K, z], \tag{5.126}$$

where, of course, $B[K, z]$ is Erlang's loss formula.[11] Thus, the server (repairman) utilization is

$$\rho = 1 - p_0 = 1 - B[K, z]. \tag{5.127}$$

Since $\rho = \lambda W_s$, we can calculate the actual arrival rate into the queueing system as

$$\lambda = \frac{\rho}{W_s}. \tag{5.128}$$

To calculate the performance statistics for the queueing system, we reason as follows. For each of the K machines a complete cycle consists of an operating period, O, followed by a queue for repair, q, and a repair (service) time, s. Thus, the average rate at which K machines break down (enter the queueing system) is given by

$$\lambda = \frac{K}{E[O] + W_q + W_s} = \frac{K}{E[O] + W}, \tag{5.129}$$

where

$$W = W_q + W_s.$$

From (5.129) we calculate

$$W = \frac{K}{\lambda} - E[O]. \tag{5.130}$$

---

[11]To bring (5.125) into the form of (5.126), multiply the numerator and denominator by $z^K/K!$

We have, of course,

$$W_q = W - W_s. \tag{5.131}$$

Little's law yields

$$L_q = \lambda W_q, \tag{5.132}$$

and

$$L = \lambda W. \tag{5.133}$$

The most difficult part of the calculations for the machine repair model is $p_0 = B[K, z]$. Fortunately, Algorithm 5.2.2 is easy to implement on a programmable calculator such as a Hewlett-Packard HP-32S or HP-28S. In fact, I wrote the HP-32S program to do this in Allen [1].

Example 5.2.6 High Tale Airfreight has 20 buffered terminals on one communication line. The terminals are used for data entry to a computer system. The average time required to key an entry into the buffer is 80 seconds; this keying time is approximately exponential. High Tale analysts have found they can model such systems using the machine repair model with $W_s = 2$ seconds. Calculate the throughput λ and the mean response time W. Repeat the calculations if 50 terminals are put on the line but everything else remains the same.

Solution Since $z = E[O]/W_s = 80/2 = 40$ for 20 terminals, we calculate

$$p_0 = B[20, 40] = 0.521307.$$

Thus,

$$\rho = 1 - p_0 = 0.478693,$$

and

$$\lambda = \frac{\rho}{W_s} = 0.2393465 \quad \text{entries per second}$$

We calculate

$$W = \frac{K}{\lambda} - E[O] = 3.56 \quad \text{seconds.}$$

For 50 terminals, we calculate

$$p_0 = B[50, 40] = 0.018690671,$$

so that $\rho = 0.981309329$, $\lambda = 0.490654665$, and $W = 21.905$ seconds. $\square$

Let q_n be the probability there are n machines in the repair facility (being repaired or awaiting repair) just before a machine breaks down; that is, that an inoperative machine finds n ahead of it when it arrives. It is not difficult to show (see Exercise 32) that

$$q_n = \frac{(K-n)p_n}{K-L}, \quad n = 0, 1, 2, \ldots, K-1.$$

To calculate the distribution function for the time a machine spends in the repair facility, $W[\cdot]$, we proceed much as we did in deriving this formula for the M/M/1 and M/M/1/K queueing systems. We let N_a be the random variable that counts the number of machines found in the repair facility by an arriving machine; that is, $P[N_a = n] = q_n$. We can write

$$
\begin{aligned}
W[t] &= \sum_{n=0}^{K-1} P[w \le t | N_a = n]\, P[N_a = n] \\
&= \sum_{n=0}^{K-1} P[w \le t | N_a = n]\, q_n.
\end{aligned}
\tag{5.134}
$$

A customer (machine) that arrives when there are n customers already there has a time in the system equal to the sum of $n+1$ independent identically distributed exponential random variables, each with mean $1/\mu$. Hence, by Theorem 3.2.6(d), the corresponding density function is given by

$$f_{w,n}(x) = \frac{\mu\,(\mu x)^n\, e^{-\mu x}}{n!}, \quad x \ge 0 \tag{5.135}$$

and

$$P[w \le t | N_a = n] = \int_0^t \frac{\mu\,(\mu x)^n\, e^{-\mu x}}{n!}\, dx. \tag{5.136}$$

Therefore,

$$
\begin{aligned}
W[t] &= \sum_{n=0}^{K-1} \left\{ \int_0^t \frac{\mu\,(\mu x)^n\, e^{-\mu x}}{n!}\, dx \right\} q_n \\
&= 1 - \sum_{n=0}^{K-1} Q[n; \mu t]\, q_n,
\end{aligned}
\tag{5.137}
$$

by Exercise 13, where

$$Q[n; \mu t] = e^{-\mu t} \sum_{k=0}^{n} \frac{(\mu t)^k}{k!}. \tag{5.138}$$

Similarly,

$$
\begin{aligned}
W_q[t] &= W_q[0] + \sum_{n=1}^{K-1} P[q \le t | N_q = n] \, q_n \\
&= q_0 + \sum_{n=1}^{K-1} q_n \int_0^t \frac{\mu(\mu x)^{n-1}}{(n-1)!} e^{-\mu x} \, dx \\
&= 1 - \sum_{n=0}^{K-2} Q[n; \mu t] \, q_{n+1}.
\end{aligned}
\tag{5.139}
$$

Note that (5.139) is exactly the same as the corresponding formulas for the M/M/1/K system! The formulas for $W[\cdot]$ agree, too. However, the q_n's in (5.139) are very different from those in the M/M/1/K system. $\square$

The formulas for $W[t]$ and $W_q[t]$ can be simplified. We note that (5.127) and (5.124) together yield

$$
p_n = \frac{\dfrac{z^{K-n}}{(K-n)!}}{\displaystyle\sum_{k=0}^{K} \frac{z^k}{k!}}, \quad n = 0, 1, 2, \ldots, K,
\tag{5.140}
$$

where $z = E[O]/W_s$. If we multiply the numerator and denominator of (5.140) by e^{-z}, we see that the random variable that counts the number of inoperative machines has a truncated Poisson distribution (see Exercise 47 of Chapter 2). Moreover,

$$
q_n = \frac{\dfrac{z^{K-n-1}}{(K-n-1)!}}{\displaystyle\sum_{k=0}^{K-1} \frac{z^k}{k!}}, \quad n = 0, 1, \ldots, K-1,
\tag{5.141}
$$

by Exercise 32. Thus, each q_n is the p_n that we would calculate if the total number of machines were reduced by one; that is, for a $M/M/1/K-1/K-1$ system. If we use the notation of (5.138) for a Poisson sum and the notation

$$
p[k; \alpha] = e^{-\alpha} \frac{\alpha^k}{k!}
$$

for a Poisson probability, we can write (5.141) as

$$
q_n = \frac{p[K-n-1; z]}{Q[K-1; z]}.
$$

Thus, (5.137) becomes

$$W[t] = 1 - \sum_{n=0}^{K-1} \frac{p[K - n - 1; z]\, Q[n; \mu t]}{Q[K - 1; z]}. \tag{5.142}$$

It is easy to show (see Exercise 33) that

$$\sum_{j=0}^{k} p[k - j; \lambda]\, Q[j; \mu] = Q[k; \lambda + \mu]. \tag{5.143}$$

Substituting (5.143) in (5.142) yields

$$W[t] = 1 - \frac{Q[K - 1; z + \mu t]}{Q[K - 1; z]}, \quad t \ge 0. \tag{5.144}$$

Similarly, we can write

$$W_q[t] = 1 - \frac{Q[K - 2; z + \mu t]}{Q[K - 1; z]}, \quad t \ge 0. \tag{5.145}$$

Recall that, in (5.144) and (5.145),

$$z + \mu t = \frac{E[O]}{W_s} + \frac{t}{W_s} = \frac{E[O] + t}{W_s}. \quad \square \tag{5.146}$$

Recall that $Q[k; y]$ is defined by the Poisson sum

$$Q[k; y] = e^{-y} \sum_{i=0}^{k} \frac{y}{i!}. \tag{5.147}$$

We ask you to show in Exercise 33 that

$$Q[k; y] = \frac{1}{k!} \int_{y}^{\infty} e^{-x} x^k \, dx. \tag{5.148}$$

But (5.148) is the "right tail" probability of a gamma random variable with parameters $\beta = k + 1$ and $\alpha = 1$; that is, if Y is such a random variable, the $Q[k; y] = P[Y > y]$. We can compute such a probability using the APL function GΔDIST or we can appeal to the definition of $Q[k; y]$ and use POISSONΔDIST.

Example 5.2.7 Consider Example 5.2.6. We found that the mean response time W was 3.56 seconds. We can use (5.144) to calculate

$$W[t] = P[w \le 5] = 1 - \frac{Q[19; 42.5]}{Q[19; 40]}.$$

Now

$$Q[19; 42.5] = e^{-42.5} \sum_{n=0}^{19} \frac{42.5^n}{n!} = 4.376854 \times 10^{-5},$$

and

$$Q[19; 40] = e^{-40} \sum_{n=0}^{19} \frac{40^n}{n!} = 0.000176303.$$

(We used POISSONΔDIST to make the calculations.)   Thus, $W[5] = 0.75174$. Similarly, by (5.145), we calculate

$$W_q[5] = 1 - \frac{Q[18; 42.5]}{Q[19; 40]} = 0.89317.$$

These calculations could also be made using the APL functions DWΔMRΔ1 and DWQΔMRΔ1, respectively. $\square$

The equations for the machine repair model M/M/1/K/K are collected in Table 10 of Appendix C.

There is a another, more general, single repairman machine repair model, the $M/G/1/K/K$ queueing system, in which the repair time has a general distribution although the up time for each machine is exponential. The $M/G/1/K/K$ system is very useful for computer system modeling. Unfortunately, it is not a birth-and-death process queueing model. The derivation of the equations to describe this model is too advanced to give here but can be found in Takács [59] and Jaiswal [25].

The only difference in the computations for the $M/G/1/K/K$ system from that of the M/M/1/K/K system is in the calculation of p_0. After we obtain p_0, we calculate ρ , λ, W_q, and W exactly as we do for the M/M/1/K/K system. Of course, the formulas for the distribution functions of q and w no longer apply. The formula for p_0 is

$$p_0 = \left[1 + \frac{K W_s}{E[O]} \sum_{n=0}^{K-1} \binom{K-1}{n} \times B_n \right]^{-1}, \qquad (5.149)$$

where

$$B_n = \begin{cases} 1 & \text{for } n = 0 \\ \displaystyle\prod_{i=1}^{n} \left(\frac{1 - W_s^*[i/E[O]]}{W_s^*[i/E[O]]} \right) & \text{for } n = 1, 2, \dots, K-1, \end{cases} \qquad (5.150)$$

and, of course, $W_s^*[\theta]$ is the Laplace–Stieltjes transform of the service (repair) time.

Table 5.2.2. Example 5.2.8

n	B_n	$\binom{3}{n}$	$B_n \times \binom{3}{n}$
0	1.00000	1	1.00000
1	0.02532	3	0.07595
2	0.00130	3	0.00389
3	0.00010	1	0.00010

		Sum	1.07994

Example 5.2.8 Richard E. Rutledge, Jr. (Dick to his friends and subordinates), a senior executive at the Blew Computer Corporation (sometimes called "Big Blew") believes he can model the word processing and electronic mail activities in his executive office suite as an $M/D/1/4/4$ queueing system. He generates so many letters, memos, and electronic mail messages each day that four secretaries ceaselessly type away at their workstations that are connected to a large computer system over a LAN. Each secretary works, on the average, for 40 seconds before he makes a request for service to the computer system. A request for service is processed in almost exactly one second. Dick has measured the mean response time using his electronic watch (he gets 1.05 seconds) and estimates the throughput as 350 requests per hour. Dick has decided to hire two additional secretaries to keep up with his prodigious productivity and will connect their workstations to the same LAN if the $M/D/1/6/6$ model indicates a mean response time of less than 1.5 seconds. Can he hire the two secretaries?

Solution Let us first check out the present system. From the definition of the Laplace–Stieltjes transform for a constant service time, we see that

$$W_S^*[\theta] = e^{-\theta}.$$

Therefore,

$$\frac{1 - W_S^*[i/E[O]]}{W_S^*[i/E[O]]} = \frac{1 - e^{-i/40}}{e^{-i/40}} = e^{i/40} - 1.$$

Thus, we can write

$$B_n = \begin{cases} 1 & \text{for } n = 0 \\ \prod_{i=1}^{n} \left(e^{i/40} - 1 \right) & \text{for } n = 1, 2, 3. \end{cases} \tag{5.151}$$

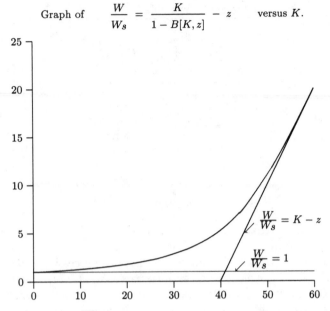

Graph of $\dfrac{W}{W_s} = \dfrac{K}{1 - B[K, z]} - z$ versus K.

$\dfrac{W}{W_s} = K - z$

$\dfrac{W}{W_s} = 1$

Figure 5.2.9. $\dfrac{W}{W_s}$ versus K for M/M/1/K/K queueing system.

We show the values needed to calculate p_0 by (5.149) in Table 5.2.2. The values in this table were calculated to ten decimal places but are shown rounded to five. Hence, by (5.149),

$$p_0 = \left[1 + \frac{4 \times 1}{40} \times 1.07994\right]^{-1} = 0.90253.$$

This could also be calculated by the APL function PZERO. Thus, $\rho = 1 - p_0 = 0.09747$, $\lambda = \rho/W_s = 0.09747$ requests per second or 350.88 requests per hour. Then

$$W = \frac{K}{\lambda} - E[O] = 1.038 \quad \text{seconds}.$$

Dick's calculations were amazingly close to the model results! We leave it as an exercise for the reader (Exercise 34) to show that with six secretaries $p_0 = 0.85390$, $\lambda = 525.95$ requests per hour and $W = 1.069$ seconds. The calculations can be made using $M\Delta D\Delta 1\Delta K\Delta K$, that uses the function PZERO to calculate p_0. These values show that Dick can add two work stations without seriously degrading the performance of his office staff. □

We can use the machine repair model (shown in Figure 5.2.7) to model some interactive computer systems. The computer terminals (or personal computers being used as terminals) correspond to the machines and the central computer system to the repair facility. We think of the service time of the computer system as having an exponential distribution. We also think of the "think time" of the terminals, that corresponds to up time of a machine, as being exponential. (Usually the time it takes to receive output from the computer, think about what to do next, and input a request to the computer is lumped together as "think time.") In Figure 5.2.9 we have a graph of such a system. We show the normalized response time, W/W_s, versus the number of active terminals, K. When there is only one active user ($K = 1$), there is no queueing for service so that $W = W_s$; the line $W/W_s = 1$ is an asymptote to the curve. As the number of active terminals grows very large (think of $K = 100,000,000,000$, that is, 100 American billion), we would expect the computer system to be busy most of the time, so $\rho \approx 1$. But then $\lambda = \rho/W_s \approx 1/W_s$ and adding more machines has little effect on the throughput. When we substitute this equation for λ into our equation for W (equation (5.130)) we obtain $W = K W_s - E[O]$ or $W/W_s = K - E[O]/W_s = K - z$. This means that the curve has the asymptote $W/W_s = K - z$ for large K. Thus, when K is large, increasing it by 1 has little effect on the throughput but increases W by approximately W_s (W/W_s by 1). The two asymptotes to the curve intersect at the point $(K^*, 1)$ where $K^* = 1 + z = (E[O] + W_s)/W_s$. Kleinrock [32, page 209, (4.66)] calls the value K^* the *saturation number*. He also gives it an interesting physical interpretation. If each terminal user has exactly $E[O]$ units of think time and uses exactly W_s units of service time per interaction, then, with perfect synchronization, K^* is the maximum number of terminals that can be supported with no mutual interference. Thus, for any real system when K is one, we have no mutual interference; when K is small, we have almost no mutual interference. However, as K increases past K^* we are certain to have some mutual interference (queueing for service) while the formula $W = K W_s - E[O]$ for a very large number of users (terminals) shows complete interference! That is, for a very large number of users, each user delays every other user by one mean service time, W_s. As we shall see later, the asymptotic formula $W = K W_s - E[O]$ is valid also for a general service time; in fact, it is true for even more general "machine repair like" queueing systems.

Figure 5.2.9 was plotted using $z = 40$ so that $K^* = 41$, but the shape of the curve will be very similar for other values of z.

Example 5.2.9 Consider Example 5.2.6 in which High Tale Airfreight had a system of 20 terminals on one communication line and was able to model it as a $M/M/1/20/20$ queueing system with $W_s = 2$ seconds and $E[O] = 80$ seconds. Suppose the nature of the application changes so that the system response time is nearly constant with the same mean. (For example, they could have attached the terminals to a lightly loaded LAN.) Let us see how this new system compares to the old. We will use the equations found in Table 11 of Appendix C. (We can check our solution using MΔDΔ1ΔKΔK.) From the definition $W_s^*[\theta] = e^{-2\theta}$, and, since $E[O] = 80$, we have

$$\frac{1 - W_s^*[i/E[O]]}{W_s^*[i/E[O]]} = \frac{1 - e^{-i/40}}{e^{-i/40}} = e^{i/40} - 1,$$

for $i = 1, 2, \ldots, 19$. Hence,

$$B_n = \begin{cases} 1 & \text{for } n = 0 \\ \prod_{i=1}^n \left(e^{i/40} - 1\right) & \text{for } n = 1, 2, \ldots, 19. \end{cases} \tag{5.152}$$

To calculate p_0, we substitute these values of B_n into (5.149) yielding

$$p_0 = \left[1 + \frac{20 \times 2}{80} \sum_{n=0}^{19} \binom{19}{n} B_n\right]^{-1} = 0.51706.$$

This calculation can also be made with the APL function PZERO. Then $\rho = 1 - p_0 = 0.48294$, $\lambda = \rho/2 = 0.24147$, and

$$W = \frac{K}{\lambda} - E[O] = 2.826 \quad \text{seconds.}$$

For this system,

$$K^* = \frac{E[O] + W_s}{W_s} = 41 \quad \text{terminals.}$$

We leave it as an exercise (Exercise 35) to show that, if High Tale had 50 terminals, then $p_0 = 0.00452$, $\rho = 0.99548$, $\lambda = 0.49774$ and $W = 20.45$ seconds. (The asymptotic formula $W = K W_s - E[O]$ yields the value $W = 20$ seconds.) $\square$

As we indicated in the footnote at the beginning of this section, it is an "abuse of notation" to use the Kendall notation M/M/1/K/K to describe the machine repair model with one repairman, because the interarrival time for customer arrival at the service center is not actually exponential. The input to the service facility is what is called *quasi-random input*. As described by Kobayashi [33, Section 2.9.2] this means that "... a finite number K of sources generate quasi-random input if (1) the probability that any particular source generates a request in an interval $(t, t + h)$ is $\alpha h + o(h)$ when the source is eligible to generate a new request at time t, and (2) all sources act independently of the states of any other sources." The system we designate by M/M/c/K/K in the next section also has quasi-random input.

5.2.7 M/M/c/K/K Queueing System (Machine Repair, Multiple Repairmen)

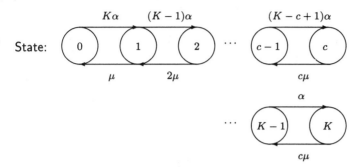

Figure 5.2.10. State-transition rate diagram for M/M/c/K/K system.

This queueing system is similar to the machine repair model considered in the last section except that we have c repairmen rather than one. We have a birth-and-death process model with the state-transition diagram shown in Figure 5.2.10. From this figure we can calculate the p_n's as

$$
p_n =
\begin{cases}
\dbinom{K}{n} \left(\dfrac{W_s}{E[O]} \right)^n p_0, & n = 0, 1, \ldots, c \\[3mm]
\dfrac{n!}{c! c^{n-c}} \dbinom{K}{n} \left(\dfrac{W_s}{E[O]} \right)^n p_0, & n = c+1, \cdots, K,
\end{cases}
\tag{5.153}
$$

where

$$
p_0 = \left[\sum_{k=0}^{c} \binom{K}{k} \left(\frac{W_s}{E[O]} \right)^k + \sum_{k=c+1}^{K} \frac{k!}{c! c^{k-c}} \binom{K}{k} \left(\frac{W_s}{E[O]} \right)^k \right]^{-1}.
\tag{5.154}
$$

We can now calculate

$$
L_q = \sum_{n=c+1}^{K} (n - c) p_n.
\tag{5.155}
$$

It is easy to see that

$$
\lambda = \frac{K}{E[O] + W},
\tag{5.156}
$$

where $W = W_q + W_s$. We can apply Little's law to obtain

$$
W_q = \frac{L_q}{\lambda} = \frac{L_q}{K} (E[O] + W_q + W_s).
\tag{5.157}
$$

Solving (5.157) for W_q yields

$$W_q = \frac{L_q}{K - L_q}(E[O] + W_s). \tag{5.158}$$

We can now substitute W_q from (5.158) into (5.156) to obtain a numerical value for λ. We can also calculate

$$W = W_q + W_s, \tag{5.159}$$

and, using Little's law,

$$L = \lambda W, \tag{5.160}$$

and

$$L_q = \lambda W_q. \tag{5.161}$$

The formulas for this model are collected in Table 12 of Appendix C. The calculation of the performance parameters of the machine repair model using the formulas of Table 12 are implemented by the APL program MACHΔREP. In Table 13 of Appendix C we give the formulas for the machine repair model $D/D/c/K/K$. See Boyse and Warn [3] for the derivations.

To derive the distribution function for q, $W_q[\cdot]$, we reason that an arriving machine (customer) must queue for repair (service) only if $n \geq c$, where n is the number of customers found in the repair system. When this is the case, the arrival must wait for the departure of $(n-c)+1$ customers. (If $n = c$, one customer must depart, if $n = c + 1$, two customers must depart, etc.) Let N_a be the number of customers an arriving machine finds in the repair facility so that $q_n = P[N_a = n]$. If $n \geq c$ and $k = n - c$, then $P[q > t|N_q = n]$ is the probability that k or fewer customers depart in an interval of length t. But this probability is given by

$$P[q > t|N_q = n] = e^{-c\mu t} \sum_{i=0}^{k} \frac{(c\mu t)^i}{i!} = Q[k; c\mu t]. \tag{5.162}$$

Formula (5.162) follows from the fact that the service facility services customers like a single server with an exponential distribution and mean service time $1/(c\mu)$. Thus, the number of customers processed in an interval of length t has a Poisson distribution with mean $c\mu t$. We have

$$P[q > t] = \sum_{n=c}^{K-1} P[q > t|N_q = n]\, q_n = \sum_{n=c}^{K-1} q_n Q[n - c; c\mu t]. \tag{5.163}$$

To complete the derivation, we need the theorem stated by Reiser [50] and proved by Sevcik and Mitrani [54]. As Reiser states the theorem,

In a closed queueing network the (stationary) state probabilities at customer arrival epochs are identical to those of the same network in long-term equilibrium with one customer removed.

In our notation this means that

$$q_n[K] = p_n[K-1], \quad n = 0, 1, \cdots, K-1, \tag{5.164}$$

where $q_n[K]$ is the steady state probability that an arriving machine finds n machines in the repair facility when there are K machines in the system, and $p_n[K-1]$ is the steady-state probability that there are n machines in the repair facility for a system with $K-1$ machines. The q_n's in (5.163) are written as $q_n[K]$ in the notation (5.164). By (5.153), when $n \geq c$, we can write, using $z = E[O]/W_s$,

$$
\begin{aligned}
p_n[K] &= \frac{n!}{c!c^{n-c}} \binom{K}{n} z^n p_0[K] \\
&= \frac{n!c^c}{c!c^n} \frac{K!z^n}{(K-n)!n!} p_0[K] \\
&= \frac{c^c}{c!} p_0[K] \frac{\dfrac{e^{-cz}(cz)^{K-n}}{(K-n)!}}{\dfrac{e^{-cz}(cz)^K}{K!}} \\
&= \frac{c^c}{c!} p_0[K] \frac{p[K-n; cz]}{p[K; cz]}. \tag{5.165}
\end{aligned}
$$

Therefore,

$$q_n[K] = p_n[K-1] = \frac{c^c}{c!} p_0[K-1] \frac{p[K-n-1; cz]}{p[K-1; cz]}. \tag{5.166}$$

Substituting the above formula for $q_n[K]$ into (5.163) yields

$$
\begin{aligned}
P[q > t] &= \sum_{n=c}^{K-1} p_n[K-1] Q[n-c; c\mu t] \\
&= \frac{c^c p_0[K-1]}{c! p[K-1; cz]} \sum_{n=c}^{K-1} p[K-n-1; cz] Q[n-c; c\mu t] \\
&= \frac{c^c p_0[K-1]}{c! p[K-1; cz]} \sum_{n=0}^{K-c-1} p[K-c-1-n; cz] Q[n; c\mu t] \\
&= \frac{c^c p_0[K-1]}{c! p[K-1; cz]} Q[K-c-1; cz + c\mu t]
\end{aligned}
$$

$$= \frac{c^c\, p_0[K-1]\, Q[K-c-1; c(E[O]+t)/W_s)]}{c!\, p[K-1; c(E[O]/W_s)]}. \qquad (5.167)$$

In the next to last step of (5.167), we used the identity

$$\sum_{j=0}^{k} p[k-j; \lambda] Q[j; \mu] = Q[k; \lambda + \mu]$$

from Exercise 33. Now we can use (5.167) to write

$$W_q[t] = 1 - \frac{c^c\, p_0[K-1] Q[K-c-1; c(E[O]+t)/W_s]}{c!\, p[K-1; c E[O]/W_s]}. \qquad (5.168)$$

The derivation of the distribution function $W[\cdot]$ is much more complicated. It is derived in the solutions manual for Kobayashi [33]. The result is

$$\begin{aligned} W[t] \;=\;\; & 1 - C_1\, e^{-t/W_s} \\ & + C_2 \times Q[K-c-1; c(E[O]+t)/W_s] \quad t \geq 0, \quad (5.169) \end{aligned}$$

where

$$C_1 = 1 + C_2 \times Q[K-c-1; cz], \qquad (5.170)$$

and

$$C_2 = \frac{c^c\, p_0[K-1]}{c!(c-1)(K-c-1)!\, p[K-1; c E[O]/W_s]}. \qquad (5.171)$$

Formula (5.154) (with K replaced by $K-1$ everywhere it appears) can be used to calculate $p_0[K-1]$ in (5.168) and (5.171).

Example 5.2.10 Slobovian Scientific has an interactive time-sharing system with 80 active terminals during the peak period of the day. They know that this system can be modeled as a machine repair queueing system with one repairman, that is, as an $M/M/1/80/80$ queueing system. Their measurements show that the average service time of the computer, $W_s = 0.18$ seconds and their average think time, $E[O] = 18$ seconds. Their performance measurements show that the average response time, $W = 0.69$ seconds, their throughput, $\lambda = 4.3$ interactions per second, and $P[w \leq 1 \text{ second}] = 0.76$. Slobovian Scientific is a growing company. They estimate that three months from now they will average 100 users during the peak period, and that within a year they will have 160 users. They want to estimate the performance of their current system with 100 users. They are considering an upgrade to their computer, that would replace the CPU with a multiprocessing system consisting of two CPUs identical to their current CPU. They want to estimate W, λ, and $P[w \leq 1 \text{ second}]$ for

(a) their current system with 100 terminals,

(b) the upgraded system with 100 terminals,

(c) the upgraded system with 160 terminals.

They want $W \leq 0.8$ seconds with $P[w \leq 1 \text{ second}] \geq .9$.

Solution

(a) The APL programs MACHΔREP and DWΔMRΔ1 show that $W = 1.474$ seconds, $\lambda = 5.13$ transactions per second, and $P[w \leq 1 \text{ second}] = 0.4219$. This is unsatisfactory.

(b) The $M/M/2/100/100$ model applies. It is necessary to correct for the fact that two CPUs operating as a multiprocessor system interfere with each other. We correct for this by assuming that each of the CPUs operates at 90% of the speed it would achieve by itself; we do this by setting $W_s = 0.18/0.9 = 0.2$ seconds. Then the APL program MACHΔREP indicates that $W = 0.28$ seconds, and $\lambda = 5.47$ interactions per second. The APL function DWΔMRΔC shows that $P[w \leq 1 \text{ second}] = 0.99326$. This performance would be very satisfactory.

(c) The $M/M/2/160/160$ queueing model applies with $W_s = 0.2$ seconds. MACHΔREP indicates that $W = 0.64925$ seconds, and $\lambda = 8.5794$ interactions per second, while DWΔMRΔC calculates $P[w \leq 1 \text{ second}] = 0.99326$. The upgrade is AOK! □

5.3 Embedded Markov Chain Systems

In Section 5.2 we showed how, in some cases, a queueing system can be modeled as a birth-and-death process. This makes it relatively easy to calculate the steady state distribution of number of customers in the system and other performance measures; these include average queueing time, average waiting time in the system, etc. One fact makes analysis of such systems straightforward; it is that the stochastic process $\{N(t), t \geq 0\}$ is Markov. This is true because of the memoryless property of the exponential service time distribution; we need not account for the service time already expended by a customer receiving service. When more general service times are allowed, not only $\{N(t), t \geq 0\}$, but also $\{R(t), t \geq 0\}$, the remaining service time for the current customer, is needed to predict future values of $N(t)$. However, for a number of queueing systems, an embedded Markov chain can be constructed, that makes it possible to compute many of the parameters of interest.

5.3.1   The M/G/1 Queueing System

We assume the queueing system has a Poisson input process with average value λ and a general service time distribution. Customers have independent service times. The other restriction on the service time that must be made in order to calculate $L = E[N]$, $W = E[w]$, and $L_q = E[q]$, is that the first and second moments of service time, $E[s]$ and $E[s^2]$, must exist. In order to calculate the standard deviations of each of the primary random values, $E[s^3]$ must also exist and be known. Let $Z(t)$ for $t \geq 0$ denote the number of customers in the queueing system at time t. Let $0 < t_1 < t_2 < \cdots < t_n < \cdots$ denote the successive times at which a customer completes service. Then the sequence $\{Z(t_n)\}$ forms a discrete time process $\{X_n\}$ where $X_n = Z(t_n)$ for $n = 1, 2, 3 \ldots$. Thus, X_n is the number of customers the nth departing customer leaves behind. We will show that $\{X_n\}$ is a Markov chain. First we note that

$$X_{n+1} = \begin{cases} X_n - 1 + A & \text{if } X_n \geq 1 \\ A & \text{if } X_n = 0, \end{cases} \tag{5.172}$$

where A is the number of customers who arrive during the service time of the $(n + 1)$st customer. The service time, s, of the $(n + 1)$st customer is independent of the service time of other customers and of the number of customers in the system. The arrival process is Poisson, which has stationary increments by Definition 4.2.1. Thus, A depends only on s and not on when the service began or the length of the queue. Since X_{n+1} depends only on the value of X_n, and the independent random variable A, and not on X_{n-1}, X_{n-2}, etc., $\{X_n\}$ is a Markov chain.

The arrival process is Poisson but the length of the service time is a random variable, so A is not a Poisson random variable.[12] However, we can write

$$P[A = n] = \int_0^\infty P[A = n | s = t] d\, W_s[t], \ n = 0, 1, \ldots, \tag{5.173}$$

by the law of total probability, where the integral is a Stieltjes integral with $W_s[\cdot]$ the distribution function of the service time. (Recall from Section 2.9 that the Stieltjes integral $\int_0^\infty h(x)\, dF(x)$ is evaluated as $\int_0^\infty h(x) f(x)\, dx$ or $\sum_{x_i} h(x_i) p(x_i)$, depending on whether the random variable X with distribution function $F(\cdot)$ is continuous with density function $f(\cdot)$ or discrete with pmf $p(\cdot)$.) By Theorem 4.2.1,

$$P[A = n | s = t] = \frac{e^{-\lambda t}(\lambda t)^n}{n!}, \quad n = 0, 1, 2, \ldots. \tag{5.174}$$

---

[12]Unless, of course, the service time is constant; then A *is* Poisson with mean λW_s.

Thus,

$$
\begin{aligned}
P_{ij} &= P[X_{n+1} = j | X_n = i] = P[A = j - i + 1] \\
&= \int_0^\infty P[A = j - i + 1] \, dW_s[t] \\
&= \begin{cases} \displaystyle\int_0^\infty e^{-\lambda t} \frac{(\lambda t)^{j-i+1}}{(j - i + 1)!} \, dW_s[t] & \text{if } j \geq i - 1 \text{ and } i \geq 1 \\[2mm] 0 & \text{if } j < i - 1 \text{ and } i \geq 1. \end{cases}
\end{aligned}
$$

Equation (5.175) allows us to calculate P_{ij} for $i \geq 1$. However, it does not tell us how to calculate the first row of the state-transition matrix P, that is, the probabilities of transition from the 0 state. We reason that, if the system is empty when a customer completes service and leaves the system, then no state transition can occur until a new customer arrives; the next transition occurs when this new customer departs. Thus, the state transition probabilities are the same for $i = 0$ as for $i = 1$; that is, the first row of P is the same as the second row. it is convenient to represent P in terms of the probabilities $\{k_n\}$, where k_n is the probability that n customers arrive during one service period. Thus,

$$
k_n = P[A = n] = \int_0^\infty e^{-\lambda t} \frac{(\lambda t)^n}{n!} \, dW_s[t], \quad n = 0, 1, 2 \ldots . \tag{5.175}
$$

We have

$$
P = \begin{bmatrix}
k_0 & k_1 & k_2 & k_3 & k_4 & \cdots \\
k_0 & k_1 & k_2 & k_3 & k_4 & \cdots \\
0 & k_0 & k_1 & k_2 & k_3 & \cdots \\
0 & 0 & k_0 & k_1 & k_2 & \cdots \\
\vdots & \vdots & \vdots & \vdots & \vdots &
\end{bmatrix}. \tag{5.176}
$$

It is intuitively clear that the queueing system should be stable if and only if the average number of customers who arrive during one service time, $E[A]$, is less than 1. But

$$
\begin{aligned}
E[A] &= \sum_{n=0}^\infty n \, k_n = \int_0^\infty e^{-\lambda t} \sum_{n=0}^\infty \frac{n(\lambda t)^n}{n!} \, dW_s[t] \\
&= \int_0^\infty e^{-\lambda t} (\lambda t) \sum_{n=0}^\infty \frac{(\lambda t)^n}{n!} \, dW_s[t] \\
&= \lambda \int_0^\infty t \, dW_s[t] = \lambda W_s = a = \rho. \tag{5.177}
\end{aligned}
$$

Hence, ρ is the average number of customers to arrive during one service time. We will now prove that, if $\rho < 1$, then the embedded Markov chain $\{X_n\}$ has a steady state probability distribution π. First we note that, since $k_n > 0$ for all n, the Markov chain is irreducible. Likewise, it is aperiodic, since $P_{ii} > 0$ for each i. So suppose $\rho < 1$ and $x_i = i/(1-\rho)$, $i = 0, 1, 2, \ldots$. Then, if $i > 0$,

$$
\begin{aligned}
\sum_{j=0}^{\infty} P_{ij} x_j &= \sum_{j=0}^{\infty} P_{ij} \frac{j}{1-\rho} = \sum_{j=i-1}^{\infty} k_{j-i+1} \left(\frac{j}{1-\rho} \right) \\
&= \frac{1}{1-\rho} \{ k_0(i-1) + k_1 i + k_2(i+1) + k_3(i+2) + \cdots \} \\
&= \frac{1}{1-\rho} \{ k_0(i-1) + k_1(i-1) + k_2(i-1) + \cdots \} \\
&\quad + \frac{1}{1-\rho} \{ k_1 + 2k_2 + 3k_3 + \cdots \} \\
&= \frac{i-1}{1-\rho} \sum_{j=0}^{\infty} k_j + \frac{1}{1-\rho} \sum_{j=1}^{\infty} j\, k_j \\
&= \frac{i-1}{1-\rho} + \frac{\rho}{1-\rho} = \frac{i}{1-\rho} - 1 = x_i - 1.
\end{aligned}
$$

Also

$$
\sum_{j=0}^{\infty} P_{0j} x_j = \sum_{j=0}^{\infty} \frac{j k_j}{1-\rho} = \frac{\rho}{1-\rho} < \infty.
$$

Hence, by Theorem 4.4.5, the embedded Markov chain $\{X_n\}$ is ergodic and thus, by Theorem 4.4.4, it has a steady state probability distribution.

Taylor and Karlin [61] show that, if $\rho = 1$, then $\{X_n\}$ is recurrent null, while, if $\rho > 1$, $\{X_n\}$ is transient. In either case $\{X_n\}$ has no steady state probability distribution. Thus, $\{X_n\}$ has a steady state probability distribution if and only if $\rho < 1$.

We assume, henceforth, that $\rho < 1$, so the embedded Markov chain $\{X_n\}$ has a steady state distribution $\pi = (\pi_0, \pi_1, \ldots)$ with $\pi_i > 0$ for each i,

$$
\sum_{i=0}^{\infty} \pi_i = 1, \quad \text{and} \quad \pi_j = \sum_{i=0}^{\infty} \pi_i P_{ij}, \quad j = 0, 1, 2 \ldots. \tag{5.178}
$$

The above series of equations, (5.178), is the component by component statement of the matrix equation $\pi = \pi P$, expressing the fact that π is a stationary distribution. In terms of (5.176), the equations are

$$\pi_i = \pi_0 k_i + \sum_{j=1}^{i+1} \pi_j k_{i-j+1}, \quad k = 0, 1, 2 \ldots. \tag{5.179}$$

We assume that X is the steady state number of customers in the system at departure instants; that is,

$$P[X = n] = \pi_n, \quad n = 0, 1, 2 \ldots.$$

We define the generating functions (see Section 2.9) of the distributions π and $K = (k_0, k_1, k_2, \ldots)$ by

$$\pi(z) = \sum_{i=0}^{\infty} \pi_i \, z^i = g_X(z), \tag{5.180}$$

and

$$K(z) = \sum_{i=0}^{\infty} k_i \, z^i = g_A(z). \tag{5.181}$$

If we multiply (5.179) by z^i, we get

$$\pi_i z^i = \pi_0 k_i z^i + \frac{1}{z} \sum_{j=0}^{i+1} \pi_j k_{i-j+1} z^{i+1} - \frac{\pi_0 k_{i+1} z^{i+1}}{z}, \tag{5.182}$$

for $i = 0, 1, 2 \ldots$. Summing (5.182) over i and recognizing

$$\sum_{j=0}^{i+1} \pi_j k_{i+1-j}$$

as a convolution (see Theorem 2.9.2(d)), we see that

$$\pi(z) = \pi_0 K(z) + \frac{1}{z}[K(z)\pi(z) - \pi_0 k_0] - \frac{\pi_0}{z}[K(z) - k_0], \tag{5.183}$$

or

$$\pi(z) = \frac{\pi_0 (1 - z) K(z)}{K(z) - z} = g_X(z). \tag{5.184}$$

Note that

$$\pi(1) = \sum_{i=0}^{\infty} \pi_i = 1 = \sum_{j=0}^{\infty} k_i = K(1),$$

and that, by Theorem 2.9.2(c),

$$K'(1) = E[A] = \sum_{n=0}^{\infty} n\, k_n = \rho.$$

Hence, applying L'Hôpital's rule to (5.184) yields

$$
\begin{aligned}
1 &= \pi(1) = \lim_{z \to 1} \pi(z) \\
&= \lim_{z \to 1} \frac{\pi_0[(1-z)K'(z) - K(z)]}{K'(z) - 1} \\
&= \frac{\pi_0 K(1)}{1 - \rho} = \frac{\pi_0}{1 - \rho},
\end{aligned}
\tag{5.185}
$$

or $\pi_0 = 1 - \rho$. That is, the steady state probability that the system is empty at departure instants is $1 - \rho$, just as it was for the M/M/1 queueing system!

We have shown that the queueing system has a steady state distribution of number of customers only at the times at which customers leave the system. However, it can be shown that what we have proven implies there exists a steady state probability distribution $P = (p_0, p_1, p_2, \ldots)$ of number of customers in the system at arbitrary points in time, furthermore $p_n = \pi_n$ for all n. That is,

$$p_n = P[N = n] = \pi_n = P[X = n].$$

These facts are proven by Gross and Harris [18] and by Kleinrock [31]. Thus,

$$\pi(z) = G_X(z) = g_N(z).$$

Hence, we can write (5.184) as

$$\pi(z) = g_N(z) = \frac{(1-\rho)(1-z)K(z)}{K(z) - z}.
\tag{5.186}$$

Since $\pi'(1) = E[N] = L$, by Theorem 2.9.2, we can calculate L from (5.186). In fact (see Exercise 42), if $\pi'(1)$ is calculated from (5.186), it yields

$$L = \pi'(1) = \rho + \frac{K''(1)}{2(1 - \rho)}.
\tag{5.187}$$

But the formula (5.187) is not very useful as it now stands because we do not know much about $K''(1)$. But, if we let $W_s^*[\theta]$ be the Laplace–Stieltjes transform of s, then, by Theorem 2.9.3,

$$W_s = E[s] = -\left.\frac{dW_s^*[\theta]}{d\theta}\right|_{\theta=0},
\tag{5.188}$$

and

$$E[s^2] = \left. \frac{d^2 W_s^*[\theta]}{d\theta^2} \right|_{\theta=0}. \tag{5.189}$$

By Theorem 2.7.2(d) and Theorem 2.9.2(c),

$$E[A^2] = \text{Var}[A] + E[A]^2 = K''(1) + K'(1). \tag{5.190}$$

But

$$
\begin{aligned}
g_A(z) &= K(z) = \sum_{n=0}^{\infty} k_n z^n = \int_0^{\infty} e^{-\lambda t} \sum_{j=0}^{\infty} \frac{(\lambda t z)^j}{j!} \, dW_s[t] \\
&= \int_0^{\infty} e^{-\lambda t} e^{\lambda t z} \, dW_s[t] = \int_0^{\infty} e^{-(\lambda - \lambda z)t} \, dW_s[t] \\
&= W_s^*[\lambda - \lambda z]. \tag{5.191}
\end{aligned}
$$

Thus, the generating function of A can be represented as the Laplace–Stieltjes transform, $W_s^*[\theta]$, of the service time, evaluated at $\theta = \lambda(1 - z)$. The formula (5.191) is a very useful relation. We can now use differentiation to solve for $E[A]$ (that we already know is ρ), $K''(1)$, and $E[A^2]$. We calculate

$$K'(z) = -\lambda W_s^{*(1)}[\lambda(1-z)] \quad \text{and} \quad K''(z) = \lambda^2 W_s^{*(2)}[\lambda(1-z)].$$

Hence, by (5.188),

$$K'(1) = -\lambda W_s^{*(1)}[0] = \lambda W_s = \rho,$$

and, by (5.189),

$$K''(1) = \lambda^2 W_s^{*(2)}[0] = \lambda^2 E[s^2]. \tag{5.192}$$

Substituting (5.192) into (5.187) yields

$$
\begin{aligned}
L &= E[N] = \rho + \frac{\lambda^2 E[s^2]}{2(1 - \rho)} = \rho + \frac{\lambda^2 \text{Var}[s] + \rho^2}{2(1 - \rho)} \\
&= \rho + \frac{\rho^2(1 + C_s^2)}{2(1 - \rho)}. \tag{5.193}
\end{aligned}
$$

The formula (5.193) in any of the forms shown is known as the *Pollaczek–Khintchine formula*.

Substituting formula (5.192) and $K'(1) = \rho$ into (5.190) yields

$$E[A^2] = \lambda^2 E[s^2] + \rho. \tag{5.194}$$

By the law of total probability,

$$p_n = \int_0^\infty A_n[t] dW[t],$$

where

$$A_n[t] = P[n \text{ arrivals during time } w | w = t],$$

and $W[\cdot]$ is the distribution function for w. But, by Theorem 4.2.1,

$$A_n[t] = e^{-\lambda t} \frac{(\lambda t)^n}{n!}.$$

Hence,

$$p_n = \pi_n = \int_0^\infty e^{-\lambda t} \frac{(\lambda t)^n}{n!} dW[t]. \tag{5.195}$$

This means that

$$
\begin{aligned}
g_N(z) &= \sum_{n=0}^\infty p_n z^n = \int_0^\infty e^{-\lambda t} \sum_{j=0}^\infty \frac{(\lambda t z)^j}{j!} dW[t] \\
&= \int_0^\infty e^{-\lambda t} e^{\lambda t z} dW[t] = \int_0^\infty e^{-\lambda(1-z)t} dW[t] \\
&= W^*[\lambda(1-z)],
\end{aligned}
\tag{5.196}
$$

where $W^*[\theta]$ is the Laplace–Stieltjes transform of the waiting time in the system, w.

We can differentiate (5.196) to get

$$
\begin{aligned}
\frac{dg_N(z)}{dz} &= \left. \frac{dW^*[v]}{dv} \frac{dv}{dz} \right|_{v=\lambda(1-z)} = -\lambda \left. \frac{dW^*[v]}{dv} \right|_{v=\lambda(1-z)} \\
&= \lambda \int_0^\infty t\, e^{-\lambda t(1-z)} dW[t].
\end{aligned}
\tag{5.197}
$$

Hence, by Theorem 2.9.2,

$$L = \left. \frac{dg_N(z)}{dz} \right|_{z=1} = \lambda \int_0^\infty t\, dW[t] = \lambda W. \tag{5.198}$$

We have derived Little's law for the special case of the M/G/1 queueing system. Proceeding as we did in deriving (5.198), we obtain

$$
\begin{aligned}
E[N(N-1)(N-2)\cdots(N-k+1)] &= \left. \frac{d^k g_N(z)}{dz^k} \right|_{z=1} \\
&= \lambda^k E[w^k],
\end{aligned}
\tag{5.199}
$$

for $k = 1, 2, 3, \ldots$. This is a generalization of Little's law. Another important formula is obtained by substituting (5.191) into (5.186) yielding

$$g_N(z) = \frac{(1 - \rho)(1 - z)W_s^*[\lambda(1 - z)]}{W_s^*[\lambda(1 - z)] - z}, \qquad (5.200)$$

that is called the *Pollaczek-Khintchine transform equation* by many authors. (We shall see two other transform equations that are also called by this name.)

If we substitute (5.196) into (5.200) we obtain

$$W^*[\lambda(1 - z)] = \frac{(1 - \rho)(1 - z)\, W_s^*[\lambda(1 - z)]}{W_s^*[\lambda(1 - z)] - z} \qquad (5.201)$$

or

$$W^*[\theta] = \frac{(1 - \rho)\theta W_s^*[\theta]}{\theta + \lambda(W_s^*[\theta] - 1)}. \qquad (5.202)$$

But, since $w = q + s$, by Theorem 2.9.3(d) we must have

$$W^*[\theta] = W_q^*[\theta]\, W_s^*[\theta]. \qquad (5.203)$$

Hence, by the uniqueness of the Laplace–Stieltjes transform,

$$W_q^*[\theta] = \frac{(1 - \rho)\theta}{\theta + \lambda(W_s^*[\theta] - 1)}. \qquad (5.204)$$

We now have three equations, each of which is sometimes called the Pollaczek-Khintchine transform equation; that is, (5.200), (5.202), and (5.204). From these equations we see that, in principle, if we know the Laplace–Stieltjes transform, $W_s^*[\theta]$, of the service time distribution, we can calculate the steady state probability distribution, $\{p_n\}$, of number of customers in the system as well as the distribution functions, $W_q[\cdot]$ and $W[\cdot]$. Of course we must be able to invert the Laplace–Stieltjes transforms in order to do this, that is, we must find the time dependent functions that have the transforms (5.200), (5.202), and (5.204). We demonstrate the procedure for the M/M/1 queueing system.

If the service time has an exponential distribution with $W_s = 1/\mu$, then, by Example 2.9.6,

$$W_s^* = \frac{\mu}{\mu + \theta}. \qquad (5.205)$$

Substituting (5.205) into (5.200) yields

$$g_N(z) = \frac{(1 - \rho)(1 - z)\dfrac{\mu}{\mu + \lambda(1 - z)}}{\dfrac{\mu}{\mu + \lambda(1 - z)} - z} = \frac{\mu(1 - \rho)(1 - z)}{\mu - z(\mu + \lambda(1 - z))}$$

$$= \frac{(1-\rho)(1-z)}{1 - z(1 + \rho(1-z))} = \frac{(1-\rho)(1-z)}{(1-z) - (1-z)\rho z}$$

$$= \frac{1-\rho}{1-\rho z} = (1-\rho) \sum_{n=0}^{\infty} (\rho z)^n \quad \text{if } |\rho z| < 1. \tag{5.206}$$

Hence, p_n can be read off as the coefficient of z^n in (5.206) or

$$p_n = (1-\rho)\rho^n, \quad n = 0, 1, 2, \ldots, \tag{5.207}$$

as we showed earlier. Substituting (5.205) into (5.202) yields

$$W^*[\theta] = \frac{(1-\rho)\theta \dfrac{\mu}{\mu+\theta}}{\theta + \lambda \left[\dfrac{\mu}{\mu+\theta} - 1 \right]} = \frac{(1-\rho)\theta\mu}{\theta(\mu+\theta) + \lambda(\mu - \mu - \theta)}$$

$$= \frac{(1-\rho)\theta}{\dfrac{\theta}{\mu}(\theta + \mu) - \rho\theta} = \frac{(1-\rho)\theta}{(1-\rho)\theta + \dfrac{\theta^2}{\mu}} = \frac{1}{1 + \dfrac{\theta}{\mu}(1-\rho)}$$

$$= \frac{\mu(1-\rho)}{\mu(1-\rho) + \theta}. \tag{5.208}$$

By Example 2.9.6, this is the Laplace–Stieltjes transform of an exponential random variable with parameter $\mu(1 - \rho)$, so we have

$$W[t] = 1 - e^{-\mu(1-\rho)t} = 1 - \exp(-t/W). \tag{5.209}$$

Similarly, as we show in Example 3.4.3, if (5.205) is substituted into (5.204), the resulting transform can be inverted to yield

$$W_q[t] = 1 - \rho \exp(-t/W). \tag{5.210}$$

Takács [60] has generated a recurrence formula for calculating moments of queueing time in terms of moments of service time. We state one of his results without proof.

Theorem 5.3.1 (*Takács Recurrence Theorem*) *Consider an M/G/1 queueing system in which $E[s^{j+1}]$ exists. Then $E[q], E[q^2], \ldots, E[q^j]$ also exist and*

$$E[q^k] = \frac{\lambda}{1-\rho} \sum_{i=1}^{k} \binom{k}{i} \frac{E[s^{i+1}]}{(i+1)} E[q^{k-i}], \quad k = 1, 2, \ldots, j, \tag{5.211}$$

where $E[q^0] = 1$.

Corollary *If the hypotheses of Theorem 5.3.1 are true, then the moments $E[w], E[w^2], \ldots, E[w^j]$ exist and*

$$E[w^k] = \sum_{i=0}^{k} \binom{k}{i} E[s^i] E[q^{k-i}], \quad k = 1, 2, \ldots, j. \tag{5.212}$$

Proof of Corollary Since q and s are independent, we can write

$$E[w^k] = E[(q + s)^k] = E\left[\sum_{i=0}^{k} \binom{k}{i} s^i q^{k-i}\right] = \sum_{i=0}^{k} \binom{k}{i} E[s^i] E[q^{k-i}]. \quad \blacksquare$$

Theorem 5.3.1 and the corollary yield the following formulas for the M/G/1 queueing system (see Exercise 39).

$$W_q = \frac{\lambda E[s^2]}{2(1 - \rho)} = \frac{\rho W_s}{1 - \rho}\left(\frac{1 + C_s^2}{2}\right). \tag{5.213}$$

$$E[q^2] = 2W_q^2 + \frac{\lambda E[s^3]}{3(1 - \rho)}. \tag{5.214}$$

$$W = W_q + W_s. \tag{5.215}$$

$$E[w^2] = E[q^2] + \frac{E[s^2]}{1 - \rho}. \tag{5.216}$$

Equation (5.213) is the most famous of these equations. It is called the *Pollaczek–Khintchine formula* or *Pollaczek's formula*. (Formula (5.193) also is called by both these names, at times, but (5.213) is what authors are more likely to mean when they refer to Pollaczek's formula.)

By Little's law,

$$L_q = \frac{\lambda^2 E[s^2]}{2(1 - \rho)}. \tag{5.217}$$

Equations (5.213)–(5.217) show that, if we know the first three moments of service time, we can calculate both the expected value and the standard deviation for the random variables q and w. (We can do the same for N and N_q by Exercises 40 and 41.) However, if we know only the first two moments of service time we must be content with average values, only. In many cases knowledge of average values, only, will not enable us to make the kind of probability calculations we desire. It is especially valuable to be able to compute percentile values, such as we did for the random variables q and w in the M/M/1 queueing system. There is no general formula to calculate percentile values of w for the M/G/1 queueing system, but James Martin [45] gives the estimates

$$\pi_w[90] \approx W + 1.3\sigma_w, \tag{5.218}$$

and

$$\pi_w[95] \approx W + 2\sigma_w. \tag{5.219}$$

(Actually, Martin gives only the second estimate, but the reasoning he gives to justify it yields (5.218), also.) Another approach is to approximate the random variable of interest with a gamma random variable or a hyperexponential random variable with the same mean and C_x^2 value, as we showed in Chapter 3.

Example 5.3.1 Four communication lines are used by a message switching system. Each has an average transmission time per message of 2.4 seconds and operates at 80% utilization during the peak period of the day. Messages for transmission arrive in a random pattern to each line; however, the message transmission time has a different distribution for each line. The transmission time for the first line is H_2 with $q_1 = 0.4$, $q_2 = 0.6$, $1/\mu_1 = 4.8$ seconds, and $1/\mu_2 = 0.8$ seconds. The service time distribution on the second line is exponential. On the third it is Erlang-3 and on the fourth it is constant. Calculate W_q and W for each line and estimate $\pi_w[90]$.

Solution The M/M/1 model applies to the second line (a special case of M/G/1) and the M/G/1 queueing model applies to the other three lines. For the first line we use the hyperexponential formulas of Chapter 3 to compute $W_s = 2.4$ seconds, $E[s^2] = 19.2$ seconds^2, and $E[s^3] = 267.264$ seconds^3. Since $\sigma_s^2 = E[s^2] - W_s^2 = 13.44$ second^2, we calculate $C_s^2 = 13.44/2.4^2 = 7/3$. No special computations are necessary for the second line, since we can use the exact M/M/1 queueing model. For the third line we use the Erlang-3 formulas from Exercise 39, Chapter 3, to calculate

$$E[s^2] = 7.68, \quad C_s^2 = 1/k = 1/3, \quad \text{and } E[s^3] = 30.72.$$

As we ask you to show in Exercise 44 (see Example 5.3.2), we can calculate the exact distribution of w for the M/H$_2$/1 queueing system of line 1, and, as we ask you to show in Exercise 52, the same is true for the M/D/1 system of line 4. Since it is trivial to calculate $\pi_w[90]$ for the M/M/1 system of line 2, this means that we must be satisfied with an approximation to $\pi_w[90]$ only for line 3. We show the results of our computations in Table 5.3.1. □

Table 5.3.1. Example 5.3.1

Line	Dist.	W_q	$\pi_w[90]$ Martin	Exact
1	H_2	16.0	44.98	45.07
2	E_1	9.6	27.60	27.63
3	E_3	6.4	18.87	—
4	D	4.8	14.40	14.43

We see by Table 5.3.1 that Martin's approximation for $\pi_w[90]$ is very good; it hardly seems worth the effort to calculate the exact values for lines 1 and 4. This example dramatically demonstrates the inimical effect of "irregularity" in service time, as measured by C_s^2. (We have all been conditioned by television ads to recognize the deleterious effects of irregularity in our personal lives.) The average queueing time is twice as large for exponential service time as for constant service time; the same is true for the 90th percentile value of w. These values are about 50% higher for the H_2 service time than they are for exponential service time.

The M/G/1 queueing model is quite a useful one because random arrival patterns are quite common although random service times are not. The M/G/1 queueing model is often used as we used it in Example 5.3.1. That is, to calculate means and estimated percentile values rather than attempting to invert the Pollaczek–Khintchine transform equations. We will show in the next example how transform methods can be used to study the steady state $M/H_2/1$ queueing system. Other examples of the use of transform methods are given in Section 3.4 and in the exercises. □

Example 5.3.2 Consider the $M/H_2/1$ queueing system. We follow Kobayashi [33, Example 3.1] in using transform methods to find the steady state distribution of the number of customers in the system. The Laplace–Stieltjes transform of s is given by

$$W_s^*[\theta] = \frac{q_1\mu_1}{\mu_1 + \theta} + \frac{q_2\mu_2}{\mu_2 + \theta}.$$

Hence,

$$
\begin{aligned}
W_s^*[\lambda(1-z)] &= \frac{q_1\mu_1}{\mu_1 + \lambda(1-z)} + \frac{q_2\mu_2}{\mu_2 + \lambda(1-z)} \\
&= \frac{q_1}{1 + \rho_1(1-z)} + \frac{q_2}{1 + \rho_2(1-z)},
\end{aligned}
$$

where $\rho_i = \lambda/\mu_i, i = 1, 2$. Substituting the above into (5.200) yields, after some simplification (see Exercise 43), the formula

$$g_N(z) = \frac{(1 - \rho)[1 + (\rho_1 + \rho_2 - \rho)(1 - z)]}{\rho_1\rho_2 z^2 - (\rho_1 + \rho_2 + \rho_1\rho_2)z + 1 + \rho_1 + \rho_2 - \rho}.$$

Assume that z_1 and z_2 are the roots of the denominator of $g_N(z)$ as written above. Then we can write $g_N(z)$ using the partial fraction representation

$$g_N(z) = C_1\frac{z_1}{z_1 - z} + C_2\frac{z_2}{z_2 - z},$$

where C_1 and C_2 are constants. Since $g_N(z)$ is the generating function of N, we know that

$$g_N(z) = \sum_{n=0}^{\infty} p_n z^n = p_0 + p_1 z + p_2 z^2 + \cdots.$$

Therefore,

$$g_N(0) = p_0 = 1 - \rho = C_1 + C_2,$$

and

$$g_N(1) = \sum_{n=0}^{\infty} p_n = 1 = C_1\frac{z_1}{z_1 - 1} + C_2\frac{z_2}{z_2 - 1}.$$

We solve the above two equations for C_1 and C_2 obtaining

$$C_1 = \frac{(z_1 - 1)(1 - \rho z_2)}{z_1 - z_2},$$

and

$$C_2 = \frac{(z_2 - 1)(1 - \rho z_1)}{z_2 - z_1}.$$

Now we can write

$$
\begin{aligned}
g_N(z) &= \frac{C_1}{1 - \dfrac{z}{z_1}} + \frac{C_2}{1 - \dfrac{z}{z_2}} \\[2mm]
&= C_1\left[1 + \frac{z}{z_1} + \left(\frac{z}{z_1}\right)^2 + \cdots + \left(\frac{z}{z_1}\right)^n + \cdots\right] \\[2mm]
&\quad + C_2\left[1 + \frac{z}{z_2} + \left(\frac{z}{z_2}\right)^2 + \cdots + \left(\frac{z}{z_2}\right)^n + \cdots\right].
\end{aligned}
$$

It follows immediately from this representation that

$$p_n = C_1 z_1^{-n} + C_2 z_2^{-n}, \quad n = 0, 1, 2, \ldots.$$

As an example of the use of this formula, let us consider the $M/H_2/1$ part of Example 5.3.1, that is, the line with an H_2 distribution with $q_1 = 0.4$, $q_2 = 0.6$, $\mu_1 = 1/4.8$, and $\mu_2 = 1/0.8$, and where $\rho = 0.8$, so that $\lambda = 0.8/2.4 = 1/3$. Hence, $\rho_1 = \lambda/\mu_1 = 1.6$, $\rho_2 = \lambda/\mu_2 = 0.8/3$ so z_1, z_2 are the roots of

$$1.28z^2 - 6.88z + 6.2 = 0.$$

The roots are $z_1 = 4.229870336$, and $z_2 = 1.145129665$. Therefore, $C_1 = 0.087843387$ and $C_2 = 0.112156613$, so we can write

$$g_N(z) = \frac{0.371566137}{4.229870336 - z} + \frac{0.128433865}{1.145129665 - z}.$$

By Theorem 2.9.2(c),

$$L = g'_N(1) = \frac{0.371566137}{(4.229870336 - 1)^2} + \frac{0.128433865}{(1.145129665 - 1)^2} = 6.133.$$

This agrees with the value we get from Little's law applied to the value $W = 18.4$ calculated in Example 5.3.1, that is, $\lambda \times 18.4 = 6.133$. $\square$

In Exercise 44 you are asked to prove that w for an $M/H_2/1$ queueing system also has a H_2 distribution. The formulas for the $M/H_2/1$ queueing system are collected together in Table 15 of Appendix C. Tables 16, 17, and 18 contain the formulas for the $M/G/1$ queueing systems in which the service time is gamma, Erlang, or constant, respectively. $\square$

5.3.2 The GI/M/1 Queueing System

The embedded Markov chain technique enables us to obtain useful results for the $GI/M/1$ queueing system. For this system we assume the interarrival times are described by independent, identically distributed, random variables. (Such an arrival pattern is called a *renewal process*.)[13] We represent the system state by the number of customers an arriving customer finds in the system. This yields a stochastic process $\{X_n\}$ where X_n is the number of customers the nth arriving customer finds in the system. By proceeding much as we did in Section 5.3.1 we can show that $\{X_n\}$ is a Markov chain and that, if $\rho = (\lambda W_s/c) < 1$, then a steady state probability distribution $\{\pi_n\}$ exists, where π_n is the probability that an arriving customer finds n customers in the system for $n = 0, 1, 2, \ldots$. The details of proving these facts are given in Takács [59], Gross and Harris [18], Kleinrock [31], and Cooper [10].

---

[13]Renewal processes are discussed in Section 4.5 of Chapter 4.

The reader should note the distinction between π_n, the probability that an arriving customer finds n customers in the system, and p_n, the probability that a random observer does. Consider a D/D/1 queueing system with $E[\tau] = 10$ minutes and $W_s = 5$ minutes. Then $\pi_0 = 1$ and $\pi_n = 0$ for $n \geq 1$, while $p_0 = 0.5$, $p_1 = 0.5$, and $p_n = 0$ for $n \geq 2$. An arriving customer never sees another customer, although the system contains one customer half the time and is empty half the time. A random observer, unlike an arriving customer, would see this "half and half" situation. Wolff [66] showed that $\pi_n = p_n$ for all n if and only if the arrival process is Poisson.

Let X be the number of customers that an arriving customer finds in a GI/M/1 queueing system. Let $\{\pi_n\}$ be the steady state distribution defined above, such that $\pi_n = P[X = n]$ for all n. The proof mentioned above also shows that X has a geometric distribution with

$$\pi_n = \pi_0(1 - \pi_0)^n, \quad n = 0, 1, 2, \ldots. \tag{5.220}$$

In addition, π_0 is the unique solution of the equation

$$1 - \pi_0 = A^*[\mu\pi_0], \tag{5.221}$$

such that $0 < \pi_0 < 1$. In equation (5.221) $A^*[\theta]$ is the Laplace–Stieltjes transform of the interarrival time, τ, and μ is the average service rate.

Since X is geometric, we have

$$E[X] = \frac{1 - \pi_0}{\pi_0}, \text{ and } \sigma_X^2 = \frac{1 - \pi_0}{\pi_0^2}. \tag{5.222}$$

Proceeding exactly as we did in deriving the distribution of q for the M/M/1 queueing system we find that

$$W_q[t] = P[q \leq t] = 1 - (1 - \pi_0) \exp(-\pi_0 t / W_s), \quad t \geq 0. \tag{5.223}$$

It is not difficult to show from (5.223) (see Exercise 59) that

$$W_q = (1 - \pi_0)\frac{W_s}{\pi_0}, \tag{5.224}$$

$$E[q^2] = 2(1 - \pi_0)\left(\frac{W_s}{\pi_0}\right)^2, \tag{5.225}$$

and

$$\sigma_q^2 = (1 - \pi_0^2)\left(\frac{W_s}{\pi_0}\right)^2. \tag{5.226}$$

For this system the probability that an arriving customer must queue for service is $1 - \pi_0$.

The same argument we used for deriving the distribution function of w for the M/M/1 queueing system shows that

$$W[t] = P[w \leq t] = 1 - \exp(\pi_0 t / W_s), \quad t \geq 0. \tag{5.227}$$

Thus, the waiting time in the system has an exponential distribution just as it did for the M/M/1 system! This remarkable fact implies that

$$W = \frac{W_s}{\pi_0} \quad \text{and} \quad \sigma_w^2 = W^2. \tag{5.228}$$

Also, since w is exponential,

$$\pi_w[r] = W \ln \left(\frac{100}{100 - r} \right). \tag{5.229}$$

We differentiate formula (5.223) to obtain f_q, the density function of q. We get

$$f_q(t) = \frac{(1 - \pi_0)\pi_0}{W_s} \exp(-\pi_0 t / W_s), \quad t > 0. \tag{5.230}$$

This means that the density function, f, of the queueing time of those who must queue, q', is given by

$$f(t) = \frac{f_q(t)}{P[q > 0]} = \frac{f_q(t)}{1 - \pi_0} = \frac{\exp(-t/W)}{W}, \quad t > 0. \tag{5.231}$$

But this is the density function of w. Hence,

$$P[q' \leq t] = P[w \leq t] = 1 - \exp(-t/W), \quad t > 0. \tag{5.232}$$

Therefore,

$$E[q'] = E[q|q > 0] = W,$$

and

$$\text{Var}[q'] = W^2. \tag{5.233}$$

Kleinrock [31] shows that

$$p_0 = 1 - \rho, \tag{5.234}$$

and

$$p_n = \rho \pi_0 (1 - \pi_0)^{n-1}, \quad n = 1, 2, \ldots. \tag{5.235}$$

Example 5.3.3 Consider the M/M/1 queueing system. Here,

$$A^*[\theta] = \frac{\lambda}{\lambda + \theta}. \tag{5.236}$$

Hence, (5.221) becomes

$$1 - \pi_0 = \frac{\lambda}{\lambda + \mu\pi_0}. \tag{5.237}$$

This equation yields

$$\pi_0 = 1 - \rho. \tag{5.238}$$

Since $p_n = \pi_n$ for all n because the arrival pattern is random, we have, by (5.220),

$$p_n = (1 - \rho)\rho^n, \quad n = 0, 1, 2, \ldots. \tag{5.239}$$

All the other formulas for the M/M/1 queueing system now agree with those from the GI/M/1 model by making the substitution $\pi_0 = 1 - \rho$. □

Example 5.3.4 Consider the E_2/M/1 queueing system. It is not difficult to show (see Exercise 64) that, if τ has an Erlang-k distribution, then

$$A^*[\theta] = \frac{k\lambda}{k\lambda + \theta}. \tag{5.240}$$

Hence, (5.221) becomes

$$1 - \pi_0 = \left(\frac{2\lambda}{2\lambda + \mu\pi_0}\right)^2.$$

This equation yields

$$\pi_0 = -2\rho + 0.5 + \sqrt{2\rho + 0.25}. \tag{5.241}$$

Hence, we can easily calculate π_0 as a function of ρ. □

Example 5.3.5 Consider the E_3/M/1 queueing system. By (5.240) and (5.221) we can write

$$1 - \pi_0 = \left(\frac{3\lambda}{3\lambda + \mu\pi_0}\right)^3,$$

that yields the cubic equation

$$\pi_0^3 + (9\rho - 1)\pi_0^2 + 9\rho(3\rho - 1)\pi_0 + 27\rho^2(\rho - 1) = 0. \tag{5.242}$$

This equation is not easy to solve with pencil and paper but can be quickly solved with a scientific calculator that has a root finder, such as a Hewlett-Packard HP-28S or any other of the HP series that has the "solve" function. It can also be solved easily using *Mathematica*. We calculated the values of π_0 for a few values of ρ in the second column of Table 20 in Appendix C. □

Example 5.3.6 Consider the U/M/1 queueing system where the interarrival times are independent random variables distributed uniformly on the interval from 0 to $2/\lambda$. Then, by Exercise 60,

$$A^*[\theta] = \frac{\lambda}{2\theta}(1 - \exp(2\theta/\lambda)).$$

We thus have the equation

$$1 - \pi_0 = \frac{\rho}{2\pi_0}(1 - \exp(-2\pi_0/\rho)). \tag{5.243}$$

This equation must be solved numerically or with the type of calculator mentioned in the previous example. □

Example 5.3.7 Consider the D/M/1 queueing system in which the arrivals are equally spaced in time. For this interarrival time distribution

$$A^*[\theta] = e^{-\theta/\lambda},$$

so (5.221) becomes

$$1 - \pi_0 = e^{-\mu\pi_0/\lambda} = e^{-\pi_0/\rho}. \tag{5.244}$$

This equation, too, must be solved numerically to yield the values in column D of Table 20. □

Example 5.3.8 Consider the $H_2/M/1$ queueing system, for which

$$A^*[\theta] = \frac{q_1\lambda_1}{\lambda_1 + \theta} + \frac{q_2\lambda_2}{\lambda_2 + \theta}.$$

Then by (5.221), the equation to determine π_0 is

$$1 - \pi_0 = \frac{q_1\lambda_1}{\lambda_1 + \mu\pi_0} + \frac{q_2\lambda_2}{\lambda_2 + \mu\pi_0}. \tag{5.245}$$

Some tedious algebra applied to (5.245) yields the quadratic equation

$$\mu^2\pi_0^2 + \mu(\lambda_1 + \lambda_2 - \mu)\pi_0 + \mu(q_1\lambda_1 + q_2\lambda_2 - \lambda_1 - \lambda_2) + \lambda_1\lambda_2 = 0 \tag{5.246}$$

The unique root of (5.246) that lies between 0 and 1 is the π_0 we seek. In the special case that Algorithm 3.2.2 is used to generate an H_2 distribution for the arrival pattern with a given $E[\tau] = 1/\lambda$ and C_τ^2, then it is easy to show (see Exercise 61) that π_0 is given by

$$\pi_0 = 0.5 - \rho + 0.5\sqrt{(1 - 2\rho)^2 + 16\rho q_1(1 - q_1)(1 - \rho)}. \tag{5.247}$$

Formula (5.247) is used to generate the last column of Table 20, Appendix C, when $C_\tau^2 = 20$. □

The values of π_0 as a function of ρ for several $GI/M/1$ queueing systems are given in Table 20, Appendix C. The detrimental effects of irregularity are evident. The probability that an arriving customer will have to queue for service is much lower for completely regular arrivals (constant inter-arrival time) than for exponential arrivals. The situation is even more dramatic for the H_2 interarrival time distribution. For the H_2 distribution shown in the last column of the table the probability an arriving customer will find the server busy is almost 19% ($1 - \pi_0 = 0.189425$) when the server utilization is only 10%.

We will now give some examples of the use of Tables 19 and 20 from Appendix C.

Example 5.3.9 Consider Example 5.3.1. Suppose the message transmission time on each of the four lines is exponential with an average value of 2.4 seconds and that each line operates at 80% utilization. Suppose further that the arrival pattern of messages to the lines is different for each line. Suppose the interarrival time for the first line is H_2 with $q_1 = 0.4$, $q_2 = 0.6$, $1/\mu_1 = 6.0$ seconds, and $1/\mu_2 = 1.0$ second; the interarrival time to the second line is exponential; it is Erlang-3 on the third line and constant on the fourth line. Find W_q, W, $\pi_q[90]$, $\pi_w[90]$, L_q, and L for each line.

Solution The $GI/M/1$ queueing system formulas of Table 19 apply with values of π_0 taken from Table 20. The results are summarized in Table 5.3.2, that should be compared to Table 5.3.1. We illustrate with the calculations for the first line. By Table 20, $\pi_0 = 0.124695$. Hence,

$$\begin{aligned}
W_q &= (1 - \pi_0)\frac{W_s}{\pi_0} = 16.85 \quad \text{seconds,} \\
W &= W_q + W_s = 19.25 \quad \text{seconds,} \\
\pi_q[90] &= W\ln(10(1 - \pi_0)) = 41.76 \quad \text{seconds,} \\
\pi_w[90] &= W\ln(10) = 44.32 \quad \text{seconds,} \\
L_q &= (1 - \pi_0)\frac{\rho}{\pi_0} = 5.62 \quad \text{messages,} \\
L &= \frac{\rho}{\pi_0} = 6.42 \quad \text{messages.}
\end{aligned}$$

It is evident from Table 5.3.2 that irregularity in the arrival process is inimical to the performance of a queueing system, just as Table 5.3.1 showed the harmful effects of lack of regularity in the service time distribution. In both Example 5.3.1 and this example, we have $E[\tau] = 3$ seconds and $W_s = 2.4$ seconds. It is interesting to compare the $M/E_2/1$ queueing system to the $E_2/M/1$, the $M/E_3/1$ queueing system to the $E_3/M/1$ system, and the $M/D/1$ system to the $D/M/1$ queueing system. They are not greatly different although not equal. □

Table 5.3.2. Results of Example 5.3.9

	Line Arrival Pattern			
	H_2	M	E_3	D
W_q	16.85	9.60	5.90	4.06
W	19.25	12.00	8.30	6.46
$\pi_q[90]$	41.76	24.95	16.28	11.88
$\pi_w[90]$	44.32	27.60	19.09	14.86
L_q	5.62	3.20	1.97	1.35
L	6.42	4.00	2.77	2.15

Example 5.3.10 Consider an $E_2/E_2/1$ queueing system with $\rho = 0.95$ and $W_s = 2$ seconds. Find an upper bound for W and $\pi_w[90]$.

Solution We can get two conservative estimates by approximating the queueing system by an $M/E_2/1$ system and by an $E_2/M/1$ system. These systems will each have performance parameters that are larger than those for the $E_2/E_2/1$ model. Using the formulas of Table 17 or the APL program MΔEKΔ1, we see that for the $M/E_2/1$ queueing system

$$W = 30.5 \text{ seconds}, \ \pi_w[90] \approx 67.24 \text{ seconds}.$$

The formulas for the $E_2/M/1$ with $\pi_0 = 0.066288$ from Table 20 yield

$$W = 30.17 \text{ seconds}, \ \pi_w[90] = 69.39 \text{ seconds}.$$

Thus, we can safely use $W = 30.17$ seconds and $\pi_w[90] = 67.24$ seconds as upper bounds for these quantities. Later we will see some better upper bounds. In Example 5.5.2 we will see that W cannot exceed 21.0 seconds. There are some rather complex algorithms that can to used for computing the performance statistics for $E_m/E_k/c$ queueing systems. Tables of values of the queue size and waiting time distributions are given in Hillier and Lo [21] and the book by Hillier and Yu [22]. For example, on page 214 of Hillier and Yu we see that, for the $E_2/E_2/1$ system of this example, with $\rho = 0.95$ and $W_s = 2$, so that $\lambda = \rho/W_s = 0.475$, we have $L_q = 8.8883$. Thus, $W_q = L_q/\lambda = 18.7122$ seconds, so that $W = W_q + W_s = 20.7122$ seconds and $L = \lambda W = 9.8383$ customers. $\square$

5.3.3 The GI/M/c Queueing System

Most of the nice properties of the GI/M/1 queueing system carry over to the GI/M/c system. Just as with the former system, if $\rho = (\lambda W_s/c) < 1$, then a steady state probability distribution $\{\pi_n\}$ exists where π_n is the probability an arriving customer finds n customers in the system for $n = 0, 1, 2, \ldots$. The details of proving these facts are given in Takács [59], Gross and Harris [18], Kleinrock [32], and Cooper [10].

The formulas for the GI/M/c queueing system are given in Table 21 of Appendix C. The reader will note that the level of computation required to calculate the performance parameters of this model increases dramatically over that for the GI/M/1 queueing model as c grows. The key parameter is ω, the unique solution of the equation

$$\omega = A^*[c\mu(1 - \omega)] \tag{5.248}$$

such that $0 < \omega < 1$.

Example 5.3.11 Consider the M/M/2 queueing system. Since $A^*[\theta] = \lambda/(\lambda + \theta)$, equation (5.248) becomes

$$\omega = \frac{\lambda}{\lambda + 2(1 - \omega)\mu} = \frac{a}{a + 2(1 - \omega)}, \tag{5.249}$$

where $a = \lambda W_s$.

The solution of (5.249) satisfying $0 < \omega < 1$ is $\omega = \rho = a/2$. We compute

$$g_j = A^*[j\mu] = \frac{\lambda}{\lambda + j\mu} = \frac{a}{a + j}, \ j = 1, 2.$$

Hence,

$$C_1 = \frac{g_1}{1 - g_1} = a,$$

and

$$C_2 = C_1 \times \frac{g_2}{1 - g_2} = \frac{a^2}{2}.$$

Therefore,

$$D = \left[\frac{1}{1 - \rho} + \frac{1}{\rho} + \frac{1}{2\rho^2}\right]^{-1} = \frac{2\rho^2(1 - \rho)}{1 + \rho}.$$

Similarly,

$$U_0 = DC_0 \left[\frac{1}{\rho} + \frac{1}{2\rho^2}\right] = (2\rho + 1) \left[\frac{1 - \rho}{1 + \rho}\right],$$

and

$$U_1 = DC_1 \left[\frac{1}{2\rho^2}\right] = \frac{Da}{2\rho^2} = 2\rho \left[\frac{1 - \rho}{1 + \rho}\right].$$

Then

$$\pi_0 = U_0 - U_1 = \frac{1 - \rho}{1 + \rho}.$$

Since $\pi_n = p_n$ for all n because the arrival process is random, we obtain

$$p_n = \begin{cases} \dfrac{1-\rho}{1+\rho}, & n = 0. \\[3mm] \dfrac{2(1-\rho)}{1+\rho}\rho^n, & n = 1, 2, \ldots. \end{cases} \tag{5.250}$$

This agrees with the formulas for p_n in Table 6 of Appendix C. Applying the formulas from Table 21 yields the other formulas in Table 6. □

The following remarkable theorem was communicated to me by the late Dr. Isaac Dukhovny, who provided the proof.

Theorem 5.3.2 *Consider a GI/M/c queueing system and a GI/M/1 queueing system each of which has the same type of distribution of interarrival time and such that $\rho < 1$, W_s, and C_T^2 are the same for both systems. Then*

$$E[q|q > 0]_{\text{GI/M/c}} = \frac{1}{c}E[q|q > 0]_{\text{GI/M/1}}. \tag{5.251}$$

Proof Let $A[\cdot]$ be the distribution function of the interarrival time for the GI/M/c system with ω the unique solution of the equation $\omega = A^*[c\mu(1-\omega)]$ such that $0 < \omega < 1$. Then, as we showed earlier,

$$W_{q'} = E[q|q > 0]_{\text{GI/M/c}} = \frac{W_s}{c(1 - \omega)}. \tag{5.252}$$

Let τ' be the interarrival time for the GI/M/1 queueing system. Since ρ for this system is the same as that for the GI/M/c system, we must have $\tau' = c\tau$ where τ is the interarrival time for the GI/M/c system. Then, if $A_1[\cdot]$ is the distribution function for τ', we must have

$$A_1[t] = P[\tau' \le t] = P[c\tau \le t] = P[\tau \le t/c] = A[t/c]. \tag{5.253}$$

This shows that

$$\begin{aligned} A_1^*[\theta] &= \int_0^\infty e^{-\theta t}\, dA_1[t] = \int_0^\infty e^{-\theta t}\, dA[t/c] \\ &= \int_0^\infty e^{-\theta cz}\, dA[z] = A^*[c\theta]. \end{aligned} \tag{5.254}$$

Therefore, the equation $\omega = A_1^*[\mu(1 - \omega)]$ becomes $\omega = A^*[c\mu(1 - \omega)]$. This is exactly the equation we solved to get the value of ω in (5.252). Since

$$E[q|q > 0]_{\text{GI/M/1}} = \frac{W_s}{1 - \omega},$$

the proof is complete. ■

Example 5.3.12 Consider the M/M/2 queueing system and compare it to the M/M/1 system. Then, since, by the last example $\omega = \rho$, we have

$$E[q|q > 0]_{\text{M/M/2}} = \frac{W_s}{2(1 - \rho)},$$

and

$$E[q|q > 0]_{\text{M/M/1}} = \frac{W_s}{1 - \rho},$$

that agrees exactly with Dukhovny's theorem. $\square$

One of the practical problems involved in using the GI/M/c queueing system is in finding the solution ω to equation (5.248). Halachmi [19] derives the following estimate of ω,

$$\hat{\omega} = \exp\left[\frac{-2(1 - \rho)}{1 + \rho^2 C_\tau^2}\right], \tag{5.255}$$

by a diffusion approximation. This can be used for ω for a quick-and-dirty estimate of the performance parameters. The estimate can also be used as a starting value for a numerical method to obtain a more exact value for ω. Of course, if you have a calculator with a root finder or have *Mathematica*, then it is almost trivial to obtain ω. Another approach is to use a numerical method such as the Newton–Raphson algorithm. Henrici [20] is a good reference for numerical methods. $\square$

Example 5.3.13 The computer performance analysts at Lingering Lead have discovered a computer subsystem that can be modeled as a D/M/2 queueing system. They want to calculate the performance parameters when $\rho = 0.9$. Then $a = 2\rho = 1.8$ erlangs. Since $A^*[\theta] = e^{-\theta/\lambda} = e^{-\theta/1.8}$, (5.248) becomes

$$\omega = \exp\left[\frac{(\omega - 1)}{0.9}\right]. \tag{5.256}$$

My HP-32S shows that the correct answer to (5.256) is 0.806899833. Halachmi's estimate is

$$\hat{\omega} = e^{-0.2} = 0.818730753,$$

that is in error by approximately 1.47%. $\square$

5.4   Priority Queueing Systems

As we mentioned in Section 5.1, all customers in a queueing system need not be treated equally, just as in most organizations, some individuals may receive preferential treatment.[14] Queueing systems in which some customers get preferential treatment are called *priority queueing systems*. The simplest queue discipline in which there are *no* priorities is the first-come, first-served assignment system, abbreviated as FCFS or FIFO (first-in, first-out). Other nonpriority queueing disciplines include last-come, first-served (LCFS or LIFO), and random-selection-for-service (RSS or SIRO). There are some whimsical queue disciplines that are part of the queueing theory folklore. These include BIFO (biggest-in, first-out).[15], FISH (first-in, still-here), and WINO (whatever-in, never-out) The reader can, no doubt, think of others to describe personal experiences with queueing systems.

In the *priority* queueing systems that we study, customers are divided into priority classes, numbered from 1 to n. We assume that the lower the priority class number, the higher the priority; that is, customers in priority class i are given preference over customers in priority class j if $i < j$. (In this case the customer in class i is said to have a *higher* priority than a customer in class j.) Customers within a given priority class are served, with respect to that class, by the FCFS queue discipline.

There are two basic control policies to resolve the situation wherein a customer of class i arrives to find a customer of class j in service, where $i < j$, called *preemptive* and *nonpreemptive* systems. In a *preemptive priority* queueing system, service is interrupted and the newly arrived customer begins service. The customer whose service was interrupted returns to the head of the jth class. As a further refinement, in a *preemptive-resume* priority queueing system, the customer whose service was interrupted begins service at the point of interruption on the next access to the service facility. There are other variations, including *preemptive-repeat*, in which the lower priority customer repeats the entire service from the beginning. In a *nonpreemptive* priority queueing system, the newly arrived customer waits until the customer in service completes service before gaining access to the service facility. This type of system is called a *head-of-the-line* system, abbreviated HOL.

Some nonpriority queue disciplines, that is, systems with no priority classes, allow preemption. One such system is the LCFS/PR (last-come, first-served, preemptive-resume) queue discipline. For this queue discipline an arriving customer preempts the customer in service.

We assume all queue disciplines are *work conserving*. This means that customers do not leave without completing service (no reneging allowed), and that the server is never idle if there are customers present requiring service.

---

[14]Sometimes this is necessary for business reasons or to optimize the use of a resource.
[15]This is actually a priority system.

The queue discipline can have important effects on the performance of a queueing system. The following example is used by Shelly Weinberg of IBM to illustrate some of the issues of queue discipline.

Example 5.4.1 Hapless Harry arrived at the office copying machine at the same time as an individual who claimed he was Leo Tolstoy. Harry wanted to copy a 10 page document, but Mr. Tolstoy wanted to reproduce his novel *War And Peace* (the 1,000 page version). Each wanted to go first; Hapless because he'd promised his boss a copy of the document and Mr. Tolstoy because his friend, Napoleon Bonaparte, would benefit from exposure to one of the greatest novels of all time. Hapless explained to Mr. Tolstoy that, assuming it took 10 seconds to copy each page, the average queueing time for the two of them, if Leo went first, would be half of 10,000 seconds (an hour and 23 minutes), but only 50 seconds if Hapless went first. Unfortunately, while Harry was stating his case, General Douglas MacArthur moved in to begin copying his 438 page book, *Reminiscences*. □

The results of Example 5.4.1 give us an intuitive feel for the proposition that W_q and W are both minimized if priority is given to customers who have the shortest required service times. This is, indeed, true. For the proof see Gelenbe and Mitrani [17, page 200].

5.4.1   M/G/1 Priority Queueing Systems

In Section 5.3.1 we developed the equations for the M/G/1 queueing system with the assumption that the queue discipline was FCFS. In the next theorem we generalize the results to include two other queue disciplines.

Theorem 5.4.1 *Consider an M/G/1 queueing system with any queue discipline that chooses customers by an algorithm that does not consider customer service times or any measure of them (such as the FCFS queue discipline). Then the steady state distribution of number in the system, N, will be the same as for FCFS. The performance parameters W, W_q, L, and L_q will also be the same as for the FCFS queue discipline. However, the distributions of w and q will depend on the particular queue discipline in effect. In particular, the second moment of q will be given by (5.257) when the queue discipline is RSS and by (5.258) when it is LCFS/NP (last-come, first-served, nonpreemptive). That is, for the RSS queue discipline,*

$$E[q^2] = \frac{2\lambda E[s^3]}{3(1-\rho)(2-\rho)} + \frac{\lambda^2 \left(E[s^2]\right)^2}{(1-\rho)^2(2-\rho)}, \qquad (5.257)$$

while for the LCFS nonpreemptive queue discipline,

$$E[q^2] = \frac{\lambda E[s^3]}{3(1-\rho)^2} + \frac{\lambda^2 \left(E[s^2]\right)^2}{2(1-\rho)^3}. \qquad (5.258)$$

For all queue disciplines independent of s, we have

$$\sigma_q^2 = E[q^2] - W_q^2, \qquad (5.259)$$

by Theorem 2.7.2(d), and

$$\sigma_w^2 = \sigma_q^2 + \sigma_s^2, \qquad (5.260)$$

since q and s are assumed independent.

Proof The main part of the theorem is proven by Kleinrock [32]. The formulas (5.257) and (5.258) are proven by Cohen [9]. (5.259) and (5.260) are true for the reasons given in the statement of the theorem. ■

The variance of q (and therefore of w, since $\sigma_w^2 = \sigma_q^2 + \sigma_s^2$) is greater for RSS than for FCFS, and for LCFS greater than RSS. To date, no one has derived explicit distribution functions for w or q for the RSS and LCFS/NP queue disciplines, even when the service time is exponential.

Example 5.4.2 Consider the $M/H_2/1$ queueing system representing the first communication line of Example 5.3.1. If the queue discipline is FCFS, the M/G/1 equations of Table 14, Appendix C, show that $W_q = 16$ seconds, $W = 18.4$ seconds, $E[w^2] = 756.48$ seconds^2 and $\sigma_w^2 = 417.92$ seconds^2. Theorem 5.4.1 shows that, if the queue discipline of this system is RSS, then W_q and W do not change, but

$$\sigma_w^2 = \sigma_q^2 + \sigma_s^2 = 844.8 + 13.44 = 858.24 \text{ seconds}^2.$$

Similarly, for the LCFS/NP queue discipline we have

$$\sigma_w^2 = \sigma_q^2 + \sigma_s^2 = 3046.4 + 13.44 = 3059.84 \text{ seconds}^2. \quad \square$$

Let us now consider M/G/1 queueing systems in which there are n priority classes. We will assume that customers of the first priority class receive the most preferential treatment, those of the second priority class the second most preferential treatment, etc. We also assume that the customers from class i arrive in a Poisson pattern with mean arrival rate $\lambda_i, i = 1, 2, \ldots, n$. Each class has its own general service time with $E[s_i] = 1/\mu_i$, and finite second and third moments $E[s_i^2]$, $E[s_i^3]$. Thus, the total arrival stream to the system has a Poisson arrival pattern with

$$\lambda = \lambda_1 + \lambda_2 + \cdots + \lambda_n. \qquad (5.261)$$

The first three moments of service time are given by

$$W_s = \frac{\lambda_1}{\lambda}E[s_1] + \frac{\lambda_2}{\lambda}E[s_2] + \cdots + \frac{\lambda_n}{\lambda}E[s_n], \qquad (5.262)$$

and, by the law of total moments,

$$E[s^2] = \frac{\lambda_1}{\lambda}E[s_1^2] + \frac{\lambda_2}{\lambda}E[s_2^2] + \cdots + \frac{\lambda_n}{\lambda}E[s_n^2], \qquad (5.263)$$

and

$$E[s^3] = \frac{\lambda_1}{\lambda}E[s_1^3] + \frac{\lambda_2}{\lambda}E[s_2^3] + \cdots + \frac{\lambda_n}{\lambda}E[s_n^3]. \qquad (5.264)$$

Equation (5.261)–(5.264) are valid not only for nonpreemptive queueing systems (HOL) and preemptive resume queueing systems but also for a system with the given customer classes but for which all customers are served on a FCFS basis. Other formulas valid for this type of system are given in Table 22 of Appendix C.

The formulas for the M/G/1 nonpreemptive priority (HOL) queueing system are given in Table 23 of Appendix C. The proofs of the validity of these formulas are given by Gelenbe and Mitrani [17] and Lavenberg [41]. The calculations for this model can be made using the APL program PΔQUEUE .

The formulas for the M/G/1 preemptive-resume priority queueing system are given in Table 24 of Appendix C. The equations for this model are adapted from Lavenberg [41]. The calculations for this model can be made using the APL program PRΔQUEUE.

We reproduce two of the equations from Table 24 for discussion. For convenience we define

$$a_j = \lambda_1 E[s_1] + \cdots + \lambda_j E[s_j], \quad j = 1, 2, \ldots, n \qquad (5.265)$$

and note that $a_n = a = \lambda W_s$. For this system the mean time in the system for a customer of class j is given by

$$W_j = \frac{1}{1 - a_{j-1}}\left[E[s_j] + \frac{\sum_{i=1}^{j}\lambda_i E[s_i^2]}{2(1 - a_j)}\right], \ a_0 = 0, \ j = 1, 2, \ldots, n. \quad (5.266)$$

Formula (5.266) shows that a customer with the highest priority (a class 1 customer) will be completely unaffected by customers of other classes. For $j = 1$ (5.266) reduces to Pollaczek's formula for an M/G/1 queueing system with $\lambda = \lambda_1$, $W_s = E[s_1]$, and $E[s^2] = E[s_1^2]$. Formula (5.266) also shows that class 2 customers are affected only by class 1 and class 2 customers, and, more generally, class j customers are affected only by those of classes 1 through j. The situation is much different for the HOL queueing discipline. For this discipline we have

$$E[q_j] = \frac{\lambda E[s^2]}{2(1 - a_{j-1})(1 - a_j)}, \quad a_0 = 0, \ j = 1, 2, \ldots, n. \tag{5.267}$$

This formula shows that priority class j customers are affected directly by all higher priority customers because of the terms $1 - a_{j-1}$ and $1 - a_j$ in the denominator, and also indirectly by customers of all classes because of the numerator, $\lambda E[s^2]$. Both λ, by (5.261), and $E[s^2]$, by (5.263), depend on all customer classes. Thus, adding a new priority class of lowest priority, say class $n + 1$, to a preemptive-resume priority queueing system will have no effect on the performance statistics for any of the existing priority classes, but could have a profound effect on the performance of all existing classes in a nonpreemptive priority system. We will illustrate this phenomenon after the following example.

Example 5.4.3 Jacques Casanova, the Chief Capacity Planner at Walla Walla Wankel, a manufacturer of exotic electronic devices, is designing a new interactive computer system. Jacques envisions three types of interactive transactions, each of which has a Poisson arrival pattern. Type 1 transactions arrive at a mean rate $\lambda_1 = 10$ transactions per second. For the processor planned by Mr. Casanova, the service time of these transactions has an Erlang-2 distribution with mean 0.04 seconds. For Type 2 transactions, $\lambda_2 = 0.5$ transactions per second with an exponential service time with $E[s_2] = 0.1$ seconds. For Type 3 transactions $\lambda_3 = 0.01$ transactions per second. For these transactions the service time is H_2 with $E[s_3] = 10$ seconds, and $E[s_3^2] = 500$ seconds^2. Jacques wants to analyze the performance of the system (a) as a FCFS queueing system, (b) as a nonpreemptive (HOL) priority with Type 1 requests getting the highest priority and Type 3 requests the lowest, and (c) as a preemptive resume priority system with priorities assigned as in (b).

Solution By (5.261), $\lambda = 10.51$ transactions per second. Using the properties of the Erlang-2 distribution, we calculate $E[s_1^2] = (1 + 1/2)E[s_1]^2 = 0.0024$ second^2. Since s_2 has an exponential distribution, we calculate $E[s_2^2] = 2! \times E[s_2]^2 = 0.02$ seconds^2. We apply (5.262) to calculate $W_s = 0.05233111$ seconds. Formula (5.263) yields $E[s^2] = 0.4789724$ seconds^2. We summarize the calculations in Table 5.4.1. They can be made using the APL programs NPΔQUEUE, PΔQUEUE, and PRΔQUEUE. Table

5.4.1 shows an overall system improvement in going from a FCFS queueing discipline to a nonpreemptive (HOL) priority system. However, the improvement has a detrimental effect on W_2 and W_3. The preemptive resume priority system provides a very dramatic improvement over the HOL queueing system. The small price paid[16] for the improvement in W, W_q, W_1, and W_2 is an increase in W_3. In Exercise 74 we ask you to investigate the effect on system performance of reversing the priority assignments. □

Let us return to the discussion just prior to Example 5.4.3. Suppose that we have two priority classes with the Type 1 and Type 2 transactions like those of Example 5.4.3. Then W_1, W_2, $E[q_1]$, and $E[q_2]$ would have the same values as those shown in Table 5.4.1 for the preemptive resume system. However, these values change for the HOL system as we now show. We calculate

$$
\begin{aligned}
\lambda &= 10.51 \text{ transactions per second} \\
W_s &= 0.0429 \text{ seconds} \\
E[s^2] &= 0.00324 \text{ seconds}^2 \\
\rho &= 0.45 \\
W_q &= 0.0294 \text{ seconds} \\
W_1 &= 0.0683 \text{ seconds} \\
W_2 &= 0.1515 \text{ seconds} \\
W &= 0.0723 \text{ seconds.}
\end{aligned}
$$

All these values are much smaller than the corresponding values in Table 5.4.1 for HOL with three priority classes. The Type 3 transactions have a tremendous negative impact on overall system performance as well as on the performance of Type 1 and Type 2 transactions!

---

[16]The observation that, "there is no free lunch," is as valid for queueing systems as everywhere else.

Table 5.4.1. Example 5.4.3

Parameter	FCFS	HOL	PR
λ	10.51	10.51	10.51
W_s	0.0523	same	same
$E[s^2]$	0.4790	same	same
$E[q_1]$	5.5933	4.195	0.020
$E[q_2]$	5.5933	7.627	0.118
$E[q_3]$	5.5933	10.17	18.35
W_1	5.6333	4.235	0.060
W_2	5.5933	7.727	0.218
W_3	15.693	20.17	28.35
W_q	5.5933	4.364	0.042
W	5.6457	4.416	0.094
L_q	58.786	45.865	0.443
L	59.336	46.415	0.993

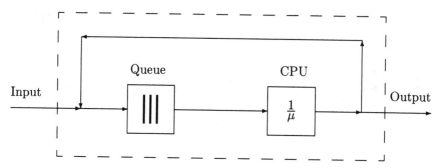

Figure 5.4.1. Round-robin system.

5.4.2 M/G/1 Processor-Sharing Priority

This model has its genesis in the round-robin algorithm for allocating the CPU resource to users. The purpose of the algorithm is to give preference to shorter jobs. The round-robin model is shown in Figure 5.4.1. Customers (user requests) are assumed to arrive at the processor in a Poisson arrival pattern. Each arriving customer enters the single CPU queue and waits in FCFS fashion for a quantum Δt of service time. The quantum size is the same for all customers. If service is completed in less than Δt time units, the customer leaves the system and the next customer in the queue starts service. If the entire quantum is expended but the customer has not completed service, the customer is forced to leave the service facility and return to the tail of the queue for another cycle. The cycle is repeated by the customer until the required service is received, whereupon the customer leaves the system. Kleinrock [32, Section 4.4] discovered that by letting the quantum Δt shrink to zero, he could produce an analytical model with much simpler expressions for the performance measures but that is a good approximation to a round-robin system with a small quantum. Kleinrock's system is called a *processor-sharing* system since, if there are k customers in the system, each receives the fraction $1/k$ of the processor capacity; that is, the customers share the processor equally. We now give a formal description of the M/G/1 processor-sharing queueing system.

The M/G/1 processor-sharing queueing system has a Poisson arrival pattern with average arrival rate λ. The service time distribution is general with average rate μ. The queue discipline is processor-sharing which means that each arriving customer immediately receives his or her share of the processor service so there is no queue. Thus, if a customer arrives when there are already $n - 1$ customers in the system, the customer receives service at the average rate μ/n. The following steady state relations hold, when $\rho = \lambda/\mu < 1$:

$$p_n = (1 - \rho)\rho^n, \quad n = 0, 1, 2, \ldots, \quad (5.268)$$

(The number of customers in the system is geometrically distributed!)

$$L = \frac{\rho}{1 - \rho}, \quad (5.269)$$

$$E[w|s = t] = \frac{t}{1 - \rho}, \quad (5.270)$$

and

$$W = \frac{W_s}{1 - \rho}. \quad (5.271)$$

Although there is no queue (and thus no queueing time), a customer requiring t units of service time suffers a delay because the full capacity of the processor is not available due to interference from other customers. The average of this delay, that we denote by $E[q|s = t]$, is the difference between $E[w|s = t]$ and t, the (full capacity) amount of processor time needed. Thus, we have

$$E[q|s = t] = \frac{\rho\, t}{1 - \rho}. \tag{5.272}$$

If we let f denote the density function of the service time we see that

$$
\begin{aligned}
W_q &= \int_0^\infty E[q|s = t]f(t)\, dt \\
&= \frac{\rho}{1 - \rho} \int_0^\infty t f(t)\, dt \\
&= \frac{\rho W_s}{1 - \rho}. \tag{5.273}
\end{aligned}
$$

The equations (5.268), (5.269), and (5.271) are exactly the same as the corresponding equations for the M/M/1 queueing system! In addition the departure stream is Poisson, as it is for the M/M/1 system. The distribution of w and q is not known in general. Ott [46] gives, in the form of Laplace–Stieltjes transforms and generating functions, the joint distribution of w and the number of customers in the system at departure for customers in the general M/G/1 processor-sharing system. Ramaswami [49] derives the mean and variance of w for the GI/M/1 processor-sharing system.

The equations for the M/G/1 processor-sharing queueing model show that this model has some very nice properties. Equation (5.270) shows that the mean conditional response time is a linear function of required service time. A customer who requires twice as much service time as another will, on the average, spend twice as much time in the system. The mean response time W, by (5.271), is independent of the service time distribution and depends only on its mean value. This contrasts with the M/G/1 queueing system in which, by Pollaczek's formula,

$$W = \frac{\lambda E[s^2]}{2(1 - \rho)} + W_s,$$

the average response time clearly depends on the second moment of the service time. Other nice properties of the model are discussed by Kleinrock [32] and by Lavenberg [41].

Example 5.4.4 Old Domination Bank is planning to install an interactive time-sharing system with a large number of terminals so that the arrival pattern to the computer center can be modeled as random (Poisson) with $\lambda = 450$ requests per second. The proposed computer center can be modeled as a single-server, processor-sharing system with a processing rate of 50,000,000 instructions per second. Suppose the average number of instructions per interaction is estimated to be 100,000, including operating system overhead. Haplas Harry Harrington, the head capacity planner at Old Domination, wants to calculate the average response time, W, the ,average number of interactions being processed by the computer, L, and the probability that 5 or more interactions are in process at once.

Solution Since $W_s = 100,000/50,000,000 = 0.002$ seconds, Haplas calculates $\rho = \lambda W_s = 0.9$. Hence, by (5.271), Haplas predicts that

$$W = \frac{W_s}{1 - \rho} = 0.02 \text{ seconds},$$

and

$$L = \lambda W = 9 \text{ interactions}.$$

By (5.268), N has the same geometric distribution as an M/M/1 system, so

$$P[N \geq n] = \rho^n, \quad n = 0, 1, 2, \ldots.$$

Therefore, he predicts that

$$P[N \geq 5] = (0.9)^5 = 0.59049. \quad \square$$

5.4.3 Multiserver Priority Systems

There is a paucity of exact results for multiserver priority systems, although there are some approximations for closed multiserver systems, that we discuss in Chapter 6. The following result is due to Cobham [8].

Theorem 5.4.2 *Consider the M/M/c nonpreemptive queueing system. Suppose there are n priority classes with each class having a Poisson arrival pattern with mean arrival rate λ_i, and that each customer has the same exponential service time requirement. Then the overall arrival pattern is Poisson with mean*

$$\lambda = \lambda_1 + \lambda_2 + \cdots + \lambda_n. \tag{5.274}$$

The performance parameters are given by

$$\rho = \frac{\lambda W_s}{c} = \frac{\lambda}{c\mu}, \tag{5.275}$$

$$E[q_1] = \frac{C[c, a]W_s}{c(1 - \lambda_1 W_s/c)}, \tag{5.276}$$

and

$$E[q_j] = \frac{C[c, a]W_s}{c\left[1 - \left(W_s \sum_{i=1}^{j-1} \lambda_i\right)/c\right]\left[1 - \left(W_s \sum_{i=1}^{j} \lambda_i\right)/c\right]}, \quad j = 2, \ldots, n.$$

$$\tag{5.277}$$

Proof See Cobham [8]. ∎

Cobham's results together with some additional useful formulas that follow either from Little's law or Theorem 3.8.1 are given in Table 26 of Appendix C.

Example 5.4.5 McQuyre Tyre has an interactive computer system for their network of stores. Phileas Fogg, Performance Czar, has successfully modeled the computer system during the peak period of the day as an M/M/3 queueing system with $\lambda = 14.7$ customers per second and $W_s = 0.2$ seconds. Using the M/M/c formulas, he calculates $W_q = 3.2$ seconds, $W = 3.4$ seconds, $L_q = 47.16$ customers, and $L = 50.1$ customers. The store managers are upset by the average response time of 3.4 seconds. Phileas offers them a service level agreement whereby they will be guaranteed a mean response time, W, of less than one second for two out of three days with the mean response time on the third day not to exceed 10 seconds. Can Phileas live up to the agreement with the present equipment?

Solution Phileas plans to use the priority queueing system described in Theorem 5.4.2. He arbitrarily assigns one third of the users to each priority class on the first day. On the second day he switches the priority classes; the first class becomes the third class, the second class becomes the first, and the third class becomes the second. On succeeding days the same switch is made so that each user spends one third of the time in each customer class. The results of the calculations described in Table 25, Appendix C, are shown in Table 5.4.2. We see that one third of the time each customer has a mean response time of 0.2953 seconds, one third of the time it is 0.4749 seconds, and it is 9.454 seconds on one third of the working days. Phileas can live up to his agreement! □

> *Razors pain you*
> *Rivers are damp;*
> *Acids stain you;*
> *And drugs cause cramp.*
> *Guns aren't lawful;*
> *Nooses give;*
> *Gas smells awful;*
> *You might as well live.*

Dorothy Parker

Table 5.4.2. Example 5.4.5

Parameter	FCFS	Nonpreemptive
λ	14.7000	14.7000
W_s	0.2000	0.2000
$E[q_1]$	3.2082	0.0953
$E[q_2]$	3.2082	0.2749
$E[q_3]$	3.2082	9.2543
W_1	3.4082	0.2953
W_2	3.4082	0.4749
W_3	3.4082	9.4543
W_q	3.2082	3.2082
W	3.4082	3.4082
L_q	47.1600	47.1600
L	50.1000	50.1000

5.5   Bounds and Approximations

In many cases when we want to use a queueing theory model to solve a problem we may find that our model isn't precisely one of the easy-to-solve models such as M/M/1, M/M/c, M/G/c, or the machine repair model. In many cases we can find the exact or approximate solution in a table in a book such as Hillier and Lo [21], Hillier and Yu [22], Kühn [37], van Hoorn [64], Sakasegawa [51], or Seelen et al. [53]. In other cases we may not need this much accuracy for quick-and-dirty, back-of-the-envelope calculations. We shall provide some of the simpler but most useful approximate formulas and bounds.

5.5.1   Heavy Traffic Approximations

We are especially interested in queueing systems as they approach overloading, that is, as the server utilization of each server approaches one. Naturally, the study of such systems is called *heavy traffic theory*.

Theorem 5.5.1 (*Heavy Traffic Approximation*) *Consider a GI/G/c queueing system. As $\rho = \lambda W_s/c$ approaches one, the distribution of queueing time approaches that of an exponential distribution with*

$$W_q = \frac{\lambda(\sigma_T^2 + \sigma_S^2/c^2)}{2(1-\rho)} = \frac{\rho W_s}{c(1-\rho)} \left\{ \frac{C_S^2 + C_T^2/\rho^2}{2} \right\}. \qquad (5.278)$$

Proof The proof for $c = 1$ was given by Kingman [28]. The general proof is given by Köllerström [35] and Kleinrock [32]. ∎

We have stated the above theorem in the form useful for approximation. Köllerström [35] gives the more mathematically precise statement of the theorem. (Unfortunately, there is a typographical error in the statement of the result on the first page of his paper.)

There is no easy way to estimate the error in using the heavy traffic approximation, but it is known that ρ must be very close to 1, say $\rho = 0.999$, for the approximation to be very accurate. We give an example below.

Example 5.5.1 Consider an $H_2/H_2/2$ queueing system where both of the hyperexponential distributions have balanced means. Suppose, also, that $C_S^2 = 2.5$, $C_T^2 = 2$, $W_S = 1$ second, and $\rho = 0.98$. Then, by the table on page 63 of Seelen *et al.* [53], the exact value of W_q is 54.498765 seconds. The heavy traffic approximation gives $W_q = 56.1352$ seconds, for an error of 3 percent. □

Suppose that ρ, C_T^2, C_S^2, and W_S are the same for two heavy-traffic queueing systems ($\rho \approx 1$): one GI/M/1 and the other GI/M/c. It is not difficult to show (see Exercise 70) that the mean queueing time for the latter system is approximately $1/c$ times the mean queueing time of the former; that is,

$$W_{q\mathrm{GI/M/c}} \approx \frac{1}{c} W_{q\mathrm{GI/M/1}}.$$

This is what intuition would suggest, but is *not* true for low values of ρ. For example, the ratio of W_q for the M/M/2 queueing system, to that for the M/M/1 queueing model is $\rho/(1 + \rho)$; this is approximately ρ for small values of ρ but approaches $1/2$ as ρ approaches 1 (the heavy traffic case). (The value of $\rho/(1 + \rho)$ is 0.09091 for $\rho = 0.1$ and 0.4975 when $\rho = 0.99$.)

5.5.2 Bounds on Queueing Systems

In the next theorem we see that both upper and lower bounds exist for W_q in a GI/G/1 queueing system; they do not require the heavy-traffic assumption.

Theorem 5.5.2 (*Bounds for W_q*) *For every GI/G/1 queueing system, the following inequalities are true:*

$$W_q \leq \frac{\rho W_S}{(1 - \rho)} \left\{ \frac{C_S^2 + C_T^2(2 - \rho)/\rho}{2} \right\}. \tag{5.279}$$

$$\max\left\{ \frac{\rho W_S C_S^2}{2(1 - \rho)} - \frac{W_S}{2}, 0 \right\} \leq W_q. \tag{5.280}$$

Proof See Stoyan [57]. ■

Daley, in a private communication, has drawn my attention to a result implicit in the unpublished work of Teunis J. Ott [47] that the lower bound (5.280) is tight, in the sense of being attained by any D/G/1 system with specified λ, W_s, and Var[s] whenever it is true that the probability is 1 that s is an integral multiple of τ. Daley also revealed that Kreinin [36] has proved the following result.

Theorem 5.5.3 (*Bounds for W_q*) *For any $E_k/G/1$ queueing system*

$$W_q \leq \frac{\rho W_s}{(1-\rho)} \left\{ \frac{C_s^2 + C_T^2}{2} \right\}, \qquad (5.281)$$

while for any $E_\gamma/G/1$ queueing system with $\gamma \leq 1$,

$$W_q \geq \frac{\rho W_s}{(1-\rho)} \left\{ \frac{C_s^2 + C_T^2}{2} \right\}, \qquad (5.282)$$

where the symbol E_γ means that the interarrival time τ has a gamma distribution with parameters $\beta = \gamma$ and $\alpha = \lambda\gamma$, so that $E[\tau] = 1/\lambda$, $\mathrm{Var}[\tau] = 1/(\lambda^2\gamma)$, and $C_T^2 = 1/\gamma \geq 1$.

Proof See Kreinin [36]. ■

Example 5.5.2 In Example 5.3.10 we considered an $E_2/E_2/1$ queueing system with $\rho = 0.95$ and $W_s = 2$ seconds. We can apply (5.281) and (5.280) to conclude that 8.5 seconds $\leq W_q \leq$ 19.0 seconds. Thus, $W \leq 19.0 + 2 = 21.0$ seconds. This is much less than the upper bound of 30.17 seconds we found in Example 5.3.10. The heavy traffic approximation yields $W_q = 20.026$ seconds or $W = 22.026$ seconds. By the heavy traffic theorem, q is approximately exponential, so $\pi_q[95] \approx W_q \ln 20 = 56.92$ seconds, if we use the upper bound 19.0 from (5.281) for W_q. It follows from Bonferroni's inequality that

$$\pi_w[90] \leq \pi_q[95] + \pi_s[95]. \qquad (5.283)$$

Since s has an Erlang-2 distribution with mean 2, it has a gamma distribution with $\beta = k = 2$ and $\alpha = k\mu = 1$, and we can use MINITAB to calculate $\pi_s[95] = 4.7439$. Therefore, by (5.283),

$$\pi_w[90] \leq 56.92 + 4.74 = 61.66 \text{ seconds}.$$

This is smaller than the 67.24 seconds we estimated in Example 5.3.10. (If we use the exact value of W_q to estimate $\pi_q[95]$, we obtain $\pi_w[90] \leq 60.80$ seconds.) □

A sharp upper bound for W_q for a GI/G/c queueing system has proven more elusive than that for the GI/G/1 queueing system. Kingman [29] proved that the value of W_q given by (5.278) for the heavy traffic case when $c = 1$ is an upper bound for W_q for all steady state systems. That is, Kingman proved that, for every GI/G/1 queueing system,

$$W_q \le \frac{\rho W_s}{(1 - \rho)} \left\{ \frac{C_s^2 + C_T^2/\rho^2}{2} \right\}, \tag{5.284}$$

for all $0 \le \rho < 1$. Daley [12, 13] improved this upper bound to obtain (5.279) which is smaller than (5.284). It would seem reasonable that, when $c > 1$, the heavy traffic W_q given by (5.278) would be an upper bound for W_q for any GI/G/c queueing system with $0 \le \rho < 1$. Brumelle [5] has shown that this is true for the GI/M/c queueing system, but we know how to solve this system to obtain the exact solution. However, since this solution requires some sophisticated computation, Brumelle's result is of some value.

Daley [13] has conjectured a slightly smaller upper bound for W_q for a GI/G/c queueing system than that given by (5.278) of the heavy traffic theorem.

Conjecture 5.5.1 (*Daley's Conjecture*) *Consider any GI/G/c queueing system with $0 \le \rho < 1$. Then*

$$W_q \le \frac{\rho W_s}{c(1 - \rho)} \left\{ \frac{C_s^2 + (2 - \rho)C_T^2/\rho}{2} \right\}. \tag{5.285}$$

See Daley [13] for a discussion of his conjecture. The formula (5.285) appears as formula (3.6) in Daley's paper.

Suzuki and Yoshida [58] have shown that (5.278) is an upper bound for W_q for a GI/G/c queueing system when $0 \le \rho < 1/c$.

In the following theorem we give the best bounds for W_q in a GI/G/c queueing system that have been proven to date. Kingman [30] used a plausibility argument to derive the upper bound

$$W_q \le \frac{\rho W_s}{c(1 - \rho)} \left\{ \frac{c\,C_s^2 + C_T^2/\rho^2}{2} \right\}. \tag{5.286}$$

Wolff [67] provides a rigorous proof of (5.286). Daley has pointed out that the proof given by Wolff is derived by showing that W_q for the GI/G/c queueing system is bounded above by W_q for the related cyclic single server queue and thus (5.286) can be strengthened to the inequality given in the following theorem.

Theorem 5.5.4 (*Bounds on W_q*) *Consider a GI/G/c queueing system with $0 \leq \rho < 1$. Then*

$$W_q \leq \frac{\rho W_s}{c(1 - \rho)} \left\{ \frac{c C_s^2 + (2 - \rho)C_T^2/\rho}{2} \right\}, \qquad (5.287)$$

and

$$\frac{\rho W_s C_s^2}{2c(1 - \rho)} - \frac{W_s[1 + (C_s^2 + 1)(c - 1)]}{2c} \leq W_q. \qquad (5.288)$$

Proof We have explained how (5.287) has been proven. To prove (5.288), we use some results of Brumelle [7]. He constructs a single-server queueing system from the original GI/G/c queueing system. This single server has the same input stream as the original system, but the single server works c times as fast as each of the original servers. That is, the service time of a customer is s/c, where s is the service time of a customer in the original system. Brumelle shows that

$$\widehat{W_q} - \frac{(C_s^2 + 1)(c - 1)W_s}{2c} \leq W_q, \qquad (5.289)$$

where $\widehat{W_q}$ refers to the constructed single server system and all the other symbols refer to the original GI/G/c queueing system. Applying the inequality (5.280) to $\widehat{W_q}$ in the above formula and some simple algebra yields (5.288). ∎

Note that inequality (5.288) provides a positive lower bound for W_q when $C_s^2[1 - c(1 - \rho)] > c(1 - \rho)$, which implies that $\rho > 1 - 1/c$. This condition is consistent with a recent result of Daley [14], who showed that, when $\rho \leq 1 - 1/c$, there is a GI/G/c queueing system with given ρ, C_s^2, and C_T^2, for which W_q is arbitrarily small.

Example 5.5.3 Computer analysts at the Fuzzy Worm Tractor Company have been able to model a planned computer subsystem as an $H_2/H_2/10$ queueing system. They estimate that $\rho = 0.98$, $C_s^2 = 4 = C_T^2$, and $W_s = 2$ seconds for the initial system. They use (5.287) and (5.288) to obtain 15 seconds $\leq W_q \leq 216.4$ seconds. According to the tables of Seelen *et al.* [53], the exact value of W_q is 37.4812 seconds. The heavy traffic approximation yields $W_q = 40.008$ seconds while Daley's conjecture (5.285) yields 40.0 seconds as an upper bound for W_q. □

5.5.3   Approximations for GI/G/c Queueing Systems

We begin by stating a very general but quite useful approximation, that was developed by John Cunneen and myself. It probably has occurred to many other people as well, but I have not seen it in the literature.

Approximation 5.5.1 (*Allen–Cunneen Approximation Formula*) *For any* $GI/G/c$ *queueing system,*

$$W_q \approx \frac{C[c,a]W_s}{c(1-\rho)} \left\{ \frac{C_T^2 + C_S^2}{2} \right\}. \tag{5.290}$$

The reader should note that, for the $M/M/c$ queueing system, (5.290) is exact. It is also exact for $M/G/1$ systems. For $M/G/c$ queueing systems, it gives Martin's estimate (see Martin [45, page 461]). The formula was developed by pattern recognition, not by any formal proof. It gives reasonably good results for many queueing systems and is easy to compute. For Example 5.3.10, it yields the estimate $W_q = 19$ seconds. The exact value is 18.7122 seconds, an error of only 1.5 percent.

Kimura [27] presents an approximation technique for the $GI/G/c$ queueing system, that utilizes several well-known approximations. His approximation is

$$W_q \approx \frac{C_T^2 + C_S^2}{A}, \tag{5.291}$$

where

$$A = \frac{1 - C_T^2}{\text{EW}(D/M/c)} + \frac{1 - C_S^2}{\text{EW}(M/D/c)} + \frac{2(C_T^2 + C_S^2 - 1)}{\text{EW}(M/M/c)} \tag{5.292}$$

where $\text{EW}(M/M/c)$ means the value of W_q for the $M/M/c$ queueing system with the same mean service time and utilization as the $GI/G/c$ system being approximated. Similar remarks apply to $\text{EW}(D/M/c)$ and $\text{EW}(M/D/c)$. The EW values can, of course, be obtained from tables such as those provided by Hillier and Lo [21], Hillier and Yu [22], Kühn [37], van Hoorn [64], or Seelen *et al.* [53]. They also can be calculated from approximation formulas. We will list some of those that Kimura found to be especially valuable. Kimura uses the following approximation formulas from Cosmetatos [11]:

$$\text{EW}(D/M/c) \approx \frac{\text{EW}(D/M/1)}{\text{EW}(M/M/1)}[1 - 4C(c,\rho)]\text{EW}(M/M/c), \tag{5.293}$$

$$\text{EW}(M/D/c) \approx \frac{1}{2}[1 + C(c,\rho)]\text{EW}(M/M/c), \tag{5.294}$$

where $C(c,\rho)$ is defined by

$$C(c,\rho) = (1-\rho)(c-1)\frac{\sqrt{4+5c}-2}{16c\rho}. \tag{5.295}$$

Kimura, of course, limits his approximation to cases where $C_S^2 \leq 1$ and $C_T^2 \leq 1$, except for the $M/H_2/c$ case.

There is an extensive literature on approximation of queueing systems. Whitt [65] provides a review of approximations for the GI/G/1 queueing system and has been a leader in the general approximation of queueing systems. Tijms [62, Chapter 4] provides an excellent review of algorithms and approximations for queueing models. Michael van Hoorn was a student of Tijms. His delightful Ph. D. dissertation [64] provides an excellent explanation of approximation methods for queueing models.

5.6   Summary

In this chapter we have introduced the reader to the fundamental ideas of queueing theory and discussed some of the basic queueing systems that are especially useful in computer science. We have illustrated the use of these systems with a number of examples. In Chapter 6 we will show how some of these basic queueing theory models can be combined into queueing network models to study more complex systems.

STUDENT SAYINGS

Roses are red;
Violets are blue
If λ is big,
Then ρ is too!

Did you say KAMAKAZY Airline or KAMAKAZY Erlang?

I have ρ-ed and ρ-ed until I'm c-sick.

MINO: "Meekest in, never-out"; a queue discipline that describes the process of being snubbed by a snooty head waiter in an exclusive restaurant.

5.7   Exercises

1. [8] Four customers per minute enter the Frugal Fast Food restaurant during lunch hour and spend an average of 4 minutes getting their food (in a queue and receiving service). After receiving their food, 40% of the customers leave the restaurant (with their food) while 60% remain to eat their food inside Frugal Fast Food. Those who stay spend an average of 25 minutes consuming their food. How many customers are inside Frugal Fast Food during the lunch hour, on the average?

2. [HM20] Consider an M/M/1 queueing system in the steady state.

   (a) Show that the probability that there are n or more customers in the system is ρ^n.

   (b) Use the result of part (a) to find the value of μ, such that, for given values of λ, n, and α, with $0 < \alpha < 1$, the probability of n or more customers in the system is α. This value of μ must be given explicitly by a formula in terms of λ, n, and α.

   (c) Use the formula developed in part (b) to find μ if $\lambda = 10$, $n = 3$, and $\alpha = 0.05$.

3. [18] Consider an M/M/1 queueing system in the steady state. Show that the following are true:

   (a)
   $$E[N_q | N_q > 0] = \frac{1}{1 - \rho}.$$

   (b)
   $$\text{Var}[N_q | N_q > 0] = \frac{\rho}{(1 - \rho)^2}.$$

   [Hint: Apply Theorem 2.9.2.]

4. [HM20] Show that, for a stable M/M/1 queueing system,
   $$\text{Var}[N_q] = \frac{\rho^2(1 + \rho - \rho^2)}{(1 - \rho)^2}.$$

   [Hint: Apply Theorem 2.9.2.]

5. [18] BRITE LITE, Inc. has production machines that break down in a Poisson pattern at the rate of three per hour during the eight hour work day. BRITE LITE is considering the repair services of I. M. Slow and I. M. Fast. Slow repairs machines with an exponential repair time distribution at an average rate of four machines per hour for a service charge of $120 per eight hour day. Fast also provides exponential repair time but with an average rate of six machines per hour; Fast charges $200 per eight hour day. Which person should be hired on a daily basis if the cost of an idle machine is $50 per hour? By "daily basis" we mean the person chosen must be paid for an eight hour day every day, even if the person is idle some of the time.

6. [15] People arrive at a telephone booth at the Fly-by-Night airline
   terminal in a random pattern with an average interarrival time of 12
   minutes. The length of phone calls from the booth, including the
   dialing time, wrong numbers, etc. is exponentially distributed with
   an average time of 4 minutes.

   (a) What is the probability that an arriving person will have to wait?

   (b) What is the average length of the waiting lines that form from
       time to time; that is, those that are not of zero length?

   (c) What is the probability that an arrival will have to wait for more
       than 10 minutes before the phone is available?

   (d) The telephone company plans to add a second booth when the
       traffic increases so much that $W_q \geq 5$ minutes. At what average
       interarrival time will $W_q = 5$ minutes occur?

7. [20] [This exercise is of difficulty 7 if you have an HP-32S or HP-28S
   calculator.] A clerk provides exponentially distributed service time
   to customers who arrive randomly at the average rate of 15 per hour.
   What average service time must the clerk provide in order that 90%
   of all customers will queue for service for a time not exceeding 12
   minutes? [Hint: A straightforward formula has never been found.
   You must use an iterative technique.]

8. [10] Show that for a steady state M/M/1 queueing system,

$$\sigma_q^2 = \frac{(2 - \rho)\,\rho\,W_s^2}{(1 - \rho)^2}.$$

   [Hint: The random variables q and s are independent so that $\text{Var}[w] = \text{Var}[q] + \text{Var}[s]$.]

9. [8] Consider a steady state M/M/1 queueing system. Prove the following two formulas from the formulas that have been proven and give
   the intuitive meaning of each of them.

   (a) $W = (L + 1)\,W_s$.

   (b) $W_q = L\,W_s$.

10. [12] In Hopeless Junction a small full service gas station is operated
    by the owner, Mirthless Snerd, by herself. On Monday mornings
    customers (cars) arrive randomly at the average rate of 15 per hour.
    Mirthless provides exponential service with a mean service time of 2.5
    minutes. Please answer the following questions.

(a) What is the mean number of customers waiting (queueing) for service?

(b) What is the mean queueing time in minutes?

(c) What is the mean time a customer spends at the station?

(d) What is the mean number of customers at the station?

(e) What is the probability that Mirthless is idle?

(f) What fraction of time does Ms. Snerd have customers waiting?

(g) What is the mean number of customers waiting for service when one or more are waiting for service? Compare with the answer to part (a).

11. [12] Los Angeles has been struck by a crime wave. Alarmed by the increasing number of bank robberies and concerned about their effect on bank customers, the Banking Upper Management Society (BUMS) adopts the following policies at each bank:

(a) A teller's window is reserved for the exclusive use of bank robbers.

(b) In order to conserve space, bank robberies may be committed only by a lone bandit.

(c) If two or more robberies occur simultaneously, the robbers are served on a first-come, first-served basis.

You are engaged as a consultant by the Bank Robbers Federation (BARF). Your job is to determine if the proposed arrangement with the BUMS is adequate. [Please keep in mind the type of overshoes you are likely to be wearing if you don't get this right.] The data you are given is:

(i) Robbers arrive at random between the hours of 9:00 a.m. and 3:00 p.m.; the average arrival rate is five robbers per hour.

(ii) Teller service time is exponential with an average value of 2 minutes (for the robber's teller). (Special robber withdrawal forms expedite service.)

(iii) The M/M/1 model seems to apply.

You are asked to determine

(1) the average time a robber must queue for service (a robbery).

(2) the average time required for a robbery (queueing time plus service time).

(3) the probability the robber's teller is busy.

(4) the average number of robbers in the bank.

(5) the probability of finding three or more robbers in the bank at the same time.

(6) the probability a robber spends more than 15 minutes in the bank.

(7) the 95th percentile of robbery time.

(The original version of this problem is due to Shelly Weinberg of IBM.)

12. [HM22] Consider an M/M/1/K queueing system. Let q_n be the probability there are n customers in the system just before a customer arrival that actually enters the system; that is, q_n is the probability that there an n customers in the system when an arrival is about to occur. Thus, $q_n = P[A_n|A]$ for $n = 0, 1, 2, \ldots, K-1$, where A_n is the event that there are n customers in the system and A is the event that an arrival is about to occur. Use Bayes' theorem (Theorem 2.4.3) to prove that

$$q_n = \frac{p_n}{1 - p_K}, \quad n = 0, 1, \ldots, K - 1.$$

[Hint: Since the arrival rate is Poisson $P[A|A_n] = \lambda h + o(h)$.]

13. [HM25] Show that

$$W[t] = 1 - \sum_{n=0}^{K-1} q_n \left(e^{-\mu t} \sum_{k=0}^{n} \frac{(\mu t)^k}{k!} \right)$$

for the M/M/1/K queueing system, where

$$q_n = \frac{p_n}{1 - p_K}.$$

[Hint: Write

$$\begin{aligned}
W[t] &= \sum_{n=0}^{K-1} \left\{ \int_0^t \frac{\mu (\mu x)^n e^{-\mu x}}{n!} \, dx \right\} q_n \\
&= \sum_{n=0}^{K-1} \left\{ 1 - \int_t^\infty \frac{\mu (\mu x)^n e^{-\mu x}}{n!} \, dx \right\} q_n \\
&= 1 - \sum_{n=0}^{K-1} q_n \int_t^\infty \frac{\mu (\mu x)^n e^{-\mu x}}{n!} \, dx.
\end{aligned}$$

Then make the change of variable $y = x - t$ in each of the integrals. By recognizing the integral form of the gamma function

$$\Gamma(t) = \int_0^\infty x^{t-1} e^{-x} \, dx, \quad t > 0,$$

and using the property of the gamma function expressed as

$$\Gamma(n+1) = n! \quad n = 0, 1, \ldots,$$

deduce that

$$\int_t^\infty \frac{\mu \left(\mu x\right)^n e^{-\mu x}}{n!} \, dx = e^{-\mu t} \sum_{k=0}^{n} \frac{(\mu t)^k}{k!},$$

for $n = 0, 1, \ldots, K - 1$.]

14. [9] Consider the cyclic queueing model of a computer system shown in Figure 5.7.1 below. It represents a computer system with a constant multiprogramming level of N jobs (programs) sharing the main memory. Server 1 is assumed to be the CPU and server 2 represents the *I/O* system.[17] Servers 1 and 2 provide exponential service with rates μ and λ, respectively. At the end of a CPU service (burst) a job requests *I/O* with probability q or leaves the system with its service complete with probability $p = 1 - q$. When a job completes service and leaves the system, it is immediately replaced by another job with identical statistics to keep the multiprogramming level at a constant N. N is called the *multiprogramming level*, abbreviated MPL. We consider the system to be a birth-and-death process with the state determined by the number of jobs (programs) at the CPU, either receiving service or in the queue. Thus, the system can be in state i for $i = 0, 1, \ldots, N$. The birth-and-death coefficients are $\lambda_i = \lambda$ for $i = 0, 1, \ldots, N - 1$, and $\mu_i = \mu q$ for $i = 1, 2, \ldots, N$. The state transition rate diagram is given in Figure 5.7.2 below. Let p_n be the probability that there are n customers at the CPU, that is, that the system is in state n. Let

$$\rho = \frac{\lambda}{\mu q}.$$

Use Equation (4.29) of Chapter 4 to show that

$$p_n = \rho^n p_0,$$

---

[17]Models as simple as this have been used to model large mainframe computers.

where

$$p_0 = \cfrac{1}{\displaystyle\sum_{n=0}^{N} \rho^n},$$

so that

$$p_0 = \begin{cases} \cfrac{1-\rho}{1-\rho^{N+1}} & \text{for } \rho \neq 1 \\[2ex] \cfrac{1}{N+1} & \text{if } \rho = 1. \end{cases}$$

The CPU utilization, ρ_1, is given by $\rho_1 = 1 - p_0$ and the I/O utilization by $\rho_2 = 1 - p_N$. To use this model to calculate mean throughput, λ, and mean turnaround time, W, we assume that each job starts with a CPU burst, that is followed by an I/O burst, after which it rejoins the CPU queue for another CPU burst, etc. After an average of m CPU bursts (m need not be an integer), it exits the system to be immediately replaced by another job; this keeps the multiprogramming level at N. Thus, each job, on the average, passes through the CPU system m times and the I/O system $m-1$ times. The probability, p, that a job leaves the system after a CPU burst is given by $p = 1/m$. To calculate the throughput, λ, that is the average rate at which jobs enter and depart the computer system, we reason that the departure rate is μp when the CPU is busy and zero otherwise, so that

$$\begin{aligned} \lambda &= \mu p \left(1 - p_0\right) + 0 \times p_0 \\ &= \mu p \rho_1. \end{aligned}$$

By Little's law, we calculate

$$W = \frac{N}{\lambda}.$$

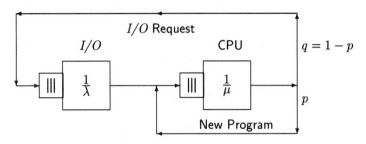

Figure 5.7.1. Cyclic queueing model.

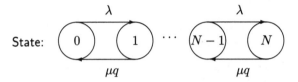

Figure 5.7.2. State-transition rate diagram.

Since the average number of visits a job makes to the CPU is m, the average CPU time used per job is

$$\frac{m}{\mu} = \frac{1}{p\mu}.$$

Similarly, since, on the average, a job makes $m - 1$ visits to the I/O facility, the average job I/O time is

$$\frac{m-1}{\lambda} = \frac{(m-1)}{m}\frac{m}{\lambda} = \frac{q}{p\lambda}.$$

Hence, the ratio of average CPU time per job to average I/O time per job is

$$\frac{\dfrac{1}{p\mu}}{\dfrac{q}{p\lambda}} = \frac{\lambda}{q\mu} = \rho.$$

Therefore, ρ provides a measure of the relative importance of CPU service and I/O service for jobs. If $\rho < 1$, the system is said to be I/O bound; if $\rho > 1$, the system is said to be *CPU bound*. Of course, if $\rho \approx 1$, the system is said to be *balanced*.

15. [12] Suppose Uptight Fawcett has a batch computer system that can be modeled by the cyclic computer model outlined in Exercise 14. Suppose the mean CPU burst time is 0.02 seconds, the mean I/O service time is 0.04 seconds, the mean number of CPU bursts required per job is 18.5, and the multiprogramming level is 10. What is

    (a) the mean throughput, λ,

    (b) mean turnaround time, W,

    (c) CPU utilization, ρ_1, and

    (d) the I/O utilization, ρ_2?

16. [HM18] Show that for an M/M/c queueing system in equilibrium, the following are true:

    (a) If $n < c$,

$$P[N \geq n] = p_0 \left\{ \sum_{k=n}^{c-1} \frac{a^k}{k!} + \frac{a^c}{c! \, (1-\rho)} \right\}.$$

    (b) If $n \geq c$,

$$P[N \geq n] = C[c, a] \, \rho^{n-c}.$$

    (c) If $c = 2$, (a) and (b) reduce to

$$P[N \geq n] = \frac{2 \rho^n}{1+\rho}, \quad n = 1, 2, \ldots.$$

    (d) If $c = 1$, (a) and (b) reduce to

$$P[N \geq n] = \rho^n, \quad n = 0, 1, 2, \ldots.$$

17. [HM20] Show that, for an M/M/c queueing system in the steady state,

$$\sigma_{N_q}^2 = \frac{\rho \, C[c, a] \, \{1 + \rho - \rho \, C[c, a]\}}{(1-\rho)^2},$$

and

$$\sigma_N^2 = \sigma_{N_q}^2 + a \, (1 + C[c, a]).$$

18. [HM20] Prove that, for an M/M/c queueing system in the steady state,

$$\sigma_q^2 = \frac{\{2 - C[c, a]\} \, C[c, a] \, W_s^2}{c^2 \, (1-\rho)^2}.$$

Hint: Use the fact that

$$\int_0^\infty x^n \, e^{-\mu x} \, dx = n! \, \mu^{-n-1}.$$

19. [20] Two analysts at Exxtol Petrol, Ltd., Log Jam and Bot L. Neck, are having an argument. They are comparing an M/M/1 system with mean arrival rate λ and mean service rate 2μ (we assume $\lambda < 2\mu$), with an M/M/2 system with mean arrival rate λ and mean service rate μ for each server. Log says the M/M/2 system is best because

$$W_{q_{M/M/2}} < W_{q_{M/M/1}}.$$

Bot responds that Log has it all wrong because

$$W_{M/M/1} < W_{M/M/2},$$

and therefore, the M/M/1 system is superior. Who is right? Note that the two systems have equal capacity; Log and Bot are comparing a single-server system to a double-server system with half-speed servers.

20. [22] Phil T. Grime, the manager of the Information Center at Gritty Soap, provides three consultants to help personal computer users solve their problems. PC users with problems arrive randomly, at an average rate of 20 per 8 hour day. The amount of time that a consultant spends with a PC user has an exponential distribution with average value of 40 minutes. Users are assigned to consultants in the order of their arrival.

    (a) What fraction of the time is each consultant busy?

    (b) What is the mean time a user spends in the queue?

    (c) What is the mean number of users waiting for a consultant?

    (d) What is the mean time a user spends in the Information Center?

    (e) What is the mean number of users in the center?

    (f) What is the probability that all the consultants are idle?

    (g) What is the probability that all the consultants are busy but no one is waiting in line?

21. [25] Customers arrive randomly (during the evening hours) at the Kittenhouse, the local house of questionable services, at an average rate of five per hour. Service time is exponential with a mean of 20 minutes per customer. There are two servers on duty.

    (a) What is the probability an arriving customer must queue?

    (b) That one or both servers are idle?

    (c) What is the average time a customer spends at the Kittenhouse?

(d) If the Kittenhouse is raided, how many customers will be caught, on the average?

(e) What is the probability that five or more customers will be caught in a raid?

(f) What is the probability that both servers are idle?

22. [C30] (See Exercise 21.) So many queueing theory students visit the Kittenhouse to collect data for this book that the proprietress, Kitty Callay (also known as the Cheshire Cat) makes some changes. She trains her kittens to provide more exotic but still exponentially distributed service and adds three more servers, for a total of five. Her captivated, titillated customers still complain that the queue is too long. Kitty commissions her most favored customer, Gnalre K. Renga to make a study of her establishment. He is to determine the mean arrival rate, λ, during the peak period, the mean service time, W_s, and to recommend the number of servers she should provide so that

(a) the mean queueing time for those who must queue will not exceed 20 minutes, and

(b) the probability that an arriving customer must wait for service will not exceed 0.25.

Mr. Renga finds that the arrival pattern is exponential with $\lambda = 9$ customers per hour. He also determines that the service time is exponential with $W_s = 30$ minutes.

(i) For the original system (with 5 servers), calculate the performance measures W_q, L_q, L, and the probability of not having to queue for service.

(ii) How many servers must be provided to satisfy the requirements (a) and (b), above?

(iii) Assume the number of servers determined in (ii) are provided. Answer the questions asked in (a)–(f) of Exercise 21 for the new Kittenhouse, where (b) now means, "What is the probability that at least one server is idle?" and (f) becomes, "What is the probability all servers are idle?"

23. [HM18] Howie Kramms, a computer science student at Ginger Tech., bragged that he could prove the following formulas for the steady

state M/M/c queueing system. He was willing to wager 10 dollars that no one else in his dormitory could. Can you call Howie's bluff?

$$
E[w^2] = \begin{cases} \dfrac{2C[c,a]W_s^2}{1 - c(1-\rho)} \left(\dfrac{1 - c^2(1-\rho)^2}{c^2(1-\rho)^2} \right) + 2W_s^2, & a \neq c - 1 \\[4mm] 4C[c,a]W_s^2 + 2W_s^2, & a = c - 1 \end{cases}
$$

Hint:
$$
\int_0^\infty t^n e^{-\mu t}\, dt = n!\, \mu^{-n-1}.
$$

24. [C20] James Martin has suggested that $\pi_w[95]$ for an M/M/c system is approximately $W + 2\sigma_w$. For the M/M/c system of part (iii) of Exercise 22,

   (a) Calculate Martin's approximation for $\pi_w[95]$.

   (b) Using (a) as a starting value, calculate $\pi_w[95]$ to at least 3 decimal places, using a numerical technique and the formula for $W[\cdot]$, the distribution function of w.

25. [20] It is shown in Section 5.2.4 that the probability that all c servers are busy in an M/M/c/c queueing system (M/M/c loss system) is given by Erlang's B formula, $B[c,a]$, defined by

$$
B[c,a] = \frac{\dfrac{a^c}{c!}}{1 + a + \dfrac{a^2}{2!} + \cdots + \dfrac{a^c}{c!}}.
$$

Consider an M/M/c queueing system in the steady state. Show that the following are true:

   (a)
$$
\frac{1}{C[c,a]} = \rho + \frac{(1-\rho)}{B[c,a]}.
$$

   (b)
$$
\frac{1}{B[1,a]} = 1 + \frac{1}{a}.
$$

   (c)
$$
\frac{1}{B[n,a]} = 1 + \frac{n}{a} \times \frac{1}{B[n-1,a]} \quad \text{for } n = 2, 3, \ldots, c.
$$

26. [P25]

   (a) Write a computer program to implement Algorithm 5.2.2.

   (b) Write a recursive program to calculate $B[c, a]$, using steps 1 and 2 of Algorithm 5.2.2.

   (c) Use the programs of (a) and (b) to calculate $B[15, 5]$. Which program runs faster?

27. [C25] JETSET Airlines, a fierce competitor of KAMAKAZY Airlines (Example 5.2.3), also is planning a new telephone reservation office. Their agents provide customers who call with an exponential service time. Like KAMAKAZY Airlines, calls that arrive when all agents are busy are held (with appropriate background music) until an agent is free. They expect a random pattern of customer calls with an average of 30 calls per hour during the peak period.

   (a) The three performance criteria are:

      1. The average time a caller waits to talk to an agent must not exceed 20 seconds.

      2. The average waiting time for customers who must queue for service must not exceed 2 minutes.

      3. Ninety-five percent of all customers must reach an agent in 30 seconds or less.

      How many agents should be provided?

   (b) If the number of agents required for part (a) is provided, calculate W_q, W, $E[q|q > 0]$, and $\pi_q[90]$ in seconds. Also compute the probability that $w \leq 3.5$ minutes and $P[N \geq 5]$.

   (c) What is the probability that all the agents are busy during the peak period? That all are idle?

28. [C20] YOUTOOLCOMPUTE has 10 portable personal computers available for rent. The average rental time is 2.5 days and is exponentially distributed. Customers arrive randomly at an average rate of five per day. If a computer is not available, a customer will go to HELL (Hewlett, Ernest, Leland, and Lial) for a computer.

   (a) What fraction of arriving customers will be lost?

   (b) What is the average number of computers on rent?

   (c) What is the probability a personal computer on rent will be on rent for more than five days?

(d) Management estimates that the profit on a computer being rented is $20 per day (they rent by the day, only). One of the computers is destroyed in an accident and cannot be replaced for 30 days. What is the maximum YOUTOOLCOMPUTE can afford to pay per day for a 30 day replacement?

29. [C20] (This exercise was contributed by Dr. Ray Bryant of IBM.) Programmers at Prolific Programming Unlimited connect to a time-sharing system over dedicated dial-up communication trunks. Arrivals of incoming programmer calls to the trunks can be modeled as a Poisson arrival process with the mean rate of 0.2 calls per minute. The length of each time-sharing session can be modeled as a uniformly distributed random variable in the interval 10 to 50 minutes; session length is independent of the load.

(a) How many dial-up lines (trunks) should there be to make sure that less than 10 percent of incoming programmer calls get busy signals?

(b) With this number of lines, what is the mean number of time-sharing sessions in progress?

(c) What is the probability that 6 or more of the lines are in use?

30. [C25] Houdini Engineering (sometimes known as Tech Type Toolers or T^3 for short) has a large room containing 50 computer workstations (called "the pit") for the use of the engineering staff. During the busiest period of the day, engineers arrive randomly at the mean rate of 49 per hour and spend an average of 30 minutes at a workstation; this latter time is exponential. Thus, the terminal room can be modeled as an M/M/50 queueing system.

(a) Calculate the performance measures W, $\pi_w[90]$ (use Martin's approximation if you can't compute it exactly), L, W_q, $\pi_q[90]$, and L_q.

(b) Approximate the values in part (a) by modeling the system as an M/M/∞ system.

31. [15] Lassettre and Scherr [40] used the machine repair queueing system to model the performance of the OS/360 time-sharing option (TSO), a system that allowed programmers to develop and run programs from a terminal in a time-sharing mode. It was, of course, the progenitor of the current TSO, that runs under the MVS operating system. (Allan Scherr received the Association of Computing Machinery's Grace

Murray Hopper award in 1975 for some earlier work (Scherr [52]), in which he used this same queueing system to model the MIT Project MAC time-sharing system called CTSS for Compatible Time-Shared System.) For their model, Lassettre and Scherr identified the parameters $E[O]$, W, W_s, and K of the machine repair system with the parameters of the TSO model as follows: $E[O]$ is the average user time, that they defined as the mean elapsed time from the request for input made by the processing program to the completion of the requested input by the user. W is the average response time that they defined as the mean length of the time interval that begins with the completion of input and ends when the program processing this input finishes and requests more input. Thus, taken together, the user time and the response time make up a complete cycle or interaction. W_s is the mean system service time, that is, the mean of the time required by the system to execute the program that processes the input entered by the user. K is the average number of users actively interacting with the system.

(a) Use Little's law to show that the mean number of interactions per second is given by

$$\lambda = \frac{K}{E[O] + W}.$$

(b) Show that the mean response time is given by

$$W = \frac{K W_s}{1 - p_0} - E[O],$$

where p_0 is the probability that there are no requests for system service pending.

(c) Show that the average number of users waiting for a response is

$$L = \frac{K W}{W + E[O]}.$$

(d) Show that the mean response time when there is a large number of users (K is large) is approximately $K W_s - E[O]$. Assume that the CPU is the computer system bottleneck.

(e) The curve W/W_s versus K has the asymptote $W/W_s = K - E[O]/W_s$ for large K. We see that for small values of K there is an asymptote to the curve, that is the horizontal line $W/W_s = 1$. The two asymptotes intersect at the point $(K^*, 1)$ where

$$K^* = \frac{W_s + E[O]}{W_s}.$$

Kleinrock [32] calls the value of K^* the *saturation number*, but Lassettre and Scherr call the point $(K^{**}, 0)$ where the line $K W_s - E[O]$ intersects the x-axis the *saturation point* and claim it is a fair approximation of the capacity of a time-sharing system. Thus,

$$K^{**} = \frac{E[O]}{W_s}.$$

Lassattre and Scherr use an $E[O]$ value of 35 seconds based on measurements for users of terminals such as Teletypes and IBM 2741s. Using this value of $E[O]$, they reported that for one TSO system with 60 active terminals the mean response time, W, was 5 seconds, with W_s equal to 0.8 seconds.[18] If CRT type terminals are acquired for this system that provide an $E[O]$ of 10 seconds and the throughput remains the same as before, what is the mean response time W? What is K^{**} for this system?

32. [HM25] Consider the M/M/c/K/K machine repair queueing system. Let p_n be the probability that n of the K machines are inoperable (either undergoing or awaiting repair) and let q_n be the probability that a machine that breaks down finds n inoperable machines in the repair facility.

(a) Prove that

$$q_n = \frac{(K-n)p_n}{K-L}, \quad n = 0, 1, \ldots, K-1.$$

(b) For an M/M/1/K/K queueing system, prove that

$$q_n = \frac{\dfrac{z^{K-n-1}}{(K-n-1)!}}{\displaystyle\sum_{k=0}^{K-1} \frac{z^k}{k!}}, \quad n = 0, 1, \ldots, K-1,$$

where $z = E[O]/W_s$, and thus, q_n has the same value as p_n for the M/M/1/K-1/K-1 system; that is, the number of machines found by arriving machines is the same as that which would be seen at a randomly chosen instant in a system with one less machine.

---

[18] A W of 5 seconds is *not* what the machine repair model gives with $K = 60$, $E[O] = 35$ seconds, and $W_s = 0.8$ seconds, but it is what they reported. Assume that the values of K, $E[O]$, and W are correct but they made a mistake in measuring W_s.

Hints: q_n is the probability that n machines are inoperable, given that a machine is about to break down and thus is equal to $P[A_n|A], n = 0, 1, \ldots, K - 1$. For part (a), use Bayes' theorem and the fact that $P[A|A_n] = (K - n)\alpha h + o(h)$ for each $h > 0$. For (b), write

$$q_n = \frac{(K - n)p_n}{K - L} = \frac{(K - n)p_n}{\displaystyle\sum_{k=0}^{K}(K - k)p_k},$$

and note that we have proven earlier that

$$p_n = \frac{\dfrac{z^{K-n}}{(K - n)!}}{\displaystyle\sum_{k=0}^{K}\dfrac{z^k}{k!}}.$$

33. [HM28] Let $p[k; \lambda]$ represent the Poisson probability

$$p[k; \lambda] = e^{-\lambda}\frac{\lambda^k}{k!}, \quad \lambda > 0, \, k = 0, 1, \ldots.$$

Let $Q[k; \mu]$ be defined by the Poisson sum

$$Q[k; \mu] = e^{-\mu}\sum_{i=0}^{k}\frac{\mu^i}{i!}, \quad \mu > 0, \, k = 0, 1, 2, \ldots.$$

(a) Prove that

$$\sum_{j=0}^{k}p[k - j; \lambda]Q[j; \mu] = Q[k; \lambda + \mu],$$

(b) Prove that

$$Q[k; y] = \int_{y}^{\infty}\frac{e^{-x}x^k}{k!}\,dx = P[Y > y],$$

where Y is a gamma random variable with parameters $\beta = k + 1$ and $\alpha = 1$.

34. [C15] Consider Example 5.2.8. Show that, for the machine repair queueing system M/D/1/6/6 with $E[O] = 40$ seconds and $W_s = 1$ second, that $p_0 = 0.85390$, $\lambda = 525.95$ requests per hour, and $W = 1.069$ seconds.

35. [C30] Consider Example 5.2.9. Suppose that High Tale puts 50 work-stations on line. Compute p_0, ρ, λ, and W for the system; that is, for the M/D/1/50/50 system with $E[O] = 80$ seconds and $W_s = 2$ seconds.

36. [C20] (The computation for this exercise can be done in about 20 minutes with a pocket calculator. The calculations are trivial if the APL functions MACHΔREP, DWQΔMRΔC, and DWΔMRΔC are available.) The English farnsworth company, Farnsworth Unlimited, Ltd., (a farnsworth is a microelectronic device for gauging the performance of farns) has a number of shops, each of which can be modeled as an M/M/2/3/3 queueing system with $E[O] = 10$ minutes and $W_s = 8$ minutes. Please do the following:

    (a) Calculate p_i for $i = 0, 1, 2, 3$, q_i for $i = 0, 1, 2$, and W, W_q, L, and L_q for each system.

    (b) Calculate the probability that an inoperative machine must queue for repair.

    (c) Calculate $E[q|q > 0]$.

    (d) Calculate $W_q[1]$ and $W[10]$ when time is measured in minutes.

37. [HM18] The M/M/c/K queueing system can be modeled as a birth-and-death process (see Sections 4.3 and 5.2).

    (a) Draw the state-transition rate diagram and from it deduce that

    $$\lambda_n = \begin{cases} \lambda & \text{for } n = 0, 1, \ldots, K - 1, \\ 0 & \text{for all other } n, \end{cases}$$

    while

    $$\mu_n = \begin{cases} n\,\mu & \text{for } n = 1, 2, \ldots, c \\ c\,\mu & \text{for } n = c + 1, \ldots, K \\ 0 & \text{otherwise.} \end{cases}$$

    (b) Show that part (a) yields

    $$p_n = \begin{cases} \dfrac{a^n}{n!}\, p_0 & \text{for } n = 1, 2, \ldots, c \\[2mm] \dfrac{a^c}{c!} \left(\dfrac{a}{c}\right)^{n-c} p_0 & \text{for } n = c + 1, \ldots, K, \end{cases}$$

    where

    $$p_0 = \left[\sum_{n=0}^{c} \frac{a^n}{n!} + \frac{a^c}{c!} \sum_{n=1}^{K-c} \left(\frac{a}{c}\right)^n \right]^{-1},$$

    and $a = \lambda W_s = \lambda/\mu$.

(c) From the formula

$$Lq = \sum_{n=c+1}^{K} (n - c)\, p_n,$$

show that

$$Lq = \frac{a^c p_0\, r \left[1 - (K - c + 1) r^{K-c} + (K - c) r^{K-c+1}\right]}{c!\,(1 - r)^2},$$

where $r = a/c$.

(d) Show that

$$L = Lq + E[N_s]$$

$$= Lq + \sum_{n=0}^{c-1} n\, p_n + c \left(1 - \sum_{n=0}^{c-1} p_n\right).$$

(e) Let q_n be the probability that an arriving customer finds n customers in the service facility. Use the same argument as that in Exercise 12 to show that

$$q_n = \frac{p_n}{1 - p_K}, \quad n = 0, 1, 2, \ldots, K - 1.$$

(f) Show that

$$E[q|q > 0] = \frac{Wq}{1 - \sum\limits_{n=0}^{c-1} q_n}.$$

(g) Let λ_a be the average arrival rate of customers who actually enter the system. Show that

$$\lambda_a = \lambda(1 - p_K).$$

38. [C22]

(a) Derive a formula for the stationary probability distribution of the number of customers in the system for an M/M/2/3 queueing system. Express p_n as a function of λ and μ.

(b) Suppose the Free K. Doubt company has modeled a computer subsystem as an M/M/2/3 queueing system with $\lambda = 15$ customers per second and $\mu = 5$ customers per second. Calculate p_n (for $n = 0, 1, 2, 3$), λ_a, ρ, Lq, L, Wq, and W for this model. (The original form of this exercise was contributed by Dr. Raymond Bryant.)

39. [18] Use Theorem 5.3.1 and its corollary to prove that the following formulas for the M/G/1 queueing system are true:

(a)
$$W_q = \frac{\lambda E[s^2]}{2(1 - \rho)}; \quad \text{(Pollaczek's formula)}$$

(b)
$$E[q^2] = 2W_q^2 + \frac{\lambda E[s^3]}{3(1 - \rho)};$$

(c)
$$E[w^2] = E[q^2] + \frac{E[s^2]}{1 - \rho}.$$

40. [20] Prove that, for the M/G/1 queueing system,

$$\sigma_N^2 = \frac{\lambda^3 E[s^3]}{3(1 - \rho)} + \left(\frac{\lambda^2 E[s^2]}{2(1 - \rho)} \right)^2$$
$$+ \frac{\lambda^2 (3 - 2\rho) E[s^2]}{2(1 - \rho)} + \rho(1 - \rho)$$

41. [18] Brumelle [6] proves that, for an M/G/1 queueing system,

$$E[N_q(N_q - 1) \cdots (N_q - k + 1)] = \lambda^k E[q^k],$$

for $k = 1, 2, 3, \ldots$.

(a) Use Brumelle's result to show that

$$\sigma_{N_q}^2 = \frac{\lambda^3 E[s^3]}{3(1 - \rho)} + \left(\frac{\lambda^2 E[s^2]}{2(1 - \rho)} \right)^2 + \frac{\lambda^2 E[s^2]}{2(1 - \rho)}.$$

(b) Use part (a) and the result of Exercise 40 to show that

$$\text{Cov}[N_q, N_s] = \frac{\lambda^2 E[s^2]}{2},$$

and thus that

$$E[N_q N_s] = \frac{\lambda^2 E[s^2]}{2(1 - \rho)}.$$

[Hint: Use Theorem 2.7.2(c) and the fact that $\sigma_{N_s}^2 = \rho(1 - \rho)$.] Note: Since $E[N_s N_q] \neq E[N_s] E[N_q]$, N_s and N_q are *not* independent random variables (see Theorem 2.7.1(d)). Of course we would not expect them to be because the number in the queue clearly depends on the number in service. However, the random variables q and s *are* independent by assumption.

42. [HM20] Show from the fact that $g_N(z)$ is given by

$$g_N(z) = \frac{(1-\rho)(1-z)K(z)}{K(z) - z}$$

that

$$L = g'_N(1) = \rho + \frac{K''(1)}{2(1-\rho)}.$$

43. [22] Use the formula

$$g_N(z) = \frac{(1-\rho)(1-z)W_s^*[\lambda(1-z)]}{W_s^*[\lambda(1-z)] - z}$$

to show that the generating function for the $M/H_2/1$ queueing system is

$$g_N(z) = \frac{(1-\rho)[1 + (\rho_1 + \rho_2 - \rho)(1-z)]}{\rho_1\rho_2 z^2 - (\rho_1 + \rho_2 + \rho_1\rho_2)z + 1 + \rho_1 + \rho_2 - \rho}$$

where $\rho_i = \lambda/\mu_i$, $i = 1, 2$. [Hint: Using the notation of Section 3.2.9, show that $q_1\rho_2 + q_2\rho_1 = \rho_1 + \rho_2 - \rho$.]

44. [HM25] Prove that w for the $M/H_2/1$ queueing system has the two-stage hyperexponential distribution function

$$W[t] = P[w \le t] = 1 - \pi_a e^{-\mu_a t} - \pi_b e^{-\mu_b t}, \ t \ge 0. \qquad (5.296)$$

In (5.296) the constants are given by

$$\pi_a = \frac{C_1 z_1}{z_1 - 1}, \qquad (5.297)$$

$$\pi_b = \frac{C_2 z_2}{z_2 - 1}, \qquad (5.298)$$

$$\mu_a = \lambda(z_1 - 1), \qquad (5.299)$$

and

$$\mu_b = \lambda(z_2 - 1), \qquad (5.300)$$

where the constants in (5.297) through (5.300) are those used in the formula for $g_N(z)$ derived in Example 5.3.2; that is,

$$g_N(z) = \frac{C_1 z_1}{z_1 - z} + \frac{C_2 z_2}{z_2 - z}. \qquad (5.301)$$

[Hint: Use the fact that $g_N(z) = W^*[\lambda(1-z)]$. Then use $\theta = \lambda(1-z)$ in (5.301) to find $W^*[\theta] = g_N(1 - \theta/\lambda)$.]

45. [HM30] Consider the $M/H_2/1$ queueing system. Invert the Laplace–Stieltjes transform of the queueing time, q, by the method of partial fractions to obtain the density function

$$f_q(t) = (1 - \rho)\delta(t) + C_3 e^{-at} + C_4 e^{-bt}, \quad t \geq 0, \qquad (5.302)$$

where $a = -z_1$ and $b = -z_2$ for the zeroes z_1 and z_2 of the polynomial

$$\theta^2 + (\mu_1 + \mu_2 - \lambda)\theta + \mu_1 \mu_2 (1 - \rho). \qquad (5.303)$$

The parameters μ_1 and μ_2 in (5.303) are the parameters for the distribution of s; that is,

$$W_s = \frac{q_1}{\mu_1} + \frac{q_2}{\mu_2}. \qquad (5.304)$$

The constants C_3 and C_4 in (5.302) are given by

$$C_3 = \frac{\lambda(1 - \rho)z_1 + \rho(1 - \rho)\mu_1 \mu_2}{z_1 - z_2}, \qquad (5.305)$$

and

$$C_4 = \frac{\lambda(1 - \rho)z_2 + \rho(1 - \rho)\mu_1 \mu_2}{z_2 - z_1}, \qquad (5.306)$$

Now integrate (5.302) to show that

$$W_q[t] = P[q \leq t] = 1 - \frac{C_3}{a}e^{-at} - \frac{C_4}{b}e^{-bt}, \quad t \geq 0. \qquad (5.307)$$

As part of deriving (5.307), you will need to show that

$$\frac{C_3}{a} + \frac{C_4}{b} = \rho. \qquad (5.308)$$

46. [HM30] Show that the density function for queueing time in the $M/M/c/K/K$ queueing system is given by

$$f_q(t) = W_q[0]\delta(t) + \frac{c^c \mu p[K - c - 1; c(t + z)]}{(c - 1)! \times p[K - 1; cz]}p_0[K - 1], \quad t \geq 0,$$

where

$$z = \frac{E[O]}{W_s}.$$

Hint: Differentiate the formula (5.309) below for $W_q[t]$. Note that q has a probability mass at the origin equal to $W_q[0] = q_0$. In differentiating (5.309), use the fact that

$$\frac{\partial}{\partial y}Q[k; y] = -p[k; y],$$

which follows from the formula

$$Q[k; y] = \int_y^\infty p[k; x]\, dx$$

of Exercise 33(b).

$$W_q[t] = P[q \le t] = 1 - \frac{c^c Q[K - c - 1; cz] p_0[K - 1]}{c! \times p[K - 1; cz]}, \quad t \ge 0, \quad (5.309)$$

where

$$Q[k; \alpha] = e^{-\alpha} \sum_{n=0}^k \frac{\alpha^n}{n!}.$$

47. [C15] Use the results of Exercise 44 to construct the distribution function of w for the first communication line of Example 5.3.1. Then calculate (a) W, (b) $W[20]$, (c) $E[w^2]$, and (d) $\pi_w[90]$.

48. [15] A favorite game played by two consenting mathematicians is called Proof or Counterexample. Player A states a theorem in the form "C implies D." Player B must either prove the theorem; that is, prove that the truth of C implies the truth of D or give a counterexample. A counterexample is an example in which C is true but D is false. You have observed that for the M/M/1 queueing system, both s and w are exponentially distributed and for the $M/H_2/1$ queueing system (see Exercise 44), both s and w have a two-stage hyperexponential distribution. You are player B in Proof or Counterexample and player A says, "For every M/G/1 queueing system, s and w have the same form of distribution." What is your response?

49. [M25] Consider the steady state $M/H_2/1$ queueing system discussed in Example 5.3.3. All parameters in this exercise are defined there. Assume $z_1 > z_2 > 1$.

(a) Prove that

$$P[N \ge n] = C_1 \frac{z_1^{-n+1}}{z_1 - 1} + C_2 \frac{z_2^{-n+1}}{z_2 - 1}, \quad n = 0, 1, \ldots.$$

(b) Prove, by using part (a), that $P[N \ge 1] = \rho$ and $P[n = 0] = 1 - \rho$.

50. [8] Consider Exercises 47 and 49. For the queueing system of Exercise 47, calculate $P[N \ge 3]$ and $P[N \ge 5]$.

51. [M20] Prove that, for the GI/G/c and GI/G/c/K/K steady state queueing systems,

$$E[q|q > 0] = \frac{W_q}{P[q > 0]},$$

where $P[q > 0]$ is the probability that an arriving customer must queue for service. Note that an arriving customer must queue for service if and only if she finds all the servers busy, but this probability is not necessarily the same as the probability that all the servers are busy; that is, the probability that a random observer finds all the servers busy. Wolff [66] has shown the two probabilities are the same only when the arrival process is Poisson. We have seen that these probabilities are different for the machine repair systems M/M/1/K/K and M/M/c/K/K.

52. [T20] Consider the steady state $M/D/1$ queueing system. Prove the following:

(a)

$$W_q[t] = \sum_{n=0}^{k-1} p_n + p_k \left(\frac{t - (k-1)W_s}{W_s} \right)$$

where

$$(k-1)W_s \le t < kW_s, \quad k = 1, 2, \ldots.$$

(b)

$$W[t] = \begin{cases} 0 & \text{for } t < W_s \\ \\ \sum_{n=0}^{k-1} p_n + p_k \left(\dfrac{t - kW_s}{W_s} \right) & \text{for } t \ge W_s, \end{cases}$$

where

$$kW_s \le t < (k+1)W_s, k = 1, 2, \ldots.$$

(In Example 3.4.6 we show how to compute the values of p_n.)

(c) Consider the M/D/1 queueing system of Example 5.3.1. Calculate $W_q[2]$, $W_q[5]$, $W[8]$, $W[14.4]$, and $W[14.43]$.

53. [12] Let b be the busy period of the server in the M/G/1 queueing system, that is, the time from a server start up to service a customer (after an idle period) until the server is again idle: Kleinrock [32] shows that

$$E[b] = \frac{W_s}{1 - \rho},$$

(note that this is the average time a customer spends in an M/M/1 system), and

$$E[b^2] = \frac{E[s^2]}{(1-\rho)^3},$$

so that

$$\sigma_b^2 = \frac{\sigma_s^2 + \rho W_s^2}{(1-\rho)^3}.$$

Find $E[b]$, $E[b^2]$, and σ_b^2 in terms of the system parameters for

(a) the M/M/1 queueing system,

(b) the M/E_k/1 queueing system, and

(c) the M/D/1 queueing system.

54. [08] (See Exercise 53.) Assume that, for (a), (b), and (c), $W_s = 2$ seconds and $\rho = 0.8$. Find the numerical values of $E[b]$, $E[b^2]$, and σ_b^2 in each case, assuming for part (b) that $k = 4$.

55. [08] Let N_b be number of customers served during one busy period of the server in the M/G/1 queueing system (see Exercise 53). Kleinrock [32] shows that

$$E[N_b] = \frac{1}{1-\rho},$$

and

$$E[N_b^2] = E[N_b] + \frac{2\rho(1-\rho) + \lambda^2 E[s^2]}{(1-\rho)^3},$$

so that

$$\sigma_{N_b}^2 = \frac{\rho(1-\rho) + \lambda^2 E[s^2]}{(1-\rho)^3}.$$

Find $E[N_b]$, $E[N_b^2]$, and $\sigma_{N_b}^2$ in terms of the system parameters for

(a) the M/M/1 queueing system,

(b) the M/E_k/1 queueing system, and

(c) the M/D/1 queueing system.

56. [05] (See Exercise 55.) Assuming that $Ws = 2$ seconds and $\rho = 0.8$ find the numerical values of $E[N_b]$, $E[N_b^2]$, and $\sigma_{N_b}^2$ for

(a) an M/M/1 queueing system,

(b) an M/E_4/1 queueing system, and

(c) an M/D/1 queueing system.

57. [10] (See Exercises 53 and 55.) Construct an H_2 probability distribution s using Algorithm 3.2.2 with $W_s = 2$ seconds and $C_s^2 = 10$. Then for the $M/H_2/1$ queueing system with the given service time and $\rho = 0.8$, calculate the numerical values of $E[b]$, $E[b^2]$, σ_{b^2}, $E[N_b]$, $E[N_b^2]$, and $\sigma_{N_b}^2$.

58. [10] Marshall [44] shows that for an M/G/1 queueing system the Laplace–Stieltjes transform of the interdeparture time is given by

$$D^*[\theta] = \frac{(\theta + \mu)\rho W_s^*[\theta]}{\theta + \lambda}.$$

Using the above result, show that the interdeparture time distribution is exponential if and only if the service time is exponential. (Disney *et al.* [15] showed that the only M/G/1 queueing system having independent, identically distributed, interdeparture times is the M/M/1 system. Such a stream is called a *renewal process*. Laslett [39] showed that the only GI/M/1 queueing system having renewal output was the M/M/1 system.)

59. [HM20] Given that, for the GI/M/1 queueing system,

$$W_q[t] = P[q \le t] = 1 - (1 - \pi_0)e^{-\pi_0 t/W_s}, \ t \ge 0,$$

prove the following:

(a)
$$W_q = (1 - \pi_0)\frac{W_s}{\pi_0}.$$

(b)
$$E[q^2] = 2(1 - \pi_0)\left(\frac{W_s}{\pi_0}\right)^2.$$

(c)
$$\sigma_q^2 = (1 - \pi_0^2)\left(\frac{W_s}{\pi_0}\right)^2.$$

60. [HM15] Show that, if the interarrival time of τ is uniformly distributed between 0 and $2/\lambda$, then

$$A^*[\theta] = \frac{\lambda}{2\theta}\left(1 - e^{-2\theta/\lambda}\right).$$

61. [HM20] Consider the $H_2/M/1$ queueing system of Example 5.3.8. If τ has the distribution generated by Algorithm 3.2.2 show that

$$\pi_0 = 0.5 - \rho + 0.5\sqrt{(1-2\rho)^2 + 16\rho q_1(1-q_1)(1-\rho)}.$$

62. [HM10] Consider the $E_k/M/1$ queueing system. Show that

$$A^*[\theta] = \left(\frac{k\lambda}{k\lambda + \theta}\right)^k.$$

63. [HM18] Consider the $H_2/M/1$ queueing system in which the parameters for τ are $q_1 = 0.4$, $\mu_1 = 0.5\lambda$, and $\mu_2 = 3\lambda$. Show that $E[\tau] = 1/\lambda$ and that the equation

$$1 - \pi_0 = A^*[\mu\pi_0]$$

reduces to the quadratic equation

$$\pi_0^2 + (3.5\rho - 1)\pi_0 + 1.5\rho(\rho - 1) = 0.$$

Show, also, that the unique value of π_0 such that $0 < \pi_0 < 1$ is given by

$$\pi_0 = 0.5 - 1.75\rho + \sqrt{1.5625\rho^2 - 0.25\rho + 0.25}.$$

64. [HM20] Show that, for the GI/M/1 queueing system,

$$\sigma_N^2 = \frac{\rho(2 - \pi_0 - \rho)}{\pi_0^2}.$$

65. [HM20] Show that, for the GI/M/1 queueing system,

$$\sigma_{N_q}^2 = \frac{\rho(1 - \pi_0)[2 - \pi_0 - \rho(1 - \pi_0)]}{\pi_0^2}.$$

66. [15] Consider Exercise 6. Suppose the telephone system at Fly-by-Night can be modeled as an $E_2/M/1$ queueing system with $E[\tau] = 10$ minutes, and $W_s = 3$ minutes. Answer the following questions:

   (a) What is the probability that an arriving customer will have to wait to use the phone?

   (b) What is the average length of a nonempty queue?

   (c) What is the probability that an arriving customer will have to wait for more than 10 minutes before the phone is available?

67. [C18] The performance analysts at Manufacturers Handover Fist have successfully modeled a computer subsystem using the D/M/1 queueing system with $E[\tau] = 0.02$ seconds and $W_s = 0.016$ seconds.

    (a) For the steady state system, calculate W_q, L_q, σ_q, L, W, $\pi_w[90]$, $W[0.1]$, $W_q[0.08]$, and $\pi_q[90]$.
    (b) Make the calculations of part (a), assuming that $E[\tau]$ has decreased to $4/245$ seconds.

68. [HM10] Show that for the GI/M/1 queueing system,

$$E[N_q|N_q > 0] = \frac{1}{\pi_0}.$$

69. [C20] Consider Example 5.3.13. Consider a D/M/2 queueing system with $W_s = 1$ second and $\rho = 0.9$ so that $\lambda = 1.8$ customers per second. Calculate the performance parameters W_q, W, L_q, L, the distribution function for w, and $\pi_w[90]$, assuming

    (a) $\omega = e^{-0.2}$ (Halachmi's approximation), and
    (b) $\omega = 0.80689933$ (the correct value).

70. [M15] Suppose ρ, C_T^2, C_S^2, and W_s are the same for two heavy-traffic queueing systems ($\rho \approx 1$); one GI/M/1 and the other GI/M/c. Show that the mean queueing time for the latter system is approximately $1/c$ times the mean queueing time of the former; that is,

$$W_{q_{GI/M/c}} \approx \frac{1}{c} W_{q_{GI/M/1}}.$$

71. [M20] Consider the H_2/M/c steady state queueing system with the Laplace–Stieltjes transform of interarrival time

$$A^*[\theta] = \frac{q\lambda_1}{\lambda_1 + \theta} + \frac{(1-q)\lambda_2}{\lambda_2 + \theta},$$

where, of course,

$$E[\tau] = \frac{q}{\lambda_1} + \frac{1-q}{\lambda_2}.$$

    (a) Show that the equation

$$\omega = A^*[c\mu(1 - \omega)] \tag{5.310}$$

    becomes, for this case,

$$c^2\mu^2\omega^2 - c\mu\omega[\lambda_1 + \lambda_2 + c\mu] + c\mu[q\lambda_1 + (1-q)\lambda_2] + \lambda_1\lambda_2 = 0. \tag{5.311}$$

(b) Show that, if Algorithm 3.2.2 is used to generate the distribution of interarrival time, then the unique solution, ω, of (5.311), with $0 < \omega < 1$, is given by

$$\omega = 0.5 + \rho - 0.5\sqrt{(1 - 2\rho)^2 + 16\rho q(1 - q)(1 - \rho)}. \quad (5.312)$$

72. [C18] Too Loose Latreck Industries is considering a computer subsystem that can be modeled as an $H_2/M/2$ queueing system. Too Loose finds that the interarrival time can be modeled as an H_2 distribution constructed by Algorithm 3.2.2 with $C_\tau^2 = 64$ and $E[\tau] = 1$ dnoces (a dnoces, that is both singular and plural, is a proprietary time unit of the company). Suppose $W_s = 1.8$ dnoces so that $\rho = 0.9$. Calculate ω, D, W_q, $W_q[300]$, and $W[300]$.

73. [10] Lili Malign, an analyst at Luigi's Contract Service, is considering the queue discipline to use for one of the main office systems. Lili modeled this system as an M/D/1 queueing system with $\lambda = 5$ customers per hour and $W_s = 9$ minutes. Help her calculate W_q, W, and σ_w assuming (a) FCFS, (b) RSS, and (c) LCFS nonpreemptive queue discipline, respectively.

74. [C15] Consider Example 5.4.3. Suppose Jacques's manager Fred Fudd tells Jacques that the system must be designed with the class priorities reversed; that is, the present type 3 customers must get the top priority and type 1 the lowest. Help Jacques compute W_1, W_2, W_3, W_q, W, L_q, and L, assuming (a) HOL and (b) preemptive resume priority.

References

[1] Arnold O. Allen, *Introduction to Computer Modeling*, Applied Computer Research, Phoenix, AZ, 1986.

[2] D. Y. Barrer, Queuing with impatient customers and indifferent clerks, *Opns. Res.*, **5**, (1957), 644–649.

[3] John W. Boyse and David R. Warn, A straightforward model for computer performance prediction, *ACM Comput. Surv.*, **7**(2), (June 1975), 73–93.

[4] E. Brockmeyer, H. L Halstrøm, and Arne Jensen, *The Life and Works of A. K. Erlang*, Acta Polytechnica Scandinavica, (287/1960).

[5] S. L. Brumelle, Bounds on the wait in a GI/M/k queue, *Manage. Sci.*, **19(7)**, (1973), 773–777.

[6] S. L. Brumelle, A generalization of $L = \lambda W$ to moments of queue length and waiting times, *Opns. Res.*, **20(6)**, (1972), 1127–1136.

[7] S. L. Brumelle, Some inequalities for parallel-server queues, *Opns. Res.*, **19(2)**, (1971), 402–413.

[8] A. Cobham, Priority assignment in waiting line problems, *Opns. Res.*, **2**, (Feb. 1954), 70–76.

[9] Jacob W. Cohen, *The Single Server Queue*, Revised ed., North-Holland, Amsterdam, 1982.

[10] Robert B. Cooper, *Introduction to Queueing Theory, Second Edition*, North-Holland, New York, 1981.

[11] George P. Cosmetatos, Approximate explicit formulae for the average queueing time in the processes (M/D/r) and (D/M/r), *INFOR*, **13(2)**, (Oct. 1975), 328–331.

[12] Darel J. Daley, Bounds on the mean waiting times in GI/G/1 queues, Research Report No. 9, School of Mathematical Sciences, University of Melbourne, 1976.

[13] Darel J. Daley, Inequalities for moments of tails of random variables, with a queueing application, *Zeitschrift für Wahrscheinlichkeitstheories und verwande Gebiete*, **41**, (1977), 139–143.

[14] Darel J. Daley, Some results for the mean waiting-time and work-load in GI/G/k queues, Statistics Department (IAS), Australian National University, July 1983.

[15] Ralph L. Disney, R. L. Farrell, and P. R. DeMorais, A characterization of queues with renewal departure processes, *Manage. Sci.*, **19(11)**, (July 1973), 1222–1228.

[16] William Feller, *An Introduction to Probability Theory and Its Applications*, Vol. II, 2nd ed. John Wiley, New York, 1971.

[17] E. Gelenbe and I. Mitrani, *Analysis and Synthesis of Computer Systems*, Academic Press, New York, 1980.

[18] Donald Gross and Carl M. Harris, *Fundamentals of Queueing Theory*, 2nd ed., John Wiley, New York, 1985.

[19] B. Halachmi, On approximating the generalized occupancy of the G/M/K queueing system, *Comput. and Opns. Res.*, **7**, 81–88.

[20] Peter Henrici, *Essentials of Numerical Analysis with Pocket Calculator Demonstrations*, John Wiley, New York, 1982.

[21] Frederick S. Hillier and F. D. Lo, Tables for multiserver queueing systems involving Erlang distributions, *Tech. Rep. 31*, Dec. 28, 1971, Stanford University, Stanford, California.

[22] Frederick S. Hillier and Oliver S. Yu, *Queueing Tables and Graphs*, North Holland, New York, 1981.

[23] Phillip Howard and Janet Butler, eds., *EDP Performance Management Handbook*, Applied Computer Research, Inc., Phoenix, AZ (updated every three months).

[24] IBM, Analysis of some queueing models in real-time systems, IBM # GF20-0007, National Accounts Division, 1133 Westchester Avenue, White Plains, NY, 10604.

[25] N. K. Jaiswal, *Priority Queues*, Academic Press, NY, 1968.

[26] David G. Kendall, Some problems in the theory of queues, *J. Roy. Statist. Soc. Ser. B*, **13**, (1951), 151–173.

[27] Toshikazu Kimura, A two-moment approximation for the mean waiting time in the GI/G/s queue, *Manage. Sci.*, **32(6)**, (June 1986), 751–763.

[28] John F. C. Kingman, On queues in heavy traffic, *J. Roy. Statist. Soc. Ser. B*, **24**, (1962), 383–392.

[29] John F. C. Kingman, Some inequalities for the queue GI/G/1, *Biometrika*, **49**, (1962), 315–324.

[30] John F. C. Kingman, Inequalities in the theory of queues, *J. Roy. Statist. Soc. Ser. B*, **32**, (1970), 102–110.

[31] Leonard Kleinrock, *Queueing Systems, Volume I: Theory*, John Wiley, New York, 1975.

[32] Leonard Kleinrock, *Queueing Systems, Volume II: Computer Applications*, John Wiley, New York, 1976.

[33] Hisashi Kobayashi, *Modeling and Analysis: An Introduction to System Performance Evaluation Methodology.* Addison-Wesley, Reading, MA, 1978.

[34] Peter Kolesar, Stalking the endangered CAT: A queueing analysis of congestion at automatic teller machines, *Interfaces*, **14**(6), (Nov.–Dec. 1984) 16–26.

[35] J. Köllerström, Heavy traffic theory for queues with several servers. I, *J. Appl. Prob.*, **11**, (1974), 544–552.

[36] A. Ya. Kreinin, Bounds for mean characteristics of systems with Erlangian input and general service times, (in Russian), *Izv. Akad. Nauk. SSSR Tehu. Kibernet*, No. 3, (1981), 187–192.

[37] Paul Kühn, *Tables on Delay Systems*, Institute of Switching and Data Technics, University of Stuttgart, 1976.

[38] Richard C. Larson, Perspectives on queues: social justice and the psychology of queueing, *Opns. Res.*, **35**(6), (1987), 895–905.

[39] G. M. Laslett, Characterizing the finite capacity GI/M/1 queue with renewal output, *Manage. Sci.*, **22**(1), (Sept. 1975), 106–110.

[40] E. R. Lassettre and Alan L. Scherr, Modeling the performance of the OS/360 time sharing option (TSO), in *Statistical Computer Performance*, (W. Freiberger, ed.), Academic Press, NY, 1972, 57–72.

[41] Stephen S. Lavenberg, ed., *Computer Performance Modeling Handbook*, Academic Press, NY, 1983.

[42] John D. C. Little, A proof of the queueing formula: $L = \lambda W$, *Opns. Res.* **9**(3), (1961), 383–387.

[43] Edward A. MacNair and Charles H. Sauer, *Elements of Practical Performance Modeling*, Prentice-Hall, Englewood Cliffs, NJ, 1985.

[44] K. T. Marshall, Some relationships between the distributions of waiting time, idle time, and interoutput time in the GI/G/1 queue, *SIAM J. Appl. Math.*, **16**(2), (March 1968).

[45] James Martin, *Systems Analysis for Data Transmission*, Prentice-Hall, Englewood Cliffs, NJ, 1972.

[46] Teunis J. Ott, The sojourn-time distribution in the M/G/1 queue with processor-sharing, *J. Appl. Prob.*, **21**, (1984), 360–378.

[47] Teunis J. Ott, Some inequalities for the D/G/1 queue, Unpublished manuscript, 1981.

[48] Emanuel Parzen, *Stochastic Processes*, Holden-Day, San Francisco, 1962.

[49] V. Ramaswami, The sojourn time in the GI/M/1 queue with processor sharing, *J. Appl. Prob.*, **21**, (1984), 437–442.

[50] Martin Reiser, Mean-value analysis and convolution method for queue-dependent servers in closed queueing networks, *Perf. Eval.*, **1**(1), (Jan. 1981), 7–18.

[51] Hirotaka Sakasegawa, *Numerical Tables of The Queueing Systems I: $E_k/E_2/s$ (New Version)*, The Institute of Statistical Mathematics, 1978.

[52] Allan L. Scherr, *An Analysis of Time-Shared Computer Systems*, MIT Press, Cambridge, MA, 1967.

[53] L. P. Seelen, Henk C. Tijms, and Michael H. van Hoorn, *Tables For Multi-Server Queues*, North-Holland, Amsterdam, 1985.

[54] Kenneth C. Sevcik and Israel Mitrani, The distribution of queueing network states at input and output instants, *JACM*, **28**(2), (April 1981), 358–371.

[55] J. Laurie Snell, *Introduction to Probability*, Random House, NY, 1988.

[56] Shaler Stidham, Jr., A last word on $L = \lambda W$, *Opns. Res.*, **22**(2), (1974), 417–421.

[57] Dietrich Stoyan (Darrel J. Daley, ed.), *Comparison Methods for Queues and Other Stochastic Models*, John Wiley, 1983.

[58] T. Suzuki and Y. Yoshida, Inequalities for the many-server queue and other queues, *J. Opns. Res. Soc. Japan*, **13**, (1970), 59–77.

[59] Lajos Takács, *Introduction to the Theory of Queues*, Oxford University Press, London and New York, 1962.

[60] Lajos Takács, A single-server queue with Poisson input, *Opns. Res.*, **10**, (1962), 388–397.

[61] Howard M. Taylor and Samuel Karlin, *A First Course in Stochastic Processes*, Second ed., Academic Press, Orlando, 1975.

[62] Henk C. Tijms, *Stochastic Modelling And Analysis: A Computational Approach*, John Wiley, Chichester, 1986.

[63] Kishor Shridharbhai Trivedi, *Probability and Statistics with Reliability, Queuing, and Computer Science Applications*, Prentice-Hall, Englewood Cliffs, NJ, 1982.

[64] Michael H. van Hoorn, *Algorithms and Approximations for Queueing Systems*, CWI Tract No. 8, CWI, Amsterdam, 1984.

65] Ward Whitt, An interpolation approximation for the mean workload in a GI/G/1 queue, *Opns. Res.*, **37(6)**, (1989), 936–952.

[66] Ronald W. Wolff, Poisson arrivals see time averages, *Opns. Res.*, **30**, (1982), 223–231.

[67] Ronald W. Wolff, *Stochastic Modeling and the Theory of Queues*, Prentice-Hall, Englewood Cliffs, NJ, 1989.

Chapter 6

Queueing Models of Computer Systems

random *adj.* **1** (*Of a number generator*) *predictable.* **2** (*Of an access method*) *unpredictable.* **3** (*Of a number*) *plucked from the drum, Tombola, by the flaky-fingered Tyche.* **4** [*From JARGON FILE.*] (*Of people, programs, systems, features*) *assorted, undistinguished, incoherent, inelegant, frivolous, fickle.*

sequential file *n A place where things can get lost in lexicographic order.*
Compare RANDOM FILE

random file *n A place where things can get lost in any sequence.*
Compare SEQUENTIAL FILE

Stan Kelly-Bootle
The Devil's DP Dictionary

6.0 Introduction

In Chapter 5 we examined basic queueing theory and applied it to selected parts of computer systems. The queueing systems we studied were essentially single resource systems; that is, there was but one service facility, although in some cases, there were multiple identical servers in the facility. Actual computer systems are multiple resource systems. Thus, we may have online terminals or workstations, communication lines, line concentrators, and communication controllers as well as the computer itself. The computer, even the simplest personal computer, has multiple resources, too, including main memory, virtual memory, coprocessors, memory and *I/O* caches, channels, input/output (*I/O*) devices, etc. There may be a queue associated with each of these resources. Thus, a computer system is a network of queues.

In this chapter most of the models studied are multiple resource models; that is, they consist of a network of the simple queueing systems we considered in Chapter 5. By "network of queues" we mean that the resources are interconnected. That is, input(s) to one queueing system may be the output(s) from one or more other queueing systems. Unfortunately very little can be done, analytically, with general queueing networks such as would be needed to model computer systems in such a way as to account for every resource (and every queue). Fortunately, however, a number of useful queueing network models do exist for modeling computer systems. The subset of queueing networks that are especially easy to evaluate are called *separable* or *product form*. The prototype of these systems is the celebrated *Jackson network* discovered by James R. Jackson [7]. In the most quoted paper of computer system modeling, Baskett, Chandy, Muntz, and Palacios [1] extended Jackson's model in several important ways.

Another breakthrough in queueing theory modeling of computer systems was the creation of the mean value analysis algorithms (MVA) by Reiser and Lavenberg [20] (see also Reiser [19]), which makes it easier to solve many queueing network models.

There are several ways of classifying queueing networks. One classification is in terms of whether or not the network accepts customers from outside the system. A queueing network is *open* if customers enter from outside the system, circulate among the service centers for service, and depart from the system. In a *closed* queueing network, such as the machine repair model, a fixed number of customers circulate indefinitely among the service centers. In such a system, a fixed number of customers contend for the resources. Still other networks have some customers who enter from outside the system and eventually leave as well as some customers who always remain in the system. Such queueing networks are called *mixed* systems.

Another useful method of classifying a queueing network model of a computer system is in terms of the computer workloads it models. A workload for a queueing network model is specified by a description of the service demands of the work to be performed. Computer workloads are classified into three basic kinds: *transaction, batch,* and *terminal.*

A transaction workload is represented by an open model, since customers are assumed to enter the system from outside, receive service, and depart. The number of customers in the system varies over time. An example of a transaction workload is a large interactive system with so many terminals or workstations that the system is not sensitive to the actual number. A typical transaction workload is processed by a data base application system, such as an airline reservations system. For a transaction workload we specify the intensity by giving the average arrival rate of transactions, λ. Thus, we may specify that $\lambda = 5,000$ transactions per hour. The primary performance measure for a transaction workload is the average response time, W, that is, average total time a transaction spends in the system.

The queueing model for a batch workload is closed, since the number of customers (jobs) is assumed fixed. We visualize a system in which a completed job leaves the system but is immediately replaced by another job from a backlog. A batch workload intensity is specified by the average number, N, of active jobs in the system. N is not required to be an integer. The primary performance measures for batch workloads include average response time, W, as well as throughput, λ. We also are interested in the utilizations of the servers throughout the system.

A terminal workload model looks like the finite population system of Figure 6.1.1. People at terminals or workstations make service requests to a central processor system. We often refer to the customers as terminals. This workload model is a closed system, since the number of customers (terminals) is fixed. Thus, a terminal workload has much in common with a batch workload. The intensity of a terminal workload is specified by the average number of active terminals, N, as well as the average think time, $T = E[t]$.

We will consider some of the key definitions that are needed to study queueing networks. A more complete table of definitions for such systems is given in Table 25 of Appendix C.

We will use K to describe the number of service centers in a queueing network and C to describe the number of customer classes. Service center k provides an average service time per visit of S_{ck} for customers of class c. A customer of class c makes an average of V_{ck} visits to node k for a total *service demand* of $D_{ck} = V_{ck} \times S_{ck}$ time units of service for class c. The total service demand at service center k is then $D_k = \sum_c D_{ck}$. If the system has average throughput λ, then $V_{ck} = \lambda_{ck}/\lambda$, where λ_{ck} is the average throughput of class c customers at node k. For a single class system, this becomes

$$V_k = \frac{\lambda_k}{\lambda}, \tag{6.1}$$

where λ_k is the average throughput of node k. Equation (6.1) is sometimes written

$$\lambda_k = \lambda V_k, \tag{6.2}$$

and called the *forced flow law* because it shows that the throughput of any node determines the throughput of all the others. (For a multiclass system (6.2) becomes $\lambda_{ck} = \lambda V_{ck}$.)

Example 6.0.1 Gimpy Gipper, the lead performance analyst at Crumbling Cookies, has made measurements on his main batch processing machine. These indicate that the average number of visits each job makes to Drive 1 is five and that the disk throughput for Drive 1 is 10 requests per second. To calculate the system throughput, λ, Gipper uses the forced flow law to conclude that

$$
\begin{aligned}
\lambda &= \frac{\lambda_1}{V_1} \\
&= \frac{10 \text{ requests/second}}{5 \text{ requests/job}} \\
&= 2 \text{ jobs/second.} \quad \square
\end{aligned}
\tag{6.3}
$$

One of the key performance concepts used in studying a computer system is the *bottleneck device* or *server*, usually referred to as the *bottleneck*. As the workload on a computer system increases some server of the system eventually becomes constantly busy. When this happens the combination of the saturated server and a randomly changing demand for that server causes response times and queue lengths to grow without bound. By *saturated server* we mean a server with a utilization of 1.0 or 100%. A system is saturated when at least one of its servers or resources is saturated. The bottleneck of a system is the first server to saturate as the load on the system is increased. We can identify the bottleneck of the system if we know the service demands D_k, for $k = 1, 2, \ldots, K$. The bottleneck device is device j where j is the integer for which $D_j = D_{\max}$, where

$$
D_{\max} = \max\{D_1, \ldots, D_K\}.
$$

It is important to note that the bottleneck is workload dependent. That is, different workloads have different bottlenecks for the same computer system. It is part of the folklore that scientific computing jobs are *CPU bound*, while business oriented jobs are *I/O bound*. That is, for scientific workloads such as CAD (computer aided design), FORTRAN compilations, etc., the CPU is usually the bottleneck. Business oriented workloads, such as data base management systems, electronic mail, payroll computations, etc., tend to have *I/O* bottlenecks. Of course, one can always find a particular scientific workload that is not CPU bound and a particular business system that is not *I/O* bound, but it is true that different workloads on the same computer system can have dramatically different bottlenecks. Since the workload on many computer systems changes during different periods of the day, so do the bottlenecks. Usually, we are most interested in the bottleneck during the peak (busiest) period of the day.

Example 6.0.2 Sandy Snodgrass, the lead performance analyst at Serendipitous Systems, measures a small batch processing computer system. She finds that the CPU has a visit ratio $V_1 = 20$ with $S_1 = 0.05$ seconds, the first I/O device has $V_2 = 11$ and $S_2 = 0.08$ seconds, while the other I/O device has $V_3 = 8$ and $S_3 = 0.04$ seconds. Hence, Sandy calculates $D_1 = 1$ second, $D_2 = 0.88$ seconds, while $D_3 = 0.32$ seconds. Sandy concludes that the bottleneck is the CPU (the system is CPU bound). □

6.1 Finite Population Models

In this section we discuss finite population queueing systems as portrayed in Figure 6.1.1. We consider a simple model for a computer system with a terminal workload and only one class of jobs. Figure 6.1.1 illustrates what is commonly known as a *finite population queueing model* of an interactive system (see Muntz [17]). The central processor system consists of one or more CPUs with the associated main memories and I/O devices. The customers (users) are interacting with the central processor system through the N terminals (personal computers or work stations). Each customer is assumed to be in exactly one of three states at any instant of time:

(1) "thinking"[1] at the terminal (this time is called *think time t* with mean $E[t] = T$),

(2) queueing for service at the central processor system,[2] or

(3) receiving service from the central processor system.

We assume that a user is not allowed to make a new request for service until the previous request has been satisfied.

In Figure 6.1.1 the customer can be represented as a token that travels around the system and that at any instant is either at a terminal, in the central processor queue (if there is one), or circulating through the central processor system. This model is closed. The particular queueing network model we use for the system in Figure 6.1.1 depends on what model we select to represent the central processor system. One of the simplest models is achieved by using a single exponential server with an associated queue to represent the central processor system. If we assume the think time is

---

[1]There may be a bit of optimism here.
[2]Sometimes there is no queue for the central processor system, so this state is omitted.

exponential this yields the machine repair model that we studied in Section 5.2.6 (the M/M/1/K/K queueing system). As Lavenberg [12] points out, the equations for the machine repair model are true when the think time has a general distribution and the central processor system has an exponential distribution.

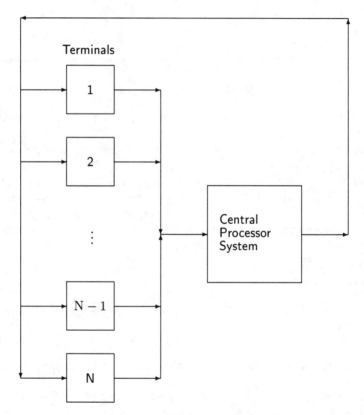

Figure 6.1.1. Finite population system.

6.1.1   Machine Repair Model

It seems incredible that the central processor system of a time-sharing system could be modeled as a simple exponential distribution, but Scherr [21] successfully used this model to analyze the Compatible Time-Sharing System (CTSS) at MIT.[3] This was an early time-sharing system in which user programs were swapped in and out of main memory with only one complete program in memory at a time. Since there was no overlap of program execution and swapping, Scherr used the sum of program execution time and swapping time as the CPU service time. The machine repair analytic model gave results that were very close to those for a simulation model and to actual measured values.

For this model, since the operating time for machines corresponds to think time, with average think time $E[t] = T$, we have, by Little's formula, the mean response time

$$W = \frac{N}{\lambda} - T. \tag{6.4}$$

But $\lambda = \rho/W_s$, so the mean response time can be written as

$$W = \frac{NW_s}{\rho} - T, \tag{6.5}$$

where

$$\rho = 1 - p_0, \tag{6.6}$$

and

$$p_0 = \left\{ \sum_{k=0}^{N} \frac{N!}{(N-k)!} \times \left(\frac{W_s}{T} \right)^k \right\}^{-1} = B[N, z], \tag{6.7}$$

where $B[\cdot, \cdot]$ is Erlang's B formula and

$$z = \frac{T}{W_s}.$$

Example 6.1.1 Slobovian Scientific has an interactive system of 20 active engineering diskless workstations connected to a file server by a LAN that can be modeled by the machine repair queueing system. The average file server service time is 2 seconds, while the mean think time is 20 seconds. Find p_0, ρ, λ, and the average response time, W. What would be the effect of adding five workstations?

---

[3]See Exercise 31 of Chapter 5 for a description of how Scherr and his colleague, Lassettre, used the machine repair model to model the performance of the OS/360 time-sharing option (TSO), a system that allowed programmers to develop and run programs from a terminal in a time-sharing mode.

Graph of $\dfrac{W}{W_s} = \dfrac{N}{1 - B[N, z]} - z$ versus N.

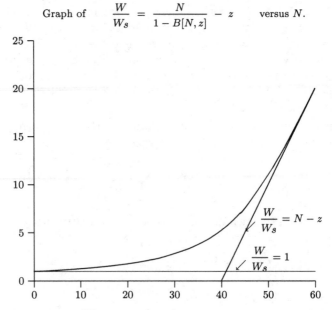

Figure 6.1.2. $\dfrac{W}{W_s}$ versus N for M/M/1/N/N queueing system.

Solution   For 20 terminals,

$$p_0 = B[20, 10.0] = 0.001869.$$

Thus,

$$\rho = 1 - p_0 = 0.998131, \quad \lambda = 0.49907 \text{ interactions/second,}$$

and

$$W = \frac{20}{0.49907} - 20 = 20.075 \text{ seconds.}$$

With 25 workstations

$$p_0 = 0.00002927, \quad \rho = 0.99997073,$$

$$\lambda = 0.499985365, \quad \text{and} \quad W = 30 \text{ seconds.}$$

Thus, the addition of 5 workstations has increased the average throughput by only 0.18%, while increasing the mean response time by 49.44%.   □

This example illustrates the concept of system saturation. Consider Figure 6.1.2, the graph of W/W_s versus N, the number of active terminals. When there is only one active user ($N = 1$), there is no queueing for service so that $W = W_s$; the line $W/W_s = 1$ is an asymptote to the curve. As the number of active terminals grows very large (think of $N = 100,000,000,000$, that is, 100 American billion) we would expect the computer system to be busy most of the time, so $\rho \approx 1$. But then $\lambda = \rho/W_s \approx 1/W_s$, and adding more machines has little effect on the throughput. When we substitute this equation for λ into our equation for W (equation (6.4)), we obtain $W = N W_s - T$ or $W/W_s = N - T/W_s = N - z$. This means that the curve has the asymptote $W/W_s = N - z$ for large N. Thus, when N is large, increasing it by 1 has little effect on the throughput but increases W by approximately W_s (W/W_s by 1). The two asymptotes to the curve intersect at the point $(N^*, 1)$ where $N^* = 1 + z = (T + W_s)/W_s$.

Kleinrock [10, page 209, (4.66)] calls the value N^* the *saturation number*. He also gives it an interesting physical interpretation. If each terminal user has exactly T units of think time and uses exactly W_s units of service time per interaction, then, with perfect synchronization, N^* is the maximum number of terminals that can be supported with no mutual interference. Thus, for any real system, when N is one we have no mutual interference; when N is small, we have almost no mutual interference. However, as N increases past N^*, we are certain to have some mutual interference (queueing for service) while the formula $W = N W_s - T$ for a very large number of users (terminals) shows complete interference! That is, for a very large number of users, each user delays every other user by one mean service time, W_s. As we shall see later, the asymptotic formula $W = N W_s - T$ is valid also for a general service time; in fact, it is true for even more general "machine repair like" queueing systems. In Example 6.1.1, $N^* = 22/2 = 11$ workstations, and the increase in W due to the change from 20 to 25 was close to $5 \times 2 = 10$ seconds (it was 9.963).

Lassettre and Scherr [11] successfully used the machine repair model to develop the OS/360 time-sharing option (TSO).

Figure 6.1.2 was plotted using $z = 40$ so that $N^* = 41$, but the shape of the curve will be very similar for other values of z.

6.1.2 Finite Processor-Sharing Model

For this model we assume the think time is exponential and the central processor system has the processor-sharing queue discipline. Kleinrock [10, Section 4.11] shows that the service time distribution is restricted only by the requirement that the Laplace–Stieltjes transform, $W_s^*[\theta]$, is rational, that is, the ratio of two polynomials in θ (with the denominator at least one degree higher than the numerator). An equivalent requirement is that the service time s be Coxian. Then the equations of the machine repair model hold. That is, equations (6.4) through (6.7) hold for this model.

This is a very useful result. It can be used to model many interactive systems with at least enough accuracy to evaluate and compare upgrade options.

Example 6.1.2 Let us consider the example on page 317 of Martin Reiser's perceptive paper, Reiser [18]. Dr. Reiser shows how to solve this problem with QNET4, a sophisticated APL program that was a precursor to the current RESQ program.[4] He considers a finite processor-sharing model with 20 active terminals, mean think time $T = 3$ seconds, a central processor service rate of 500,000 instructions per second, and an average interaction requirement of 100,000 instructions. Therefore, $W_s = 100,000/500,000 = 0.2$ seconds. Hence, $z = T/W_s = 15$, and $p_0 = B[20, 15] = 0.045593216$. By (6.5) we have

$$W = \frac{NW_s}{1 - p_0} - T = 1.191 \quad \text{seconds,}$$

which agrees with Reiser's solution, as does the average throughput

$$\lambda - \frac{\rho}{W_s} = 4.772 \quad \text{interactions per second.}$$

The average number in the central processor system

$$\lambda W = 5.6835, \quad \text{and} \quad \rho = 0.9544$$

also agrees with Reiser's solution. □

6.2 Jackson Networks

James R. Jackson in his papers [7, 8] defined some queueing networks with remarkable properties. The network of Jackson [7] is described in the following theorem. The type of network defined in Jackson [8] is described by Kleinrock [9, Section 4.8].

Theorem 6.2.1 (*Jackson's Theorem*) *Suppose a queueing network consists of K nodes satisfying the following three conditions:*

1. *Each node consists of c_k identical exponential servers, each with average service rate μ_k.*

2. *Customers arriving at node k from outside the system arrive in a Poisson pattern with the average arrival rate λ_k. (Customers also arrive at node k from other nodes within the network.)*

---

[4]In their book, MacNair and Sauer [14] show how to use RESQ for modeling computer systems.

3. Once served at node k, a customer goes (instantly) to node j ($j = 1, 2, \ldots, m$) with probability p_{kj}; or leaves the network with probability $1 - \sum_{j=1}^{K} p_{kj}$.

Then, for each node k, the average arrival rate to the node, Λ_k, is given by

$$\Lambda_k = \lambda_k + \sum_{j=1}^{K} p_{jk} \Lambda_j. \tag{6.8}$$

In addition, if we let $p(n_1, n_2, \ldots, n_K)$ denote the steady state probability that there are n_k customers in the kth node for $k = 1, 2, \ldots, K$, and if $\Lambda_k < c_k \mu_k$ for $k = 1, 2, \ldots, K$ (so there is a steady state distribution), then

$$p(n_1, n_2, \ldots, n_m) = p_1(n_1)\, p_2(n_2) \cdots p_K(n_K), \tag{6.9}$$

where $p_k(n_k)$ is the steady state probability that there are n_k customers in the kth node if it is treated as an $M/M/c_k$ queueing system with an average arrival rate Λ_k and average service time $1/\mu_k$ for each of the c_k servers. Furthermore, each node k behaves as if it were an independent $M/M/c_k$ queueing system with average arrival rate Λ_k.

Proof See Jackson [7]. ■

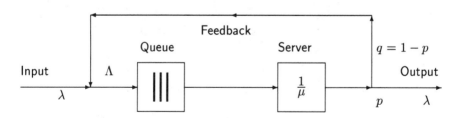

Figure 6.2.1. M/M/1 queue with feedback.

Example 6.2.1 Consider the simple $M/M/1$ feedback queueing system of Figure 6.2.1. Suppose the $M/M/1$ queueing system represents a message switching facility that transmits messages to the required destination. We assume the time to transmit a message and receive an acknowledgment of correct receipt is exponential. (We assume an error detecting code is used.)

The probability that a message is received correctly is p; with probability $q = 1 - p$ the message must be retransmitted. Then, by Jackson's theorem,

$$\Lambda = \lambda + q\Lambda \quad \text{or} \quad \Lambda = \frac{\lambda}{p}, \quad \text{and } \rho = \Lambda W_s = \frac{\lambda}{p\mu}.$$

Since we have an M/M/1 system,

$$L = \frac{\rho}{1 - \rho} = \frac{\lambda}{p\mu - \lambda}, \text{ and } W = \frac{W_s}{1 - \rho} = \frac{p}{p\mu - \lambda}.$$

Consider now the message switching center of Example 5.2.1, where $\lambda = 4$ messages per second and $W_s = 0.22$ seconds. Suppose it is the basis for our feedback queueing system and that the probability, p, of correctly transmitting a message is 0.99. Then we have $\Lambda = 4.0404$ messages per second, $\rho = 0.8889$, $L = 8$ messages, and $W = 1.98$ seconds. $\square$

As the following example shows, the arrival stream to the feedback queueing system of the last example is *not* Poisson.

Example 6.2.2 Consider the M/M/1 queueing system with Bernoulli feedback shown in Figure 6.2.1 and discussed in Example 6.2.1. Trivedi [25, Exercise 3, pages 422–423] shows that the input stream with average rate Λ has a two-stage hyperexponential distribution with

$$q_1 = \frac{q\mu}{\mu - \lambda},$$

$\mu_1 = \mu$, $\mu_2 = \lambda$, and $q_2 = 1 - q_1$. However, he also shows that the output stream is Poisson with average rate λ. $\square$

> *I've researched supermarkets*
> *And now I can be vocal;*
> *The Express line, I have found*
> *Behaves more like a local.*

> Mimi Kay

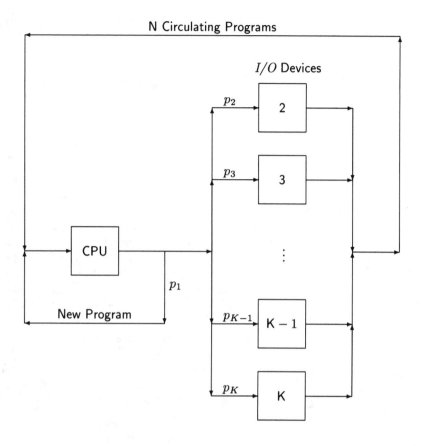

Figure 6.2.2. Central server system.

6.2.1 Central Server Model (Single Class)

This queueing network was used by Buzen [3] to model a multiprogramming computer system. Thus, it models one of the most difficult resources to model in a computer system; that is, the main memory of the computer. The central server model, shown in Figure 6.2.2, is a closed Jackson queueing network model. The central server is, of course, the CPU. The model is closed because it contains a fixed number of programs (has a constant multiprogramming level), N. The programs can be thought of as markers or tokens that cycle around the system interminably. However, each time a program makes the cycle from the CPU directly back to the end of the CPU queue, we assume a program execution has been completed and a new program enters the system. Thus, there must be a backlog of jobs ready to enter the computer system at all times. There are $K - 1$ of the I/O devices, each with its own queue, and each exponentially distributed with average service rate μ_k $(k = 2, 3, \ldots, K)$. The CPU was assumed to have an exponential distribution in Buzen's models, but Lavenberg and Sauer in Lavenberg [12, Section 3.4.2] show that the CPU can also have a general distribution with a processor-sharing queue discipline. The queue discipline is assumed to be FCFS for the I/O servers. Upon completion of a CPU service, a job returns to the CPU queue (completes execution) with probability p_1 or enters service at I/O device k with probability p_k, $k = 2, 3, \ldots, K$. Upon completion of an I/O service, the job returns to the CPU queue for another cycle. If we let $(n_1, n_2, \ldots, n_K)$ represent the state of the system, where n_k is the number of jobs at the kth service center, Buzen [4] provides an algorithm that allows one to calculate $p(n_1, n_2, \ldots, n_K)$ and other performance measures, such as the average job turnaround times and the utilization of all resources. His algorithm is called the *convolution algorithm*. Buzen's algorithm is efficient but not at all intuitive. We provide a mean value analysis (MVA) algorithm, below, that is more intuitive and that can be solved using the APL program CENT and the *Mathematica* program **cent**.

Algorithm 6.2.1 does not deal directly with the average service time, $S_k = 1/\mu_k$, for each visit but rather with $D_k = V_k S_k$, so that D_k, the *service demand*, is the total amount of service that a job needs at the kth service center. We will not need to use the branching probabilities directly, either. In Algorithm 6.2.2 we show how to compute each D_k from the collection $\{S_k, p_k, k = 1, 2, \ldots, K\}$.

Algorithm 6.2.1 *Consider the central server system of Figure* 6.2.2. *Suppose we are given the mean total resource requirement, D_k for each of the K resources and the multiprogramming level N. Then we calculate the performance measures of the system as follows.*

Step 1 [Initialize] *Set $L_k[0] = 0$, $k = 1, 2, \ldots, K$.*

Step 2 [Iterate] *For $n = 1, 2, \ldots, N$, calculate*

$$W_k[n] = D_k(1 + L_k[n-1]), \quad k = 1, 2, \ldots, K. \tag{6.10}$$

$$W[n] = \sum_{k=1}^{K} W_k[n]. \tag{6.11}$$

$$\lambda[n] = \frac{n}{W[n]}. \tag{6.12}$$

$$L_k[n] = \lambda(n)W_k[n], \quad k = 1, 2, \ldots, K. \tag{6.13}$$

$$\tag{6.14}$$

Step 3 [Compute performance measures]
   Set job response time (turnaround time)

$$W = W[N].$$

Set throughput

$$\lambda = \lambda[N].$$

Set server utilization

$$\rho_k = \lambda D_k, \quad k = 1, 2, \ldots, K. \quad \blacksquare$$

In Step 2, $W_k[n]$ is the average total time a customer spends at the kth node for V_k visits (not just the time spent in one visit), and $W[n]$ is the average total time a customer spends in the central server system. The key formula in Algorithm 6.2.1 is formula (6.10) for $W_k[n]$. It says that the average time an arriving customer will spend at server k when there are n customers in the central server system is the average service required for a customer at the server, D_k, times the total number of customers who must be served, including the arriving customer. This number is one plus the average number of customers the arriving customer finds in the system. The fundamental principle of mean value analysis is that the average number of customers an arriving customer finds at the server is $L_k[n-1]$. As Reiser [19], one of the discoverers of MVA, puts it,

> Mean-value analysis is based on the arrival theorem [10-11] that states: *In a closed queueing network the (stationary) state probabilities at customer arrival epochs are identical to those of the same network in long-term equilibrium with one customer removed.*

We state the arrival theorem formally for reference.

Theorem 6.2.2 (*The Arrival Theorem*) *In a closed queueing network the (stationary) state probabilities at customer arrival epochs are identical to those of the same network in long-term equilibrium with one customer removed.*

Proof For the proof see Sevcik and Mitrani [23].

   ∎

Note that there is a boot strapping technique involved in Step 2. For $n = 1$ all the $L_k[0]$ values are zero, so that each $W_k[1] = D_k$, by formula (6.10). Then we calculate $\lambda[1]$ using formulas (6.11) and (6.12); the latter is

Little's law. We can then calculate $L_k[1] = \lambda[1]W_k[1]$ for all k, and we are ready for the next iteration ($n = 2$). We are on our way! Let us consider a simple example.

Example 6.2.3 Suppose we have a computer system that consists of a CPU and one *I/O* device and that the multiprogramming level is 2. Thus, $K = N = 2$. Suppose, also, that the average CPU time needed per job is 0.4 seconds while an average of 0.6 seconds of *I/O* service is required ($D_1 = 0.4$ seconds and $D_2 = 0.6$ seconds). We also suppose that both service times are exponential. Let us step through Algorithm 6.2.1. In Step 1 we set $L_1[0] = L_2[0] = 0$. In Step 2, for $n = 1$, we compute $W_1[1] = D_1 = 0.4$ seconds and $W_2[1] = D_2 = 0.6$ seconds. Therefore, $W[1] = 1.0$ second and $\lambda[1] = 1/1 = 1$ job per second. We compute $L_1[1] = 0.4$ and $L_2[1] = 0.6$. Now we can set $n = 2$ and compute

$$W_1[2] = 0.4(1 + 0.4) = 0.56 \text{ seconds},$$

and

$$W_2[2] = 0.6(1 + 0.6) = 0.96 \text{ seconds}.$$

Hence, $W[2] = 0.56 + 0.96 = 1.52$ seconds and

$$\lambda = \lambda[2] = \frac{2}{1.52} = 1.3158 \text{ jobs per second},$$

$\rho_1 = \lambda D_1 = 0.5263$, and $\rho_2 = 0.7895$. $\square$

The calculations for the example could be done using the APL function CENT or the *Mathematica* program **cent**. Now let us look at the algorithm needed to convert from branching probabilities to service demands.

Algorithm 6.2.2 *Consider the central server system of Figure* 6.2.2. *This algorithm will construct the parameters needed to use Algorithm* 6.2.1.

Step 1 *Set the visit ratio V_1 for the CPU to*

$$V_1 = \frac{1}{p_1}.$$

Step 2 *For $n = 2, 3, \ldots, K$ calculate*

$$V_n = p_n V_1.$$

Step 3 *[Calculate the demands, D_k] Set*

$$D_k = V_k \times S_k, \quad k = 1, 2, \ldots, K. \tag{6.15}$$

Proof See Denning and Buzen [5, page 237]. ■

The APL program FIX implements Algorithm 6.2.2. It is used by CENTP. Thus, CENTP can use as input the branching probabilities and the values of S_k for $k = 1, 2, \ldots, K$. FIX handles the conversion so that CENTP can call CENT to calculate the performance parameters using Algorithm 6.2.1.

The central server model was designed to model the contention for the processor and I/O resources between programs in a multiprogramming computer system. It has been used with some success but, like all models, it has both strengths and weaknesses. Among its strengths is the fact that, unlike the machine repair model, it allows I/O devices of different speeds. It also permits the modeling of multiple programs simultaneously occupying the main memory. Thus, one can model the improvement that would occur if more main memory were obtained, so that the multiprogramming level can be increased. Modeling of improvements due to more or faster I/O devices or a faster CPU can also be obtained. There are several weaknesses in the model, as well. It does not represent the overlap between CPU processing and I/O activity for a program. It assumes that each I/O device can handle service for only one program at a time and that different I/O devices do not interfere with each other. It also assumes random routing of I/O requests to I/O devices and that there is no priority system in effect for either the CPU or for I/O devices. Finally, the model does not represent the time it takes the operating system to allocate memory to incoming programs. These weaknesses are necessary to make the model analytically tractable. In spite of the weaknesses the model has produced useful results. In the next example, based on Ferrari *et al.* [6, Case 9.6, pages 423–437], we demonstrate how the central server model can be used. The tricky part of the example is that the paging rate of the paging drum depends upon the multiprogramming level, N. Ferrari *et al.* solved this problem by determining empirically the paging rate as a function of the size of the program under execution.

Example 6.2.4 In Case 9.6 Ferrari *et al.* [6] study a multiprogrammed, virtual memory system, that is running a batch workload. They model the system as a central server model with a CPU and three I/O devices. The I/O devices consist of the paging drum to handle the paging requests of the virtual storage system, a file drum to service some of the program I/O requests, and a file disk to service the remainder of the I/O requirements. The main memory is large enough to support 400 page frames, of which 300 are available for user programs. The 300 page frames are divided evenly among the programs. Thus, if the multiprogramming level $N = 3$, then each program has 100 page frames available. For the workload in the study, the CPU speed is 0.7 MIPS (millions of instructions per second) and a typical program executes 2.1 million instructions, so the average total CPU time per program is $2.1/0.7 = 3$ seconds. (This is D_1 in our notation.) A program requires an average of 150 file drum accesses and 250 file disk accesses. The average individual file access for either the file drum or the paging drum requires $1/80.455 = 0.01242931$ seconds. To apply the central server model we need to know the average number of paging drum accesses per program. This number depends upon the number of page frames assigned to a program. Since there are just 300 page frames for all the programs in main memory, the number of page frames assigned to a program is $300/N$. Ferrari et al. provide a lifetime curve (Figure 9.28 of their book) for the computer system. This curve gives the average CPU time between page faults as a function of the number of resident page frames occupied by a program. (Professor Ferrari has assured me that such a curve can be constructed for a computer system by extensive tracing operations.) When the multiprogramming level is three, so that the number of resident page frames is 100, the mean time between page faults is 0.013 seconds. Thus, the mean number of page faults per program is $V_2 = 3/0.013 = 230$. This means that the visit ratio, V_1, for the CPU is $V_1 = 230 + 150 + 250 + 1 = 631$. ($V_1 = V_2 + V_3 + \cdots + V_K + 1$, since the job always leaves the system after receiving a CPU burst.) We show the values of S_k, V_k, and D_k when the multiprogramming level $N = 3$ in Table 6.2.1. It is clear that the file disk is the bottleneck for this system. Table 6.2.2 displays the results from running the APL program CENT for this model. Table 6.2.3 displays the output of the Myriad program Central Server for this example. (The *Mathematica* program **cent** agrees with the Myriad results a little more closely than CENT does.)

When $N = 4$, the lifetime curve shows that the mean time between page faults is approximately 0.012, so that $V_2 = 3/0.012 = 250$ and $V_1 = 250 + 150 + 250 + 1 = 651$. Only three values in Table 6.2.1 need be changed for multiprogramming level 4; V_1, V_2 and $D_2 = 3.10733$ seconds. We show the model results from the *Mathematica* program **cent** for this case in Table 6.2.4. Increasing the multiprogramming level from 3 to 4 had a desirable effect and an undesirable one as well. The throughput increased from 0.1168 jobs per second to 0.12503 jobs per second, an increase of 7.05%, while the response time increased from 25.69 seconds to 31.99 seconds, or 24.52%. Ferrari *et al.* show that increasing the multiprogramming level beyond 5 decreases the throughput and increases the response time, because the paging drum becomes the bottleneck. In fact, for $N = 10$ the throughput drops to 0.0643 jobs per second and the response time increases to 155.4 seconds. This very poor performance is caused by thrashing—the paging drum is constantly bringing in new page frames and very little processing is accomplished.

Table 6.2.1. Model Input

k	V_k	S_k	D_k
1	631	$3/V_1$	3.00000
2	230	$1/80.455$	2.85874
3	150	$1/80.455$	1.86440
4	250	$1/33.734$	7.41092

Table 6.2.2. Model Output

$$W = 25.690 \qquad \lambda = 0.117$$
$$\rho_1 = 0.350 \qquad \rho_2 = 0.334$$
$$\rho_3 = 0.218 \qquad \rho_4 = 0.865$$
$$L_1 = 0.475 \qquad L_2 = 0.447$$
$$L_3 = 0.263 \qquad L_4 = 1.814$$

Table 6.2.3. Myriad Output Example 6.2.2

```
05:13:47        Myriad v 2.0 A (c) Pallas Int. Corp., 1986-1990    02/03/90
                      -- Global Model Parameters --
Peripheral servers:                      3
CPUs (central server):                   1
Users in system:                         3
                      -- Server 0  Parameters --
Average service time per visit:        3/631
Relative end of job frequency:           1
                      -- Server 1  Parameters --
Average service time per visit:        1/80.455
Relative visit frequency:               230
                      -- Server 2  Parameters --
Average service time per visit:        1/80.455
Relative visit frequency:               150
                      -- Server 3  Parameters --
Average service time per visit:        1/33.734
Relative visit frequency:               250


                -- Central Server Analysis Results --
   Server    Total time   Util.(%)   Throughput   Queue time  Nb at servr
   System      25.6888                 0.1168       0.0000       3.0000
   Central      0.0065     35.035     73.6896       0.0017       0.4755
   Periph  1    0.0166     33.385     26.8599       0.0042       0.4466
   Periph  2    0.0150     21.773     17.5173       0.0026       0.2630
   Periph  3    0.0622     86.546     29.1955       0.0325       1.8149
```

Table 6.2.4. Model Output

$$W = 31.990 \qquad \lambda = 0.1250$$
$$\rho_1 = 0.375 \qquad \rho_2 = 0.3885$$
$$\rho_3 = 0.233 \qquad \rho_4 = 0.9266$$
$$L_1 = 0.551 \qquad L_2 = 0.5793$$
$$L_3 = 0.294 \qquad L_4 = 2.5761$$

For lower multiprogramming levels, the file disk is the bottleneck. By moving some files from the file disk to the file drum Ferrari *et al.* were able to reduce V_4 from 250 to 118 while increasing V_3 from 150 to 282. This yields the model input shown in Table 6.2.5; the output from **cent** is shown in Table 6.2.6. This system is now quite well tuned, with $D_1 \approx D_2$ and $D_3 \approx D_4$. Note that the throughput has increased to 0.17384 jobs per second while the response time has decreased to 23 seconds![5] $\square$

Table 6.2.5. Model Input

i	V_i	S_i	D_i
1	671	$3/V_1$	3.00000
2	250	1/80.455	3.10733
3	282	1/80.455	3.50506
4	118	1/33.734	3.49795

Table 6.2.6. Final Output

$$W = 23.010 \qquad \lambda = 0.174$$
$$\rho_1 = 0.522 \qquad \rho_2 = 0.540$$
$$\rho_3 = 0.609 \qquad \rho_4 = 0.608$$
$$L_1 = 0.867 \qquad L_2 = 0.915$$
$$L_3 = 1.111 \qquad L_4 = 1.107$$

---

[5] Some of the data for this example came from Case 9.6, pages 423–437 of the book *Measurement and Tuning of Computer Systems* by Domenico Ferrari, Giuseppe Serazzi, and Alessandro Zeigner, Prentice Hall, Englewood Cliffs, NJ, 1983 with the permission of Professor Ferrari and Prentice Hall, Inc., Englewood Cliffs, NJ.

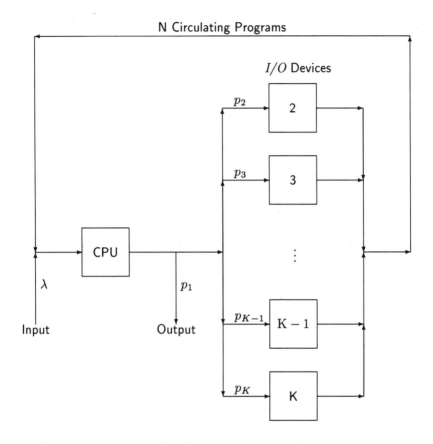

Figure 6.2.3. Open central server system.

In Example 3.26 of Lavenberg [12], Lavenberg and Sauer use the central server queueing model to study memory management of a multiprogrammed computer system. Their example is somewhat different from the preceding example, because they are trying to optimize the performance of the operating system rather than improve the performance of an existing system. In both examples measurement is needed to determine the parameters for the central server model.

An open model version of the central server model is often used because it is much simpler and therefore requires less computation. It can be used to model a system, such as a transaction processing system in which the system is not limited by the multiprogramming level. The model, shown in Figure 6.2.3, is a modified form of the central server model shown in

Figure 6.2.2. The multiprogramming level is no longer fixed, so the "New Program" branch is replaced by a simple outgoing arrow to indicate where customers exit the system. The branching probability remains p_1. An incoming arrow leading to the CPU queue is added to indicate the arriving traffic with average rate λ. The arrival pattern is assumed to have an exponential interarrival time. We assume the system is not overloaded, so the throughput is also λ. Jackson's theorem (Theorem 6.2.1) can be applied to the system, as stated in the following algorithm.

Algorithm 6.2.3 *Consider the open central server system described above. Suppose we are given the average arrival rate λ, the average service time at each device, S_k, $k = 2, 3, \ldots, K$, and the branching probabilities p_k, $k = 2, 3, \ldots, K$. Then we use Jackson's theorem to calculate the performance measures of the system as follows:*

Step 1 [Calculate demands] *If the demands are not known, use Algorithm 6.2.2 to calculate*

$$D_k, \quad k = 1, 2, \ldots, K.$$

Step 2 [Calculate device performance] *For $k = 1, 2, \ldots, K$, calculate*

$$\rho_k = \lambda D_k \quad L_k = \frac{\rho_k}{1 - \rho_k}, \quad and \quad W_k = \frac{D_k}{1 - \rho_k}.$$

Step 3 [Compute system performance measures] *Set average response time (average time in system) to*

$$W = \sum_{k=1}^{K} W_k.$$

Set average number of customers in the system to

$$L = \lambda W.$$

The bottleneck device is device j, where j is the integer for which $D_j = D_{\max}$, where

$$D_{\max} = max\{D_1, \ldots, D_K\}.$$

The maximum possible throughput is given by

$$\lambda_{\max} = \frac{1}{D_{\max}}.$$

Proof The algorithm follows immediately from Theorem 6.2.1. ∎

Example 6.2.5 Rick Rockyfeller, a performance analyst at Sleek Slimming Salons, investigates a transaction processing computer system that can be modeled as an open central server model with one CPU and three disks. Rick finds that the average transaction arrival rate is $1/5$ transactions per second with an average CPU service requirement of 3 seconds per transaction. The average I/O service requirements are 1, 2, and 4 seconds, respectively. Rick uses the equations of Algorithm 6.2.3 to construct Table 6.2.7. From the table he calculates the average transaction response time

$$W = W_1 + W_2 + W_3 + W_4 = 32.08 \text{ seconds,}$$

and the average number of transactions being processed is

$$L = \lambda W = 6.42.$$

The bottleneck is the third disk drive, so the maximum possible throughput is $1/4 = 0.25$ transactions per second.

Rick discovers that he has inadvertently solved the example on pages 8–9 of Lazowska *et al.* [13]. □

Table 6.2.7

k	ρ_k	W_k
1	0.6	7.50
2	0.2	1.25
3	0.4	3.33
4	0.8	20.00

Table 6.2.8

k	ρ_k	L_k
1	0.6835	1.656
2	0.2278	0.290
3	0.4557	0.779
4	0.9113	3.274

Example 6.2.6 For Example 6.2.5 consider a model with the same parameters Rick used, but that is now a closed central server model—thus, the throughput λ is no longer one of the given parameters but is one of the performance measures calculated for the model. Suppose the multiprogramming level $N = 6$, the closest integer to the value $L = 6.42$ calculated by Rick. Then CENT yields the values in Table 6.2.8. It also shows that $W = 26.3359$ seconds and $\lambda = 0.2278$. The model corresponds to a computer system in which there is always a queue of jobs (customers) waiting to

enter the computer memory as soon as the previous job is completed, and in which the multiprogramming level is kept at 6. Note that, for this closed model, the throughput is greater than for the open model used by Rick, and the mean response time, W, is smaller. However, the closed model does not account for the waiting time outside the system and assumes there is always a queue of jobs waiting to enter the system. □

The open central server model can also be used to represent the central processor system in Figure 6.1.1, provided the multiprogramming level need not be modeled. However, such a model is no longer a Jackson network model because each terminal cannot be modeled as an exponential server. Each terminal can be modeled as an infinite server, that is, as an M/M/∞ system, often called a "delay system", but then the model is no longer a Jackson network but rather is a BCMP queueing network. We discuss such systems in the next section.

6.3   BCMP Queueing Networks

In their classic paper Baskett, Chandy, Muntz, and Palacios [1] generalize the Jackson queueing network to allow different classes of customers, each with different service requirements, as well as service time distributions other than exponential. They also allow open, closed, and mixed networks of queues. Finally, they allow customers to change classes after completing service at a service center before going to another service center. The number of service centers is K and the number of customer classes is C. There are four types of service centers, each with a different queueing discipline. The four types of service centers are:

Type 1

The queue discipline is FCFS, and each customer has the same exponential service requirement. The service rate can be load dependent with $\mu(j)$ representing the service rate when there are j customers at the center.

Type 2

The queue discipline is processor-sharing, with each customer class allowed to have a different Coxian service time distribution.

Type 3

There are an infinite number of servers so that each arriving customer begins service immediately. Each customer class is allowed to have a different Coxian service time distribution. We will sometimes describe a type 3 service center as a *delay center*.

Type 4

The queue discipline is LCFS preemptive repeat with a single server. The distribution of service time is Coxian with each customer class allowed a different distribution.

For open networks there are two possible Poisson arrival patterns. In the first pattern, the arrival rate from outside has a mean rate, $\lambda(N)$, that depends upon the number of customers, N, in the system. In the second pattern, there are C Poisson arrival streams, one for each customer class. The arrival rate to the jth class depends upon the number already present in the class.

In our discussion of BCMP queueing networks, we will consider important special cases. We will not consider these networks in their full generality. For example, we do not allow customer class switching.

6.3.1 Single Class BCMP Queueing Networks

The open single class BCMP queueing network model we consider is that of Figure 6.2.3, in which we allow not only type 1 service centers but also type 2 (processor-sharing) and type 3 (delay) service centers. The following algorithm is the MVA algorithm for this model.

Algorithm 6.3.1 *Consider the open central server system shown in Figure 6.2.3 in which BCMP types of service centers are allowed. Suppose we are given the average arrival rate, λ, the average service time at each device, S_k, $k = 2, 3, \ldots, K$, and the branching probabilities, p_k, $k = 2, 3, \ldots, K$. Then we calculate the performance measures of the system as follows:*

Step 1 [Calculate demands] *If the demands are not known, use Algorithm 6.2.2 to calculate*

$$D_k, \quad k = 1, 2, \ldots, K.$$

Step 2 [Calculate device performance] *For $k = 1, 2, \ldots, K$, calculate:*

$$\rho_k = \lambda D_k \tag{6.16}$$

for all service centers. Calculate

$$L_k = \begin{cases} \rho_k & \text{for delay centers (type 3)} \\[2mm] \dfrac{\rho_k}{1 - \rho_k} & \text{for queueing centers (type 1 and 2)} \end{cases} \tag{6.17}$$

and

$$W_k = \begin{cases} D_k & \text{for delay centers} \\ \\ \dfrac{D_k}{1 - \rho_k} & \text{for queueing centers} \end{cases} \tag{6.18}$$

Step 3 [Compute system performance Measures] *Set average response time (average time in system) to*

$$W = \sum_{k=1}^{K} W_k. \tag{6.19}$$

Set average number of customers in the system to

$$L = \lambda W.$$

The bottleneck device is device j, where j is the integer for which $D_j = D_{max}$, *where*

$$D_{max} = max\{D_1, \ldots, D_K\}.$$

The maximum possible throughput is given by

$$\lambda_{max} = \frac{1}{D_{max}}.$$

Proof See Lazowska *et al.* [13]. ∎

Since delay centers are used primarily to represent customers at terminals, delay centers rarely appear in open models.

Our closed BCMP model for one class is shown in Figure 6.3.1 and is constructed by replacing the central processor system in Figure 6.1.1 by the open central server model of Figure 6.2.3. The N terminals are type 3 servers (delay centers), the CPU can be considered either a type 1 (exponential) or type 2 (PS) server and the I/O devices are all type 1 servers. The dashed box in Figure 6.3.1 encloses the central computer system. The main memory of this system is not modeled. That is, there is no queue for requests to enter main computer memory. This means that no limit is imposed on the number of requests that can, simultaneously, reside in the central computer system. Other models, that we will discuss later, must be used to model systems in which the multiprogramming level (MPL) is specifically modeled. The MVA algorithm for calculating the performance measures of the computer system of Figure 6.3.1 follows.

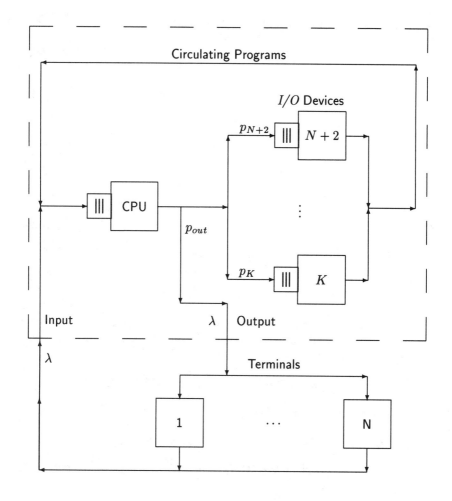

Figure 6.3.1. Closed BCMP computer system.

Algorithm 6.3.2 *Consider the closed BCMP system of Figure 6.3.1. Suppose the mean think time $E[t] = T$ for each of the N terminals. Each terminal is a type 3 server. The CPU (device number $N+1$) is either a type 1 or type 2 server with service demand $D = D_{N+1}$ given. We are also given the service demands D_k, $k = N+2, N+3, \ldots, K$ for the I/O devices. Then calculate the performance measures of the system as follows.*

Step 1 [Initialize] *Set $L_k[0] = 0$, $k = N+1, N+2, \ldots, K$,*

Step 2 [Iterate] *For $n = 1, 2, \ldots, N$, calculate*

$$W_k[n] = D_k(1 + L_k[n-1]), \quad k = N+1, N+2, \ldots, K, \quad (6.20)$$

$$W[n] = \sum_{k=N+1}^{K} W_k[n], \quad (6.21)$$

$$\lambda[n] = \frac{n}{W[n] + T}, \quad (6.22)$$

$$L_k[n] = \lambda(n)W_k[n], \quad k = N+1, N+2, \ldots, K. \quad (6.23)$$

Step 3 [Compute performance measures] *Set system throughput to*

$$\lambda = \lambda[N]. \quad (6.24)$$

Set job response time (turnaround time) to

$$W = \frac{N}{\lambda} - T. \quad (6.25)$$

[We have already calculated $W = W[N]$ in Step 2, but we want to show the dependence of W on λ, N, and T, explicitly.] Set server utilizations to

$$\rho_k = \lambda D_k, \quad k = N+1, N+2, \ldots, K. \quad (6.26)$$

Set server throughputs to

$$\lambda_k = \lambda V_k, \quad k = N+1, N+2, \ldots, K. \quad (6.27)$$

We calculated $L_k[N]$ and $W_k[N]$ for each server in the last iteration of Step 2.

Proof Formula (6.19) is the fundamental principle of MVA. Formula (6.20) merely says that the total time a job spends in the system is the sum of the times the job spends at each service center. Formulas (6.21)–(6.26) are all implementations of Little's law. ∎

Example 6.3.1 Mellow Marmalade has an interactive computer system consisting of 50 active terminals connected to a central computer system as in Figure 6.3.1. The company performance analysts find that they can model the system by the queueing model described in Algorithm 6.3.2 with three *I/O* devices. They find that the average think time is 15 seconds, the mean CPU service demand per interaction (job) is 0.15 seconds, and the mean total service demand per interaction on the three *I/O* devices is 0.02, 0.04, and 0.05 seconds, respectively. The calculations of Algorithm 6.3.2 are shown, in part, in Table 6.3.1. The entries in the table are rounded from those obtained using the APL program BCMP. Using the unrounded results, we find that $\lambda = 3.24485$ interactions per second, $W = 0.40904$ seconds, $\rho_{CPU} = 0.4867$ (the CPU is the bottleneck device), and the average number in the central computer system $L = 1.327$. □

Table 6.3.1. Calculations of Example 6.3.1

	k	$n = 0$	$n = 1$	$n = 2$	$n = 50$
L_k	51	0	0.00983	0.01985	0.91652
L_k	52	0	0.00131	0.00262	0.06929
L_k	53	0	0.00262	0.00526	0.14865
L_k	54	0	0.00328	0.00657	0.19279
W_k	51	-	0.15000	0.15147	0.28246
W_k	52	-	0.02000	0.02003	0.02136
W_k	53	-	0.04000	0.04010	0.04581
W_k	54	-	0.05000	0.05016	0.05941
λ		-	0.06553	0.13105	3.24485

As we previously noted, the closed BCMP system of Figure 6.3.1, as implemented by Algorithm 6.3.2, has the major weakness that it assumes there are no memory constraints, that is, that there is always room in main memory to handle each transaction (job) as it arrives. This is a reasonable assumption for a lightly loaded system, such as the one we considered in Example 6.3.1. In practice every computer system must limit the multiprogramming level. For a lightly loaded batch or interactive system we can use the model described by Algorithm 6.3.2 (for a batch system, N is the fixed number of batch jobs in the system at all times and the mean think time is zero). For a very heavily loaded system with fixed multiprogramming level (it could be fixed at the maximum level thought feasible), the closed central server model of Algorithm 6.2.1 can be used to calculate the throughput λ; then the response time formula $W = N/\lambda - T$ can be used to calculate the mean response time.

The remaining problem, that we will now address, is the system in which there is an upper limit, say m, set on the multiprogramming level but for which the average is much lower. While the computer is in operation we assume that the actual MPL (multiprogramming level) will fluctuate from zero during the rare times when the computer is doing nothing to m when the maximum allowed level has been reached and there is a queue of transactions or jobs waiting to enter main memory. Thus, to model the effect of memory on the performance of the computer system, we must take the fluctuating MPL into account.

A technique has been developed for handling memory management modeling in two stages. In the first stage we replace the entire central computer system shown in the dashed box in Figure 6.3.1 plus the queue (the queue that develops when the multiprogramming level m has been reached), by a *flow equivalent service center*, abbreviated FESC. This flow equivalent service center can be thought of as a black box, that, when given the workload of the system, provides the same throughput and response time as the real system. The FESC is a *load dependent server*. This means that throughput depends on the number of customers in the FESC. When there is only one transaction being processed (the MPL is one), the throughput and the mean response time are much lower than when the computer is operating with the maximum MPL. The second stage is to model the system of Figure 6.3.1 with the FESC replacing the central computer system.

The entire modeling process is somewhat complex. The following algorithm provides enough detail to write the computer code to implement the model we have described. The computational load of this modeling process requires a computer, even for very simple systems.

Algorithm 6.3.3 *Consider the closed BCMP system of Figure 6.3.1. We assume that all the information about the computer system mentioned in Algorithm 6.3.2 is available. We also assume that the central computer system has a maximum allowed multiprogramming level (MPL) of m transactions (jobs). Then calculate the performance measures of the system as follows.*

Step 1 [Calculate preliminary parameters.] *Use Algorithm 6.2.1 to calculate the average service rate, $\mu[j] = \lambda[j]$, ($\lambda[j]$ is the average throughput), for multiprogramming level $j = 1, 2, \ldots, m$.*

Step 2 [Initialize] *Set $q[0] = 1$ and create the vector $\boldsymbol{\mu}$ of dimension N where $\mu[j]$ is defined in Step 1 for $j = 1, 2, \ldots, m$ and $\mu[j] = \mu[m]$ for $j = m + 1, \ldots, N$.*

Step 3 [Iterate] *For $n = 1, 2, \ldots, N$ repeat Steps 4–6 in sequence.*

Step 4 [Compute W for n customers in FESC] *Set $W = 0$ and repeat the following calculation for $j = 1, 2, \ldots, n$.*

$$W = W + \frac{j}{\mu[j]} \, q[j-1].$$

Step 5 [Compute $\lambda[n]$] *Set*

$$\lambda[n] = \frac{n}{W + T}.$$

Step 6 [Compute parameters] *Set $s = 0$, and for $j = n, n - 1, n - 2, \ldots, 1$, calculate*

$$q[j] = \frac{\lambda[n]}{\mu[j]} q[j - 1],$$

and

$$s = s + q[j].$$

Then set $q[0] = 1 - s$.

Step 7 [Compute performance measures] *Set system throughput to*

$$\lambda = \lambda[N]. \tag{6.28}$$

Set job response time (turnaround time) to W. Set server utilizations to

$$\rho_k = \lambda D_k, \quad k = N + 1, N + 2, \ldots, K. \tag{6.29}$$

Set server throughputs to

$$\lambda_k = \lambda V_k, \quad k = N + 1, N + 2, \ldots, K. \tag{6.30}$$

Finally,

$$q[j] = P[j \text{ transactions are in process}] \quad j = 0, 1, 2, \ldots, N. \tag{6.31}$$

Proof See Lazowska *et al.* [13]. ■

Note that, in Step 2 we take note of the fact that the throughput of the central computer system can never exceed that which occurs when the maximum multiprogramming level, m, is in effect. Step 3 means that, on the first pass, Step 4 is executed with $n = 1$, then Step 5 is executed with $n = 1$, and, finally, Step 6 is executed with $n = 1$. This completes one pass. On the second pass the first pass is repeated with $n = 2$, etc. After Step 6 is executed with $n = N$, Step 7 is executed. This is the algorithm that Myriad uses for its Interactive System model; that is, it kicks out exactly the same numbers as the *Mathematica* program **online**. The value of $q[j]$ is, of course, conditioned on there being n customers in the entire computer system. Thus, $q[j]$ is the probability there are j customers in the FESC, given that there are n customers in the model, that is, that there are n active terminals.

Example 6.3.2 Let us apply the Algorithm 6.3.3 model to the Mellow Marshmallow interactive computer system of Example 6.3.1. Let us use $m = 3$. The *Mathematica* program **subcent** shows that $\mu[1] = 3.84615$, $\mu[2] = 5.49683$, and $\mu[3] = 6.19921$. Then **online** shows that Algorithm 6.3.3 predicts $\lambda = 3.24223$ transactions per second and $W = 0.421502$ seconds. These values are very close to those we obtained using the Algorithm 6.3.2 model. We would expect this, since the system is lightly loaded. Moreover, **online** provides the distribution of active customers X, that is, those that have obtained access to main memory. We find that $P[X = 0] = 0.343824$ (this is the probability that the central system is idle), $P[X = 1] = 0.177085$, $P[X = 2] = 0.297981$, and $P[X = 3] = 0.1811$. From these probabilities, we calculate the average number of active customers $E[X] = 1.19548$. By Little's law, the average time a transaction spends in the central system, once a memory partition has been obtained, is 0.3687 seconds. Subtracting this number from W, we see that the average time a transaction queues for main memory is 0.05278 seconds. Although our new model doesn't improve the accuracy of λ and W very much for this lightly loaded case, it does give us more detailed information. Suppose now that the load is doubled to 100 active users. Then **online** predicts that $\lambda = 5.9397$ transactions per second, $W = 1.83587$ seconds, $E[X] = 2.811$, and the mean time an active customer spends in the central computer system is 0.4733 seconds. The approximate solution generated by Algorithm 6.3.2 is $\lambda = 6.124$ transactions per second and $W = 1.329$ seconds. □

In the last example the added complexity of the FESC model, that takes the limitations of main memory into account, was needed only for heavily loaded systems. Let us consider another example, that illustrates a situation where the Algorithm 6.3.3 model is really needed.

Example 6.3.3 Puerile Publications, a publisher of children's books, has 30 diskless workstations connected to a computer system that acts as a file server on a LAN. Peter Piper, the analyst for their computer systems, uses the Algorithm 6.3.3 model for the system with $m = 4$. He has measured the system, which has two disk drives on the file server. He finds that $D_1 = .3$ seconds, $D_2 = 0.4$ seconds, and $D_3 = 0.2$ seconds. The mean think time is 10 seconds. By using **subcent** and **online**, Peter finds that $\lambda = 2.065$ transactions per second and $W = 4.53$ seconds. He also finds that, on the average, 9.35 workstations have transactions in progress, either queueing for the server or receiving service from it. He finds that the probability distribution for customers active in the server memory is $p_0 = 0.00539$, $p_1 = 0.01455$, $p_2 = 0.02579$, $p_3 = 0.03743$, and $p_4 = 0.91684$, so that the average number of active customers in main memory is 3.85. The average time they spend there is 1.86 seconds, and the average time spent queueing to get in is 2.67 seconds. The Algorithm 6.3.2 model predicts $\lambda = 2.25$ transactions per second, $W = 3.33$ seconds, and $L = 7.5$ workstation users. $\square$

6.3.2 Multiple Job Class Models

Very few computer systems are dedicated to one application so completely that the system can be modeled very accurately as a single job class model. To do so requires a great deal of averaging over the actual classes, which can lead to inaccurate performance projections for upgraded systems.

6.3.2.1 Open Multiclass Models

An open multiclass model has only open classes. Open class models are very easy to solve. For this reason they are somewhat popular, although most open class models represent only an approximation of reality and can cause modeling problems for mixed systems with both open and closed workload classes. For this reason some experienced modelers avoid using mixed models.

We assume there are C open classes in the model and that each class c has an arrival rate λ_c. Each class also has a service demand D_{ck} for each service center k. We assume that a steady state solution exists. For this to be true, we must have

$$\max_k \left\{ \sum_{c=1}^C \lambda_c D_{ck} \right\} < 1. \tag{6.32}$$

This condition guarantees that no service center will receive more service requests than it can handle.

We can calculate the utilization of each service center by summing the utilizations due to each customer class. Thus,

$$\rho_k = \sum_{c=1}^C \lambda_c D_{ck} \quad k = 1, 2, \ldots, K. \tag{6.33}$$

By the arrival theorem, we can write

$$W_{ck} = \begin{cases} D_{ck} & \text{for delay centers} \\ \dfrac{D_{ck}}{1 - \displaystyle\sum_{j=1}^C \rho_{jk}} & \text{for queueing centers,} \end{cases} \tag{6.34}$$

where

$$\rho_{jk} = \lambda_j D_{jk}$$

for all j and k. Little's law gives us

$$L_{ck} = \lambda_c W_{ck}. \tag{6.35}$$

The mean response time for class c customers is given by

$$W_c = \sum_{k=1}^{K} W_{ck}, \tag{6.36}$$

while, by Little's law,

$$L_c = \lambda_c W_c. \tag{6.37}$$

6.3.2.2 Closed Multiclass Models

These models are much more difficult to solve, both in terms of complexity of the algorithms and in computer time required. Many open class models can be solved in a few minutes with a pocket calculator, but a closed model with only a few classes and only a few workloads in each class can make massive demands on a computer as well as a significant investment in programming effort.

A closed, multiple class workload model has C customer classes, each of which has a fixed population N_c. We write the vector $\vec{N} = (N_1, N_2, \ldots, N_C)$ to indicate the overall population. Like all MVA algorithms this model depends on Little's law and the arrival theorem. Little's law is used to calculate the throughput $\lambda_c[\vec{N}]$ for each class by the formula

$$\lambda_c[\vec{N}] = \frac{N_c}{T_c + W_c[\vec{N}]} \tag{6.38}$$

where

$$W_c[\vec{N}] = \sum_{k=1}^{K} W_{ck}[\vec{N}] \tag{6.39}$$

and each $W_{ck}[\vec{N}]$ is calculated by (6.41) below. We apply Little's law to each service center k to calculate the average number of class c customers there as follows

$$L_{ck}[\vec{N}] = \lambda_c[\vec{N}] W_{ck}[\vec{N}], \tag{6.40}$$

and sum the values to obtain

$$L_k[\vec{N}] = \sum_{c=1}^{K} L_{ck}[\vec{N}]. \tag{6.41}$$

The heart of the algorithm is the equation

$$W_{ck}[\vec{N}] = \begin{cases} D_{ck} & \text{for delay centers} \\ \\ D_{ck}[1 + A_{ck}[\vec{N}]] & \text{for queueing centers} \end{cases} \qquad (6.42)$$

where $A_{ck}[\vec{N}]$ is the number of customers at service center k seen by an arriving class c customer. The arrival theorem implies that

$$A_{ck}[\vec{N}] = L_k[\overrightarrow{N - 1_c}], \qquad (6.43)$$

where $\overrightarrow{N - 1_c}$ is the population $\vec{N}$ with one class c customer removed. What makes the multiclass MVA algorithm difficult to implement (and even to describe in detail) is the following: We start with the empty population

$$\vec{0} = \overbrace{(0, 0, \ldots, 0)}^{k \ 0\text{'s}} \qquad (6.44)$$

by setting

$$L_k[\vec{0}] = 0, \quad \text{for } k = 1, 2, \ldots K. \qquad (6.45)$$

Then we can use (6.42) together with (6.37)–(6.41) iteratively to calculate solutions for increasing populations $\vec{n}$ until we reach $\vec{N}$. However, the solution for each feasible population $\vec{n} = (n_1, n_2, \ldots, n_c)$ (to be feasible we must have $n = n_1 + n_2 + \cdots + n_C$, where $n_c \leq N_c$, $c = 1, 2 \ldots, C$) requires as input the C solutions $L_k[\overrightarrow{n - 1_c}]$, $c = 1, 2, \ldots, C$. Thus, the following concerns are the two major procedural difficulties in implementing the multiclass MVA algorithm:

1. Generating all the feasible values of $\vec{n}$ satisfying the conditions (i) $n = n_1 + n_2 + \cdots + n_C$ and (ii) $n_c \leq N_c$, $c = 1, 2 \ldots, C$.

2. Saving the solutions $L_k[\overrightarrow{n - 1_c}]$, $c = 1, 2, \ldots, C$, in such a way that they can be retrieved as needed.

Making sure that all the feasible values of $\vec{n}$ are actually used isn't trivial, either. In addition to the procedural problems, there is quite a heavy computation and storage requirement for the computer system that implements the algorithm.

We state the exact multiclass MVA algorithm in a concise form. It is described more completely by the *Mathematica* program **Exact**, that implements the exact multiclass MVA modeling algorithm. We will use it as our detailed statement of the algorithm. In the algorithm below we do not number the terminals but number the service centers in the central computer system starting with $k = 1$ for the CPU.

Algorithm 6.3.4 (*Exact Closed Multiclass MVA Algorithm*) *Consider the closed BCMP system of Figure 6.3.1. We assume there are C customer classes, N_c customers in each class, and that each customer class c has the service demand D_{ck}, $k = 1, 2, \ldots, K$. Then we calculate the performance measures of the system as follows.*

Step 1 [Loop] *Repeat Steps 2, 3, and 4 for all feasible customer populations, $\vec{n}$.*

Step 2 [Calculate values of $W_{ck}[\vec{n}]$] *For $k = 1, 2, \ldots, K$, and $c = 1, 2, \ldots, C$, calculate*

$$W_{ck}[\vec{n}] = \begin{cases} D_{ck} & \text{for delay centers} \\ \\ D_{ck}[1 + L_k[\overrightarrow{n - 1_c}]] & \text{for queueing centers} \end{cases} \tag{6.46}$$

Step 3 [Calculate values of $\lambda_c[\vec{n}]$] *For $c = 1, 2, \ldots, C$, calculate*

$$\lambda_c[\vec{n}] = \frac{N_c}{T_c + W_c[\vec{n}]}, \tag{6.47}$$

where

$$W_c[\vec{n}] = \sum_{k=1}^{K} W_{ck}[\vec{n}]. \tag{6.48}$$

Step 4 [Calculate values of $L_k[\vec{n}]$] *For $k = 1, 2, \ldots, K$, calculate*

$$L_k[\vec{n}] = \sum_{c=1}^{C} \lambda_c[\vec{n}] W_{ck}[\vec{n}]. \tag{6.49}$$

Table 6.3.2

c	k	D_{ck}
1	1	0.50 seconds
1	2	0.04 seconds
1	3	0.06 seconds
2	1	0.40 seconds
2	2	0.20 seconds
2	3	0.30 seconds
3	1	1.20 seconds
3	2	0.05 seconds
3	3	0.06 seconds

Example 6.3.4 The performance analysts at Notorious Northgate have found they can model one of their small computer systems using Algorithm 6.3.4 with three customer classes, two terminal and one batch, with $\vec{N} = \{5, 10, 5\}$. There are 3 service centers with service demands D_{ck} as given in Table 6.3.2. The mean think time for the first terminal class is 20 seconds; it is 30 seconds for the second terminal class. The results from running **Exact** for this system are shown in Table 6.3.3. All times are in seconds. The CPU utilization is actually slightly less than 1; the exact value is 0.999999620457793. □

Table 6.3.3. Results

c/k	λ_c	W_c	ρ_k	L_k
1	0.210	3.81	1.0000	6.5867
2	0.298	3.56	0.1003	0.1111
3	0.647	7.73	0.1408	0.1628

The program **Exact** is reasonably efficient. Example 6.3.4 was modeled in 25.93 seconds on an IBM compatible PC with a 33 MHz Intel 80386 microprocessor. However, the exact multiclass MVA modeling algorithm is not practical for modeling large multiclass systems. Therefore, we will describe an approximate algorithm. We use the same numbering system in this algorithm as in Algorithm 6.3.4.

Algorithm 6.3.5 (*Approximate Closed Multiclass MVA Algorithm*) *Consider the closed BCMP system of Figure 6.3.1. We assume there are C customer classes, N_c customers in each class, and that each customer class c has the service demand D_{ck}, $k = 1, 2, \ldots, K$. Let ϵ be an input to specify our error criterion. Then we calculate the performance measures of the system as follows.*

Step 1 [Calculate approximate values of $L_{ck}[\vec{N}]$] *Set*

$$L_{ck}^{(1)}[\vec{N}] = \frac{N_c}{K} \qquad (6.50)$$

for all c and k.

Step 2 [Approximate $A_{ck}[\vec{N}]$] *Calculate*

$$A_{ck}[\vec{N}] = \frac{N_c - 1}{N_c} L_{ck}^{(1)}[\vec{N}] + \sum_{\substack{j=1 \\ j \neq c}}^{C} L_{jk}^{(1)}[\vec{N}] \qquad (6.51)$$

for all c and k.

Step 3 [Calculate $L_{ck}[\vec{N}]$] *Using the estimates for $A_{ck}[\vec{N}]$ from Step 2, calculate*

$$W_{ck}[\vec{N}] = \begin{cases} D_{ck} & \text{for delay centers} \\ \\ D_{ck}[1 + A_{ck}[\vec{N}]] & \text{for queueing centers} \end{cases} \tag{6.52}$$

for all c and k. Then we calculate

$$W_c[\vec{N}] = \sum_{k=1}^{K} W_{ck}[\vec{N}], \tag{6.53}$$

$$\lambda_c[\vec{N}] = \frac{N_c}{T_c + W_c[\vec{N}]}, \tag{6.54}$$

for $c = 1, 2, \ldots, C$, and

$$L_{ck}^{(2)}[\vec{N}] = \lambda_c[\vec{N}] \, W_{ck}[\vec{N}] \tag{6.55}$$

for $c = 1, 2, \ldots, C$, and $k = 1, 2, \ldots, K$.

Step 4 [Test] *If*

$$|L_{ck}^{(1)}[\vec{N}] - L_{ck}^{(2)}[\vec{N}]| < \epsilon, \tag{6.56}$$

for all c and k go to Step 5. Otherwise return to Step 2 after replacing $L_{ck}^{(1)}[\vec{N}]$ by $L_{ck}^{(2)}[\vec{N}]$ for all c and k.

Step 5 [Compute performance measures] *Set $W_{ck}[\vec{N}]$, $W_c[\vec{N}]$, and $\lambda_c[\vec{N}]$ to the values calculated in Step 3 for all values of c and k. If $L_k[\vec{N}]$ is desired, calculate it using (6.40).*

Proof See Lazowska *et al.* [13]. ∎

Example 6.3.5 Suppose we compute the performance statistics for the computer system of Example 6.3.4 using Algorithm 6.3.5. If we use $\epsilon = 0.001$, we obtain the values in Table 6.3.4. The computation requires 2.03 seconds on an IBM compatible PC with a 33MHz Intel 80386 microprocessor. If we use $\epsilon = 10^{-13}$, the output is very little different, and the calculation takes 11.75 seconds. Note that, although the algorithm converges to a solution very quickly, the solution it produces is *not* the exact solution, no matter how small an ϵ is used. □

Table 6.3.4. Results

c/k	λ_c	W_c	ρ_k	L_k
1	0.210	3.83	0.9686	6.5979
2	0.298	3.57	0.0990	0.1092
3	0.620	8.06	0.1391	0.1603

6.3.2.3 Mixed Multiclass Models

Mixed multiclass models have both open (transaction) and closed (terminal and batch) workload classes. Although Baskett *et al.* [1] assure us that mixed multiclass models of BCMP queueing network models of computer systems have solutions, the solutions tend to be "kludgy" approximate solutions. We outline the solution technique given by Lazowska *et al.* [13]. The idea is to reduce the solution to the solution of a closed multiclass model, that we can solve using Algorithm 6.3.4 or Algorithm 6.3.5. The influence of the open classes on the closed classes is introduced by "inflating" the service demand of each closed class at each service center by dividing each D_{ck} for a closed class at service center k by one minus the total utilization of service center k by all open classes. That is,

$$D_{ck}^* = \frac{D_{ck}}{1 - \rho_{\{O\},k}}, \tag{6.57}$$

where $\rho_{\{O\},k}$ is the total utilization of service center k by all open classes. Thus, open classes effectively slow down the service centers for closed classes. We follow Lazowska *et al.* [13] by letting $c \in \{O\}$ mean that class c is an open workload class and similarly for $c \in \{C\}$.

Algorithm 6.3.6 *Consider a mixed multiclass BCMP computer system. Suppose the closed classes have service demands D_{ck}, for all $c \in \{C\}$ and $k = 1, 2, \ldots, K$. An approximate solution is given by the following steps:*

1. For each center k, calculate

$$\rho_{ck} = \lambda_c D_{ck}, \quad \text{for all } c \in \{O\}, \tag{6.58}$$

and the total utilization by all open classes

$$\rho_{\{O\},k} = \sum_{c \in \{O\}} \rho_{ck}. \tag{6.59}$$

2. Use Algorithm 6.3.4 to solve the closed model with the open classes deleted, where the service demand for closed class c at service center k is given by

$$D_{ck}^* = \frac{D_{ck}}{1 - \rho_{\{O\},k}}. \tag{6.60}$$

Then the number at each service center, throughputs, and response times from this model are the performance measures for the closed classes in the original mixed class model. The utilization, ρ_{ck}, for the closed class c at service center k is given by

$$\rho_{ck} = \lambda_c D_{ck}. \tag{6.61}$$

3. The mean response time, W_{ck}, and the average number of open class c customers at service center k, L_{ck}, is computed in terms of performance measures of the closed classes as follows:

$$W_{ck} = \frac{D_{ck}\left[1 + L_{\{C\},k}\right]}{1 - \rho_{\{O\},k}} \quad \text{for all } c \in \{O\}, \tag{6.62}$$

$$L_{ck} = \lambda_c W_{ck} \quad \text{for all } c \in \{O\}, \tag{6.63}$$

where $\rho_{\{O\},k}$ was calculated in Step 1 and $L_{\{C\},k}$ is the average total number of closed class customers at service center k calculated from the solution of the closed class model in Step 2.

Let us consider a simple example.

Table 6.3.5

c	k	D_{ck}
1	1	0.20 seconds
1	2	0.50 seconds
2	1	0.50 seconds
2	2	0.80 seconds
3	1	0.80 seconds
3	2	1.25 seconds

Example 6.3.6 Suppose we have a computer system with one open (transaction) and two closed (batch) workload classes. Suppose the computer system has two devices, one CPU and one I/O device, and that each batch class has one member. We also assume that $\lambda_1 = 1$ transaction per second, and the service demands are given in Table 6.3.5. Furthermore, we assume there is one customer in each closed class. We see that $\rho_{\{O\},1} = \lambda_1 D_{11} = 0.2$ and that $\rho_{\{O\},2} = \lambda_1 D_{12} = 0.5$ Hence, by (6.59), $D_{21}^* = 0.5/(1 - 0.2) = 0.625$, $D_{22}^* = 0.8/(1-0.5) = 1.6$, $D_{31}^* - 0.8/(1-0.2) = 1$, and $D_{32}^* = 1.25/0.5 = 2.5$, where all service demands are in seconds. Solving the closed model of Step 2 in the algorithm yields the numbers in Table 6.3.6. Then, by (6.61), we calculate

$$W_{11} = \frac{0.2[1 + 0.4562]}{1 - 0.2} = 0.364 \text{ seconds,}$$

and

$$W_{12} = \frac{0.5[1 + 1.5438]}{1 - 0.5} = 2.5438 \text{ seconds,}$$

so that $W_1 = 2.908$ seconds. Finally, we see that the utilization of the CPU is $\rho_1 = \sum_{c=1}^{3} \lambda_c D_{c1} = 0.4844$, and the utilization of the I/O device is $\rho_2 = \sum_{c=1}^{3} \lambda_c D_{c2} = 0.9497$. $\square$

Table 6.3.6. Results

c/k	λ_c	W_c	L_k
2	0.2820	3.546	0.4562
3	0.1793	5.579	1.5438

It is evident that there is both good news and bad news about Algorithm 6.3.6. The good news is that it is relatively easy to implement. (We leave it as an exercise for the reader to revise the code for the *Mathematica* program **Exact** to implement this algorithm.) The algorithm has been used to model some computer systems. The bad news is that there are some less than desirable features. For example, it is clear that the algorithm has the effect of giving all open classes priority over all closed classes, although this might not be true of the system that is to be modeled. The algorithm creates this effect by the way it inflates the service demands of closed classes in equation (6.59). The other, more troublesome, problem is that there is a double level of approximation because an open (transaction) workload is always an approximation of an actual workload.

6.4 Summary

In this chapter we have discussed a number of queueing network models of computer systems. We have limited the discussion to those that are relatively easy to implement. The more complex queueing network models are very difficult to implement and require very complex software. The standard reference on queueing network models of computer systems is the excellent book by Lazowska, Zahorjan, Graham, and Sevcik [13]. Schwartz [22] provides some advice on how to fit the model provided by a sophisticated queueing network modeling package to an actual computer system.

It is the mark of an instructed mind to rest satisfied with the degree of precision which the nature of the subject admits, and not to seek exactness when only an approximation of the truth is possible.
Aristotle

6.5 Exercises

1. [6] Measurements of an interactive computer system at Piper's Pickles show that the average response time is 1.5 seconds, the number of active users is 100, the CPU utilization is 75%, and the average CPU

time used per interaction was 0.3 seconds. What was the average think time?

2. [5] An interactive computer system at Anchor Anchovies has 70 active terminals with a mean think time of 30 seconds. The paging disk averages 5 accesses per interaction with the average access time of 0.05 seconds. The paging disk has an average utilization of 0.5. What is the average system response time?

3. [8] During a period when the performance of a computer system at Weezl Words was being measured a particular disk was busy 30 percent of the time. If each transaction required 25 accesses to the disk, on the average, each of which takes 25 milliseconds, what was the average throughput of the system?

4. [8] An interactive system at Harvey Wallbangers was measured during a period in which 10 terminals were active with an average think time of 10 seconds. If each interaction required half a second of CPU processing and the CPU utilization was 40 percent, what was the average response time?

5. [10] The interactive system at Myth and Smesson was measured when 50 terminals were active, the average think time was 15 seconds, and the average response time was 1.5 seconds. What was the mean number of transactions active in the central subsystem?

6. [10] Slobovian Scientific of Example 6.1.1 decides to upgrade their system so that the average file server service time is 0.5 seconds; all the other parameters are unchanged. Calculate p_0, ρ, λ, and W, assuming the system can be modeled as a machine repair queueing system.

7. [05] Consider Example 6.2.1. Suppose λ and W_s are as in the example but that $p = 0.9999$. Calculate Λ, ρ, L, and W.

8. [10 if you have *Mathematica* and can use **cent**; 22 if you must write your own code.] The analysts at Image Power believe they can model their small batch computer system as a central server model using Algorithm 6.2.1. They have a CPU and two I/O devices, with the total service demands of 2, 1, and 0.5 seconds, respectively. If the MPL (multiprogramming level) is 5, find λ, W, ρ_1, ρ_2, and ρ_3.

9. [12] Consider Example 6.2.5. Suppose Rick upgrades the computer system so that the service demands are 1.5, 0.5, 1, and 2 seconds, respectively, for the CPU and the three I/O devices. If $\lambda = 1/5$, find W and L.

10. [15 if you have *Mathematica*; 25 if you must write your own code.] Suppose Peter Piper of Puerile Publications, Example 6.3.3, discovers that he must add 10 more workstations to their system (business is booming at Puerile). Use his model to do the following:

   (a) Calculate λ and W for the current system with 10 more workstations, that is, $N = 40$.

   (b) Calculate λ and W with $N = 40$ for an upgraded file server that has $D_1 = 0.1$, $D_2 = 0.4$, and $D_3 = 0.2$ (all values in seconds).

   (c) Calculate λ and W with $N = 40$ for an upgraded file server consisting of the original CPU but a single *I/O* device, that is much faster because of caching and a faster disk drive so that $D_1 = 0.3$ seconds and $D_2 = 0.02$ seconds.

11. [10] Interactive Systems has a transaction processing computer system that processes two kinds of transactions. The analysts at Interactive feel they can model their system as an open multiclass model described in Section 6.3.2.1 with one CPU and two *I/O* devices. The service demands and arrival rates are described in the table below. Calculate W_1, W_2, L_1, and L_2.

c	k	D_{ck}	λ_c
1	1	0.2	1.0
1	2	0.3	—
1	3	0.4	—
2	1	0.4	1.5
2	2	0.3	—
2	3	0.2	—

12. [15 if you have *Mathematica*; 25 if you must write your own code.] Consider Example 6.3.4. Suppose it is discovered that the think times have been recorded incorrectly and should be 40 seconds for the first terminal class and 60 seconds for the second. Assume the other values are correct and use Algorithm 6.3.4 to calculate the correct values for Table 6.3.3.

13. [15 if you have *Mathematica*; 20 if you must write your own code.] Calculate the approximate values for Exercise 12 using Algorithm 6.3.5 with $\epsilon = 0.001$.

So naturalists observe, a flea
Hath smaller fleas that on him prey.
And these have smaller still to bite 'em
And so proceed ad infinitum.

Jonathan Swift

It is much easier to be critical than to be correct.
Benjamin Disraeli

The public is the only critic whose opinion is worth anything at all.
Mark Twain

A cynic is a blackguard whose faulty vision sees things as they are, and
not as they ought to be.
Ambrose Bierce

As everyone knows, he who joins a waiting line is sure to wait for an
abnormally long time, and similar bad luck follows us on all occasions.
How much can probability theory contribute towards an explanation?
... We have here a new confirmation for the persistence of bad luck.
Assuredly Peter has reason for complaint if he has to wait three times as
long as Paul, but the distribution (5.2) attributes to this event probability
$\frac{1}{4}$. *It follows that, on the average, in one out of two cases either Paul or*
Peter has reason for complaint. The observed frequency increases in
practice because very short waiting times naturally pass unnoticed.
William Feller
An Introduction to Probability and Its Applications, Vol. II

References

[1] Forest Baskett, K. Mani Chandy, Richard R. Muntz, and Fernando G. Palacios, Open, closed, and mixed networks of queues with different classes of customers, *JACM*, **22(2)**, (April 1975), 248–260.

[2] Gunter Bolch, *Leistungsbewertung von Rechensystemen*, B. G. Teubner, Stuttgart, 1989.

[3] Jeffrey P. Buzen, Queueing network models of multiprogramming, Ph.D. dissertation, Division of Engineering and Applied Physics, Harvard University, Cambridge, MA, May 1971. Available from NTIS, Springfield, VA, as AD 731575, (August 1971).

[4] Jeffrey P. Buzen, Computational algorithms for closed queueing networks with exponential servers, *Comm. ACM*, **16(9)**, (September 1973), 527–531.

[5] Peter J. Denning and Jeffrey P. Buzen, The operational analysis of queueing network models, *ACM Computing Surveys*, **10(3)**, (Sept. 1978), 225–261.

[6] Domenico Ferrari, Giuseppe Serazzi, and Alessandro Zeigner, *Measurement and Tuning of Computer Systems*, Prentice-Hall, Englewood Cliffs, NJ, 1983.

[7] James R. Jackson, Networks of waiting lines, *Operations Res.*, **5(4)**, (August 1957), 518–521.

[8] James R. Jackson, Jobshop-like queueing systems, *Management Science*, **10(1)**, (October 1963), 131–142.

[9] Leonard Kleinrock, *Queueing Systems, Volume I: Theory*, John Wiley, New York, 1975.

[10] Leonard Kleinrock, *Queueing Systems, Volume II: Computer Applications*, John Wiley, New York, 1976.

[11] E. R. Lassettre and Allen L. Scherr, Modelling the performance of the OS/360 time sharing option (TSO), in *Statistical Computer Performance Evaluation*, (W. Freiberger, ed.), Academic Press, 1972, 57–72.

[12] Stephen S. Lavenberg, ed., *Computer Performance Modeling Handbook*, Academic Press, New York, 1983.

[13] Edward D. Lazowska, John Zahorjan, G. Scott Graham, and Kenneth C. Sevcik, *Quantitative System Performance: Computer System Analysis Using Queueing Network Models*, Prentice-Hall, Englewood Cliffs, NJ, 1984.

[14] Edward A. MacNair and Charles H. Sauer, *Elements of Practical Performance Modeling*, Prentice-Hall, Englewood Cliffs, NJ, 1985.

[15] James Martin, *Systems Analysis for Data Transmission*, Prentice-Hall, Englewood Cliffs, NJ, 1972.

[16] Michael K. Molloy, *Fundamentals of Performance Modeling*, Macmillan, New York, 1989.

[17] Richard R. Muntz, Analytic modeling of interactive systems, *Proc. IEEE*, **63**, (June 1975), 946–953.

[18] Martin Reiser, Interactive modeling of computer systems, *IBM Systems J.*, **15(4)**, (1976), 309–327.

[19] Martin Reiser, Mean-Value analysis and convolution method for queue-dependent servers in closed queueing networks, *Perf. Eval.*, **1**(1), (Jan. 1981), 7–18.

[20] Martin Reiser and Stephen S. Lavenberg, Mean value analysis of closed multichain queueing networks, *JACM*, **27(2)**, (April 1980), 313–322.

[21] Allen L. Scherr, *An Analysis of Time-Shared Computer Systems*, MIT Press, Cambridge, MA, 1967.

[22] Jeffry A. Schwartz, Techniques for improving the accuracy of your queueing network models, *CMG '89 Proceedings*, CMG, Chicago, IL, 1989.

[23] Kenneth C. Sevcik and Israel Mitrani, The distribution of queueing network states at input and output instants, *JACM*, **28(2)**, (April 1981), 358–371.

[24] Henk C. Tijms, *Stochastic Modelling and Analysis: A Computational Approach*, John Wiley, Chichester, 1986.

[25] Kishor Shridharbhai Trivedi, *Probability and Statistics with Reliability, Queuing, and Computer Science Applications*, Prentice-Hall, Englewood Cliffs, NJ, 1982.

Part Three:

Statistical Inference

Preface to Part Three: Statistical Inference

Some people hate the very name of statistics, but I find them full of beauty and interest. Whenever they are not brutalized, but delicately handled by the higher methods, and are warily interpreted, their power of dealing with complicated phenomena is extraordinary. They are the only tools by which an opening can be cut through the formidable thicket of difficulties that bars the path of those who pursue the Science of man.

Francis Galton

Francis Galton (1822-1911) was a renowned British biologist and scientist who is often credited with founding or at least making respectable the field of regression analysis. His work on fingerprints made possible their use in human identification. The above Galton quote beautifully summarizes what statistics can and cannot do.

Statistics has gotten some bad press. The popular quote, "There are lies, damned lies, and statistics,"[6] strikes a responsive chord in many of us. Part of this attitude may be instilled in some by excellent books such as the ever popular *How to Lie with Statistics* by Darrel Huff [3]. However, Huff does *not* tell us how to lie with statistics but rather how some advertisers have misled the public with statements or graphs that *appear* to be valid statistics. In fact, Otto Frisch [2] shows that the widely believed statement, "You can prove anything with statistics," is not true; that is, it is impossible to use valid statistics and valid data to prove anything that isn't so. Of course this does not prevent some people from trying. Statistics has passed the test of time as being one of the most useful of all intellectual disciplines. The outstanding book, Tanur *et al.* [7], provides for the lay reader a number of interesting essays on how statistics has solved many real-world problems.

---

[6]Attributed by Mark Twain to Benjamin Disraeli.

You may be asking yourself the following two cosmic questions:[7]

- What is statistics?

- How can it help me better understand that mysterious place, the real world?

To help answer these two questions we quote from the article "STATIS-TICS: The Field," published in Kruskal and Tanur [4]. William H. Kruskal says:

> My description of statistics is, of course, a personal one, but one that many statisticians would generally agree with. Almost any characterization of statistics would include the following general functions:
>
> 1. to help in summarizing and extracting relevant information from data, that is, from observed measurements, whether numerical, classificatory, ordinal, or whatever;
> 2. to help in finding and evaluating patterns shown by the data, but obscured by inherent random variability;
> 3. to help in the efficient design of experiments and surveys;
> 4. to help communication between scientists (if a standard procedure is cited, many readers will understand without need of detail).
>
> There are some other roles that activities called "statistical" may, unfortunately, play. Two such misguided roles are
>
> 1. to sanctify or provide seals of approval (one hears, for example, of thesis advisers or journal editors who insist on certain formal statistical procedures, whether or not they are appropriate);
> 2. to impress, obfuscate, or mystify (for example, some research papers contain masses of undisguised formulas that serve no purpose except that of indicating what a bright fellow the author is).

These two misguided roles are what Huff writes about in his popular book. It is a valuable book to read to prevent specious statistics from misleading you.

Wallis and Roberts [8] provide the crisp, enlightening definition, "Statistics is a body of methods for making wise decisions." That is what we will attempt to provide in Part Three.

---

[7]And then again, you may not be.

Some Statistical Computer Systems We Use in Part Three

There are a number of valuable statistical computer systems available for assistance in making statistical calculations and for displaying data in various formats. These systems are especially useful for performing exploratory data analysis. We have chosen three of them to use in this book, not because everyone agrees they are the best, but because they are available to the author and will probably be available to most readers. The systems are discussed in the preface and are listed again here:

1. MINITAB [5].

2. the EXPLORE programs of Doane [1].

3. SAS/STAT for IBM PC's and compatibles [6].

Cited References

[1] David P. Doane, *Exploring Statistics With The IBM PC, Second Edition*, Addison-Wesley, Reading, MA, 1988.

[2] Otto R. Frisch, You can prove anything with statistics, in *The Encyclopedia of Delusions*, edited by Ronald Duncan and Miranda Weston-Smith, Simon & Schuster, New York, 1979.

[3] Darrel Huff (with illustrations by I. Geis), *How to Lie with Statistics*, W. W. Norton, New York, 1954.

[4] William H. Kruskal and Judith M. Tanur (eds.), *International Encyclopedia of Statistics, Vols. 1 & 2*, The Free Press, New York, 1978.

[5] *MINITAB Reference Manual, Release 7*, Minitab Inc., State College, PA, 1989.

[6] *SAS Language Guide for Personal Computers, Release 6.03 Edition*, SAS Institute, Cary, NC, 1988.

[7] Judith M. Tanur, *et al.*, *Statistics: A Guide to The Unknown, 3rd Ed.*, Wadsworth & Brooks/Cole, Pacific Grove, CA, 1989.

[8] W. Allen Wallis and Harry V. Roberts, *Statistics: A New Approach*, The Free Press, New York, 1956.

Chapter 7

Estimation and Data Analysis

> *"Would you tell me, please, which way I ought to go from here?"*
> *"That depends a good deal on where you want to get to," said the Cat.*
> *"I don't much care where—" said Alice.*
> *"Then it doesn't matter which way you go," said the cat.*
>
> Lewis Carroll
> *Alice's Adventures in Wonderland*

7.0 Introduction

Heretofore we have assumed in all our probability models that we knew the exact probability distribution of each random variable under consideration. That is, we assumed a knowledge of both the form of the probability distribution and the values of the parameters of the distribution. In that mythical place often called the *real world* we are sure of neither. (I am, of course, excepting the rare individual who has direct communication with the Supreme Being. If you are one of these you have no need for the remainder of this book.) For most of us, our information about a particular random variable must be based on a sampling of observed values. Nearly everyone uses this technique to make judgments about such things as the quality of food and service at a restaurant, the entertainment value of a TV series, the talent of an actress or actor, etc.

Part Three of this book is part of a subject area called *statistical inference*. Statistical inference is based on a sample from the population of all items under consideration.

We will usually be concerned with obtaining a sample $x_1, x_2, \ldots, x_n$ of values from the *population* of all possible values of a random variable X. For the sample to have desirable mathematical properties it should be what is called a *random sample*. We can visualize the process of obtaining a random sample as a step-by-step procedure in which a series of observations is obtained in such a way that (a) each observed or selected value is independent of the others, and (b) at each step the selected value has the same probability of being chosen as any other element in the population. This can be conceptualized as a sequence $X_1, X_2, \ldots, X_n$ of independent random variables, each with the same distribution as X. We will therefore define a *random sample of size n* to be a sequence of independent identically distributed random variables $X_1, X_2, \ldots, X_n$. Once a random sample has been taken (the random variables have assumed values), we indicate the sample by $x_1, x_2, \ldots, x_n$. Thus, we take note of the fact that the values of two different random samples of size n from the same population are usually different; one random sample of five response times may be 1.2, 0.85, 0.35, 0.87, 0.98 seconds, while another random sample may yield 0.76, 0.45, 0.92, 1.18, 0.54 seconds. There will usually be some observed "randomness" between the values of two different random samples.

For computer science applications it is usually not too difficult to obtain random samples. However, in the area of political sampling, this can be a major problem. One of the most celebrated cases of lack of randomness in a sample is *Literary Digest*'s presidential poll of 1936. The *Digest* predicted on the basis of their poll that the Republican nominee, Governor Alf Landon, would defeat the incumbent Franklin D. Roosevelt by a margin of 3 to 2. Roosevelt won by one of the most one-sided landslides in American political history, obtaining 62% of the popular vote and 46 of 48 states. Bryson [6] explains why the sample obtained by the poll was not random and discusses some of the myths often quoted about the poll. The poll was *not* a telephone poll taken of *Literary Digest* subscribers as is often stated.

A number of questions concerning the use of a random sample have probably occurred to the reader. We list some of the most common concerns as a series of "cosmic questions" below. We will not be able to answer fully all of these questions in this book, but we shall attack each of them vigorously, if not rigorously.

(The reader should note that we will often use μ to denote the average or expected value of a random variable, although the symbol was reserved in Chapters 5 and 6 for average service rate of a server.)

Some Cosmic Questions

Given the values of a random sample $x_1, x_2, \ldots, x_n$ from a population determined by the random variable X (Thus, we assume that $X_1, X_2, \ldots, X_n$ are independent with the same probability distribution as X), the following cosmic questions may arise.

1. How do we estimate the values of the parameters of X such as $\mu = E[X]$, $\sigma^2 = \text{Var}[X]$, and σ, the standard deviation of X?

2. How do we make probability judgments about the accuracy of these estimates?

3. Assuming that the technique for estimating a parameter θ of X is to use a random variable $\hat{\theta}$ ($\hat{\theta}$ is pronounced "theta hat"), called an *estimator of θ*, which depends upon the random sample,[1] what are some desirable properties of estimators?

We will consider these cosmic questions and other important matters in this chapter.

> *Do what you can, with what you have, where you are.*
> Theodore Roosevelt

7.1 Estimators

An *estimator* $\hat{\theta}$ of a parameter θ of a random variable X is a random variable, which depends upon a random sample $X_1, X_2, \ldots, X_n$. The two most common estimators are the *sample mean*, also known as the *arithmetic mean*, $\overline{X}$, defined by

$$\overline{X} = \frac{1}{n} \sum_{i=1}^{n} X_i, \tag{7.1}$$

and the sample variance S^2 defined by[2]

$$S^2 = \frac{1}{n-1} \sum_{i=1}^{n} (X_i - \overline{X})^2. \tag{7.2}$$

We define the *sample standard deviation* S, of course, as the positive square root of S^2.

---

[1] We indicate this fact by writing $\hat{\theta} = \hat{\theta}(X_1, X_2, \ldots, X_n)$.

[2] We ask you to answer the question, "Why $n-1$ rather than n in the denominator of (7.2)?" in Exercise 15.

As the names suggest, $\overline{X}$ is an estimator of $\mu = E[X]$, S^2 of $\sigma^2 = \text{Var}[X]$, and S of the standard deviation σ.

There is a certain awkwardness of notation in statistics, that occurs in other areas of mathematics as well. This concerns the distinction between a function, which is a mapping from one set to another, and a particular value of the function. In Chapter 3 we tried to be consistent about indicating the distribution function of a random variable X by the symbol F, or sometimes $F(\cdot)$, but reserved the notation $F(x)$ to represent the value of F at the point x, such as in the formula

$$F(x) = 1 - \exp(-x/E[X]), \quad x > 0 \tag{7.3}$$

for the distribution function of an exponential random variable. When we indicate a random variable, which is a function, we usually use a capital letter such as X; we use a small x to indicate a particular value of X. (We consistently violated this convention in Chapters 5 and 6.) Thus, we indicate a random sample, which is a collection of functions, by $X_1, X_2, \ldots, X_n$, while we indicate a particular random sample that has been selected by $x_1, x_2, \ldots, x_n$. Similarly, when we talk about the sample mean, as in (7.1), we use a capital letter; we do the same for the sample variance in (7.2). An actual calculated value of the sample mean would be written as

$$\overline{x} = \frac{x_1 + x_2 + \cdots + x_n}{n}, \tag{7.4}$$

where the $x_1, x_2, \ldots, x_n$ are the values of the sample. Similarly, the actual calculated value of the sample variance would be written as

$$s^2 = \frac{\left((x_1 - \overline{x})^2 + (x_2 - \overline{x})^2 + \cdots + (x_n - \overline{x})^2\right)}{n - 1}, \tag{7.5}$$

where $\overline{x}$ was calculated by (7.4).

The next theorem would receive five stars if theorems were rated with stars as restaurants are. This theorem will help us answer the cosmic questions that were raised in the introduction of this chapter.

Theorem 7.1.1 (*The Sampling Theorem*) *Let $X_1, X_2, \ldots, X_n$ be a random sample of size n from a population determined by the random variable X that has finite mean $\mu = E[X]$ and finite variance $\sigma^2 = \text{Var}[X]$. Let $\overline{X}$ be defined by (7.1) and S^2 by (7.2). Then*

(a) $E[\overline{X}] = \mu$.

(b) $E[S^2] = \sigma^2$.

(c) $\text{Var}[\overline{X}] = \frac{\sigma^2}{n}$.

(d) For large n, the random variable

$$Z = \frac{\overline{X} - \mu}{\frac{\sigma}{\sqrt{n}}} \qquad (7.6)$$

has approximately the standard normal distribution.

(e) If X has a normal distribution, then the random variable

$$Y = \frac{\overline{X} - \mu}{\frac{S}{\sqrt{n}}} \qquad (7.7)$$

has the Student's t distribution with n − 1 degrees of freedom.

Proof We omit the proof. It can be found in Kreyszig [21]. ■

The standard deviation of $\overline{X}$, that, by the sampling theorem is $\sigma/\sqrt{n}$, is often called the *standard error*. Since we usually do not know the value of σ, the approximate value of the standard error is taken to be $s/\sqrt{n}$. Note that, if Z in (7.6) has a standard normal distribution, then $\overline{X}$ is normally distributed with mean μ and standard deviation $\sigma/\sqrt{n}$. Before we show how this theorem can be applied let us discuss some desirable properties of estimators. We write $\hat{\theta} = \hat{\theta}(X_1, X_2, \ldots, X_n)$ for an estimator of the parameter θ to emphasize the fact that the value of the estimator depends upon the value of a random sample.

An estimator $\hat{\theta}$ of θ is *unbiased* if $E[\hat{\theta}] = \theta$. Intuitively this means that the estimated values of θ will cluster about θ. Theorem 7.1.1 tells us that both the sample mean and the sample variance are unbiased estimators. (The reason we divided by $n - 1$ in (7.2) rather than by n was to make S^2 an unbiased estimator. We ask you to prove that S^2 is unbiased in Exercise 15.)

An estimator $\hat{\theta}$ having the property that, for each $\epsilon > 0$,

$$\lim_{n \to \infty} P[|\hat{\theta} - \theta| < \epsilon] = 1, \qquad (7.8)$$

is called a *consistent estimator of θ*. We also say that $\hat{\theta}$ *converges in probability to θ*. An estimator lacking this property, that is, an inconsistent estimator, could be said to "miss the point." Cramér [11] shows that the sample mean and the sample variance are consistent estimators.

Let us consider some consequences of properties of the sample mean $\overline{X}$ as an estimator of the mean μ of X. Suppose we took k random samples, each of size n, and calculated the corresponding k values of the sample mean, say $\overline{x}_1, \overline{x}_2, \ldots, \overline{x}_k$. Then these k numbers would cluster about the value μ because $\overline{X}$ is unbiased. The variance of $\overline{X}$ (by Theorem 7.1.1 $\text{Var}[\overline{X}] = \sigma^2/n$) is a measure of the tightness of the clustering about μ. By Chebyshev's inequality, at least three-fourths of the values $\overline{x}_1, \overline{x}_2, \ldots, \overline{x}_k$ are within two standard deviations of μ, that is, not farther than $2\sigma/\sqrt{n}$ from μ. Actually, about 95% of the sample means are this close to μ, if n is large enough that Z in (7.6) is approximately normal. A widely used rule of thumb for normality of Z and Thus, of $\overline{X}$ is that n be at least 30. The size that n must be depends a great deal upon the distribution of X. In fact, if X is normally distributed, then $\overline{X}$ is normal for $n = 1$. If the distribution of X is very different from normal, then values of n much greater than 30 may be required, perhaps as large as 100. Doane [12] has an excellent EXPLORE program MONTE that will let you explore the distribution of $\overline{X}$ on your personal computer.

Let us consider an example.

Example 7.1.1 Rick Rivets, a performance analyst at Fast Fasteners, collected the random sample of 40 think times shown in Table 7.1.1. He calculates the sample mean (9.32345 seconds) and the sample standard deviation (10.7977 seconds). Rick believes the think time has an exponential distribution. In Chapter 8 we will learn how he can test this hypothesis. It is tedious to calculate the above estimates with a four-function calculator. Fortunately, most scientific and business calculators have built-in functions to make such calculations. The EXPLORE program ANALYZ of Doane [12] can easily make these calculations. So, of course, can the MINITAB command DESCRIBE [26] and the UNIVARIATE procedure of SAS/STAT [27]. We will give examples of the use of these facilities in Section 7.3. □

Table 7.1.1. Random Sample

0.080	0.171	0.672	0.715	0.932
1.661	1.851	2.705	2.784	2.797
2.824	3.052	3.243	3.538	3.754
3.857	3.871	3.975	4.246	4.573
5.293	6.292	6.349	6.406	6.539
7.787	8.723	8.771	9.571	10.051
11.117	13.356	13.943	15.187	23.860
27.031	29.188	30.366	33.524	48.283

Two more desirable properties of estimators should be mentioned. The first of these concerns efficiency. Clearly, if we compare two unbiased estimators,

$\hat{\theta}_1$ and $\hat{\theta}_2$, the one with the smaller variance will tend to be more efficient for a given sample size. We say that an unbiased estimator $\hat{\theta}$ is the *minimum variance unbiased estimator of* θ if $\text{Var}[\hat{\theta}] < \text{Var}[\hat{\theta}_1]$ when $\hat{\theta}_1$ is any other unbiased estimator of θ. The minimum variance unbiased estimator is said to be *efficient*.

Theorem 7.1.2 *Let* $X_1, X_2, \ldots, X_n$ *be a random sample from a population determined by* X. *Then the following hold:*

(a) *If* X *is normally distributed with mean* μ *and variance* σ^2, *then the estimators* $\overline{X}$ *and* S^2 *are unbiased, consistent, minimum variance estimators of the parameters* μ *and* σ^2, *respectively.*

(b) *If* X *is Poisson with parameter (expected value)* α, *then* $\overline{X}$ *is an unbiased, consistent, estimator of* α. *It is also the minimum variance unbiased estimator of* α.

(c) *If* X *is Bernoulli with parameter* p, *then* k/n, *where* k *is the number of successes observed in* n *independent trials, is the maximum likelihood estimator of* p. (*We will define maximum likelihood estimator in Section 7.1.2.*) *It is also unbiased, consistent, and the minimum variance unbiased estimator of* p.

Proof See Hogg and Craig [18]. ■

Although unbiased estimators are desirable in many respects, there is not always one available for a particular estimate. In such cases we may consider a consistent estimator $\hat{\theta}$ with minimum mean-squared error, where we define the *mean-squared error of* $\hat{\theta}$ to be

$$E[(\hat{\theta} - \theta)^2] = \text{Var}[\hat{\theta}] + (E[\hat{\theta}] - \theta)^2. \tag{7.9}$$

The term $E[\hat{\theta}] - \theta$ is called the *bias* of $\hat{\theta}$ and, of course, is zero for unbiased estimators. It is possible for a biased estimator $\hat{\theta}$ to have a smaller mean-squared error than any unbiased estimator, if $\text{Var}[\hat{\theta}]$ is small.

We have discussed some desirable properties of estimators and have shown that, for some special populations, the sample mean $\overline{X}$ and the sample variance S^2 have many of these properties. We have, however, not given any general methods for constructing estimators. In the next two subsections we consider the two most popular techniques for constructing estimators.

A mathematician named Klein
Thought the Moebius band was divine.
  Said he, "If you glue
  The edges of two,
You'll get a weird bottle like mine."

7.1.1 Method of Moments Estimation

Suppose we are given the values $x_1, x_2, \ldots, x_n$ of a random sample taken from a population determined by a random variable X. Suppose, also, that X is characterized by k parameters $\theta_1, \theta_2, \ldots, \theta_k$, that we wish to estimate. We define the jth *sample moment* by

$$M_j = \frac{1}{n} \sum_{i=1}^{n} x_i^j, \tag{7.10}$$

for $j = 1, 2, \ldots, k$. We then equate the k sample moments and the population moments $E[X^j]$ (defined in Chapter 2), giving

$$M_j = E[X^j], \tag{7.11}$$

for $j = 1, 2, 3, \ldots, k$. The values $\hat{\theta}_1, \hat{\theta}_2, \ldots, \hat{\theta}_k$ obtained by solving the k simultaneous equations of the form (7.11) are the method of moment estimates of the parameters. We illustrate this method with some examples.

Example 7.1.2 Suppose the processing time X for an inquiry for the interactive computer system developed at Barnaby Brass has been found by Big Brass, its founder, to have a gamma distribution with parameters β and α (see Chapter 3) and that n random values of processing time $x_1, x_2, \ldots, x_n$ have been observed. We calculate M_1 and M_2 using (7.10). Then we set

$$M_1 = \mu = \frac{\beta}{\alpha}, \tag{7.12}$$

and

$$M_2 - M_1^2 = \sigma^2 = \frac{\beta}{\alpha^2}. \tag{7.13}$$

If we let $\hat{\beta}$, $\hat{\alpha}$ denote the solution to these equations, we see that

$$\hat{\beta} = \frac{M_1^2}{(M_2 - M_1^2)}, \tag{7.14}$$

and

$$\hat{\alpha} = \frac{M_1}{(M_2 - M_1^2)}. \quad \square \tag{7.15}$$

Example 7.1.3 The pattern of arrivals to the main office of Bigbucks Financial during the busiest period of the day has a Poisson distribution. The number of arrivals for each of n randomly selected 10-minute intervals has been collected yielding the values $x_1, x_2, \ldots, x_n$. Doug Dinglehoffer, the lead performance analyst, wants to estimate the mean arrival rate per 10-minute interval, $\lambda = E[X]$, as well as the standard deviation of X. Doug uses the method of moments. Therefore, he sets

$$\hat{\lambda} = M_1 = \bar{x}.$$

He also sets

$$M_2 = E[X^2] = \sigma^2 + E[X]^2, \tag{7.16}$$

so the method of moments estimate of σ is given by[3]

$$\hat{\sigma} = \sqrt{M_2 - M_1^2}. \tag{7.17}$$

The formulas we developed in Example 7.1.2 are the general formulas for estimating the mean and variance by the method of moments. That is, we always use

$$\hat{\mu} = M_1 \tag{7.18}$$

and

$$\widehat{\sigma^2} = M_2 - M_1^2 \tag{7.19}$$

for method of moments estimates of μ and σ^2. $\square$

We used the method of moments method in Example 3.2.7 to fit an Erlang-k distribution to an observed message length distribution.

The method of moments technique has the twin virtues of being intuitively satisfying as well as easy to apply in most cases. However, the method of maximum likelihood, that will be discussed next, is even more intuitively appealing. In addition it has a deeper theoretical foundation. In many cases the two methods yield the same estimators.

7.1.2 Maximum Likelihood Estimation

The idea of maximum likelihood estimation of the parameters $\theta_1, \theta_2, \ldots, \theta_k$, that characterize a random variable X, is to choose the parameter value or values that makes the observed sample values $x_1, x_2, \ldots, x_n$ most probable. We illustrate with an example before we set up the formal procedures.

---

[3]We ask you to show in Exercise 6 that

$$M_2 - M_1^2 = \left(\frac{n-1}{n}\right) S^2.$$

Thus, (7.17) becomes $\hat{\sigma} = S\sqrt{\frac{n-1}{n}}$.

Example 7.1.4 The scientists at the Big Defense Company have many application programs collected from a number of sources. Fineas Foog, the program librarian, carefully tabulated the programs and discovered that proportion p of the programs are written in the C language. Fineas lost the paper on which he had written the number p; he is sure that either $p = 0.6$ or $p = 0.8$. He decides to estimate p from a random sample. Fifteen programs are selected randomly and eight are found to be C programs. Fineas uses the maximum likelihood estimate for p. Since X, the number of the fifteen programs written in C, has a binomial distribution with parameters 15 and p, we can calculate the probability of the observed result if $p = 0.6$, and if $p = 0.8$. If $p = 0.6$, the probability we will observe eight C programs is

$$\binom{15}{8}(0.6)^8(0.4)^7 = 0.17708,$$

while, if $p = 0.8$, the probability is

$$\binom{15}{8}(0.8)^8(0.2)^7 = 0.0138.$$

Hence, Fineas estimates that $p = 0.6$, since this value has the greater probability of yielding the observed sample. $\square$

Suppose now that X is a random variable, discrete or continuous, whose distribution depends upon a single parameter θ. Let $x_1, x_2, \ldots, x_n$ be an observed random sample. If X is discrete, the probability that a random sample consists of exactly these values is given by

$$l(\theta) = p(x_1)p(x_2)\cdots p(x_n), \tag{7.20}$$

where $p(\cdot)$ is the probability mass function of X. The function l defined by (7.20) is called the *likelihood function* and is a function of θ; that is, the value of (7.20) depends both upon the selected sample values and the choice of θ. If X is continuous with density function $f(\cdot)$, then the likelihood function $l(\cdot)$ is defined by

$$l(\theta) = f(x_1)(x_2)\cdots f(x_n), \tag{7.21}$$

where f is the density function of X. The *maximum likelihood estimate of* θ is the value of θ that maximizes the likelihood function (7.20) or (7.21). If l is a differentiable function of θ, then a necessary condition for l to have a maximum value is that

$$\frac{\partial l}{\partial \theta} = 0. \tag{7.22}$$

We indicate a partial derivative in (7.22) because l depends both on θ and the sample values $x_1, x_2, \ldots, x_n$. Thus, to find the maximum likelihood estimate of θ we solve (7.22) to find the value of θ that maximizes l. If we replace the values $x_1, x_2, \ldots, x_n$ in the solution by the random sample $X_1, X_2, \ldots, X_n$, we obtain the random variable $\hat{\theta}$, that is the *maximum likelihood estimator of* θ.

If the distribution of X involves several parameters $\theta_1, \theta_2, \ldots, \theta_k$, then to find the values that maximize the likelihood function, we can solve the system of equations

$$\frac{\partial l}{\partial \theta_1} = 0, \frac{\partial l}{\partial \theta_2} = 0, \ldots, \frac{\partial l}{\partial \theta_k} = 0, \tag{7.23}$$

to determine maximum likelihood estimates for the parameters. Some care must be taken to ensure that the solution of the k simultaneous equations (7.23) maximizes l; a minimum point is also characterized by zero partial derivatives.

In many cases it is more convenient to work with $L = \ln l$, the logarithmic likelihood function. Since the logarithm function $\ln$ is a monotonically increasing function, a maximum of L is a maximum of l and vice versa. In this case we replace (7.23) by

$$\frac{\partial L}{\partial \theta_1} = 0, \frac{\partial L}{\partial \theta_2} = 0, \ldots, \frac{\partial L}{\partial \theta_k} = 0. \tag{7.24}$$

Example 7.1.5 Consider Example 7.1.4. Suppose Fineas now reports that perhaps the proportion p of C programs was *not* 0.6 or 0.8. Fineas decides to make a maximum likelihood estimate on the basis of the data. If the sample is of size n, then we can visualize the sample as a sequence of Bernoulli trials with probability p of success on each trial. Thus, if the observed number of C programs is k, then the likelihood function l is, by (7.20),

$$l(p) = p^k (1 - p)^{n-k}.$$

Thus,

$$L = \ln l = k \times \ln p + (n - k) \times \ln(1 - p),$$

and (7.24) becomes

$$\frac{\partial L}{\partial p} = \frac{k}{p} - \frac{n - k}{1 - p} = 0.$$

Solving for p yields

$$\hat{p} = \frac{k}{n}.$$

This is the result of Theorem 7.1.2(c). In this example, $n = 15$ and $k = 8$, so

$$\hat{p} = \frac{8}{15} = 0.5333. \quad \Box$$

Example 7.1.6 Pourtnoy's Complaint Service receives requests for service in a Poisson pattern and wants to estimate the average arrival rate λ from the random sample $k_1, k_2, \ldots, k_n$ of arrivals per one-minute interval. Thus,

$$l(\lambda) = \left(e^{-\lambda}\frac{\lambda^{k_1}}{k_1!}\right) \cdots \left(e^{-\lambda}\frac{\lambda^{k_n}}{k_n!}\right) = \frac{1}{k_1! \cdots k_n!}e^{-(n\lambda)}\lambda^{n\overline{k}}, \quad (7.25)$$

where

$$\overline{k} = \frac{k_1 + k_2 + \cdots k_n}{n}. \quad (7.26)$$

Thus,

$$L = \ln l = -\ln(k_1! \times \cdots \times k_n!) - n \times \lambda + n \times \overline{k} \times \ln(\lambda), \quad (7.27)$$

and (7.24) is

$$\frac{\partial L}{\partial \lambda} = -n + \frac{n \times \overline{k}}{\lambda} = 0. \quad (7.28)$$

Solving for λ yields

$$\hat{\lambda} = \overline{k}. \quad (7.29)$$

Thus, the sample mean is the maximum likelihood estimate for λ. This is the same solution we got by the method of moments in Example 7.1.3. $\quad \Box$

Example 7.1.7 Suppose the random variable X has a normal distribution with parameters μ and σ^2. If we have obtained a random sample $x_1, x_2, \ldots, x_n$, then the likelihood function is

$$\begin{aligned} l(\mu, \sigma^2) &= \left(\frac{1}{\sigma\sqrt{2\pi}}\right)\exp\left(\frac{-(x_1 - \mu)^2}{2\sigma^2}\right) \\ &\quad \times \left(\frac{1}{\sigma\sqrt{2\pi}}\right)\exp\left(\frac{-(x_2 - \mu)^2}{2\sigma^2}\right) \\ &\quad \times \cdots \times \left(\frac{1}{\sigma\sqrt{2\pi}}\right)\exp\left(\frac{-(x_n - \mu)^2}{2\sigma^2}\right) \\ &= \left(\frac{1}{2\pi\sigma^2}\right)^{n/2}\exp\left(-\sum_{i=1}^{n}\frac{(x_i - \mu)^2}{2\sigma^2}\right). \end{aligned} \quad (7.30)$$

Taking the natural logarithm of (7.30) we obtain

$$L(\mu, \sigma^2) = -\frac{n}{2} \times \ln 2\pi - \frac{n}{2} \times \ln\sigma^2 - \sum_{i=1}^{n}\frac{(x_i - \mu)^2}{2\sigma^2}. \quad (7.31)$$

Setting the partial derivatives equal to zero yields

$$\frac{\partial L}{\partial \mu} = \sum_{i=1}^{n} \frac{(x_i - \mu)}{\sigma^2} = 0, \tag{7.32}$$

and

$$\frac{\partial L}{\partial \sigma^2} = \frac{-n}{2\sigma^2} + \sum_{i=1}^{n} \frac{(x_i - \mu)^2}{2\sigma^4} = 0. \tag{7.33}$$

Solving (7.32) we obtain

$$\hat{\mu} = M_1 = \bar{x}. \tag{7.34}$$

Substituting this value for $\hat{\mu}$ into (7.33) yields

$$\widehat{\sigma^2} = \sum_{i=1}^{n} \frac{(x_i - \bar{x})^2}{n} = M_2 - M_1^2 = \frac{n-1}{n} \times S^2, \tag{7.35}$$

where the second and third equalities are proven in Exercise 6. As we previously remarked, these are the solutions we would get using the method of moments. Although $\hat{\mu}$ is unbiased, $\widehat{\sigma^2}$ is a biased estimator. $\square$

Example 7.1.8 A method has been developed to estimate the size of an animal (or fish) population by performing a capture/recapture experiment. Suppose the actual size of the population we want to estimate is N. We first capture and tag r of the animals. The r animals are then released and allowed to mix into the general population. Later, n animals are captured (or recaptured) and the number of tagged animals, k, is counted. The probability $p_k(N)$ that the second set of captured animals contains exactly k tagged animals is given by

$$P_k(N) = \frac{\binom{r}{k}\binom{N-r}{n-k}}{\binom{N}{n}}, \tag{7.36}$$

since this capture/recapture experiment can be modeled exactly by a hyper-geometric random variable with parameters n, N, and r. (See the definition of the hypergeometric random variable preceding Exercise 11 in Chapter 3.) The parameters n, N, r of our experiment correspond to the number of elements chosen without replacement, the total number of elements in the set, and the number of red elements in the set, respectively. For given values of r, k, n the value of N that maximizes $p_k(N)$ is the maximum likelihood estimate of N, designated $\widehat{N}$. To find $\widehat{N}$ we consider the ratio

$$\frac{p_k(N)}{p_k(N-1)} = \frac{(N-r)(N-n)}{N(N-r-n+k)}. \tag{7.37}$$

This ratio is larger than one if and only if $Nk < rn$ (see Exercise 7). Thus, as N increases $p_k(N)$ increases as long as $N < rn/k$; then it decreases. The value of $p_k(N)$ reaches its maximum when N is the largest integer less than or equal to rn/k, that is, when $\widehat{N} = [rn/k]$. (Recall that $[x]$ is the symbol for "the largest integer less than or equal to x.") Suppose the population of fish in a small lake is to be estimated. Fifty fish are caught, tagged, and released. Later 40 fish are caught and 4 of them are found to be tagged. The maximum likelihood estimate of the number of fish in the lake is

$$\widehat{N} = \left[50 \times \frac{40}{4} \right] = 500. \ \square$$

Feller [15, pages 45–46] discusses the technique of Example 7.1.8 in greater depth. He shows how to estimate a confidence interval for a special case. We should note that the maximum likelihood estimate $\widehat{N}$ has a simple intuitive interpretation. It is approximately true that $\widehat{N} = rn/k$ or the fraction k/n of tagged animals in the sample is equal to the fraction $r/\widehat{N}$ of tagged animals in the population.

> *And new Philosophy calls all in doubt,*
> *The Element of fire is quite put out,*
> *The Sun is lost, and the'earth and no mans wit*
> *Can well direct him, where to looke for it.*

John Donne

7.2 Confidence Intervals

In Section 7.1 we talked about some nice properties of estimators such as being unbiased, consistent, the minimum variance unbiased estimator, etc. (The reader may feel like adding "trustworthy, loyal, ..., reverent.") However, not one of these desirable properties is of any help in making a probability judgment about the quality or accuracy of the estimate delivered. The confidence interval, as you probably suspected from the title of this section, is what enables us to do that.

The idea of a confidence interval is very similar to that of an error limit in numerical analysis. If we calculate a value x and know that the error in the calculation does not exceed δ (where $\delta > 0$), then we know the true value lies between $x - \delta$ and $x + \delta$. In the case of an estimator, we are dealing with a random variable, so we cannot predict with certainty that the true value of the parameter, θ, is within any finite interval. We can, however, construct a *confidence interval*, so there is a specified confidence or probability that the true value θ lies within the interval. The 95% and 99% intervals are particularly popular, corresponding to α values of 0.05 (five percent) and 0.01 (one percent). For a given confidence level, of course, the shorter the interval, the more accurate the estimate.

The sequence of theorems that follow indicate how to calculate confidence intervals for the parameters of some random variables of interest. The theorems will be interspersed with partial proofs and examples.

Theorem 7.2.1 *Let $x_1, x_2, \ldots, x_n$ be the values of a random sample from a population determined by the random variable X that has finite mean μ and finite variance σ^2. Suppose further that either:*

(a) X is normally distributed, or

(b) n is large enough that, by Theorem 7.1.1, $\overline{X}$ is approximately normally distributed.

Then, if we assume σ is known, the $100(1 - \alpha)$ percent confidence interval for μ is given by

$$\overline{x} \pm E, \tag{7.38}$$

where

$$E = z_{\alpha/2} \times \frac{\sigma}{\sqrt{n}}. \tag{7.39}$$

To ensure that the width of the confidence interval does not exceed w, choose

$$n \geq \left(\frac{2 \times z_{\alpha/2} \times \sigma}{w} \right)^2. \tag{7.40}$$

Proof If the stated conditions are true, $(\overline{X} - \mu)/(\sigma/\sqrt{n})$ has (at least approximately) a standard normal distribution so that

$$P[-z_{\alpha/2} \leq \frac{\overline{x} - \mu}{\frac{\sigma}{\sqrt{n}}} \leq z_{\alpha/2}] = 1 - \alpha, \tag{7.41}$$

by the symmetry of the normal distribution. A little manipulation of (7.41) yields

$$P[\overline{x} - z_{\alpha/2} \times \frac{\sigma}{\sqrt{n}} \leq \mu \leq \overline{x} + z_{\alpha/2} \times \frac{\sigma}{\sqrt{n}}] = 1 - \alpha, \tag{7.42}$$

that proves (7.38) and (7.39). To prove (7.40) set

$$w = 2 \times E,$$

and solve for n. ∎

(Recall that by the definition in Chapter 3, z_α is the value of z such that $P[Z > z_\alpha] = \alpha$, where Z is the standard normal random variable.) The value $z_{\alpha/2}$ can be picked out of Table 3 in Appendix A or chosen from Table 7.2.1. Of course it can also be calculated by the APL function IN, the MINITAB command INVCDF with the subcommand NORMAL, or the SAS/STAT function PROBIT. It can even be calculated by the HP-21S or similar calculator. The reader is cautioned that to use Table 3, one sets $z_{\alpha/2}$ to the value of z that yields the value $1 - \alpha/2$ in the table. Thus, for a 95 percent confidence interval, $\alpha = 0.05$ and $z_{\alpha/2} = z_{0.025} = 1.96$, since this is the z value corresponding to the Table 3 value of $1 - 0.025 = 0.975$.)

Table 7.2.1. Values of $z_{\alpha/2}$

$1 - \alpha$	0.900	0.950	0.990	0.999
$z_{\alpha/2}$	1.645	1.960	2.576	3.291

The main problem with putting Theorem 7.2.1 into practice is that sometimes we know *neither* the mean μ nor the variance σ^2. In these cases, when n is fairly large, we can use s in place of σ in (7.39). When n is small, but X is normally distributed, we can use Theorem 7.2.2, that requires only numbers that can be calculated from the sample.

Example 7.2.1 A random sample of 225 interactive response times measured at user terminals at Bonanza Banana yields $\bar{x} = 7$ jerks (a jerk is a proprietary time unit whose value in seconds is known only to Barry Blast, Chief Systems Analyst) with a sample standard deviation of 3 jerks. Find a 95% confidence interval for the interactive response time.

Solution $\overline{X}$ is approximately normal because the sample size is relatively large; s is a good approximation to σ. As we showed above, $z_{\alpha/2} = z_{0.025} = 1.96$. Hence, by (7.39),

$$E = 1.96 \times \frac{3}{15} = 0.392 \text{ jerks.}$$

Therefore, the 95% confidence interval is 7 ± 0.392 jerks. $\square$

Example 7.2.2 Consider Example 7.2.1. Suppose Barry does not want the length of the 95% confidence interval to exceed 0.5 jerks. How large should his sample be? (Barry can enlarge the random sample of 225 values he has now. The current length of the interval is $2 \times 0.392 = 0.784$.)

Solution Barry uses (7.40) to obtain

$$n = \left(\frac{2 \times 1.96 \times 3}{0.5}\right)^2 = 553.19.$$

Thus, Barry needs a sample of size 554; he must randomly select 329 more response times. □

Theorem 7.2.2 *Suppose $x_1, x_2, \ldots, x_n$ are the values of a random sample from a population determined by a normally distributed random variable X with unknown mean and variance. Then the $100(1 - \alpha)\%$ confidence interval for the mean of X is given by*

$$\bar{x} \pm E, \tag{7.43}$$

where

$$E = t_{n-1,\alpha/2} \times \frac{s}{\sqrt{n}}, \tag{7.44}$$

and $t_{n-1,\alpha/2}$ is defined by $P[T > t_{n-1,\alpha/2}] = \alpha/2$, where T has a Student's t distribution with $n - 1$ degrees of freedom.

Proof The theorem follows immediately from the fact that

$$(\overline{X} - E[X])/(S/\sqrt{n})$$

has a Student's t distribution with $n - 1$ degrees of freedom. This was part of Theorem 7.1.1. ■

The value of $t_{n-1,\alpha/2}$ can be obtained by calculation using MINITAB, SAS/STAT, or the HP-21S or similar calculator. It can also be obtained from Table 5 in Appendix A. To find $t_{n-1,\alpha/2}$ using Table 5, look under the column for α using the value of $\alpha/2$, and in the row labeled with an n value of $n - 1$. This is illustrated in Example 7.2.3.

Example 7.2.3 Claude Chandon, the Chief Performance Seer at Tipplers Vineyards, believes X, the number of interactive message buffers in use during the peak period on the computer he is studying, has a normal distribution. He takes a random sample of 9 values, which yields $\bar{x} = 120$ and $s = 10$. Assuming Claude is right about the normality of X, find a 99% confidence interval for the mean number of buffers in use.

Solution For a 99% confidence interval $\alpha = 0.01$. By Table 5, $t_{8,0.005} = 3.355$. Hence, by (7.44),

$$E = 3.355 \times \frac{10}{3} = 11.18 \text{ buffers,}$$

so the 99% confidence interval for mean number of buffers in use is 120 ± 11.18 buffers. □

Theorem 7.2.2 is much more satisfying than Theorem 7.2.1, because we need deal only with entities that we can measure and calculate; for Theorem 7.2.1 we had to assume a knowledge of σ that we may not have. It is common when applying Theorem 7.2.2 to have a large confidence interval corresponding to a small sample. Uncertainty is to be expected if we have little information to base our estimate upon.

Figure 7.2.1 illustrates the relationship between 10 typical large sample 95% confidence intervals for a mean μ and the true value of the mean. Note that one of the confidence intervals fails to contain the mean.

The next theorem tells us how to construct a confidence interval for the variance of a normally distributed random variable.

$$\mu - 1.96\frac{\sigma}{\sqrt{n}} \quad \mu \quad \mu + 1.96\frac{\sigma}{\sqrt{n}}$$

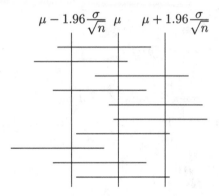

Figure 7.2.1. Ten confidence intervals.

Theorem 7.2.3 *Let $X_1, X_2, \ldots, X_n$ be a random sample from a population determined by a normally distributed random variable X with mean μ and variance σ^2. Then the random variable $Y = (n-1)S^2/\sigma^2$ has a chi-square distribution with $n-1$ degrees of freedom. Consequently, the $100(1-\alpha)\%$ confidence interval for σ^2 is given by*

$$\frac{(n-1) \times s^2}{\chi^2_{n-1,\alpha/2}} \leq \sigma^2 \leq \frac{(n-1) \times s^2}{\chi^2_{n-1,1-\alpha/2}}. \tag{7.45}$$

$\chi^2_{n-1,\alpha/2}$ and $\chi^2_{n-1,1-\alpha/2}$ can be determined from Table 4 in Appendix A. They can also be calculated using MINITAB, SAS/STAT, or the HP-21S or similar calculator.

Proof We omit the proof that can be found in Kreyszig [21]. ∎

Example 7.2.4 Hazunga Enterprises has collected the mean number of lines of code per programmer-day for 30 large programming projects. The number of lines per programmer-day, X, has a normal distribution. If $\bar{x} = 75$ and $s^2 = 90$, find 95% confidence intervals for μ and σ^2.

Solution By Table 4 of Appendix A,

$$\chi^2_{29,0.975} = 16.047,$$

and

$$\chi^2_{29,0.025} = 45.722.$$

Hence, by (7.45), the 95% confidence interval for σ^2 is given by

$$57.984 \le \sigma^2 \le 162.647.$$

By Theorem 7.2.2, the 95% confidence interval for μ is given by $\bar{x} \pm E$, where

$$E = t_{n-1,\alpha/2} \times \frac{s}{\sqrt{n}} = 3.542.$$

Hence, the 95% confidence interval for μ is given by

$$71.458 \le \mu \le 78.542. \quad \square$$

This large confidence interval for σ^2 is typical of the results given by Theorem 7.2.3. It takes a very large sample to produce a narrow confidence interval for σ^2.

Another important parameter we may want to estimate is the probability, p, of success in one Bernoulli trial. Equivalently, p is sometimes called the *true proportion of a population*. In the latter case, we have in mind a population in which each member may or may not have a certain attribute. We call those members that have the attribute "successes" and those that don't, "failures." From a random sample of size n, we estimate p by counting the number of successes, k. By Example 7.1.5 we know the maximum likelihood estimate $\hat{p}$ is k/n. The next theorem shows how to construct a confidence interval for p.

Theorem 7.2.4 (*Large-sample confidence interval for p*). *An approximate $100(1 - \alpha)\%$ confidence interval for the Bernoulli parameter p (true proportion of a population) is given by*

$$\hat{p} - z_{\alpha/2}\sqrt{\frac{\hat{p}(1 - \hat{p})}{n}} < p < \hat{p} + z_{\alpha/2}\sqrt{\frac{\hat{p}(1 - \hat{p})}{n}}, \qquad (7.46)$$

where

$$\hat{p} = \frac{k}{n}.$$

(Here k is the number of successes in a random sample of size n.) To ensure the confidence interval has width not exceeding w, choose

$$n \geq \frac{4z_{\alpha/2}^2 \times \hat{p}(1 - \hat{p})}{w^2}. \tag{7.47}$$

Proof By the central limit theorem (Theorem 3.3.1),

$$\frac{\hat{p} - p}{\sqrt{\frac{\hat{p}(1 - \hat{p})}{n}}} \tag{7.48}$$

has approximately a standard normal distribution. Hence,

$$P[-z_{\alpha/2} < \frac{\hat{p} - p}{\sqrt{\frac{\hat{p}(1 - \hat{p})}{n}}} < z_{\alpha/2}] = 1 - \alpha. \tag{7.49}$$

A little algebra applied to (7.49) yields (7.46). By (7.46), the width of the confidence interval w is given by

$$w = 2 \times z_{\alpha/2} \sqrt{\frac{\hat{p}(1 - \hat{p})}{n}}. \tag{7.50}$$

Solving (7.50) for n yields (7.47). ∎

Example 7.2.5 NSPSP (the National Society for Professional Scientific Programmers) obtains a random sample of 100 professional scientific programmers that contains 18 who consider themselves primarily C programmers. Find the 95% confidence interval for the true proportion of C programmers among all professional scientific programmers. How large a sample is required so that the confidence interval will not exceed 0.05 in length?

Solution Since $\hat{p} = 0.18$, (7.46) yields the confidence interval $0.1047 \leq p \leq 0.2553$ for p. Unfortunately, the width of this confidence interval is 0.1506, almost as large as $\hat{p}$ itself. To decrease the width of the confidence interval to 0.05 requires, by (7.47), $n \geq 907.23$, that is, a total sample size of at least 908. Hence, NSPSP should increase their sample size by 808. □

We summarize our confidence interval calculations in Table 7.2.2.

Table 7.2.2. Confidence Intervals

Parameter	Assumptions	Endpoints
μ	$N(\mu, \sigma^2)$ or n large σ^2 known	$\overline{x} \pm z_{\alpha/2} \dfrac{\sigma}{\sqrt{n}}$
μ	$N(\mu, \sigma^2)$ σ^2 unknown	$\overline{x} \pm t_{n-1, \alpha/2} \dfrac{s}{\sqrt{n}}$
$\mu_1 - \mu_2$	Independent Distributions σ_1^2, σ_2^2 known n_1, n_2 large	$\overline{x} - \overline{y} \pm z_{\alpha/2} \sqrt{\dfrac{\sigma_1^2}{n_1} + \dfrac{\sigma_2^2}{n_2}}$
$\mu_1 - \mu_2$	Independent Normal Distributions σ_1^2, σ_2^2 unknown but equal	$\overline{x} - \overline{y} \pm t_{n_1 + n_2 - 2, \alpha/2}$ $\times \sqrt{\dfrac{(n_1 - 1)s_x^2 + (n_2 - 1)s_y^2}{n_1 + n_2 - 2} \left(\dfrac{1}{n_1} + \dfrac{1}{n_2} \right)}$
p	Binomial $B(n, p)$ n large	$\dfrac{k}{n} \pm z_{\alpha/2} \sqrt{\dfrac{(k/n)(1 - k/n)}{n}}$

standard deviation *n. A sexual activity formerly considered perverted
but now universally practiced and accepted.*
Stan Kelly-Bootle
The Devil's DP Dictionary

*All the problems become smaller if you don't dodge them, but confront
them. Touch a thistle timidly, and it pricks you; grasp it boldly, and its
spines crumble.*
Admiral William Halsey

You can observe a lot just by watching.
Yogi Berra

7.3   Exploratory Data Analysis

> *Exploratory data analysis (EDA) seeks to reveal structure, or simple*
> *descriptions, in data. We look at numbers or graphs and try to find*
> *patterns. We pursue leads suggested by background information,*
> *imagination, patterns perceived, and experience with other*
> *data analyses.*
>
> Persi Diaconis[4]

So far in this chapter we proceeded as though we knew exactly what we wanted to do. We assumed we had a random sample, $x_1, x_2, \ldots, x_n$, determined by X, that had a known distribution type (exponential, normal, gamma, etc.); all we needed to do was estimate μ and σ. However, our sample is probably not completely random. In addition, we may have great doubts about the distribution of X. This may lead us into what is often called *exploratory data analysis* (abbreviated EDA) after the title of John Tukey's book, Tukey [30]. It is also known as *preliminary data analysis* by Cox and Snell [10] and *the initial examination of data* by Chatfield [8]. Whatever we call it, the purpose of EDA is to answer some new cosmic questions (not to be confused with the cosmic questions at the beginning of the chapter).

More Cosmic Questions

4. How do we screen the sample (data) for errors, outliers, and missing observations?

5. How can we use descriptive statistics to analyze the data?

6. What visual methods are available to help answer cosmic questions 4 and 5?

These three cosmic questions inevitably lead us to the following cosmic question:

7. What is a histogram, a stem-and-leaf plot, and a box-and-whiskers plot; how are they used?

My favorite visual display for analyzing data is the histogram. However, to be effective it must be supplemented by other displays and by descriptive statistics. It is best to use all the EDA display techniques in concert, especially if you have a computer available to do the computations. Statistics

---

[4]Quoted with permission from *Exploring Data Tables, Trends, and Shapes*, edited by David C. Hoaglin, Frederick Mosteller & John W. Tukey, John Wiley, New York, 1985.

books tend to make histograms seem more complicated than necessary with rules about how many classes you should have, how to choose class boundaries, etc. Therefore, it is convenient to have a good computer program to make choices for you. The better programs allow you to override the computer choices with your own if you are not satisfied with the initial results. Doane's statistical system EXPLORE, Doane [12], has a routine ANALYZ, that produces useful descriptive statistics as well as beautiful histograms. The histogram algorithm uses Sturges' rule (usually), that says that the ideal number of classes is given by

$$1 + \log_2(n). \tag{7.51}$$

For its histogram setup procedure ANALYZ produces the screen shown in Figure 7.3.1. If Option 1 is chosen, ANALYZ produces a histogram automatically. In Figure 7.3.2 we show the result of choosing Option 2 assuming the data under analysis is that of Table 7.3.1. If one chooses the options recommended by ANALYZ, then the histogram of Figure 7.3.2 is produced, just as though Option 1 had originally been selected.

We have found that an effective way to get pleasing histograms is to first choose Option 1. If not satisfied with the histogram produced automatically by ANALYZ, then use Option 2 to improve the first histogram. Option 2 can be applied over and over again until a satisfactory histogram is found; rarely are more than two iterations required.

A histogram is used to display information about a sample of data by dividing the data into classes based on the magnitude of the data elements. Typically, the number of classes ranges from 5 to 20. The class intervals determine to which class each data element belongs. The height of the histogram over a class interval represents either the *frequency* (number of data elements in the class) or the *relative frequency* (number of elements divided by n). For the histogram of Figure 7.3.2 ANALYZ chose 6 classes, although Sturges' rule indicates that 7 classes should be used. The 6 classes are each of width 2 with the class boundaries falling at $0, 2, \ldots, 12$. MINITAB (Figure 7.3.3) for the same data chose 11 class intervals, each of width 1, with centers at $0, 2, \ldots, 10$. SAS/STAT (Figures 7.3.6 and 7.3.7) chose 6 class intervals (the same number chosen by ANALYZ), each of width 2, with centers at $0, 2, \ldots, 10$. Unfortunately, the point 10.23 fails to be registered in Figure 7.3.7, although it is counted in Figure 7.3.6.

Example 7.3.1 Kari Grant, a senior performance consultant, is helping Gerard's Gigantic Gilded Guernsey Dairy improve the interactive performance of their central computer system. Kari collected the random sample of 50 response times measured at user terminals shown in Table 7.3.1. Kari has access to the MINITAB statistical system on a mainframe; she also has the EXPLORE programs of Doane [12] and SAS/STAT available on her personal computer. Figure 7.3.2 is the histogram Kari made of her data using the EXPLORE ANALYZ program. Figure 7.3.4 contains a set of descriptive statistics that Ms. Grant produced using ANALYZ with the same data. Figure 7.3.5 is a similar printout of descriptive statistics provided as part of the output of the SAS/STAT procedure UNIVARIATE. Figure 7.3.3 is the histogram she made using MINITAB. The histogram of Figure 7.3.2 clearly shows the characteristic shape of an exponential distribution. Figure 7.3.3 looks like it could be exponential, but the cluster of values with midpoint 5 is surprising. Kari decided to generate a histogram using SAS/STAT. SAS/STAT has a procedure CHART, that can generate an ordinary histogram of the type seen in most statistics books with vertical bars like those in a prison. SAS calls it a "frequency bar chart." Kari used CHART to generate the conventional histogram shown in Figure 7.3.7. The procedure CHART can also produce another type of frequency bar chart that looks like most computer generated histograms; that is, it has horizontal bars like venetian blinds. Ms. Grant produced such a frequency bar chart for the data of Table 7.3.1 using CHART; it appears in Figure 7.3.6. This latter histogram provides more information than the vertical histogram of Figure 7.3.7. It is elegant, with the midpoints explicitly listed and some extra useful information appended. Somehow, though, it does not look as exponential as Figure 7.3.2. (The collection of data in Table 7.3.1 *is* a sample from an exponential distribution.)

Kari decides to make some stem-and-leaf plots, too. Figure 7.3.8 is the stem-and-leaf plot she generated using MINITAB. Figure 7.3.9 is the stem-and-leaf plot she got from the SAS/STAT UNIVARIATE procedure. It does not provide as much information as the MINITAB plot, but a rather anemic looking box-and-whiskers plot is thrown in. Kari decides that the sample is probably exponential, but withholds final judgment until she can perform some of the goodness-of-fit tests we will discuss in Chapter 8. □

HISTOGRAM SET-UP
To construct the histogram, shall the computer:
1. make all decisions automatically
2. offer advice, but leave decisions to you
3. leave decisions to you entirely
? 1
Figure 7.3.1. Histogram set up by ANALYZ.

Table 7.3.1. Response Time

3.93	1.63	0.40	1.13	0.30
0.21	0.19	2.60	2.24	0.24
10.23	4.69	3.47	0.16	6.54
0.29	0.80	0.65	0.72	7.15
0.22	1.61	2.38	4.69	0.30
4.79	4.93	0.24	1.52	2.41
0.24	4.75	1.39	2.22	0.79
0.63	1.82	3.45	0.22	0.94
5.21	0.26	3.74	1.28	0.50
3.25	1.25	0.13	2.87	0.38

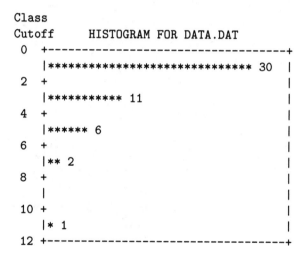

```
Class
Cutoff       HISTOGRAM FOR DATA.DAT
  0   +----------------------------------+
      |**************************** 30   |
  2   +                                  |
      |*********** 11                    |
  4   +                                  |
      |****** 6                          |
  6   +                                  |
      |** 2                              |
  8   +                                  |
      |                                  |
 10   +                                  |
      |* 1                               |
 12   +----------------------------------+
```

Figure 7.3.2. Histogram produced by ANALYZ.

```
MTB > histogram c1

Histogram of C1   N = 50

Midpoint   Count
      0      15   ***************
      1      11   ***********
      2       8   ********
      3       5   *****
      4       2   **
      5       6   ******
      6       0
      7       2   **
      8       0
      9       0
     10       1   *
```

Figure 7.3.3. Histogram produced by MINITAB.

GENERAL FACTS: DISPERSION:

File = DATA.DAT   Variance = 4.842021 (4.745181)
Observations =   50   St. Dev. = 2.200459 (2.178344)
Minimum = .13   Coef. of Var. = 103.8 (102.8)
Maximum = 10.23   Avg. Dev. About Mean = 1.72592
Range= 10.1   Avg. Dev. About Median = 1.6208

CENTRAL TENDENCY: SKEWNESS AND KURTOSIS:

Mean = 2.1196   2nd Moment = 4.74518
Median = 1.335   3rd Moment = 15.32232
Geom. Mean = 1.139498   4th Moment = 118.5946
1st Quartile = .3   Skewness =1.482 (Pearson Beta 1)
2nd Quartile = 1.335   Kurtosis =5.267 (Pearson Beta 2)
3rd Quartile = 3.45   1 Outlier(s) Detected

Note: Standard deviation using N instead of N-1 shown in
parentheses.

Figure 7.3.4. Statistics produced by ANALYZ.

                              Moments

       N   50   Sum Wgts   50
       Mean   2.1196   Sum   105.98
       Std Dev   2.200459   Variance   4.84202
       Skewness   1.528577   Kurtosis   2.641456
       USS   461.8942   CSS   237.259
       CV   103.8148   Std Mean   0.311192
       T:Mean=0   6.811231   Prob>|T| 0.0001
       Sgn Rank   637.5   Prob>|S| 0.0001
       Num ^= 0   50

Figure 7.3.5. Statistics produced by UNIVARIATE.

                          FREQUENCY OF TIME

        TIME   CUM   CUM
      MIDPOINT   FREQ   FREQ   PERCENT   PERCENT
            |
         0 |********************** 22   22   44.00   44.00
         2 |**************   14   36   28.00   72.00
         4 |**********   10   46   20.00   92.00
         6 |**   2   48   4.00   96.00
         8 |*   1   49   2.00   98.00
        10 |*   1   50   2.00   100.00
            ----+----+----+----+--
            5   10   15   20

Figure 7.3.6. SAS/STAT horizontal frequency bar chart.

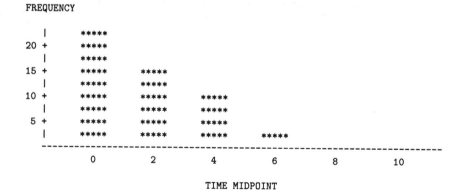

Figure 7.3.7. SAS/STAT vertical frequency bar chart.[5]

---

[5]I am unable to persuade SAS/STAT to plot all the points.

The stem-and-leaf plot or display originated as a kind of "poor man's histogram." It was originally used to hand order a collection of data into a display that looks somewhat like a histogram; it also makes the data easy to analyze. To construct a stem-and-leaf plot proceed as follows:

1. Split each data value into two sets of digits. The leading (first) set of digits is the *stem*, and the trailing (second) set of digits is the *leaf.*

2. List all stem digits from lowest to highest. (SAS/STAT (see Figure 7.3.9) lists them in the opposite order but that is uncommon.)

3. For each data element write the leaf numbers on the line labeled by the appropriate stem number. The leaves should be listed in increasing order.

Note that the stem can consist of two or more digits. It is a good idea to list the leaf unit above the display as MINITAB did in Figure 7.3.8. In that way one can tell the size of the numbers in the plot. It is common to list the number of leaves on each line as SAS/STAT does. Instead, MINITAB lists the *depth* of the line on each line. The depth is the total number of leaves either on that line or on the other lines toward the nearest end of the plot (unless the depth is in parentheses). The line with parentheses around the depth contains the middle observation, if n is odd, and the middle two values, if n is even. The depth listed on this line is merely the number of observations on the line. The 8 in parentheses on line 2 indicates that the second line contains the middle two observations (observations 25 and 26); also indicated is the fact that there are 8 values on this line. The 9 on the fifth line indicates there are 9 items total on the fifth line and on higher numbered lines. (Actually, there are 5 items on line 5 and one each on lines 6, 7, 8, and 11.) It is also common to drop the trailing digits of each leaf, as is done in Figure 7.3.8 and Figure 7.3.9. The stem-and-leaf plot has some advantages over the histogram. Like the histogram it provides the "shape" of a distribution; in fact, a stem-and-leaf display looks a lot like a histogram. However the stem-and-leaf display provides additional information, such as:

- Where the values are concentrated.

- How wide the range of values is.

- How symmetrical or nonsymmetrical the data is.

- Whether there are gaps in the data.

- Whether there are outliers.

- In some cases the actual data values (when the trailing digits of the leaf are not dropped).

For large collections of data, stem-and-leaf plots become complicated and unwieldy, but for small sets of data a stem-and-leaf plot is superb.

```
MTB > stem-and-leaf of c1

Stem-and-leaf of C1        N  = 50
Leaf Unit = 0.10


     22     0 1112222222233345667789
     (8)    1 12235668
     20     2 223468
     14     3 24479
      9     4 66779
      4     5 2
      3     6 5
      2     7 1
      1     8
      1     9
      1    10 2
```

Figure 7.3.8. MINITAB stem-and-leaf plot.

```
Stem Leaf                                  #        Boxplot
  10 2                                     1           0
   8
   6 52                                    2           |
   4 778892                                6           |
   2 22446935579                          11        +--+--+
   0 122222222333344567788913345668       30        *-----*
     ----+----+----+----+----+----+
```

Figure 7.3.9. SAS/STAT stem-and-leaf plot.

Example 7.3.1 should give you an idea of some things you can do easily with computer statistical systems.

The most important activity of professionals is *pattern recognition*. A great performance analyst can quickly examine the measurements taken at a computer installation and detect patterns that tell her how well the installation is operating; she immediately knows if changes are needed and, if so, *what* changes are necessary. Similarly, a statistician with some knowledge of the field under study can deduce a lot just by looking at the data; particularly by using histograms, stem-and-leaf plots, and box-and-whiskers plots. The patterns stand out *if you know what you are looking for.*

The final exploratory analysis display tool we will discuss is the box-and-whiskers plot, often called the *boxplot*. I prefer the former designation because it has more pizzazz (sizzle). But then I would *never* call garbanzo beans (how romantic!) chick peas (ughh!). The box-and-whiskers plot provides more information than the histogram and the stem-and-leaf plot. However, in order to understand this information we need some new definitions.

In the definitions to follow, we assume we are talking about a sample of n values, say $x_1, x_2, \ldots, x_n$.

> *The race is not always to the swift,*
> *nor the battle to the strong,*
> *but that's the way to bet.*
> Damon Runyon

7.3.1   Measures of Central Tendency

Recall from the first part of this chapter that the *sample mean*, $\bar{x}$, sometimes called the *arithmetic mean*, is the sum of the sample values divided by the number of values; that is, by

$$\bar{x} = \frac{1}{n} \sum_{i=1}^{n} x_i. \tag{7.52}$$

For the sample of Table 7.3.1 we see by Figure 7.3.4 that $\bar{x} = 2.1196$. Formula (7.52) generates the infamous class average that we have all been concerned about when we took an exam.

Exploratory data analysts like John Tukey tend to emphasize the *sample median* (the word "sample" is often left out) that, roughly speaking, is the middle observation; that is, the observation that is greater than half the observations and less than the other half. To pin it down exactly, we must consider two cases. If the size of the sample is odd, the median is the middle measurement when the sample is listed in increasing order. Thus, if the sample is 25, 27, 457, the median is 27. (The median is then the $[n/2 + 1/2]$th observation from either end.) If the size of the sample is even, the median is the mean of the two middle values when the sample is listed in increasing order. If the sample is 25, 27, 29, 457, then the median is $(27 + 29)/2 = 28$. (When n is even, the median is the average of the $n/2$th and the $[n/2 + 1]$st values.)

The *mode* is another measure of central tendency that is not used very much in science and engineering. The mode of a sample is the observation that occurs most often. It is not necessarily unique. For example, if the sample consists of 25 *1*'s and 25 *2*'s, it has two modes. It is *bimodal*. If there were three modes it would be *trimodal*, etc. The set of data in Table 7.3.1 is *unimodal* with a mode of 0.24.

The sample mean, $\bar{x}$, is the preferred measure of central tendency for most statisticians because it is the best estimator of the population mean, μ, and because more useful theory has been developed for dealing with it than the other measures. EDA people prefer the median because it is more resistant to outliers. An *outlier* is an extreme value (much larger or much smaller than the remainder of the observations). Outliers must be checked to see if they are correct or an error was made in collecting the data. It is not always clear whether or not an outlier *could* be a bona fide observation. Entire books have been written on the subject. (See, for example, Barnett and Lewis [3].) Statisticians tend to suspect that any data element more than 3 standard deviations from the mean is an outlier. Snedecor and Cochran [29] list several statistical measures that can be used to determine whether or not a sample observation from normally distributed data is an outlier.

Two other important measures of central tendency, the harmonic mean and the geometric mean, are discussed later in this chapter.

> *The power of imagination makes us infinite.*
> John Muir

7.3.2   Measures of Spread or Dispersion

The mean, median, or mode doesn't by itself tell the whole story. For example, consider two very different samples that have the same mean and median: the first consists of the single observation 1000 and the second consists of the two observations 500 and 1500. Both have a mean that is equal to the median and both have a mean of 1000 but the samples are very different. We need some measure of the spread of the distribution about the mean or median. The simplest measure is the range. The *range* of a sample is the difference between the largest and smallest values. The range for the data of Table 7.3.1 is $10.23 - 0.13 = 10.1$.

Another important characterization of spread is given in terms of *percentiles*, that we discussed in Chapter 3 for random variables. The *p*th *percentile* observation of a sample is the number such that *p*% of the values do not exceed it and $(100 - p)$% are greater. The 90th percentile value of a sample is the observation that divides the top 10% from the remaining 90%. Percentile values are commonly used to report the results of aptitude tests. The 25th, 50th, and 75th percentiles are also known as the *lower quartile*, *middle quartile* (median), and *upper quartile*, respectively. We will see them marked in our box-and-whiskers plots.

The *interquartile range* is the distance between the lower (first) quartile and the upper (third) quartile. It is an important measure of spread in terms of quartiles and is easy to see on a box-and-whiskers plot.

We have frequently used the *sample variance* s^2 of a sample, which is the sum of the squared deviations from the sample mean divided by $n - 1$; that is,

$$s^2 = \frac{1}{n-1} \sum_{i=1}^{n} (x_i - \overline{x})^2. \tag{7.53}$$

Naturally, as we have stated earlier in this chapter, s is a good estimate of the population standard deviation, σ. It has been learned empirically, that is, by trial and error, that a ballpark approximation of s is given by

$$s \approx \frac{\text{range}}{4}. \tag{7.54}$$

For the data of Table 7.3.1 (7.54) yields $s \approx 10.1/4 = 2.525$, which is not far from the correct value of 2.200459.

The sample standard deviation is the most useful measure of spread for performing traditional statistical tests.

The final measure of spread is the *sample coefficient of variation*. It is defined as the positive square root of the *sample squared coefficient of variation*, that is defined by

$$\widehat{C^2} = \frac{s^2}{\overline{x}^2}, \tag{7.55}$$

where, of course, we assume $\overline{x} \neq 0$. The value of the coefficient of variation (often written CV) is usually given in percent by statistical packages. (Figure 7.3.4 shows that ANALYZ labels the coefficient of variation "Coef. of Var.", and gives its value as 103.8 percent. The SAS/STAT procedure UNIVARIATE output in Figure 7.3.5 gives it the value 103.8148 percent and labels it CV.) The squared coefficient of variation is a favorite statistic of certain queueing theory aficionados. Some of them attribute mystical powers to this statistic.

Most people are about as happy as they make up their minds to be.
Abraham Lincoln

7.3.3 Measures of Shape

Let us now consider the shape of the distribution represented by our sample. Statisticians (and almost everyone else) tend to think of the bell-shaped normal distribution as the *normal* distribution; that is, the goodness of any distribution is measured relative to how much it looks like the normal distribution. The density function of a normal distribution is perfectly symmetrical about the mean (the point right under the middle of the bell). Any distribution that is not symmetrical about the mean is said to be *skewed*. Figure 7.3.10 displays a distribution that is skewed to the right. A distribution is skewed to the right if the mean is larger than the median and to the left if the mean is smaller than the median. The *coefficient of skewness*, often called, simply, *skewness* measures the lack of symmetry of a distribution or of a sample from a distribution. Naturally, the normal distribution has coefficient of skewness zero.

The coefficient of skewness of a random variable is defined by

$$\frac{E[(X - \mu)^3]}{\sigma^3}, \tag{7.56}$$

where $\mu = E[X]$. The ANALYZ program estimate of the skewness for the data of Table 7.3.1 is 1.482 (shown in Figure 7.3.4). The SAS/STAT estimate provided by UNIVARIATE is 1.528577 (shown in Figure 7.3.5). (We ask you to prove in Exercise 12 that an exponential random variable has skewness two.[6]) A positively skewed distribution has a positive skewness coefficient just as a negative skewness coefficient indicates a distribution that is skewed to the left; naturally, the skewness of a symmetrical distribution is zero.

Another measure of shape is *kurtosis* or the *kurtosis coefficient*. Many statisticians use the following definition for the kurtosis of a random variable X:

$$\frac{E[(X - \mu)^4]}{\sigma^4}. \tag{7.57}$$

---

[6]Skewness estimators are very inaccurate for skewed data. I used SAS/STAT to generate 5 independent random samples from an exponential population, each of size 3,200. The SAS/STAT procedure UNIVARIATE was applied to each sample. The skewness values provided ranged from 1.689568 to 2.072886. The estimate closest to the population skewness was 2.025028. The kurtosis estimates were between 3.584992 and 6.461421; the estimate closest to the population kurtosis used by SAS/STAT for exponential distributions was 6.333024. (SAS/STAT uses the convention that an exponential random variable has kurtosis 6.)

Snedecor and Cochran [29] define the kurtosis of a random variable X to be

$$\frac{E[(X - \mu)^4]}{\sigma^4} - 3, \tag{7.58}$$

so that the kurtosis of a normal distribution is zero. SAS/STAT uses the same convention.

A normal distribution has kurtosis zero according to (7.58) but three according to (7.57). One must be careful in interpreting the kurtosis obtained from a computer statistical system. SAS/STAT estimated the kurtosis for the data of Table 7.3.1 as 2.641456; the EXPLORE program ANALYZ estimated it as 5.267. As we mentioned earlier, the random sample in Table 7.3.1 is from an exponential population with kurtosis nine by (7.57) but six by (7.58). You may wonder exactly what kurtosis measures. According to the *SAS Procedures Guide for Personal Computers* [27] and A. S. C. Ehrenberg [14] it measures "heaviness of tails" or differences from the normal distribution in the proportion of the values that fall a long way from the mean. Doane [12] provides a practical way of making that determination from reasonably large samples in Appendix G of his book.

The density of the Erlang-2 distribution is skewed to the right.
The mean is to the right of the median.

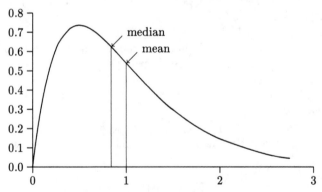

Figure 7.3.10. Positively Skewed Distribution

Box-and-Whiskers Plots

Example 7.3.1 (continued) Now that you have been exposed to all the necessary definitions, you may want to look at some box-and-whiskers plots. Kari Grant of Example 7.3.1 made the MINITAB box-and-whiskers plot of Figure 7.3.11 (she also used the DESCRIBE command to make the information more complete). Ms. Grant constructed the box-and-whiskers plot of Figure 7.3.12 using the EXPLORE program BOXPLOT. She also provided the table from the SAS/STAT UNIVARIATE Procedure shown in Figure 7.3.13 to complement the somewhat sparse box-and-whiskers plot that was part of Figure 7.3.9. The box in a box-and-whiskers plot is constructed by drawing a box between the lower quartile and the upper quartile. A line is drawn across the box at the median so one can tell at a glance which way the sample is skewed. The two whiskers point to the extreme values. Thus, a cursory examination of a box-and-whiskers plot will tell you

1. the upper and lower quartile values, Q_3 and Q_1.

2. the interquartile range $Q_3 - Q_1$.

3. the most extreme values (lowest and highest). They are pointed to by the whiskers.

4. the symmetry or asymmetry of the sample.

The MINITAB box-and-whiskers plot of Figure 7.3.11 is an extended form of the standard boxplot that was developed by Velleman and Hoaglin[7] and is discussed in their book, Velleman and Hoaglin [31]. It does not use whiskers to mark the extreme elements.

```
MTB > describe  c1

            N      MEAN    MEDIAN    TRMEAN    STDEV    SEMEAN
C1         50     2.120     1.335     1.854    2.200     0.311

           MIN       MAX        Q1        Q3
C1       0.130    10.230     0.300     3.455

MTB > boxplot c1
```

```
        ----------------
    -I      +         I-------------------                 *
        ----------------
--+---------+---------+---------+---------+---------+---------+----C1
0.0       2.0       4.0       6.0       8.0      10.0
```

Figure 7.3.11. MINITAB box-and-whiskers plot.

---

[7]They actually wrote the code for the MINITAB box-and-whiskers plot.

S U M M A R Y   D E S C R I P T I V E   S T A T I S T I C S

File = DATA.DAT	Mean = 2.1196	Skewness = 1.482
Cases = 50	St. Dev. = 2.200459	Kurtosis = 5.267
Low = .13	1st Quartile = .3	
High = 10.23	2nd Quartile = 1.335	
Range = 10.1	3rd Quartile = 3.45	

B O X P L O T   (File = DATA.DAT)

LOQ1   Q2   Q3   HI

```
        +-----+-------------+
    |-|      |             |----------------------------------------|
        +-----+-------------+
```

Figure 7.3.12. BOXPLOT box-and-whiskers plot.

Quantiles(Def=5)

	100% Max	10.23	99%	10.23
	75% Q3	3.45	95%	6.54
	50% Med	1.335	90%	4.86
	25% Q1	0.3	10%	0.22
	0% Min	0.13	5%	0.19
			1%	0.13
	Range	10.1		
	Q3-Q1	3.15		
	Mode	0.24		

Figure 7.3.13. SAS/STAT UNIVARIATE table.

benchmark *v. trans. To subject (a system) to a series of tests in order to obtain prearranged results not available on competitive systems. See also*
MENDACITY SEQUENCE.
mendacity sequence *n An ISO standard sorting sequence allowing the F's in a truth table to be ordered by degree of falsehood.*

⇒ *The basic sequence, in ascending order, is: lies; damn lies; statistics; damn statistics; benchmarks; delivery promises; DP dictionary entries. Further refinements can be expected.*

Stan Kelly-Bootle
The Devil's DP Dictionary

7.3.4 The Harmonic Mean

In this section and the next we will examine two means that are seldom discussed in statistics books but that are useful for some applications, especially those in computer science. The first is called the *harmonic mean* and the second the *geometric mean*.

The harmonic mean is particularly useful for analyzing the results of benchmark tests. A *benchmark* is a program or collection of programs whose running time on a computer system is used to measure its performance relative to other computer systems for performing the same task.

Dongarra, Martin, and Worlton [13] provide an excellent overview of computer benchmarking and discuss several standard benchmarks that are widely used. This what they say about standard benchmarks:

WIDELY USED BENCHMARKS

Livermore Loops: computational routines extracted from programs at Lawrence Livermore National Laboratory; used to test scalar and vector floating-point performance.

Linpack: linear algebra software library; used to test scalar and vector floating-point performance.

Whetstones: synthesized benchmark of basic arithmetic; used to test performance of midsize and small computers.

Dhrystones: synthesized benchmark for system-programming operations; used to test nonnumeric performance of midsize and smaller computers.

Each of the above benchmarks is useful for deciding what computer will perform best for you *only* if your workload is similar to that measured by the benchmark. Recently the debit-credit benchmark has a become a de facto standard for measuring online transaction processing performance (OLTP) (see [2]).

To give you an idea of the relative size of a couple of these benchmarks, I ran them on the four IBM PC AT compatible computers that I used in writing this book and indicate the results in Table 7.3.2. Computer A is the 8 MHz IBM PC AT (the real thing) that I started this book with and that has an Intel 80286 microprocessor. All the others are IBM PC AT compatibles with Intel microprocessors and floating-point coprocessors. Computer B has a 20 MHz 80386 microprocessor. Computer C, my office machine, has a 16 MHz 80386 microprocessor. Computer D, my current personal home computer, has a 33 MHz 80386 microprocessor and a hardware cache between the CPU and main memory. Another widely quoted PC benchmark is the SI benchmark that is part of the Norton Utilities. Peter Norton claims it measures the computing power of an IBM compatible computer relative to the IBM PC XT. The SI rating of Computer B is 21.2 and of Computer D is 40.6. That is, Peter Norton, who has reached a pinnacle of fame never before achieved by a computer professional by appearing in an advertisement for a premium scotch whiskey, claims my current home computer has 40.6 times the computing power of an IBM XT! The manufacturer also claims it has an 8 MIPS rating but doesn't explain how the rating was measured.

Table 7.3.2. Benchmark Results

Computer	Dhrystones	Whetstones
A	1,517	181.2K
B	4,552	1,059.0K
C	3,251	816.4K
D	11,379	2,588.4K

Example 7.3.2 Suppose you drive your car one mile at 20 miles per hour and a second mile at 60 miles per hour. What is your average speed for the two miles?

Solution You may be tempted to add 20 to 60 and divide by 2 to obtain 40 miles per hour, but you already know that is not the correct answer; otherwise the question would not have been asked the way it was. It *would* be the correct answer if we had specified that you drove for one hour at 20 miles per hour and for one hour at 60 miles per hour. The average speed is the fixed speed that will move the car over the two miles in exactly the same time that you drove the two miles. Since 60 miles per hour is 1 mile per minute, it took you 1 minute to drive the second mile, and Thus, 3 minutes to drive the first mile, for a total time of 4 minutes. Thus, the average speed is $2/4 = 1/2$ miles per minute or 30 miles per hour. This is what the harmonic mean h would have given you.

The *harmonic mean* h of the nonzero numbers $x_1, x_2, \ldots, x_n$ is defined by

$$h = \frac{n}{\dfrac{1}{x_1} + \dfrac{1}{x_2} + \cdots + \dfrac{1}{x_n}}. \tag{7.59}$$

For the question as given,

$$h = \frac{2}{\dfrac{1}{20} + \dfrac{1}{60}} = 30 \quad \text{miles per hour.} \quad \square$$

Example 7.3.3 Program A and program B each require the execution of 1,000,000 instructions when they are executed on computer X. Program A has been written in an optimized form for the hardware of computer X so that it executes at the rate of 2,000,000 instructions per second; program B has not been optimized and therefore executes at the rate of 500,000 instructions per second. If program A and program B are each executed once on computer X, what is the average instruction execution rate?

Solution Program A executes in 0.5 seconds and program B in 2 seconds. To obtain the average instruction execution rate we need to weight each execution speed by the fraction of time the computer is executing at that speed. Therefore, the average instruction execution rate is

$$\frac{0.5}{2.5} \times 2,000,000 + \frac{2}{2.5} \times 500,000 = 800,000 \text{ instructions per second.}$$

It is simpler to make this calculation using the harmonic mean. We calculate

$$\frac{2}{\dfrac{1}{500,000} + \dfrac{1}{2,000,000}} = \frac{2,000,000}{2.5} = 800,000 \text{ instructions per second.}$$

$\square$

The unit of measure commonly used for instruction execution rate is millions of instructions per second, abbreviated MIPS.[8] Thus, the computer in Example 7.3.3 is executing at 0.8 MIPS. Note that there is no 1 MIP computer, but there might be a 1 MIPS computer.

One of the problems of measuring the performance of a computer system by running benchmarks occurs when several different benchmarks are run. We would like to generate a single number to summarize the performance. Smith [28] considers this problem and agrees with me that

> Harmonic mean should be used for summarizing performance expressed as a rate. It corresponds accurately with computation time that will actually be consumed by running real programs. Harmonic mean, when applied to a rate, is equivalent to calculating the total number of operations divided by the total time.

In the following proposition, we summarize an important property of the harmonic mean that is useful in analyzing benchmarks and for other applications as well.

---

[8]When the MIPS rate is quoted by a computer manufacturer for a particular computer a certain mix of instructions is assumed. The mix is what the manufacturer believes to be typical for users of the machine.

Proposition 7.3.1 *Consider the following situation.*

Case 1: Suppose each of $s_1, s_2, \ldots, s_n$ represents the average speed of a car, ship, etc., each maintained over the same distance d.

Case 2: Suppose each of $s_1, s_2, \ldots, s_n$ represents the average computer instruction execution rate for a computer executing a computer program that executes d instructions.

Then the harmonic mean h of $s_1, s_2, \ldots, s_n$ is the average speed in the case of a car or ship; that is, the speed that would cover the distance $n \times d$ in the same time as was achieved by the speeds $s_1, s_2, \ldots, s_n$ in succession. In the case of a computer, the harmonic mean is the average instruction execution rate over the time period that the programs are run.

Proof We will give the proof for Case 1. It will be clear that the same proof applies to Case 2. For $n = 1$ there is nothing to prove, so assume that $n \geq 2$. Let t_1 be the time to make the first traversal of the distance d, t_2 the time for the second traversal, $\ldots$, t_n the time for the nth traversal. Let

$$T = t_1 + t_2 + \ldots + t_n. \tag{7.60}$$

Since we need to weight the speeds by the average time spent at that speed, the required average, $\bar{s}$, is given by

$$\bar{s} = \frac{t_1}{T} \times s_1 + \frac{t_2}{T} \times s_2 + \cdots + \frac{t_n}{T} \times s_n. \tag{7.61}$$

Since

$$d = t_i \times s_i, \tag{7.62}$$

or

$$t_i = \frac{d}{s_i}, \tag{7.63}$$

for $i = 1, 2, \ldots, n$, we can substitute (7.63) into (7.61) for each i. Then the ith term in (7.61) is

$$\frac{\frac{d}{s_i} \times s_i}{\frac{d}{s_1} + \frac{d}{s_2} + \cdots + \frac{d}{s_n}} = \frac{d}{\frac{d}{s_1} + \frac{d}{s_2} + \cdots + \frac{d}{s_n}} = \frac{1}{\frac{1}{s_1} + \frac{1}{s_2} + \cdots + \frac{1}{s_n}}. \tag{7.64}$$

Since there are n terms identical to (7.64) in (7.61) the sum (7.61) *is* the harmonic mean. ■

You are probably thinking the following cosmic question.

A Cosmic Question

8. I understand how you used the harmonic mean to solve Example 7.3.2 and Example 7.3.3, but what if I run a whole suite of programs (benchmarks) of path lengths (computer instructions) $l_1, l_2, \ldots, l_n$, respectively, that run at the average instruction rates $s_1, s_2, \ldots, s_n$? How can I use the harmonic mean, or any other technique, to obtain the overall average instruction rate?

Answer to a Cosmic Question

You use the *generalized harmonic mean* H defined by[9]

$$H = \frac{l_1 + l_2 + \cdots + l_n}{\dfrac{l_1}{s_1} + \dfrac{l_2}{s_2} + \cdots + \dfrac{l_n}{s_n}}. \tag{7.65}$$

Proposition 7.3.2 *Suppose a sequence of programs of path lengths $l_1, l_2, \ldots, l_n$ instructions, respectively, run at the respective average instruction rates $s_1, s_2, \ldots, s_n$. Then the average instruction rate for running all the programs in sequence is given by the generalized harmonic mean, H.*

Proof For $n = 1$ there is nothing to prove, so assume that $n \geq 2$. The time it takes to run program i is given by

$$t_i = \frac{l_i}{s_i}, \tag{7.66}$$

for $i = 1, 2, \ldots, n$. Let us set

$$T = t_1 + t_2 + \cdots + t_n. \tag{7.67}$$

Weighting each speed by the average time the computer runs at that speed yields the average speed

$$\bar{s} = \frac{t_1}{T} \times s_1 + \frac{t_2}{T} \times s_2 + \cdots + \frac{t_n}{T} \times s_n. \tag{7.68}$$

Each term in (7.68) is of the form

$$\frac{\dfrac{l_i}{s_i} \times s_i}{T} = \frac{l_i}{T}. \tag{7.69}$$

---

[9]Sometimes H is called the *weighted harmonic mean*. See (7.82).

Hence, (7.68) becomes

$$\bar{s} = \frac{l_1 + l_2 + \cdots + l_n}{T}. \tag{7.70}$$

Since

$$T = \frac{l_1}{s_1} + \frac{l_2}{s_2} + \cdots + \frac{l_n}{s_n},$$

we have shown that $\bar{s}$ is given by (7.65), and the proof is complete.

Example 7.3.4 Underground Services, a money laundering company, has a Model XYZ computer from Little Blue in Beta Test. Random Numbers, the Little Blue performance analyst studying the system, has measured the performance of the computer and constructed Table 7.3.3. In Table 7.3.3, f_i is the average number of times program i is run each day, l_i is the average number of XYZ instructions executed when program i is run (measured in millions of instructions), and s_i is the average rate at which instructions are executed (measured in millions of instructions per second, MIPS). Random wants to calculate the average instruction rate of the computer per day in units of MIPS (counted only when the computer is running programs, of course.)

<u>Table 7.3.3</u>

f_i	l_i	s_i
10	30	3
20	40	4
15	30	5

Solution Random uses the *generalized harmonic mean*, H. Since $f_i \times l_i$ is the number of instructions executed at the rate s_i, he calculates:

$$H = \frac{f_1 \times l_1 + f_2 \times l_2 + f_3 \times l_3}{\dfrac{f_1 \times l_1}{s_1} + \dfrac{f_2 \times l_2}{s_2} + \dfrac{f_3 \times l_3}{s_3}} = \frac{1,550}{390} = 3.974 \text{ MIPS.} \quad \square$$

We assume in Example 7.3.4 that the average instruction rates as well as the path lengths are known for the programs under consideration. This is something that must be measured. Computer manufacturers design computers with a number of different instructions, not all of which execute at the same speed. Designers of RISC (reduced instruction set computers) use simple instructions designed to complete execution in one machine cycle, with very few exceptions. This means that more of these instructions must be executed (at a faster instruction rate) to perform a given task than would be true with a CISC (complex instruction set computer). To determine the average instruction rate to perform a given task on a uniprocessor computer system of any architecture, the generalized harmonic mean (7.65) can be used. The task to be measured is performed and a monitor counts l_i, the number of times instruction i is executed for $i = 1, 2, \ldots, n$, where n is the number of instructions the computer can perform. We assume that s_i for each instruction is known. Of course the instruction rate determined by (7.65) is valid only for one particular task. Determining a benchmark or suite of benchmarks that is *representative* of the computer workload of potential users of the computer system is an extremely difficult task.

7.3.5 The Geometric Mean

The *geometric mean* is defined for a sequence $x_1, x_2, \ldots, x_n$ of nonnegative numbers by the formula

$$g = \sqrt[n]{x_1 \times x_2 \times \cdots \times x_n}. \qquad (7.71)$$

In some cases we need to use the *weighted geometric mean* defined by

$$G = x_1^{w_1} \times x_2^{w_2} \times \cdots \times x_n^{w_n}, \qquad (7.72)$$

where the $w_1, w_2, \ldots, w_n$ are nonnegative *weights* such that

$$w_1 + w_1 + \cdots + w_n = 1. \qquad (7.73)$$

The geometric mean, like the harmonic mean, appears in very few statistics books[10] but does have some important applications. The main application of the geometric mean is in averaging ratios of numbers. For this application it functions exactly like the harmonic mean does in averaging speeds.

Fleming and Wallace [17] discuss its use in analyzing benchmarking results. They believe that benchmarking results should be normalized to a particular processor, say X, by dividing the time taken for each benchmark to run by the time it took to run the same benchmark on processor X. Then, for each processor, the normalized times are averaged to obtain the overall rating of the processor. Fleming and Wallace prove that "the geometric mean is the only correct average of normalized measurements." They do

---

[10]Wallis and Roberts [32] is a rare exception. This very readable book discusses both the harmonic and geometric means with examples of how each is used.

not indicate *why* you should use normalized measurements. It is not clear to this author exactly what normalization between two processors means because different processors with different architectures may accomplish a task in completely different ways. It would seem that the normalization discussed by the authors would be appropriate only for processors in a family of very similar machines from one manufacturer.

I agree with Smith [28], who says

> The geometric mean has the property of performance relationships consistently maintained regardless of the computer that is used as the basis for normalization. The geometric mean does provide a consistent measure in this context, but it is consistently wrong. The solution to the problem of normalizing with respect to a given computer is not to use geometric mean, as suggested in [Fleming and Wallace [17]] but to always normalize results *after* the appropriate aggregate measure is calculated, not before.

We follow Smith [28] in using the numbers from Table IX in the Fleming and Wallace paper to show that their technique of using the geometric mean of normalized numbers leads to incorrect results. The results are shown in Table 7.3.4. The weighted arithmetic means of the times to run the two benchmarks show that processor X is the fastest, processor Z is 14 percent slower than X, and processor Y is 36 percent slower than processor X. This is the correct way to analyze the results in terms of the time it takes to run the different benchmarks. However, if we calculate the normalized geometric mean, as advocated by Fleming and Wallace, we would conclude that processor Y is not the *slowest* machine but rather the *fastest*. This clearly is in error. The problem is that of normalizing the numbers. (The same result is obtained by normalizing a geometric mean after calculating the geometric mean as by normalizing first and then calculating the geometric mean.) To be fair to Fleming and Wallace, they do say

> RULE 3: Use the Sum (or arithmetic mean) of Raw, Unnormalized Results whenever This "Total" Has Some Meaning

Example 7.3.5 Suppose the Information Systems Group at Grapeshot Graphite has budgets for the years 1983–1987 as shown in Table 7.3.5. (The budget values are in millions of dollars.) What is the average value of the ratio of this year's to last year's budget?

Table 7.3.4. Benchmark Tests

Benchmark	Weight	Processor		
		X	Y	Z
1	0.6	20	10	40
2	0.4	40	80	20
Weighted arithmetic mean		28	38	32
Normalized to X		1.00	1.36	1.14
Weighted geometric mean		26.4	23.0	30.3
Normalized to X		1.00	0.87	1.15

Solution You are probably thinking that it should be the geometric mean because that's the section of the book you are reading; you are right. The geometric mean is designed to average ratios. For the ratios in Table 7.3.5 the geometric mean is 1.195457673. It is the correct average ratio, since

$$2.425 \times 1.195457673^4 = 4.95277.$$

The arithmetic mean of the ratios in the table is 1.2, but

$$2.425 \times 1.2^4 = 5.02848,$$

a value with an error of 1.529%. □

Table 7.3.5. Budget

Year	Budget	Ratio
1983	2.42500	—
1984	2.54625	1.05
1985	2.97911	1.17
1986	3.96222	1.33
1987	4.95277	1.25

You are probably thinking the following cosmic question.

A Cosmic Question

9. What relationship, if any, is there between the arithmetic mean, the harmonic mean, and the geometric mean?

Answer to a Cosmic Question

The answer is given in the following proposition.

Proposition 7.3.3 *Consider a sequence of positive numbers, $a_1, a_2, \ldots, a_n$. Recall that the arithmetic mean, A, (also called the sample mean, $\bar{a}$, when $a_1, a_2, \ldots, a_n$ is a sample) is defined by*

$$A = \frac{a_1 + a_2 + \cdots + a_n}{n}. \tag{7.74}$$

The harmonic mean, h, is defined by

$$h = \frac{n}{\frac{1}{a_1} + \frac{1}{a_2} + \cdots + \frac{1}{a_n}} = \left[\frac{1}{n} \times \left(\frac{1}{a_1} + \frac{1}{a_2} + \cdots + \frac{1}{a_n} \right) \right]^{-1}. \tag{7.75}$$

The geometric mean, g, is defined by

$$g = \sqrt[n]{a_1 \times a_2 \times \cdots \times a_n}. \tag{7.76}$$

Then we have

$$h \leq g \leq A \tag{7.77}$$

with equality if and only if all the a_i's are equal.

Proof We present a proof of the arithmetic-geometric mean inequality provided by Ross Honsberger [19], that he attributes to George Polya. We begin by noting that it is easy to show by elementary calculus (we ask you to do so in Exercise 14) that the function

$$e^x - 1 - x$$

has a unique minimum value of zero when $x = 0$, which means that for all real x, it is true that

$$e^x \geq 1 + x.$$

Let $a_1, a_2, \ldots, a_n$ denote positive real numbers. Letting x assume the values $(a_i/A) - 1$ for $i = 1, 2, \ldots, n$, we obtain the n inequalities

$$e^{(a_1/A)-1} \geq \frac{a_1}{A},$$

$$e^{(a_2/A)-1} \geq \frac{a_2}{A},$$

$$\cdots$$

$$e^{(a_n/A)-1} \geq \frac{a_n}{A}.$$

Multiplying these inequalities together yields

$$\exp\left(\frac{a_1 + a_2 + \cdots + a_n}{A} - n\right) \geq \frac{a_1 \times a_2 \times \cdots \times a_n}{A^n},$$

which is

$$e^{n-n} \geq \frac{g^n}{A^n}$$

or

$$1 \geq \frac{g^n}{A^n},$$

from which it follows that

$$A \geq g.$$

Note that $A = g$ only if equality holds in all n relations. This means we must have

$$\frac{a_i}{A} - 1 = 0,$$

for $i = 1, 2, \ldots, n$. This shows that each a_i is equal to A. This completes the proof of the arithmetic-geometric mean inequality. To prove the geometric-harmonic mean inequality, let us apply the arithmetic-geometric mean inequality to the numbers $a_1^{-1}, a_2^{-1}, \ldots, a_n^{-1}$ to obtain

$$\frac{(a_1^{-1} + a_2^{-1} + \cdots + a_n^{-1})}{n} \geq (a_1^{-1} \times a_2^{-1} \times \cdots \times a_n^{-1})^{1/n}. \tag{7.78}$$

From the definitions of g and h, we see that (7.78) can be interpreted as

$$h^{-1} \geq g^{-1},$$

that is equivalent to

$$h \leq g. \tag{7.79}$$

Furthermore, there is equality if and only if

$$a_1^{-1} = a_2^{-1} = \cdots = a_n^{-1},$$

that is equivalent to

$$a_1 = a_2 = \cdots = a_n.$$

This completes the proof. ∎

Weighted Means

Recall the weighted geometric mean G defined by formula (7.72), that we repeat as

$$G = a_1^{w_1} \times a_2^{w_2} \times \cdots \times a_n^{w_n}, \tag{7.80}$$

where the w_i are nonnegative and

$$w_1 + w_2 + \cdots + w_n = 1. \tag{7.81}$$

The generalized harmonic mean H, defined by (7.65), can also be written in the form of a *weighted harmonic mean*

$$H = \frac{1}{W}, \tag{7.82}$$

where

$$W = \frac{w_1}{s_1} + \frac{w_2}{s_2} + \cdots + \frac{w_n}{s_n}. \tag{7.83}$$

We assume that

$$w_i = \frac{l_i}{L}$$

for $i = 1, 2, \ldots, n$, where

$$L = \sum_{i=1}^{n} l_i.$$

Thus, each w_i is positive and (7.81) holds. We can define the *weighted arithmetic mean*, WA, by

$$\text{WA} = w_1 \times a_1 + w_2 \times a_2 + \cdots + w_n \times a_n, \tag{7.84}$$

where, of course, we assume each w_i is nonnegative and that (7.81) holds. The next proposition shows that the weighted means behave just like the unweighted means.

Proposition 7.3.4 *Consider a sequence of positive numbers, $a_1, a_2, \ldots, a_n$. Suppose the weighted arithmetic mean, WA, is defined by (7.84), the weighted geometric mean, G, is defined by (7.80) and the weighted harmonic mean, H, is defined by (7.82). Then the following inequalities hold with equality in all cases if and only if all the a_i are equal.*

$$H \leq G \leq \text{WA}. \tag{7.85}$$

Proof. Using the fact that $-\ln t$ is convex for positive t, Boas [4] proves that

$$G \leq \text{WA}, \tag{7.86}$$

with equality if and only if

$$a_1 = a_2 = \cdots = a_n. \tag{7.87}$$

We can apply (7.86) to obtain

$$\frac{w_1}{a_1} + \frac{w_2}{a_2} + \cdots + \frac{w_n}{a_n} \geq \frac{1}{a_1^{w_1}} \times \frac{1}{a_2^{w_2}} \times \cdots \times \frac{1}{a_n^{w_n}}. \tag{7.88}$$

Formula (7.88) can be interpreted as

$$H^{-1} \geq G^{-1}, \tag{7.89}$$

or

$$H \leq G. \tag{7.90}$$

It follows from previous results that we have equality in (7.90) if and only if

$$a_1 = a_2 = \cdots = a_n. \tag{7.91}$$

> *Yet what are all such gaieties to me*
> *Whose thoughts are full of indices and surds?*

$$x^2 + 7x + 53$$

$$= \frac{11}{3}.$$

Lewis Carroll

7.4 Exercises

1. [HM15] Suppose the random variable X has the density function

$$f(x) = \begin{cases} (1+\lambda)x^\lambda, & 0 < x < 1 \\ 0 & \text{otherwise.} \end{cases}$$

Show that the maximum likelihood estimate of λ based on a given random sample of size n is given by

$$\hat{\lambda} = -\left(1 + \frac{n}{\displaystyle\sum_{i=1}^{n} \ln x_i}\right).$$

2. [00] Consider a uniform random variable defined on the interval $0 \leq x \leq \beta$, where β is unknown. Use the method of moments to find an estimate of β based on a given random sample of size n.

3. [C15] If the $n = 20$ values of processing time mentioned in Example 7.1.2 are 16.39, 25.09, 16.31, 20.94, 17.58, 19.06, 17.21, 18.48, 16.88, 15.51, 25.87, 17.63, 29.13, 21.34, 11.14, 26.03, 23.28, 21.13, 18.46, 14.25, find the method of moments estimates of β and α.

4. [C10] If the $n = 20$ values of number of customer arrivals in a 10 minute period mentioned in Example 7.1.3 are 25, 32, 34, 22, 27, 29, 23, 22, 30, 31, 33, 21, 28, 25, 24, 35, 27, 30, 34, 26, find the method of moments estimates of α and σ.

5. [10] Bortkiewicz [5] in 1898 fitted the Poisson distribution to the number of deaths from horse kicks in the Prussian cavalry per corps-year for each of 200 corps-years. His data is given in the table. Estimate the Poisson parameter α using the method of maximum likelihood and construct a 95% confidence interval for it using $\hat{\sigma} = \sqrt{\hat{\alpha}}$.

Horse Kick Deaths by Prussian Cavalry

Number of Deaths	Observed Number Corps-Years
0	109
1	65
2	22
3	3
4	1
≥ 5	0
Total	200

6. [M15] Consider Example 7.1.3.

(a) Prove that

$$\sum_{i=1}^{n}(X_i - \overline{X})^2 = \sum_{i=1}^{n} X_i^2 - n(\overline{X})^2. \qquad (7.92)$$

(b) Using (a) show that

$$M_2 - M_1^2 = \left(\frac{n-1}{n}\right) S^2. \qquad (7.93)$$

7. [M15] Consider the ratio (7.37) in Example 7.1.8. Show that this ratio is larger than unity (that is, that $p_k(N) > p_k(N-1)$) if and only if $Nk < rn$.

8. [M10] The farm experts at Fanny Farmers Farm plan to lay out a square plot of land with side μ by using a long rod of length μ so the area of the plot is μ^2. Unfortunately, the length of the rod is not known exactly, so n independent measurements by n independent farmers are taken, yielding the values $x_1, x_2, \ldots, x_n$. We assume that each X_i has mean μ and variance σ^2.

(a) Show that $(\overline{X})^2$ is *not* an unbiased estimator of μ^2, the area of the field.

(b) For what value of k is the estimator $(\overline{X})^2 - k \times S^2$, where S^2 is the sample variance, an unbiased estimator for μ^2?

9. [M10] Consider the Pascal distribution described in Chapter 3 preceding Exercise 20. It describes a sequence of Bernoulli trials, that continues until r successes occur. The probability that k failures occur before the rth success is given by

$$p(k; r, p) = \binom{r + k - 1}{k} \times p^k \times q^k,$$

for $k = 0, 1, \ldots$, where $q = 1 - p$.

(a) Suppose $r \geq 2$. To estimate p, suppose trials are continued until r successes are achieved. Suppose k failures occur before the rth success. Then let $\hat{p} = (r-1)/(k + r - 1)$. Show that $\hat{p}$ is an unbiased estimator of p.

(b) The Vice President of Information Systems wants to interview five Information Center users who are satisfied with the service. An assistant polls a number of users. The sequence of responses is SSFFSFFFSS, where S means the user is satisfied and F that he or she is not. Estimate the true proportion of Information Center users who are satisfied.

10. [C8] Elvin Prellvin, the chief statistician for Heartbreak Hotels, is provided with a random sample from a normal population with $\bar{x} = 12.9$ and $s = 3.2$.

    (a) If $n = 121$, what should Elvin obtain as a 95% confidence interval for μ and for σ^2?

    (b) Answer (a), if $n = 9$ rather than 121.

11. [10] Renaissance Resistors made a random check of 50 of their professional employees who had been provided with either a personal computer or a computer workstation and found that 22 of them used their computers for more than 6 hours per day.

    (a) Find the maximum likelihood estimate of the true proportion of professionals who use their computers at least 6 hours per day.

    (b) Find the 95% confidence interval for the estimate of part (a).

    (c) How large should the random sample be to ensure that the width of the 95% confidence interval of part (b) does not exceed 0.1?

12. [M10] Prove that an exponential random variable has skewness two.

13. [M10] Prove that an exponential random variable has kurtosis nine according to the definition used by Doane [12], (six according to SAS/STAT [27]).

14. [HM08] Prove that the function g defined for all real x by $g(x) = e^x - 1 - x$ has a unique minimum value of zero when $x = 0$.

15. [HM12] Prove that the sample variance S^2 defined by (7.2) is unbiased. Hint: Use the formula proved in Exercise 6(a).

References

[1] Milton Abramowitz and Irene A. Stegun, *Handbook of Mathematical Functions With Formulas, Graphs, and Mathematical Tables*, National Bureau of Standards, Washington DC, 1964. Also published by Dover Publications, New York.

[2] Anon *et al.*, A measure of transaction processing power, *Datamation*, April 1, 1985.

[3] V. Barnett and T. Lewis, *Outliers in Statistical Data*, John Wiley, New York, 1978.

[4] Ralph P. Boas, Jr., *A Primer of Real Functions*, 3rd ed., The Mathematical Association of America, Washington DC, 1981.

[5] Ladislaus von Bortkiewicz, *Das Gesetz der Kleinen Zahlen*, Teubner, Leipzig, 1898.

[6] Maurice C. Bryson, The literary digest poll: making a statistical myth, *The Amer. Stat.* **30(4)**, (Nov. 1976), 184–185.

[7] M. G. Bulmer, *Principles of Statistics*, 2nd ed., MIT Press, Cambridge, MA, 1967. Now published by Dover Publications, New York.

[8] Chris Chatfield, The initial examination of data (with discussion), *Journal of the Royal Statistical Society A*, **148**, 1985, 214–253.

[9] Chris Chatfield, Exploratory data analysis, *European Journal of Operations Research*, **23**, 1986, 5–13.

[10] D. R. Cox and E. J. Snell, *Applied Statistics*, Chapman and Hall, London, 1981.

[11] Harald Cramér, *Mathematical Methods of Statistics*, Princeton University Press, Princeton, 1946.

[12] David P. Doane, *Exploring Statistics With The IBM PC*, 2nd ed., Addison-Wesley, 1988. (Comes with a diskette containing the EXPLORE statistical system and a data diskette.)

[1] Jack Dongarra, Joanne L. Martin, and Jack Worlton, Computer benchmarking: paths and pitfalls, *IEEE Spectrum*, July 1987, 38–43.

[14] A. S. C. Ehrenberg, *A Primer in Data Reduction*, John Wiley, London, 1982.

[15] William Feller, *An Introduction to Probability Theory and Its Applications, Volume I*, 3rd ed., Wiley, New York, 1968.

[16] Domenico Ferrari, *Computer Systems Performance Evaluation*, Prentice-Hall, Englewood Cliffs, NJ, 1978.

[17] Philip J. Fleming and John J. Wallace, How not to lie with statistics: the correct way to summarize benchmark results, *CACM*, **29(3)**, March 1986, 218–221.

[18] Robert V. Hogg and Allen T. Craig, *Introduction to Mathematical Statistics*, 3rd ed., Macmillan, New York, 1970.

[19] Ross Honsberger, *Mathematical Morsels*, The Mathematical Association of America, Washington DC, 1978.

[20] Murray S. Klamkin, *International Mathematical Olympiads 1979–1985*, Mathematical Association of America, Washington DC, 1986.

[21] Erwin Kreyszig, *Introductory Mathematical Statistics*, Wiley, New York, 1970.

[22] Donald P. Minassian, The arithmetic-geometric mean inequality revisited: elementary calculus and negative numbers, *Amer. Math. Month.* **94(10)** (Dec. 1987), 977–978.

[23] Frederick Mosteller and John W. Tukey, *Data Analysis and Regression: A Second Course in Statistics*, Addison-Wesley, Reading, MA, 1977.

[24] Paul Newbold, *Statistics For Business And Economics*, Prentice-Hall, Englewood Cliffs, NJ, 1984.

[25] Sheldon M. Ross, *Introduction to Probability and Statistics for Engineers and Scientists*, John Wiley, New York, 1987.

[26] Barbara F. Ryan, Brian L. Joiner, and Thomas A. Ryan, Jr., *MINITAB Handbook*, 2nd ed., Duxbury Press, Boston, 1985.

[27] *SAS Procedures Guide for Personal Computers, Version 6 Edition*, SAS Institute, Cary, NC, 1985.

[28] James E. Smith, Characterizing computer performance with a single number, *CACM*, (Oct. 1988), 1202–1206.

[29] George W. Snedecor and William G. Cochran, *Statistical Methods*, 7th ed., The Iowa State University Press, Ames, 1980.

[30] John W. Tukey. *Exploratory Data Analysis*, Addison-Wesley, Reading, MA, 1977.

[31] Paul F. Velleman and David C. Hoaglin, *Applications, Basics, and Computing of Exploratory Data Analysis*, Duxbury Press, Boston, MA, 1981.

[32] W. Allen Wallis and Harry V. Roberts, *Statistics: A New Approach*, The Free Press, Chicago, 1956.

Chapter 8

Hypothesis Testing

All knowledge resolves itself into probability.
David Hume

Chance favors the prepared mind.
Louis Pasteur

8.0 Introduction

In this chapter we continue our study of statistical inference. Recall that
statistical inference is the process of drawing conclusions about a population on the basis of a random sample. We assume that the population
is determined by a random variable; that is, that the population consists
of the possible values of a random variable. In Chapter 7 we examined
the problem of estimating the values of the parameters of the random variable. We were concerned not only with making the estimates, but also with
making probability judgments about the quality of our estimates.

Hypothesis testing is a procedure for determining, from information con-
tained in a random sample from a population, whether to accept or reject
a certain statement (hypothesis) about the random variable determining
the population. A statistical hypothesis is usually stated as a proposition
concerning the distribution of this random variable. It may be a statement
about the values of one or more of the parameters of a given distribution;
it also may concern the form of the distribution. Examples of statistical
hypotheses follow. In some cases they may not *seem* to be statistical hypotheses because we haven't made explicit what the random variable under
consideration is.

1. The average response time at the terminals used by the systems programmers does not exceed 0.5 seconds during the time period from 10 a.m. till 12 a.m.

2. The programmers in Department Able are more productive (in terms of lines of code per programmer-day) on the average than the programmers in Department Baker.

3. The arrival pattern of requests from the interactive system used by the application programmers to the central computer complex is Poisson.

A statistical hypothesis test is a formal, step-by-step procedure described below. It is based upon the intuitively appealing idea that, "an event with a low probability of occurrence does not happen very often." That is, if a hypothesis implies a certain event to have a low probability, its occurrence is evidence *against* the hypothesis.

> *When you're away, I'm restless, lonely*
> *Wretched, bored, dejected, only*
> *Here's the rub, my darling dear,*
> *I feel the same when you are here.*

> Samuel Hoffenstein
> *Poems in Praise of Practically Nothing*

8.1 Hypothesis Test Procedure

We define a hypothesis test procedure formally as follows:

Procedure 8.1.1 (*Procedure For Testing a Hypothesis*)

Step 1 *Decide upon a null hypothesis H_0 and an alternative hypothesis H_1. (Although, by convention, H_1 is called the alternative hypothesis, H_1 is usually the hypothesis we want to test.)*

Step 2 *Select a test statistic; that is, a formula for calculating a number based upon the random sample, say $t(X_1, X_2, \cdots, X_n)$. (Common test statistics are the sample mean $\overline{X}$ and the sample variance S^2.)*

Step 3 *Choose α, the* level of significance *of the test. It is usually chosen to be either 0.05 or 0.01, (5% or 1%), but any value between 0 and 1 can be selected. (We will see below what α means.)*

Step 4 *Choose a* rejection region (*often called a* critical region) *for the test. That is, choose a set of possible test statistic values such that if H_0 is true, then the probability that the value of the test statistic will fall in the rejection region is α. A critical region is often chosen to be all the numbers greater (or smaller) than a critical value. (This is a* one-tailed test.) *Another popular choice is the set of all numbers either less than a left critical value or greater than a right critical value. (Naturally, this is a* two-tailed test.) *[The complement of the rejection region is called the* acceptance region (*what else?*).]

Step 5 *Calculate the test statistic of a random sample from the population. If this value falls in the critical region, reject H_0 and accept H_1, otherwise accept H_0.*

Step 6 *Optionally, calculate the p-value of the test. The p-value is the probability that, if H_0 is true, a value at least as extreme as the observed test statistic will be observed. The decision rule for accepting or rejecting then becomes: reject H_0 if the p-value is less than α. Whether H_0 is accepted or rejected, the p-value provides useful information.*

We illustrate the procedure in the following example.

Example 8.1.1 Symple Symon Software has a large software development group. Gabriella Gauss, their Supreme Statistician, has verified that X, the number of lines of code per programmer week, has a normal distribution with mean 300 and standard deviation 20. Symple programmers all write code in their proprietary programming language, Symply Super. Six months ago they adopted a new programming paradigm advocated by Super Software Sycophants. Ms. Gauss decides to use a statistical test to determine whether the new paradigm has led to better programmer productivity. She obtains a random sample from 100 programmers to determine the average number of lines of code they each produced in a recent week. The sample yields $\bar{x} = 310$ lines per week. Gabriella wants to determine, at the 5 % level of significance, whether the mean of X has increased.

Solution As is customary (and wise) for problems of this sort, Gabriella chooses the null hypothesis that there has been no change. Thus, she specifies H_0: $\mu = \mu_0 = 300$ lines per week. Since she is hoping that μ has increased, she chooses H_1: $\mu > 300$ lines per week and $\sigma = 20$ lines per week. Clearly, a large value for $\bar{x}$ is evidence that μ has increased. But how large is large? If H_0 is true, then, by the sampling theorem (Theorem

7.2.1), $\overline{X}$ is normally distributed with mean 300 and standard deviation

$$\frac{\sigma}{\sqrt{n}} = \frac{20}{10} = 2. \qquad (8.1)$$

Therefore,

$$\frac{\overline{X} - 300}{2} \qquad (8.2)$$

is a standard normal random variable. Since $z_{0.05} = 1.645$, she chose the rejection region to be all $\overline{x}$ such that

$$\frac{\overline{x} - 300}{2} > 1.645, \qquad (8.3)$$

or $\overline{x} > 303.29$. Since $\overline{x}$ is 310, Gabriella rejects H_0 and concludes that $\mu > 300$. The procedure of this example is illustrated in Figure 8.1.1. $\square$

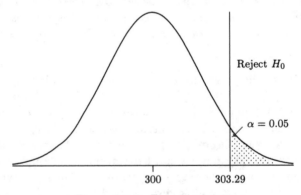

Reject H_0

$\alpha = 0.05$

300   303.29

Figure 8.1.1. Example 8.1.1.

You have probably noted from this example that statistical hypothesis testing is not foolproof. In Example 8.1.1 it is possible by pure chance that $\mu = 300$ but for the sample of 100 programmer weeks, $\overline{x}$ is greater than 303.29. We would reject H_0 and conclude that $\mu > 300$. This would be a *Type I error*; that is, we would have rejected H_0 when it was true. The probability of a Type I error is α because of the way we set up the statistical test. The other possible error is accepting H_0 when it is false. This is known (imaginatively) as a *Type II error*. We denote its probability by β. The concepts of Type I and Type II errors are illustrated in Table 8.1.1. (P in Table 8.1.1 is the probability of the indicated result. Thus, the probability of accepting H_0 when H_1 is true is β.) In applying a statistical test, we can make only these errors. Statisticians, however, have identified a couple more errors made by *other* statisticians. A Type III error is assuming the wrong distribution for X. A Type IV error is solving the wrong problem. These latter types of errors are part of the statistical folklore but not of the discipline. Suppose we are comparing statistical tests of a null hypothesis H_0 with the same level of significance α. We would like a test that will best avoid our making a Type II error. This characteristic is measured by the *power* of the test. It is defined to be $1 - \beta$. It is sometimes difficult to measure the power of a test. It is especially difficult if H_1 is a *composite hypothesis* rather than a *simple hypothesis*. For a simple hypothesis not only the form of the distribution but also its parameters are specified. Suppose, in Example 8.1.1, H_1 was "X has a normal distribution with mean 305 and standard deviation 20." Then H_1 is simple and the test can be visualized as in Figure 8.1.2. Here β, the probability of accepting H_0 when H_1 is true, is the area under the density function for $\overline{X}$ (when H_1 is true) to the left of 303.29. Since for this calculation we assume H_1 is true, β is also the area of the tail of a standard normal density to the right of the value

$$z = \frac{305 - 303.29}{2} = 0.855. \qquad (8.4)$$

Thus, we find by consulting Table 3 of Appendix A that β is 0.1963. Hence, the power of this test is 0.8037.

Let us illustrate the relationship between α and β. Suppose we reduce α to 0.01 (one percent). The critical region becomes all $\overline{x}$ greater than

$$300 + 2 \times 2.326 = 304.652, \qquad (8.5)$$

since $z_{0.01} = 2.326$. This means that β is the area under the standard normal density function to the right of

$$\frac{305 - 304.652}{2} = 0.174, \qquad (8.6)$$

so that $\beta = 0.43093$. Thus, making α smaller (moving the critical value to the right) increases β while making α larger (moving the critical value to the left) decreases β. The level of significance chosen for a test is a compromise.

As we mentioned earlier, the desirability of a statistical test for a fixed α and H_0 is measured by the power of the test, which is

$$1 - \beta = P[\text{rejecting } H_0 | H_1 \text{ is true}]. \qquad (8.7)$$

In the sequel, whenever possible we shall compute the power of the tests discussed. In some cases it can be proven, mathematically, that a particular test has the maximum power of any test of a simple hypothesis H_0 at the α level of significance against an alternative composite hypothesis; such a test is called a *uniformly most powerful test.*

The true density function is on the left, if $\mu = 300$, and on the right, if $\mu = 305$.

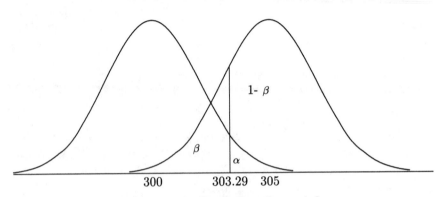

Figure 8.1.2. Illustration of α and β.

Statisticians do not regard accepting the null hypothesis H_0 as being equivalent to believing it to be true. When we accept H_0 as the result of a statistical test, we merely believe that there is not sufficient evidence to reject it or that H_0 may be only approximately true. Suppose, for example, that the null hypothesis is "X is normally distributed with mean $\mu = 30$ and standard deviation $\sigma = 10$," while H_1 is "X is normally distributed with standard deviation $\sigma = 10$ but $\mu > 30$." Then, if the test fails to reject H_0 at a particular level of significance, say 5%, we still may believe that μ is not *exactly* 30; it may be 30.001 or 29.998. We are, however, reasonably sure that μ is not 35.6 or 40.2. That is, μ does not significantly exceed 30. In fact, the terminology used by statisticians is to say the result of a test is *significant* if the value of the test statistic falls in the critical (rejection) region.

Table 8.1.1. Type I and Type II Errors

Decision	Unknown Truth	
	H_0 true	H_1 true
Accept H_0	True decision $P = 1 - \alpha$	Type II error $P = \beta$
Accept H_1	Type I error $P = \alpha$	True decision $P = 1 - \beta$

> *Houston, Tranquility Base here. The Eagle has landed.*
> Neil A. Armstrong

8.2   Tests of Means

There are two basic kinds of hypothesis tests about means. They are called *one-sample tests* and *two-sample tests*. The first type is a test of the null hypothesis that the mean of a population (that is, of the random variable X, that determines the population) has some specified value μ_0 against some appropriate alternative hypothesis. For example, we may wish to test the null hypothesis that the programmers at Spacey Sputniks write an average of 50 lines of code per day against the alternative hypothesis that the average is higher. Example 8.1.1 illustrated a one-sample test of a mean.

A two-sample test compares the means from two different populations. Thus, we might want to compare the average lines of code per day produced by different programming groups.

In the next section we consider one-sample tests and in the following section we consider two-sample tests.

8.2.1   One-Sample Tests of Means

In order to be able to make a means test of a population we must have (1) a sample large enough to ensure that $\overline{x}$ is normally distributed or (2) a normal population.

For the first alternative (as we have previously mentioned) an often used rule of thumb is that $n \geq 30$, although this is not an absolute guarantee that $\overline{x}$ is normal. We state the one-sample means test as an algorithm.

Algorithm 8.2.1 (*Test of Value of Mean for Large Sample or Small Normal Sample*) *Given a random sample $x_1, x_2, \ldots, x_n$ from a population determined by a random variable X, this algorithm will determine at the α level of significance whether to accept or reject the null hypothesis $H_0 : \mu = \mu_0$ against one of the alternative hypotheses (a) $H_1 : \mu > \mu_0$, (b) $H_1 : \mu < \mu_0$, or (c) $H_1 : \mu \neq \mu_0$.*

Step 1 [Calculate the test statistic] *For the large sample case ($n \geq 30$) calculate the test statistic*

$$z = \frac{(\overline{x} - \mu_0)}{\frac{s}{\sqrt{n}}}. \tag{8.8}$$

If σ is known it should be used in place of the sample standard deviation s in (8.8).

For the small sample case in which X is assumed to be normal (or, at least, approximately normal), calculate the test statistic

$$t = \frac{(\overline{x} - \mu_0)}{\frac{s}{\sqrt{n}}}. \tag{8.9}$$

Step 2 [Find the critical region] *Let us consider the large-sample case first. The critical region for alternative hypothesis (a) is the set of all z such that $z > z_\alpha$; for alternative hypothesis (b), it is the set of all z such that $z < z_\alpha$. For alternative hypothesis (c), the critical region comprises the two tails consisting of all z such that $z < -z_{\alpha/2}$ and all z such that $z > z_{\alpha/2}$. The critical regions for the small-sample case are very similar. We replace z by t and z_α by $t_{n-1,\alpha}$, where $t_{n-1,\alpha}$ refers to a Student's t distribution with $n-1$ degrees of freedom.*

Step 3 [Accept or reject H_0] *If the test statistic calculated in Step 1 falls in the critical region determined in Step 2, reject H_0 and accept H_1; otherwise accept H_0.*

Proof We will give an indication of why Algorithm 8.2.1 should be true. Kreyszig [13] provides a formal proof. For the large-sample case, by the sampling theorem, Theorem 7.2.1, $(\overline{X} - \mu_0)/(\sigma/\sqrt{n})$ has approximately a standard normal distribution. If σ is not known, then s should be at least a fair approximation to it. Hence, each of the critical regions defined in Step 1 does have approximately probability α. Similarly, by the same theorem, $(\overline{X} - \mu_0)/(s/\sqrt{n})$ has a Student's t distribution with $n-1$ degrees of freedom. Again, the critical regions determined in Step 2 have probability α. Kreyszig [13] proves that if X is normally distributed with σ known, then for both of the one-tailed alternatives, Algorithm 8.2.1 provides the uniformly most powerful tests. ∎

Figures 8.2.1(a) and (b) illustrate Algorithm 8.2.1. The figure to illustrate the test for the alternative hypothesis $\mu < \mu_0$ is similar to Figure 8.2.1(a) except that H_0 is rejected when $z < -z_\alpha$. Let us consider some examples of the use of Algorithm 8.2.1.

Example 8.2.1 A random sample of 400 response times taken during the peak period of the interactive inquiry system at Futile Finance yields a sample mean $\overline{x}$ of 21 time units[1] with a sample standard deviation s of 12

---

[1]It is rumored that 20 time units is equal to 0.8 seconds. Futile has subsecond response time.

time units. The service level agreement between the MIS Department and the users of the inquiry system is that the mean response time for the peak period should not exceed 20 time units. A test is to be made each day at the 1% confidence level ($\alpha = 0.01$). Does the system pass today?

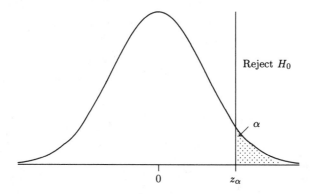

Figure 8.2.1(a). Alternative hypothesis $\mu > \mu_0$.

Solution The null hypothesis is that $\mu = 20$; the alternative hypothesis is that $\mu > 20$. In Step 1 of Algorithm 8.2.1, we calculate $z = (21 - 20)/(12/\sqrt{400}) = 20/12 = 5/3 = 1.6667$. In Step 2 we find the critical region is the set of all z's greater than $z_{0.01} = 2.326$. Thus, we cannot reject H_0 at the 1% level of significance. Since $z_{0.05} = 1.645$, we could reject H_0 at the 5% level. $\square$

Example 8.2.2 The mean time for a clerk to service a customer at Salacious Savings is 3 minutes with a standard deviation of 1 minute; the service time is normally distributed. Sunny Solono, Vice President of Service, decides to test the feasibility of installing an improved interactive computer system. He trains Super Sally to use the proposed system. (Sally is considered to be an average clerk.) He then runs a test in which she uses a prototype of the proposed system at the benchmarking center of the manufacturer. Sally processes 16 customers during the test. The time it takes her to process each customer (in seconds) is shown in Table 8.2.1. Sunny believes the proposed system will be cost effective if the mean clerk service time does not exceed 110 seconds. At the 5% level of significance, does the test indicate that the service time criterion is satisfied?

Solution We assume the clerk service time remains normally distributed. For the test H_0 is $\mu = 110$ seconds; H_1 is $\mu > 110$ seconds. In Step 1 of

Table 8.2.1. Data

156	73	100	110
123	101	125	83
92	101	64	109
56	179	165	182

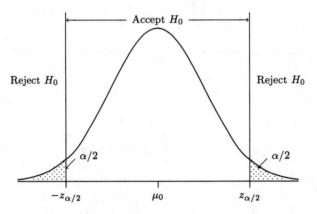

Figure 8.2.1(b). Alternative hypothesis $\mu \neq \mu_0$.

```
MTB > note: Example 8.2.2
MTB > ttest mu=110 'lines';
SUBC> alternative=+1.

 TEST OF MU = 110.000 VS MU G.T. 110.000

             N      MEAN     STDEV    SE MEAN       T    P VALUE
 lines      16   113.687    39.140     9.785     0.38      0.36
```

Figure 8.2.2. MINITAB TTEST Example 8.2.2.

Algorithm 8.2.1, we calculate $t = (113.687 - 110)/(39.14/4) = 0.38$. In Step 2, by Table 5 of Appendix A, the critical region is the set of all $t > t_{15,0.05} = 1.753$. Hence, we cannot reject the null hypothesis at the 5% level of significance. Figure 8.2.2 shows the MINITAB TTEST for this example. Since the p-value is 0.36 we cannot reject the null hypothesis. □

Sample Sizes Needed

We have discussed the size of the Type I error (α) and the size of the Type II error (β) but have not emphasized the fact that the sample size n is important in both types of error; that is, both α and β decrease as n increases. In some cases we can choose the required sample size to give ourselves the desired power in a hypothesis test. For example, suppose X is approximately normal with σ unknown, and the alternative hypothesis is that $\mu > \mu_0$. Suppose, further, that we want the power to be $1 - \beta$ when the true mean $\mu = \mu_0 + \delta$. Then we have

$$
\begin{aligned}
1 - \beta &= P\left[\frac{\overline{x} - \mu_0}{\frac{\sigma}{\sqrt{n}}} > z_\alpha | \mu = \mu_0 + \delta\right] \\
&= P\left[\frac{\overline{x} - (\mu_0 + \delta)}{\frac{\sigma}{\sqrt{n}}} > z_\alpha - \frac{\delta}{\frac{\sigma}{\sqrt{n}}} | \mu = \mu_0 + \delta\right]. \quad (8.10)
\end{aligned}
$$

But, if $\mu = \mu_0 + \delta$, then

$$
\frac{\overline{X} - (\mu_0 + \delta)}{\frac{\sigma}{\sqrt{n}}} \quad (8.11)
$$

has approximately a standard normal distribution. This implies that

$$
1 - \beta = P[z > z_\alpha - (\delta\sqrt{n}/\sigma)], \quad (8.12)
$$

or

$$
- z_\beta = z_\alpha - \left(\frac{\delta\sqrt{n}}{\sigma}\right). \quad (8.13)
$$

(See Figure 8.2.3 where $z = \mu_0 + z_\alpha\sigma/\sqrt{n}$.) Hence, we must have

$$
n = (z_\alpha + z_\beta)^2 \times \frac{\sigma^2}{\delta^2}. \quad (8.14)
$$

By symmetry the same size n is required for the alternative hypothesis $\mu < \mu_0$, if $\mu = \mu_0 - \delta$. Let us consider an example.

Example 8.2.3 Consider Example 8.2.2. Suppose Sunny wants $1 - \beta$ to be at least 0.8 when $\delta = 20$, that is, when $\mu = 130$. How large should n be?

Solution If $1 - \beta = 0.8$, then $\beta = 0.2$. Sunny can use $\hat{\sigma} = s = 39.14$. Hence, since to 4 decimal places, $z_{0.05} = 1.6449$ and $z_{0.2} = 0.8416$, formula (8.14) yields 23.68. Hence, Sunny should use $n \geq 24$. $\square$

The true density function is on the left, if $\mu = \mu_0$,
and on the right, if $\mu = \mu_0 + \delta$.

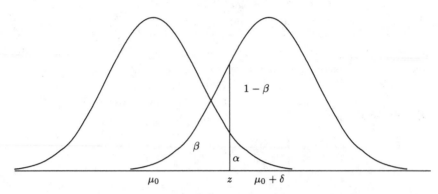

Figure 8.2.3. Graph to illustrate sample sizes needed.

Posten [22] shows that Algorithm 8.2.1 is robust as long as the population has a distribution that is not skewed. That is, the test gives good results even when the population is nonnormal provided it is not strongly skewed.

We will now consider two-sample tests of means.

8.2.2   Two-Sample Tests of Means

Statistics can be used for the most common of all decision making processes: that of comparing the performance (or the lack thereof) of two entities. Thus, we may compare the productivity of two groups, the difference in effectiveness of two software systems or two programming languages, the difference in down time of two systems, the difference in the yield of good solid-state chips from different processes, etc. The test that is commonly used for all these comparisons is the two-sample test of means, also called the *t-test* (or *ttest*, because that is the name of the procedure in some computer statistical systems). This test is accomplished by taking independent (usually) samples from the two populations and comparing the sample means. Although knowledge of hypothesis testing is not a substitute for knowledge of the systems under test, good statistical testing can prevent egregious errors. The mathematical basis for some of the manipulations required by a two-sample test of means is given in the following theorem, which is stated without proof.

Theorem 8.2.1 *Suppose two independent random variables X and Y have the means μ_X and μ_Y and the variance σ_X^2 and σ_Y^2. Then the distribution* *of* $X - Y$ *has the mean* $\mu_X - \mu_Y$ *and the variance* $\sigma_X^2 + \sigma_Y^2$. *Furthermore, if a random sample of size n is taken from the population determined by* X, *and a random sample of size m is taken from the population determined by* Y, *then the following formulas are true:*

$$E[\,\overline{X} - \overline{Y}\,] = \mu_X - \mu_Y, \tag{8.15}$$

and

$$\mathrm{Var}[\,\overline{X} - \overline{Y}\,] = \frac{\sigma_X^2}{n} + \frac{\sigma_Y^2}{m}. \;\blacksquare \tag{8.16}$$

We state the two-sample test of means first for the case of large samples in which we do *not* assume that the variances of the two populations are equal.

Algorithm 8.2.2 (*Two-Sample Test of Means, Large Samples*) *Given a random sample* $x_1, x_2, \ldots, x_n$ *from a population determined by* X, *and a random sample* $y_1, y_2, \cdots, y_m$ *from a population determined by* Y, *where* X *and* Y *are independent, this algorithm will determine, at the* α *level of significance, whether to accept or reject the null hypothesis* $H_0 : \mu_X - \mu_Y = d_0$ *against one of the alternative hypotheses* (a) $H_1 : \mu_X - \mu_Y > d_0$, (b)$H_1 : \mu_X - \mu_Y < d_0$, *or* (c) $H_1 : \mu_X - \mu_Y \neq d_0$. *It is assumed that* $n \geq 30$, $m \geq 30$, *and that either* (a) σ_X *and* σ_Y *are known or* (b) s_X *and* s_Y *are good estimates of the respective standard deviations.*

Step 1 [Calculate the test statistic] *Compute the test statistic*

$$z = \frac{(\overline{x} - \overline{y}) - d_0}{\sqrt{(\sigma_X^2/n) + (\sigma_Y^2/m)}}. \tag{8.17}$$

If σ_X *and* σ_Y *are not known, substitute* s_X *and* s_Y *for them.*

Step 2 [Determine the critical region] *For alternative hypothesis* (a), *the critical region is the set of all* $z > z_\alpha$; *for alternative hypothesis* (b), *it is the set of all* $z < -z_\alpha$; *and for alternative* (c), *it comprises the two tails consisting of all* z *such that* $z < -z_{\alpha/2}$ *and all* z *such that* $z > z_{\alpha/2}$, *which can be written as the set of all* z *such that* $|z| > z_{\alpha/2}$.

Step 3 [Accept or reject H_0**]** *If the test statistic of Step 1 falls in the critical region determined in Step 2, reject* H_0; *otherwise accept* H_0.

Example 8.2.4 Insular Insurance has two separate interactive computer systems; System I and System II. Insular attempts to keep the two systems equally loaded so that the mean response time is the same for each system.

A sample of 100 response times[2] from System I yields $\overline{x} = 20.24$ time units
with $s_X = 5.6$ time units. A sample of 120 response times from System II
yields $\overline{y} = 18.72$ time units with $s_Y = 4.2$ time units. Low Pockets, the
Supreme Performance Analyst, wants to test the hypothesis, at the 2.5%
level of significance, that $\mu_X > \mu_Y$.

Solution Low decides to use Algorithm 8.2.2 with $d_0 = 0$ and the null
hypothesis $H_0 : \mu_X = \mu_Y$. The alternative hypothesis is that $\mu_X > \mu_Y$.
Low calculates the statistic

$$z = \frac{20.24 - 18.72}{\sqrt{5.6^2/100 + 4.2^2/120}} = 2.24. \tag{8.18}$$

Since $z_{0.025} = 1.96$, Low rejects the null hypothesis in favor of the hypothesis that the mean response time is higher on System I. He decides to switch
some of the users from System I to System II. Note that the hypothesis
that Low wanted to test was *not* made the null hypothesis. This is the way
such tests are usually made. □

Sometimes we want to compare means from two different populations
when only small samples are available. The next algorithm shows how to do
this when the two populations are normal even though the variances may
not be the same. The solution given is called *Welch's approximate solution
to the Behrens-Fisher problem* and is discussed in Best and Rayner [4]. In
fact, Best and Rayner believe that Welch's approximation should be used
even when the variances may be equal, provided $\nu \geq 5$ in (8.20). Ryan,
Joiner, and Ryan [24] and other authors agree with this assessment.

Algorithm 8.2.3 (*Two-Sample Test of Means, Small Samples, Normal
Population, Variances Not Assumed Equal*) *Given that $x_1, x_2, \ldots, x_n$ is a
random sample from a normally distributed population determined by X,
and that $y_1, y_2, \cdots, y_m$ is a random sample from a normally distributed population determined by Y, where X and Y are independent with variances
that are unknown and not necessarily the same, this algorithm will determine, at the α level of significance, whether to accept or reject the null
hypothesis $H_0 : \mu_X - \mu_Y = d_0$ against one of the alternative hypotheses (a)
$H_1 : \mu_X - \mu_Y > d_0$, (b) $H_1 : \mu_X - \mu_Y < d_0$, or (c) $H_1 : \mu_X - \mu_Y \neq d_0$.
It is assumed that at least one of n and m is less than 30 for otherwise we
could use Algorithm 8.2.2.*

Step 1 [Calculate the test statistic] *Compute*

$$t = \frac{(\overline{x} - \overline{y}) - d_0}{\sqrt{\dfrac{s_X^2}{n} + \dfrac{s_Y^2}{m}}}. \tag{8.19}$$

---

[2]The size of the time unit is proprietary but widely believed to be 50 milliseconds.

Step 2 [Determine the critical region] *Compute*

$$\nu = \frac{(s_X^2/n + s_Y^2/m)^2}{(s_X^2/n)^2/(n-1) + (s_Y^2/m)^2/(m-1)}. \tag{8.20}$$

The test statistic is Student's t with ν degrees of freedom. The critical region for alternative hypothesis (a) is the set of all $t > t_{\nu,\alpha}$; for alternative hypothesis (b), it is the set of all $t < -t_{\nu,\alpha}$; and for alternative hypothesis (c), it is the two tails $|t| > t_{\nu,\alpha/2}$.

Step 3 [Accept or reject H_0] *If the test statistic calculated in Step 1 falls in the critical region found in Step 2 reject H_0; otherwise accept it.*

The power of this test is discussed in Best and Rayner [4]. One problem with applying this test is that the value of ν calculated by formula (8.20) may not be an integer. The SAS/STAT function PROBT and the MINITAB command CDF with the subcommand T are able to calculate Student's t distribution for fractional degrees of freedom. However, MINITAB rounds down to the next smaller integer for ν in the TTEST procedure. If you are making this calculation with a pocket calculator, you can round ν down to the next lower integer value. The corresponding critical value of t will be a reasonable approximation to the correct value and will err on the conservative side. That is, the p-value will be larger than the exact value, so that it will be more difficult to reject the null hypothesis.

Pearson and Please [21] show that Algorithm 8.2.3 is quite robust to the assumption that the populations are normal. The test is particularly robust when the two sample sizes are equal. If the two populations are symmetric, this increases the robustness.

Example 8.2.5 Hoopers Hoops has used two different self-study courses, Course A and Course B, to teach its beginning programmers good programming techniques. The success of each course is evaluated by scores the students achieve on the Hoopers Hula Hoop Programmers Test. Nine students using Course A achieved an average test score of 89.6 with a sample standard deviation of 3.6. The seven programmers who took Course B got an average score of 81.9 with a sample standard deviation of 12.7. Assuming test scores are normally distributed, test the null hypothesis $\mu_A = \mu_B$ against the alternative hypothesis that $\mu_A > \mu_B$ at the 10% level of significance, using Algorithm 8.2.3.

Solution We calculate the test statistic

$$t = \frac{89.6 - 81.9}{\sqrt{(3.6^2/9) + (12.7^2/7)}} = 1.556. \tag{8.21}$$

We calculate the number of degrees of freedom ν to be

$$\nu = \frac{(3.6^2/9 + 12.7^2/7)^2}{(3.6^2/9)^2/8 + (12.7^2/7)^2/6} = 6.7536. \tag{8.22}$$

We use $\nu = 6$ and find from Table 5 that $t_{6,0.1} = 1.44$, so we must reject H_0 at the 10% level of significance and conclude that Course A is the most effective. The SAS/STAT function TINV shows that $t_{6.7536,0.1} = 1.42029$, while the MINITAB command INVCDF with the subcommand T yields $t_{6.7536,0.1} = 1.4203$. The SAS/STAT function PROBT yields the exact p-value of 0.082606. The MINITAB command CDF with subcommand T yields the p-value 0.0826. My HP-21S yields the p-value 0.08536 using $\nu = 6$. $\square$

For the next algorithm we assume that the two populations are normal and have the *same* variance. When this is true, the test is exact. In addition the true value of the variance can be estimated more accurately because the values from the two samples are pooled for the estimate. The SAS/STAT procedure TTEST provides a statistical test of the hypothesis that the variances are the same. It also calculates the p-values for Algorithms 8.2.3 and 8.2.4, and allows the user to decide which value to use. The EXPLORE program TWOSAM does most of what TTEST does except that no p-values are calculated. The MINITAB command TTEST assumes the variances are unequal unless you specify that you want it to assume they are equal. In the next section we will discuss the two-sample test for equal variances.

Algorithm 8.2.4 (*Two-Sample Test of Means, Small Samples, Normal Population, Variances Equal*) *Given a random sample $x_1, x_2, \ldots, x_n$ from a normally distributed population determined by X, and a random sample $y_1, y_2, \cdots, y_m$ from a normally distributed population determined by Y, where X and Y are independent with equal but unknown variances, this algorithm will determine, at the α level of significance, whether to accept or reject the null hypothesis $H_0 : \mu_X - \mu_Y = d_0$ against one of the alternative hypotheses (a) $H_1 : \mu_X - \mu_Y > d_0$, (b) $H_1 : \mu_X - \mu_Y < d_0$, or (c) $H_1 : \mu_X - \mu_Y \neq d_0$. It is assumed that at least one of n and m is less than 30, for otherwise we could use Algorithm 8.2.2.*

Step 1 [Calculate the test statistic] *Compute*

$$t = \frac{(\overline{x} - \overline{y}) - d_0}{s_p\sqrt{\frac{1}{n} + \frac{1}{m}}}, \tag{8.23}$$

where

$$s_p = \left[\frac{(n-1)s_X^2 + (m-1)s_Y^2}{n+m-2}\right]^{1/2} \tag{8.24}$$

Step 2 [Determine the critical region] *The test statistic is Student's t with $n+m-2$ degrees of freedom. The critical region for alternative hypothesis (a) is the set of all $t > t_{n+m-2,\alpha}$; for alternative hypothesis (b), it is the set of all $t < -t_{n+m-2,\alpha}$; and for alternative hypothesis (c), it is the two tails $|t| > t_{n+m-2,\alpha/2}$.*

Step 3 [Accept or reject H_0] *If the test statistic calculated in Step 1 falls in the critical region found in Step 2, reject H_0; otherwise accept it.*

Proof The proof is given by Larsen and Marx [15]. ∎

Example 8.2.6 The Hotstuff Chili Company has two teams of application programmers, Team Able and Team Baker, independently develop the software for a new interactive system. (A team of experts will decide which of the systems to implement.) Table 8.2.2 shows the mean number of lines of code each of the 20 programmers on Team Able produced per day while Table 8.2.3 shows the corresponding mean numbers for Team Baker. The average for Team Able is 97.38 and for Team Baker is 87.13 lines per programmer day. Sam Cool (often called Cool Sam), the manager of MIS, decides to do a statistical analysis to determine whether or not Team Able is more productive than Team Baker—Team Able averages 10.25 lines of code per day more than Team Baker. Sam uses the 5% level of significance.

Solution Sam decides to use the null hypothesis that the mean number of lines of code per programmer day is the same for the two teams. The alternative hypothesis is that the mean is larger for Team Able. Figure 8.2.4 shows the MINITAB TWOSAMPLE solution. The subcommand ALTERNATIVE=+1 tells MINITAB that H_1 is $\mu_1 > \mu_2$. Since we did not use the subcommand POOLED, TWOSAMPLE assumes the variances are unequal and makes Welch's Behrens–Fisher correction to arrive at 34 degrees of freedom for the test. The value of the test statistic is 0.96 with a p-value of 0.17, so the null hypothesis cannot be rejected. Figure 8.2.5 provides the solution by the EXPLORE program TWOSAM. TWOSAM provides the data to test the hypothesis of equal variances as well as the value of the test statistic and the number of degrees of freedom of the test statistic. The SAS/STAT TTEST solution of Figure 8.2.6 provides even more information. It shows that the p-value of the F-test for equal variances is 0.1415, which is strong but not overwhelming evidence in favor of equality of the variances. The F-test for equal variances is not a robust test. That is, it depends strongly upon the normality of both populations. We discuss the lack of redeeming features of this test in more detail at the end of Section 8.3.2.   The p-value 0.3431, given by the SAS/STAT procedure TTEST for the equal variances case, assumes that the alternative hypothesis is $\mu_1 \neq \mu_2$, so it is twice the correct value for the alternative hypothesis $\mu_1 > \mu_2$ that we are testing. $\square$

Table 8.2.2.   Team Able

121.2	64.3	79.1	135.7	105.6	92.7	114.8	138.27	87.2	63.7
119.6	142.7	81.4	58.9	175.3	20.7	47.8	112.70	53.4	132.6

Table 8.2.3.   Team Baker

53.798	42.832	100.318	68.122	99.532	76.452	76.585	115.624
139.437	66.254	103.859	91.770	93.435	63.412	89.562	60.232
81.448	144.096	116.115	59.682				

```
MTB > TWOSAMPLE 'able' 'baker';
 SUBC> ALTERNATIVE=+1.
 TWOSAMPLE T FOR able VS baker
           N       MEAN     STDEV    SE MEAN
 able     20       97.4      39.0      8.7
 baker    20       87.1      27.6      6.2
 95 PCT CI FOR MU able - MU baker: (-11.5, 32.0)
 TTEST MU able = MU baker (VS GT): T= 0.96  P=0.17  DF=  34
```

Figure 8.2.4.   MINITAB solution to Example 8.2.6.

> *Any clod can have the facts, but having opinions is an art.*
> Charles McCae

> *I have a feeling we're not in Kansas anymore, Toto.*
> Dorothy
> *The Wizard of Oz*

```
TEST FOR DIFFERENCES BETWEEN TWO SAMPLES: ABLE VS. BAKER
1. Test of Two Means: Assuming Equal Population Variances
        Student's t =    0.960  with d.f.= 38
2. Test of Two Means: Assuming Unequal Population Variances
        Student's t =    0.960  with d.f.= 38 ( d.f. =  34 using
                                  Welch's Behrens-Fisher correction)
3. Test For Equality of Population Variances
        F =    1.994 with d.f.= 19 for numerator
                       and d.f.= 19 for denominator
4. Pooled Estimate of Population Variance
        Estimated Pooled Variance = 1140.511
        Estimated Pooled Std. Dev.= 33.77145
```

Figure 8.2.5. EXPLORE TWOSAM solution to Example 8.2.6.

```
Ttest Procedure
Variable: LINES
ID      N        Mean      Std Dev     Std Error     Minimum       Maximum
------------------------------------------------------------------------------
a      20  97.38000000  38.97641122  8.71539050  20.70000000  175.3000000
b      20  87.12825000  27.60183268  6.17195742  42.83200000  144.0960000


Variances        T       DF     Prob>|T|
-----------------------------------------
Unequal     0.9599    34.2     0.3438
Equal       0.9599    38.0     0.3431

For HO: Variances are equal, F'=    1.99 with 19 and 19 DF
        Prob > F'= 0.1415
```

Figure 8.2.6. SAS/STAT solution to Example 8.2.6.

We have met the enemy, and they is us.
Pogo
(Walt Kelly)

What we see depends mainly on what we look for.
John Lubbock

8.3 Tests of Variances

Unfortunately, the tests available concerning variance are not as extensive as those concerning means, and the conditions under which the tests are

valid are more restrictive. We must rejoice in what *is* available. Just as
with means there are two basic kinds of hypothesis tests of variances. You
guessed it: they are called *one-sample tests* and *two-sample tests*. The
first type is a test of the null hypothesis that the variance of a population
(that is, of the random variable X, which determines the population) has
some specified value σ_0^2 against some appropriate alternative hypothesis.
A two-sample test compares the variances from two different populations.
We might want to test this hypothesis so that we know whether or not to
pool our estimates of the variance; that is, choose whether to use Algorithm
8.2.3 or Algorithm 8.2.4.

In the next section we consider one-sample tests and in the following
section we consider two-sample tests.

> *God does not throw dice.*
> Albert Einstein

> *The die is cast.*
> Julius Caesar

8.3.1   One-Sample Test of Variance

Unfortunately, we can test the variance only from a normal population. We
state the test available for a normal population as an algorithm.

Algorithm 8.3.1 (*Test of Value of Variance, Normal Population*) *Given
a random sample* $x_1, x_2, \ldots, x_n$ *from a population determined by a normal
random variable* X, *this algorithm will determine at the* α *level of significance whether to accept or reject the null hypothesis* $H_0 : \sigma^2 = \sigma_0^2$ *against
one of the alternative hypotheses* (a) $H_1 : \sigma^2 > \sigma_0^2$, (b) $H_1 : \sigma^2 < \sigma_0^2$, *or*
(c) $H_1 : \sigma^2 \neq \sigma_0^2$.

Step 1 [Calculate the test statistic] *Calculate the test statistic*

$$\chi^2 = (n-1) \times \frac{s^2}{\sigma_0^2}. \tag{8.25}$$

(χ^2 *has approximately a chi-square distribution with* $n-1$ *degrees of
freedom.*)

Step 2 [Find the critical region] *The critical region for alternative
hypothesis* (a) *is the set of all* $\chi^2 > \chi_{n-1,\alpha}^2$; *for alternative hypothesis*
(b), *it is the set of all* $\chi^2 < \chi_{n-1,1-\alpha}^2$ *For alternative hypothesis* (c),
the critical region consists of the two tails $\chi^2 < \chi_{n-1,1-\alpha/2}^2$ *and* $\chi^2 >
\chi_{n-1,\alpha}^2$.

Step 3 [Accept or reject H_0] *If the test statistic calculated in Step 1 falls in the critical region determined in Step 2, reject H_0 and accept H_1; otherwise accept H_0.*

Proof The proof follows immediately from Theorem 7.3.3. ∎

Power of Variance Test

It is relatively easy to calculate the power of the hypothesis test of Algorithm 8.3.1 for any particular value of the true σ^2, say σ_1^2. Calculation of the power of a number of assumed values allows one to construct a *power curve* for the test.

Suppose that the null and alternative hypotheses are $\sigma^2 = \sigma_0^2$, and $\sigma^2 > \sigma_0^2$, respectively. If $\sigma^2 = \sigma_1^2 > \sigma_0^2$, the power of the test is given by

$$
\begin{aligned}
1 - \beta &= P\left[(n-1)\frac{s^2}{\sigma_0^2} > \chi_{n-1,\alpha}^2 | \sigma^2 = \sigma_1^2\right] \\
&= P\left[(n-1)\frac{s^2}{\sigma_1^2} > \left(\frac{\sigma_0^2}{\sigma_1^2}\right)\chi_{n-1,\alpha}^2 | \sigma^2 = \sigma_1^2\right]. \quad (8.26)
\end{aligned}
$$

But if $\sigma^2 = \sigma_1^2$, then $(n-1)s^2/\sigma_1^2$ has a chi-square distribution with $n-1$ degrees of freedom; we can replace this expression in (8.26) by the symbol χ_{n-1}. Thus, (8.26) becomes

$$
1 - \beta = P\left[\chi_{n-1}^2 > \left(\frac{\sigma_0^2}{\sigma_1^2}\right)\chi_{n-1,\alpha}^2\right]. \quad (8.27)
$$

We can use (8.27) to calculate the power for any particular values of β and σ_1^2.

Let us consider an example of the use of Algorithm 8.3.1.

Example 8.3.1 For years, the scores achieved by persons taking the Programmer Aptitude Test at Brownstain International were normally distributed with mean 82.6 and variance 19.78. With the advent of new programming techniques two years ago, the scores seem to have changed. The test scores for the 150 tests observed in the last two years yield a sample mean of 83.2 and a sample variance of 27.3. At the 5% level of significance, does Algorithm 8.3.1 indicate that the variance has increased? What is the power of the test when the actual variance is 26.0?

Solution The null hypothesis is $H_0 : \sigma^2 = 19.78$ and the alternate hypothesis is $H_1 : \sigma^2 > 19.78$. The test statistic is

$$
\chi^2 = 149 \times \frac{27.3}{19.78} = 205.647. \quad (8.28)
$$

Table 4 of Appendix A is, unfortunately, not much help in calculating $\chi^2_{149, 0.05}$. The SAS/STAT function CINV yields the value 178.48535277, the MINITAB command INVCDF with subcommand CHISQUARE yields 178.4848, and the HP-21S yields 178.4854. Therefore, we reject H_0 and conclude that $\sigma^2 > 19.78$. By (8.27), the power of the test when $\sigma^2 = 26$ is

$$P[\chi^2_{149} > 135.79]. \qquad (8.29)$$

The EXPLORE program AREA yields the value 0.773 for (8.29) as does the SAS/STAT program PROBCHI, the MINITAB command CDF with sub-command CHISQUARE, the HP-21S, and the APL function CHISQUAREΔ-DIST. □

> *When you win nothing hurts.*
> Joe Namath

8.3.2   Two-Sample Test of Variance

We state the two-sample test of variance as an algorithm. We assume in the following test that we have renamed the random variables, if necessary, so that

$$s^2_X \geq s^2_Y, \qquad (8.30)$$

so that only a right-tailed test is necessary for both alternative hypotheses.

Algorithm 8.3.2 (*Two-Sample Test of Variances, Normal Population*) *Given a random sample $x_1, x_2, \ldots, x_n$ from a normal population determined by X, and a random sample $y_1, y_2, \cdots, y_m$ from a normal population determined by Y, where X and Y are independent, this algorithm will determine, at the α level of significance, whether to accept or reject the null hypothesis $H_0 : \sigma^2_X = \sigma^2_Y$ against one of the alternative hypotheses (a) $H_1 : \sigma^2_X \neq \sigma^2_Y$ or (b) $H_1 : \sigma^2_X > \sigma^2_Y$. It is assumed that the variables are renamed, if necessary, so that*

$$s^2_X \geq s^2_Y. \qquad (8.31)$$

Step 1 [Calculate the test statistic] *Compute the test statistic*

$$f = \frac{s^2_X}{s^2_Y}. \qquad (8.32)$$

Step 2 [Determine the critical region] *The test statistic has a Snedecor-F distribution with $\nu_1 = n - 1$ and $\nu_2 = m - 1$ degrees of freedom. Hence, for alternative hypothesis (a), the critical region is the set of all $f > f_{\alpha/2}(n - 1, m - 1)$. For alternative hypothesis (b), it is the set of all $f > f_\alpha(n - 1, m - 1)$.*

Step 3 [Accept or reject H_0] *If the test statistic of Step 1 falls in the critical region determined in Step 2, reject H_0; otherwise accept H_0.*

Proof By Theorem 7.3.3, $(n-1)S_X^2/\sigma_X^2$ and $(m-1)S_Y^2/\sigma_Y^2$ are independent chi-square random variables with $n-1$ and $m-1$ degrees of freedom, respectively. Therefore, $(S_X^2/\sigma_X^2)/(S_Y^2/\sigma_Y^2)$ has an F distribution with parameters $n-1$ and $m-1$ by the definition in Section 3.2.8. Hence, when H_0 is true, the same is true of S_X^2/S_Y^2, so that

$$P[f_{1-\alpha/2}(n-1, m-1) < \frac{S_X^2}{S_Y^2} < f_{\alpha/2}] = 1 - \alpha, \qquad (8.33)$$

and

$$P[\frac{S_X^2}{S_Y^2} > f_\alpha(n-1, m-1)] = \alpha. \qquad (8.34)$$

Thus, the proof is complete. ∎

Example 8.3.2 Use Algorithm 8.3.2 to determine, at the 5% level of significance, whether it is reasonable to suppose that the variance of the number of lines of code per programmer day is the same for the two groups of programmers in Example 8.2.6.

Solution The null hypothesis is that $\sigma_X^2 = \sigma_Y^2$ and the alternative hypothesis is $\sigma_X^2 \neq \sigma_Y^2$. The test statistic is

$$f = \frac{s_X^2}{s_Y^2} = \left(\frac{38.9764}{27.6018}\right)^2 = 1.994. \qquad (8.35)$$

The test statistic has an F distribution with $\nu_1 = \nu_2 = 19$. We cannot read $f_{0.025}(19, 19)$ directly from Table 6 of Appendix A. However, we can read $f_{0.025}(20, 19) = 2.51$, which shows that $f_{0.025}(19, 19) \geq 2.51$, so that we cannot reject the null hypothesis. (According to the SAS/STAT function FINV, $f_{0.025}(19, 19) = 2.52645$. It is 2.5264 according to the MINITAB command INVCDF with subcommand F, and 2.52645 by my HP-21S.) We can see from Figure 8.2.6 that SAS/STAT has determined the p-value of this test to be 0.1415, which agrees with the value calculated with my HP-21S. □

Pearson and Please [21] show that F-tests of variance, such as those of Algorithms 8.3.1 and 8.3.2, are *not* robust to the assumption of normality. In fact, Moore and McCabe [19, pages 568–570], on the basis of the Pearson and Please paper (which they discuss in detail) claim that inferences about variances are "so lacking in robustness as to be of little use in practice."

This agrees with my experience. This also adds credibility to the practice of assuming the variances are not equal when performing the two-sample t-test of means, that is, to using Algorithm 8.2.3 rather than Algorithm 8.2.4 for performing the two-sample t-test of differences between means.

> *Don't be afraid to take a big step when one is indicated. You can't cross a*
> *chasm in two small jumps.*
> David Lloyd George

> *When in doubt tell the truth.*
> Mark Twain

> *Always do right. This will surprise some people and astonish the rest.*
> Mark Twain

> **documentation** *n.* [*Latin documentum "warning."*] **1** *The promised*
> *literature that fails to arrive with the supporting hardware.* **2** *A single,*
> *illegible, photocopied page of proprietary caveats and suggested infractions.*
> **3** *The detailed, unindexed description of a superseded package.*
> Stan Kelly-Bootle
> *The Devil's DP Dictionary*

8.4   Bernoulli Tests

The Bernoulli random variable is not only the simplest but also one of the most useful random variables because several other useful random variables can easily be formulated in terms of it. Examples include the binomial, the geometric, and the Pascal. Theorem 7.3.4 explained how to calculate a confidence interval for p. We will discuss hypothesis tests concerning the parameter p of a Bernoulli random variable.

8.4.1   One-Sample Bernoulli Tests

As mentioned earlier, p is sometimes known as the true proportion of successes in a population. In the latter case, we have in mind a population in which each member may or may not have a certain attribute. We call those members that have the attribute *successes* and those that don't *failures*, even though this sounds judgmental. From a random sample of size n, we estimate p by counting the number of successes, k. By Example 7.2.5 the maximum likelihood estimate of p is $\hat{p} = k/n$. In Algorithm 8.4.1 we assume that the size of the sample is large enough that we can treat $\hat{p}$ as a normal random variable. We use the common rule of thumb that $n \times p_0 \geq 5$ and

$n \times (1 - p_0) \geq 5$, where p_0 is the value of p assumed in the null hypothesis. The calculations then are rather simple. For smaller samples we must use Algorithm 8.4.2. It requires more computation than Algorithm 8.4.1 but is more accurate. In fact, it can be used for the large sample case as well.

We state the large-sample, one-sample Bernoulli test as an algorithm.

Algorithm 8.4.1 (*Test of Value of Bernoulli Parameter p, Large Sample*) *Suppose a random sample of size n from a population yields k successes when the true proportion of successes is p. Thus, the maximum likelihood estimate of p is $\hat{p} = k/n$. This algorithm will determine at the α level of significance whether to accept or reject the null hypothesis $H_0 : p = p_0$ against one of the alternative hypotheses (a) $H_1 : p > p_0$, (b) $H_1 : p < p_0$, or (c) $H_1 : p \neq p_0$. We assume that $n \times p_0 \geq 5$ and $n \times (1 - p_0) \geq 5$.*

Step 1 [Calculate the test statistic]

$$z = \frac{(\hat{p} - p_0)}{\sigma_p}, \tag{8.36}$$

where

$$\sigma_p = \frac{\sqrt{\hat{p}(1 - \hat{p})}}{\sqrt{n}}. \tag{8.37}$$

Step 2 [Find the critical region] *The critical region for alternative hypothesis (a) is the set of all z such that $z > z_\alpha$. For alternative hypothesis (b), it is the set of all z such that $z < z_\alpha$. For alternative hypothesis (c), the critical region comprises the two tails consisting of all z such that $z < -z_{\alpha/2}$ and all z such that $z > z_{\alpha/2}$.*

Step 3 [Accept or reject H_0] *If the test statistic calculated in Step 1 falls in the critical region determined in Step 2, reject H_0 and accept H_1; otherwise accept H_0.*

Proof Since $E[X] = p$, Algorithm 8.4.1 is a special case of Algorithm 8.2.1. ∎

Let us consider an example of the use of this algorithm.

Example 8.4.1 Super Solid State (sometimes known as S-cubed) produces a popular microprocessor called the Super A. Let p be the fraction of the Super A microprocessors that are what Super Solid State calls *super grade*. Until recently p was 0.67. However, Super Sandy, the resident production genius, has developed a new production process, that she claims will dramatically increase p. Using the new method, S-cubed produced 200 Super A microprocessors of which 150 are super grade. Super Sandy decides to test the statistical hypothesis, at the 5% level, that $p > 0.67$.

Solution Super Sandy uses the null hypothesis $H_0 : p = 0.67$ with the alternative hypothesis $H_1 : p > 0.67$. She calculates

$$\hat{p} = \frac{150}{200} = 0.75. \tag{8.38}$$

Hence, by (8.36), $z = 2.6128$. Since $z_{0.05} = 1.645$, Sandy can reject the null hypothesis and conclude that $p > 0.67$. The p-value of the test is 0.0045; there can be little doubt that the new production method is superior to the old. $\square$

Example 8.4.2 Boob Tubes, a manufacturer of television picture tubes, claims that 90% or more of its tubes last at least three years. Couch Potatoes, a consumer advocate and publisher, challenges this claim. Couch collects a random sample of 200 TV tubes sold by Boob Tubes and finds that at least 40 of them failed within 3 years of their purchase. At the one percent level of significance, what does Couch Potatoes conclude?

Solution Couch lets p be the fraction of defective tubes sold by Boob, that is, those that fail to function for at least 3 years. The null hypothesis chosen is $H_0 : p = 0.1$ with the alternative hypothesis $H_1 : p > 0.1$. From the data Couch Potatoes calculates $\hat{p} = 40/200 = 0.2$,

$$\sigma_p = \sqrt{\frac{0.2 \times 0.8}{200}} = 0.028284271, \tag{8.39}$$

so that

$$z = \frac{0.1}{\sigma_p} = 3.536. \tag{8.40}$$

Since $z_{0.01} = 2.3263$, Couch decides that the null hypothesis must be rejected; it appears that Boob Tubes' tubes are not as long-lived as claimed. $\square$

Algorithm 8.4.2 (*Test of Value of Bernoulli Parameter p, Small Sample*) *Suppose a random sample of size n yields k successes when the true proportion of successes is p. Thus, $\hat{p} = k/n$. This algorithm will determine at the α level of significance whether to accept or reject the null hypothesis $H_0 : p = p_0$ against one of the alternative hypotheses (a) $H_1 : p > p_0$, (b) $H_1 : p < p_0$, or (c) $H_1 : p \neq p_0$. Let Y be the random variable that counts the number of successes in the sample. Thus, under H_0, Y has a binomial distribution with parameters n and p_0.*

Step 1 [Calculate the p-value of the test]

For case (a), calculate the p-value using the formula

$$p\text{-}value = P[Y \geq k] = \sum_{i=k}^{n} \binom{n}{i} \times p_0^i \times (1 - p_0)^{n-i}. \qquad (8.41)$$

For case (b), calculate the p-value using the formula

$$p\text{-}value = P[Y \leq k] = \sum_{i=0}^{k} \binom{n}{i} \times p_0^i \times (1 - p_0)^{n-i}. \qquad (8.42)$$

For case (c), the p-value depends upon whether $k \leq n \times p_0$ or $k > n \times p_0$. If the former is true, the p-value is twice the p-value calculated by (8.42). If $k > n \times p_0$, then the p-value is twice the p-value calculated by (8.41).

Step 2 [Accept or reject H_0]

If the test p-value calculated in Step 1 is less than α, reject H_0 and accept H_1; otherwise accept H_0.

Proof Algorithm 8.4.2 follows immediately from the definition of the binomial distribution. ∎

Example 8.4.3 Let us apply Algorithm 8.4.2, a more exact test, to Example 8.4.1.

Solution Since the alternative hypothesis is that $p > 0.67$, we must use (8.41) to calculate the p-value of the test. The APL function BINSUM shows that the p-value is 0.00875. Hence, there is no doubt that the null hypothesis should be rejected in favor of the alternative hypothesis $H_1 : p > 0.67$. □

Example 8.4.4 In the Chips, a manufacturer of solid-state memory chips, claims that not more than one percent of its chips are defective. Computer Clones decides to check this claim by testing a random selection of 50 chips and finds that two are defective. Let p be the fraction of defective chips. At the 10 percent level of significance, can Computer Clones reject the null hypothesis $H_0 : p = 0.01$ in favor of the alternative hypothesis $H_1 : p > 0.01$?

Solution Since the alternative hypothesis is $H_1 : p > 0.01$, we calculate the p-value using (8.41). Thus,

$$
\begin{aligned}
p\text{-value} \;&=\; \sum_{i=2}^{50} \binom{50}{i}(0.01)^i \times (0.99)^{50-i} \\[2mm]
&=\; 1 - \binom{50}{0}(0.01)^0 \times (0.99)^{50} - \binom{50}{1} \times 0.01 \times (0.99)^{49} \\[2mm]
&=\; 0.089435313.
\end{aligned}
\tag{8.43}
$$

Since the p-value is smaller than 0.1, Computer can reject the null hypothesis at the 10 percent level. It could not, however, be rejected at the five percent level. $\square$

We will now consider two-sample Bernoulli tests.

> *A burden in the bush is worth two in your hands.*
> James Thurber

8.4.2   Two-Sample Bernoulli Tests

In this section we will compare the probability of success for two independent Bernoulli random variables X and Y. We assume that we have a random sample of size n from a population determined by X, that contains k_X successes and thus yields the estimate $\hat{p}_X = k_X/n$. Similarly, we have a random sample of size m from a population determined by Y, that contains k_Y successes and thus yields the estimate $\hat{p}_Y = k_Y/m$. In Algorithm 8.4.3 we consider the large-sample case in which the sampling distribution of $\widehat{p_X - p_Y}$ is approximately normal. If the null hypothesis is $H_0 : p_X - p_Y = d_0 \neq 0$, the hypothesis test procedure is slightly different than it is with the null hypothesis $H_0 : p_X = p_Y$. In the former case, the test statistic is

$$
z = \frac{(\hat{p}_X - \hat{p}_Y) - d_0}{s_{p_X - p_Y}},
\tag{8.44}
$$

where

$$
s_{p_X - p_Y} = \sqrt{\frac{\hat{p}_X(1 - \hat{p}_X)}{n} + \frac{\hat{p}_Y(1 - \hat{p}_Y)}{m}}.
\tag{8.45}
$$

In the latter case we can improve our estimate of the common value of the probability of success by pooling our estimate, that is, by using

$$
\hat{p} = \frac{k_X + k_Y}{n + m} = \frac{n\hat{p}_X + m\hat{p}_Y}{n + m}.
\tag{8.46}
$$

Substituting (8.46) into (8.45) yields

$$s_p = \sqrt{\hat{p}(1 - \hat{p}) \left(\frac{1}{n} + \frac{1}{m} \right)}. \tag{8.47}$$

Algorithm 8.4.3 (*Two-Sample Bernoulli Test, Large Samples*) *Suppose a random sample of size n from a population yields k_X successes while a random sample of size m from another independent population yields k_Y successes. We estimate the fraction of successes in the first population by $\hat{p}_X = k_X/n$ and in the second by $\hat{p}_Y = k_Y/m$. This algorithm will determine, at the α level of significance, whether to accept or reject the null hypothesis $H_0 : p_X - p_Y = d_0$ against one of the alternative hypotheses (a) $H_1 : p_X - p_Y > d_0$, (b) $H_1 : p_X - p_Y < d_0$, or (c) $H_1 : p_X - p_Y \neq d_0$. Alternatively, the null hypothesis may be $H_0 : p_X = p_Y$ against one of the null hypotheses (a) $H_1 : p_X > p_Y$, (b) $H_1 : p_X < p_Y$, or (c) $H_1 : p_X \neq p_Y$. It is assumed that $n \times \hat{p}_X \geq 5$, $n \times (1 - \hat{p}_X) \geq 5$, $m \times \hat{p}_Y \geq 5$, and $m \times (1 - \hat{p}_Y) \geq 5$.*

Step 1 [Calculate the test statistic] *For the null hypothesis $p_X - p_Y = d_0$, compute the test statistic*

$$z = \frac{(\hat{p}_X - \hat{p}_Y) - d_0}{s_{p_X - p_Y}}. \tag{8.48}$$

For the null hypothesis $p_X = p_Y$, the test statistic is

$$z = \frac{\hat{p}_X - \hat{p}_Y}{s_p}. \tag{8.49}$$

Step 2 [Determine the critical region] *For alternative hypothesis (a), the critical region is the set of all $z > z_\alpha$; for alternative hypothesis (b), it is the set of all $z < -z_\alpha$; and for alternative (c), it comprises the two tails consisting of all z such that $z < -z_{\alpha/2}$ and all z such that $z > z_{\alpha/2}$.*

Step 3 [Accept or reject H_0] *If the test statistic of Step 1 falls in the critical region determined in Step 2, reject H_0; otherwise accept H_0.*

Example 8.4.5 In 1954 a Salk polio vaccine field trial was held to determine the effectiveness of the vaccine. Of the 401,826 children involved in the test 200,712 (the treatment group) were given the Salk vaccine and 201,114 (the control group) were given a placebo. (This was a double-blind experiment in which neither the children nor the administrators of

the vaccine knew which children were getting the Salk vaccine and which
the placebo.) Only 33 of the children who received the Salk vaccine con-
tracted polio while 115 of the control group did. We let p be the fraction
of children contracting polio. The null hypothesis is $H_0 : p_X = p_Y$ where
p_X is the proportion of children in the treated group who got polio and p_Y
is the proportion in the control group. The alternative hypothesis is $H_1 :$
$p_X < p_Y$. From the data, we see that $\hat{p}_X = 33/200,712 = 0.000164415$ and
$\hat{p}_Y = 115/201,114 = 0.000571815$. We also calculate

$$\hat{p} = \frac{33 + 115}{401,826} = 0.000368319. \tag{8.50}$$

Hence, (8.47) yields $s_p = 0.00006054$, so the test statistic $z = -6.729$. The
p-value of the test is thus almost zero; the APL function NDIST produces
the value 8.59×10^{-12} and the SAS/STAT function PROBNORM yields
8.54×10^{-12}, which is confirmed by my HP-21S. We conclude that the Salk
vaccine *is* effective. $\Box$

For the small-sample Bernoulli test it is convenient to set up our hypothesis
test procedure using the binomial distribution. Suppose X is binomial with
parameters n and p_X, while Y is binomial with parameters m and p_Y.
Suppose also that X and Y are independent. We wish to test the null
hypothesis $H_0 : p_X = p_Y$ against, say, $H_1 : p_X > p_Y$. Let $k = k_X + k_Y$ be
the number of successes in $n + m$ trials. If H_0 is true so that $p_X = p_Y = p$,
then

$$P[X + Y = k] = \binom{n + m}{k} p^k \times (1 - p)^{n+m-k}. \tag{8.51}$$

Now

$$P[X = k_X] = \binom{n}{k_X} p^{k_X} \times (1 - p)^{n-k_X}, \tag{8.52}$$

and

$$P[Y = k_Y] = \binom{m}{k_Y} p^{k_Y} \times (1 - p)^{m-k_Y}. \tag{8.53}$$

Therefore,

$$\begin{aligned} P[X = k_X | X + Y = k] &= \frac{P[X = k_X \text{ and } Y = k - k_X]}{P[X + Y = k]} \\ &= \frac{\binom{n}{k_X} \times \binom{m}{k - k_X}}{\binom{n + m}{k}}, \end{aligned} \tag{8.54}$$

for $k_X = 0, 1, \cdots, n$. Let us define a random variable U by

$$P[U = j] = P[X = j | X + Y = k], \tag{8.55}$$

for $j = 0, 1, \ldots, k$. Hence, by (8.54), U is a hypergeometric random variable with parameters $n = k$, $N = n + m$, and $r = k$. (We defined *hypergeometric random variable* in Chapter 3 just before Exercise 11.) Note that under H_0, the conditional distribution of X given $X + Y = k$ is independent of the common value of $p_X = p_Y$. Let us consider how to use (8.54) to test $H_0 : p_X = p_Y$ against $H_1 : p_X > p_Y$. For a given value of k, it is clear that a small value of k_X tends to support H_0, while a large value tends to support H_1. We can therefore reject H_0 at the α level if, assuming H_0, $k_X > c$, where c is the smallest integer such that

$$P[X \geq c | X + Y = k] \leq \alpha. \tag{8.56}$$

This is the Fisher–Irwin test. The p-value of this test is

$$P[X \geq k_X | X + Y = k] = \sum_{j=k_X}^{k} P[U = j]. \tag{8.57}$$

Further tests for other alternative hypotheses are given in Algorithm 8.4.4.

Algorithm 8.4.4 (*Two-Sample Bernoulli Test, The Fisher–Irwin Test*)
Let X be the binomial random variable that counts the number of successes in a random sample of size n from a population with the proportion p_X of successes. Let Y be a binomial random variable that counts the number of successes in a random sample of size m from a population with proportion p_Y of successes. Assume that X and Y are independent. This algorithm will determine, at the α level of significance, whether to accept or reject $H_0 : p_X = p_Y$ against one of the alternative hypotheses (a) $H_1 : p_X > p_Y$, (b) $H_1 : p_X < p_Y$, or (c) $H_1 : p_X \neq p_Y$. Let $k = k_X + k_Y$ be the total number of successes in the two samples and let U be defined by (8.55).

Step 1 [Calculate the test statistic] *Calculate the number of successes k_X in the sample from the first population. Calculate $k = k_X + k_Y$, the total number of successes in the two populations.*

Step 2 [Determine the critical region] *For alternative hypothesis (a), compute c, the smallest integer such that under H_0,*

$$P[X \geq c | X + Y = k] \leq \alpha. \tag{8.58}$$

Then the critical region is the set of all $k_X > c$. The p-value of this test is

$$P[X \geq k_X | X + Y = k] = \sum_{j=k_X}^{k} P[U = j]. \tag{8.59}$$

For alternative hypothesis (b), compute c, the largest integer such that

$$P[X \leq c | X + Y = k] \leq \alpha. \tag{8.60}$$

Then the critical region is the set of all $k_X < c$. The p-value of this test is

$$P[X \leq k_X | X + Y = k] = \sum_{j=0}^{k_X} P[U = j]. \tag{8.61}$$

For alternative hypothesis (c) the critical region consists of the set of all $k_X > c_1$ and all $k_X < c_2$ where c_1 and c_2 are determined as above with probability $\alpha/2$ in each tail. The p-value of this test is two times the smaller tail probability, that is, two times the minimum of the two numbers obtained from (8.59) and (8.61).

Step 3 [Accept or reject H_0] *If the test statistic calculated in Step 1 falls in the critical region found in Step 2 or if the p-value of the test is less than α, reject H_0; otherwise accept it.*

Example 8.4.6 The IRS audited the tax returns of 30 families in Boondocks Center. Twenty of the returns were from lower-income families and ten from high-income families. Four low-income families and five high-income families had underpaid their taxes according to the IRS auditors. At the five percent level of significance, are the proportions of families who underpaid the same?

Solution The null hypothesis is $H_0 : p_X = p_Y$ and the alternative hypothesis is $H_1 : p_X \neq p_Y$. Using the APL program HYPERGΔDIST with parameters $n = 9$, $N = 30$, and $r = 20$ (or the *Mathematica* program **hypergdist**), we calculate

$$P[X \geq 4 | X + Y = 9] = \sum_{j=4}^{9} P[U = j] = 0.9816, \tag{8.62}$$

and

$$P[X \leq 4 | X + Y = 9] = \sum_{j=0}^{4} P[U = j] = 0.1037278564, \tag{8.63}$$

where U is defined by (8.55). Therefore, the p-value of the test is $2 \times 0.1037278564 = 0.2074557$. (The innovative BASIC program 6-6-1 of Ross [23] provides the same answer. Ross's program always uses the alternative hypothesis (c), so it cannot be used to solve Exercise 8.) Hence, we cannot reject the null hypothesis that the proportions are the same. $\square$

> *The conduct of e*
> *Is abhorrent to me.*
> *He is (not to enlarge on his disgrace)*
> *More than a little base.*

J. A. Lindon

8.5 Chi-Square Tests

The chi-square test is a versatile test that is easy to perform and requires little calculation. It takes two basic forms: one for testing goodness-of-fit and one for testing independence in contingency tables. Algorithm 8.5.1 describes how to perform a goodness-of-fit test and Algorithm 8.5.2 describes the chi-square test of independence for a contingency table. It is the *only* goodness-of-fit test discussed in many, if not most, statistics books. By *goodness-of-fit test*, we mean a method of testing whether a given population is determined by a particular random variable, such as exponential, normal, Poisson, Erlang-k, gamma, etc. We will, of course, have some clues from the sample parameters, such as $\overline{x}$ or s. In addition, we may have used some of the exploratory data analyses techniques of Section 7.4 to determine what type of random variable we are dealing with. For example, if $\overline{x} = 10$ and $s = 30$, we know that the underlying random variable X is not likely to be exponential, since the mean is equal to the standard deviation for an exponential distribution, but it could be gamma or hyperexponential.

Once we have established what type of random variable we believe determines the population, we can apply a goodness-of-fit test to make a probability judgment about our choice. Such a test is a special class of hypothesis test in which the null hypothesis is that the population is determined by a particular type of random variable (normal, gamma, etc.) and the alternative hypothesis is that it is not. Unfortunately, when a goodness-of-fit test leads to the rejection of the null hypothesis, no conclusion is obtained as to what type of random variable might fit better.

We will now consider the chi-square goodness-of-fit test. In the next section we will consider some more powerful goodness-of-fit tests. The more powerful tests, as might be expected, require more computation and are more difficult to apply.

Algorithm 8.5.1 (*Chi-Square Goodness-of-Fit Test*) *Each element of a given random sample $x_1, x_2, \ldots, x_n$, determined by the random variable X, falls into exactly one of k categories or cells (category and cell will be used interchangeably) $C_1, C_2, \cdots, C_k$. This test will determine, at the α level* *of significance, whether it is reasonable to suppose that the observed distribution of the n sample values into categories or cells is consistent with the null hypothesis that X has a given distribution.*

Step 1 [Count O_i] *Count the number, O_i, of observed elements in category C_i, for $i = 1, 2, \cdots, k$.*

Step 2 [Calculate the E_i] *On the basis of the null hypothesis that X has a given distribution, calculate E_i, the expected number of elements in category C_i, for $i = 1, 2, \cdots, k$.*

Step 3 [Calculate χ^2] *Calculate the chi-square statistic*

$$\chi^2 = \sum_{i=1}^{k} \frac{(O_i - E_i)^2}{E_i}. \tag{8.64}$$

Step 4 [Calculate the number of degrees of freedom m of the underlying chi-square distribution] *Set $m = k - 1$. Then subtract one from m for each independent parameter that is estimated from the data to generate the E_i values in Step 2.*

Step 5 [Find critical value] *Find the critical value $\chi^2_{m,\alpha}$ such that the probability that a chi-square random variable with m degrees of freedom will exceed $\chi^2_{m,\alpha}$ is α. (Table 4 of Appendix A gives these values.)*

Step 6 [Accept or reject H_0] *If $\chi^2 \geq \chi^2_{m,\alpha}$, reject H_0; otherwise accept H_0.*

The χ^2 statistic provides an intuitively satisfying measure of the deviation between what is observed and what is expected. The deviation for each cell is first measured by $(O_i - E_1)^2$ and then scaled by dividing by the E_i term. Finally, we add up the contributions from all the cells. The mathematical justification for the chi-square test is that the distribution of the χ^2 statistic approaches that of a chi-square distribution with m degrees of freedom as $n \to \infty$. (This is proven in Cramér [8].) Thus, we can associate a probability with the result. For small values of n, the distribution of the χ^2 statistic may not be closely approximated by a chi-square distribution, and the chi-square test will not yield good results. It has been discovered, empirically, that the chi-square test works best when all the E_i are at least 5, for if E_i is small, the division by E_i in the term $(O_i - E_i)^2/E_i$ can cause a large error in the value of χ^2. We sometimes pool categories or cells to make each $E_i \geq 5$. Another, less demanding rule of thumb is that each $E_i \geq 1$ and at least 80 percent of the E_i exceed 5. Let us consider some examples to illustrate the test of Algorithm 8.5.1.

Example 8.5.1 Let us consider the horse-kick data of Exercise 5, Chapter 7. This set of data, shown in Table 8.5.1, is one of the most famous in statistics. In his widely referenced book Bortkiewicz [6] analyzed the data concerning deaths caused by horse kicks in the 19th century Prussian cavalry. Ten Prussian cavalry corps were monitored over a period of 20 years. The random variable X measured the number of fatal horse kicks per corps-year. Bortkiewicz fitted the Poisson distribution to the data. Let us use Algorithm 8.5.1 to determine whether or not X appears to have a Poisson distribution at the 5 percent level of significance. As we asked you to determine in the above exercise, the maximum likelihood estimate of α, the average number of deaths per corps-year, is 0.61 (the total number of deaths, 122, divided by the number of corps-years, 200, $= 122/200$). We calculate

$$P[X = i] = e^{-0.61} \times \frac{0.61^i}{i!},$$

for $i = 0, 1, 2, 3$. We also calculate

$$P[X \geq 4] = 1 - \sum_{i=0}^{3} P[X = i].$$

(The maximum number of observed fatalities in a corps-year was 4.) Then we calculate

$$E_i = 200 \times P[X = i]$$

for each i. The results are shown in Table 8.5.1. Since $E_4 = P[X \geq 4] = 0.72$ and $E_3 = 4.11$ are both less than 5, we form a new class called "≥ 3" in the revised table, Table 8.5.2. For this class, the expected value is very close to 5. We now apply Step 3 of Algorithm 8.5.1 to calculate

$$\chi^2 = \frac{(109 - 108.67)^2}{108.67} + \frac{(65 - 66.28)^2}{66.28} + \frac{(22 - 20.22)^2}{20.22} + \frac{(4 - 4.83)^2}{4.83} = 0.325.$$

Since we estimated the mean of X from the data, Step 5 yields $m = 2$. By Table 4 of Appendix A, $\chi^2_{2,0.05} = 5.9915$. Since 0.325 is much smaller than this number, we accept the null hypothesis that X has a Poisson distribution.[3] In Figures 8.5.1–8.5.3, we show how Doane's EXPLORE

---

[3]In fact the fit is so good that we should be a little suspicious of the data. See Freedman, Pisani and Purves [12, pages 425–427] for an account of how the great statistician R. A. Fisher showed that Gregor Mendel's pea experiment data were fudged. Winsor [34], in his review of Bortkiewicz's book, gives him high marks. He says that Bortkiewicz was not proclaiming the merits of the Poisson distribution but rather, "He was giving a name to a phenomenon of statistics which he had, or thought he had, discovered." Bishop *et al.* [5] also discuss Bortkiewicz's results as well as Mendel's pea experiment.

program GOODFT makes these calculations. GOODFT combines cells in the same manner that we did. It is a truly effortless way to apply Algorithm 8.5.1 to the Poisson case. □

Table 8.5.1. Horse Kick Deaths by Cavalry Corps-Year

Number of Deaths i	Observed Number O_i	Estimated Number E_i
0	109	108.67
1	65	66.28
2	22	20.22
3	3	4.11
≥ 4	1	0.72
Totals	200	200

Table 8.5.2. Horse Kick Deaths by Cavalry Corps-Year (Modified)

Number of Deaths i	Observed Number O_i	Estimated Number E_i
0	109	108.67
1	65	66.28
2	22	20.22
≥ 3	4	4.83
Totals	200	200

Example 8.5.2 Every year the World Series of professional baseball features a series of games between an American League and a National League team. The series ends when one of the teams has won 4 games. (This team is declared the winner, of course.) Thus, a World Series consists of 4 through 7 games. Table 8.5.3 shows the number of games of each series for the 50 years between 1926 and 1975. There are cynics who feel that the preponderance of seven-game series shows that economics has a strong impact on the series; the owners as well as the players make more money if more games are played. What can a goodness-of-fit test tell us?

Solution Let us assume that the teams are evenly matched and that the games are independent. Then, for each team, the series can be thought of as a Bernoulli sequence of trials with the probability of success on each trial equal to 0.5. The Pascal distribution described preceding Exercise 20 in Chapter 3 seems to fit here. For each team, winning the series in n games is achieved if $n = r + k$ with $r = 4$ and $k = n - 4 = 0, 1, 2, 3$. This probability is given by

$$p(k; 4, 0.5) = \begin{pmatrix} 4 + k - 1 \\ k \end{pmatrix} \times 0.5^4 \times 0.5^k$$

for $k = 0, 1, 2, 3$. Let X be the number of games of the series. Since the above formula yields the probability that one of the teams wins in n games, say the National League team, the probability that the series is completed in n games is given by

$$P[X = n] = 2 \times p(n - 4; 4, 0.5), \quad n = 4, 5, 6, 7, \quad (8.65)$$

since either team can win. (We note that (8.65) yields $P[X = 7] = P[X = 6] = 0.3125$, $P[X = 5] = 0.25$, and $P[X = 4] = 0.125$.) Using (8.65), we obtain the last column in Table 8.5.3. We calculate

$$\chi^2 = \frac{(9 - 6.25)^2}{6.25} + \frac{(11 - 12.5)^2}{12.5} + \frac{(8 - 15.625)^2}{15.625} + \frac{(22 - 15.625)^2}{15.625} = 7.712.$$

Since $\chi^2_{3, 0.05} = 7.8147$, we accept the null hypothesis at the 5 percent level of significance. The p-value of the test is 0.052354. We would reject the hypothesis at the 10 percent level. In Exercise 13 we ask you to consider the situation in which one team is always favored over the other. For more (and deeper) discussions of the World Series problem, see Woodside [35], Tannenbaum [33], and Brunner [7]. □

```
Enter the value of m? 4
  X    f(o)
 ---   ----
  0    ? 109
  1    ? 65
  2    ? 22
  3    ? 3
  4    ? 1
```

Figure 8.5.1. GOODFT data entry for Example 8.5.1.

```
... POISSON GOODNESS OF FIT TEST ...
Mean =    0.61 events per unit of time/space
          (rounded to nearest .01)
   X            P(X)        f(o)        f(e)
------        ------        ----        ------
   0          0.5434        109         108.7
   1          0.3314         65          66.3
   2          0.1011         22          20.2
   3          0.0206          3           4.1
4 and over    0.0036          1           0.7
------        ------        ----        ------
Totals:       1.0000        200         200.0
```

Figure 8.5.2. GOODFT screen for Example 8.5.1.

```
DETAILS OF POISSON TEST STATISTIC CALCULATION
                                2
   X                    [f(o) - f(e)] /f(e)
-----                   -----------------
   0                          0.001
   1                          0.025
   2                          0.157
3 and more                    0.140
-----                   -----------------
Total:                        0.324
Chi-Square =                  0.324 with d.f. =  2
```

Figure 8.5.3. Last GOODFT screen for Example 8.5.1.

Table 8.5.3. World Series Games

Number of Games	Number of Years	Estimated Number
i	O_i	E_i
4	9	6.250
5	11	12.500
6	8	15.625
7	22	15.625
Totals	50	50

8.5.1 Contingency Tables

An important application of the chi-square test is for testing the independence of the rows and columns of *contingency tables*. A contingency table is one in which data have a two-way classification—one by row and one by column. Suppose, for example, that Dr. Nancy Nipps, the chief admissions officer at Histate University, wants to know whether there is a difference in the graduation rate of male and female students who begin their academic careers at Histate. A random sample of entering freshman at the university in September 1980 is taken. The students are classified by sex and by whether or not they graduated within five years. This data are shown in Table 8.5.4 in a *two by two contingency table*. In all contingency tables the goal is to determine whether the row classification is independent of the column classification. In Table 8.5.4 the question is, "For the entering students, is the proportion graduating within 5 years the same for female students as it is for male students?"

Table 8.5.4. Data on entering freshmen

Sex	Graduated	Did Not Graduate	Totals
Male	15	35	50
Female	12	18	30
Totals	27	53	80

Contingency Table Notation

p_{ij} The probability an observation is in row i and column j.

$\hat{p}_{ij}$ The estimate of p_{ij}.

$p_{i.}$ The marginal probability for row i: $p_{i.} = \sum_j p_{i,j}$.

$\hat{p}_{i.}$ The estimated marginal probability for row i: $\hat{p}_{i.} = \sum_j \hat{p}_{ij}$.

$p_{.j}$ The marginal probability for column j. $p_{.j} = \sum_i p_{ij}$.

$\hat{p}_{.j}$ The estimated marginal probability for column j. $\hat{p}_{.j} = \sum_i \hat{p}_{ij}$.

n The number of observations.

O_{ij} The observed cell count of the cell in row i, column j.

The probability p_{ij} is estimated by

$$\hat{p}_{ij} = \frac{O_{ij}}{n}.$$

The marginal probabilities are estimated from the observed row totals R_i and column totals C_j by the formulas

$$\hat{p}_{i.} = \frac{R_i}{n}, \tag{8.66}$$

and

$$\hat{p}_{.j} = \frac{C_j}{n}. \tag{8.67}$$

The null hypothesis is that the row and column classifications are independent; that is,

$$p_{ij} = p_{i.} \times p_{.j}. \tag{8.68}$$

The expected cell count under the assumption of independence is

$$E_{ij} = n \times \hat{p}_{i.} \times \hat{p}_{.j} = n \times \frac{R_i}{n} \times \frac{C_j}{n} = \frac{R_i \times C_j}{n}. \tag{8.69}$$

A standard $r \times c$ contingency table of r rows and c columns is set up as shown in Table 8.5.5.

Table 8.5.5. General Contingency Table

	Column				
	1	2	$\cdots$	c	Totals
Row 1	O_{11}	O_{12}	$\cdots$	O_{1c}	R_1
Row 2	O_{21}	O_{22}	$\cdots$	O_{2c}	R_2
$\cdots$	$\cdots$	$\cdots$	$\cdots$	$\cdots$	$\cdots$
Row r	O_{r1}	O_{r2}	$\cdots$	O_{rc}	R_r
Totals	C_1	C_2	$\cdots$	C_c	n

The algorithm for the independence test of a contingency table should be stated slightly differently than the way we stated it for a goodness-of-fit test in Algorithm 8.5.1.

Algorithm 8.5.2 (*Chi-Square Test of Independence for a Contingency Table*) *It is assumed that a contingency table as shown in Table 8.5.5 has been formed from a random sample. Each observation described in the table is classified independently of other observations. This assumption is satisfied if there is only one sample that is classified by two criteria (rows and columns), or if each row is a random sample classified by column. This test will determine, at the α level of significance, whether it is reasonable to suppose that the row classification is independent of the column classification. The mathematical formulation of the null hypothesis is $H_0 : p_{i,j} = p_{i.} \times p_{.j}$ for all i and j.*

Step 1 [Calculate the expected cell counts]

$$E_{ij} = \frac{R_i \times C_j}{n}. \tag{8.70}$$

Step 2 [Calculate the χ^2 test statistic]

$$\chi^2 = \sum_{i,j} \frac{(O_{ij} - E_{ij})^2}{E_{ij}}. \tag{8.71}$$

If the table is two-by-two the χ^2 test statistic is adjusted by the Yates continuity correction, and (8.71) becomes

$$\chi^2 = \sum_{i,j} \frac{(\text{maximum}(0, (|O_{ij} - E_{ij}| - 0.5)))^2}{E_{ij}}. \tag{8.72}$$

Step 3 [Find the critical value] *Set $m = (r-1)(c-1)$ and find the critical value $\chi^2_{m,\alpha}$.*

Step 4 [Accept or reject H_0] *If $\chi^2 \geq \chi^2_{m,\alpha}$, reject H_0; otherwise accept H_0.*

Recently D'Agostino *et al.* [10] have recommended that the Yates continuity correction *not* be applied for the two-by-two table, although this has been traditional. The authors claim the Yates correction makes the test too conservative. In the next example we apply Algorithm 8.4.2 to the Nancy Nipps problem. We make the calculations with and without the Yates correction in the example.

Example 8.5.3 Consider the two-by-two contingency table displayed in Table 8.5.4. Dr. Nipps wishes to determine whether the proportion of male and female students who graduate from Histate is the same. The raw data from Table 8.5.4 shows that the proportion for men is $15/50 = 0.3$ and for women it is $12/30 = 0.4$; the proportions do not appear to be the same. However, the difference may be due to sampling error. Algorithm 8.5.2 will help Nancy decide. Applying Step 1 of the algorithm, we calculate the expected cell counts using (8.70). We obtain

$$E_{11} = \frac{50 \times 27}{80} = 16.875, \tag{8.73}$$

$$E_{12} = \frac{50 \times 53}{80} = 33.125, \tag{8.74}$$

$$E_{21} = \frac{30 \times 27}{80} = 10.125, \tag{8.75}$$

and

$$E_{22} = \frac{30 \times 53}{80} = 19.875. \tag{8.76}$$

Since we have a two-by-two table (the table may *appear* to have four rows and four columns but only the cells containing observation counts are counted in making the assessment of r and c), we use (8.72) to calculate the χ^2 test statistic. Fortunately, in making the Yates correction, we find that $|O_{ij} - E_{ij}| = 1.875$ for all i and j. Hence, we obtain

$$\chi^2 = 1.375^2 \times \sum_{i,j} \frac{1}{E_{ij}} = 0.4510.$$

If we do not apply the Yates correction, as recommended by D'Agostino *et al.* [10], we find that χ^2 is only 0.83857. Since $\chi^2_{1,0.05} = 3.8415$, we accept the null hypothesis at the five percent level of significance with or without the Yates correction. The fraction who graduate seems to be the same for men and women. Using the SAS/STAT function CHISQUARE, the MINITAB command CDF with subcommand CHISQUARE, the HP-21S, or the EXPLORE program AREA, it is easy to show that the p-value of this test is 0.502 with the Yates correction and 0.3598 without it. □

Statistical packages remove a great deal of the labor (and chance for error) in statistical tests for independence.

> *Fear not the atom in its fission*
> *The cradle will outwit the hearse.*
> *Man on this earth has a mission*
> *To survive and keep on getting worse.*

Samuel Hoffenstein
Pencil in Air

8.6   EDF Tests

In this section we will discuss some goodness-of-fit tests that are both more powerful and more sophisticated than the chi-square tests of the last section. These tests are called EDF tests because they are based on a comparison of the *empirical distribution function*, $S_n(\cdot)$ (which we will define below) to the hypothesized distribution function, $F(\cdot)$. The function $S_n(\cdot)$ is defined in terms of a random sample from a population determined by the random variable X.

Given a random sample of size n, $x_1, x_2, \ldots, x_n$, we first sort it into ascending order. Thus, after sorting and renumbering (if necessary), $x_1 \leq x_2 \leq \cdots \leq x_n$. Then we define the empirical distribution function, $S_n(\cdot)$, by

$$S_n(x) = \frac{i}{n}, \tag{8.77}$$

where i is the number of sample values $\le x$. Thus, $S_n(\cdot)$ is a step function, that is zero for x less than the smallest x_i, has a jump of $1/n$ at each x_i (unless, of course, $x_i = x_{i+1} = \cdots = x_{i+j}$ for some $j > 0$ in which case the jump is $(j+1)/n$), and is 1 for all x that are greater than or equal to the largest x_i.

The first EDF test that we will consider is the Kolmogorov–Smirnov test. This test sounds like it could be a sobriety test or a test taken in a bar.[4] Kolmogorov and Smirnov are both well-known statisticians and their test is widely used. The Kolmogorov–Smirnov (K–S) test compares the vertical distance between $S_n(x)$ and the assumed distribution function $F(x)$ at each point x. There are three common measures of this distance (statistics). They are:

$$D^+ = \sup_x \{S_n(x) - F(x)\}, \tag{8.78}$$

$$D^- = \sup_x \{F(x) - S_n(x)\}, \tag{8.79}$$

and the most common measure D defined by

$$D = \max\{D^+, D^-\}. \tag{8.80}$$

Because $S_n(\cdot)$ is a nondecreasing step function and $F(\cdot)$ is also nondecreasing (because it is a distribution function) it suffices to check (8.78) and (8.79) at the finite set of points $x_1, x_2, \ldots, x_n$. In fact,

$$D^+ = \max_i \{\frac{i}{n} - F(x_i)\}, \tag{8.81}$$

and

$$D^- = \max_i \{F(x_i) - \frac{(i-1)}{n}\}. \tag{8.82}$$

The above three statistics are called *supremum statistics*. There is another class of EDF statistics called *quadratic statistics*. We shall limit ourselves to the Anderson–Darling [2] quadratic statistic, A^2, defined by

$$A^2 = n \int_{-\infty}^{+\infty} (S_n(x) - F(x))^2 F(x)(1 - F(x))^{-1} dF(x). \tag{8.83}$$

---

[4]Some statisticians prefer to perform the test in a bar, but this is not required.

Stephens in D'Agostino and Stephens [9, Chapter 4] found the quadratic statistical tests to be more powerful than the Kolmogorov-Smirnov test. He claims that the A^2 test is one of the most powerful of the quadratic tests.

One disadvantage of EDF goodness-of-fit tests, compared to chi-square tests, is that a table of critical values is required for each probability distribution family, such as exponential, normal, etc. The one exception is the special case when all parameters are completely specified. In the most common situation, in which the parameters must be estimated from the data, a table of critical values is required for each distribution. In Algorithm 8.6.1 we consider the completely specified case.

Algorithm 8.6.1 (*Kolmogorov-Smirnov and A^2 Tests with Parameters Known*) *Given a random sample $x_1, x_2, \ldots, x_n$ of size n that is sorted in ascending order, this test will determine at the α level of significance whether it is reasonable to suppose that the population distribution function is F, where the parameters of F are assumed known.*

Step 1 [Calculate the test statistics] *Calculate*

$$z_i = F(x_i), \tag{8.84}$$

for $i = 1, 2, \ldots, n$,

$$D^+ = \max_i\{\frac{i}{n} - z_i\}, \tag{8.85}$$

$$D^- = \max_i\{z_i - \frac{(i-1)}{n}\}, \tag{8.86}$$

and

$$D = \max\{D^+, D^-\}. \tag{8.87}$$

Then calculate the modified value, D^*, using the formula

$$D^* = D\left(\sqrt{n} + 0.12 + \frac{0.11}{\sqrt{n}}\right). \tag{8.88}$$

The same factor is used with D^+ and D^- if they are to be used in the test. This gives all the (K–S) statistics. Calculate A^2 by the formula

$$A^2 = -\frac{1}{n}\sum_{i=1}^{n}(2i-1)\{ln(z_i) + ln(1 - z_{n+1-i})\} - n. \tag{8.89}$$

Step 2 [Accept or reject the null hypothesis] *If the modified D is greater than the critical value in Table 12 of Appendix A, reject H_0; otherwise accept it by the Kolmogorov–Smirnov test. If A^2 is greater than the critical value in Table 12 of Appendix A, reject H_0; otherwise accept it by the A^2 test.*

Proof See Stephens in D'Agostino and Stephens [9, Section 4.4]. ∎

D^+ is used to test the hypothesis that $F(x) \leq S_n(x)$ for all x while D^- is used to test the hypothesis that $S_n(x) \leq F(x)$ for all x.

Table 8.6.1 Example 8.6.1.

1.52	4.65	5.36	6.28
15.75	20.92	40.73	40.89
60.72	93.74	95.41	106.36
125.69	151.18	200.86	268.75
459.86	827.67	840.33	1087.33

Let us consider an example of the use of Algorithm 8.6.1.

Example 8.6.1 The Grand Viceroy of Statistics at Kolmogorov Krusty Krumpets is given the data of Table 8.6.1, which he's been told comes from a normal population with $\mu = 220$ and $\sigma = 300$. Grand decides to use both the Kolmogorov–Smirnov and the A^2 tests at the five percent level of significance to test the null hypothesis that the population is $N(220, 300^2)$ as claimed.

Solution Grand uses the null hypothesis that the population is normally distributed with a mean of 220 and standard deviation of 300. The alternative hypothesis is that the distribution is different from that assumed in H_0. Grand uses the APL program KSΔSPEC to find that the modified D value D^* is 1.34218. Since the critical value in Table 12 of Appendix A is 1.358, he cannot reject the null hypothesis using the Kolmogorov–Smirnov test. The APL program AΔ2ΔSPEC, however, shows that the A^2 value is 2.569975. Since the critical value in Table 12 of Appendix A is 2.492, the A^2 test shows Grand should reject the null hypothesis at the five percent level. The A^2 test has more power! □

The details of the Kolmogorov–Smirnov test are shown in Table 8.6.2. From the third column we see that $D^+ = 0.2907$ and the last column shows that $D^- = 0.2332$. Hence, $D = 0.2907$ and D^* is

$$D^* = D(\sqrt{20} + 0.12 + \frac{0.11}{\sqrt{20}}) = 4.6167D = 1.3421. \qquad (8.90)$$

This is slightly different from the value obtained by KSΔSPEC, which carried more decimal places.

When you have eliminated the impossible, whatever remains, however
improbable, must be the truth.
Sherlock Holmes

Vision is the art of seeing thing invisible.
Jonathan Swift

Table 8.6.2. Kolmogorov–Smirnov Test of Example 8.6.1

x_i	$z_i = \Phi\left(\dfrac{x_i - 220}{300}\right)$	$\dfrac{i}{30} - z_i$	$z_i - \dfrac{i-1}{30}$
1.52	0.2332	-0.1832	0.2332
4.65	0.2364	-0.1364	0.1864
5.36	0.2372	-0.0872	0.1372
6.28	0.2381	-0.0381	0.0881
15.75	0.2480	0.0020	0.0480
20.92	0.2535	0.0465	0.0035
40.73	0.2751	0.0749	-0.0249
40.89	0.2752	0.1248	-0.0748
60.72	0.2977	0.1523	-0.1023
93.74	0.3369	0.1631	-0.1131
95.41	0.3390	0.2110	-0.1610
106.36	0.3524	0.2476	-0.1976
125.69	0.3766	0.2734	-0.2234
151.18	0.4093	0.2907	-0.2407
200.86	0.4746	0.2754	-0.2254
268.75	0.5645	0.2355	-0.1855
459.86	0.7880	0.0620	-0.0120
827.67	0.9786	-0.0786	0.1286
840.33	0.9807	-0.0307	0.0867
1087.33	0.9981	0.0019	0.0481

We will now consider the EDF tests for exponentiality when the mean must be estimated from the data.

Algorithm 8.6.2 (*The Kolmogorov–Smirnov and A^2 Tests for Exponentiality (Mean Estimated)*) *Given a random sample $x_1, x_2, \ldots, x_n$ of size n, sorted in ascending order, this test will determine at the α level of significance whether it is reasonable to suppose that the population is exponential.*

Step 1 [Calculate the test statistics] *Calculate*

$$\overline{x} = \frac{1}{n} \times \sum_{i=1}^{n} x_i, \qquad (8.91)$$

$$w_i = \frac{x_i}{\overline{x}}, \qquad (8.92)$$

and

$$z_i = 1 - e^{-w_i} \tag{8.93}$$

for $i = 1, 2, \ldots, n$. Then calculate

$$D^+ = \max_i \{\frac{i}{n} - z_i\}, \tag{8.94}$$

$$D^- = \max_i \{z_i - \frac{(i-1)}{n}\}, \tag{8.95}$$

and

$$D = \max\{D^+, D^-\}. \tag{8.96}$$

Calculate the modified form, D^*, using the formula

$$D^* = (D - \frac{0.2}{n})\left(\sqrt{n} + 0.26 + \frac{0.5}{\sqrt{n}}\right). \tag{8.97}$$

Calculate A^2:

$$A^2 = -\frac{1}{n}\sum_{i=1}^{n}(2i - 1)\{ln(z_i) + ln(1 - z_{n+1-i})\} - n, \tag{8.98}$$

and the modified form, A^* defined by:

$$A^* = A^2 \left(1 + \frac{0.6}{n}\right). \tag{8.99}$$

Step 2 [Accept or reject the null hypothesis] *Reject the null hypothesis that the population is exponential, if the modified value of D exceeds the critical value in Table 13 of Appendix A; otherwise accept H_0 by the Kolmogorov–Smirnov test. Reject the null hypothesis that the population is exponential if the modified A^2 exceeds the critical value in Table 13 of Appendix A; otherwise accept H_0 by the A^2 test.*

Proof See Stephens in D'Agostino and Stephens [9, Section 4.9.3]. ∎

Table 8.6.3. Exponential Sample

3.93	1.63	0.40	1.13	0.30
0.21	0.19	2.60	2.24	0.24
10.23	4.69	3.47	0.16	6.54
0.29	0.80	0.65	0.72	7.15
0.22	1.61	2.38	4.69	0.30
4.79	4.93	0.24	1.52	2.41
0.24	4.75	1.39	2.22	0.79
0.63	1.82	3.45	0.22	0.94
5.21	0.26	3.74	1.28	0.50
3.25	1.25	0.13	2.87	0.38

Example 8.6.2 The numbers in Table 8.6.3 are the numbers from Table 7.3.1 and comprise a random sample from an exponential distribution with a mean of 2. Let us apply Algorithm 8.6.2 to the data. We used the APL function EXPONΔT to make both the Kolmogorov–Smirnov test calculations and the A^2 calculations. It calculated a modified D value of 0.9179967. By Table 13 of Appendix A, the critical value is 1.094, so we cannot reject the null hypothesis by the Kolmogorov–Smirnov test. EXPONΔT also yields $A^2 = 1.0337149$ and a modified A^2 value equal to 1.046119. Since the critical value from Table 13 is 1.321, the A^2 test does not reject the null hypothesis, either. The data also passed the chi-square test made by the APL function CHISQΔEXPON. This function divides the sample into four classes based on the sample mean and the assumption of exponentiality. Thus, the critical value is determined by a chi-square distribution with 2 degrees of freedom. The critical value, $\chi^2_{2,0.05}$ is equal to 5.9915, and the calculated value from the test is 2.16. We expect 10 of the 40 observations to be in each class. The numbers observed in the 4 classes were 16, 10, 10, 14, respectively. The three quartiles of the data according to the null hypothesis are 0.60977, 1.46919, and 2.93839. The calculated p-value of the test is 0.339596. This calculated value was based on the assumption that χ^2 has a chi-square distribution with 2 degrees of freedom.) We ask you to verify these numbers in Exercise 12. $\square$

The exponential distribution is so important in applied probability that special tests have been devised to determine whether an observed distribution is exponential. Spinelli and Stephens [31] examined a number of tests for exponentiality. They find, for example, that the test by Shapiro and Wilk [28] has a major weakness in that it is not consistent; that is, the power of the test will not necessarily approach 1 as the sample size increases. Spinelli and Stephens claim the A^2 test is the best overall test for exponentiality. They provide a general test for a shifted exponential that tests for an ordinary exponential as a special case. Algorithm 8.6.2 is the test recommended by Stephens in his chapter, "Tests for the Exponential Distribution," in D'Agostino and Stephens [9] for testing for an ordinary (nonshifted) exponential.

No discussion of goodness-of-fit tests would be complete without a test for the normal distribution. It is, of course, the paramount probability distribution in statistics, theoretical or applied. D'Agostino in his chapter, "Tests for the Normal Distribution," in D'Agostino and Stephens [9] recommends a number of tests including the Anderson–Darling EDF A^2 test. In Algorithm 8.6.3 we present the Kolmogorov–Smirnov as well as the Anderson–Darling EDF A^2 test for normality. The A^2 test is the more powerful of the two tests, although the Kolmogorov–Smirnov test is more frequently presented in textbooks. In fact, the Anderson–Darling test appears in extremely few textbooks. D'Agostino also says:

> For testing for normality, the Kolmogorov–Smirnov test is only a historical curiosity. It should never be used. It has poor power in comparison to the above procedures.

> For testing for normality, when a complete sample is available, the chi-square test should not be used. It does not have good power when compared to the above tests.

My experience is certainly in agreement with Professor D'Agostino's comments. However, one should be aware of the chi-square and Kolmogorov–Smirnov tests, since they appear in most statistics textbooks.

Algorithm 8.6.3 (*The Kolmogorov–Smirnov and A^2 Tests for Normality (Parameters Estimated)*) *Given a random sample $x_1, x_2, \ldots, x_n$ of size n, that is sorted in ascending order, this test will determine at the α level of significance whether it is reasonable to suppose that the population is normal.*

Step 1 [Calculate the test statistics] *Calculate*

$$\overline{x} = \frac{1}{n} \times \sum_{i=1}^{n} x_i, \tag{8.100}$$

$$s^2 = \frac{1}{n-1} \sum_{i=1}^{n} (x_1 - \overline{x})^2, \tag{8.101}$$

$$w_i = \frac{x_i - \overline{x}}{s}, \tag{8.102}$$

and

$$z_i = \Phi(w_i) \tag{8.103}$$

for $i = 1, 2, \ldots, n$. Then calculate

$$D^+ = \max_i \{\frac{i}{n} - z_i\}, \tag{8.104}$$

$$D^- = \max_i \{z_i - \frac{(i-1)}{n}\}, \tag{8.105}$$

and

$$D = max\{D^+, D^-\}. \tag{8.106}$$

Calculate the modified form of D using the formula

$$D^* = D\left(\sqrt{n} - 0.01 + \frac{0.85}{\sqrt{n}}\right). \tag{8.107}$$

Calculate A^2

$$A^2 = -\frac{1}{n}\sum_{i=1}^{n}(2i-1)\{ln(z_i) + ln(1 - z_{n+1-i})\} - n, \tag{8.108}$$

and the modified form defined by

$$A^* = A^2\left(1 + \frac{0.75}{n} + \frac{2.25}{n^2}\right). \tag{8.109}$$

Step 2 [Accept or reject the null hypothesis] *Reject the null hypothesis that the population is normal if the modified value of D exceeds the critical value in Table 14 of Appendix A; otherwise accept H_0 by the Kolmogorov–Smirnov test. Reject the null hypothesis that the population is normal if the modified A^2 exceeds the critical value in Table 14 of Appendix A; otherwise accept H_0 by the A^2 test.*

Proof See Stephens in D'Agostino and Stephens [9, Section 4.8]. ∎

Example 8.6.3 Fred Fripple, Performance Vizier at Macho Motors, believes the multiprogramming level for one of their computers has a normal distribution. A sample of the multiprogramming level for 15 randomly selected times is given in Table 8.6.4. Fred wants to use the Kolmogorov–Smirnov and A^2 tests at the five percent level to check his hypothesis.

Solution The APL function NORMALΔT calculates $D^* = 1.23296$ and $A^* = 2.042969$. By Table 14 of Appendix A, the critical value of D^* is 0.895, so the Kolmogorov–Smirnov test rejects the null hypothesis. The critical value of A^* is 0.752, so the A^2 test rejects the null hypothesis, too. □

Although we used the asymptotic tables (Tables 12, 13, and 14 of Appendix A) for Kolmogorov–Smirnov tests, Lilliefors [16,17] and others have provided tables for Kolmogorov–Smirnov tests that can be entered with both D and n. Tables 7–9 of Appendix A are of this type. We chose to use the asymptotic tables because they can be adapted more easily to computer programs. The direct tables, however, can be used to construct confidence intervals for the true distribution function from the empirical distribution function and the critical value D_α. For example, consider the empirical distribution function S_{40} from Example 8.6.2. By Table 8 of Appendix A, we can be 95% confident that the true distribution function F satisfies the inequality

$$|F(x) - S_{40}(x)| < \frac{1.06}{\sqrt{40}} = 0.168 \tag{8.110}$$

for all real x. We can also write

$$S_{40}(x) - 0.168 < F(x) < S_{40}(x) + 0.168 \tag{8.111}$$

with 95% confidence (and obvious changes to the upper and lower bounds for some values of x to prevent probability values greater than one or less than zero).

Neave's outstanding book and statistical tables, [20] provides K–S tables for the normal and completely specified cases. He also provides some excellent examples of their use. It is difficult to praise Neave's little book too highly. He goes to extraordinary pains to make the tables foolproof to the user and supplies a number of insightful examples of the use of the tables.

As with most things in life, there is good news and bad news about EDF goodness-of-fit tests compared to chi-square goodness-of-fit tests. The good news is that EDF tests are more powerful and, in many cases, easier to implement on a computer than chi-square tests. On the other hand, EDF tests tend to require more computation and thus are labor-intensive if only a calculator or even a programmable calculator is available. Most statistical packages don't directly support EDF goodness-of-fit tests, so the user must write some code to use the techniques. In addition, tables of critical values are available for only a few probability distributions. We provide them for the completely specified case (all distributions) and for the important cases of normal and exponential when the parameters must be estimated. Stephens, in his chapter "Tests Based on EDF Statistics," in D'Agostino and Stephens [9] provides tests and tables for several more continuous distributions including the Weibull, gamma, and Cauchy. In addition, he discusses EDF tests for discrete distributions. The discrete tests, however, are limited to completely specified distributions with k classes. Naturally, the table for this test must be somewhat limited, too.

We therefore recommend that EDF tests, particularly the A^2 test, be used for testing all completely specified continuous distributions as well as the normal and exponential distributions. For discrete distributions we recommend the chi-square test.

Table 8.6.4. Example 8.6.3

1.75	2.10	2.11	2.50	2.60
2.91	3.10	3.20	3.31	4.62
5.40	8.50	13.10	19.00	25.00

Comparisons do ofttime great grievance.
John Lydgate, c. 1440

8.7   Analysis of Variance

The analysis of variance is traditionally abbreviated ANOVA. In fact, the whole name is rarely written out in statistical publications. It is a powerful tool of statistics. In this section we look at its simplest form, called *one-way analysis of variance*. This form can be used to compare a number of population means, simultaneously. Thus, we can avoid making a large number of two-sample tests.

The language often used in ANOVA derives from its roots in agriculture and medicine. Thus, different populations are often called "treatments." From this point of view, we can imagine that we have k different treatments where the result of applying treatment i is a normal random variable with mean μ_i and variance σ^2. We assume there are k independent samples from k treatments (populations) of size $n_1, n_2, \cdots, n_k$, respectively. We assume that $N = n_1 + n_2 + \cdots + n_k$ is the total number of observations. We let x_{ij} be the jth observation in the ith population or treatment. Then the overall sample mean, sometimes called the *grand sample mean*, is given by

$$\overline{x} = \frac{1}{N} \sum_j \sum_i x_{ij}. \tag{8.112}$$

The *total sum of squares*, TSS, is given by

$$\text{TSS} = \sum_j \sum_i (x_{ij} - \overline{x})^2. \tag{8.113}$$

TSS measures the total sum of squares of the deviation from the overall sample mean. Some of this variability is due to the differences *within* the population (treatment) samples and some is due to the differences *between* the sample means of the different populations. The *sum of squares for treatment* (SST), often called the *sum of squares between*, is defined by

$$\text{SST} = \sum_{i=1}^{k} n_i(\overline{x}_i - \overline{x})^2. \tag{8.114}$$

It measures the weighted sum of squares difference *between* the population sample means and the grand sample mean. Thus, a large value of SST is an indication that the means of the populations are different. The *error sum of squares* (SSE), sometimes called the *sum of squares within*, is defined by

$$\text{SSE} = \sum_{i=1}^{k} \sum_{j=1}^{n_i} (x_{ij} - \overline{x}_i)^2. \tag{8.115}$$

The SSE measures the sum of the square of variability within the populations (treatments). A large value of SSE is evidence *against* the equality of the means.

The theory of ANOVA depends upon the *partitioning* of the total sum of squares. For one-way analysis of variance, we have

$$\text{TSS} = \text{SST} + \text{SSE}, \tag{8.116}$$

where SST has $k - 1$ degrees of freedom and SSE has $N - k$ degrees of freedom. Equation (8.116) is an algebraic identity.

We now state the one-way analysis of variance algorithm.

Algorithm 8.7.1 (*One-Way* ANOVA *Test of k Means from Normal Populations with Equal Variances*) *Given k random samples, each from an independent normal population with the same variance, this algorithm will determine at the α level of significance whether to accept the null hypothesis,*

$$H_0: \mu_1 = \mu_2 = \cdots = \mu_k,$$

or the alternative hypothesis

$$H_1 : \textit{Not all the population means are the same.}$$

Step 1 Calculate the test statistic] *Compute*

$$F = \frac{\text{MST}}{\text{MSE}}, \tag{8.117}$$

where

$$\text{MST} = \frac{\text{SST}}{k - 1}, \tag{8.118}$$

and

$$\text{MSE} = \frac{\text{SSE}}{N - k}. \tag{8.119}$$

(SST *is defined by* (8.114) *and* SSE *by* (8.115).)

Step 2 [Determine the critical region] *The test statistic F has a Snedecor-F distribution with $k - 1$ and $N - k$ degrees of freedom. The critical region is the set of all $F > f_{k-1, N-k, \alpha}$.*

Step 3 [Accept or reject H_0] *If the test statistic calculated in Step 1 falls in the critical region found in Step 2, reject H_0; otherwise accept H_0.*

Proof See Chapter 8 of Ross [23]. Ross essentially shows that MST has a χ^2 distribution with $k - 1$ degrees of freedom and that MSE has a χ^2 distribution with $N - k$ degrees of freedom, so that F has an F-distribution with parameters (degrees of freedom) $k - 1$ and $N - k$. Ross also shows that the expected value of MSE is σ^2 and that if the null hypothesis is false, then the expected value of MST is greater than σ^2. ∎

It is customary, when making an ANOVA calculation, to display the partition of the total sum of squares (TSS) and the other elements of the calculation in an *analysis of variance table* in the form shown in Table 8.7.1.

Table 8.7.1. An Analysis of Variance Table

SOURCE	DF	SS	MS	F-Value	$Pr > F$
Treatments	$k - 1$	SST	MST	F	p-value
Error	$N - k$	SSE	MSE		
Total	$N - 1$	TSS			

All computer statistical systems with ANOVA routines provide an analysis of variance table as part of the output. However, some of them do not provide the p-value of the test, so you must look up the critical value of F to complete the test. The SAS/STAT procedure ANOVA and the

MINITAB[5] command AOVONEWAY provide the p-value; the EXPLORE program ANOVA does not.

Let us consider an example of the use of Algorithm 8.7.1.

Example 8.7.1 Recall Example 8.2.6. The Hotstuff Chili Company also has a team of application programmers, Team Charly, in the Engineering Department. Chi Chivari (sometimes known as χ^2), the manager of the team, has trained his programmers to use some programming techniques that utilize the power of their computers to help them program more effectively. Table 8.7.2 shows the average number of lines of code per programmer-day written by each of 20 randomly selected programmers from Team Charly. Let us apply Algorithm 8.7.1 to test the null hypothesis that Teams Able, Baker, and Charly produce the same number of lines of code per programmer-day, on the average.

Solution Here $N = 60$, $k = 3$, and $n_i = 20$ for $i = 1, 2, 3$. We know from Example 8.2.6 that $\bar{x}_1 = 97.38$ and $\bar{x}_2 = 87.12825$ lines of code per programmer day, respectively. We compute $\bar{x}_3 = 137.0273$, and, thus, $\bar{x} = 107.1785$. Therefore,

$$\text{SST} = 20 \times \sum_{i=1}^{3} (\bar{x}_i - \bar{x})^2 = 27,779.48, \tag{8.120}$$

and

$$\text{SSE} = \sum_{i=1}^{3} (19)s_i^2 = 51,559.01. \tag{8.121}$$

We easily calculate MST$= 13,889.74$ and MSE$= 904.54$. Hence,

$$F = \text{MST}/\text{MSE} = 15.36.$$

By Table 6 of Appendix A, we estimate

$$f_{2,57,0.05} \approx 3.162. \tag{8.122}$$

(SAS/STAT, MINITAB, and the HP-21S provide the more accurate value 3.159.) Since $F = 15.36$, we must reject the null hypothesis. □

Figure 8.7.1 displays the analysis of variance table provided by the SAS/STAT routine ANOVA for this example and the top part of Figure

---

[5]Version 7 of MINITAB as well as the Student Edition of MINITAB provide the p-value. Some earlier versions of MINITAB, such as those used in the examples of Ryan et al. [24] did not.

8.7.2 provides the analysis of variance table provided by the MINITAB command AOVONEWAY. Notice that the sum of squares numbers are slightly different from the ones we calculated. Note also that SAS/STAT provides the nominal p-value of the test as 0.0001. A simple calculation with the HP-21S shows that the actual value is approximately 4.6×10^{-6}.

You may have noticed that although we used many numbers that were obtained from Example 8.2.6 to make the calculations, the calculations were not trivial. A computer is needed for any serious ANOVA computation. Some computer statistical systems also do the pairwise comparisons of the means for you, although Algorithm 8.7.1 does not require this. One technique for making the comparisons is to construct a confidence interval for each mean and compare these intervals to judge the relative size of the means. The estimate usually used for σ^2 is MSE, so the $100(1 - \alpha)\%$ confidence interval for μ_i is the interval from

$$\overline{x}_i - t_{N-k,\alpha/2}\sqrt{\frac{\text{MSE}}{N - k}} \tag{8.123}$$

to

$$\overline{x}_i + t_{N-k,\alpha/2}\sqrt{\frac{\text{MSE}}{N - k}}. \tag{8.124}$$

The MINITAB command AOVONEWAY does this calculation for you and provides graphs as output. The bottom part of Figure 8.7.2 shows the graphs provided by AOVONEWAY for the confidence intervals in Example 8.7.1. From the graphs we see that $\mu_1 < \mu_3$ and $\mu_2 < \mu_3$ at the 5 percent level of significance but that μ_1 and μ_2 appear to be equal at this level of significance. The SAS/STAT procedure ANOVA provides a table with the same information as the graphs of Figure 8.7.2.

Table 8.7.2. Team Charly

53.798	42.832	100.318	68.122
99.532	76.452	76.585	115.624
139.437	66.254	103.859	91.770
93.435	63.412	89.562	60.232
81.448	144.096	116.115	59.682

Analysis of Variance Procedure

Dependent Variable: LINES

Source	DF	Sum of Squares	Mean Square	F Value	Pr > F
Model	2	27779.47978	13889.73989	15.36	0.0001
Error	57	51558.98880	904.54366		
Corrected Total	59	79338.46858			

Figure 8.7.1. SAS/STAT ANOVA table for Example 8.7.1.

```
ANALYSIS OF VARIANCE
SOURCE    DF        SS       MS       F       p
FACTOR     2     27779    13890    15.36   0.000
ERROR     57     51559      905
TOTAL     59     79338
                                 INDIVIDUAL 95 PCT CI'S FOR MEAN
                                 BASED ON POOLED STDEV
LEVEL      N      MEAN    STDEV   -+---------+---------+---------+-----
able      20     97.38    38.98        (----*----)
baker     20     87.13    27.60   (-----*----)
charly    20    137.03    20.80                       (-----*----)
                                 -+---------+---------+---------+-----
POOLED STDEV =    30.08            75       100       125       150
```

Figure 8.7.2. MINITAB ANOVA table for Example 8.7.1.

In the next section we will see that, for the Kruskal–Wallis test, we can weaken the requirement that each population is normal with a common variance.

8.8   Nonparametric Tests

The tests of this section are called *nonparametric* to distinguish them from most of the tests we have previously considered, that is, from *parametric tests*. The name is a bit of a misnomer, since nonparametric tests *do* usually depend upon some parameter—a population proportion or median, for example. They never depend upon a normal distribution, however, so the term *distribution-free* is more accurate, although not as widely used. Many parametric tests depend upon the assumption that the population from which the sample was taken is normal. For sufficiently large samples, the central limit theorem can be invoked to avoid the normality assumption.

However, if the distribution of the population is badly skewed or otherwise significantly nonnormal, parametric statistical tests will not be correct for small sample sizes. Nonparametric tests obviate this difficulty.

Another requirement for parametric tests is that sample statistics such as the sample mean and variance can be computed from the sample and used to estimate the corresponding population parameters. Data that are purely nominal in nature (such as "approve", "disapprove", and "have no opinion") or which are given in rank order only (such as ranking the engineers in a department from 1 to n) do not yield such meaningful results.

Nonparametric tests can be used to test hypotheses about data which are not normal or are not measured on an interval scale. Since these tests do not depend on the shape of the distribution, they are sometimes called *distribution-free tests*. The reason they are more often called *nonparametric tests* is because they do not depend upon the mean or the variance.

Nonparametric methods, as we shall see, often are computationally simpler than many parametric methods and tend to be easier to understand. Since we know "there ain't no free lunch," we must be prepared for some disadvantages of nonparametric methods, too. One disadvantage of some nonparametric tests is that they waste information, since exact numerical data are often reduced to a qualitative form. Nonparametric tests also tend to be less sensitive than their parametric counterparts and thus require stronger evidence to reject a null hypothesis. For this reason, if the requirements for a parametric test are met, it is best to use it rather than a nonparametric test. On the other hand, if we want to test data to see if they are randomly selected, the only available test is a nonparametric test, the runs test.

8.8.1   The Sign Test

The sign test is one of the most versatile of the nonparametric tests and yet one of the simplest to apply. The first version of the test we will consider is the one-sample sign test.

Algorithm 8.8.1 (*One-Sample Sign Test*) *Given a random sample $x_1, x_2,$ $\ldots, x_n$ from a population with median m, this algorithm will determine at the α level of significance whether to accept or reject the null hypothesis $H_0 : m = m_0$ against one of the alternative hypotheses (a) $H_1 : m > m_0$, (b) $H_1 : m < m_0$, or (c) $H_1 : m \neq m_0$.*

Step 1 [Calculate the test statistics] *Compute $x_+ = $ the number of sample values greater than m_0, $x_- = $ the number of values less than m_0, and $x_= = $ the number of values equal to m_0. Then let the $x = \max\{x_+, x_-\}$, and $m = n - x_=$.*

Step 2 [Determine the p-value of the test] *Let X be a binomial random variable with parameters m and 0.5. For alternative hypothesis (a), the p-value is given by*

$$p\text{-}value = P[X \geq x_+]. \tag{8.125}$$

For alternative hypothesis (b), the p-value is given by

$$p\text{-}value = P[X \geq x_-]. \tag{8.126}$$

For alternative hypothesis (c), the p-value is given by

$$p\text{-}value = 2 \times P[X \geq x]. \tag{8.127}$$

Step 3 [Accept or reject H_0] *If the p-value calculated in Step 2 is less than α, reject H_0; otherwise accept it.*

Proof If the null hypothesis is true, then the probability that any element of the random sample is greater than m_0 is 0.5. If we let X be the number of elements of the sample that are greater than m_0, then X has a binomial distribution with parameters n and 0.5. But the same argument shows that Y, the number of elements of the sample smaller than m_0, has the same distribution as X. This shows that the algorithm is valid. ∎

Table 8.8.1

75.2	65.1	42.6	39.7
100.8	63.5	72.3	49.6
45.3	92.7	79.3	32.7
76.3	88.1	87.3	48.7
74.7	47.6	43.2	89.8

Example 8.8.1 Table 8.8.1 represents the average number of lines of code per programmer-day produced by 20 randomly selected programmers at Helpful Heuristics. Harry Houdini, the lead chief programmer, wants to test the hypothesis that the median of the mean number of lines of code per day of each programmer is more than 50. The null hypothesis is that the median is 50; the alternative hypothesis is that the median is greater than 50. For the sign test, Harry notes that $x_+ = 12$ and $n = 20$, since no value in the table is equal to 50. Therefore, the p-value of the test is given by

$$\left(\frac{1}{2}\right)^{20} \times \sum_{k=12}^{20} \binom{20}{k} = 0.251723. \tag{8.128}$$

If Harry uses $\alpha = 0.05$, he cannot reject the null hypothesis that the median is 50. □

MINITAB has the command STEST, that will do the calculations for a sign test. It yields the same p-value that Harry Houdini calculated. The calculations can become a trifle daunting if you do not have computer assistance available. The calculation in (8.125) can be made using the normal approximation for large values of n. It is a two step process using

$$z = \frac{(x_+ - 0.5) - 0.5n}{0.5\sqrt{n}}, \tag{8.129}$$

if $x_+ > 0.5n$ or

$$z = \frac{(x_+ + 0.5) - 0.5n}{0.5\sqrt{n}}, \tag{8.130}$$

if $x_+ < 0.5n$. Then the p-value is calculated as

$$p\text{-value} = P[Z > z], \tag{8.131}$$

where Z is a standard normal variable. Formula (8.131) yields the excellent approximation 0.25117 as the p-value for Example 8.8.1. One would not expect such a good approximation when $n = 20$. A similar approximation can be used for the other alternative hypotheses when n is large.

Algorithm 8.8.1 can be used in a slightly modified form to perform the sign test on paired samples. The testing of two paired samples—either before-and-after observations or matched pairs is a very common procedure. For before-and-after observations of elements of the same sample, we may be interested in determining the effectiveness of a class, a different procedure, treatment for a disease, etc. For matched pairs we consider two populations in which each element of the first is matched with an element of the second. When the populations consist of people, the matching may be on the basis of sex, age, IQ, health, or other factors.

Algorithm 8.8.2 (*Sign Test for Differences Between Paired Samples*) *Given n pairs of observations from two randomly obtained populations with medians m_1 and m_2, respectively, this algorithm will determine at the α level of significance whether to accept or reject the null hypothesis H_0 : $m_1 = m_2$ against one of the alternative hypotheses (a) H_1 : $m_1 > m_2$, (b)H_1 : $m_1 < m_2$, or (c) H_1 : $m_1 \neq m_2$.*

Step 1 [Calculate the test statistics] *Compute x_+ which is the number of paired sample values in which the value in the second population is greater than that of the first; x_- which is the number of paired sample values in which the observed value in the second population is less than that of the first population; and $x_=$ which is the number of paired sample values that are equal. Then let the $x = \max\{x_+, x_-\}$, and $m = n - x_=$.*

Step 2 [Determine the p-value of the test] *Let X be a binomial random variable with parameters m and 0.5. For alternative hypothesis (a). the p-value is given by*

$$p\text{-}value = P[X \geq x_+]. \tag{8.132}$$

For alternative hypothesis (b), the p-value is given by

$$p\text{-}value = P[X \geq x_-]. \tag{8.133}$$

For alternative hypothesis (c), the p-value is given by

$$p\text{-}value = 2 \times P[X \geq x]. \tag{8.134}$$

Step 3 [Accept or reject H_0] *If the p-value calculated in Step 2 is less than α, reject H_0; otherwise accept it.*

Proof This test can be reduced to the one-sample sign test (Algorithm 8.8.1) by creating a new sample as follows: subtract each element of the first sample from the corresponding (paired) element of the second sample to create a new sample. Then apply the one-sample sign test to the resulting sample with $m_0 = 0$. ∎

Table 8.8.2. Data For Example 8.8.2

	Acci	dents	
Before	After	Before	After
5	3	3	1
2	0	0	2
4	3	1	3
6	4	4	1
1	0	3	0
3	2	6	4

Example 8.8.2 Bigg Stuff, the mayor of Rosefull, is persuaded by Crash Bangg, the Rosefull Highway Czar, to install a new traffic control system at the 12 most dangerous intersections within the city. Table 8.8.2 shows the number of accidents during the six weeks before and the six weeks after installation of the new systems. Crash decides to use Algorithm 8.8.2 to test at the 5% level of significance the null hypothesis that the traffic control system is not effective against the alternative hypothesis that it is $(m_1 > m_2)$. Since $m_- = 10$ and $n = 12$, the p-value of the test is

$$\left(\frac{1}{2}\right)^{12} \sum_{k=10}^{12} \binom{12}{k} = 0.01929.$$

Hence, Crash can reject the null hypothesis and conclude that the new traffic system is effective. □

> When angry count four; when very angry, swear.
>
> Mark Twain

8.8.2   The Kruskal–Wallis Test

The *Kruskal–Wallis Test*, often called the *Kruskal–Wallis H Test*, is used in place of the one-way analysis of variance test when the populations cannot be assumed normal. Thus, it can be used to test whether or not the means from k independent samples are equal when the populations are not normal. The Kruskal–Wallis H test is due to Kruskal and Wallis [14]. It uses a ranking method. Let n_i be the number of observations in the ith sample. The k samples are combined and the $n = n_1 + n_2 + \cdots + n_k$ observations ranked from smallest to largest, substituting the appropriate rank from $1, 2, \ldots, n$ for each observation. Observations with the same values are given the average of their ranks. The sum of the ranks for each sample are then computed. R_i is the sum of the ranks from the ith sample. Kruskal and Wallis denoted by H the random variable

$$H = \frac{12}{n(n+1)} \sum_{i=1}^{k} \frac{R_i^2}{n_i} - 3(n+1). \tag{8.135}$$

When the samples are from the same distribution (H_0 is true) and each sample consists of at least 5 observations, then H can be approximated by a chi-square distribution with $k - 1$ degrees of freedom. For small samples the approximation is not very good, so Kruskal and Wallis provide tables of the exact distribution for the case of $k = 3$, $n_i \leq 5$. We state the test as a formal algorithm.

Algorithm 8.8.3 (*Kruskal–Wallis Test*) *Suppose we have k independent samples from k populations. We wish to test the null hypothesis*

H_0: the samples are from identical populations

against the alternative hypothesis

H_1: the populations are not identical

at the α level of significance.

Step 1 [Compute h] *Calculate*

$$h = \frac{12}{n(n+1)} \sum_{i=1}^{k} \frac{R_i^2}{n_i} - 3(n+1).$$

Step 2 [Accept or reject H_0] *If $h > \chi^2_{k-1,\alpha}$, reject H_0; otherwise accept* H_0.

Proof See Kruskal and Wallis [14] ∎

Example 8.8.3 Consider Example 8.7.1. Let us use the Kruskal–Wallis test to test the null hypothesis that the three programming teams, Able, Baker, and Charly, have the same average productivity in lines of code per programmer day. In Table 8.8.3 we indicate the rank of each observation from the three teams. We see that $R_1 = 513$, $R_2 = 415$, and $R_3 = 902$. (We checked these values using the SAS/STAT procedure NPAR1WAY.) Then we calculate $h = 21.75377$. Since $\chi^2_{2,0.05} = 5.9915$, we must reject the null hypothesis as before. The p-value of the test is 0.00001889. The SAS/STAT procedure NPAR1WAY reports that $h = 21.754$ with the p-value 0.0001, while the MINITAB command KRUSKAL-WALLIS reports that $h = 21.75$ with p-value 0.000. □

Table 8.8.3. Example 8.8.3

Lines	Team	Rank	Lines	Team	Rank
121.200	Able	36	64.300	Able	11
79.100	Able	16	135.700	Able	44
105.600	Able	29	92.700	Able	22
114.800	Able	31	138.200	Able	46
87.200	Able	19	63.700	Able	10
119.600	Able	35	142.700	Able	51
81.400	Able	17	58.900	Able	6
175.300	Able	60	20.700	Able	1
47.800	Able	3	112.700	Able	30
53.400	Able	4	132.600	Able	42
53.798	Baker	5	42.832	Baker	2
100.318	Baker	27	68.122	Baker	13
99.532	Baker	26	76.452	Baker	14
76.585	Baker	15	115.624	Baker	32
139.437	Baker	50	66.254	Baker	12
103.859	Baker	28	91.770	Baker	21
93.435	Baker	23	63.412	Baker	9
89.562	Baker	20	60.232	Baker	8
81.448	Baker	18	144.096	Baker	52
116.115	Baker	33	59.682	Baker	7

Table 8.8.3. (continued) Example 8.8.3

Lines	Team	Rank	Lines	Team	Rank
122.430	Charly	38	158.459	Charly	56
131.325	Charly	41	163.751	Charly	57
96.136	Charly	24	153.743	Charly	55
138.884	Charly	47	148.112	Charly	53
139.404	Charly	49	149.350	Charly	54
135.342	Charly	43	167.758	Charly	58
97.156	Charly	25	121.453	Charly	37
128.118	Charly	40	136.361	Charly	45
139.313	Charly	48	171.223	Charly	59
124.637	Charly	39	117.591	Charly	34

8.9   Summary Table

Table 8.9.1 provides a summary of some of the simpler hypothesis tests from the beginning of the chapter. Unfortunately, we were not able to find a table format that would allow us to provide a table for the more complex tests, such as those for goodness-of-fit and analysis of variance.

Table 8.9.1 Hypothesis Tests

Hypothesis	Rejection Region of H_0
H_0: $\mu = \mu_0$ H_1: $\mu > \mu_0$	$\dfrac{\bar{x} - \mu_0}{\sigma/\sqrt{n}} \geq z_\alpha$
H_0: $\mu = \mu_0$ H_1: $\mu > \mu_0$	$\dfrac{\bar{x} - \mu_0}{s/\sqrt{n}} \geq t_{n-1,\alpha}$
H_0: $\mu_1 = \mu_2$ H_1: $\mu_1 > \mu_2$	$\dfrac{\bar{x} - \bar{y}}{\sqrt{\sigma_1^2/n_1 + \sigma_2^2/n_2}} \geq z_\alpha$
H_0: $\mu_1 = \mu_2$ H_1: $\mu_1 > \mu_2$	$\dfrac{\bar{x} - \bar{y}}{\sqrt{\dfrac{(n_1 - 1)s_x^2 + (n_2 - 1)s_y^2}{n_1 + n_2 - 2}\left(\dfrac{1}{n_1} + \dfrac{1}{n_2}\right)}} \geq t_{n_1+n_2-2,\alpha}$
H_0: $p = p_0$ H_1: $p > p_0$	$\dfrac{\dfrac{k}{n} - p_0}{\sqrt{\dfrac{p_0(1 - p_0)}{n}}} \geq z_\alpha$
H_0: $p_1 = p_2$ H_1: $\mu_1 > \mu_2$	$\dfrac{\dfrac{k_1}{n_1} + \dfrac{k_2}{n_2}}{\sqrt{\left(\dfrac{k_1 + k_2}{n_1 + n_2}\right)\left(1 - \dfrac{k_1 + k_2}{n_1 + n_2}\right)\left(\dfrac{1}{n_1} + \dfrac{1}{n_2}\right)}} \geq z_\alpha$

Far better it is to dare mighty things, to win glorious triumphs,
even though checkered by failure, than to take rank with those
poor spirits who neither enjoy much nor suffer much, because
they live in the gray twilight that knows not victory nor defeat.

Theodore Roosevelt

8.10   Exercises

1. [C10] The number of terminals in use by Group Alfa during the lunch hour at Fairlady Aircraft is normally distributed. A random sample of 40 observed values of the number of active terminals is shown in the table. Assume $\sigma = 5$. At the 5% level of significance, test the null hypothesis that the mean number of active terminals $\mu = 20$ against the alternative hypothesis that $\mu > 20$.

Random Sample

22.5585	16.7495	19.8800	19.8130
22.3250	23.2865	15.7555	11.8815
24.5805	22.1430	30.7650	24.0120
23.1480	16.2845	21.1555	22.1095
19.0270	18.8885	24.2645	21.9145
26.7180	27.4775	22.8960	14.3475
18.3270	10.4450	27.1350	11.1425
23.0950	26.8640	18.0160	9.9325
21.5260	27.2705	21.5315	20.2230
9.0565	18.7445	24.9890	17.7345

2. [C10] Fowler Heir Mining has selected the random sample of response times observed at the terminal used by the Vice President for Canaries. The values, in seconds, are shown in the table. Assuming the response times are normally distributed, test the null hypothesis that the mean response time $\mu = 0.5$ seconds against the alternative hypothesis that $\mu > 0.5$ seconds. Use the 5% level of significance.

Random Sample at Fowler Heir

0.53	0.60	0.57	0.58	0.44
0.68	0.49	0.55	0.51	0.59

3. [8] [5 if a computer statistical system is used] The 200 programmers at Division A of Mickeysoft Corporation have averaged 76.21 lines of code per programmer-day with a sample standard deviation of 10.37. The 150 programmers of Division B have averaged 72.72 lines with a sample standard deviation of 10.07. At the one percent level of significance, does it appear that the programmers of Division A are more productive than those of Division B?

4. [C10] [5 if a computer statistical system is used] Consider the situation of Exercise 3. Suppose it is discovered that the statistics gathered at Mickeysoft are based on only 6 programmers at Division A and 5 at division B. (The numbers are the same, though.) Assume the distribution of lines of code per programmer-day at both divisions is normally distributed with the same (unknown) mean. At the one percent level of significance, does it now appear that the programmers of Division A are more productive than those of Division B?

5. [C10] [5 if a computer statistical system is used] Consider Exercise 3. Assuming the distributions of lines of code per programmer-day are normal, test at the 10% level of significance the null hypothesis that the variances of lines of code per programmer-day are the same at the two divisions. Use the alternative hypothesis that they are different.

6. [C10] Consider Exercise 2. Assuming the response time distribution is normal, test at the 10% level of significance the null hypothesis that the variance $\sigma^2 = 0.004$. Use the alternative hypothesis that $\sigma^2 > 0.004$.

7. [C15] Badyere Tire claims that 90% of its Randy Radials last for more than 50,000 miles. Diligent Dealer wear tests ten Randy Radials and finds that two of them fail before 50,000 miles. Use Algorithm 8.4.2 to test the Badyere Tire claim at the 10% level of significance.

8. [C20] Richilan Radials advertises that Richilan radial tires are better than Randy Radials. As proof, they tested 20 of their radials and 20 Randy Radials. Richilan claims that only 2 of their tires failed to reach 50,000 miles, but 7 Randy Radials failed. Use the Fisher-Irwin test at the 5% level of significance to test Richilan's claim. The null hypothesis is that there is no difference in the tires. The alternative hypothesis is that the fraction of good Richilan tires is higher than the fraction of good Randy tires.

9. [C10] Parsimonious Peripherals has developed a new *I/O* device. The company claims that the table contains the retrieval times of 30 randomly selected *I/O* requests. The times are in milliseconds. Parsimonious claims the processing times are normally distributed with a mean of 10 milliseconds and a standard deviation of 5 milliseconds. Test this claim using the chi-square test as follows. Normalize the data by the formula $z_i = (x_i - 10)/5$. Partition the real line into 5 intervals such that if H_0 is true, then one-fifth of the $z_i's$ would be expected in each interval. Apply the chi-square test with these intervals as cells. Use the 5% level of significance.

Parsimonious Sample

14.560	4.102	7.514	8.873	10.373
7.582	10.651	5.424	6.470	7.589
3.022	8.666	1.966	17.381	15.564
17.715	13.655	14.860	12.037	10.475
5.298	9.750	-1.371	7.274	14.901
18.845	16.881	7.229	16.833	9.440

10. [C15] Whoopdedoo Fashions has taken a random sample of 15 communication line times of their interactive order entry system to yield the values in the table. (The units of time are druds.) Use the Kolmogorov–Smirnov test and the A^2 test to test the hypothesis that line time is exponential. Use the 10% level of significance.

Whoopdedoo Fashions Sample

2.31	17.29	26.23	79.83	30.35
3.59	1.29	0.58	4.81	15.87
28.73	3.87	18.99	2.81	62.46

11. [C15] Consider the random sample of response times in Exercise 2. Use the Kolmogorov–Smirnov and A^2 tests at the 5% level to test the hypothesis that the response time is normally distributed. Assume the mean and variance are not known.

12. [10] Without using the APL function CHISQΔEXPON, verify that the numbers for the chi-square test for exponentiality at the 5 percent level of significance quoted in Example 8.6.2 are correct.

13. [15] Consider Example 8.5.2. Suppose that for each World Series one team is more likely to win each game than the other. Assume the probability of winning each game is 0.6 for one team and 0.4 for the other team. Analyze the 50 games that we analyzed in Example 8.5.2, at the 5 percent level of significance.

14. [C15] Silicon Valley Doodads has 4 computers in the central computer center that are run 3 shifts per day. Andy Exponential, the manager of the computer center, makes up the following contingency table, by computer and shift, of the number of times that the operators of a particular computer have been forced to do an IPL (initial program load) during the shift. Andy wants to test at the 5 percent level of significance whether the need for an IPL on a computer is independent of the computer and the shift. Please make the calculation for Andy.

	Computer				
	A	B	C	D	Totals
Shift 1	5	3	2	7	17
Shift 2	7	12	9	16	44
Shift 3	1	2	4	2	9
Totals	13	17	15	25	70

15. [20] A consumer testing organization tests three brands of pickup trucks by crashing them into a barrier and recording the repair cost in dollars. Six vehicles of each type were tested. The results are recorded in the table below. Use one-way analysis of variance to test the hypothesis that the average damage to each brand of pickup is the same. Use the five percent level of significance.

	Pickup	
Gorilla	Warrior	Gladiator
200	75	120
50	470	570
150	20	600
75	140	450
100	220	700
250	210	350

16. [20] Solve Exercise 15 using the Kruskal–Wallis test.

References

[1] Milton Abramowitz and Irene A. Stegun, *Handbook of Mathematical Functions With Formulas, Graphs, and Mathematical Tables*, National Bureau of Standards, Washington, DC, 1964. Also published in paperback by Dover Publications, New York.

[2] T. W. Anderson and D. A. Darling, A test for goodness-of-fit, *J. Amer. Statist. Assoc.*, **49** (1954), 300–310.

[3] T. W. Anderson and Barrett P. Eynon, *MINITAB Guide to The Statistical Analysis of Data*, The Scientific Press, Palo Alto, 1986.

[4] D. J. Best and J. C. W. Rayner, Welch's approximate solution to the Behrens–Fisher problem, *Technometrics*, **29**(2), May 1987, 205–210.

[5] Yvonne M. M. Bishop, Stephen E. Fienberg, and Paul W. Holland, *Discrete Multivariate Analysis: Theory and Practice*, The MIT Press, Cambridge, MA, 1975.

[6] Ladislaus von Bortkiewicz, *Das Gesetz der Kleinen Zahlen*, Teubner, Leipzig, 1898.

[7] Jim Brunner, Absorbing Markov Chains and the number of games in a World Series, *The UMAP Journal*, **8(2)**, (Summer 1987), 99–108.

[8] Harald Cramér, *Mathematical Methods of Statistics*, Princeton University Press, Princeton, 1946.

[9] Ralph B. D'Agostino and Michael A. Stephens, eds., *Goodness-of-Fit Techniques*, Marcel Dekker, New York, 1986.

[10] Ralph B. D'Agostino, Warren Chase, and Albert Belanger, The appropriateness of some common procedures for testing the equality of two independent binomial populations, *The American Statistician*, **42(3)**, (August 1988), 198–202.

[11] David P. Doane, *Exploring Statistics With The IBM PC, Second Edition*, Addison-Wesley, Reading, MA, 1988. (Comes with a diskette containing the EXPLORE statistical system and a data diskette.)

[12] David Freedman, Robert Pisani, and Roger Purves, *Statistics*, W. W. Norton, New York, 1978.

[13] Erwin Kreyszig, *Introductory Mathematical Statistics*, Wiley, New York, 1970.

[14] William H. Kruskal and W. Allen Wallis, Use of ranks in one-criterion analysis of variance, *Journal of the American Statistical Association*, **47**, (1952), 583–621.

[15] Richard J. Larsen and Morris L. Marx, *An Introduction to Mathematical Statistics and Its Applications*, Prentice-Hall, Englewood Cliffs, NJ, 1986.

[16] Hubert W. Lilliefors, On the Kolmogorov–Smirnov test for normality with mean and variance unknown, *J. Amer. Statist. Assoc.*, **62** (1967), 399–402.

[17] Hubert W. Lilliefors, On the Kolmogorov–Smirnov test for the exponential distribution with mean unknown, *J. Amer. Statist. Assoc.*, **64** (1969), 387–389.

[18] *MINITAB Reference Manual Release 7*, Minitab, State College, PA, 1989.

[19] David S. Moore and George P. McCabe, *Introduction to the Practice of Statistics*, Freeman, New York, 1989.

[20] Henry R. Neave, *Elementary Statistics Tables*, George Allen & Unwin Ltd., London, 1981.

[21] E. S. Pearson and N. W. Please, Relation between the shape of population distribution and the robustness of four simple test statistics, *Biometrika*, **62(2)**, (1975), 223–241.

[22] Harry O. Posten, The robustness of the one-sample *t*-test over the Pearson system, *J. Statist. Comput. and Simul.*, **9**, (1979), 133–149.

[23] Sheldon M. Ross, *Introduction to Probability and Statistics for Engineers and Scientists*, John Wiley, New York, 1987.

[24] Barbara F. Ryan, Brian L. Joiner, and Thomas A. Ryan, Jr, *Minitab Handbook, Second Edition*, Duxbury Press, Boston, 1985.

[25] *SAS/STAT Guide for Personal Computers, Version 6 Edition*, SAS Institute, Cary, NC, 1987.

[26] *SAS Language Guide for Personal Computers, Release 6.03 Edition*, SAS Institute, Cary, NC, 1988.

[27] Robert L. Schaefer and Richard B. Anderson, *The Student Edition of MINITAB*, Addison-Wesley, Benjamin/Cummings, 1989.

[28] S. S. Shapiro and M. B. Wilk, An analysis of variance test for normality, *Biometrika*, **52** (1965), 591–612.

[29] S. S. Shapiro and M. B. Wilk, An analysis of variance test for the exponential distribution (complete samples), *Technometrics*, **14(2)** (May 1972), 355–370.

[30] George W. Snedecor and William G. Cochran, *Statistical Methods, Seventh Edition*, The Iowa State University Press, Ames, Iowa, 1980.

[31] John J. Spinelli and Michael A. Stephens, Tests for exponentiality when origin and scale parameters are unknown, *Technometrics*, **29**(4) (Nov. 1987), 471–476.

[32] Michael A. Stephens, EDF statistics for goodness of fit and some comparisons, *J. Amer. Stat. Assoc.*, **69** (1974), 730–737.

[33] Peter Tannenbaum, Handicapping a championship series with elementary symmetric polynomials, *The UMAP Journal*, **10(2)** (Summer 1989), 115–136.

[34] C. P. Winsor, Quotations "Das Gesetz der Kleinen Zahlen," *Human Biology*, **19** (1947), 154–161.

[35] William Woodside, Winning streaks, shutouts, and the length of the World Series, *The UMAP Journal*, **10(2)** (Summer 1989), 99–113.

Chapter 9

Regression and Correlation Analysis

An optimist is someone who believes the future is uncertain.
Edward Teller

Never make forecasts; especially about the future.
Samuel Goldwyn

9.0 Introduction

In this chapter we study the relationship between two random variables, say X and Y. It is customary to call X the *predictor* or *independent random variable* and Y the *response* or *dependent random variable*. The two procedures commonly used to study the relationship between the random variables X and Y are *regression analysis* and *correlation analysis*.

The primary measurement used in correlation analysis is the *correlation (coefficient)* of X and Y, written $\rho(X, Y)$, which we defined in Chapter 2 by the formula

$$\rho(X, Y) = \frac{\text{Cov}(X, Y)}{(\text{Var}[X]\text{Var}[Y])^{1/2}}, \tag{9.1}$$

provided both variances are nonzero. As we noted in Chapter 2, $|\rho(X, Y)| \leq 1$ with equality if and only if $P[Y = aX + b] = 1$ for some a and b. For a proof of this result, see Rice [18, pages 125–127]. Thus, ρ can be used to measure the strength of the linear association between X and Y. We will find that correlation analysis is a useful tool in linear regression analysis.

For regression analysis, a key definition is the *curve of regression of Y on X*, which we defined in Chapter 3 by the equation

$$E[Y|X = x] = \int_{-\infty}^{\infty} y f_{Y|X}(y|x) dy. \tag{9.2}$$

The *curve of regression of Y on X* is used to describe the statistical relationship between X and Y. It is particularly useful when the curve turns out to be a straight line. The result of Theorem 3.2.5(c) is that if X and Y have a bivariate normal distribution, then the curve of regression of Y on X is the straight line

$$E[Y|X = x] = \mu_{y|x} = \mu_Y + \frac{\rho \sigma_Y}{\sigma_X}(x - \mu_X), \tag{9.3}$$

and for each $x, Y = Y_x$ is a normal random variable with mean

$$E[Y_x] = \mu_Y + \frac{\rho \sigma_Y (x - \mu_X)}{\sigma_X} \tag{9.4}$$

and standard deviation

$$\sigma_{Y_x} = \sigma_Y \sqrt{1 - \rho^2}. \tag{9.5}$$

Similar remarks apply to the curve of regression X on Y.

When we apply regression analysis to real life situations, of course, we do not know the exact parameters of the random variables X and Y, so we cannot calculate the curve of regression of Y on X. What is usually done is to observe the value of the random variable Y for each of n different values of x, say $x_1, x_2, \ldots, x_n$. When the n independent experiments have been performed, we have n pairs of numbers $(x_1, y_1), (x_2, y_2), \ldots, (x_n, y_n)$. We can then plot these n pairs of points to form a *scatter diagram*. The scatter diagram is a useful tool to make an assessment about the possible relation between X and Y. A curve can often be fitted to the data to define a relationship between the two random variables. The relationship that is estimated is $E[Y|X = x]$ versus x, that is, the regression of Y on X.

It would seem that *curve-fitting* might be a better choice of words to describe this kind of procedure than *regression analysis*, but the expression is rooted in history. Regression is the word used by Sir Francis Galton to describe some anomalies discovered in his pioneering work in predicting the heights of children from the heights of their parents. He found that parents who were taller than average also had children who were taller than average but that their children tended to be shorter than their parents. Galton called this phenomenon "regression to mediocrity." The word *regression* has been adopted by statisticians to represent all techniques in which one random variable is predicted from the values of another random variable or from the values of several other random variables.

9.1 Simple Linear Regression

For this model it is assumed that the curve of regression of Y on X is of the form

$$E[Y|X = x] = \beta_0 + \beta_1 x \quad \text{for all } x, \tag{9.6}$$

where β_0 and β_1 are constants, that is, where the curve is a straight line. Simple linear regression is applied after the known constants $x_1, x_2, \ldots, x_n$ are chosen by the investigator for the values of x at which Y is observed to yield the set of pairs $(x_1, y_1), (x_2, y_2), \ldots, (x_n, y_n)$. The following additional mathematical assumptions are made for what is called *the standard statistical model*:

1. The random variables Y_i, $i = 1, 2, \ldots, n$ are of the form

$$Y_i = \beta_0 + \beta_1 x_i + \epsilon_i, \quad i = 1, 2, \ldots, n. \tag{9.7}$$

2. The random variables $\epsilon_1, \epsilon_2, \ldots, \epsilon_n$ in (9.7) are errors that create the deviations about the linear relationship $\beta_0 + \beta_1 x$, $i = 1, 2, \ldots, n$, respectively. The errors are independent and normally distributed with

$$E[\epsilon_i] = 0 \quad \text{and} \quad \text{Var}[\epsilon_i] = \sigma^2, \quad i = 1, 2, \ldots, n. \tag{9.8}$$

It follows from these two assumptions that each Y_i has a normal distribution with

$$E[Y_i] = E[Y|X = x_i] = E[\beta_0 + \beta_1 x_i + \epsilon_i] = \beta_0 + \beta_1 x_i = \mu_i, \tag{9.9}$$

and

$$\text{Var}[Y_i] = \text{Var}[Y|X = x_i] = \text{Var}[\beta_0 + \beta_1 x_i + \epsilon_i] = \text{Var}[\epsilon_i] = \sigma^2. \tag{9.10}$$

The random variables $Y_1, Y_2, \ldots, Y_n$ are mutually independent because of the mutual independence of $\epsilon_1, \epsilon_2, \ldots, \epsilon_n$.

In Table 9.1.1 we show the number of pages and the price of 21 books reviewed in the November 1989 issue of *Technometrics*. (Four additional books were reviewed with no price given.)

Table 9.1.1. Data for Books Reviewed

Pages	Price	Pages	Price	Pages	Price
264	45.00	405	45.00	307	37.50
427	45.00	53	7.95	88	25.00
273	36.00	130	39.95	163	45.00
546	51.00	243	39.95	265	59.40
328	59.95	188	39.95	370	84.25
296	47.50	595	48.64	475	55.00
63	8.75	96	9.95	393	51.95

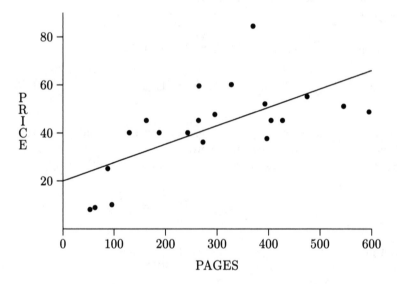

Figure 9.1.1. Scatter diagram for data in Table 9.1.1.

The scatter diagram for the data in Table 9.1.1 is shown in Figure 9.1.1. The data look slightly linear, but it would be difficult to fit a straight line to the data by eye. In the next section we will show how we fit the straight line to the data using the *least squares* method.

9.1.1   Estimation of Parameters

We must estimate three parameters from the data for the simple linear regression model. The three parameters to be estimated are β_0, β_1, and σ^2. The first two parameters could be estimated if we could find a straight line that fits the observed points $(x_1, y_1), (x_2, y_2), \ldots, (x_n, y_n)$ in the scatter diagram. The most popular method in use is the least squares method, in which each of the deviations $y_i - \beta_0 - \beta_1 x_i$ is squared and added together to form the sum of squares

$$S[\beta_0, \beta_1] = \sum_{i=1}^{n} [y_i - \beta_0 - \beta_1 x_i]^2. \tag{9.11}$$

To apply the method of least squares, we choose the slope, β_1, and intercept, β_0, of the straight line to minimize $S[\beta_0, \beta_1]$. The minimizing values are designated by $\hat{\beta}_0$ and $\hat{\beta}_1$. To find $\hat{\beta}_0$ and $\hat{\beta}_1$, we calculate

$$\frac{\partial S}{\partial \beta_0} = -2 \sum_{i=1}^{n} [y_i - \beta_0 - \beta_1 x_i], \tag{9.12}$$

and

$$\frac{\partial S}{\partial \beta_1} = -2 \sum_{i=1}^{n} x_i [y_i - \beta_0 - \beta_1 x_i]. \tag{9.13}$$

Setting these partial derivatives equal to zero, we obtain the following equations for the parameters $\hat{\beta}_0$ and $\hat{\beta}_1$.

$$\hat{\beta}_0 n + \hat{\beta}_1 \sum_{i=1}^{n} x_i = \sum_{i=1}^{n} y_i, \tag{9.14}$$

and

$$\hat{\beta}_0 \sum_{i=1}^{n} x_i + \hat{\beta}_1 \sum_{i=1}^{n} x_i^2 = \sum_{i=1}^{n} x_i y_i. \tag{9.15}$$

Solving for $\hat{\beta}_0$ and $\hat{\beta}_1$, we obtain

$$\hat{\beta}_1 = \frac{n \sum_{i=1}^{n} x_i y_i - \left(\sum_{i=1}^{n} x_i \right) \left(\sum_{i=1}^{n} y_i \right)}{n \sum_{i=1}^{n} x_i^2 - \left(\sum_{i=1}^{n} x_i \right)^2}, \tag{9.16}$$

and

$$\hat{\beta}_0 = \bar{y} - \hat{\beta}_1 \bar{x}. \tag{9.17}$$

An equivalent formula for $\hat{\beta}_1$ is given by

$$\hat{\beta}_1 = \frac{\sum_{i=1}^{n} (x_i - \bar{x})(y_i - \bar{y})}{\sum_{i=1}^{n} (x_i - \bar{x})^2} = \frac{\sum_{i=1}^{n} (x_i - \bar{x}) y_i}{\sum_{i=1}^{n} (x_i - \bar{x})^2} = \sum_{i=1}^{n} c_i y_i. \tag{9.18}$$

Thus, $\hat{\beta}_1$ is a linear combination of the $y_i's$. We ask you to prove, in Exercise 2, that (9.18) is equivalent to (9.16).

A shorthand notation for some of the parameters used in the derivations of $\hat{\beta}_0$ and $\hat{\beta}_1$ is widely used (with slight variations). This notation makes it much easier to write down some of the formulas. Following Mosteller, Fienberg, and Rourke [15], we define

$$\text{SSx} = \sum_{i=1}^{n}(x_i - \overline{x})^2 = \sum_{i=1}^{n} x_i^2 - n\overline{x}^2, \qquad (9.19)$$

$$\text{SSy} = \sum_{i=1}^{n}(y_i - \overline{y})^2 = \sum_{i=1}^{n} y_i^2 - n\overline{y}^2, \qquad (9.20)$$

and

$$\text{Sxy} = \sum_{i=1}^{n}(x_i - \overline{x})(y_i - \overline{y}) = \sum_{i=1}^{n} x_i y_i - n\overline{x}\,\overline{y}. \qquad (9.21)$$

Draper and Smith [6] use the above notation with x and y written in capital letters but use S_{XX} in place of SSx and S_{YY} in place of SSy. Montgomery and Peck [14] use the same notation as Draper and Smith except that they do not capitalize x and y.

Using the above notation, we can write (9.18) as

$$\hat{\beta}_1 = \frac{\text{Sxy}}{\text{SSx}} = \sum_{i=1}^{n} c_i y_i, \qquad (9.22)$$

where

$$c_i = \frac{x_i - \overline{x}}{\text{SSx}}. \qquad (9.23)$$

Example 9.1.1 For the data of Table 9.1.1, my HP-42S shows that

$$\sum_{i=1}^{21} x_i = 5,968 \qquad \sum_{i=1}^{21} x_i^2 = 2,180,968$$

$$\sum_{i=1}^{21} y_i = 882.69 \qquad \sum_{i=1}^{21} y_i^2 = 43,621.3521$$

$$\sum_{i=1}^{21} x_i y_i = 288,500.5 \qquad \overline{x} = 284.190476 \qquad \overline{y} = 42.03286$$

$$\text{Sxy} = 32,648.391688 \quad \text{and} \quad \text{SSx} = 484,919.24037.$$

Hence, by (9.22),

$$\hat{\beta}_1 = \frac{\text{Sxy}}{\text{SSx}} = 0.0776385, \qquad (9.24)$$

and, by (9.17),

$$\hat{\beta}_0 = 42.03286 - 0.0776385 \times 284.190476 = 19.9687.$$

Hence, the *least squares line* or *sample regression of Y on X* is given by

$$\hat{y} = 19.9687 + 0.0776385x, \qquad (9.25)$$

where x is the number of pages. Thus, a book costs, on the average, $19.9687 plus $0.0776 per page. Figure 9.1.2 shows the MINITAB statistics for this regression example produced by the MINITAB command REGRESS. The first table of these statistics shows that $\hat{\beta}_0 = 19.969$ and $\hat{\beta}_1 = 0.07764$. We will explain the other statistics later. Using (9.25) to calculate the estimated price for the first edition of the book you are reading, we obtained $50.25, although the list price was $39.95. □

Most serious pocket calculators (both scientific and business), where *serious* means a purchase price of about $30 or more, have a built-in facility for simple linear regression analysis. Most will provide the slope and intercept of the least squares regression line, as well as the sample correlation coefficient, r, that we will discuss below. In addition, intermediate quantities, such as $\sum x_i$, $\sum y_i$, $\sum x_i y_i$, $\sum x^2$, can be obtained. Many will also calculate estimated values for y given x; that is, $\hat{y} = \hat{\beta}_0 + \hat{\beta}_1 x$. It is frightfully dull work to calculate these numbers with a four-function calculator.

It is important when doing simple linear regression to look at the scatter diagram to see whether the data appear to be linear. Anscombe [1] has provided four data sets, each containing eleven points, that have identical least squares lines and identical regression statistics but very different scatter diagrams. We show the scatter diagrams in Figures 9.1.3–9.1.6.[1]

Clearly, only the first data set has a reasonable straight line fit. The second data set looks like a curve such as a parabola would fit the points, but a straight line does not. The third data set looks like a straight line with one point that popped out—probably an error in the data. The measurement for that data point may have been improperly recorded. The last data set has all the points on a vertical line except for one unusual point. Linear regression is certainly not called for if the unusual point really belongs to

---

[1] Adapted by permission from Professor Francis J. Anscombe, "Graphs in Statistical Analysis," *The American Statistician*, **27(1)**, (February 1973), 17–21.

```
The regression equation is
Price = 20.0 + 0.0776 Pages
```

Predictor	Coef	Stdev	t-ratio	p
Constant	19.969	6.367	3.14	0.005
Pages	0.07764	0.01976	3.93	0.001

```
s = 13.76      R-sq = 44.8%      R-sq(adj) = 41.9%
```

Analysis of Variance

SOURCE	DF	SS	MS	F	p
Regression	1	2923.0	2923.0	15.44	0.001
Error	19	3596.4	189.3		
Total	20	6519.4			

Unusual Observations

Obs.	Pages	Price	Fit	Stdev.Fit	Residual	St.Resid
15	370	84.25	48.69	3.45	35.56	2.67R

R denotes an obs. with a large st. resid.

Figure 9.1.2. MINITAB statistics for Example 9.1.1.

the data set. Although the MINITAB command REGRESS provides almost identical statistics for the four data sets, it does issue a warning that there is an outlier in Data Set 3. It does the same for Data Set 4. The Anscombe data sets are delivered with the MINITAB system.

Anscombe's first data set.   Anscombe's second data set.

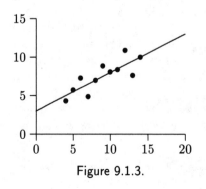

Figure 9.1.3.

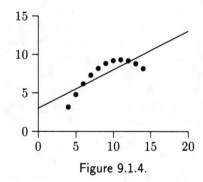

Figure 9.1.4.

Anscombe's third data set.   Anscombe's fourth data set.

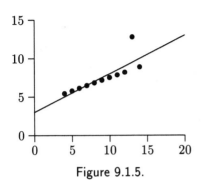 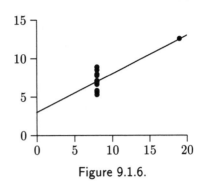

Figure 9.1.5.   Figure 9.1.6.

Although we have provided an algorithm for constructing the least squares line, we have not provided any method besides looking at the scatter diagram to judge how well a straight line represents the relationship between the random variables X and Y. That is our next objective.

In Theorem 9.1.1 we provide some useful properties of the estimators β_0 and β_1.

Theorem 9.1.1 *Let $\hat{\beta}_0$ and $\hat{\beta}_1$ be the least squares estimators for the simple linear regression model with the standard statistical assumptions. Then $\hat{\beta}_0$ and $\hat{\beta}_1$ are normally distributed with the following parameters:*

(a) $E[\hat{\beta}_0] = \beta_0$

(b) $E[\hat{\beta}_1] = \beta_1$

(c) $\mathrm{Var}[\hat{\beta}_0] = \dfrac{\displaystyle\sum_{i=1}^{n} x_i^2}{n\,\mathrm{SSx}}\sigma^2$

(d) $\mathrm{Var}[\hat{\beta}_1] = \dfrac{\sigma^2}{\mathrm{SSx}}.$

Finally, each $\hat{\beta}_i$ is the maximum likelihood estimator of β_i and has minimum variance among all unbiased linear estimators of β_i. (The fact that $\hat{\beta}_i$ has the minimum variance among all unbiased linear estimators is known as the Gauss–Markov theorem and does not require the assumption that the errors ϵ_i's are normally distributed.)

Proof By (9.18) we have

$$\hat{\beta}_1 = \sum_{i=1}^{n} c_i Y_i, \tag{9.26}$$

where

$$c_i = \frac{x_i - \bar{x}}{\text{SSx}} \quad i = 1, 2, \ldots, n. \tag{9.27}$$

Hence, $\hat{\beta}_1$ is a linear combination of the n independent normal random variables $Y_1, Y_2, \ldots, Y_n$, so that by Theorem 3.2.4, $\hat{\beta}_1$ is normal with mean

$$E[\hat{\beta}_1] = \sum_{i=1}^{n} c_i E[Y_i] = \sum_{i=1}^{n} c_i(\beta_0 + \beta_1 x_i)$$

$$= \beta_0 \sum_{i=1}^{n} c_i + \beta_1 \sum_{i=1}^{n} c_i x_i = \beta_1, \tag{9.28}$$

because

$$\sum_{i=1}^{n} c_i = 0 \quad \text{and} \quad \sum_{i=1}^{n} c_i x_i = 1,$$

as we ask you to prove in Exercise 3. We calculate

$$\text{Var}[\hat{\beta}_1] = \sum_{i=1}^{n} c_i^2 \text{Var}[Y_i] = \frac{\sum_{i=1}^{n}(x_i - \bar{x})^2}{\left[\sum_{i=1}^{n}(x_i - \bar{x})^2\right]^2}\sigma^2 = \frac{\sigma^2}{\text{SSx}}. \tag{9.29}$$

This proves that $\hat{\beta}_1$ is normally distributed and that (b) and (d) are true. The remainder of the proof is similar. See Montgomery and Peck [14] for the remainder of the proof, except the proof that each $\hat{\beta}_i$ is the maximum likelihood and minimum variance unbiased estimator of β_i. These facts are proved by Neter *et al.* [17]. ■

Since $\hat{\beta}_0$ and $\hat{\beta}_1$ are normally distributed, if we had an estimate of σ^2 we could calculate confidence limits for β_0 and β_1, as well as perform other useful statistical tests. The parameter σ^2 is a measure of the expected vertical spread about the least squares line in the scatter diagram and thus a measure of how well this line represents the relation between X and Y.

The *i*th *predicted (fitted) value*, written $\hat{y}_i$, is $\hat{y}_i = \hat{\beta}_0 + \hat{\beta}_1 x_i$ ($i = 1, 2, \ldots, n$), and the *i*th *residual* is $e_i = y_i - \hat{y}_i$. Thus, in the scatter diagram the residuals are the vertical distances between the plotted points and the least squares line. We define the *error sum of squares* by

$$\text{SSE} = \sum_{i=1}^{n} e_i^2 = \sum_{i=1}^{n}(y_i - \hat{y}_i)^2. \tag{9.30}$$

The following theorem gives us the estimate usually used for σ^2 as well as other useful information.

Theorem 9.1.2 *Suppose the hypotheses of Theorem 9.1.1 are true. Then the mean square error $\hat{\sigma}^2$ defined by*

$$\hat{\sigma}^2 = s^2 = \frac{\text{SSE}}{n-2} \equiv \text{MSE} \tag{9.31}$$

is an unbiased estimator of σ^2. The random variable

$$\frac{(n-2)\text{MSE}}{\sigma^2} \tag{9.32}$$

has a chi-square distribution with $n-2$ degrees of freedom so the $100(1-\alpha)\%$ confidence interval for σ^2 is given by

$$\frac{(n-2)\text{MSE}}{\chi^2_{n-2,\alpha/2}} \leq \sigma^2 \leq \frac{(n-2)\text{MSE}}{\chi^2_{n-2,1-\alpha/2}}. \tag{9.33}$$

By Theorem 9.1.1, the estimated variance of $\hat{\beta}_0$ is given by

$$s^2_{\hat{\beta}_0} = \frac{\sum\limits_{i=1}^{n} x_i^2}{n\,\text{SSx}} s^2, \tag{9.34}$$

and the estimated variance of $\hat{\beta}_1$ is given by

$$s^2_{\hat{\beta}_1} = \frac{s^2}{\text{SSx}}. \tag{9.35}$$

Furthermore, each of the normalized random variables

$$\frac{\hat{\beta}_i - \beta_i}{s_{\hat{\beta}_i}}, \quad i = 0, 1, \tag{9.36}$$

has a Student's t distribution with $n-2$ degrees of freedom. Hence, the $100(1-\alpha)\%$ confidence interval for the intercept β_0 is given by

$$\hat{\beta}_0 \pm t_{n-2,\alpha/2} \times s_{\hat{\beta}_0}, \tag{9.37}$$

and the $100(1-\alpha)\%$ confidence interval for the slope β_1 is given by

$$\hat{\beta}_1 \pm t_{n-2,\alpha/2} \times s_{\hat{\beta}_1}. \tag{9.38}$$

Proof See Montgomery and Peck [14, pages 16–28]. ∎

MSE is also called the *residual mean square*. The number $s = \sqrt{\text{MSE}}$ is called the *standard error*, the *standard error of the estimate* or the *standard error of regression* and is often written as s_e.

The relation (9.36) for $i = 1$ is used to test the null hypothesis, H_0: $\beta_1 = 0$, against the alternate hypothesis, H_1: $\beta_1 \neq 0$, as a basic measure of the effectiveness of the regression equation. Failure to reject this null hypothesis implies that there is no linear relationship between X and Y. Similarly, using (9.36) for $i = 0$ is the basis for the test of H_0: $\beta_0 = 0$ versus H_1: $\beta_0 \neq 0$. However, this test is not quite as important in determining the effectiveness of the regression study. We present the following algorithms for testing the significance of β_1 and β_0.

Algorithm 9.1.1 (*Test of significance of the parameter β_1*) *This algorithm will determine at the α level of significance whether to accept or reject the null hypothesis, H_0: $\beta_1 = 0$, against the alternative hypothesis, H_1: $\beta_1 \neq 0$.*

Step 1 [Calculate the test statistic] *Calculate*

$$t = \frac{\hat{\beta}_1}{s_{\hat{\beta}_1}}, \tag{9.39}$$

using the formulas in Theorem 9.1.2.

Step 2 [Accept or reject H_0] *If*

$$|t| > t_{n-2,\alpha/2},$$

reject H_0 and accept H_1; otherwise accept H_0.

Algorithm 9.1.2 (*Test of significance of the parameter β_0*) *This algorithm will determine at the α level of significance whether to accept or reject the null hypothesis, H_0: $\beta_0 = 0$, against the alternative hypothesis, H_1: $\beta_0 \neq 0$.*

Step 1 [Calculate the test statistic] *Calculate*

$$t = \frac{\hat{\beta}_0}{s_{\hat{\beta}_0}}, \tag{9.40}$$

using the formulas in Theorem 9.1.2.

Step 2 [Accept or reject H_0] *If*

$$|t| > t_{n-2,\alpha/2},$$

reject H_0 and accept H_1; otherwise accept H_0.

Proof The proofs of the two algorithms are immediate from Theorem 9.1.2. ∎

The above algorithms are executed in standard statistical packages such as MINITAB and SAS/STAT. For example, the results of the tests are given by MINITAB in Figure 9.1.2 for the data of Table 9.1.1. The value of $|t| = |\hat{\beta}_1/s_{\hat{\beta}_1}|$ is 3.93, given in the column *t-ratio* and row *Pages*. Since the p-value is 0.001, we can reject the null hypothesis at the 0.1% level. Similarly, we can reject the null hypothesis that $\beta_0 = 0$ in favor of the alternative hypothesis that $\beta_0 \neq 0$, at the 0.5% level of significance.

Example 9.1.2 Consider Example 9.1.1. Let us calculate a 95% confidence interval for σ^2, β_0, and for β_1. From Figure 9.1.2 we see that MSE= 189.3. Hence, (9.33) yields

$$109.48 \leq \sigma^2 \leq 403.828,$$

since $\chi^2_{19,0.025} = 32.8523268$ and $\chi^2_{19,0.975} = 8.906516$. We also see by Figure 9.1.2 that $s_{\hat{\beta}_0} = 6.367$ and $s_{\hat{\beta}_1} = 0.01976$. (In Exercise 4 we ask you to verify that these are the values that are yielded by formulas (9.34) and (9.35), respectively.) Since $t_{19,0.25} = 2.0930$, by (9.37) the 95% confidence interval for β_0 is 19.9687 ± 13.3261 or 6.6426 to 33.2948. Similarly, by (9.38) the 95% confidence for β_1 is 0.07764 ± 0.04136 or 0.03628–0.1190. $\square$

9.1.2 Analysis of Variance in Linear Regression

In Theorem 9.1.2 we saw that the error sum of squares, SSE, can be used to estimate σ^2 by the formula

$$\hat{\sigma}^2 = \frac{\text{SSE}}{n-2} = \text{MSE}. \tag{9.41}$$

SSE can also be interpreted as a measure of how much variation in Y is left unexplained by the linear regression model. The *total sum of squares*, SST, is defined by

$$\text{SST} = \sum_{i=1}^{n} (y_i - \overline{y})^2. \tag{9.42}$$

Note that from the definition of SS_y, we have $\text{SS}_y = \text{SST}$. The following proposition provides some information about the total sum of squares.

Proposition 9.1.1 *The total sum of squares,* SST, *can be written as*

$$\text{SST} = \text{SSR} + \text{SSE}, \tag{9.43}$$

where

$$\text{SSE} = \sum_{i=1}^{n} (y_i - \hat{y}_i)^2, \tag{9.44}$$

and

$$\text{SSR} = \sum_{i=1}^{n} (\hat{y}_i - \bar{y})^2. \tag{9.45}$$

Furthermore,

$$\text{SSR} = \hat{\beta}_1 S_{xy}, \tag{9.46}$$

and

$$\text{SSE} = \text{SS}_y - \hat{\beta}_1 S_{xy}. \tag{9.47}$$

Proof The equation (9.43) follows immediately from the following equation, which we will prove.

$$\sum_{i=1}^{n} (y_i - \bar{y})^2 = \sum_{i=1}^{n} (\hat{y}_i - \bar{y})^2 + \sum_{i=1}^{n} (y_i - \hat{y}_i)^2. \tag{9.48}$$

To prove (9.48), we write

$$
\begin{aligned}
\sum_{i=1}^{n} (y_i - \bar{y})^2 &= \sum_{i=1}^{n} [(\hat{y}_i - \bar{y}) + (y_i - \hat{y}_i)]^2 \\
&= \sum_{i=1}^{n} (\hat{y}_i - \bar{y})^2 + \sum_{i=1}^{n} (y_i - \hat{y}_i)^2 \\
&\quad + 2 \sum_{i=1}^{n} (\hat{y}_i - \bar{y})(y_i - \hat{y}_i).
\end{aligned} \tag{9.49}
$$

But

$$
\begin{aligned}
\sum_{i=1}^{n} (\hat{y}_i - \bar{y})(y_i - \hat{y}_i) &= \sum_{i=1}^{n} e_i(\hat{y}_i - \bar{y}) \\
&= \sum_{i=1}^{n} e_i(\hat{\beta}_0 + \hat{\beta}_1 x_i) - \bar{y} \sum_{i=1}^{n} e_i \\
&= \hat{\beta}_0 \sum_{i=1}^{n} e_i + \hat{\beta}_1 \sum_{i=1}^{n} e_i x_i - \bar{y} \sum_{i=1}^{n} e_i = 0, \tag{9.50}
\end{aligned}
$$

because the residuals satisfy the two equations

$$\sum_{i=1}^{n} e_i = 0 \quad \text{and} \quad \sum_{i=1}^{n} e_i x_i = 0, \tag{9.51}$$

as can be seen by setting

$$0 = \frac{\partial S}{\partial \beta_0} = -2 \sum_{i=1}^{n} [y_i - \beta_0 - \beta_1 x_i] = -2 \sum_{i=1}^{n} e_i, \qquad (9.52)$$

and

$$0 = \frac{\partial S}{\partial \beta_1} = -2 \sum_{i=1}^{n} x_i [y_i - \beta_0 - \beta_1 x_i] = -2 \sum_{i=1}^{n} e_i x_i. \qquad (9.53)$$

Equation (9.50) substituted into (9.49) yields the proof of (9.48), that is equivalent to (9.43).

In Exercise 5 we ask you to prove that $(\overline{x}, \overline{y})$ is on the regression line so that $\overline{y} = \hat{\beta}_0 + \hat{\beta}_1 \overline{x}$. Therefore, we can write

$$
\begin{aligned}
\text{SSR} \quad &= \quad \sum_{i=1}^{n} (\hat{y}_i - \overline{y})^2 \\
&= \quad \sum_{i=1}^{n} [\hat{\beta}_0 + \hat{\beta}_1 x_i - (\hat{\beta}_0 + \hat{\beta}_1 \overline{x})]^2 = \sum_{i=1}^{n} [\hat{\beta}_1 (x_i - \overline{x})]^2 \\
&= \quad (\hat{\beta}_1)^2 \sum_{i=1}^{n} (x_i - \overline{x})^2 = (\hat{\beta}_1)^2 \text{SS}_x = \hat{\beta}_1 S_{xy}, \qquad (9.54)
\end{aligned}
$$

since $S_{xy} = \hat{\beta}_1 \text{SS}_x$ by (9.22). This proves (9.46). Formula (9.47) follows from (9.46), since $\text{SS}_y = \text{SST}$. ∎

Hogg and Tanis [11, pages 492–493] show the validity of the ANOVA table (Table 9.1.2) for simple linear regression.

Table 9.1.2. ANOVA Table for Simple Linear Regression

SOURCE	DF	SS	MS	*F*-Value	*Pr > F*
Regression	1	SSR	MSR	*F*	*p*-value
Error	$n - 2$	SSE	MSE		
Total	$n - 1$	SST			

In the table

$$\text{MSR} = \frac{\text{SSR}}{1},$$

since SSR has only one degree of freedom. The F-value is

$$F = \frac{\text{MSR}}{\text{MSE}}. \tag{9.55}$$

The value of F provides the test of the null hypothesis that $\beta_1 = 0$, against the alternative hypothesis that $\beta_1 \neq 0$. The null hypothesis is rejected if $F > F_{1,n-2;\alpha}$, since F has an F-distribution with 1 and $n - 2$ degrees of freedom. This turns out to be equivalent to the t-test that we described earlier (Algorithm 9.1.1), since

$$\left(\frac{\hat{\beta}_0}{s_{\hat{\beta}_0}}\right)^2 = \frac{\text{MSR}}{\text{MSE}},$$

and the square of a Student's t distribution with $n - 2$ degrees of freedom is an F distribution with 1 and $n - 2$ degrees of freedom. Most computer statistical systems provide both tests for the significance of β_1. For example, the SAS/STAT procedure REG claims the t-value of the t-test for the data of Table 9.1.1 is 3.93 with a p-value of 0.0009. It gives an F value of $15.422 = 3.929631^2$ with a p-value of 0.0009. See Figures 9.1.11–9.1.12. MINITAB provides roughly the same numbers in Figure 9.1.2 but with slightly less precision. The ANOVA table that is generated by the MINITAB command REGRESS for Example 9.1.1 appears in Figure 9.1.2, while that given by the SAS/STAT procedure REG appears in Figure 9.1.12.

9.1.3   Correlation Analysis

In Chapter 2 we defined the *correlation (coefficient) of X and Y*, written $\rho(X, Y)$, by

$$\rho(X, Y) = \frac{\text{Cov}(X, Y)}{(\text{Var}[X]\text{Var}[Y])^{1/2}}, \tag{9.56}$$

provided both variances are nonzero. We mentioned that $|\rho(X, Y)| \leq 1$ with equality if and only if $P[Y = aX + b] = 1$ for some a and b. Thus, the *correlation coefficient* ρ is a measure of the strength of linear association between two variables. A correlation coefficient with absolute value near 1 does not necessarily mean there is a predictive relationship between the two variables. For example, it was once noticed that there was a strong correlation between the number of stork nests and the number of babies born in English villages. This does not mean, however, that storks bring babies or that babies attract storks. The reason for the correlation is that both variables are partially determined by the size of a village. A large village tends to have lots of stork nests and babies and a small village a paucity of both.

9.1.3.1 The Coefficient of Determination

Suppose the least squares line fits the data in the scatter diagram very closely. Then

$$\text{SSE} = \sum_{i=1}^{n}(y_i - \hat{y}_i)^2 = \text{SSy} - \hat{\beta}_1\text{Sxy} \qquad (9.57)$$

will be small. Since SSR = SST − SSE, this means that SSR will be large. The regression sum of squares, SSR, is interpreted as the amount of total variation that *is* explained by the regression model, that is, attributed to an approximate linear relationship between Y and X. Hence, the *coefficient of determination*, R^2, defined by

$$R^2 = \frac{\text{SSR}}{\text{SST}} = \frac{\text{SST} - \text{SSE}}{\text{SST}} = 1 - \frac{\text{SSE}}{\text{SST}}, \qquad (9.58)$$

is the ratio of the explained variation to the total variation. It is clear from the definition of R^2 that $0 \leq R^2 \leq 1$, and the closer it is to one, the better a straight line fits the data in the scatter diagram. It is one when all the points in the scatter diagram lie on a straight line, as in Figures 9.1.7-9-.1.8. As the symbol suggests, R^2 is the square of the sample regression coefficient, r, that we will discuss below. For the least squares line of Figure 9.1.1, $R^2 = 0.448$, while in Figures 9.1.3–9.1.6, $R^2 = 0.667$. An R^2 value of 0.667 is considered evidence that the values of y can be predicted from those of x using the regression line. For scientific or engineering problems, however, we like to see R^2 values of at least 0.8. The R^2 value of 0.448 in Example 9.1.1 shows that the number of pages is not very useful for predicting the price of a book by using the least squares line.

9.1.3.2 The Sample Coefficient of Correlation

When we perform a simple linear regression study we do not know the value of the population correlation coefficient $\rho(X, Y)$, so we estimate it by the *sample correlation coefficient, r,* that is defined by

$$r = \frac{\sum_{i=1}^{n}(x_i - \overline{x})(y_i - \overline{y})}{\sqrt{\sum_{i=1}^{n}(x_i - \overline{x})^2 \sum_{i=1}^{n}(y_i - \overline{y})^2}} = \frac{\text{Sxy}}{\sqrt{[\text{SSx}][\text{SSy}]}}. \qquad (9.59)$$

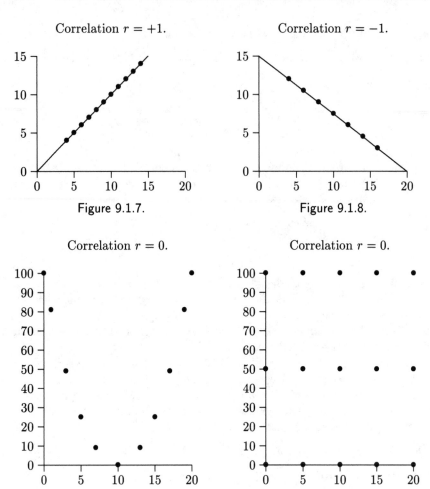

Correlation $r = +1$.

Figure 9.1.7.

Correlation $r = -1$.

Figure 9.1.8.

Correlation $r = 0$.

Figure 9.1.9.

Correlation $r = 0$.

Figure 9.1.10.

Kass [12] shows that r can be written as

$$r = \frac{X \cdot Y}{|U||V|} = \cos(U, V), \tag{9.60}$$

where U and V are the elements of n-dimensional Euclidian space defined by $U = (u_1, u_2, \ldots, u_n)$ where $u_i = x_i - \bar{x}$, and $V = (v_1, v_2, \ldots, v_n)$ where $v_i = y_i - \bar{y}$, $i = 1, 2, \ldots, n$. Thus, r has a geometric interpretation.

If all the points in the scatter diagram fall on the least squares line, then $r = 1$ if, as in Figure 9.1.7, the slope is positive so that increasing values of x correspond to increasing values of y. If all points of the scatter diagram fall on a straight line with negative slope, as in Figure 9.1.8, then $r = -1$. The sample correlation r for Figures 9.1.9–9.1.10 is zero. The points in Figure 9.1.9 fall on the parabola $y = (x - 10)^2$ but certainly not on a line. One would expect r to be zero for Figure 9.1.10 because there is no way to fit a straight line to the points in a satisfactory way. The least squares line, however, is $y = 50$.

There is a close relationship between the sample correlation coefficient r, and $\hat{\beta}_1$. From the relationships

$$r = \frac{\text{Sxy}}{\sqrt{[\text{SSx}][\text{SSy}]}}, \tag{9.61}$$

and

$$\hat{\beta}_1 = \frac{S_{xy}}{SS_x}, \tag{9.62}$$

we conclude that

$$\hat{\beta}_1 = \left(\frac{SS_y}{SS_x}\right)^{1/2} r. \tag{9.63}$$

Although (9.63) shows that r and $\hat{\beta}_1$ are closely related, they measure different things. The sample correlation coefficient r measures the degree of linear association between Y and X; $\hat{\beta}_1$ is the estimated slope of the regression line, that is, the estimated change in y for a unit change in x.

We calculate

$$\begin{aligned} r^2 &= (\hat{\beta}_1)^2 \frac{SS_x}{SS_y} = \hat{\beta}_1 \frac{S_{xy}}{SS_y} \\ &= \frac{\text{SSR}}{\text{SST}} = R^2, \end{aligned} \tag{9.64}$$

which shows that the coefficient of determination R^2 is the square of r.

Algorithm 9.1.3 *Given two random variables X and Y with sample correlation coefficient r, this algorithm will test at the α level of significance the null hypothesis, H_0: $\rho = 0$, against the alternative hypothesis, H_1: $\rho \neq 0$.*

Step 1 Calculate the test statistic

$$t_0 = \frac{r\sqrt{n - 2}}{\sqrt{1 - r^2}}. \tag{9.65}$$

Step 2 Accept or reject H_0 *If*

$$|t_0| > t_{n-2, \alpha/2},$$

reject the null hypothesis and accept H_1.

Proof Hogg and Craig [10, pages 339–342] show that if the null hypothesis is true, then t_0 has a Student's t distribution with $n-2$ degrees of freedom. The proof follows from this fact. ∎

Snedecor and Cochran [21, Section 10.5] prove that this test is equivalent to the t-test we gave in Algorithm 9.1.1.

Example 9.1.3 The sample correlation coefficient for the data of Table 9.1.1 is 0.66959. Since $n = 21$, $t_0 = 3.92965$. Table 5 of Appendix A indicates that $t_{19,0.025} = 2.093$, so we can reject the null hypothesis at the 5% level of significance. The p-value of the test is 0.00089971. Hence, we can reject the null hypothesis at the 0.09% level of significance. These numbers agree with those produced by SAS/STAT in Figure 9.1.11. □

We exhibit some output produced by several SAS/STAT procedures in applying simple linear regression to the data of Table 9.1.1. SAS/STAT provides the same data we obtained from the MINITAB command REGRESS and displayed in Figure 9.1.2. SAS/STAT provides slightly more decimal digits of precision. (The MINITAB command REGRESS, however, produced a warning that the point $(370, 84.25)$ appeared to be an outlier while the SAS/STAT procedure REG produced no such warning.) Figure 9.1.11 shows part of the output from the SAS/STAT Procedure REG. It provides the estimates of the parameters β_0 and β_1 together with the t-tests to determine whether they are significant.

Figure 9.1.12 provides the ANOVA table for Example 9.1.1. Slightly more information is provided than the MINITAB ANOVA table shown in Figure 9.1.2.

Price Versus Pages

Variable	DF	Parameter Estimate	Standard Error	T for H0: Parameter=0	Prob > |T|
INTERCEP	1	19.968731	6.36704128	3.136	0.0054
PAGES	1	0.077639	0.01975708	3.930	0.0009

Figure 9.1.11. SAS/STAT parameter estimates for Example 9.1.1.

Dependent Variable: PRICE

Analysis of Variance

Source	DF	Sum of Squares	Mean Square	F Value	Prob>F
Model	1	2922.96646	2922.96646	15.442	0.0009
Error	19	3596.40297	189.28437		
C Total	20	6519.36943			

Root MSE	13.75807	R-square	0.4484	
Dep Mean	42.03286	Adj R-sq	0.4193	
C.V.	32.73169			

Figure 9.1.12. SAS/STAT ANOVA table for Example 9.1.1.

Figure 9.1.13 shows the statistics generated by the SAS/STAT procedure CORR. (There is a similar procedure in MINITAB.) It shows that $r = 0.66959$ for Example 9.1.1. It also gives the mean, standard deviation, the sum of the values of x_i and y_i, and the p-value of the test that $\rho = 0$.

CORRELATION ANALYSIS

Simple Statistics

Variable	N	Mean	Std Dev	Sum
PAGES	21	284.19048	155.71115	5968
PRICE	21	42.03286	18.05460	882.69000

Simple Statistics

Variable	Minimum	Maximum
PAGES	53.00000	595.00000
PRICE	7.95000	84.25000

Price Versus Pages

CORRELATION ANALYSIS

Pearson Correlation Coefficients / Prob > |R| under Ho: Rho=0 / N = 21

	PAGES	PRICE
PAGES	1.00000	0.66959
	0.0	0.0009
PRICE	0.66959	1.00000
	0.0009	0.0

Figure 9.1.13. SAS/STAT correlation analysis for Example 9.1.1.

9.1.4   Interval Estimation of Predicted Values

One of the major reasons for using regression analysis is for estimating values of the response variable from the predictor variable. Thus, for a value of x, say x_0, we estimate the corresponding y value to be $\hat{y}_0 = \hat{\beta}_0 + \hat{\beta}_1 x_0$. In Example 9.1.1 we were interested in estimating the cost of a book of 390 pages. The value \$50.25 that we calculated is the estimate of the average cost of a statistics book of 390 pages. In general, when we fix x_0 and calculate $\hat{y}_0 = \hat{\beta}_0 + \hat{\beta}_1 x_0$, we are estimating $E[Y|X = x_0] = \beta_0 + \beta_1 x_0$. Montgomery and Peck [14, pages 29–31] show that the estimator $\hat{y}_0 = \hat{\beta}_0 + \hat{\beta}_1 x_0$ for $E[Y|X = x_0]$ is unbiased and has a normal distribution with variance

$$\sigma^2 \left[\frac{1}{n} + \frac{(\bar{x} - x_0)^2}{\text{SSx}} \right]. \tag{9.66}$$

They also show that the $100(1 - \alpha)$ confidence interval for $E[Y|X = x_0]$ is

$$\hat{y}_0 - t_{n-2,\alpha/2} \sqrt{\text{MSE} \left[\frac{1}{n} + \frac{(\bar{x} - x_0)^2}{\text{SSx}} \right]}$$

$$\leq E[Y|X = x_0] \leq \hat{y}_0 + t_{n-2,\alpha/2} \sqrt{\text{MSE} \left[\frac{1}{n} + \frac{(\bar{x} - x_0)^2}{\text{SSx}} \right]}. \tag{9.67}$$

This confidence interval is narrowest when $x_0 = \bar{x}$, as one would expect.

Of even greater importance than the estimation of $E[Y|X = x_0]$ is the prediction of y when $x = x_0$. Montgomery and Peck[14, pages 31–33] show that $\hat{y}_0 = \hat{\beta}_0 + \hat{\beta}_1 x_0$ is also an estimator of y_0. They show that the $100(1 - \alpha)\%$ prediction interval for the predicted value y_0 at x_0 is given by

$$\hat{y}_0 - t_{n-2,\alpha/2} \sqrt{\text{MSE} \left[1 + \frac{1}{n} + \frac{(\bar{x} - x_0)^2}{\text{SSx}} \right]}$$

$$\leq y_0 \leq \hat{y}_0 + t_{n-2,\alpha/2} \sqrt{\text{MSE} \left[1 + \frac{1}{n} + \frac{(\bar{x} - x_0)^2}{\text{SSx}} \right]}. \tag{9.68}$$

Let us consider an example.

Example 9.1.4 Consider Example 9.1.1. Based on a simple linear regression applied to the data of Table 9.1.1, the average price for a statistics book reviewed in Technometrics with 390 pages is \$50.25. By (9.67), the 95% confidence interval for the average price of a 390 page book is \$42.59

to \$57.91 since $t_{19,0.025} = 2.093$, SSx= $484,919.24037$, $\bar{x} = 284.190476$, MSE= 189.3 and

$$t_{n-2,\alpha/2} \sqrt{\text{MSE} \left[\frac{1}{n} + \frac{(\bar{x} - x_0)^2}{\text{SSx}}\right]} = 7.657286.$$

Similarly, the 95% prediction interval for the price of a book of 390 pages is \$20.44 to \$80.05. In Figure 9.1.14 we show the MINITAB command and subcommand that will make this calculation for us, assuming that c1 is the "pages" data of Table 9.1.1 and c2 is the "price" data. We have left out several pages of additional output generated by the MINITAB command **regress**. By "95% C. I." MINITAB means the 95% confidence interval for the mean of the predicted value and by "95% P. I." MINITAB means the 95% prediction interval of the predicted value. $\square$

```
MTB > regress c2 1 c1;
 SUBC> predict 390.
    Fit  Stdev.Fit        95% C.I.          95% P.I.
   50.25      3.66    ( 42.59, 57.91)   ( 20.44, 80.05)
```

Figure 9.1.14. MINITAB output for Example 9.1.3.

9.1.5 Errors and Diagnostics

Regression analysis is one of the best developed parts of statistics. There is no doubt that if the hypotheses of the standard statistical model are true, the theory can yield very good results. Nevertheless, more egregious statistics has probably been perpetuated using linear regression than all the other statistical methods put together. We have provided some of the standard measures of goodness of fit of the least squares line for the simple linear regression model. However, as the MINITAB output for Anscombe's fourth data set (the scatter plot is shown in Figure 9.1.6) shown in Figure 9.1.16 demonstrates, a bad set of data can give good statistics. In fact, the statistics shown in Figure 9.1.16 are almost identical to those for Anscombe's first data set, whose scatter diagram appears in Figure 9.1.3. How can one doubt a model with $R^2 = 0.667$, $s = 1.236$, and good numbers in all the other right places? Fortunately, MINITAB does provide the diagnostic in the last line, that warns the user that the point at $(19, 12.5)$ has a large influence because of the large value of x. There are really two parts to the problem of getting good results with regression analysis. The first is determining whether the hypotheses of the standard statistical model are valid, that is, whether X and Y jointly have a bivariate normal distribution. The second problem is determining points that are outliers or have unusual influence. An *outlier* is a point that doesn't seem to fit in with the remainder of the data. A point whose removal causes major changes in the analysis is called *influential*. The two problems are not completely independent, of course. Fortunately, diagnostic techniques have been developed and added to statistical packages such as MINITAB and SAS/STAT to make it easier for practitioners to use regression analysis with more confidence.

The parameters, such as R^2, r, s, and the ANOVA table, do not tell us whether the hypotheses of the standard statistical model are valid. The key to this problem seems to lie with the analysis of the residuals. Montgomery and Peck [14, pages 57–70] show how to analyze the residuals with different kinds of plots. The detailed analysis of residuals is beyond the scope of this book but is covered in Atkinson [2], Cook and Weisberg [4], and in chapter 5 of Weisberg [22].

```
MTB > regress c6 1 c5 'resids' 'fits'
 The regression equation is
 Y4 = 3.00 + 0.500 X4
  Predictor        Coef       Stdev     t-ratio        p
 Constant         3.002       1.124        2.67    0.026
 X4               0.4999      0.1178       4.24    0.002
  s = 1.236      R-sq = 66.7%     R-sq(adj) = 63.0%
 Analysis of Variance
  SOURCE        DF          SS          MS          F        p
 Regression      1      27.490      27.490      18.00    0.002
 Error           9      13.742       1.527
 Total          10      41.232
 Unusual Observations
 Obs.       X4         Y4       Fit Stdev.Fit  Residual   St.Resid
   8      19.0     12.500     12.500     1.236     0.000       * X
 X denotes an obs. whose X value gives it large influence.
MTB > note The above was Anscombe's fourth data set.
```

Figure 9.1.16. MINITAB output for fourth Anscombe data set.

The second problem is more tractable and most statistical packages provide help. We will follow Hoaglin [9] in developing some of the concepts. Hoaglin shows that we can write the fitted y values, $\hat{y}_i$, in the form

$$\hat{y}_i = \sum_{j=1}^{n} h_{ij} y_j \quad i = 1, 2, \ldots, n, \tag{9.69}$$

where h_{ij} indicates how changing y_j affects $\hat{y}_i$ and is defined by

$$h_{ij} = \frac{1}{n} + \frac{(x_i - \overline{x})(x_j - \overline{x})}{\displaystyle\sum_{k=1}^{n}(x_k - \overline{x})^2}. \tag{9.70}$$

The number h_{ij} shows how changing y_j affects $\hat{y}_i$. The symmetric $n \times n$ matrix $H = (h_{ij})$ is called the *hat matrix* because it takes the vector of observed y-values $(y_1, y_2, \ldots, y_n)^T$ into the vector of fitted y-values $(\hat{y}_1, \hat{y}_2, \ldots, \hat{y}_n)^T$. Usually, we are concerned with the observed and fitted y-values at the same data point. In linear regression, the *leverage* of observation i is h_{ii}, the corresponding diagonal element of the hat matrix. For the simple regression line,

$$h_{ii} = \frac{1}{n} + \frac{(x_i - \overline{x})^2}{\displaystyle\sum_{k=1}^{n}(x_k - \overline{x})^2}. \tag{9.71}$$

The hat matrix and its diagonal elements have the following basic properties:

1. H is a *projection matrix*, that is, $H^2 = H$.

2. $0 \le h_{ii}$.

3. In simple linear regression, $\sum_{i=1}^{n} h_{ii} = 2$ (unless all the x_i are equal), so the average size of an h_{ii} is $2/n$.

4. If $h_{ii} = 0$ or $h_{ii} = 1$, then $h_{ij} = 0$ for all $j \ne i$.

The extreme cases $h_{ii} = 0$ and $h_{ii} = 1$ can be interpreted as follows. If $h_{ii} = 0$, then $\hat{y}_i = 0$; it is not affected by y_i or any other y_j. An example of this phenomenon occurs with $x = 0$ when fitting a straight line through the origin. If $h_{ii} = 1$, then $\hat{y}_i = y_i$, and the regression model always fits this data point exactly.

Equation (9.71) shows that for simple linear regression, the observations farthest from $\overline{x}$ are the ones that have the highest leverage. Thus, in Figure 9.1.6, the rightmost point has high leverage. Sure enough, as previously noted, in Figure 9.1.16 we see that MINITAB says "X denotes an obs. whose X value gives it large influence" about this point.

It is customary to use h_i in place of h_{ii} to simplify the notation. Then each residual e_i has the standard deviation $\sigma\sqrt{1 - h_i}$, that is estimated by $s\sqrt{1 - h_i}$. Dividing e_i by its estimated standard deviation yields the *standardized residual*

$$\frac{e_i}{s\sqrt{1 - h_i}}. \tag{9.72}$$

MINITAB and other statistical systems use the standardized residuals to identify outliers. Thus, in Figure 9.1.2, MINITAB identifies the point $(370, 84.25)$ as being one with a large standardized residual value of 2.67. MINITAB identifies every data point with a standardized residual greater than 2. Many statisticians use the rule of thumb that any observation with $|y_i - \hat{y}_i| \geq 3s_e$ should be considered an outlier.

One key technique in evaluating the influence of an observation is to delete the observation from the regression analysis and see how various results change. To denote a quantity calculated from the data without observation i it is customary to append (i) which is read as *not i* or *i omitted*. Thus, $\hat{y}_i(i)$ is the predicted y-value at x_i when the regression is fitted to the data without observation i. The measure most commonly used to measure the influence of observation i on the fitted value $\hat{y}_i$ is DFITS$_i$, defined to be the result of dividing the difference of $\hat{y}_i$ and $\hat{y}_i(i)$ by an estimate of the standard error of $\hat{y}_i$. Hoaglin [9] derives the formula

$$\text{DFITS}_i = \frac{\sqrt{h_i}\, e_i}{s(i)(1 - h_i)}. \tag{9.73}$$

The MINITAB command REGRESS has a subcommand that will calculate DFITS.

Another measure used to detect outliers is the *studentized residual* defined by

$$e_i^* = \frac{e_i}{s(i)\sqrt{1 - h_i}}. \tag{9.74}$$

The MINITAB command REGRESS has a subcommand that will generate the studentized residuals. Hoaglin shows that the following relation holds between DFITS and the studentized residual e_i^*:

$$\text{DFITS}_i = \left(\frac{h_i}{1 - h_i}\right)^{1/2} e_i^*. \tag{9.75}$$

Hoaglin uses a real data set showing the number of employees and the cost of public affairs for 12 U.S. government agencies to demonstrate the use of regression diagnostics. We provide a demonstration in Table 9.1.3, that displays the fitted value $\hat{y}_i$, the standard residual, DFITS$_i$, and the studentized residual e_i^* for each x_i in Anscombe's third data set, whose graph is shown in Figure 9.1.5. The point $(13.0, 12.74)$ is clearly an outlier. The values in Table 9.1.4 were created using the MINITAB command REGRESS with subcommands DFITS and TRESIDUAL.

Table 9.1.3. Anscombe's Third Data Set

x_i	$\hat{y}_i$	Std. Res.	DFITS$_i$	e_i^*
10	7.99973	-0.46018	-0.146	-0.44
8	7.00027	-0.19633	-0.062	-0.19
13	9.49891	2.99999	676.904	1216.69
9	7.50000	-0.33085	-0.099	-0.31
11	8.49945	-0.59695	-0.219	-0.57
14	9.99864	-1.13497	-0.790	-1.16
6	6.00082	0.07042	0.030	0.07
4	5.00136	0.38070	0.247	0.36
12	8.99918	-0.75518	-0.336	-0.74
7	6.50055	-0.06974	-0.025	-0.07
5	5.50109	0.21188	0.111	0.20

9.1.6 Regression through the Origin

In some cases the nature of the random variables means that we would expect that $Y = 0$ when $X = 0$ so that the curve of regression of Y on X must go through the origin. Then, if the curve of regression of Y on X is linear,

$$Y = \beta_1 X + \epsilon, \tag{9.76}$$

where ϵ is $N(0, \sigma^2)$. (We are assuming the standard statistical model.) Montgomery and Peck [14, pages 38–43] show that

$$\hat{\beta}_1 = \frac{\sum\limits_{i=1}^{n} x_i y_i}{\sum\limits_{i=1}^{n} x_i^2}. \tag{9.77}$$

They also show that the residual mean square is given by

$$\hat{\sigma}^2 = \frac{\text{SSE}}{n-1} \equiv \text{MSE}. \tag{9.78}$$

The MINITAB command REGRESS has a subcommand NOCONSTANT that calculates the regression line that goes through the origin. Similarly, the SAS/STAT procedure REG has an option NOINT that suppresses the intercept term.

Example 9.1.5 Andy Allright, Lead Performance Wizard at Big Britches, is studying a computer system dedicated to batch processing. He discovers that there are two random variables associated with each job: X, the number of disk I/O operations, and Y, the CPU time per job. He wants to predict Y from X. He made the measurements in Table 9.1.4 and decides to build a simple regression line through the origin with X the number of disk I/O operations, measured in thousands, and Y the CPU time, measured in seconds. Figure 9.1.17 shows the MINITAB commands and output for the model. The regression line is thus $y = 14.2575\,x$, where y is CPU seconds and x is the number of I/O operations in thousands for the job. Thus, a job that requires 1,000 I/O operations will require 14.2575 seconds of CPU time, according to the model. MINITAB does not provide R^2 for this model, because it is difficult to interpret for a regression line through the origin. However, the t-ratio of 71.44 for $\hat{\beta}_1$ shown in the first table indicates that Andy has a good fit. The ANOVA table also indicates a good fit. Just to be on the safe side, however, Andy runs MINITAB with no subcommand, thus allowing a constant term. This shows a great fit, too! The R^2 value is 0.982 and the standard error of regression $s = 3.201$, that is smaller than that for the through-the-origin line. However, the t-value for the constant term is only 1.15 with a p-value of 0.283, so the null hypothesis that $\beta_0 = 0$ cannot be rejected. The through-the-origin line appears to be the best. We show the scatter diagram with both regression curves in Figure 9.1.18. It is clear from the scatter diagram that both lines provide a good fit to the data. □

Table 9.1.4. Big Britches Data

Disk I/O Ops. (Thousands)	CPU Time Seconds	Disk I/O Ops. (Thousands)	CPU Time Seconds
x	y	x	y
4.2	58.3	6.7	95.3
5.3	72.6	7.1	99.9
3.6	55.0	2.7	35.2
4.1	61.7	5.2	78.3
3.1	48.9	7.2	99.8

```
MTB > regress c2 1 c1 'resids' 'fits';
SUBC> NOCONSTANT.

The regression equation is
CPU = 14.3 {\it I/O}

Predictor        Coef        Stdev      t-ratio        p
NOCONSTANT
I/O            14.2575      0.1996       71.44      0.000

s = 3.258
Analysis of Variance

SOURCE       DF        SS          MS         F         p
Regression   1       54189       54189    5104.36    0.000
Error        9          96          11
Total       10       54285
```

Figure 9.1.17. MINITAB solution to Example 9.1.5.

```
MTB > regress c2 1 c1

The regression equation is
CPU = 3.84 + 13.5 I/O

Predictor        Coef        Stdev      t-ratio        p
Constant        3.840       3.338        1.15      0.283
I/O            13.5487      0.6465       20.96      0.000

s = 3.201      R-sq = 98.2%      R-sq(adj) = 98.0%

Analysis of Variance

SOURCE       DF        SS          MS         F         p
Regression   1       4500.3      4500.3    439.15     0.000
Error        8         82.0        10.2
Total        9       4582.3
```

Figure 9.1.18. MINITAB solution with constant term.

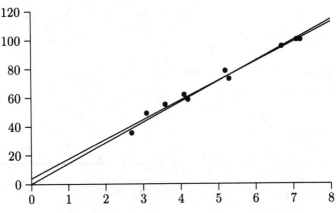

Figure 9.1.19. Curves for Example 9.1.5.

9.2   Nonlinear Regression

So far we have considered only the case where the curve of regression of Y on X is linear, that is, where $E[Y|X = x] = \beta_0 + \beta_1 x$. In this section we shall consider cases in which the regression curve is nonlinear but for which the methods of Section 9.1 can be applied, by applying a transformation to the sample data to "straighten out the curve." We shall consider polynomial regression, where $E[Y|X = x] = \beta_0 + \beta_1 x + \beta_2 x^2 + \cdots + \beta_k x^k$, as part of the section on multiple linear regression.

9.2.1   Regression with Transformed Variables

Often a nonlinear model relating Y to X can be transformed on X or Y (or both), such that the transformed random variable Y' can be written as

$$Y' = \beta_0 + \beta_1 X'. \tag{9.79}$$

In Table 9.2.1 we show some of the common transformations that are used to obtain a relation of the form (9.79).

Table 9.2.1. Useful Transformations

Original Relationship	Transformation	Linear Relationship
(a) $Y = \beta_0 + \dfrac{\beta_1}{X}$	$X' = \dfrac{1}{X}$	$Y' = \beta_0 + \beta_1 X'$
(b) $\dfrac{1}{Y} = \beta_0 + \beta_1 X$	$Y' = \dfrac{1}{Y}$	$Y' = \beta_0 + \beta_1 X$
(c) $\dfrac{1}{Y} = \beta_0 + \dfrac{\beta_1}{X}$	$X' = \dfrac{1}{X}, \quad Y' = \dfrac{1}{Y}$	$Y' = \beta_0 + \beta_1 X'$
(d) $Y = \alpha e^{\beta X}$	$Y' = \ln(Y)$	$Y' = \ln(\alpha) + \beta X$
(e) $Y = \alpha x^\beta$	$Y' = \ln(Y), \quad X' = \ln(X)$	$Y' = \ln(\alpha) + \beta X'$
(f) $Y = \alpha + \beta \ln(X)$	$X' = \ln(X)$	$Y = \alpha + \beta X'$

The major advantage of making one of transformations in Table 9.2.1 is that the normal equations of the simple linear regression model can be used to estimate the two parameters of the transformed equations. Thus, standard statistical packages, such as MINITAB and SAS/STAT, can be used to calculate the estimates. The least squares method can be used with the untransformed data to calculate the parameters of a nonlinear curve fit to the data, assuming one *can* decide what curve to use, but the calculations tend to be very difficult.

Table 9.2.2. Tire Wear Data

Miles Driven (Thousands) x	Percentage Usable y	$\ln(y)$
5	82.5	4.41280
10	65.6	4.18358
15	55.0	4.00733
25	35.5	3.56953
35	24.6	3.20275
45	15.2	2.72130
55	10.9	2.38876

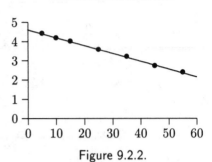

Figure 9.2.1.   Figure 9.2.2.

Example 9.2.1 The relationship described by transformation (d) in Table 9.2.1 is called an exponential curve (not to be confused with the exponential distribution). We illustrate the least squares curve fit of exponential data by the data in Table 9.2.2, that was produced by making tests on Turbo Tire premium tires. The second column describes the percentage of tires that still can be used after being driven the number of miles shown in column one. For example, the first row shows that after 5000 miles, 82.5% of the tires can still be used; 17.5% of them have failed. We apply the transformations to y specified in Table 9.2.1 by taking the natural logarithm of y to get the value in column 3 of Table 9.2.2. We use the MINITAB REGRESS command to find that the least squares line for the transformed data is $\hat{y}' = 4.60606 - 0.040823\,x$. The linear fit is excellent with $R^2 = 99.9\%$. Since $\ln(\alpha) = 4.60606$, our estimate of α is $\exp(4.60606) = 100.089$, and our estimate of β is -0.040823. Hence, our final equation for $\hat{y}$ is $\hat{y} = 100.089\exp(-0.040823\,x)$. The scatter plot of the original observations is displayed in Figure 9.2.1; the scatter plot of the of the transformed data is shown in Figure 9.2.2. My HP-42S has an exponential curve fitting routine, that yields $\hat{y} = 100.0890269\exp(-0.040823031\,x)$. □

You may have the following cosmic questions in mind:

1. I can often see from the scatter diagram of the original observations that the curve of regression should be a curve rather than a straight line, but how do I decide *what curve* might fit? That is, how do I decide which of the transformations in Table 9.2.2 to use?

2. If I successfully transform the data and get a linear fit, how do I calculate confidence intervals on the resulting parameters and on the estimates of y values?

There are several answers to the first question. In some cases one knows from physical considerations that the form of the relationship between y and x is exponential, or one of the other forms in Table 9.2.1 but the values of the parameters are not known. Applying the transformation and fitting a straight line using simple linear regression will yield an estimate of the parameters, as we saw in Example 9.2.1. In other cases we don't know which transformation will be successful. However, by using a statistical package such as MINITAB or SAS/STAT, we can apply test transformations and make scatter diagrams until we find one that looks approximately linear. Ryan *et al.* [19, pages 254–257] provide some helpful examples of this technique. We recommend that you use the MINITAB command GPLOT (a high resolution plotting routine) rather than the ordinary plotting command PLOT. PLOT, like the SAS/STAT procedure of the same name, has a rather low resolution, so the plot of points on an actual straight line may not look like a straight line. When using SAS for this type of plot, it is better to use the high resolution routines in the SAS package SAS/GRAPH, although this package requires a large amount of hard disk space.

In earlier days, engineers plotted the scatter diagram on graph paper with special scaling such as semilog paper that has logarithmic scaling on one axis, or log–log paper, that has logarithmic scaling on both axes. Using this method, if the points fall on a straight line, the form of the relationship between y and x is clear. For example, if semilog paper is used and a linear relationship is apparent, then transformation (d) of Table 9.2.1 should be used. Similarly, if log–log paper is used, a linear relationship in the plot would indicate that transformation (e) should be used. If none of the transformations in Table 9.2.1 are effective, it may be necessary to use polynomial curve fitting, the subject of a later section.

The answer to the second question is that unless some special hypotheses are true, one cannot make such predictions with much certainty. There are some special cases in which this can be done, but the assumptions are rather stringent. Some statisticians make interval estimates on the transformed data if the transformed error ϵ' appears to be normally distributed. Tests must be performed on the residuals to see if this is a reasonable assumption. Draper and Smith [6, Chapter 5] discuss regression with transformed variables in great depth.

> *Lives of great men all remind us*
> *We can make our lives sublime.*
> *And, departing, leave behind us*
> *Footprints on the sands of time.*

> Henry Wadsworth Longfellow

9.3   Multiple Linear Regression

For this model it is assumed that the curve of regression of Y on $X = (X_1, X_2, \ldots, X_k)$ is of the form

$$E[Y|X = (x_1, x_2, \ldots, x_k)] = \beta_0 + \beta_1 x_1 + \beta_2 x_2 + \cdots + \beta_k x_k + \epsilon_i, \qquad (9.80)$$

for all $(x_1, x_2, \ldots, x_k)$, where $\beta_0, \beta_1, \ldots, \beta_k$ are constants.

This model is called the *multiple linear regression model* with k explanatory variables (or regressors) $x_1, x_2, \ldots, x_k$.

The data in multiple linear regression consist of n $(k+1)$-tuples of the form $(x_{1i}, x_{2i}, \ldots, x_{ki}, y_i)$, $i = 1, 2, \ldots, n$, where the x_{ji} values are assumed to be known without error, but the y_i values are values of random variables.

The following additional mathematical assumptions are made for what is called *the standard statistical model*:

1. The random variables Y_i, $i = 1, 2, \ldots, n$ are of the form

$$Y_i = \beta_0 + \beta_1 x_{1i} + \beta_2 x_{2i} + \cdots + \beta_k x_{ki} + \epsilon_i, \quad i = 1, 2 \ldots, n. \quad (9.81)$$

2. The random variables $\epsilon_1, \epsilon_2, \ldots, \epsilon_n$ in (9.81) are errors that create the deviations about the linear relationship

$$\beta_0 + \beta_1 x_{1i} + \beta_2 x_{2i} + \cdots + \beta_k x_{ki}, \qquad (9.82)$$

for $i = 1, 2, \ldots, n$, respectively. The errors are independent and normally distributed with

$$E[\epsilon_i] = 0 \quad \text{and} \quad \text{Var}[\epsilon_i] = \sigma^2, \quad i = 1, 2, \ldots, n. \qquad (9.83)$$

It follows from these assumptions that each Y_i has a normal distribution with

$$
\begin{aligned}
E[Y_i] &= E[Y|X = (x_{1i}, x_{2i}, \ldots, x_{ki})] \\
&= E[\beta_0 + \beta_1 x_{1i} + \beta_2 x_{2i} + \cdots + \beta_k x_{ki} + \epsilon_i] \\
&= \beta_0 + \beta_1 x_{1i} + \beta_2 x_{2i} + \cdots + \beta_k x_{ki}
\end{aligned}
\qquad (9.84)
$$

and

$$
\begin{aligned}
\text{Var}[Y_i] &= \text{Var}[Y|X = (x_{1i}, x_{2i}, \ldots, x_{ki})] \\
&= \text{Var}[\beta_0 + \beta_1 x_{1i} + \beta_2 x_{2i} + \cdots + \beta_k x_{ki} + \epsilon_i] \\
&= \text{Var}[\epsilon_i] = \sigma^2.
\end{aligned}
\qquad (9.85)
$$

The random variables $Y_1, Y_2, \ldots, Y_n$ are mutually independent because of the mutual independence of $\epsilon_1, \epsilon_2, \ldots, \epsilon_n$.

9.3.1 Estimation of Parameters

Just as we did for simple linear regression, we fit the multiple regression model (9.80) by the method of least squares. To do this we minimize

$$S[\beta_0, \beta_1, \beta_2, \ldots, \beta_k] = \sum_{i-1}^{n} (y_i - \beta_0 - \beta_1 x_{1i} - \beta_2 x_{2i} - \cdots - \beta_i x_{ki})^2. \quad (9.86)$$

To do this we set to zero the $k + 1$ partial derivatives

$$\frac{\partial S}{\partial \beta_0} = -2 \sum_{i=1}^{n} [y_i - \beta_0 - \beta_1 x_{1i} - \cdots - \beta_k x_{ki}],$$

$$\frac{\partial S}{\partial \beta_j} = -2 \sum_{i=1}^{n} [y_i - \beta_0 - \beta_1 x_{1i} - \cdots - \beta_k x_{ki}](x_{ji}), \quad j = 1, 2, \ldots, k,$$

to obtain the $k + 1$ *normal equations*

$$\hat{\beta}_0 n + \hat{\beta}_1 \sum_{i=1}^{n} x_{1i} + \hat{\beta}_2 \sum_{i=1}^{n} x_{2i} + \cdots + \hat{\beta}_k \sum_{i=1}^{n} x_{ki} = \sum_{i=1}^{n} y_i$$

$$\hat{\beta}_0 \sum_{i=1}^{n} x_{1i} + \hat{\beta}_1 \sum_{i=1}^{n} x_{1i}^2 + \hat{\beta}_2 \sum_{i=1}^{n} x_{1i} x_{2i} + \cdots + \hat{\beta}_k \sum_{i=1}^{n} x_{1i} x_{ki} = \sum_{i=1}^{n} x_{1i} y_i,$$

$$\hat{\beta}_0 \sum_{i=1}^{n} x_{2i} + \hat{\beta}_1 \sum_{i=1}^{n} x_{2i} x_{1i} + \hat{\beta}_2 \sum_{i=1}^{n} x_{2i}^2 + \cdots + \hat{\beta}_k \sum_{i=1}^{n} x_{2i} x_{ki} = \sum_{i=1}^{n} x_{2i} y_i,$$

$$\vdots \qquad \qquad \vdots \qquad \qquad \vdots$$

$$\hat{\beta}_0 \sum_{i=1}^{n} x_{ki} + \hat{\beta}_1 \sum_{i=1}^{n} x_{ki} x_{1i} + \hat{\beta}_2 \sum_{i=1}^{n} x_{ki} x_{2i} + \cdots + \hat{\beta}_k \sum_{i=1}^{n} x_{ki}^2 = \sum_{i=1}^{n} x_{ki} y_i.$$

We solve the above normal equations to obtain the least squares estimates $\hat{\beta}_0$, $\hat{\beta}_1$, $\hat{\beta}_2$, ..., $\hat{\beta}_k$ of the parameters β_0, β_1, β_2, ..., β_k. For the standard statistical model, these estimates are also the maximum likelihood estimates of the parameters. Unfortunately, when $k \geq 2$ we no longer have simple formulas for the estimates. Fortunately, we can turn the computation over to a personal computer statistical system, such as MINITAB, SAS/STAT, or EXPLORE.

Table 9.3.1. Tree Data

Diameter	Height	Volume	Diameter	Height	Volume
8.3	70	10.3	8.6	65	10.3
8.8	63	10.2	10.5	72	16.4
10.7	81	18.8	10.8	83	19.7
11.0	66	15.6	11.0	75	18.2
11.1	80	22.6	11.2	75	19.9
11.3	79	24.2	11.4	76	21.0
11.4	76	21.4	11.7	69	21.3
12.0	75	19.1	12.9	74	22.2
12.9	85	33.8	13.3	86	27.4
13.7	71	25.7	13.8	64	24.9
14.0	78	34.5	14.2	80	31.7
14.5	74	36.3	16.0	72	38.3
16.3	77	42.6	17.3	81	55.4
17.5	82	55.7	17.9	80	58.3
18.0	80	51.5	18.0	80	51.0
20.6	87	77.0			

Let us consider an example.

Example 9.3.1 The data in Table 9.3.1 have been collected for 31 trees at Loggers Lumber. The analysts at Loggers feel that the trees are representative of the trees of interest and want a simple formula to calculate the volume of a tree from the diameter and height. Suppose they try linear regression. Let X_1 be the diameter of a tree, X_2 the height of a tree, and Y the volume. Then the normal equations become

$$\hat{\beta}_0 n + \hat{\beta}_1 \sum_{i=1}^{n} x_{1i} + \hat{\beta}_2 \sum_{i=1}^{n} x_{2i} = \sum_{i=1}^{n} y_i$$

$$\hat{\beta}_0 \sum_{i=1}^{n} x_{1i} + \hat{\beta}_1 \sum_{i=1}^{n} x_{1i}^2 + \hat{\beta}_2 \sum_{i=1}^{n} x_{1i}x_{2i} = \sum_{i=1}^{n} x_{1i}y_i$$

$$\hat{\beta}_0 \sum_{i=1}^{n} x_{2i} + \hat{\beta}_1 \sum_{i=1}^{n} x_{1i}x_{2i} + \hat{\beta}_2 \sum_{i=1}^{n} x_{2i}^2 = \sum_{i=1}^{n} x_{2i}y_i$$

If we substitute the values of the sums in the above normal equations (*Mathematica* made it easy to calculate the sums), we obtain the following set of three simultaneous equations:

$$31\hat{\beta}_0 + 410.7\hat{\beta}_1 + 2356\hat{\beta}_2 = 935.3$$

$$410.7\hat{\beta}_0 + 5736.55\hat{\beta}_1 + 31524.7\hat{\beta}_2 = 13887.9$$
$$2356\hat{\beta}_0 + 31524.7\hat{\beta}_1 + 180274\hat{\beta}_2 = 72962.6$$

The *Mathematica* Solve function yields the solution:

$$\hat{\beta}_0 = -57.9865 \quad \hat{\beta}_1 = 4.70835 \quad \hat{\beta}_2 = 0.339204.$$

The MINITAB solution is shown in Figure 9.3.1. The fit of the linear regression appears to be excellent, except for the tree with diameter 20.6 which appears to be an outlier. □

```
MTB > regress c3 2 c1 c2

The regression equation is
VOLUME = - 58.0 + 4.71 DIAMETER + 0.339 HEIGHT

Predictor       Coef      Stdev    t-ratio        p
Constant     -57.988      8.638      -6.71    0.000
DIAMETER      4.7082     0.2643      17.82    0.000
HEIGHT        0.3393     0.1302       2.61    0.014

s = 3.882      R-sq = 94.8%      R-sq(adj) = 94.4%

Analysis of Variance

SOURCE       DF        SS        MS        F        p
Regression    2    7684.2    3842.1   254.97    0.000
Error        28     421.9      15.1
Total        30    8106.1

SOURCE       DF     SEQ SS
DIAMETER      1     7581.8
HEIGHT        1      102.4


Unusual Observations
Obs.DIAMETER     VOLUME       Fit Stdev.Fit  Residual  St.Resid
  31     20.6     77.000    68.515     1.850     8.485      2.49R

R denotes an obs. with a large st. resid.
```

Figure 9.3.1. MINITAB output for Example 9.3.1.

You may be thinking, "What's this nonsense? Don't you remember the formula we learned in grade school that the volume of a cylinder is given by $\pi r^2 h$, where r is the radius and h the length?" However, if this formula is applied to the data in Table 9.3.1, the calculated volume is 2.5–3 times too large. Trees are not shaped like a cylinder but taper from the bottom to the top. The data on the trees we give in Table 9.3.1 can be found on page 329 of Ryan et al. [19]. The data set is an actual sample of black cherry trees in Allegheny National Forest, Pennsylvania. The tree data were obtained from the book *Forest Mensuration* by H. Arthur Meyer, Penns Valley Publishers, State College, Pennsylvania, 1953. It appears in the MINITAB worksheet TREES and can be accessed by the MINITAB command RETRIEVE. In Table 9.3.1 the diameter of a tree is given in inches, the height in feet, and the volume in cubic feet. The diameter is measured 4.5 feet above the ground. The true volume of a tree is obtained by cutting the tree down, submerging it in a tank of water, and measuring the amount of water displaced, according to Archimedes' principle. Mosteller, Fienberg, and Rourke [14, Example 7, pages 392–395] discuss the problem of estimating the volume of a tree from the diameter and height. They point out that the volume of a cone, that is one-third the volume of a cylinder, might be more appropriate. They show how to use the nonlinear estimator $V = Cr^2h$. They linearize it by using linear regression on $y = \ln(V)$. The result is very sensitive to the data. It appears that the linear relationship we developed in Example 9.3.1 would work well for the purpose of estimating the total volume of lumber in a tract.

9.3.2   Analysis of Variance for Multiple Regression

Most of the procedures that we use in multiple linear regression are simple extensions of those used in simple linear regression analysis. We test the usefulness of the multiple regression equation as a predictor by constructing an analysis of variance table and performing an F test. We can also determine confidence and prediction intervals, but the equations are much more complex and require a computer for implementation.

We can partition the total sum of squares (SST) of the response variable into the sum of squares due to regression (SSR) and the sum of squares due to error (SSE); that is,

$$\text{SST} = \text{SSR} + \text{SSE}.$$

Just as with simple linear regression, we have $\text{SST} = \sum(y - \bar{y})^2$, $\text{SSR} = \sum(\hat{y} - \bar{y})^2$, and $\text{SSE} = \sum(y - \hat{y})^2$. In simple linear regression there are two population parameters, β_0 and β_1, so SSE has $n - 2$ degrees of freedom. For multiple linear regression there are $k + 1$ population parameters, $\beta_0, \beta_1, \ldots, \beta_k$, so SSE has $n - (k + 1)$ degrees of freedom. Similarly, SSR has k degrees of freedom rather than 1. Then SST must have $n - (k + 1) + k = n - 1$ degrees of freedom. Statistical packages, such as SAS/STAT and MINITAB, present the analysis of variance results in a table like Table 9.3.2. For this table we have

$$\text{MSR} = \frac{\text{SSR}}{k},$$

and

$$\text{MSE} = \frac{\text{SSE}}{n - k - 1}.$$

Table 9.3.2. ANOVA Table for Multiple Linear Regression

SOURCE	DF	SS	MS	F-Value	$Pr > F$
Regression	k	SSR	MSR	F	p-value
Error	$n - k - 1$	SSE	MSE		
Total	$n - 1$	SST			

Just as for simple linear regression, the F-value is given by

$$F = \frac{\text{MSR}}{\text{MSE}}.$$

The value of F provides the test of the null hypotheses that $\beta_1 = \beta_2 = \cdots \beta_k = 0$, against the alternative hypothesis that at least one β_i is not zero. If the null hypothesis is true, then changes in predictor variables have no effect on the response variable. Thus, if the null hypothesis is true, the regression equation has no value. Since F has an F-distribution with k and $n - k - 1$ degrees of freedom, the null hypothesis is rejected if $F > F_{k,n-k-1,\alpha}$. MINITAB and SAS/STAT provide the p-value of the test, that is, the probability that, if the null hypothesis is true, a value of F larger than that actually observed will occur. Thus, the p-value is given by

$$p\text{-value} = P[F_{k,n-k-1} > F].$$

If the p-value is small, the null hypothesis can be rejected. In Table 9.3.1 MINITAB indicates that the p-value is zero. My HP-21S indicates that the p-value actually is 1.0712×10^{-18}, which is rather small.

Just as for the simple linear regression case, MSE is the unbiased estimator of σ^2 usually used, and $s_e = \sqrt{\text{MSE}}$ is the standard error of the estimate.

We define the *coefficient of multiple determination*, R^2, by the formula

$$R^2 = \frac{\text{SSR}}{\text{SST}} = 1 - \frac{\text{SSE}}{\text{SST}}.$$

Just as in the simple linear regression case it measures how much of the observed variation is due to the regression equation. We have $0 \le R^2 \le 1$. If R^2 is close to one, the regression model is a good predictor while if it is near zero the regression model is a poor predictor. Figure 9.3.1 shows that $R^2 = 0.948$ for Example 9.3.1. MINITAB and SAS/STAT provide the adjusted R^2 value, R_a^2, as well as R^2 in their regression procedures. R_a^2 is defined by

$$R_a^2 = 1 - \frac{\text{SSE}/(n - k - 1)}{\text{SST}/(n - 1)}. \tag{9.87}$$

It can be shown that

$$R_a^2 \le R^2.$$

Some statisticians feel that the adjusted value is a better indicator of the success of the regression model than R^2.

The F value from the analysis of variance table and R^2 are related by the simple formula

$$F = \frac{R^2}{1 - R^2} \left(\frac{n - k - 1}{k} \right). \tag{9.88}$$

9.3.3   Inferences in Multiple Regression

Most books on linear regression put everything into matrix notation. We won't bother to do that, here, but it will be convenient to state a few results in terms of the matrix $C = (X'X)^{-1}$, where X is the n by $k + 1$ matrix given by

$$X = \begin{bmatrix} 1 & x_{11} & x_{12} & \cdots & x_{1k} \\ 1 & x_{21} & x_{22} & \cdots & x_{2k} \\ \vdots & \vdots & \vdots & \cdots & \vdots \\ 1 & x_{n1} & x_{n2} & \cdots & x_{nk} \end{bmatrix} \tag{9.89}$$

and X' is the transpose of X obtained by interchanging the rows and columns to form a $k + 1$ by n matrix. Thus, C is a $k + 1$ by $k + 1$ square matrix that is written as

$$C = \begin{bmatrix} C_{00} & C_{01} & C_{02} & \cdots & C_{0k} \\ C_{10} & C_{11} & C_{12} & \cdots & C_{1k} \\ \vdots & \vdots & \vdots & \cdots & \vdots \\ C_{k0} & C_{k1} & C_{k2} & \cdots & C_{kk} \end{bmatrix}, \qquad (9.90)$$

to make it easier to state some of the results of the following theorem:

Theorem 9.3.1 *Let $\hat{\beta}_0$, $\hat{\beta}_1$, $\hat{\beta}_2$, ..., $\hat{\beta}_k$ be the least squares estimators of the parameters β_0, β_1, β_2, ..., β_k for the multiple linear regression model with the standard statistical assumptions. Then each of the $\hat{\beta}_i$ is normally distributed with mean β_i and variance $\sigma^2 C_{ii}$, where C_{ii} is the ith diagonal element of the matrix $C = (X'X)^{-1}$. Furthermore, each of the normalized random variables*

$$\frac{\hat{\beta}_i - \beta_i}{\sqrt{\text{MSE } C_{ii}}}, \quad i = 0, 1, \ldots, k \qquad (9.91)$$

has a Student's t distribution with $n - (k + 1)$ degrees of freedom. Hence, the $100(1 - \alpha)\%$ confidence interval for β_i is given by

$$\hat{\beta}_i \pm t_{n-k-1, \alpha/2} \times \sqrt{\text{MSE } C_{ii}}, \quad i = 0, 1, \ldots, k. \qquad (9.92)$$

Finally, each $\hat{\beta}_i$ is the maximum likelihood estimator of β_i and has minimum variance among all unbiased linear estimators of β_i. (The fact that $\hat{\beta}_i$ has the minimum variance among all unbiased linear estimators is known as the Gauss–Markov theorem and does not require the assumption that the errors, ϵ_i, are normally distributed.)

Proof See Montgomery and Peck [14, Chapter 4] for the proof of everything except the proof that each $\hat{\beta}_i$ is the maximum likelihood and minimum variance unbiased estimator of β_i, which is proved by Neter *et al.* [17]. ∎

The quantity

$$s_{\hat{\beta}_i} = \sqrt{\text{MSE } C_{ii}} \qquad (9.93)$$

is called the *standard error* of the regression coefficient $\hat{\beta}_i$. Although we could use the formula (9.93) to compute the standard error, for multiple linear regression we would use a statistical system such as MINITAB or SAS/STAT to make the calculation for us because of the labor involved in hand calculation. Once we find by the F test or by the value of R^2 that there appears to be a significant regression relationship, we will want to assess the significance of the individual regression coefficients $\hat{\beta}_i$. The above theorem shows that each $\hat{\beta}_i$ is an unbiased estimator of β_i and that $(\hat{\beta}_i - \beta_i)/s_{\hat{\beta}_i}$ has a Student's t distribution with $n-k-1$ degrees of freedom. Hence, we can calculate confidence intervals for the β_i and test hypotheses concerning their values. To test the significance of an individual regression coefficient, we test H_0: $\beta_i = 0$ versus H_1: $\beta_i \neq 0$. For this test we use the t-ratio, $t_{\hat{\beta}_i} = \hat{\beta}_i/s_{\hat{\beta}_i}$. If $|t_{\hat{\beta}_i}| > t_{n-k-1,\alpha/2}$, we reject the null hypothesis and accept H_1. The t-ratios and p-values of the tests are given by the MINITAB command REGRESS and by the SAS/STAT procedure REG. Thus, in Table 9.3.1 we see that the p-values for the constant term and the coefficient of DIAMETER are zero while the p-value for the coefficient of HEIGHT is 0.014.

Example 9.3.2 Consider the regression equation

$$\text{VOLUME} = -57.988 + 4.7082\,\text{DIAMETER} + 0.3393\,\text{HEIGHT}, \qquad (9.94)$$

which was developed in Example 9.3.1. Let us construct a 95% confidence interval for β_1, the coefficient of DIAMETER in (9.94). From the MINITAB output in Figure 9.3.1, we see that $s_{\hat{\beta}_1} = 0.2643$. Since $t_{\hat{\beta}_1}$ has a Student's t distribution with $31 - 2 - 1 = 28$ degrees of freedom and $t_{28,0.025} = 2.0484$, the 95% confidence interval for β_1 is given by

$$\hat{\beta}_1 \pm t_{28,0.025}s_{\hat{\beta}_1} \quad \text{or} \quad 4.1668 \leq \beta_1 \leq 5.2496. \quad \square$$

Table 9.3.3.

x_1	x_2	y
0.9	10	35
1.2	13	40
1.5	14	45
1.7	16	46
2.1	17	50
2.4	21	65

Example 9.3.3 Harry Hewdy of Kashmir Kravits decides to use multiple linear regression with the data in Table 9.3.3 to predict y from x_1 and x_2. He constructed the matrix X described by equation (9.89) as

$$X = \begin{pmatrix} 1 & 0.9 & 10 \\ 1 & 1.2 & 13 \\ 1 & 1.5 & 14 \\ 1 & 1.7 & 16 \\ 1 & 2.1 & 17 \\ 1 & 2.4 & 21 \end{pmatrix}.$$

Using *Mathematica* he calculated the matrix C to be

$$C = \begin{pmatrix} 8.30389 & 8.65537 & -1.46864 \\ 8.65537 & 15.3208 & -2.22062 \\ -1.46864 & -2.22062 & 0.335977 \end{pmatrix}.$$

He found that using $s = \sqrt{\text{MSE}} = 2.894$ from Figure 9.3.2, part of the output from the MINITAB command REGRESS for the data in Table 9.3.3, the formula (9.93)

$$s_{\hat{\beta}_i} = \sqrt{\text{MSE } C_{ii}}, \tag{9.95}$$

with the values of C_{ii} from the matrix C, he obtains the same values for the standard errors as are given in Figure 9.3.2. That is, he obtains $s_{\hat{\beta}_0} = 2.894\sqrt{8.30389} = 8.339$, $s_{\hat{\beta}_1} = 2.894\sqrt{15.3208} = 11.328$, and $s_{\hat{\beta}_2} = 2.894\sqrt{0.335977} = 1.677$. (Recall that we assume the rows and columns are numbered 0, 1, 2.) □

```
The regression equation is
C3 = 4.73 - 2.8 C1 + 3.08 C2

Predictor      Coef      Stdev    t-ratio        p
Constant      4.731      8.339       0.57    0.610
C1            -2.83      11.33      -0.25    0.819
C2            3.081      1.677       1.84    0.164

s = 2.894      R-sq = 95.3%     R-sq(adj) = 92.1%
```

Figure 9.3.2. Output from REGRESS for Example 9.3.2.

Just as for simple linear regression, multiple linear regression allows the calculation of confidence intervals for the estimated mean of the prediction variable and confidence limits for the predicted values. Recall that you should use confidence intervals for the mean predicted value if you want the limits to show the region that should contain the population regression curve. You should use confidence limits on the predicted values if you want the limits to show the region that should contain most of the population of all possible observations. Thus, the latter interval is expected to be much wider. Both types of intervals are very difficult to compute by hand for multiple regression models but can be obtained with ease from a statistical package such as MINITAB, or SAS/STAT. . Therefore, we shall not give the exact formulas. They can be found in all advanced linear regression books, such as Draper and Smith [6] or Neter *et al.* [17].

9.3.4   Selection of Independent (Predictor) Variables

There are some special opportunities as well as pitfalls in multiple linear regression. It often happens that there are a number of independent variables that could be used in the model. We would like to use only those that are important, that is, those that make a difference. From a computational point of view, it would also be desirable to use a small number of variables. The goal is to determine the best subset of variables to include in the regression model. One difficulty is that the variables are rarely truly independent in the probabilistic sense. Thus, the obvious strategy of starting with all the variables in the model and then eliminating variables one at a time, eliminating the variable with the smallest p-value at each stage, will often *not* yield the best subset. The p-value usually depends upon the other variables in the model. Fortunately, statistical computer systems can aid the selection process. If there are not too many variables to consider, one can have the computer try all possible subsets of independent variables.

Example 9.3.4 The MINITAB worksheet REALEST.MTW in the student edition contains information on each home sold in Oxford, Ohio, during 1987. A description of the data is given in Table 9.3.4.

Table 9.3.4. Description of REALEST.MTW

Column	Name	Description
C1	SELL $	Selling price of house
C2	ASK $	Asking price of house
C3	BATHS	Number of bathrooms
C4	BEDROOMS	Number of bedrooms
C5	AGE	Age of house (years)
C6	BASEMENT	Basement?
C7	CARS	Size of garage (cars)

The BASEMENT variable is an indicator variable that tells us whether there is a basement and will not be considered in the analysis. Our initial trial is to use all five of the other variables, as we show in the MINITAB REGRESS output of Figure 9.3.3. The overall fit is certainly good, with the p-value for the F-test essentially zero. Surprisingly, all the individual coefficients appear to be significant except for the number of bathrooms and the number of bedrooms, especially the latter. These are the two values most often used to categorize a house! Since we are operating intuitively, let us drop both the number of bedrooms and the number of bathrooms for our next trial, shown in Figure 9.3.4. Our statistics indicate that we are able to predict almost as well with only three variables as we did with five and all of the coefficients are highly significant.

For our next trial we used the MINITAB command BREG, that tries all possible subsets and reports all those that look promising. The output from this trial is shown in Figure 9.3.5. The output shows that we made the best three variable choice. Our choice can be improved only slightly by adding the variable BATHS. This four variable model has the same R^2 value as the full five variable example, but has a slightly smaller standard error. The command BREG uses a very efficient algorithm so it took very little computer time to generate the results shown in Figure 9.3.5. SAS/STAT has a procedure called RSQUARE, that always identifies the model with the largest R^2 for each number of variables considered. It also reports the C_p value. It can be used to investigate all subsets of variables. RSQUARE is somewhat more accurate than BREG but also much slower. It apparently uses double precision for all calculations while BREG on my IBM PC compatible with an Intel 386 microprocessor seems to use single precision arithmetic. □

```
The regression equation is
SELL $ = 6.36 + 0.712 ASK $ + 2.79 BATHS + 0.22 BEDROOMS - 0.111 AGE
           + 3.37 CARS

43 cases used 8 cases contain missing values

Predictor      Coef       Stdev     t-ratio        p
Constant      6.364       3.269       1.95      0.059
ASK $        0.71242     0.03433     20.75      0.000
BATHS         2.786       1.774       1.57      0.125
BEDROOMS      0.217       1.068       0.20      0.840
AGE         -0.11065     0.02948     -3.75      0.001
CARS          3.3746      0.8318       4.06      0.000

s = 4.045      R-sq = 96.5%      R-sq(adj) = 96.0%

Analysis of Variance

SOURCE       DF         SS          MS         F         p
Regression    5      16694.1      3338.8     204.09    0.000
Error        37        605.3        16.4

Total        42      17299.4
```

Figure 9.3.3. MINITAB REGRESS output.

```
The regression equation is
SELL $ = 9.39 + 0.737 ASK $ - 0.125 AGE + 3.95 CARS

43 cases used 8 cases contain missing values

Predictor        Coef       Stdev    t-ratio        p
Constant        9.391       2.418       3.88    0.000
ASK $         0.73692     0.02994      24.61    0.000
AGE          -0.12525     0.02586      -4.84    0.000
CARS          3.9503      0.7666        5.15    0.000

s = 4.089      R-sq = 96.2%     R-sq(adj) = 95.9%

Analysis of Variance

SOURCE         DF          SS         MS        F        p
Regression      3      16647.3     5549.1   331.89    0.000
Error          39        652.1       16.7
Total          42      17299.4
```

Figure 9.3.4. MINITAB REGRESS output.

```
MTB > breg 'sell $' c2 c3 c4 c5 c7

Best Subsets Regression of SELL $

43 cases used 8 cases contain missing values.

                                          B
                                          E
                                          D
                                      A   B   R
                                      S   A   O     C
                                      K   T   O  A  A
                         Adj.             H   M  G  R
     Vars  R-sq  R-sq   C-p      s    $   S   S  E  S

        1  91.9  91.7   46.5   5.8395  X
        1  42.3  40.9  571.3  15.605       X
        2  94.1  93.8   25.9   5.0722  X   X
        2  94.0  93.7   26.8   5.1097  X          X
        3  96.2  95.9    4.9   4.0890  X       X  X
        3  95.0  94.6   17.6   4.6992  X   X      X
        4  96.5  96.1    4.0   3.9934  X   X   X  X
        4  96.3  95.9    6.5   4.1221  X       X  X  X
        5  96.5  96.0    6.0   4.0447  X   X   X  X  X
```

Figure 9.3.5. MINITAB REGRESS output.

```
MTB > stepwise 'sell $' c2 c3 c4 c5 c7

    STEPWISE REGRESSION OF  SELL $  ON  5 PREDICTORS, WITH N =   43
    N(CASES WITH MISSING OBS.) =   8 N(ALL CASES) =   51

      STEP         1         2         3         4         5
    CONSTANT    6.4188    0.6128   -0.5208    6.6762    9.3907

    ASK $        0.819     0.736     0.744     0.715     0.737
    T-RATIO      21.59     18.66     20.29     22.32     24.61

    BATHS                    7.1       5.3       2.9
    T-RATIO                 3.79      2.88      1.70

    CARS                             2.57      3.39      3.95
    T-RATIO                          2.76      4.14      5.15

    AGE                                      -0.108    -0.125
    T-RATIO                                   -4.00     -4.84

    S            5.84      5.07      4.70      3.99      4.09
    R-SQ        91.92     94.05     95.02     96.50     96.23
```

Figure 9.3.6. MINITAB output.

For comparison, we ran the most common procedure used by statistical computer systems. This technique, called *stepwise regression*, starts with the first variable listed and adds new variables only if they meet certain criteria. Variables can be dropped, too, as we see in Step 5. We will discuss the procedure in more detail below. Figure 9.3.6 shows the MINITAB command to initiate the procedure, with the variables considered in the order listed. Note that the final choice is the three variable model that we got by intuition and that the routine never gave the output of using all five variables. □

Recall that we assume we have observed each of n cases for the k predictors (independent variables) $X_1, X_2, \ldots, X_k$ and a response Y. We use the convention in the following discussion that p is the number of predictors in a selected subset, usually including an intercept. Thus, for the full subset $p = k + 1$.

Although we have not specifically noted what criteria are used by MINITAB and other statistical systems to evaluate the goodness of fit of a subset of the predictors, we see from Figure 9.3.5 that the MINITAB command BREG uses R^2, R_a^2, $s = s_e$, and Mallow's C_p. Mallow's C_p is defined by

$$C_p = \frac{\text{SSE}_p}{\hat{\sigma}^2} + 2p - n, \qquad (9.96)$$

where SSE_p is the error sum of squares for the p variable subset and $\hat{\sigma}^2$ is the mean squared error, MSE, for the full set of variables $X_1, X_2, \ldots, X_k$. It is a little difficult to use C_p as a criterion for comparing the goodness of different subsets although Mallows claims that a good subset should have $C_p \approx p$. Computer algorithms using C_p to select good subsets usually choose those with the smallest values of C_p. Thus, in the above example, the model using ASK $, AGE, and CARS would be rated better on the basis of C_p than the model using ASK $, BATHS, and CARS, for two reasons. The former has a much smaller C_p value, 4.9 versus 17.6. In addition, 4.9 is much closer to 4 than 17.6 is.

Since statistical packages tend to use similar algorithms for selecting variables, we will describe the algorithm for testing all subsets of variables by quoting from the MINITAB manual [13], which describes the algorithm used for the command BREG:

Computational Method

BREG employs a procedure called the Hamiltonian Walk, that is a method for "visiting" all possible subsets in the same number of steps; one step for each subset. That is, all $2^m - 1$ subsets are visited in $2^m - 1$ steps, and a different subset regression is evaluated at each step. Each subset in the Hamiltonian Walk differs from the preceding subset by the addition or deletion of only one variable. The method used to perform the regression calculations is the sweep operator described in Goodnight's paper [8]. The sweep operator "sweeps" a variable in or out of the regression on each step of the Hamiltonian Walk, and calculates the SSE for each subset. [For more information, see [7, 20, and 8].]

The stepwise procedures are the most popular selection methods. However, Weisberg [22, page 214] warns:

The stepwise methods are easy to explain, inexpensive to compute, and widely used. The comparative simplicity of the results from stepwise regression seems to appeal to many analysts. But stepwise methods must be used with caution. The model selected in a stepwise fashion need not optimize any reasonable criterion function for choosing a model. The apparent ordering of the predictors is an artifact of the method and need not reflect relationships of substantive interest. Finally, stepwise regression may seriously overstate significance of results.

If the number of variables is small, it is a good idea to test all subsets of the variables. Since the algorithms used by most computer statistical systems are quite fast, this will not take much computer time and finding the best regression model is more likely.

The basic selection method calculates an F-statistic for each variable in the model. We use the notation $\text{SSE}[X_1, \ldots, X_p]$ for the sum of squares error for the multiple regression model that uses the variables $X_1, X_2, \ldots, X_p$ with the corresponding mean square error $\text{MSE}[X_1, \ldots, X_p]$ while $\text{SSE}[X_1, \ldots, X_{i-1}, X_{i+1}, \ldots, X_p]$ is the sum of squares error for the same system with the variable X_i removed. Then, if the variables $X_1, \ldots, X_p$ are in the model, the F-statistic for X_i is

$$\frac{\text{SSE}[X_1, \ldots, X_{i-1}, X_{i+1}, \ldots, X_p] - \text{SSE}[X_1, \ldots, X_p]}{\text{MSE}[X_1, \ldots, X_p]}, \qquad (9.97)$$

and has 1 and $n - p - 1$ degrees of freedom. If the F-statistic for any variable is less than the parameter F-out (this parameter is called SLSTAY in SAS and FREMOVE in MINITAB), then the variable with the smallest F-statistic is removed from the model. The regression equation is calculated for this smaller model and the procedure goes to a new step. The square root of the F-statistic is the familiar t-ratio or t-statistic, that is printed out by most statistical systems, rather than the F-statistic.

If no variable can be removed, the basic selection procedure attempts to add a new variable. An F-statistic is calculated for each variable not yet in the model. If the model now contains $X_1, \ldots, X_p$, then the F-statistic for the new variable X_{p+1} is

$$\frac{\text{SSE}[X_1, \ldots, X_p] - \text{SSE}[X_1, \ldots, X_p, X_{p+1}]}{\text{MSE}[X_1, \ldots, X_p, X_{p+1}]}. \qquad (9.98)$$

The variable with the largest F-statistic is added provided it is larger than the parameter F-Add (this parameter is called FENTER in MINITAB and SLENTRY for SAS). Thus, the new variable that most reduces the error sum of squares is added. The regress equation is then calculated and the procedure goes to a new step. If no variable can be added, the stepwise procedure ends.

There are several modifications of the basic selection procedure, including *forward selection* and *backwards elimination*.

The forward selection procedure adds variables in the same manner as the basic selection procedure, but once a variable is entered, it is never removed. The forward selection procedure ends when no variable not in the equation can be entered.

The backwards elimination procedure starts with the model containing all the variables. The procedure then removes variables, one at a time, using the same method as the basic selection procedure. No variable, once removed, is allowed to reenter the model. The backwards elimination procedure ends when no variable in the model can be deleted.

Although a computer statistical system is a great aid to an analyst in choosing the best regression model, there is no algorithm that can guarantee the best possible model. The greatest asset to the analyst is a deep knowledge of the area under study and of each of the variables, including expected sign and magnitude of the coefficient.

Example 9.3.5 Sally Selling, the most successful realtor in Oxford, Ohio, in early 1988, examines the results of Example 9.3.4. She knows that the selling price is in units of $1000 (a grand), as is the asking price. Sally knows that most people ask more for a house than they can sell it for but that, on the average, they get about 75 percent of what they ask for. She knows that the number of bathrooms has little effect upon the price of houses in Oxford because almost all houses have an adequate number of bathrooms. Similarly, the number of bedrooms has little effect upon housing prices because most houses have at least three. Garages are a different matter. Because of the inclement weather, everyone wants to keep all cars inside; a garage brings a premium of about $4000 per car space. The age of a house, in years, has little effect unless the house is quite old, when it could have a substantial negative effect. Sally likes the regression formula

$$\text{SELL \$} = 9.391 + 0.73692\,\text{ASK \$} - 0.12525\,\text{AGE} + 3.9503\,\text{CARS}. \quad \square$$

9.3.5   Polynomial Regression

For polynomial regression it is assumed that the curve of regression of Y on X is of the form

$$E[Y|X = x] = \beta_0 + \beta_1 x + \beta_2 x^2 + \cdots + \beta_k x^k, \qquad (9.99)$$

so that the regression model can be represented as

$$Y_i = \beta_0 + \beta_1 x_i + \beta_2 x_i^2 + \cdots + \beta_k x_i^k + \epsilon_i, \quad i = 1, 2, \ldots, n. \qquad (9.100)$$

Polynomial regression is used to fit curves to scatter diagrams in which a straight line does not provide a good fit because a curve is required to fit the trend, that is, when the plot looks curved. This model is actually a special case of the multiple linear regression model with $x_1 = x$, $x_2 = x^2, \ldots, x_k = x^k$. Hence, the methodology for the multiple regression model can be applied directly to the polynomial model. This may seem strange, but although the model is nonlinear in x, it *is* linear in the regression coefficients $\beta_0, \beta_1, \ldots, \beta_k$.

The polynomial model is of degree k, corresponding to the degree of its highest order term. As one would expect, the higher the degree of the polynomial, the more complex a curve it can represent. To keep the analysis simpler, the goal is to choose the polynomial of lowest order that will adequately represent the trend in the scatter diagram. The first degree polynomial is a straight line. The second degree polynomial is called a quadratic and has a curve with just one bend. The cubic or third degree polynomial has a curve with two bends. In general, the kth degree polynomial has a curve with at most $k - 1$ bends.

Table 9.3.5.

x	y	x	y
1	24.08	2	17.21
3	20.47	4	20.63
5	18.48	6	21.45
7	26.10	8	32.74
9	43.00	10	41.99

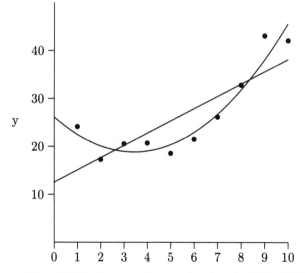

Figure 9.3.7. Scatter diagram for data in Table 9.3.5.

Example 9.3.6 Mickey Mike of Funnyland wants to fit a curve to the data in Table 9.3.5. He would prefer a simple linear curve but thinks he may have to consider a polynomial. He constructs the scatter diagram of Figure 9.3.7. (This figure also shows the least squares linear regression line and the least squares quadratic curve for the data.) Mickey uses the REGRESS command of MINITAB to fit a straight line to the data and gets the output of Figure 9.3.8. The fit is not too bad. The p-value for the F test is only 0.004 and both coefficients are significant at a reasonable level. However, R^2 is only 67.3% so Mickey uses REGRESS again to fit a quadratic to the data and obtains the output of Figure 9.3.9. The new regression curve is

$$\hat{y} = 26.177 - 4.261\,x + 0.6202\,x^2,$$

that has an R^2 value of 92.5% with very small p-values on all the tests. The value of the standard error, s, has declined from 5.735 for the linear model to 2.930 for the quadratic model. Figure 9.3.7, too, seems to indicate that the quadratic fit is superior to the linear fit. Mickey is convinced that he should use the quadratic model. $\square$

```
The regression equation is
y = 12.5 + 2.56 x

Predictor        Coef       Stdev     t-ratio        p
Constant       12.533       3.918        3.20    0.013
x              2.5603      0.6314        4.05    0.004

s = 5.735      R-sq = 67.3%      R-sq(adj) = 63.2%

Analysis of Variance

SOURCE        DF          SS          MS         F        p
Regression     1      540.80      540.80     16.44    0.004
Error          8      263.16       32.89
Total          9      803.96
```

Figure 9.3.8. MINITAB output Example 9.3.6.

```
The regression equation is
y = 26.2 - 4.26 x + 0.620 xsquared

Predictor       Coef      Stdev    t-ratio        p
Constant       26.177     3.446       7.60    0.000
x              -4.261     1.439      -2.96    0.021
xsquared        0.6202    0.1275      4.86    0.000

s = 2.930      R-sq = 92.5%      R-sq(adj) = 90.4%

Analysis of Variance

SOURCE       DF         SS         MS        F        p
Regression    2      743.86     371.93    43.32    0.000
Error         7       60.09       8.58
Total         9      803.96

SOURCE       DF     SEQ SS
x             1     540.80
xsquared      1     203.06
```

Figure 9.3.9. MINITAB output Example 9.3.6.

There are often computational difficulties with polynomial regression because of the correlation between X, X^2, and higher-power terms. A transformation used to improve accuracy is $x'_i = (x_i - \overline{x})/s_x$, where s_x is the sample standard deviation of the x_i's. When we use this transformation, we are using normalized coordinates. In Exercise 19 we ask you to try the normalizing transformation.

Polynomial regression can also employ more than one independent variable. For example, a model employing a second order polynomial in two variables could be written as

$$Y_i = \beta_0 + \beta_1 x_{1i} + \beta_2 x_{2i} + \beta_{11} x_{1i}^2 + \beta_{22} x_{2i}^2 + \beta_{12} x_{1i} x_{2i} + \epsilon_i, \quad (9.101)$$

for $i = 1, \ldots, n$, where $\beta_{12} x_{1i} x_{2i}$ represents the cross-product term. Naturally, the regression of Y on $X = (X_1, X_2)$ is given by

$$E[Y|X = (x_1, x_2)] = \beta_0 + \beta_1 x_1 + \beta_2 x_2 + \beta_{11} x_1^2 + \beta_{22} x_2^2 + \beta_{12} x_1 x_2. \quad (9.102)$$

Table 9.3.6. Example 9.3.7.

Lines	CPU time (seconds)	Cost dollars
2000	42	15.03
1500	35	10.81
1250	37	11.33
1000	20	6.36
2500	57	21.68
2750	60	23.40
3000	68	25.81
750	30	7.63
800	31	8.84
1120	26	7.37

Example 9.3.7 Simply Scintillating has a central computer system that runs a certain batch program for Department Widget many times per day. (The program is started remotely from the workstations on the Scintillating LAN, but the printout is made at the computer center and delivered by messenger.) Relentless Rudy, the Chief Analyst, decides to model the cost per run of the program using the regression model

$$y = \beta_0 + \beta_1 x_1 + \beta_2 x_2 + \beta_{12} x_1 x_2 + \epsilon,$$

where x_1 is the number of lines printed and x_2 is the CPU time used by the program. He uses the 10 measurements of Table 9.3.6 to construct his model. The output from the MINITAB command REGRESS is shown in Figure 9.3.10. Relentless finds, as is often the case, that there are some good things about the model and some problems, too. (The old good news, bad news.) The good news is that $R^2 = 99.6$ percent, the standard error s is only 0.5804, and the p-value of the F-test is very small. However, three out of four of the coefficients are not significant at the 5% level of significance. We ask you to help Relentless in Exercises 20 and 21. □

```
The regression equation is
y = - 1.90 + 0.00223 x1 + 0.256 x2 +0.000020 x1.x2

Predictor        Coef       Stdev     t-ratio        p
Constant       -1.903       1.843       -1.03    0.342
x1           0.002228    0.001227        1.82    0.119
x2            0.25620     0.06209        4.13    0.006
x1.x2       0.00002038  0.00002348        0.87    0.419


s = 0.5804      R-sq = 99.6%     R-sq(adj) = 99.4%


Analysis of Variance

SOURCE        DF          SS          MS         F        p
Regression     3      472.37      157.46    467.46    0.000
Error          6        2.02        0.34
Total          9      474.39
```

Figure 9.3.10. REGRESS output Example 9.3.7.

After we have applied a computer procedure to select a list of possible regression models to use for our study, we need to make the final choice. Part of this final choice must involve the use of diagnostic procedures to detect possible errors. Multiple linear regression has all the error opportunities of simple linear regression, plus some new ones that are due to problems of interaction of the regressors. Fortunately, all the techniques we discussed in Section 9.1.5 are available for multiple regression, too. Many of the advanced techniques depend upon a detailed analysis of the residuals, which is an advanced topic that we won't cover here, but that is discussed in the more advanced books listed in Section 9.1.5.

One special problem of multiple linear regression is *multicollinearity*; that is high correlation between the regressor variables. An ideal situation would be that in which each regressor is strongly correlated with the predictor variable but weakly correlated with each other. Multicollinearity means that some of the regressors are strongly correlated with each other. If only two regressors are strongly correlated, the multicollinearity is easy to detect. More complex linear dependence is more difficult to detect.

One indication of multicollinearity is when none or almost none of the individual β's are significant, even though the overall F-test is significant. We saw this in Example 9.3.7. When this happens we can usually remove one of the regressors from the model.

Another indication of multicollinearity is obtaining radically different values for some of the $\hat{\beta}_i$'s when only minor changes are made to the model by adding or deleting data.

The best way to remove multicollinearity is to drop regressors selectively, if they are found to be highly correlated with other regressors.

*Give us the fortitude to endure the things which cannot be changed, and
the courage to change the things which should be changed, and the wisdom
to know one from the other.*
Oliver J. Hart

9.4   Exercises

For these exercises, if you do not have a statistical computer system available, you should have a calculator with linear regression built in, such as the HP-21S. Otherwise, some of the exercises will be very tedious.

1. [15] See Example 9.1.1. In the table below we show the number of pages and the price of 20 books reviewed in the November 1988 issue of *Technometrics*. (Two other books were reviewed with no price given.)

Books Reviewed Data

Pages	Price	Pages	Price	Pages	Price
311	34.50	610	49.95	384	49.95
415	80.00	278	34.95	232	32.50
408	34.50	492	39.25	435	29.95
699	33.95	687	42.50	540	99.75
416	25.00	447	36.95	614	72.95
307	34.95	429	69.75	260	34.95
171	17.95	162	19.95	-	-

Let X be the number of pages and Y the price of the books.

(a) Draw the scatter diagram for the data. Does it look like a straight line would fit the data?

(b) Find the least squares regression line. How well does it seem to fit the data? Are there any outliers?

(c) Find the average number of pages per book and average price per book.

(d) Use the equation of the least squares regression line to estimate the price of a book with 390 pages. How does this compare to the estimate we made in Example 9.1.1?

2. [15] Prove that the formula (9.18) for $\hat{\beta}_1$ is equivalent the formula (9.16).

3. [20] Consider the proof of Theorem 9.1.1. Prove that $\sum_{i=1}^{n} c_i = 0$ and $\sum_{i=1}^{n} c_i x_i = 1$.

4. [10] Consider Example 9.1.2. Prove that, for this example, (9.34) yields $s_{\hat{\beta}_0} = 6.367$ and (9.35) yields $s_{\hat{\beta}_1} = 0.01976$. You may assume that $s^2 = 189.3$ and that you are given the values calculated in Example 9.1.1.

5. [5] Show that the fitted linear regression line $\hat{y} = \hat{\beta}_0 + \hat{\beta}_1 x$ goes through the point $(\overline{x}, \overline{y})$.

6. [7] Analysts at Blithering Boats have constructed the linear regression line $\hat{y} = 15 + 2.3x$.

   (a) What is the estimated change in the mean value of y when x increases by 2?

   (b) If $\overline{y} = 42.62$, determine $\overline{x}$.

7. [5] Suppose an analyst at Dogwood's Doughnuts is constructing a simple linear regression line with 41 points. She has calculated $S_{xy} = -540$, $SS_x = 130$, $\overline{y} = 35.7$, and $\overline{x} = 12.3$. Find the least squares regression line.

8. [12] Wimpering Willie is constructing a simple linear regression line with 25 points. Willie has calculated the value of SSE to be 374.2.

   (a) What is the standard error, s_e?

   (b) For $x = 75$, $\hat{y} = 126.3$. If the observation $(75, 140.2)$ is part of the original data set, should it be considered an outlier according to the oft quoted rule of thumb that any point for which $|y_i - \hat{y}| \geq 3s_e$ is an outlier?

   (c) If the observation $(75, 140.2)$ has $h_i = 0.07$, what is the standard residual of the point?

   (d) Would MINITAB flag the observation of part (c) as "an observation with a large standard residual?"

9. [10] Consider Example 9.1.1. Using the results calculated in the example and the parameters calculated by the MINITAB command REGRESS and shown in Figure 9.1.2, do the following:

(a) Find the 95% confidence interval for β_1.

(b) Test the null hypothesis $H_0 : \beta_1 = 0.07$ against $H_1 : \beta_1 \neq 0.07$ at the 5% level of significance.

10. [10] The sample correlation coefficient, r, between two random variables is 0.625.

   (a) If $n = 21$, test the null hypothesis H_0: $\rho = 0$ against the alternative hypothesis H_1: $\rho \neq 0$ at the 5% level of significance.

   (b) What is the p-value of the test?

11. [C10] If $n = 17$ and the intermediate values for a simple linear regression model are

$$\sum x = 136 \quad \sum y = 552.212 \quad \sum xy = 5081.33$$

$$\sum x^2 = 1496 \quad \sum y^2 = 19031.7,$$

do the following:

   (a) Calculate Sxy, SSx, and SSy.

   (b) Find the sample regression of Y on X.

   (c) Construct the ANOVA table.

12. [10] Professor Ruddy Redback checks the chlorine residual in his swimming pool at the times shown in the following table, after his pool service has treated the pool with a water purifier.

Hours (after treatment) x	Chlorine Residual (parts per million) y
06	1.86
12	1.73
18	1.61
24	1.50
30	1.39
36	1.30

Use the method of least squares to fit an exponential curve of the form $y = \alpha \exp(\beta x)$ to the data for Professor Redback's pool.

13. [12] The Perfect Product Prediction Company has collected the data shown in the table for the sales of Golden Jelly Belly jelly beans in ten market areas. The price is in cents for 5 beans and the sales are in thousands of cartons.

Price	Sales
x	y
14	97.46
15	90.56
16	84.54
17	79.26
18	74.57
19	70.40
20	66.66
21	63.28
22	60.22
23	57.44

Use the method of least squares to fit a curve of the form $y = \alpha x^{\beta}$ to the data.

14. [8] Henry Hunk, the head statistician at the Hunky Dory Boat Company, uses the multiple linear regression model

$$Y = \beta_0 + \beta_1 x_1 + \cdots + \beta_5 x_5 + \epsilon,$$

that seems to satisfy the standard statistics model assumptions. His least squares calculations from 30 data points yield SSE = 0.42 and $R^2 = 0.93$. Test the null hypothesis, H_0: $\beta_1 = \beta_2 = \cdots = \beta_5 = 0$, against the alternative hypothesis, H_1: At least one of the parameters $\beta_1, \beta_2, \ldots, \beta_5$ is not zero. Use $\alpha = 0.05$.

15. [10] Sweet Suzy at Candy Canes has developed the linear multiple regression model

$$\hat{y} = 18.7 + 3.2\, x_1 + 0.015\, x_2,$$

that seems to satisfy the standard statistics model assumptions. Her least squares calculations from 30 data points also yield

$$s_{\hat{\beta}_2} = 0.0036 \quad \text{and} \quad R^2 = 0.78.$$

(a) Test the null hypothesis, $H_0: \beta_1 = \beta_2 = 0$, against the alternative hypothesis, H_1: At least one of the parameters β_1, β_2 is not zero. Use $\alpha = 0.05$.

(b) Test the null hypothesis, $H_0: \beta_2 = 0$, against the alternative hypothesis, $H_1: \beta_2 \neq 0$. Use $\alpha = 0.05$.

16. [08] Show that for the multiple linear regression model, the F-value from the analysis of variable table, and R^2 are related by the simple formula (9.88),

$$F = \frac{R^2}{1 - R^2} \left(\frac{n - k - 1}{k} \right).$$

17. [10] In a multiple regression study at Carnal Cruises, part of the ANOVA table is as follows:

SOURCE	DF	SS
Regression	4	48
Error	36	36

(a) What is the sample size?

(b) How many predictor variables are there?

(c) Test the null hypothesis that all of the predictor variable coefficients are zero, against the alternative hypothesis that at least one of the coefficients is not zero. Use $\alpha = 0.05$.

(d) Calculate the standard error of the estimate.

18. [10] George Grep found the following, partially filled in, ANOVA table from a multiple regression study with 3 predictor variables. Help George by filling in the five blank spaces in the table (don't forget MSE).

SOURCE	DF	SS	MS
Regression			170.30
Error			
Total	9	538.24	

(a) Calculate R^2.

(b) Test the null hypothesis that all of the predictor variable coefficients are zero, against the alternative hypothesis that at least one of the coefficients is not zero. Use $\alpha = 0.05$.

19. [20] You will need a computer system, such as MINITAB, SAS/STAT, or EXPLORE to do this exercise. Sam Spade of Hilbert Hackers has discovered the following table that relates x to y. He wants to fit a cubic to the data and is concerned about the numerical stability of the calculations. Help Sam by doing the following:

    (a) Fit the regression function $y = \beta_0 + \beta_1 x + \beta_2 x^2 + \beta_3 x^3$ to the data. Record the values of $\hat{\beta}_0$, $\hat{\beta}_1$, $\hat{\beta}_2$, $\hat{\beta}_3$, the estimated standard deviation of $\hat{\beta}_3$, R^2, and s_e. Is β_3 significant at the five percent level?

    (b) Standardize x by the formula $x' = (x - \bar{x})/s_x$, and fit the regression function $y = \beta_0^* + \beta_1^* x' + \beta_2^* (x')^2 + \beta_3^* (x')^3$ to the data. Then record the same values that you recorded for (a) and make the same test.

Data

x	y	x	y	x	y
10	7	10	8	10	6
15	12	15	15	15	13
20	10	20	11	20	7
25	14	25	16	25	17

20. [15] You will need a computer for this exercise and the next. Fit the regression equation $y = \beta_0 + \beta_1 x_1 + \beta_2 x_2$ to the data of Table 9.3.6. Compare the fit to that of the model fitted in Example 9.3.7.

21. [25] Consider Example 9.3.7. Consider the regressors x_1, x_2, and $x_3 = x_1 x_2$ used in that example, plus the regressors $x_4 = x_1^2$, and $x_5 = x_2^2$. Then find the best subsets of regressors of size 1, 2, 3, and 4.

Alan Turing
Found alluring
Machines whose only fault
Was that they would not halt.

Karl David
Wells College

The dice of God are always loaded.
Ralph Waldo Emerson

I shall never believe that God plays dice with the world.
Albert Einstein

God not only plays dice. He also sometimes throws the dice where
they cannot be seen.
Stephen Hawking

We figured the odds as best we could, and then we rolled the dice.
Jimmy Carter

References

[1] Francis J. Anscombe, Graphs in statistical analysis, *The American Statistician*, **27(1)**, (February 1973), 17–21.

[2] A. C. Atkinson, *Plots, Transformations and Regression: An Introduction to Graphical Methods of Diagnostic Regression Analysis*, Oxford University Press, New York, 1985.

[3] Samprit Chatterjee and Ali S. Hadi, Influential observations, high leverage points, and outliers in linear regression, *Statistical Science*, **1(3)**, (August 1986), 379–416 (with discussion).

[4] R. Dennis Cook and Sanford Weisberg, *Residuals and Influence in Regression*, Chapman and Hall, New York, 1982.

[5] David P. Doane, *Exploring Statistics With The IBM PC, Second Edition*, Addison-Wesley, Reading, MA, 1988. (Comes with a diskette containing the EXPLORE statistical system and a data diskette.)

[6] Norman Draper and Harry Smith, *Applied Regression Analysis*, 2nd ed., John Wiley, New York, 1981.

[7] M. J. Garside, Some computational procedures for the best subset problem, *Applied Statisticsm* **20**, (1971), 8–15.

[8] James H. Goodnight, A tutorial on the sweep operator, *The American Statistician*, **33**, (1979), 149–158.

[9] David C. Hoaglin, Using leverage and influence to introduce regression diagnostics, *The College Mathematics Journal*, **19(5)**, (November 1985), 387–401.

[10] Robert V. Hogg and Allen T. Craig, *Introduction to Mathematical Statistics*, 3rd ed., Macmillan, New York, 1970.

[11] Robert V. Hogg and Elliot A. Tanis, *Probability and Statistical Inference*, 3rd ed., Macmillan, New York, 1988.

[12] Seymour Kass, An eigenvalue characterization of the correlation coefficient, *The American Mathematical Monthly*, **96(10)**, (December 1989), 910–911.

[13] *MINITAB Reference Manual, Release 7*, Minitab, Inc., State College, PA, 1989.

[14] Douglas C. Montgomery and Elizabeth A. Peck, *Introduction to Linear Regression Analysis*, John Wiley, New York, 1982.

[15] Frederick Mosteller, Stephen E. Fienberg, and Robert E. K. Rourke, *Beginning Statistics with Data Analysis*, Addison-Wesley, Reading, MA, 1983.

[16] Frederick Mosteller and John W. Tukey, *Data Analysis and Regression: A Second Course in Statistics*, Addison-Wesley, Reading, MA, 1977.

[17] John Neter, William Wasserman, and Michael H. Kutner, *Applied Linear Statistical Models*, 2nd ed., Richard D. Irwin, Homewood, IL, 1985.

[18] John A. Rice, *Mathematical Statistics and Data Analysis*, Wadsworth & Brooks/Cole, Belmont, CA, 1988.

[19] Barbara F. Ryan, Brian L. Joiner, and Thomas A. Ryan, Jr., *Minitab Handbook, Second Edition*, Duxbury Press, Boston, 1985.

[20] M. Schatzoff, R. Tsao, and S. Fienberg, Efficient calculation of all possible regressions, *Technometrics*, **10**, (1968), 769–779.

[21] George W. Snedecor and William G. Cochran, *Statistical Methods, Seventh Edition*, The Iowa State University Press, Ames, 1980.

[22] Sanford Weisberg, *Applied Linear Regression*, 2nd ed., John Wiley, New York, 1985.

Appendix A

Statistical Tables

A.1   Discrete Random Variables

Table 1A. Properties of Some Common Discrete Random Variables[1]

Random Variable	Parameters	$p(\cdot)$
Bernoulli	$0 < p < 1$	$p(k) = p^k q^{1-k}$ $k = 0, 1$
Binomial	n $0 < p < 1$	$p(k) = \binom{n}{k} p^k q^{n-k},$ $k = 0, 1, \ldots, n$
Multinomial	n, r, p_i, k_i $\displaystyle\sum_{i=1}^{r} p_i = 1$ $\displaystyle\sum_{i=1}^{r} k_i = n,$	$p(\overline{k}) = \dfrac{n!}{k_1! k_2! \cdots k_r!} p_1^{k_1} p_2^{k_2} \cdots p_r^{k_r}$ where $\overline{k} = (k_1, k_2, \ldots, k_r)$

---

[1] $q = 1 - p.$

Table 1A. (continued)

Random Variable	Parameters	$p(\cdot)$
Hypergeometric	$N > 0$	$p(k) = \dfrac{\dbinom{r}{k}\dbinom{N-r}{n-k}}{\dbinom{N}{n}},$
	$n, k \geq 0$	$k = 0, 1, \ldots, n,$ where $k \leq r$ and $n - k \leq N - r.$
Multivariate Hypergeometric	$\displaystyle\sum_{i=1}^{l} r_i = N$	$p(k_1, k_2, \ldots, k_l) = \dfrac{\dbinom{r_1}{k_1}\dbinom{r_2}{k_2}\cdots\dbinom{r_l}{k_l}}{\dbinom{N}{n}},$ for $k_i \in \{0, 1, \ldots, n\}$, $k_i \leq r_i \; \forall i$ and $\displaystyle\sum_{i=1}^{l} k_i = n.$
Geometric	$0 < p < 1$	$p(k) = q^k p, \quad k = 0, 1, \ldots.$
Pascal (negative binomial)	$0 < p < 1$ r positive integer	$p(k) = \dbinom{r + k - 1}{k} p^r q^k,$ $k = 0, 1, \cdots$
Poisson	$\alpha > 0$	$p(k) = e^{-\alpha}\dfrac{\alpha^k}{k!}, \quad k = 0, 1, \cdots$

Table 1B. Properties of Some Common Discrete Random Variables[2]

Random Variable	z-transform $g[z]$	$E[X]$	$\text{Var}[X]$
Bernoulli	$q + pz$	p	pq
Binomial	$(q + pz)^n$	np	npq
Multinomial	$(p_1 z_1 + p_2 z_2 + \cdots + p_r z_r)^n$	$E[X_i] = np_i$	$\text{Var}[X_i] = np_i q_i$
Hypergeometric	—	$\dfrac{nr}{N}$	$\dfrac{nr(N-r)(N-n)}{N^2(N-1)}$
Multivariate Hypergeometric	—	—	—
Geometric	$\dfrac{p}{1-qz}$	$\dfrac{q}{p}$	$\dfrac{q}{p^2}$
Pascal (negative binomial)	$p^r(1-qz)^{-r}$	$\dfrac{rq}{p}$	$\dfrac{rq}{p^2}$
Poisson	$e^{\alpha(z-1)}$	α	α

---

[2] $q_i = 1 - p_i$.

A.2   Continuous Random Variables

Table 2A. Properties of Some Common Continuous Random Variables

Random Variable	Parameters	Density $f(\cdot)$
Uniform	$a < b$	$\dfrac{1}{b-a},\ a \le x \le b,\quad 0 \text{ otherwise}$
Exponential	$\alpha > 0$	$f(x) = \alpha e^{-\alpha x},\ x > 0,\quad 0 \text{ if } x \le 0$
Gamma	$\beta, \alpha > 0$	$f(x) = \dfrac{\alpha(\alpha x)^{\beta-1}}{\Gamma(\beta)} e^{-\alpha x},\ x > 0$ $0,\ x \le 0$
Erlang-k	$k > 0$ $\mu > 0$	$f(x) = \dfrac{\mu k(\mu k x)^{k-1}}{(k-1)!} e^{-\mu k x},\ x > 0$ $0,\quad x \le 0$
H_k [3]	$q_i, \mu_i > 0$ $\displaystyle\sum_{i=1}^{k} \dfrac{q_i}{\mu_i} = \dfrac{1}{\mu}$	$f(x) = \displaystyle\sum_{i=1}^{k} q_i \mu_i e^{-\mu_i x},\ x > 0$ $0,\quad x \le 0$
Chi-square	$n > 0$	$f(x) = \dfrac{x^{((n/2)-1)} e^{-x/2}}{2^{n/2}\Gamma(n/2)},\ x > 0,\quad 0 \text{ if } x \le 0$
Normal	$\sigma > 0$	$f(x) = \dfrac{1}{\sigma\sqrt{2\pi}} \exp\left(-\dfrac{1}{2}\dfrac{(x-\mu)^2}{\sigma^2}\right)$
Student's t	n	$f(x) = \dfrac{\Gamma[(n+1)/2]}{\sqrt{n\pi}\,\Gamma(n/2)}\left(1+\dfrac{x^2}{n}\right)^{-(n+1)/2}$
F	n, m	$f(x) = \dfrac{(n/m)^{n/2}\Gamma[(n+m)/2]x^{((n/2)-1)}}{\Gamma(n/2)\Gamma(m/2)(1+(n/m)x)^{(n+m)/2}},\ x > 0$

---

[3]Hyperexponential with k stages.

Table 2B. Properties of Some Common Continuous Random Variables

Random Variable	$E[X]$	$\text{Var}[X]$	Laplace–Stieltjes Transform $X^*[\theta]$
Uniform	$\dfrac{a+b}{2}$	$\dfrac{(b-a)^2}{12}$	$\dfrac{e^{-b\theta}-e^{-a\theta}}{\theta(a-b)}$
Exponential	$\dfrac{1}{\alpha}$	$\dfrac{1}{\alpha^2}$	$\dfrac{\alpha}{\alpha+\theta}$
Gamma	$\dfrac{\beta}{\alpha}$	$\dfrac{\beta}{\alpha^2}$	$\left(\dfrac{\alpha}{\alpha+\theta}\right)^{\beta}$
Erlang-k	$\dfrac{1}{\mu}$	$\dfrac{1}{k\mu^2}$	$\left(\dfrac{k\mu}{k\mu+\theta}\right)^{k}$
H_k [4]	$\dfrac{1}{\mu}$	$\left(2\displaystyle\sum_{i=1}^{k}\dfrac{q_i}{\mu_i^2}\right)-\dfrac{1}{\mu^2}$	$\displaystyle\sum_{i=1}^{k}\dfrac{q_i\mu_i}{\mu_i+\theta}$
Chi-square	n	$2n$	$\left(\dfrac{1}{1+2\theta}\right)^{n/2}$
Normal	μ	σ^2	$\exp\left(-\theta\mu-\tfrac{1}{2}\theta^2\sigma^2\right)$
Student's t	0 for $n>1$	$\dfrac{n}{n-2}$ for $n>2$	does not exist
F	$\dfrac{m}{m-2}$ if $m>2$	$\dfrac{m^2(2n+2m-4)}{n(m-2)^2(m-4)}$ if $m>4$	does not exist

---

[4]Hyperexponential with k stages.

A.3   Statistical Tables

Table 3

The Normal Distribution Functions $\Phi(z) = \int_{-\infty}^{z} \dfrac{e^{-t^2/2}}{\sqrt{2\pi}}\, dt$

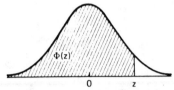

z	0.00	0.01	0.02	0.03	0.04	0.05	0.06	0.07	0.08	0.09
0.0	.50000	.50399	.50798	.51197	.51595	.51994	.52392	.52790	.53188	.53586
0.1	.53983	.54380	.54776	.55172	.55567	.55962	.56356	.56749	.57142	.57535
0.2	.57926	.58317	.58706	.59095	.59483	.59871	.60257	.60642	.61026	.61409
0.3	.61791	.62172	.62552	.62930	.63307	.63683	.64058	.64431	.64803	.65173
0.4	.65542	.65910	.66276	.66640	.67003	.67364	.67724	.68082	.68439	.68793
0.5	.69146	.69497	.69847	.70194	.70540	.70884	.71226	.71566	.71904	.72240
0.6	.72575	.72907	.73237	.73565	.73891	.74215	.74537	.74857	.75175	.75490
0.7	.75804	.76115	.76424	.76730	.77035	.77337	.77637	.77935	.78230	.78524
0.8	.78814	.79103	.79389	.79673	.79955	.80234	.80511	.80785	.81057	.81327
0.9	.81594	.81859	.82121	.82381	.82639	.82894	.83147	.83398	.83646	.83891
1.0	.84134	.84375	.84614	.84849	.85083	.85314	.85543	.85769	.85993	.86214
1.1	.86433	.86650	.86864	.87076	.87286	.87493	.87698	.87900	.88100	.88298
1.2	.88493	.88686	.88877	.89065	.89251	.89435	.89617	.89796	.89973	.90147
1.3	.90320	.90490	.90658	.90824	.90988	.91149	.91308	.91466	.91621	.91774
1.4	.91924	.92073	.92220	.92364	.92507	.92647	.92785	.92922	.93056	.93189
1.5	.93319	.93448	.93574	.93699	.93822	.93943	.94062	.94179	.94295	.94408
1.6	.94520	.94630	.94738	.94845	.94950	.95053	.95154	.95254	.95352	.95449
1.7	.95543	.95637	.95728	.95818	.95907	.95994	.96080	.96164	.96246	.96327
1.8	.96407	.96485	.96562	.96638	.96712	.96784	.96856	.96926	.96995	.97062
1.9	.97128	.97193	.97257	.97320	.97381	.97441	.97500	.97558	.97615	.97670
2.0	.97725	.97778	.97831	.97882	.97932	.97982	.98030	.98077	.98124	.98169
2.1	.98214	.98257	.98300	.98341	.98382	.98422	.98461	.98500	.98537	.98574
2.2	.98610	.98645	.98679	.98713	.98745	.98778	.98809	.98840	.98870	.98899
2.3	.98928	.98956	.98983	.99010	.99036	.99061	.99086	.99111	.99134	.99158
2.4	.99180	.99202	.99224	.99245	.99266	.99286	.99305	.99324	.99343	.99361
2.5	.99379	.99396	.99413	.99430	.99446	.99461	.99477	.99492	.99506	.99520
2.6	.99534	.99547	.99560	.99573	.99585	.99598	.99609	.99621	.99632	.99643
2.7	.99653	.99664	.99674	.99683	.99693	.99702	.99711	.99720	.99728	.99736
2.8	.99744	.99752	.99760	.99767	.99774	.99781	.99788	.99795	.99801	.99807
2.9	.99813	.99819	.99825	.99831	.99836	.99841	.99846	.99851	.99856	.99861
3.0	.99865	.99869	.99874	.99878	.99882	.99886	.99889	.99893	.99896	.99900
3.1	.99903	.99906	.99910	.99913	.99916	.99918	.99921	.99924	.99926	.99929
3.2	.99931	.99934	.99936	.99938	.99940	.99942	.99944	.99946	.99948	.99950
3.3	.99952	.99953	.99955	.99957	.99958	.99960	.99961	.99962	.99964	.99965
3.4	.99966	.99968	.99969	.99970	.99971	.99972	.99973	.99974	.99975	.99976
3.5	.99977	.99978	.99978	.99979	.99980	.99981	.99981	.99982	.99983	.99983
3.6	.99984	.99985	.99985	.99986	.99986	.99987	.99987	.99988	.99988	.99989
3.7	.99989	.99990	.99990	.99990	.99991	.99991	.99992	.99992	.99992	.99992
3.8	.99993	.99993	.99993	.99994	.99994	.99994	.99994	.99995	.99995	.99995

Table 4

Critical Values of the Chi-Square Distribution[a]

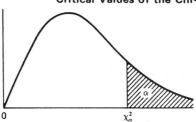

n[b] $\diagdown$ α	0.995	0.990	0.975	0.950	0.05	0.025	0.010	0.005
1	0.0⁴393[c]	0.0³157[c]	0.0⁹982[c]	0.0²393[c]	3.8415	5.0239	6.6349	7.8794
2	0.0100	0.0201	0.0506	0.1026	5.9915	7.3778	9.2103	10.597
3	0.0717	0.1148	0.2158	0.3518	7.8147	9.3484	11.345	12.838
4	0.2070	0.2971	0.4844	0.7107	9.4877	11.143	13.277	14.860
5	0.4117	0.5543	0.8312	1.1455	11.071	12.833	15.086	16.750
6	0.6757	0.8721	1.2373	1.6354	12.592	14.449	16.812	18.548
7	0.9893	1.2390	1.6899	2.1674	14.067	16.013	18.475	20.278
8	1.3444	1.6465	2.1797	2.7326	15.507	17.535	20.090	21.955
9	1.7350	2.0879	2.7004	3.3251	16.920	19.023	21.666	23.589
10	2.1559	2.5582	3.2470	3.9403	18.307	20.483	23.209	25.188
11	2.6032	3.0535	3.8158	4.5748	19.675	21.920	24.725	26.757
12	3.0738	3.5706	4.4038	5.2260	21.026	23.337	26.217	28.300
13	3.5650	4.1069	5.0087	5.8919	22.362	24.736	27.688	29.819
14	4.0747	4.6604	5.6287	6.5706	23.685	26.119	29.141	31.319
15	4.6009	5.2294	6.2621	7.2609	24.996	27.488	30.578	32.801
16	5.1422	5.8122	6.9077	7.9616	26.296	28.845	32.000	34.267
17	5.6972	6.4078	7.5642	8.6718	27.587	30.191	33.409	35.719
18	6.2648	7.0149	8.2308	9.3905	28.869	31.526	34.805	37.156
19	6.8440	7.6327	8.9066	10.117	30.144	32.852	36.191	38.582
20	7.4339	8.2604	9.5908	10.851	31.410	34.170	37.566	39.997
21	8.0337	8.8972	10.283	11.591	32.671	35.479	38.932	41.401
22	8.6427	9.5425	10.982	12.338	33.924	36.781	40.289	42.796
23	9.2604	10.196	11.689	13.091	35.173	38.076	41.638	44.181
24	9.8862	10.856	12.401	13.848	36.415	39.364	42.980	45.559
25	10.520	11.524	13.120	14.611	37.653	40.647	44.314	46.928
26	11.160	12.198	13.844	15.379	38.885	41.923	45.642	48.290
27	11.808	12.879	14.573	16.151	40.113	43.194	46.963	49.645
28	12.461	13.565	15.308	16.928	41.337	44.461	48.278	50.993
29	13.121	14.257	16.047	17.708	42.557	45.722	49.588	52.336
30	13.787	14.954	16.791	18.493	43.773	46.980	50.892	53.672
40	20.707	22.164	24.433	26.509	55.759	59.342	63.691	66.766
50	27.991	29.707	32.357	34.764	67.505	71.420	76.154	79.490
60	35.535	37.485	40.482	43.188	79.082	83.298	88.380	91.952
70	43.275	45.442	48.758	51.739	90.531	95.023	100.425	104.215
80	51.172	53.540	57.153	60.392	101.879	106.629	112.329	116.321
90	59.196	61.754	65.647	69.126	113.145	118.136	124.116	128.299
100	67.328	70.065	74.222	77.930	124.342	129.561	135.807	140.169
z_α	−2.5758	−2.3263	−1.9600	−1.6449	+1.6449	+1.9600	+2.3263	+2.5758

[a] Adapted from *Biometrika Tables for Statisticians*, (E. S. Pearson and H. O. Hartley, eds.), Vol. 1, 4th ed. Cambridge University Press, Cambridge, 1966, by permission of Biometrika Trustees.

[b] For $n > 100$ use

$$\chi_\alpha^2 = n\left\{1 - \frac{2}{9n} + z_\alpha \sqrt{\frac{2}{9n}}\right\}^3$$

where z_α is given on the bottom line of the table.

[c] The expression 0.0⁴393 means 0.0000393, etc.

Table 5

Critical Values of the Student-t Distribution[a]

n \ α	0.10	0.05	0.025	0.01	0.005
1	3.078	6.314	12.706	31.821	63.657
2	1.886	2.920	4.303	6.965	9.925
3	1.638	2.353	3.182	4.541	5.841
4	1.533	2.132	2.776	3.747	4.604
5	1.476	2.015	2.571	3.365	4.032
6	1.440	1.943	2.447	3.143	3.707
7	1.415	1.895	2.365	2.998	3.499
8	1.397	1.860	2.306	2.896	3.355
9	1.383	1.833	2.262	2.821	3.250
10	1.372	1.812	2.228	2.764	3.169
11	1.363	1.796	2.201	2.718	3.106
12	1.356	1.782	2.179	2.681	3.055
13	1.350	1.771	2.160	2.650	3.012
14	1.345	1.761	2.145	2.624	2.977
15	1.341	1.753	2.131	2.602	2.947
16	1.337	1.746	2.120	2.583	2.921
17	1.333	1.740	2.110	2.567	2.898
18	1.330	1.734	2.101	2.552	2.878
19	1.328	1.729	2.093	2.539	2.861
20	1.325	1.725	2.086	2.528	2.845
21	1.323	1.721	2.080	2.518	2.831
22	1.321	1.717	2.074	2.508	2.819
23	1.319	1.714	2.069	2.500	2.807
24	1.318	1.711	2.064	2.492	2.797
25	1.316	1.708	2.060	2.485	2.787
26	1.315	1.706	2.056	2.479	2.779
27	1.314	1.703	2.052	2.473	2.771
28	1.313	1.701	2.048	2.467	2.763
29	1.311	1.699	2.045	2.462	2.756
30	1.310	1.697	2.042	2.457	2.750
40	1.303	1.684	2.021	2.423	2.704
60	1.296	1.671	2.000	2.390	2.660
120	1.289	1.658	1.980	2.358	2.617
∞	1.282	1.645	1.960	2.326	2.576

[a] Adapted from *Biometrika Tables for Statisticians* (E. S. Pearson and H. O. Hartley, eds.), Vol. 1, 4th ed. Cambridge University Press, Cambridge, 1966, by permission of Biometrika Trustees.

Table 6

Critical Values of the F Distribution*

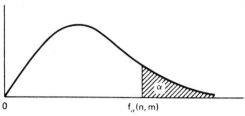

$f_\alpha(n, m)$

Denom-inator	Numer-ator	$f_{0.05}(n, m)$								
m	n	1	2	3	4	5	6	7	8	9
1		161.40	199.50	215.70	224.60	230.20	234.00	236.80	238.90	240.50
2		18.51	19.00	19.16	19.25	19.30	19.33	19.35	19.35	19.38
3		10.13	9.55	9.28	9.12	9.01	8.94	8.89	8.85	8.81
4		7.71	6.94	6.59	6.39	6.26	6.16	6.09	6.04	6.00
5		6.61	5.79	5.41	5.19	5.05	4.95	4.88	4.82	4.77
6		5.99	5.14	4.76	4.53	4.39	4.28	4.21	4.15	4.10
7		5.59	4.74	4.35	4.12	3.97	3.87	3.79	3.73	3.68
8		5.32	4.46	4.07	3.84	3.69	3.58	3.50	3.44	3.39
9		5.12	4.26	3.86	3.63	3.48	3.37	3.29	3.23	3.18
10		4.96	4.10	3.71	3.48	3.33	3.22	3.14	3.07	3.02
11		4.84	3.98	3.59	3.36	3.20	3.09	3.01	2.95	2.90
12		4.75	3.89	3.49	3.26	3.11	3.00	2.91	2.85	2.80
13		4.67	3.81	3.41	3.18	3.03	2.92	2.83	2.77	2.71
14		4.60	3.74	3.34	3.11	2.96	2.85	2.76	2.70	2.65
15		4.54	3.68	3.29	3.06	2.90	2.79	2.71	2.64	2.59
16		4.49	3.63	3.24	3.01	2.85	2.74	2.66	2.59	2.54
17		4.45	3.59	3.20	2.96	2.81	2.70	2.61	2.55	2.49
18		4.41	3.55	3.16	2.93	2.77	2.66	2.58	2.51	2.46
19		4.38	3.52	3.13	2.90	2.74	2.63	2.54	2.48	2.42
20		4.35	3.49	3.10	2.87	2.71	2.60	2.51	2.45	2.39
21		4.32	3.47	3.07	2.84	2.68	2.57	2.49	2.42	2.37
22		4.30	3.44	3.05	2.82	2.66	2.55	2.46	2.40	2.34
23		4.28	3.42	3.03	2.80	2.64	2.53	2.44	2.37	2.32
24		4.26	3.40	3.01	2.78	2.62	2.51	2.42	2.36	2.30
25		4.24	3.39	2.99	2.76	2.60	2.49	2.40	2.34	2.28
26		4.23	3.37	2.98	2.74	2.59	2.47	2.39	2.32	2.27
27		4.21	3.35	2.96	2.73	2.57	2.46	2.37	2.31	2.25
28		4.20	3.34	2.95	2.71	2.56	2.45	2.36	2.29	2.24
29		4.18	3.33	2.93	2.70	2.55	2.43	2.35	2.28	2.22
30		4.17	3.32	2.92	2.69	2.53	2.42	2.33	2.27	2.21
40		4.08	3.23	2.84	2.61	2.45	2.34	2.25	2.18	2.12
60		4.00	3.15	2.76	2.53	2.37	2.25	2.17	2.10	2.04
120		3.92	3.07	2.68	2.45	2.29	2.17	2.09	2.02	1.96
∞		3.84	3.00	2.60	2.37	2.21	2.10	2.01	1.94	1.88

Table 6

Critical Values of the *F* Distribution *(Continued)*

Denom- inator *m*	Numer- ator *n*	$f_{0.05}(n, m)$									
		10	12	15	20	24	30	40	60	120	∝
1		241.9	243.9	245.9	248.0	249.1	250.1	251.1	252.2	253.3	254.3
2		19.40	19.41	19.43	19.45	19.45	19.46	19.47	19.48	19.49	19.50
3		8.79	8.74	8.70	8.66	8.64	8.62	8.59	8.57	8.55	8.53
4		5.96	5.91	5.86	5.80	5.77	5.75	5.72	5.69	5.66	5.63
5		4.74	4.68	4.62	4.56	4.53	4.50	4.46	4.43	4.40	4.36
6		4.06	4.00	3.94	3.87	3.84	3.81	3.77	3.74	3.70	3.67
7		3.64	3.57	3.51	3.44	3.41	3.38	3.34	3.30	3.27	3.23
8		3.35	3.28	3.22	3.15	3.12	3.08	3.04	3.01	2.97	2.93
9		3.14	3.07	3.01	2.94	2.90	2.86	2.83	2.79	2.75	2.71
10		2.98	2.91	2.85	2.77	2.74	2.70	2.66	2.62	2.58	2.54
11		2.85	2.79	2.72	2.65	2.61	2.57	2.53	2.49	2.45	2.40
12		2.75	2.69	2.62	2.54	2.51	2.47	2.43	2.38	2.34	2.30
13		2.67	2.60	2.53	2.46	2.42	2.38	2.34	2.30	2.25	2.21
14		2.60	2.53	2.46	2.39	2.35	2.31	2.27	2.22	2.18	2.13
15		2.54	2.48	2.40	2.33	2.29	2.25	2.20	2.16	2.11	2.07
16		2.49	2.42	2.35	2.28	2.24	2.19	2.15	2.11	2.06	2.01
17		2.45	2.38	2.31	2.23	2.19	2.15	2.10	2.06	2.01	1.96
18		2.41	2.34	2.27	2.19	2.15	2.11	2.06	2.02	1.97	1.92
19		2.38	2.31	2.23	2.16	2.11	2.07	2.03	1.98	1.93	1.88
20		2.35	2.28	2.20	2.12	2.08	2.04	1.99	1.95	1.90	1.84
21		2.32	2.25	2.18	2.10	2.05	2.01	1.96	1.92	1.87	1.81
22		2.30	2.23	2.15	2.07	2.03	1.98	1.94	1.89	1.84	1.78
23		2.27	2.20	2.13	2.05	2.01	1.96	1.91	1.86	1.81	1.76
24		2.25	2.18	2.11	2.03	1.98	1.94	1.89	1.84	1.79	1.73
25		2.24	2.16	2.09	2.01	1.96	1.92	1.87	1.82	1.77	1.71
26		2.22	2.15	2.07	1.99	1.95	1.90	1.85	1.80	1.75	1.69
27		2.20	2.13	2.06	1.97	1.93	1.88	1.84	1.79	1.73	1.67
28		2.19	2.12	2.04	1.96	1.91	1.87	1.82	1.77	1.71	1.65
29		2.18	2.10	2.03	1.94	1.90	1.85	1.81	1.75	1.70	1.64
30		2.16	2.09	2.01	1.93	1.89	1.84	1.79	1.74	1.68	1.62
40		2.08	2.00	1.92	1.84	1.79	1.74	1.69	1.64	1.58	1.51
60		1.99	1.92	1.84	1.75	1.70	1.65	1.59	1.53	1.47	1.39
120		1.91	1.83	1.75	1.66	1.61	1.55	1.50	1.43	1.35	1.25
∝		1.83	1.75	1.67	1.57	1.52	1.46	1.39	1.32	1.22	1.00

Table 6

Critical Values of the *F* Distribution *(Continued)*

Denom- inator	Numer- ator	$f_{0.01}(n, m)$								
m	*n*	1	2	3	4	5	6	7	8	9
1		4052	4999.5	5403	5625	5764	5859	5928	5982	6022
2		98.50	99.00	99.17	99.25	99.30	99.33	99.36	99.37	99.39
3		34.12	30.82	29.46	28.71	28.24	27.91	27.67	27.49	27.35
4		21.20	18.00	16.69	15.98	15.52	15.21	14.98	14.80	14.66
5		16.26	13.27	12.06	11.39	10.97	10.67	10.46	10.29	10.16
6		13.75	10.92	9.78	9.15	8.75	8.47	8.26	8.10	7.98
7		12.25	9.55	8.45	7.85	7.46	7.19	6.99	6.84	6.72
8		11.26	8.65	7.59	7.01	6.63	6.37	6.18	6.03	5.91
9		10.56	8.02	6.99	6.42	6.06	5.80	5.61	5.47	5.35
10		10.04	7.56	6.55	5.99	5.64	5.39	5.20	5.06	4.94
11		9.65	7.21	6.22	5.67	5.32	5.07	4.89	4.74	4.63
12		9.33	6.93	5.95	5.41	5.06	4.82	4.64	4.50	4.39
13		9.07	6.70	5.74	5.21	4.86	4.62	4.44	4.30	4.19
14		8.86	6.51	5.56	5.04	4.69	4.46	4.28	4.14	4.03
15		8.68	6.36	5.42	4.89	4.56	4.32	4.14	4.00	3.89
16		8.53	6.23	5.29	4.77	4.44	4.20	4.03	3.89	3.78
17		8.40	6.11	5.18	4.67	4.34	4.10	3.93	3.79	3.68
18		8.29	6.01	5.09	4.58	4.25	4.01	3.84	3.71	3.60
19		8.18	5.93	5.01	4.50	4.17	3.94	3.77	3.63	3.52
20		8.10	5.85	4.94	4.43	4.10	3.87	3.70	3.56	3.46
21		8.02	5.78	4.87	4.37	4.04	3.81	3.64	3.51	3.40
22		7.95	5.72	4.82	4.31	3.99	3.76	3.59	3.45	3.35
23		7.88	5.66	4.76	4.26	3.94	3.71	3.54	3.41	3.30
24		7.82	5.61	4.72	4.22	3.90	3.67	3.50	3.36	3.26
25		7.77	5.57	4.68	4.18	3.85	3.63	3.46	3.32	3.22
26		7.72	5.53	4.64	4.14	3.82	3.59	3.42	3.29	3.18
27		7.68	5.49	4.60	4.11	3.78	3.56	3.39	3.26	3.15
28		7.64	5.45	4.57	4.07	3.75	3.53	3.36	3.23	3.12
29		7.60	5.42	4.54	4.04	3.73	3.50	3.33	3.20	3.09
30		7.56	5.39	4.51	4.02	3.70	3.47	3.30	3.17	3.07
40		7.31	5.18	4.31	3.83	3.51	3.29	3.12	2.99	2.89
60		7.08	4.98	4.13	3.65	3.34	3.12	2.95	2.82	2.72
120		6.85	4.79	3.95	3.48	3.17	2.96	2.79	2.66	2.56
∞		6.63	4.61	3.78·	3.32	3.02	2.80	2.64	2.51	2.41

Table 6

Critical Values of the F Distribution *(Continued)*

Denom-inator m	Numer-ator n	$f_{0.01}(n, m)$									
		10	12	15	20	24	30	40	60	120	∞
1		6056	6106	6157	6209	6235	6261	6287	6313	6339	6366
2		99.40	99.42	99.43	99.45	99.46	99.47	99.47	99.48	99.49	99.50
3		27.23	27.05	26.87	26.69	26.60	26.50	26.41	26.32	26.22	26.13
4		14.55	14.37	14.20	14.02	13.93	13.84	13.75	13.65	13.56	13.46
5		10.05	9.89	9.72	9.55	9.47	9.38	9.29	9.20	9.11	9.02
6		7.87	7.72	7.56	7.40	7.31	7.23	7.14	7.06	6.97	6.88
7		6.62	6.47	6.31	6.16	6.07	5.99	5.91	5.82	5.74	5.65
8		5.81	5.67	5.52	5.36	5.28	5.20	5.12	5.03	4.95	4.86
9		5.26	5.11	4.96	4.81	4.73	4.65	4.57	4.48	4.40	4.31
10		4.85	4.71	4.56	4.41	4.33	4.25	4.17	4.08	4.00	3.91
11		4.54	4.40	4.25	4.10	4.02	3.94	3.86	3.78	3.69	3.60
12		4.30	4.16	4.01	3.86	3.78	3.70	3.62	3.54	3.45	3.36
13		4.10	3.96	3.82	3.66	3.59	3.51	3.43	3.34	3.25	3.17
14		3.94	3.80	3.66	3.51	3.43	3.35	3.27	3.18	3.09	3.00
15		3.80	3.67	3.52	3.37	3.29	3.21	3.13	3.05	2.96	2.87
16		3.69	3.55	3.41	3.26	3.18	3.10	3.02	2.93	2.84	2.75
17		3.59	3.46	3.31	3.16	3.08	3.00	2.92	2.83	2.75	2.65
18		3.51	3.37	3.23	3.08	3.00	2.92	2.84	2.75	2.66	2.57
19		3.43	3.30	3.15	3.00	2.92	2.84	2.76	2.67	2.58	2.49
20		3.37	3.23	3.09	2.94	2.86	2.78	2.69	2.61	2.52	2.42
21		3.31	3.17	3.03	2.88	2.80	2.72	2.64	2.55	2.46	2.36
22		3.26	3.12	2.98	2.83	2.75	2.67	2.58	2.50	2.40	2.31
23		3.21	3.07	2.93	2.78	2.70	2.62	2.54	2.45	2.35	2.26
24		3.17	3.03	2.89	2.74	2.66	2.58	2.49	2.40	2.31	2.21
25		3.13	2.99	2.85	2.70	2.62	2.54	2.45	2.36	2.27	2.17
26		3.09	2.96	2.81	2.66	2.58	2.50	2.42	2.33	2.23	2.13
27		3.06	2.93	2.78	2.63	2.55	2.47	2.38	2.29	2.20	2.10
28		3.03	2.90	2.75	2.60	2.52	2.44	2.35	2.26	2.17	2.06
29		3.00	2.87	2.73	2.57	2.49	2.41	2.33	2.23	2.14	2.03
30		2.98	2.84	2.70	2.55	2.47	2.39	2.30	2.21	2.11	2.01
40		2.80	2.66	2.52	2.37	2.29	2.20	2.11	2.02	1.92	1.80
60		2.63	2.50	2.35	2.20	2.12	2.03	1.94	1.84	1.73	1.60
120		2.47	2.34	2.19	2.03	1.95	1.86	1.76	1.66	1.53	1.38
∞		2.32	2.18	2.04	1.88	1.79	1.70	1.59	1.47	1.32	1.00

Table 6

Critical Values of the F Distribution *(Continued)*

Denom-inator	Numer-ator	$f_{0.025}(n, m)$								
m	n	1	2	3	4	5	6	7	8	9
1		647.8	799.5	864.2	899.6	921.8	937.1	948.2	956.7	963.3
2		38.51	39.00	39.17	39.25	39.30	39.33	39.36	39.37	39.39
3		17.44	16.04	15.44	15.10	14.88	14.73	14.62	14.54	14.47
4		12.22	10.65	9.98	9.60	9.36	9.20	9.07	8.98	8.90
5		10.01	8.43	7.76	7.39	7.15	6.98	6.85	6.76	6.68
6		8.81	7.26	6.60	6.23	5.99	5.82	5.70	5.60	5.52
7		8.07	6.54	5.89	5.52	5.29	5.12	4.99	4.90	4.82
8		7.57	6.06	5.42	5.05	4.82	4.65	4.53	4.43	4.36
9		7.21	5.71	5.08	4.72	4.48	4.32	4.20	4.10	4.03
10		6.94	5.46	4.83	4.47	4.24	4.07	3.95	3.85	3.78
11		6.72	5.26	4.63	4.28	4.04	3.88	3.76	3.66	3.59
12		6.55	5.10	4.47	4.12	3.89	3.73	3.61	3.51	3.44
13		6.41	4.97	4.35	4.00	3.77	3.60	3.48	3.39	3.31
14		6.30	4.86	4.24	3.89	3.66	3.50	3.38	3.29	3.21
15		6.20	4.77	4.15	3.80	3.58	3.41	3.29	3.20	3.12
16		6.12	4.69	4.08	3.73	3.50	3.34	3.22	3.12	3.05
17		6.04	4.62	4.01	3.66	3.44	3.28	3.16	3.06	2.98
18		5.98	4.56	3.95	3.61	3.38	3.22	3.10	3.01	2.93
19		5.92	4.51	3.90	3.56	3.33	3.17	3.05	2.96	2.88
20		5.87	4.46	3.86	3.51	3.29	3.13	3.01	2.91	2.84
21		5.83	4.42	3.82	3.48	3.25	3.09	2.97	2.87	2.80
22		5.79	4.38	3.78	3.44	3.22	3.05	2.93	2.84	2.76
23		5.75	4.35	3.75	3.41	3.18	3.02	2.90	2.81	2.73
24		5.72	4.32	3.72	3.38	3.15	2.99	2.87	2.78	2.70
25		5.69	4.29	3.69	3.35	3.13	2.97	2.85	2.75	2.68
26		5.66	4.27	3.67	3.33	3.10	2.94	2.82	2.73	2.65
27		5.63	4.24	3.65	3.31	3.08	2.92	2.80	2.71	2.63
28		5.61	4.22	3.63	3.29	3.06	2.90	2.78	2.69	2.61
29		5.59	4.20	3.61	3.27	3.04	2.88	2.76	2.67	2.59
30		5.57	4.18	3.59	3.25	3.03	2.87	2.75	2.65	2.57
40		5.42	4.05	3.46	3.13	2.90	2.74	2.62	2.53	2.45
60		5.29	3.93	3.34	3.01	2.79	2.63	2.51	2.41	2.33
120		5.15	3.80	3.23	2.89	2.67	2.52	2.39	2.30	2.22
∞		5.02	3.69	3.12	2.79	2.57	2.41	2.29	2.19	2.11

Table 6

Critical Values of the F Distribution (Continued)

Denom- inator m	Numer- ator n	\multicolumn{10}{c}{$f_{0.025}(n, m)$}									
		10	12	15	20	24	30	40	60	120	∞
1		968.6	976.7	984.9	993.1	997.2	1001	1006	1010	1014	1018
2		39.40	39.41	39.43	39.45	39.46	39.46	39.47	39.48	39.49	39.50
3		14.42	14.34	14.25	14.17	14.12	14.08	14.04	13.99	13.95	13.90
4		8.84	8.75	8.66	8.56	8.51	8.46	8.41	8.36	8.31	8.26
5		6.62	6.52	6.43	6.33	6.28	6.23	6.18	6.12	6.07	6.02
6		5.46	5.37	5.27	5.17	5.12	5.07	5.01	4.96	4.90	4.85
7		4.76	4.67	4.57	4.47	4.42	4.36	4.31	4.25	4.20	4.14
8		4.30	4.20	4.10	4.00	3.95	3.89	3.84	3.78	3.73	3.67
9		3.96	3.87	3.77	3.67	3.61	3.56	3.51	3.45	3.39	3.33
10		3.72	3.62	3.52	3.42	3.37	3.31	3.26	3.20	3.14	3.08
11		3.53	3.43	3.33	3.23	3.17	3.12	3.06	3.00	2.94	2.88
12		3.37	3.28	3.18	3.07	3.02	2.96	2.91	2.85	2.79	2.72
13		3.25	3.15	3.05	2.95	2.89	2.84	2.78	2.72	2.66	2.60
14		3.15	3.05	2.95	2.84	2.79	2.73	2.67	2.61	2.55	2.49
15		3.06	2.96	2.86	2.76	2.70	2.64	2.59	2.52	2.46	2.40
16		2.99	2.89	2.79	2.68	2.63	2.57	2.51	2.45	2.38	2.32
17		2.92	2.82	2.72	2.62	2.56	2.50	2.44	2.38	2.32	2.25
18		2.87	2.77	2.67	2.56	2.50	2.44	2.38	2.32	2.26	2.19
19		2.82	2.72	2.62	2.51	2.45	2.39	2.33	2.27	2.20	2.13
20		2.77	2.68	2.57	2.46	2.41	2.35	2.29	2.22	2.16	2.09
21		2.73	2.64	2.53	2.42	2.37	2.31	2.25	2.18	2.11	2.04
22		2.70	2.60	2.50	2.39	2.33	2.27	2.21	2.14	2.08	2.00
23		2.67	2.57	2.47	2.36	2.30	2.24	2.18	2.11	2.04	1.97
24		2.64	2.54	2.44	2.33	2.27	2.21	2.15	2.08	2.01	1.94
25		2.61	2.51	2.41	2.30	2.24	2.18	2.12	2.05	1.98	1.91
26		2.59	2.49	2.39	2.28	2.22	2.16	2.09	2.03	1.95	1.88
27		2.57	2.47	2.36	2.25	2.19	2.13	2.07	2.00	1.93	1.85
28		2.55	2.45	2.34	2.23	2.17	2.11	2.05	1.98	1.91	1.83
29		2.53	2.43	2.32	2.21	2.15	2.09	2.03	1.96	1.89	1.81
30		2.51	2.41	2.31	2.20	2.14	2.07	2.01	1.94	1.87	1.79
40		2.39	2.29	2.18	2.07	2.01	1.94	1.88	1.80	1.72	1.64
60		2.27	2.17	2.06	1.94	1.88	1.82	1.74	1.67	1.58	1.48
120		2.16	2.05	1.94	1.82	1.76	1.69	1.61	1.53	1.43	1.31
∞		2.05	1.94	1.83	1.71	1.64	1.57	1.48	1.39	1.27	1.00

Table 7

Critical Values of D in the Kolmogorov–Smirnov Test with Parameters of F Known[a]

Sample size n	Level of a significance for $D = \max\lvert F(x) - S_n(x) \rvert$			
	$\alpha = 0.20$	$\alpha = 0.10$	$\alpha = 0.05$	$\alpha = 0.01$
1	0.9000	0.9500	0.9750	0.9950
2	0.6838	0.7764	0.8419	0.9293
3	0.5648	0.6360	0.7076	0.8290
4	0.4927	0.5652	0.6239	0.7342
5	0.4470	0.5095	0.5633	0.6685
6	0.4104	0.4680	0.5193	0.6166
7	0.3815	0.4361	0.4834	0.5758
8	0.3583	0.4096	0.4543	0.5418
9	0.3391	0.3875	0.4300	0.5133
10	0.3226	0.3687	0.4093	0.4889
11	0.3083	0.3524	0.3912	0.4677
12	0.2958	0.3382	0.3754	0.4491
13	0.2847	0.3255	0.3614	0.4325
14	0.2748	0.3142	0.3489	0.4176
15	0.2659	0.3040	0.3376	0.4042
16	0.2578	0.2947	0.3273	0.3920
17	0.2504	0.2863	0.3180	0.3809
18	0.2436	0.2785	0.3094	0.3706
19	0.2374	0.2714	0.3014	0.3612
20	0.2316	0.2647	0.2941	0.3524
25	0.2079	0.2377	0.2640	0.3166
30	0.1903	0.2176	0.2417	0.2899
35	0.1766	0.2019	0.2243	0.2690
Over 35	$1.07/\sqrt{n}$	$1.22/\sqrt{n}$	$1.36/\sqrt{n}$	$1.63/\sqrt{n}$

[a] Adapted from Table 1 of L. H. Miller, Table of percentage points of Kolmogorov statistics, *J. Amer. Statist. Assoc.* **51** (1956), 113, and Table 1 of F. J. Massey, Jr., The Kolmogorov–Smirnov test for goodness-of-fit, *J. Amer. Statist. Assoc.* **46** (1951), 70, with permission of the authors and publisher.

Table 8

Critical Values of D in the Kolmogorov–Smirnov Test for the Exponential Distribution with Mean Unknown[a]

Sample size n	Level of significance for $D = \max \lvert F(x) - S_n(x) \rvert$				
	$\alpha = 0.20$	$\alpha = 0.15$	$\alpha = 0.10$	$\alpha = 0.05$	$\alpha = 0.01$
3	0.451	0.479	0.511	0.551	0.600
4	0.396	0.422	0.449	0.487	0.548
5	0.359	0.382	0.406	0.442	0.504
6	0.331	0.351	0.375	0.408	0.470
7	0.309	0.327	0.350	0.382	0.442
8	0.291	0.308	0.329	0.360	0.419
9	0.277	0.291	0.311	0.341	0.399
10	0.263	0.277	0.295	0.325	0.380
11	0.251	0.264	0.283	0.311	0.365
12	0.241	0.254	0.271	0.298	0.351
13	0.232	0.245	0.261	0.287	0.338
14	0.224	0.237	0.252	0.277	0.326
15	0.217	0.229	0.244	0.269	0.315
16	0.211	0.222	0.236	0.261	0.306
17	0.204	0.215	0.229	0.253	0.297
18	0.199	0.210	0.223	0.246	0.289
19	0.193	0.204	0.218	0.239	0.283
20	0.188	0.199	0.212	0.234	0.278
25	0.170	0.180	0.191	0.210	0.247
30	0.155	0.164	0.174	0.192	0.226
Over 30	$0.86/\sqrt{n}$	$0.91/\sqrt{n}$	$0.96/\sqrt{n}$	$1.06/\sqrt{n}$	$1.25/\sqrt{n}$

[a] Adapted from Table 1 of H. W. Lilliefors, On the Kolmogorov–Smirnov test for the exponential with mean unknown, *J. Amer. Statist. Assoc.* **64** (1969), 388, with permission of the author and publisher.

Table 9

Critical Values of D in the Kolmogorov–Smirnov Test for Normality with Mean and Variance Unknown[a]

Sample size n	Level of significance for $D = \max \lvert \Phi(z) - z \rvert				
	$\alpha = 0.20$	$\alpha = 0.15$	$\alpha = 0.10$	$\alpha = 0.05$	$\alpha = 0.01$
4	0.300	0.319	0.352	0.381	0.417
5	0.285	0.299	0.315	0.337	0.405
6	0.265	0.277	0.294	0.319	0.364
7	0.247	0.258	0.276	0.300	0.348
8	0.233	0.244	0.261	0.285	0.331
9	0.223	0.233	0.249	0.271	0.311
10	0.215	0.224	0.239	0.258	0.294
11	0.206	0.217	0.230	0.249	0.284
12	0.199	0.212	0.223	0.242	0.275
13	0.190	0.202	0.214	0.234	0.268
14	0.183	0.194	0.207	0.227	0.261
15	0.177	0.187	0.201	0.220	0.257
16	0.173	0.182	0.195	0.213	0.250
17	0.169	0.177	0.189	0.206	0.245
18	0.166	0.173	0.184	0.200	0.239
19	0.163	0.169	0.179	0.195	0.235
20	0.160	0.166	0.174	0.190	0.231
25	0.142	0.147	0.158	0.173	0.200
30	0.131	0.136	0.144	0.161	0.187
Over 30	$0.736/\sqrt{n}$	$0.768/\sqrt{n}$	$0.805/\sqrt{n}$	$0.886/\sqrt{n}$	$1.031/\sqrt{n}$

[a] Adapted from Table 1 of H. W. Lilliefors, On the Kolmogorov–Smirnov test for normality with mean and variance unknown, *J. Amer. Statist. Assoc.* **62** (1967), 400, with permission of the author and publisher. Correction for $n = 25$ provided by Professor Lilliefors.

A.4   The Laplace–Stieltjes Transform

Table 10. Laplace Transform Properties and Identities[5]

	Function	Transform
1.	$f(t)$	$f^*[\theta] = \int_{0^-}^{\infty} e^{-\theta t} f(t)\, dt$
2.	$af(t) + bg(t)$	$af^*[\theta] + bg^*[\theta]$
3.	$f\left(\frac{t}{a}\right), \quad a > 0$	$af^*[a\theta]$
4.	$f(t - a)$ for $t \ge a$	$e^{-a\theta} f^*[\theta]$
5.	$e^{-at} f(t)$	$f^*[\theta + a]$
6.	$t f(t)$	$-\dfrac{df^*[\theta]}{d\theta}$
7.	$t^n f(t)$	$(-1)^n \dfrac{d^n f^*[\theta]}{d\theta^n}$
8.	$\int_0^t f(u) g(t - u)\, du$	$f^*[\theta] g^*[\theta]$
9.	$\dfrac{df(t)}{dt}$	$\theta\, f^*[\theta] - f(0)$
10.	$\dfrac{d^n f(t)}{dt^n}$	$\theta^n\, f^*[\theta] - \displaystyle\sum_{i=1}^{n} \theta^{n-i} f^{(i-1)}(0)$
11.	$\int_0^t f(x)\, dx$	$\dfrac{f^*[\theta]}{\theta}$
12.	$\dfrac{\partial f(t)}{\partial a} \quad a$ a parameter	$\dfrac{\partial f^*[\theta]}{\partial a}$

[5] All functions f are assumed to be piecewise continuous and of exponential order. That is, there exist positive constants M and a such that $|f(t)| \le M e^{at}$ for $t \ge 0$.

Table 11. Laplace Transform Pairs

	Function	Transform
1.	$f(t)$	$f^*[\theta] = \int_{0-}^{\infty} e^{-\theta t} f(t)\, dt$
2.	$f(t) = c$	$\dfrac{c}{\theta}$
3.	$t^n, \quad n = 1, 2, 3, \cdots$	$\dfrac{n!}{\theta^{n+1}}$
4.	$t^a, \quad a > 0$	$\dfrac{\Gamma(a+1)}{\theta^{a+1}}$
5.	e^{at}	$\dfrac{1}{\theta - a}, \quad \theta > a$
6.	te^{at}	$\dfrac{1}{(\theta - a)^2}, \quad \theta > a$
7.	$t^n e^{at}$	$\dfrac{n!}{(\theta - a)^{n+1}}, \quad \theta > a$
8.[6]	$\delta(t)$	1
9.	$\delta(t - a)$	$e^{-a\theta}$
10.[7]	$U(t - a)$	$\dfrac{e^{-a\theta}}{\theta}$
11.	$f(t - a)U(t - a)$	$e^{-a\theta} f^*[\theta]$

---

[6]The Dirac delta function $\delta(\cdot)$ is defined by $\delta(t) = 0$ for $t \neq 0$ but $\int_{a-\epsilon}^{a+\epsilon} \delta(t - a) f(t)\, dt = f(a)$ for each f and each $\epsilon > 0$.

[7]The unit step function $U(\cdot)$ is defined by

$$U(t - a) = \begin{cases} 0 & \text{for } t < a \\ 1 & \text{for } t \geq a. \end{cases}$$

A.4.1   Goodness-of-Fit Tables

Table 12. Upper Percentage Points for Kolmogorov-Smirnov and A^2 Tests with Parameters Known[8]

| Statistic | | Upper-tail significance level α | | | |
| | | Significance Level α | | | |
T	Modified T	0.10	0.05	0.025	0.01
$D^+(D^-)$	$D^+(\sqrt{n} + 0.12 + \frac{0.11}{\sqrt{n}})$	1.073	1.224	1.358	1.518
D	$D(\sqrt{n} + 0.12 + \frac{0.11}{\sqrt{n}})$	1.224	1.358	1.480	1.628
A^2	for all $n \geq 5$	1.933	2.492	3.070	3.857

Table 13. Upper Percentage Points for Kolmogorov-Smirnov and A^2 Exponentiality Tests, Mean Estimated[8]

| Statistic | | Upper-tail significance level α | | | |
| | | Significance Level α | | | |
T	Modified T	0.10	0.05	0.025	0.01
D	$(D - \frac{0.2}{n})(\sqrt{n} + 0.26 + \frac{0.5}{\sqrt{n}})$	0.995	1.094	1.184	1.298
A^2	$A^2(1.0 + \frac{0.6}{n})$	1.062	1.321	1.591	1.959

Table 14. Upper Percentage Points for Kolmogorov–Smirnov and A^2 Normality Tests, Parameters Estimated[8]

| Statistic | | Upper-tail significance level α | | | |
| | | Significance Level α | | | |
T	Modified T	0.10	0.05	0.025	0.01
D	$D(\sqrt{n} - 0.01 + \frac{0.85}{\sqrt{n}})$	0.819	0.895	0.995	1.035
A^2	$A^2(1.0 + \frac{0.75}{n} + \frac{2.25}{n^2})$	0.631	0.752	0.873	1.035

---

[8]Table 12 is adapted from Table 4.2, Table 13 from Table 4.11, and Table 14 from Table 4.7 of *Goodness-of-Fit Techniques* by R. B. D'Agostino and M. A. Stephens, Marcel Dekker, New York, 1986 with permission.

Appendix B

APL Programs

All the APL programs mentioned in the book are listed here. The programs were written using the APL★PLUS system from STSC, Inc. As in the first edition of this book the programs are "naive" in the sense that with a few exceptions, no attempt has been made to optimize the efficiency of execution. The queueing theory programs are, for the most part, direct translations of the equations of Appendix C. No attempt has been made to organize the calculations to minimize round-off error or to avoid numerical instabilities. Therefore, no guarantee can be made that a given program will produce correct results. I have tested the programs extensively, however, and know of no errors in the programs.

There are a number of excellent APL textbooks so I will not attempt to introduce the language here. My favorite introduction is the excellent book by Jerry R. Turner [5]. Ramsey and Musgrave [4] provide an elementary introduction to statistical computing in APL. The book by Gilman and Rose [1] is a classic and is current with PC versions of APL as well as APL2, the latest IBM version of APL. The short handbook by Dave Macklin [3] provides some useful public domain utility functions. It has a very practical format, so that one can see exactly how an included function works. I found Lee's *Quick Reference Guide* [2] indispensable when writing APL★PLUS code.

Cited References

[1] Leonard Gilman and Allen J. Rose, *APL: An Interactive Approach*, 3rd ed., John Wiley, New York, 1984.

[2] Christopher H. Lee, *Quick Reference Guide APL★PLUS System for the*

PC, STSC, Rockville, MD, 1988.

[3] Dave Macklin, *The APL Handbook of Techniques*, IBM, White Plains, NY, 1978.

[4] James B. Ramsey and Gerald L. Musgrave, *APL-STAT: A Do-It-Yourself Guide to Computational Statistics Using APL*, Lifetime Learning Publications, Wadsworth, Inc. Belmont, CA, 1981.

[5] Jerry R. Turner, *APL Is Easy!*, John Wiley, New York, 1987.

B.1   APL Programs

```
    ∇ ALPHA A∆2∆SPEC Z
[1]    ⍝THIS FUNCTION PERFORMS THE ANDERSON-DARLING TEST
[2]    ⍝FOR SPECIFIED DISTRIBUTION FUNCTION AS DEFINED IN
[3]    ⍝MY ALGORITHM 8.6.1.
[4]    A∆2←((-÷N)×+/(((2×I←⍳N←⍴,Z)-1)×((⍟Z)+⍟(1-ΦZ))))-N
[5]    'THE VALUE OF THE ANDERSON-DARLING STATISTIC A-SQUARED IS'
[6]    ⍕A∆2
[7]    →((ALPHA=0.01),(ALPHA=0.025),(ALPHA=0.05))/S1,S2,S3
[8]    →TEST
[9]    S1:→(A∆2>3.857)↑END
[10]   □←'NULL HYPOTHESIS APPEARS TO BE TRUE FOR ALPHA= ',⍕ALPHA
[11]   →0
[12]   S2:→(A∆2>3.07)↑END
[13]   □←'NULL HYPOTHESIS APPEARS TO BE TRUE FOR ALPHA= ',⍕ALPHA
[14]   →0
[15]   S3:→(A∆2>2.492)↑END
[16]   □←'NULL HYPOTHESIS APPEARS TO BE TRUE FOR ALPHA= ',⍕ALPHA
[17]   →0
[18]   TEST:
[19]   →((ALPHA=0.1),(ALPHA=0.15))/S4,S5
[20]   □←'ALPHA IS NOT ONE OF THE ALLOWED VALUES'
[21]   →0
[22]   S4:→(A∆2>1.933)↑END
[23]   □←'NULL HYPOTHESIS APPEARS TO BE TRUE FOR ALPHA= ',⍕ALPHA
[24]   →0
[25]   S5:→(A∆2>1.61)↑END
[26]   □←'NULL HYPOTHESIS APPEARS TO BE TRUE FOR ALPHA= ',⍕ALPHA
[27]   →0
[28]   END:□←'THE NULL HYPOTHESIS IS REJECTED FOR ALPHA= ',⍕ALPHA
    ∇


    ∇ Z←C BBCA A;B;N
[1]    ⍝BBCA CALCULATES ERLANG'S B FORMULA WHERE
[2]    ⍝C IS THE NUMBER OF SERVERS AND A IS THE
[3]    ⍝TRAFFIC INTENSITY OR OFFERED LOAD
[4]    Z←A÷1+A
[5]    N←1
[6]    START:→0×⍳C<N←N+1
[7]    →START,Z←B÷N+B←A×Z
    ∇
```

B.2   APL Programs

```
      ∇ N∆T BCMP D
[1]   ⍝THIS PROGRAM EXECUTES ALGORITHM 6.3.1
[2]   ⍝THE CALL IS N∆T BCMP D WHERE N∆T IS THE CATENATION
[3]   ⍝OF THE NUMBER OF TERMINALS WITH THE MEAN THINK TIME AND
[4]   ⍝D IS THE VECTOR OF SERVICE DEMAND TIMES, STARTING WITH
[5]   ⍝THAT FOR THE CPU.
[6]   N←N∆T[1]
[7]   T←N∆T[2]
[8]   K←⍴D
[9]   L←K⍴0
[10]  I←0
[11]  LOOP:→PERF×⍳N<I←I+1
[12]  W←D×(1+L)
[13]  WN←+/W
[14]  LAMBDA←I÷WN+T
[15]  L←LAMBDA×W
[16]  →LOOP
[17]  PERF:'THE THROUGHPUT IS ',⍕LAMBDA
[18]  'THE RESPONSE TIME IS ',⍕WN
[19]  'THE SERVER UTILIZATIONS ARE '
[20]  ⍕LAMBDA×D
[21]  'THE NUMBER AT EACH SERVER IS  '
[22]  ⍕L
[23]  'THE RESPONSE TIME AT EACH SERVER IS '
[24]  ⍕W
[25]  'THE AVERAGE NUMBER OF CUSTOMERS IN COMPUTER SYSTEM'
[26]  'IS  ',⍕+/L
      ∇


      ∇ Z←C2 BH2 EX;Q1;Q2;MU1;MU2
[1]   ⍝BH2 CALCULATES THE PARAMETERS FOR A TWO STAGE HYPEREXPONENTIAL
[2]   ⍝RANDOM VARIABLE WITH BALANCED MEANS THAT HAS A GIVEN SQUARED
[3]   ⍝COEFFICIENT OF VARIATION, C2, AND MEAN, EX.  THE CALLING
[4]   ⍝SEQUENCE IS 'C2 BH2 EX'.  THE OUTPUT IS THE VECTOR Q1∆Q2∆MU1∆MU2
[5]   ⍝NEEDED AS THE LEFT PARAMETER BY THE TWO STAGE HYPEREXPONENTIAL
[6]   ⍝DISTRIBUTION FUNCTION H2∆DIST.
[7]   Q1←0.5×(1-(((C2-1)÷(C2+1))*0.5))
[8]   Q2←1-Q1
[9]   MU1←2×Q1×÷EX
[10]  MU2←2×Q2×÷EX
[11]  Z←Q1,Q2,MU1,MU2
      ∇


      ∇ Z←NP BINOMIAL K;X;Y;N;P;Q
[1]   ⍝BINOMIAL COMPUTES THE PROBABILITY THAT A BINOMIAL RANDOM VARIABLE
[2]   ⍝WITH PARAMETERS N AND P ASSUMES THE VALUE K.  NP IS THE VECTOR
[3]   ⍝FORMED BY CATENATING N AND P.
[4]   N←NP[1]
[5]   P←NP[2]
[6]   Q←1-P
[7]   Z←*(((K×(⍟P))-(⁻1↑X←+\⍟⍳K))+(((N-K)×(⍟Q))-(⁻1↑Y←+\⍟⍳(N-K)))+(+/⍟⍳N))
      ∇
```

B.3 APL Programs

```
    ∇ Z←I BINSUM P∆N
[1]   ⍝THIS FUNCTION IS ESSENTIALLY A TRANSLITERATION OF THE FUNCTION
[2]   ⍝PROGRAM 3-1 OF SHELDON ROSS'S BOOK ''INTRODUCTION TO PROBABILITY
[3]   ⍝AND STATISTICS FOR ENGINEERS AND  SCIENTISTS''.  IF X IS A BINOMIAL
[4]   ⍝RANDOM VARIABLE WITH PARAMETERS N AND P, AND I IS A NUMBER BETWEEN
[5]   ⍝0 AND N, BINSUM WILL CALCULATE THE PROBABILITY THAT X≤I.  THE
[6]   ⍝CALL IS 'I BINSUM P,N '.  THIS FUNCTION USES THE FUNCTION IF.
[7]   ⍝LAST CHANGE 1/31/88
[8]   P←1↑P∆N
[9]   N←¯1↑P∆N
[10]  S←(1-P)⋆N
[11]  →L1 IF S=0
[12]  A←P÷1-P
[13]  T←S
[14]  →END IF I=0
[15]  Z←((N×(Iρ1))-((⍳I)-1))÷⍳I
[16]  Z←A×Z
[17]  Z[1]←S×Z[1]
[18]  Z←+/×\Z
[19]  T←T+Z
[20]  Z←T
[21]  →END
[22]  L1:J←I
[23]  →JUMP IF(N×P)≥J
[24]  J←⌊N×P
[25]  JUMP:U1←⍟(((N+1)×((J)ρ1))-(⍳J))
[26]  U2←⍟⊕⌽⍳J
[27]  L←+/(U1-U2)
[28]  L←L+(J×⍟P)+(N-J)×⍟1-P
[29]  L←⋆L
[30]  B←(1-P)÷P
[31]  T←+/×\((B×(⌽⍳J))÷((N-J)+(⍳J)))
[32]  →ABEND IF J=I
[33]  C←÷B
[34]  T←T++/×\((C×((-(J+⍳(I-J)))+(N+1)))÷(J+⍳(I-J)))
[35]  ABEND:Z←(T+1)×L
[36]  END:→0
    ∇


    ∇ Z←BRANCH X
[1]   ⍝THIS FUNCTION DEMONSTRATES HOW THE FUNCTION ''IF'' WORKS.
[2]   ⍝IF WAS BORROWED FROM THE APL HANDBOOK OF TECHNIQUES.
[3]   →L1 IF X=0
[4]   →L2 IF X≠0
[5]   L1:Z←0
[6]   →0
[7]   L2:Z←1
    ∇
```

B.4 APL Programs

```
      ∇ K∆N CENT D
[1]   K←K∆N[1]
[2]   N←K∆N[2]
[3]   L←0×ιK
[4]   M←0
[5]   START:→END×ι(N<M←M+1)
[6]   W←D×1÷L
[7]   WN←+/W
[8]   LAMBDAN←M÷WN
[9]   L←LAMBDAN×W
[10]  →START
[11] END:'THE COMPUTER RESPONSE TIME IS    ',⍕WN
[12]  'THE THROUGHPUT IS  ',⍕LAMBDAN
[13]  'THE UTILIZATION OF EACH  DEVICE IS '
[14]  LAMBDAN×D
[15]  'THE NUMBER OF CUSTOMERS AT EACH DEVICE IS'
[16]  L
      ∇


      ∇ K∆N CENTP P∆S;K;D
[1]   ACENTP CALCULATES THE STATISTICS FOR THE
[2]   ACENTRAL SERVER MODEL.  IT USES THE FUNCTION
[3]   AFIX TO PROVIDE THE PARAMETER NEEDED FOR THE
[4]   APROGRAM CENT WHICH MAKES THE CALCULATIONS FOR
[5]   ATHE MODEL USING THE EXACT MVA ALGORITHM
[6]   K←1↑K∆N
[7]   D←K FIX P∆S
[8]   K∆N CENT D
      ∇


      ∇ Z←T CHISQUARE∆DIST N;BETA;ALPHA
[1]   ACHISQUARE∆DIST IS THE DISTRIBUTION FUNCTION FOR A CHISQUARE
[2]   ARANDOM VARIABLE WITH N DEGREES OF FREEDOM.
[3]   BETA←N÷2
[4]   ALPHA←0.5
[5]   BETA∆ALPHA←BETA,ALPHA
[6]   Z←T G∆DIST BETA∆ALPHA
      ∇


      ∇ ALPHA CHISQ∆EXPON Y;X
[1]   ATHIS FUNCTION PERFORMS THE CHISQUARE TEST ON THE SAMPLE, Y, TO
[2]   ADETERMINE, AT THE ALPHA LEVEL OF SIGNIFICANCE, WHETHER OR NOT
[3]   AY IS A RANDOM SAMPLE FROM AN EXPONENTIAL POPULATION.
[4]   X←SORT Y
[5]   XBAR←(+/X)÷N←ρ,X
[6]   ACALCULATE QUARTILES
[7]   X25←-XBAR×⍟0.75
[8]   X50←-XBAR×⍟0.5
[9]   X75←-XBAR×⍟0.25
[10]  O←(+/(X<X25)),(+/((X25≤X)∧(X<X50))),(+/((X50≤X)∧(X<X75))),+/(X75≤X)
[11]  E←4ρN÷4
[12]  CHISQ←+/((O-E)*2)÷E
[13]  Q←CHISQ CHISQUARE∆DIST 2
[14]  P←1-Q
[15]  'THE P VALUE IS ',⍕P
[16]  →(P<ALPHA)↑END
[17]  'THE POPULATION APPEARS TO BE EXPONENTIAL FOR ALPHA= ',⍕ALPHA
[18]  →0
[19] END:'THE NULL HYPOTHESIS MUST BE REJECTED FOR ALPHA= ',⍕ALPHA
      ∇
```

B.5 APL Programs

```
      ∇ CHIΔN Z
[1]    N←ρ,Z
[2]    Z1←IN 0.2
[3]    Z2←IN 0.4
[4]    Z3←IN 0.6
[5]    Z4←IN 0.8
[6]    E←5ρ(N÷5)
[7]    O←ι5
[8]    O[1]←+/(Z<Z1)
[9]    O[2]←+/(Z1≤Z)∧(Z<Z2)
[10]   O[3]←+/(Z2≤Z)∧(Z<Z3)
[11]   O[4]←+/(Z3≤Z)∧(Z<Z4)
[12]   O[5]←+/(Z4≤Z)
[13]   CHISQ←+/((((O-E)*2)÷(N÷5))
[14]   PVAL←1-CHISQ CHISQUAREΔDIST 4
[15]   'THE VALUE OF CHISQUARE IS ',⍕CHISQ
[16]   'THE P-VALUE OF THE TEST IS ',⍕PVAL
      ∇
```

```
      ∇ P CONDEXPECT EXΔYEX2ΔY
[1]    ⍝CONDEXPECT CALCULATES THE MEAN ,EX, THE SECOND MOMENT, EX2,
[2]    ⍝THE VARIANCE , VARX, AND THE STANDARD DEVIATION ,SIGX,
[3]    ⍝WHERE THE FIRST AND SECOND MOMENTS OF X GIVEN Y ARE KNOWN.
[4]    ⍝TO CALL TYPE ' P CONDEXPECT EXΔYEX2ΔY ' WHERE P IS THE SET
[5]    ⍝OF VALUES OF THE PROBABILITY MASS FUNCTION OF Y AND
[6]    ⍝EXΔYEX2ΔY IS THE VECTOR OF CONDITIONAL EXPECTATIONS OF
[7]    ⍝X GIVEN Y CATENATED WITH THE VECTOR OF THE SECOND CON -
[8]    ⍝DITIONAL MOMENTS OF X GIVEN Y.
[9]    N←ρP
[10]   EX←+/P×N↑EXΔYEX2ΔY
[11]   EX2←+/P×N↓EXΔYEX2ΔY
[12]   VARX←EX2-EX*2
[13]   SIGX←VARX*0.5
[14]   'THE MEAN OF X IS  ',⍕EX
[15]   'THE SECOND MOMENT OF X  IS  ',⍕EX2
[16]   'THE VARIANCE OF X IS  ',⍕VARX
[17]   'THE STANDARD DEVIATION IS  ',⍕SIGX
      ∇
```

```
      ∇ Z←C CΔCΔA A;RHO;B
[1]    B←C WBCA A
[2]    RHO←A÷C
[3]    Z←B÷((RHO×B)+(1-RHO))
      ∇
```

```
      ∇ LAMBDA DMΔMΔ1ΔK ESΔKΔT;ES;K;T;A;P0;P;Q
[1]    ⍝THIS FUNCTION COMPUTES THE DISTRIBUTION FUNCTION
[2]    ⍝FOR BOTH QUEUEING TIME AND SYSTEM TIME FOR
[3]    ⍝THE M/M/1/K QUEUEING SYSTEM.  THE CALL IS
[4]    ⍝LAMBDA DMΔMΔ1ΔK ES, K, T.  THE FUNCTION
[5]    ⍝PARTIALΔSUM IS USED BY THIS FUNCTION.
[6]    ES←ESΔKΔT[1]
[7]    K←ESΔKΔT[2]
[8]    T←ESΔKΔT[3]
[9]    A←LAMBDA×ES
[10]   P0←(1-A)÷1-A*K+1
[11]   P←P0×A*(⁻1+ιK+1)
[12]   Q←P[ιK]÷(1-P[K+1])
[13]   MUΔT←T÷ES
[14]   WT←1-+/Q×(MUΔT PARTIALΔSUM K-1)
[15]   WQT←1-+/(1↓Q)×(MUΔT PARTIALΔSUM K-2)
[16]   'THE PROBABILITY THE WAITING TIME IN THE SYSTEM,'
[17]   'IS NOT GREATER THAN T, WT, IS  ',⍕WT
[18]   'THE PROBABILITY THE QUEUEING TIME, Q, '
[19]   'IS NOT GREATER THAN T, WQT, IS  ',⍕WQT
      ∇
```

B.6   APL Programs

```
    ∇ Z←K DWQΔMRΔ1 EOΔESΔT;EO;ES;T;NUMERATOR;DENOMINATOR
[1]  ⍝THIS FUNCTION IS THE DISTRIBUTION FUNCTION FOR THE
[2]  ⍝QUEUEING TIME, WQ, FOR THE MACHINE REPAIR MODEL WITH
[3]  ⍝ONE REPAIRMAN.  THE CALL IS K DWQΔMRΔ1 EO, ES, T.  IN THE
[4]  ⍝CALL K IS NUMBER OF MACHINES, EO THE AVERAGE UP TIME
[5]  ⍝FOR A MACHINE, AND T IS THE TIME.  THUS
[6]  ⍝20 DWQΔMRΔ1 80,2,5 YIELDS 0.8931653946.
[7]    EO←EOΔESΔT[1]
[8]    ES←EOΔESΔT[2]
[9]    T←EOΔESΔT[3]
[10]   NUMERATOR←((EO+T)÷ES)POISSONΔDIST K-2
[11]   DENOMINATOR←(EO÷ES)POISSONΔDIST K-1
[12]   Z←1-NUMERATOR÷DENOMINATOR
    ∇
```

```
    ∇ Z←KΔC DWQΔMRΔC EOΔESΔT;K;C;EO;ES;T;P0;Q1;P
[1]  ⍝THIS IS THE DISTRIBUTION FUNCTION FOR THE QUEUEING TIME
[2]  ⍝OF A MACHINE  IN A MACHINE REPAIR MODEL WITH K MACHINES
[3]  ⍝AND C REPAIRMEN.
[4]    K←KΔC[1]
[5]    C←KΔC[2]
[6]    EO←EOΔESΔT[1]
[7]    ES←EOΔESΔT[2]
[8]    T←EOΔESΔT[3]
[9]    P0←(K,C)PO(EO,ES)
[10]   Q1←(C×(EO+T)÷ES)POISSONΔDIST K-C+1
[11]   P←(C×EO÷ES)POISSON K-1
[12]   Z←1-(C⋆C)×P0×Q1÷(!C)×P
    ∇
```

```
    ∇ Z←C DWQΔMΔMΔC AΔESΔT;A;ES;T
[1]  ⍝THIS THIS IS THE DISTRIBUTION FUNCTION FOR QUEUEING
[2]  ⍝TIME FOR THE M/M/C QUEUEING SYSTEM.  THE CALL IS
[3]  ⍝A DWQΔMΔMΔC A,ES,T.
[4]    A←AΔESΔT[1]
[5]    ES←AΔESΔT[2]
[6]    T←AΔESΔT[3]
[7]    CCA←C CΔCΔA A
[8]    Z←1-CCA×⋆-(C-A)×T÷ES
    ∇
```

```
    ∇ Z←K DWΔMRΔ1 EOΔESΔT;EO;ES;T;NUMERATOR;DENOMINATOR
[1]  ⍝THIS FUNCTION IS THE DISTRIBUTION FUNCTION FOR THE
[2]  ⍝SYSTEM TIME, W, FOR THE MACHINE REPAIR MODEL WITH
[3]  ⍝ONE REPAIRMAN.  THE CALL IS K DWΔMRΔ1 EO, ES, T.
[4]  ⍝K IS THE NUMBER OF MACHINES, EO THE AVERAGE UP TIME
[5]  ⍝FOR A MACHINE, AND T IS THE TIME.  THUS
[6]  ⍝20 DWΔMRΔ1 80,2,5 YIELDS 0.751742371.
[7]    EO←EOΔESΔT[1]
[8]    ES←EOΔESΔT[2]
[9]    T←EOΔESΔT[3]
[10]   NUMERATOR←((EO+T)÷ES)POISSONΔDIST K-1
[11]   DENOMINATOR←(EO÷ES)POISSONΔDIST K-1
[12]   Z←1-NUMERATOR÷DENOMINATOR
    ∇
```

B.7 APL Programs

```
    ∇ Z←K∆C DW∆MR∆C EO∆ES∆T;K;C;EO;ES;T;P0;Q1;Q2;P;C1;C2
[1]   ⍝THIS IS THE DISTRIBUTION FUNCTION FOR THE TOTAL TIME A
[2]   ⍝,MACHINE IS DOWN IN A MACHINE REPAIR MODEL WITH K MACHINES
[3]   ⍝AND C REPAIRMEN.
[4]   K←K∆C[1]
[5]   C←K∆C[2]
[6]   EO←EO∆ES∆T[1]
[7]   ES←EO∆ES∆T[2]
[8]   T←EO∆ES∆T[3]
[9]   P0←(K,C)PO(EO,ES)
[10]  Q1←(C×(EO+T)÷ES)POISSON∆DIST K-C+1
[11]  Q2←(C×EO÷ES)POISSON∆DIST K-C+1
[12]  P←(C×EO÷ES)POISSON K-1
[13]  C2←(C*C)×P0÷(!C)×(C-1)×(!K-C+1)×P
[14]  C1←1+(C2×Q2)
[15]  Z←(1-C1×*-T÷ES)+C2×Q1
    ∇


    ∇ Z←C DW∆M∆M∆C A∆ES∆T;CCA;A;ES;T;C1;C2
[1]   ⍝THIS IS THE DISTRIBUTION FUNCTION FOR TIME IN THE
[2]   ⍝SYSTEM, W, FOR AN M/M/C QUEUEING SYSTEM.  THE CALL
[3]   ⍝IS C DW∆M∆M∆C A,ES,T.
[4]   A←A∆ES∆T[1]
[5]   ES←A∆ES∆T[2]
[6]   T←A∆ES∆T[3]
[7]   CCA←C C∆C∆A A
[8]   RHO←A÷C
[9]   →EXCEPT×⍳A=C-1
[10]  C1←(A+1-(C+CCA))÷C-(A+1)
[11]  C2←CCA÷A+1-C
[12]  Z←1+(C1×*(-T÷ES))-C2×*(-C×T×(1-RHO)÷ES)
[13]  →0
[14] EXCEPT:Z←1-(1+CCA×T÷ES)×*(-T÷ES)
    ∇


    ∇ Z←T ERLANG K∆EX;Y
[1]   ⍝FINDS THE PROBABILITY THAT AN ERLANG-K RANDOM VARIABLE X WITH MEAN EX
[2]   ⍝ASSUMES A VALUE LESS THAN OR EQUAL TO T; THAT IS, IT IS THE
[3]   ⍝DISTRIBUTION FUNCTION OF X.
[4]   Y←K∆EX[1],K∆EX[1]÷K∆EX[2]
[5]   Z←T G∆DIST Y
    ∇
```

B.8   APL Programs

```
     ∇ ALPHA EXPONΔT X
[1]    ⍝THIS FUNCTION CALCULATES THE EDF STATISTICS D, WΔSQUARED, UΔSQUARED,
[2]    ⍝AND AΔSQUARED FOR TESTING FOR EXPONENTIALITY AS DESCRIBED BY
[3]    ⍝SPINELLI AND STEPHENS, TECHNOMETRICS, NOVEMBER 1987.
[4]    ⍝THE FUNCTION ALSO TESTS THE MODIFIED ANDERSON+DARLING (AΔSQUARED)
[5]    ⍝AT SEVERAL LEVELS OF SIGNIFICANCE USING THE TABLE ON PAGE 135
[6]    ⍝OF THE STEPHEN'S BOOK.  WE HAVE MODIFIED THE CACLCULATION HERE
[7]    ⍝FOR THE NONSHIFTED EXPONENTIAL ONLY AS DESCRIBED BY STEPHENS IN
[8]    ⍝HIS BOOK ON PAGES 134 AND 101
[9]    X←X[⍋X]
[10]   XBAR←(+/X)÷(N←⍴,X)
[11]   'XBAR IS   ',⍕XBAR
[12]   W←X÷XBAR
[13]   Z←1-*-W
[14]   I←⍳N
[15]   DPLUS←⌈/((I÷N)-Z)
[16]   DMINUS←⌈/(Z-(I-1)÷N)
[17]   D←DPLUS⌈DMINUS
[18]   V←DPLUS+DMINUS
[19]   'THE VALUE OF THE KOLMOGOROV-SMIRNOV STATISTIC, D, IS ',⍕D
[20]   MD←(D-0.2÷N)×((SN)+0.26+0.5÷SN←N*0.5)
[21]   'THE MODIFIED VALUE OF D IS   ',⍕MD
[22]   MV←(V-0.2÷SN)×(SN+0.24+0.35÷SN)
[23]   'THE VALUE OF V IS   ',⍕V
[24]   'THE MODIFIED VALUE OF V IS   ',⍕MV
[25]   WΔ2←+/(Z-(((2×I)-1)÷2×N))*2
[26]   WΔ2←WΔ2+÷12×N
[27]   'THE VALUE OF THE CRAMER-VON MISES STATISTIC, WΔSQUARED, IS ',⍕WΔ2
[28]   ZBAR←(+/Z)÷N
[29]   UΔ2←WΔ2-N×((ZBAR-0.5)*2)
[30]   'THE VALUE OF THE WATSON STATISTIC, UΔSQUARED, IS ',⍕UΔ2
[31]   AΔ2←((-÷N)×+/(((2×I)-1)×((⍟Z)+⍟(1-ΦZ))))-N
[32]   'THE VALUE OF THE ANDERSON-DARLING STATISTIC, AΔSQUARED, IS   ',⍕AΔ2
[33]   MΔAΔ2←AΔ2×(1+(0.6÷N))
[34]   'THE VALUE OF THE MODIFIED ANDERSON-DARLING STATISTIC IS   ',⍕MΔAΔ2
[35]   →((ALPHA=0.01),(ALPHA=0.025),(ALPHA=0.05))/S1,S2,S3
[36]   →TEST
[37]   S1:→(MΔAΔ2>1.959)↑END
[38]     ⎕←'POPULATION APPEARS TO BE EXPONENTIAL FOR  ALPHA= ',⍕ALPHA
[39]   →0
[40]   S2:→(MΔAΔ2>1.591)↑END
[41]     ⎕←'POPULATION APPEARS TO BE EXPONENTIAL FOR  ALPHA= ',⍕ALPHA
[42]   →0
[43]   S3:→(MΔAΔ2>1.321)↑END
[44]     ⎕←'POPULATION APPEARS TO BE EXPONENTIAL FOR  ALPHA= ',⍕ALPHA
[45]   →0
[46]   TEST:
[47]   →((ALPHA=0.1),(ALPHA=0.15),(ALPHA=0.25))/S4,S5,S6
[48]     ⎕←'ALPHA IS NOT ONE OF THE ALLOWED VALUES'
[49]   →0
[50]   S4:→(MΔAΔ2>1.062)↑END
[51]     ⎕←'POPULATION APPEARS TO BE EXPONENTIAL FOR  ALPHA= ',⍕ALPHA
[52]   →0
[53]   S5:→(MΔAΔ2>0.916)↑END
[54]     ⎕←'POPULATION APPEARS TO BE EXPONENTIAL FOR  ALPHA= ',⍕ALPHA
[55]   →0
[56]   S6:→(MΔAΔ2>0.736)↑END
[57]     ⎕←'POPULATION APPEARS TO BE EXPONENTIAL FOR  ALPHA= ',⍕ALPHA
[58]   →0
[59]   END:⎕←'THE POPULATION IS NOT EXPONENTIAL FOR ALPHA= ',⍕ALPHA
```

B.9 APL Programs

```
      ∇ ALPHA EXPON∆TEST X
[1]   ⍝THIS FUNCTION CALCULATES THE EDF STATISTICS D, W∆SQUARED, U∆SQUARED,
[2]   ⍝AND A∆SQUARED FOR TESTING FOR EXPONENTIALITY AS DESCRIBED BY
[3]   ⍝SPINELLI AND STEPHENS, TECHNOMETRICS, NOVEMBER 1987.
[4]   ⍝THE FUNCTION ALSO TESTS THE MODIFIED ANDERSON+DARLING (A∆SQUARED)
[5]   ⍝AT SEVERAL LEVELS OF SIGNIFICANCE USING THE TABLE ON PAGE 473
[6]   ⍝OF THE REFERENCED PAPER.
[7]   X←SORT X
[8]   XBAR←(+/X)÷N←ρ,X
[9]   'XBAR IS  ',⍕XBAR
[10]  BHAT←N×(XBAR-X[1])÷N-1
[11]  'BHAT IS  ',⍕BHAT
[12]  AHAT←X[1]-BHAT÷N
[13]  'AHAT IS  ',⍕AHAT
[14]  W←(X-AHAT)÷BHAT
[15]  Z←1-*-W
[16]  I←⍳N
[17]  DPLUS←⌈/((I÷N)-Z)
[18]  DMINUS←⌈/(Z-(I-1)÷N)
[19]  D←DPLUS⌈DMINUS
[20]  V←DPLUS+DMINUS
[21]  'THE VALUE OF THE KOLMOGOROV-SMIRNOV STATISTIC, D, IS ',⍕D
[22]  'THE VALUE OF SQUARE-ROOT OF N TIMES D IS ',⍕(N*0.5)×D
[23]  'THE VALUE OF SQUARE-ROOT OF N TIMES V IS ',⍕(N*0.5)×V
[24]  W∆2←(+/(Z-(((2×I)-1)÷2×N))*2)+÷12×N
[25]  'THE VALUE OF THE CRAMER-VON MISES STATISTIC, W∆SQUARED, IS ',⍕W∆2
[26]  ZBAR←(+/Z)÷N
[27]  U∆2←W∆2-N×((ZBAR-0.5)*2)
[28]  'THE VALUE OF THE WATSON STATISTIC, U∆SQUARED, IS ',⍕U∆2
[29]  A∆2←((-÷N)×+/(((2×I)-1)×((⍟Z)+⍟(1-ΦZ))))-N
[30]  'THE VALUE OF THE ANDERSON-DARLING STATISTIC, A∆SQUARED, IS  ',⍕A∆2
[31]  M∆A∆2←A∆2×(1+(5.4÷N)-(11÷N×N))
[32]  'THE VALUE OF THE MODIFIED ANDERSON-DARLING STATISTIC IS  ',⍕M∆A∆2
[33]  →((ALPHA=0.01),(ALPHA=0.025),(ALPHA=0.05))/S1,S2,S3
[34]  →TEST
[35]  S1:→(M∆A∆2>1.959)↑END
[36]  □←'POPULATION APPEARS TO BE EXPONENTIAL FOR  ALPHA= ',⍕ALPHA
[37]  →0
[38]  S2:→(M∆A∆2>1.591)↑END
[39]  □←'POPULATION APPEARS TO BE EXPONENTIAL FOR  ALPHA= ',⍕ALPHA
[40]  →0
[41]  S3:→(M∆A∆2>1.321)↑END
[42]  □←'POPULATION APPEARS TO BE EXPONENTIAL FOR  ALPHA= ',⍕ALPHA
[43]  →0
[44]  TEST:
[45]  →((ALPHA=0.1),(ALPHA=0.15),(ALPHA=0.25))/S4,S5,S6
[46]  □←'ALPHA IS NOT ONE OF THE ALLOWED VALUES'
[47]  →0
[48]  S4:→(M∆A∆2>1.062)↑END
[49]  □←'POPULATION APPEARS TO BE EXPONENTIAL FOR  ALPHA= ',⍕ALPHA
[50]  →0
[51]  S5:→(M∆A∆2>0.916)↑END
[52]  □←'POPULATION APPEARS TO BE EXPONENTIAL FOR  ALPHA= ',⍕ALPHA
[53]  →0
[54]  S6:→(M∆A∆2>0.736)↑END
[55]  □←'POPULATION APPEARS TO BE EXPONENTIAL FOR  ALPHA= ',⍕ALPHA
[56]  →0
[57]  END:□←'THE POPULATION IS NOT EXPONENTIAL FOR ALPHA= ',⍕ALPHA
      ∇
```

B.10   APL Programs

```
     ∇ D←K FIX P∆S;P;S;VCPU;TRUNCP;VEND;V
[1]   ⍝THIS FUNCTION IS NEEDED BY THE PROGRAM CENTP
[2]   ⍝TO SET UP THE PARAMETERS FOR THE FUNCTION CENT
[3]      P←K↑P∆S
[4]      S←K↓P∆S
[5]      VCPU←÷P[1]
[6]      TRUNCP←1↓P
[7]      VEND←TRUNCP×VCPU
[8]      V←VCPU,VEND
[9]      D←V×S
     ∇
```

```
     ∇ Z←C2 GH2 EX;Q1;Q2;MU1;MU2
[1]   ⍝GH2 CALCULATES THE PARAMETERS Q1, Q2, MU1, AND MU2 FOR THE TWO
[2]   ⍝STAGE HYPEREXPONENTIAL DISTRIBUTION WITH THE GAMMA NORMALIZATION.
[3]   ⍝THE CALLING SEQUENCE IS 'C2 GH2 EX' WHERE C2 IS THE DESIRED SQUARED
[4]   ⍝COEFFICIENT OF VARIATION AND EX IS THE DESIRED MEAN.
[5]   ⍝THE OUTPUT IS THE VECTOR Q1∆Q2∆MU1∆MU2 THAT IS NEEDED AS THE LEFT
[6]   ⍝PARAMETER OF THE TWO STAGE HYPEREXPONENTIAL DISTRIBUTION FUNCTION
[7]   ⍝H2∆DIST
[8]      MU1←(2÷EX)×1+((C2-0.5)÷(C2+1))*0.5
[9]      MU2←(4÷EX)-MU1
[10]     Q1←MU1×((MU2×EX)-1)÷MU2-MU1
[11]     Q2←1-Q1
[12]     Z←Q1,Q2,MU1,MU2
     ∇
```

```
     ∇ PI0 GI∆M∆1 LAMBDA∆ES
[1]   ⍝THIS FUNCTION COMPUTES THE PERFORMANCE STATISTICS
[2]   ⍝FOR THE GI/M/1 QUEUEING SYSTEM. THE CALL IS
[3]   ⍝PI0 GI∆M∆1 LAMBDA ES WHERE PI0 IS THE PROBABILITY
[4]   ⍝AN ARRIVING CUSTOMER WILL FIND THE SYSTEM EMPTY.
[5]      LAMBDA←LAMBDA∆ES[1]
[6]      ES←LAMBDA∆ES[2]
[7]      RHO←LAMBDA×ES
[8]      'SERVER UTILIZATION, ρ, IS   ',⍕RHO
[9]      W←ES÷PI0
[10]     WQ←W-ES
[11]     VARQ←(1-PI0*2)×(ES÷PI0)*2
[12]     SIGQ←VARQ*0.5
[13]     C2Q←VARQ÷WQ*2
[14]     'THE AVERAGE QUEUEING TIME, WQ , IS   ',⍕WQ
[15]     'WITH STANDARD DEVIATION, SIGQ, ',⍕SIGQ
[16]     'C-SQUARED FOR Q, C2Q, IS   ',⍕C2Q
[17]     '90TH PERCENTILE QUEUEING TIME, PIQ90, IS'
[18]     ⎕←PIQ90←(W×⍟10×(1-PI0))⌈0
[19]     '95TH PERCENTILE QUEUEING TIME, PI95, IS'
[20]     ⎕←PIQ95←(W×⍟20×(1-PI0))⌈0
[21]     'THE DISTRIBUTION OF QUEUEING TIME FOR THOSE WHO'
[22]     'MUST IS THE SAME AS W.  THAT IS, EXPONENTIAL.'
[23]     'THE AVERAGE WAITING TIME IN THE SYSTEM, W, IS'
[24]     W
[25]     'WITH STANDARD DEVIATION ',⍕W
[26]     'C-SQUARED FOR W IS 1, SINCE W IS EXPONENTIAL   '
[27]     '90TH PERCENTILE WAITING TIME IN THE SYSTEM, PIW90,'
[28]     ' IS ',⍕W×⍟10
[29]     '95TH PERCENTILE WAITING TIME IN THE SYSTEM, PIW95,'
[30]     ' IS ',⍕W×⍟20
[31]     LQ←LAMBDA×WQ
[32]     L←LAMBDA×W
[33]     VARN←RHO×(2-(PI0+RHO))÷PI0*2
[34]     VARNQ←RHO×(1-PI0)×((2-PI0)-RHO×(1-PI0))÷PI0*2
```

B.11 APL Programs

```
[35]  SIGN←VARN*0.5
[36]  SIGNQ←VARNQ*0.5
[37]  'THE AVERAGE NUMBER IN THE QUEUE, LQ, IS'
[38]  LQ
[39]  'THE STANDARD DEVIATION OF NQ, SIGNQ, IS'
[40]  SIGNQ
[41]  'THE AVERAGE NUMBER IN THE SYSTEM, L, IS'
[42]  L
[43]  'THE STANDARD DEVIATION OF N, SIGN, IS'
[44]  SIGN
      ∇
```

```
      ∇ X←T GΔDIST BETAΔALPHA
[1]   ⍝GΔDIST IS THE DISTRIUTION FUNCTION FOR A GAMMA RANDOM VARIABLE WITH
[2]   ⍝PARAMETERS BETA AND ALPHA, THE NOTATION WE USE IN THE SECOND EDITION.
[3]   ⍝THE CORRESPONDING NOTATION IN THE FIRST EDITION IS ALPHA AND LAMBDA.
[4]   ⍝IN THE SECOND EDITION NOTATION TO CALCULATE THE PROBABILITY THAT SUCH
[5]   ⍝A RANDOM VARIABLE ASSUMES A VALUE ≤T, TYPE 'T GΔDIST BETA,ALPHA'
[6]   ⍝FOR EXAMPLE, TO CALCULATE THE PROBABILITY REQUESTED IN THE FIRST PART
[7]   ⍝OF EXAMPLE 3.25, TYPE '7.2 GΔDIST 2.5 O.5' AND GET THE ANSWER 0.79381408.
[8]   ⍝WE USED PART OF THE FUNCTION IGF FROM ANSCOMBE'S BOOK.
[9]   A←BETAΔALPHA[1]
[10]  Z←BETAΔALPHA[2]×T
[11]  X←A IGF Z
      ∇
```

```
      ∇ Z←Q1ΔQ2ΔMU1ΔMU2 H2ΔDIST X
[1]   ⍝H2ΔDIST IS THE DISTRIBUTION FUNCTION OF A TWO-STAGE HYPEREXPONENTIAL
[2]   ⍝RANDOM VARIABLE, U.  THE CALLING SEQUENCE IS 'Y H2ΔDIST X'. Y IS A VECTOR
[3]   ⍝OBTAINED BY CATENATING Q1, Q2, MU1, AND MU2 AND X IS THE VALUE IN THE
[4]   ⍝FORMULA 'P[U≤X]' WHERE U IS THE HΔ2 DISTRIBUTION IN QUESTION.
[5]   Q1←Q1ΔQ2ΔMU1ΔMU2[1]
[6]   Q2←Q1ΔQ2ΔMU1ΔMU2[2]
[7]   MU1←Q1ΔQ2ΔMU1ΔMU2[3]
[8]   MU2←Q1ΔQ2ΔMU1ΔMU2[4]
[9]   Z←1-((Q1×*-MU1×X)+(Q2×*-MU2×X))
      ∇
```

```
      ∇ Z←C HAM A;RHO
[1]   Z←B÷(RHO×B←C BBCA A)+1-RHO←A÷C
      ∇
```

```
      ∇ Z←K HYPERG MΔNΔR
[1]   M←MΔNΔR[1]
[2]   N←MΔNΔR[2]
[3]   R←MΔNΔR[3]
[4]   Z←(K!R)×((M-K)!(N-R))÷(M!N)
[5]   ⍝THIS FUNCTION CALCULATES THE PROBABILITIES FOR A HYPERGEOMETRIC
[6]   ⍝RANDOM VARIABLE.  IN THE NOTATION GIVEN JUST BEFORE EXERCISE 11
[7]   ⍝IN CHAPTER 3 OF THE SECOND EDITION M (WHICH WE CALLED LITTLE N)
[8]   ⍝IS THE SIZE OF THE SAMPLE, N IS THE SIZE OF THE POPULATION, AND
[9]   ⍝R IS THE NUMBER OF RED ELEMENTS.  THIS FUNCTION CALCULATES THE
[10]  ⍝PROBABILITY THAT EXACTLY K RED ELEMENTS ARE FOUND IN THE SAMPLE
[11]  ⍝OF SIZE M.  THE CALL IS K HYPERG M,N,R.
      ∇
```

B.12   APL Programs

```
     ∇ Z←J HYPERGΔDIST MΔNΔR;I
[1]    Z←0
[2]    I←¯1
[3]    START:→0×ιJ<I←I+1
[4]    Z←Z+I HYPERG MΔNΔR
[5]    →START
[6]    ⍝THIS FUNCTION IS THE DISTRIBUTION FUNCTION FOR A
[7]    ⍝HYPERGEOMETRIC DISTRIBUTION WITH PARAMETERS M, N, AND R.
[8]    ⍝IT CALLS THE FUNCTION HYPERG.  SEE THAT FUNCTION FOR MORE
[9]    ⍝DISCUSSION.  THE CALL IS J HYPERGΔDIST M,N,R.  IT RETURNS
[10]   ⍝THE PROBABILITY THAT THE RANDOM VARIABLE ASSUMES THE VALUE
[11]   ⍝NOT EXCEEDING J.
     ∇


     ∇ Z←A IF B
[1]    ⍝IF IS A VERY USEFUL FUNCTION BORROWED (IN THE TOM SAWYER SENSE)
[2]    ⍝FROM THE APL HANDBOOK OF TECHNIQUES, IBM ORDER NUMBER
[3]    ⍝S320+5996.  IT IS USED IN MANY OF MY FUNCTIONS.  SEE, FOR
[4]    ⍝EXAMPLE, BRANCH, WHICH WAS WRITTEN TO TEST IT.
[5]    Z←B/A
     ∇


     ∇ S←A IGF Z;J;K;R
[1]    ⍝THIS IS THE INCOMPLETE GAMMA FUNCTION ADAPTED FROM THE VERSION IN
[2]    ⍝THE FRANCIS ANSCOMBE BOOK 'COMPUTING IN STATISTICAL SCIENCE THROUGH
[3]    ⍝APL.  IN THE NOTATION I USE IN THE SECOND EDITION OF MY BOOK, THE
[4]    ⍝CALLING SEQUENCE IS 'A IGF Z' WHERE A IS BETA AND Z IS ALPHA×T.
[5]    ⍝THE VALUE CALCULATED IS THE VALUE OF THE DISTRIBUTION FUNCTION
[6]    ⍝OF A GAMMA RANDOM VARIABLE WITH PARAMETERS BETA AND ALPHA AT
[7]    ⍝THE POINT T.
[8]    →□LC+1+(∧/,A>0)∧(0=ρρA)∧ρρZ←,Z⌈0
[9]    →0,ρ□←'NO GO.',S←''
[10]   →(A=⌊A)/L2
[11]   S←Z≠Z
[12]   →(0=K←+/J←Z≥7.107)/L1
[13]   S[J/ιρS]←1-(*-J/Z)×((J/Z)∘.*A-ιR)+.÷(1+R=ιR)×!A-ιR←⌈7.107+A
[14]   L1:→(0=K←+/J←(Z>0)∧~J)/0
[15]   →0,S[J/ιρS]←(*-J/Z)×(((J/Z)∘.*¯1+A+ιR)×1+(÷2×¯1+(A+R)÷J/Z)∘.×R=ιR)+.÷!¯1+A
[16]   L2:S←1-(*-Z)×(Z∘.*¯1+ιA)+.÷!¯1+ιA
     ∇


     ∇ X←IN P;T
[1]    ⍝THIS IS ESSENTIALLY ANSCOMBE'S FUNCTION INIF.  HE SAYS IT IS THE
[2]    ⍝ODEY-EVANS APPROXIMATION.  ''IN'' IS THE INVERSE STANDARD NORMAL
[3]    ⍝FUNCTION.
[4]    T←P⌈1-P
[5]    T←(¯2×⍟T)*0.5
[6]    Y← 0.09934846266 0.588581570495 0.531103462366
[7]    Y←Y, 0.10353775285 3.8560700634E¯3
[8]    X←(×P-0.5)×T-((T∘.*0,ι4)+.× 0.322232431088 1 0.342242088547 0.020423121024
[9]
     ∇


     ∇ ALPHA KS XΔY
[1]    N←(ρXΔY)÷2
[2]    X←N↑XΔY
[3]    Y←N↓XΔY
[4]    FJ←(ιN)÷N
[5]    FJ1←((ιN)-1)÷N
[6]    DPLUS←⌈/FJ-Y
[7]    DMINUS←⌈/Y-FJ1
[8]    D←DPLUS⌈DMINUS
[9]    MD←D×((N*0.5)+0.12+(0.11÷N*0.5))
[10]   'MODIFIED D IS ',⍕MD
     ∇
```

B.13 APL Programs

```
      ∇ ALPHA KS∆EXPON Y
[1]    ⍝THIS FUNCTION PERFORMS THE KOLMOGOROV-SMIRNOV TEST
[2]    ⍝FOR EXPONENTIAL DISTRIBUTION (UNKNOWN MEAN) AS DEFINED IN
[3]    ⍝SCHEAEFFER AND MCCLAVE ON PAGE 329.
[4]    FJ←(⍳N)÷N←ρY
[5]    FJ1←((⍳N)-1)÷N
[6]    YBAR←(+/Y)÷N
[7]    FY←1-*-Y÷YBAR
[8]    DPLUS←⌈/FJ-FY
[9]    DMINUS←⌈/FY-FJ1
[10]   D←DPLUS⌈DMINUS
[11]   MD←(D-0.2÷N)×((N*0.5)+0.26+(0.5÷N*0.5))
[12]   'MODIFIED D IS ',⍕MD
[13]   →((ALPHA=0.01),(ALPHA=0.025),(ALPHA=0.05))/S1,S2,S3
[14]   →TEST
[15]   S1:→(MD>1.308)↑END
[16]   ⎕←'NULL HYPOTHESIS APPEARS TO BE TRUE FOR ALPHA= ',⍕ALPHA
[17]   →0
[18]   S2:→(MD>1.19)↑END
[19]   ⎕←'NULL HYPOTHESIS APPEARS TO BE TRUE FOR ALPHA= ',⍕ALPHA
[20]   →0
[21]   S3:→(MD>1.094)↑END
[22]   ⎕←'NULL HYPOTHESIS APPEARS TO BE TRUE FOR ALPHA= ',⍕ALPHA
[23]   →0
[24]   TEST:
[25]   →((ALPHA=0.1),(ALPHA=0.15))/S4,S5
[26]   ⎕←'ALPHA IS NOT ONE OF THE ALLOWED VALUES'
[27]   →0
[28]   S4:→(MD>0.99)↑END
[29]   ⎕←'NULL HYPOTHESIS APPEARS TO BE TRUE FOR ALPHA= ',⍕ALPHA
[30]   →0
[31]   S5:→(MD>0.926)↑END
[32]   ⎕←'NULL HYPOTHESIS APPEARS TO BE TRUE FOR ALPHA= ',⍕ALPHA
[33]   →0
[34]   END:⎕←'THE NULL HYPOTHESIS IS REJECTED FOR ALPHA= ',⍕ALPHA
      ∇

      ∇ ALPHA KS∆NORMAL Y
[1]    ⍝THIS FUNCTION PERFORMS THE KOLMOGOROV-SMIRNOV TEST
[2]    ⍝FOR NORMAL DISTRIBUTION (UNKNOWN MEAN AND VARIANCE)
[3]    ⍝AS DEFINED BY SCHEAEFFER AND MCCLAVE ON PAGE 329.
[4]    FJ←(⍳N)÷N←ρY
[5]    FJ1←((⍳N)-1)÷N
[6]    YBAR←(+/Y)÷N
[7]    S←((+/(Y-YBAR)*2)÷N-1)*0.5
[8]    FY←NDIST(Y-YBAR)÷S
[9]    DPLUS←⌈/FJ-FY
[10]   DMINUS←⌈/FY-FJ1
[11]   D←DPLUS⌈DMINUS
[12]   MD←D×(((N*0.5)-0.01)+(0.85÷N*0.5))
[13]   'MODIFIED D IS ',⍕MD
[14]   →((ALPHA=0.01),(ALPHA=0.025),(ALPHA=0.05))/S1,S2,S3
[15]   →TEST
[16]   S1:→(MD>1.035)↑END
[17]   ⎕←'NULL HYPOTHESIS APPEARS TO BE TRUE FOR ALPHA= ',⍕ALPHA
[18]   →0
[19]   S2:→(MD>0.955)↑END
[20]   ⎕←'NULL HYPOTHESIS APPEARS TO BE TRUE FOR ALPHA= ',⍕ALPHA
[21]   →0
[22]   S3:→(MD>0.895)↑END
[23]   ⎕←'NULL HYPOTHESIS APPEARS TO BE TRUE FOR ALPHA= ',⍕ALPHA
[24]   →0
[25]   TEST:
[26]   →((ALPHA=0.1),(ALPHA=0.15))/S4,S5
[27]   ⎕←'ALPHA IS NOT ONE OF THE ALLOWED VALUES'
[28]   →0
[29]   S4:→(MD>0.819)↑END
```

B.14   APL Programs

```
[30]  □←'NULL HYPOTHESIS APPEARS TO BE TRUE FOR ALPHA= ',⍕ALPHA
[31]  →0
[32]  S5:→(MD>0.775)↑END
[33]  □←'NULL HYPOTHESIS APPEARS TO BE TRUE FOR ALPHA= ',⍕ALPHA
[34]  →0
[35]  END:□←'THE NULL HYPOTHESIS IS REJECTED FOR ALPHA= ',⍕ALPHA
     ∇
     ∇ ALPHA KSΔSPEC FY
[1]   ⍝THIS FUNCTION PERFORMS THE KOLMOGOROV-SMIRNOV TEST
[2]   ⍝FOR SPECIFIED DISTRIBUTION FUNCTION AS DEFINED IN
[3]   ⍝SCHEAEFFER AND MCCLAVE ON PAGE 329.
[4]   FJ←(⍳N)÷N←⍴FY
[5]   FJ1←((⍳N)-1)÷N
[6]   DPLUS←⌈/FJ-FY
[7]   DMINUS←⌈/FY-FJ1
[8]   D←DPLUS⌈DMINUS
[9]   MD←D×((N*0.5)+0.12+(0.11÷N*0.5))
[10]  'MODIFIED D IS ',⍕MD
[11]  →((ALPHA=0.01),(ALPHA=0.025),(ALPHA=0.05))/S1,S2,S3
[12]  →TEST
[13]  S1:→(MD>1.626)↑END
[14]  □←'NULL HYPOTHESIS APPEARS TO BE TRUE FOR ALPHA= ',⍕ALPHA
[15]  →0
[16]  S2:→(MD>1.48)↑END
[17]  □←'NULL HYPOTHESIS APPEARS TO BE TRUE FOR ALPHA= ',⍕ALPHA
[18]  →0
[19]  S3:→(MD>1.358)↑END
[20]  □←'NULL HYPOTHESIS APPEARS TO BE TRUE FOR ALPHA= ',⍕ALPHA
[21]  →0
[22]  TEST:
[23]  →((ALPHA=0.1),(ALPHA=0.15))/S4,S5
[24]  □←'ALPHA IS NOT ONE OF THE ALLOWED VALUES'
[25]  →0
[26]  S4:→(MD>1.224)↑END
[27]  □←'NULL HYPOTHESIS APPEARS TO BE TRUE FOR ALPHA= ',⍕ALPHA
[28]  →0
[29]  S5:→(MD>1.138)↑END
[30]  □←'NULL HYPOTHESIS APPEARS TO BE TRUE FOR ALPHA= ',⍕ALPHA
[31]  →0
[32]  END:□←'THE NULL HYPOTHESIS IS REJECTED FOR ALPHA= ',⍕ALPHA
     ∇
     ∇ X MACHΔREP Y
[1]   ⍝THIS FUNCTION CALCULATE THE STATISTICS FOR THE MACHINE REPAIR
[2]   ⍝QUEUEING SYSTEM WITH K MACHINES AND C REPAIRMEN.
[3]   ⍝CALLING SEQUENCE IS X MACHΔREP Y WHERE X IS THE VECTOR
[4]   ⍝K, C AND Y IS THE VECTOR EO, ES WHERE EO IS THE AVERAGE TIME
[5]   ⍝A MACHINE IS IN OPERATION AND ES IS THE AVERAGE REPAIR TIME.
[6]   K←X[1]
[7]   C←X[2]
[8]   EO←Y[1]
[9]   ES←Y[2]
[10]  Z1←÷Z←EO÷ES
[11]  PNΔPO←1,(J!K)×Z1*(J←⍳C)
[12]  PNΔPO←PNΔPO,(!J)×(÷!C)×(÷(C*(J-C)))×(J!K)×Z1*(J←C+⍳K-C)
[13]  P←PNΔPO×PO←÷+/PNΔPO
[14]  ⍝P IS THE VECTOR OF PROBABILITES THAT THERE ARE N
[15]  ⍝MACHINES DOWN FOR N=0, 1, ... , K.
[16]  LQ←+/P[(C+1)+⍳K-C]×⍳K-C
[17]  WQ←LQ×(EO+ES)÷(K-LQ)
[18]  LAMBDA←K÷ETB←EO+WQ+ES
[19]  RHO←LAMBDA×ES÷C
[20]  W←WQ+ES
[21]  L←LAMBDA×W
[22]  Q←K↑(K-(¯1+⍳K+1))×P÷K-L
[23]  D←+/P[C+⍳K-C-1]
[24]  WQΔD←WQ÷D
[25]  N2←+/P×(¯1+⍳K+1)*2
[26]  SIGN←(VARN←N2-L*2)*0.5
[27]  'THE UTILIZATION OF EACH REPAIRMAN IS     ',⍕RHO
[28]  'LAMBDA, THE AVERAGE THROUGHPUT OF THE QUEUEING'
```

B.15   APL Programs

```
[29]    'SYSTEM IS ',⍕LAMBDA
[30]    'THE AVERAGE RESPONSE TIME, W, IS  ',⍕W
[31]    'THE AVERAGE QUEUEING TIME, WQ, IS  ',⍕WQ
[32]    'THE AVERAGE NUMBER OF MACHINES DOWN, L, IS  ',⍕L
[33]    'THE AVERAGE NUMBER OF MACHINES WAITING TO BE'
[34]    'REPAIRED, LQ, IS  ',⍕LQ
[35]    'THE AVERAGE QUEUEING TIME FOR THOSE WHICH MUST'
[36]    'WAIT, WQ∆D, IS ',⍕WQ∆D
[37]    'THE PROBABILITY A MACHINE MUST WAIT FOR SERVICE IS  ',⍕D
      ∇
      ∇ ALPHA MCHISQ∆EXPON Y;X
[1]     ⍝THIS FUNCTION PERFORMS THE CHISQUARE TEST ON THE SAMPLE, Y, TO
[2]     ⍝DETERMINE, AT THE ALPHA LEVEL OF SIGNIFICANCE, WHETHER OR NOT
[3]     ⍝Y IS A RANDOM SAMPLE FROM AN EXPONENTIAL POPULATION.
[4]     X←SORT Y
[5]     XBAR←(+/X)÷N←ρ,X
[6]     ⍝CALCULATE TENTILES
[7]     X10←-XBAR×⍟0.9
[8]     X20←-XBAR×⍟0.8
[9]     X30←-XBAR×⍟0.7
[10]    X40←-XBAR×⍟0.6
[11]    X50←-XBAR×⍟0.5
[12]    X60←-XBAR×⍟0.4
[13]    X70←-XBAR×⍟0.3
[14]    X80←-XBAR×⍟0.2
[15]    X90←-XBAR×⍟0.1
[16]    O←(+/(X<X10)),(+/((X10≤X)∧(X<X20))),(+/((X20≤X)∧(X<X30)))
[17]    O←O,(+/((X30≤X)∧(X<X40))),(+/((X40≤X)∧(X<X50)))
[18]    O←O,(+/((X50≤X)∧(X<X60))),(+/((X60≤X)∧(X<X70)))
[19]    O←O,(+/((X70≤X)∧(X<X80))),(+/((X80≤X)∧(X<X90)))
[20]    O←O,(+/(X90≤X))
[21]    E←10ρN÷10
[22]    CHISQ←+/((O-E)*2)÷E
[23]    Q←CHISQ CHISQUARE∆DIST 8
[24]    P←1-Q
[25]    'THE P VALUE IS ',⍕P
[26]    →(P<ALPHA)↑END
[27]    'THE POPULATION APPEARS TO BE EXPONENTIAL FOR ALPHA= ',⍕ALPHA
[28]    →0
[29]    END:'THE NULL HYPOTHESIS MUST BE REJECTED FOR ALPHA= ',⍕ALPHA
      ∇
      ∇ Z←MEAN X
[1]     Z←(+/X)÷N←ρ,X
      ∇
      ∇ MOMENTS Y
[1]     ⍝MOMENTS CALCULATES THE FIRST THREE MOMENTS OF A TWO STAGE
[2]     ⍝HYPEREXPONENTIAL RANDOM VARIABLE.  THE CALL IS 'MOMENTS Y'
[3]     ⍝WHERE Y IS THE VECTOR FORMED BY CATENATING Q1, Q2, MU1, AND MU1.
[4]     Q1←Y[1]
[5]     Q2←Y[2]
[6]     MU1←Y[3]
[7]     MU2←Y[4]
[8]     ES←(Q1÷MU1)+Q2÷MU2
[9]     ES2←2×(Q1÷MU1*2)+Q2÷MU2*2
[10]    ES3←6×(Q1÷MU1*3)+Q2÷MU2*3
[11]    'THE MEAN, ES, IS ',⍕ES
[12]    'THE SECOND MOMENT, ES2, IS  ',⍕ES2
[13]    'THE THIRD MOMENT, ES3, IS  ',⍕ES3
      ∇
      ∇ K M∆D∆1∆K∆K EO∆ES;Z;EO;ES;RHO;W;P0;LAMBDA
[1]     ⍝THIS FUNCTION CALCULATES THE STATISTICS FOR THE MACHINE
[2]     ⍝REPAIR MODEL WITH CONSTANT REPAIR TIME.
[3]     EO←EO∆ES[1]
[4]     ES←EO∆ES[2]
[5]     P0←K PZERO Z←EO÷ES
[6]     RHO←1-P0
[7]     LAMBDA←RHO÷ES
[8]     W←(K÷LAMBDA)-EO
[9]     'THE PROBABILITY THE REPAIRMAN IS IDLE IS  ',⍕P0
[10]    'THE SERVER (REPAIRMAN) UTILIZATION IS  ',⍕RHO
```

B.16   APL Programs

```
[11]    'THE MEAN TIME WAITING FOR REPAIRS IS   ',⍕WQ←W-ES
[12]    'THE MEAN NUMBER WAITING FOR REPAIRS IS   ',⍕LAMBDA×WQ
[13]    'THE MEAN TIME A MACHINE IS DOWN, W, IS   ',⍕W
[14]    'THE MEAN NUMBER OF MACHINES DOWN, W, IS   ',⍕LAMBDA×W
[15]    'THE THROUGHPUT, LAMBDA, IS   ',⍕LAMBDA
        ∇

        ∇ K M∆EK∆1 X
[1]     LAMBDA←X[1]
[2]     ES←X[2]
[3]     RHO←LAMBDA×ES
[4]     R1←1-RHO
[5]     ES2←(1+÷K)×ES*2
[6]     ES3←(1+2÷K)×ES2×ES
[7]     LQ←(ES2×LAMBDA*2)÷2×R1
[8]     WQ←LQ÷LAMBDA
[9]     WQ∆Q←WQ÷RHO
[10]    EQ2←((÷3×R1)×LAMBDA×ES3)+(÷2)×(C1←LAMBDA×ES2÷R1)*2
[11]    VARQ←EQ2-WQ*2
[12]    SIGQ←VARQ*0.5
[13]    L←LQ+RHO
[14]    W←L÷LAMBDA
[15]    EW2←EQ2+ES2÷R1
[16]    VARW←EW2-W*2
[17]    SIGW←VARW*0.5
[18]    VARN←((÷3×R1)×ES3×LAMBDA*3)+(0.25×(LAMBDA*2)×C1*2)
[19]    VARN←VARN+((3-2×RHO)×C1×LAMBDA÷2)+RHO×R1
[20]    SIGN←VARN*0.5
[21]    'THE AVERAGE NUMBER QUEUEING,LQ, IS   ',⍕LQ
[22]    'THE AVERAGE TIME WAITING IN QUEUE, WQ, IS   ',⍕WQ
[23]    'WITH STANDARD DEVIATION, SIGQ,   ',⍕SIGQ
[24]    'THE AVERAGE QUEUEING TIME FOR THOSE WHO MUST'
[25]    'WAIT, IS   ',⍕WQ∆Q
[26]    'THE AVERAGE NUMBER IN THE SYSTEM, L, IS   ',⍕L
[27]    'WITH STANDARD DEVIATION   ',⍕SIGN
[28]    'THE AVERAGE TIME IN THE SYSTEM,W, IS   ',⍕W
[29]    'WITH STANDARD DEVIATION   ',⍕SIGW
[30]    'THE MARTIN RULE ESTIMATE OF THE 90TH PERCENTILE'
[31]    'TIME IN THE SYSTEM IS   ',⍕W90←W+1.3×SIGW
[32]    'THE C-SQUARED VALUE FOR WAITING TIME IS   ',⍕C2Q←VARQ÷WQ*2
[33]    'THE C-SQUARED VALUE FOR SYSTEM TIME IS   ',⍕C2W←VARW÷W*2
[34]    'THE C-SQUARED VALUE FOR NUMBER IN THE SYSTEM IS   ',⍕C2N←VARN÷L*2
        ∇

        ∇ X M∆G∆1 ES;PIW90;PIW95
[1]     ⍝ M∆G∆1 CALCULATES THE USUAL STATISTICS FOR THE
[2]     ⍝ M/G/1 MODEL USING THE POLLACZEK-KHINTCHINE
[3]     ⍝ EQUATIONS.  THE CALLING SEQUENCE IS ''LAMBDA SIG M∆G∆1 ES''
[4]     ⍝ WHERE LAMBDA IS THE AVERAGE ARRIVAL RATE, SIG IS THE STANDARD
[5]     ⍝ DEVIATION OF SERVICE TIME, AND ES IS AVERAGE SERVICE TIME.
[6]     LAMBDA←X[1]
[7]     SIG←X[2]
[8]     RHO←LAMBDA×ES
[9]     LQ←(((LAMBDA×SIG)*2)+RHO*2)÷2×(1-RHO)
[10]    L←LQ+RHO
[11]    WQ←LQ÷LAMBDA
[12]    W←WQ+ES
[13]    'THE SERVER UTILIZATION, RHO, IS   ',⍕RHO
[14]    'THE AVERAGE NUMBER IN THE SYSTEM IS   ',⍕L
[15]    'THE AVERAGE NUMBER WAITING FOR SERVICE IS   ',⍕LQ
[16]    'THE AVERAGE TIME IN THE SYSTEM IS   ',⍕W
[17]    'THE AVERAGE WAITING TIME IS   ',⍕WQ
[18]    'THE AVERAGE WAITING TIME FOR CUSTOMERS DELAYED IS   ',⍕WQ÷RHO
        ∇

        ∇ LAMBDA M∆G∆1∆EXT X
[1]     ⍝ M∆G∆1∆EXT CALCULATES THE USUAL STATISTICS FOR THE
[2]     ⍝ M/G/1 MODEL, BUT ALSO  CALCULATES THE STANDARD DEVIATIONS
[3]     ⍝ AND THE C-SQUARED VALUES FOR MANY OF THE VARIABLES SO
[4]     ⍝ THAT ERLANG APPROXIMATIONS CAN BE MADE.  THE CALLING
[5]     ⍝ SEQUENCE IS '' LAMBDA M∆G∆1∆EXT ES ES2 ES3 ''
[6]     ⍝ WHERE ES ES2 ES3 ARE THE FIRST THREE MOMENTS OF
[7]     ⍝ THE SERVICE TIME DISTRIBUTION.
```

B.17   APL Programs

```
[8]     ES←X[1]
[9]     ES2←X[2]
[10]    ES3←X[3]
[11]    RHO←LAMBDA×ES
[12]    LQ←(LAMBDA*2)×ES2÷2×(1-RHO)
[13]    L←LQ+RHO
[14]    VARN←(LAMBDA*3)×ES3÷3×(1-RHO)
[15]    VARN←VARN+(LAMBDA*4)×(ES2*2)÷4×(1-RHO)*2
[16]    VARN←VARN+(LAMBDA*2)×(3-2×RHO)×ES2÷2×(1-RHO)
[17]    VARN←VARN+RHO×(1-RHO)
[18]    SIGN←VARN*0.5
[19]    WQ←LQ÷LAMBDA
[20]    EQ2←(LAMBDA×ES3÷3×(1-RHO))+2×WQ*2
[21]    VARQ←EQ2-WQ*2
[22]    SIGQ←VARQ*0.5
[23]    W←WQ+ES
[24]    EW2←EQ2+ES2÷1-RHO
[25]    EW2←EW2
[26]    VARW←EW2-W*2
[27]    SIGW←VARW*0.5
[28]    'SERVER UTILIZATION,RHO, IS  ',⍕RHO
[29]    'THE AVERAGE NUMBER IN THE SYSTEM, L, IS ',⍕L
[30]    'WITH STANDARD DEVIATION ',⍕SIGN
[31]    'C-SQUARED FOR NUMBER IN THE SYSTEM IS  ',⍕C2N←VARN÷L*2
[32]    'THE AVERAGE NUMBER QUEUEING ,LQ, IS ',⍕LQ
[33]    'THE AVERAGE TIME IN THE SYSTEM,W, IS ',⍕W
[34]    'WITH STANDARD DEVIATION ',⍕SIGW
[35]    'C-SQUARED FOR TIME IN THE SYSTEM IS  ',⍕C2W←VARW÷W*2
[36]    'THE AVERAGE QUEUEING TIME,WQ, IS ',⍕WQ
[37]    'WITH STANDARD DEVIATION ',⍕SIGQ
[38]    'C-SQUARED FOR QUEUEING TIME IS   ',⍕C2Q←VARQ÷WQ*2
[39]    ' THE MARTIN ESTIMATES OF 90TH AND 95TH'
[40]    ' PERCENTILES OF TIME IN THE SYSTEM ARE'
[41]    ⎕←W+1.3×SIGW
[42]    ⎕←W+2×SIGW
      ∇
      ∇ LAMBDA M∆M∆1 ES;RHO;L;SIGN;W;WQ;SIGQ;W90;W95;Q90;Q95
[1]    ⍝THIS FUNCTION CALCULATES THE STATISTICS FOR THE CLASSICAL
[2]    ⍝M/M/1 QUEUEING SYSTEM.   THE CALL IS IS GIVEN ON LINE 0
[3]    ⍝WHERE LAMBDA IS THE AVERAGE ARRIVAL RATE AND ES IS THE
[4]    ⍝AVERAGE SERVICE TIME.
[5]    RHO←LAMBDA×ES
[6]    'THE SERVER UTILIZATION, ρ, IS  ',⍕RHO
[7]    'THE MEAN NUMBER IN THE SYSTEM, L, IS'
[8]    ⎕←L←RHO÷1-RHO
[9]    'WITH STANDARD DEVIATION, SIGN, '
[10]   ⎕←SIGN←(L÷(1-RHO))*0.5
[11]   'THE MEAN NUMBER IN THE QUEUE, LQ, IS'
[12]   ⎕←LQ←RHO×L
[13]   'WITH STANDARD DEVIATION, SIGNQ, '
[14]   ⎕←SIGNQ←((L*2)×(1+RHO-RHO*2))*0.5
[15]   'THE MEAN TIME IN THE SYSTEM, W, IS'
[16]   ⎕←W←ES÷1-RHO
[17]   'WITH STANDARD DEVIATION  ',⍕W
[18]   'THE MEAN TIME IN THE QUEUE, WQ, IS'
[19]   ⎕←WQ←RHO×W
[20]   'WITH STANDARD DEVIAITON, SIGQ, '
[21]   ⎕←SIGQ←SIGN×ES×(2-RHO)*0.5
[22]   'THE MEAN QUEUEING TIME FOR THOSE WHO MUST QUEUE IS W'
[23]   ⎕←W
[24]   'THE MEAN NUMBER QUEUEING WHEN THE QUEUE IS NOT EMTPY'
[25]   'IS ',⍕÷1-RHO
[26]   '90TH PERCENTILE TIME IN THE SYSTEM IS  ',⍕W×⍟10
[27]   '95TH PERCENTILE TIME IN THE SYSTEM IS  ',⍕W×⍟20
[28]   '90TH PERCENTILE QUEUEING TIME IS  ',⍕(W×⍟10×RHO)⌈0
[29]   '95TH PERCENTILE QUEUEING TIME IS  ',⍕(W×⍟20×RHO)⌈0
      ∇
      ∇ LAMBDA M∆M∆1∆K X
[1]    ⍝ M∆M∆1∆K CALCULATES THE STATISTICS FOR THE M∆M∆1
[2]    ⍝ MODEL RESTRICTED SO THAT NO MORE THAN K CUSTOMERS
```

B.18   APL Programs

```
[3]    ⍝ ARE ALLOWED IN THE SYSTEM.
[4]    ES←X[1]
[5]    K←X[2]
[6]    U←LAMBDA×ES
[7]    P0←(1-U)÷XN1←1-U⋆K+1
[8]    P←P0×U⋆(¯1+⍳K+1)
[9]    PK←P0×U⋆K
[10]   'THE PROBABILITY, PSUBZERO, THAT THE SYSTEM IS EMPTY IS'
[11]   P0
[12]   'THE PROBABILITY THAT AN ARRIVING CUSTOMER IS TURNED'
[13]   'AWAY, PSUBK, IS  ',⍕PK
[14]   'THE TRAFFIC INTENSITY, U, IS  ',⍕U
[15]   'THE SERVER UTILIZATION, RHO, IS  ',⍕(1-PK)×U
[16]   L←(U×1+((K×U)-(K+1))×U⋆K)÷(XU←1-U)×XN1
[17]   EN2←+/P×(¯1+⍳K+1)⋆2
[18]   VARN←EN2-L⋆2
[19]   C2N←VARN÷L⋆2
[20]   SIGN←VARN⋆0.5
[21]   LQ←L-(1-P0)
[22]   ENQ2←+/P[2+⍳K-1]×(⍳K-1)⋆2
[23]   VARNQ←ENQ2-LQ⋆2
[24]   C2NQ←VARNQ÷LQ⋆2
[25]   SIGNQ←VARNQ⋆0.5
[26]   LAMBDAA←LAMBDA×(1-PK)
[27]   WQ←LQ÷LAMBDAA
[28]   W←WQ+ES
[29]   WQΔQ←WQ÷(1-P0)
[30]   'THE AVERAGE NUMBER IN THE SYSTEM, L, IS  ',⍕L
[31]   'WITH STANDARD DEVIATION  ',⍕SIGN
[32]   'C-SQUARED FOR NUMBER IN THE SYSTEM, N, IS  ',⍕C2N
[33]   'THE AVERAGE NUMBER WAITING, LQ, IS  ',⍕LQ
[34]   'WITH STANDARD DEVIATION  ',⍕SIGNQ
[35]   'C-SQUARED FOR NUMBER IN THE QUEUE, NQ, IS  ',⍕C2NQ
[36]   'THE AVERAGE TIME IN THE SYSTEM, W, IS  ',⍕W
[37]   'THE AVERAGE WAITING TIME, WQ, IS  ',⍕WQ
[38]   'THE AVERAGE WAITING TIME FOR THOSE WHO MUST WAIT IS  ',⍕WQΔQ
       ∇
    ∇ C MΔMΔC AΔES;C;ES;RHO;LAMBDA;CCA;LQ;VARNQ;SIGNQ;WQ;W
[1]    ⍝THIS FUNCTION CALCULATES THE STATISTICS FOR THE M/M/C
[2]    ⍝QUEUEING SYSTEM.   THE CALL IS ON LINE 0.
[3]    A←AΔES[1]
[4]    ES←AΔES[2]
[5]    RHO←A÷C
[6]    LAMBDA←A÷ES
[7]    'THE AVERAGE ARRIVAL RATE IS  ',⍕LAMBDA
[8]    CCA←C CΔCΔA A
[9]    'THE PROBABILITY ALL SERVERS ARE BUSY IS  ',⍕CCA
[10]   'THE SERVER UTILIZATION, ρ, IS  ',⍕RHO
[11]   LQ←RHO×CCA÷1-RHO
[12]   'THE AVERAGE NUMBER IN THE QUEUE IS  ',⍕LQ
[13]   VARNQ←LQ×(1+RHO-RHO×CCA)÷1-RHO
[14]   'WITH STANDARD DEVIATION  ',⍕SIGNQ←VARNQ⋆0.5
[15]   VARN←VARNQ+A×(1+CCA)
[16]   WQ←LQ÷LAMBDA
[17]   'THE AVERAGE QUEUEING TIME IS  ',⍕WQ
[18]   VARQ←(2-CCA)×CCA×(ES⋆2)÷(((1-RHO)⋆2)×C⋆2)
[19]   'WITH STANDARD DEVIATION  ',⍕VARQ⋆0.5
[20]   'THE AVERAGE QUEUEING TIME FOR CUSTOMERS DELAYED IS'
[21]   ⎕←WQ÷CCA
[22]   W←WQ+ES
[23]   L←LAMBDA×W
[24]   'THE AVERAGE NUMBER IN THE SYSTEM IS  ',⍕L
[25]   'WITH STANDARD DEVIATION  ',⍕VARN⋆0.5
[26]   'THE AVERAGE TIME IN THE SYSTEM IS  ',⍕W
[27]   →LΔAΔCΔ1×⍳A=C-1
[28]   EW2←(2×CCA×ES⋆2)÷(A+1-C)
[29]   EW2←EW2×(1-(C-A)⋆2)÷(C-A)⋆2
[30]   →LΔSIGΔW,EW2←EW2+2×ES⋆2
[31] LΔAΔCΔ1:EW2←((4×CCA)+2)×ES⋆2
[32] LΔSIGΔW:VARW←EW2-W⋆2
```

B.19   APL Programs

```
[33]    SIGW←VARW*0.5
[34]    'WITH STANDARD DEVIATION   ',⍕SIGW
[35]    K←ES÷C-A
[36]    WQ90←0⌈K×⍟10×CCA
[37]    WQ95←0⌈K×⍟20×CCA
[38]    'THE 90TH PERCENTILE QUEUEING TIME IS   ',⍕WQ90
[39]    'THE 95TH PERCENTILE QUEUEING TIME IS   ',⍕WQ95
[40]    'THE MARTIN ESTIMATE OF 90TH PERCENTILE SYSTEM TIME IS'
[41]    ⎕←EST90←W+1.3×SIGW
[42]    X1←W
[43]    X2←W+2×SIGW
[44]    'THE 90TH PERCENTILE SYTEM TIME BY FUNCTION PIW90 IS'
[45]    ⎕←CPIW90←C PIW90 A,ES,X1,X2
[46]    'THE MARTIN ESTIMATE OF 95TH PERCENTILE SYSTEM TIME IS'
[47]    ⎕←EST95←W+2×SIGW
[48]    CPIW95←C PIW95 A,ES,X1,W+2×SIGW
[49]    'THE 95TH PERCENTILE SYSTEM TIME BY FUNCTION PIW95 IS'
[50]    CPIW95
      ∇
      ∇ P←NDIST T;S;R;Z
[1]   ⍝ ' NIDST T ' EVALUATES THE STANDARD NORMAL DISTRIBUTION
[2]   ⍝ FUNCTION AT THE POINT OR VERCTOR T, USING FORMULA 26.2.17
[3]   ⍝ IN ABRAMOWITZ AND STEGUN.
[4]     R←⍴T
[5]     S←(T<0)/⍳⍴T←,T
[6]     T[S]←-T[S]
[7]     P←(÷(○2)*0.5)×*-(T*2)÷2
[8]     T←÷1+0.2316419×T
[9]     Z←⁻0.356563782+T×1.781477937+T×⁻1.821255978+T×1.330274429
[10]    P←1-P×T×0.31938153+T×Z
[11]    P[S]←1-P[S]
[12]    P←R⍴P
      ∇
      ∇ X NEXPON∆TEST ALPHA
[1]   ⍝THIS FUNCTION CALCULATES THE EDF STATISTICS D, W∆SQUARED, U∆SQUARED,
[2]   ⍝AND A∆SQUARED FOR TESTING FOR EXPONENTIALITY AS DESCRIBED BY
[3]   ⍝STEPHENS IN HIS BOOK ON PAGES 134 AND 101.  IT TESTS FOR THE
[4]   ⍝UNSHIFTED EXPONENTIAL.
[5]   ⍝THE FUNCTION ALSO TESTS THE MODIFIED ANDERSON+DARLING (A∆SQUARED)
[6]   ⍝AT SEVERAL LEVELS OF SIGNIFICANCE USING THE TABLE ON PAGE 473
[7]   ⍝OF THE SPINELLI/STEPHENS PAPER IN NOV 1987 TECHNOMETRICS.
[8]     X←SORT X
[9]     XBAR←(+/X)÷N←,⍴X
[10]    'XBAR IS  ',⍕XBAR
[11]    W←X÷XBAR
[12]    Z←1-*-W
[13]    I←⍳N
[14]    DPLUS←⌈/((I÷N)-Z)
[15]    DMINUS←⌈/(Z-(I-1)÷N)
[16]    D←DPLUS⌈DMINUS
[17]    V←DPLUS+DMINUS
[18]    'THE VALUE OF THE KOLMOGOROV-SMIRNOV STATISTIC, D, IS ',⍕D
[19]    'THE VALUE OF SQUARE-ROOT OF N TIMES D IS ',⍕(N*0.5)×D
[20]    'THE VALUE OF SQUARE-ROOT OF N TIMES V IS ',⍕(N*0.5)×V
[21]    W∆2←+/(Z-(((2×I)-1)÷2×N))*2
[22]    'THE VALUE OF THE CRAMER-VON MISES STATISTIC, W∆SQUARED, IS ',⍕W∆2
[23]    ZBAR←(+/Z)÷N
[24]    U∆2←W∆2-N×((ZBAR-0.5)*2)
[25]    'THE VALUE OF THE WATSON STATISTIC, U∆SQUARED, IS ',⍕U∆2
[26]    A∆2←((-÷N)×+/(((2×I)-1)×((⍟Z)+⍟(1-⌽Z))))-N
[27]    'THE VALUE OF THE ANDERSON-DARLING STATISTIC, A∆SQUARED, IS   ',⍕A∆2
[28]    M∆A∆2←A∆2×(1+(5.4÷N)-(11÷N×N))
[29]    'THE VALUE OF THE MODIFIED ANDERSON-DARLING STATISTIC IS   ',⍕M∆A∆2
[30]    →((ALPHA=0.01),(ALPHA=0.025),(ALPHA=0.05))/S1,S2,S3
[31]    →TEST
[32]    S1:→(M∆A∆2>1.959)↑END
[33]    ⎕←'POPULATION APPEARS TO BE EXPONENTIAL FOR  ALPHA= ',⍕ALPHA
[34]    →0
[35]    S2:→(M∆A∆2>1.591)↑END
[36]    ⎕←'POPULATION APPEARS TO BE EXPONENTIAL FOR  ALPHA= ',⍕ALPHA
```

B.20   APL Programs

```
[37]    →0
[38] S3:→(M△A△2>1.321)↑END
[39]    □←'POPULATION APPEARS TO BE EXPONENTIAL FOR   ALPHA= ',▼ALPHA
[40]    →0
[41] TEST:
[42]    →((ALPHA=0.1),(ALPHA=0.15),(ALPHA=0.25))/S4,S5,S6
[43]    □←'ALPHA IS NOT ONE OF THE ALLOWED VALUES'
[44]    →0
[45] S4:→(M△A△2>1.062)↑END
[46]    □←'POPULATION APPEARS TO BE EXPONENTIAL FOR   ALPHA= ',▼ALPHA
[47]    →0
[48] S5:→(M△A△2>0.916)↑END
[49]    □←'POPULATION APPEARS TO BE EXPONENTIAL FOR   ALPHA= ',▼ALPHA
[50]    →0
[51] S6:→(M△A△2>0.736)↑END
[52]    □←'POPULATION APPEARS TO BE EXPONENTIAL FOR   ALPHA= ',▼ALPHA
[53]    →0
[54] END:□←'THE POPULATION IS NOT EXPONENTIAL FOR ALPHA= ',▼ALPHA
[55]
[56]
[57]
        ▽

        ▽ ALPHA NORMAL△T X
[1]     �)THIS FUNCTION CALCULATES THE EDF STATISTICS D, W△SQUARED, U△SQUARED,
[2]     ▵AND A△SQUARED FOR TESTING FOR NORMALITY AS DESCRIBED BY
[3]     ▵STEPHENS IN HIS BOOK.
[4]     ▵THE FUNCTION ALSO TESTS THE MODIFIED ANDERSON-DARLING (A△SQUARED)
[5]     ▵AT SEVERAL LEVELS OF SIGNIFICANCE USING THE TABLE ON PAGE 123
[6]     ▵OF THE STEPHEN'S BOOK.  WE HAVE MODIFIED THE CACLCULATION HERE
[7]     ▵FOR THE NONSHIFTED EXPONENTIAL ONLY AS DESCRIBED BY STEPHENS IN
[8]     ▵HIS BOOK ON PAGES 122 AND 101
[9]     X←X[♠X]
[10]    XBAR←(+/X)÷(N←ρ,X)
[11]    'XBAR IS ',▼XBAR
[12]    S←((+/(X-XBAR)*2)÷N-1)*0.5
[13]    W←(X-XBAR)÷S
[14]    Z←NDIST W
[15]    I←ιN
[16]    DPLUS←⌈/((I÷N)-Z)
[17]    DMINUS←⌈/(Z-(I-1)÷N)
[18]    D←DPLUS⌈DMINUS
[19]    'THE VALUE OF THE KOLMOGOROV-SMIRNOV STATISTIC, D, IS ',▼D
[20]    MD←D×(SN+(0.85÷SN+N*0.5)-0.01)
[21]    'THE MODIFIED VALUE OF D IS   ',▼MD
[22]    A△2←((-÷N)×+/(((2×I)-1)×((●Z)+●(1-ΦZ))))-N
[23]    'THE VALUE OF  A△SQUARED, IS ',▼A△2
[24]    M△A△2←A△2×(1+(0.75÷N)+(2.25÷N*2))
[25]    'THE VALUE OF THE MODIFIED A△SQUARED STATISTIC IS   ',▼M△A△2
[26]    →((ALPHA=0.01),(ALPHA=0.025),(ALPHA=0.05))/S1,S2,S3
[27]    →TEST
[28] S1:→(M△A△2>1.035)↑END
[29]    □←'POPULATION APPEARS TO BE NORMAL FOR   ALPHA= ',▼ALPHA
[30]    →0
[31] S2:→(M△A△2>0.873)↑END
[32]    □←'POPULATION APPEARS TO BE NORMAL FOR   ALPHA= ',▼ALPHA
[33]    →0
[34] S3:→(M△A△2>0.752)↑END
[35]    □←'POPULATION APPEARS TO BE NORMAL FOR   ALPHA= ',▼ALPHA
[36]    →0
[37] TEST:
[38]    →((ALPHA=0.1),(ALPHA=0.15),(ALPHA=0.25))/S4,S5,S6
[39]    □←'ALPHA IS NOT ONE OF THE ALLOWED VALUES'
[40]    →0
[41] S4:→(M△A△2>0.631)↑END
[42]    □←'POPULATION APPEARS TO BE NORMAL FOR   ALPHA= ',▼ALPHA
[43]    →0
[44] S5:→(M△A△2>0.561)↑END
[45]    □←'POPULATION APPEARS TO BE NORMAL FOR   ALPHA= ',▼ALPHA
[46]    →0
```

B.21 APL Programs

```
[47]  S6:→(M∆A∆2>0.47)↑END
[48]  ⎕←'POPULATION APPEARS TO BE NORMAL FOR  ALPHA= ',⍕ALPHA
[49]   →0
[50]  END:⎕←'THE POPULATION IS NOT NORMAL FOR ALPHA= ',⍕ALPHA
    ∇

    ∇ Z←MU PARTIAL∆SUM K;N
[1]   ⍝THIS FUNCTION IS NEEDED FOR THE FUNCTION
[2]   ⍝DM∆M∆1∆K.  IT COMPUTES A SEQUENCE OF POISSON
[3]   ⍝SUMS.
[4]   N←0
[5]   Z←*-MU
[6]   START:→0×⍳K<N←N+1
[7]   →START,Z←Z,MU POISSON∆DIST N
    ∇

    ∇ Z←C PIW90 A∆ES∆X1∆X2;A;ES
[1]   ⍝THIS FUNCTION ESTIMATES THE 90TH PERCENTILE VALUE
[2]   ⍝OF W FOR THE M/M/C QUEUEING SYSTEM.  X1 AND
[3]   ⍝X2 BRACKET THE VALUE.
[4]   A←A∆ES∆X1∆X2[1]
[5]   ES←A∆ES∆X1∆X2[2]
[6]   X1←A∆ES∆X1∆X2[3]
[7]   X2←A∆ES∆X1∆X2[4]
[8]   N←0
[9]   TEST:→OUT IF 50<N←N+1
[10]  T←(X1+X2)÷2
[11]  →HIGH IF(C DW∆M∆M∆C(A,ES,T))>0.9
[12]  E←|D←(C DW∆M∆M∆C(A,ES,T))-0.9
[13]  →OUT IF E<1E¯4
[14]  →TEST,X1←(X1+X2)÷2
[15]  HIGH:→TEST,X2←(X1+X2)÷2
[16]  OUT:Z←T
    ∇

    ∇ Z←C PIW95 A∆ES∆X1∆X2;A;ES
[1]   ⍝THIS FUNCTION ESTIMATES THE 95TH PERCENTILE VALUE
[2]   ⍝OF W FOR THE M/M/C QUEUEING SYSTEM.  X1 AND
[3]   ⍝X2 BRACKET THE VALUE.
[4]   A←A∆ES∆X1∆X2[1]
[5]   ES←A∆ES∆X1∆X2[2]
[6]   X1←A∆ES∆X1∆X2[3]
[7]   X2←A∆ES∆X1∆X2[4]
[8]   N←0
[9]   TEST:→OUT IF 50<N←N+1
[10]  T←(X1+X2)÷2
[11]  →HIGH IF(C DW∆M∆M∆C(A,ES,T))>0.95
[12]  E←|D←(C DW∆M∆M∆C(A,ES,T))-0.95
[13]  →OUT IF E<1E¯4
[14]  →TEST,X1←(X1+X2)÷2
[15]  HIGH:→TEST,X2←(X1+X2)÷2
[16]  OUT:Z←T
    ∇

    ∇ Z←X PO Y;K;C;EO;ES;Z1;PN∆P0
[1]   ⍝THIS FUNCTION CALCULATES THE PROBABILITY THAT ALL MACHINES ARE UP
[2]   ⍝FOR A MACHINE REPAIR MODEL WITH K-1 MACHINES, C REPAIRMEN, AVERAGE
[3]   ⍝UP TIME PER MACHINE EO AND AVERAGE REPAIR TIME ES.
[4]   ⍝THE CALL IS X PO Y WHERE X = K, C AND Y = EO, ES.
[5]   K←X[1]-1
[6]   C←X[2]
[7]   EO←Y[1]
[8]   ES←Y[2]
[9]   Z1←÷Z←EO÷ES
[10]  PN∆P0←1,(J!K)×Z1*(J←⍳C)
[11]  PN∆P0←PN∆P0,(!J)×(÷!C)×(÷(C*(J-C)))×(J!K)×Z1*(J←C+⍳K-C)
[12]  Z←P0←÷+/PN∆P0
    ∇

    ∇ Z←ALPHA POISSON K;X
[1]   ⍝ ' ALPHA POISSON K ' COMPUTES THE PROBABILITY THAT
[2]   ⍝ A POISSON RANDOM VARIABLE WITH MEAN ALPHA ASSUMES
[3]   ⍝ THE VALUE K.
[4]   Z←(*-ALPHA)×*(K×(⍟ALPHA))-(¯1↑X←+\⍟⍳K)
    ∇
```

B.22   APL Programs

```
      ∇ Z←ALPHA POISSON∆DIST N
[1]   ⍝ ' ALPHA POISSON∆DIST N ' COMPUTES THE PROBABILITY
[2]   ⍝ THAT A POISSON RANDOM VARIABLE WITH MEAN ALPHA
[3]   ⍝ ASSUMES A VALUE ≤ N; THAT IS, THE DISTRIBUTION
[4]   ⍝ FUNCTION OF THE VARIABLE.
[5]      Z←(*-ALPHA)×+/1,*((⍟ALPHA)×⍳N)-+\⍟⍳N
      ∇

      ∇ M POLL N
[1]   ⍝POLL ASSUMES A LINE WITH M TERMINALS, N OF
[2]   ⍝WHICH ARE READY TO TRANSMIT.  POLL CALCULATES
[3]   ⍝THE PROBABILITY THAT 1,2,...,M-N+1 POLLS ARE
[4]   ⍝REQUIRED TO FIND THE FIRST READY TERMINAL.
[5]   ⍝THE AVERAGE, EX, VARIANCE, VARX, AND THE
[6]   ⍝STANDARD DEVIATION, SIGX, ALSO ARE COMPUTED.
[7]      P←((N-1)!M-⍳(M-N)+1)÷(N!M)
[8]      X←⍳(M-N)+1
[9]      EX←+/X×P
[10]     VARX←+/((X-EX)*2)×P
[11]     SIGX←VARX*0.5
[12]     'THE AVERAGE NUMBER OF POLLS REQUIRED IS:'
[13]     EX
[14]     'THE STANDARD DEVIATION, SIGX, IS   ';SIGX
[15]     'TO SEE THE PROBABILITY THAT 1, 2, ..., M+N+1'
[16]     'POLLS ARE REQUIRED, TYPE P.'
      ∇

      ∇ Z←M POLL2M N;P;X
[1]   ⍝ THIS FUNCTION IS USED BY MPOLL.
[2]      P←((N-1)!M-⍳(M-N)+1)÷(N!M)
[3]      X←⍳(M-N)+1
[4]      Z←+/P×X×X
      ∇

      ∇ Z←M POLLM N;P;X
[1]   ⍝ THIS FUNCTION IS USED BY MPOLL.
[2]      P←((N-1)!M-⍳(M-N)+1)÷(N!M)
[3]      X←⍳(M-N)+1
[4]      Z←+/P×X
      ∇

      ∇ Z←X POWER N
[1]      I←0
[2]      Z←X
[3]   START:→0×⍳(N≤I←I+1)
[4]      Z←Z+.×X
[5]      →START
      ∇

      ∇ PL∆PES PR∆QUEUE PES2
[1]   ⍝PR∆QUEUE IS A PROGRAM FOR CALCULATING THE STATISTICS FOR
[2]   ⍝AN M/G/1 PREEMPTIVE-RESUME QUEUEING SYSTEM.
[3]   ⍝THE CALL IS LINE 0 WHERE PL∆PES IS THE CATENATION OF
[4]   ⍝PL, THE VECTOR OF MEAN ARRIVAL RATES OF PRIORITY CLASSES,
[5]   ⍝WITH ES, THE VECTOR OF MEAN SERVICE TIME OF CLASSES.
[6]   ⍝PES2 IS THE VECTOR OF SECOND MOMENTS OF SERVICE TIME.
[7]      N←⍴PES2
[8]      PLAMBDA←N↑PL∆PES
[9]      PES←(-N)↑PL∆PES
[10]     LAMBDA←+/PLAMBDA
[11]     PROB←PLAMBDA÷LAMBDA
[12]     ES←+/PROB×PES
[13]     ES2←+/PROB×PES2
[14]     PU←0,+\PLAMBDA×PES
[15]     DIV1←N↑(1-PU)
[16]     DIV2←(-N)↑(1-PU)
[17]     PV←+\PLAMBDA×PES2
[18]     PW←(PES+PV÷2×DIV2)÷DIV1
[19]     PL←PLAMBDA×PW
[20]     PWQ←PW-PES
```

B.23   APL Programs

```
[21]    PLQ←PLAMBDA×PWQ
[22]    WQ←+/PROB×PWQ
[23]    LQ←LAMBDA×WQ
[24]    W←+/PROB×PW
[25]    L←LAMBDA×W
[26]    'ES= ',⍕ES
[27]    'ES2= ',⍕ES2
[28]    'LAMBDA= ',⍕LAMBDA
[29]    'RHO= ',⍕RHO
[30]    'WQ= ',⍕WQ
[31]    'W= ',⍕W
[32]    'LQ= ',⍕LQ
[33]    'L= ',⍕L
[34]    'THE AVERAGE QUEUEING TIMES FOR THE '
[35]    'THE RESPECTIVE PRIORITY CLASSES ARE:'
[36]    PWQ
[37]    'THE CORRESPONDING SYSTEM TIMES ARE:'
[38]    PW
[39]    'THE AVERAGE NUMBER IN THE SYSTEM FROM THE'
[40]    'RESPECTIVE CLASSES IS:'
[41]    PL
[42]    'THE CORRESPONDING MEAN NUMBER QUEUEING IS:'
[43]    PLQ
     ∇

     ∇ P←K PZERO Z
[1]    ⍝THIS IS A FUNCTION WHICH CALCULATES THE PROBABILITY THERE
[2]    ⍝ARE NO MACHINES DOWN FOR THE M/G/1/K/K MACHINE REPAIR SYSTEM.
[3]    ⍝K IS THE NUMBER OF MACHINES AND Z=EO÷ES.
[4]    P←÷1+(K÷Z)×+/((⁻1+⍳K)!J)×1,×\(*((⍳J←(K-1))÷Z))-1
     ∇

     ∇ PLΔPES PΔQUEUE PES2
[1]    ⍝PΔQUEUE IS A PROGRAM FOR CALCULATING THE STATISTICS FOR
[2]    ⍝AN M/G/1 NONPREEMPTIVE QUEUEING SYSTEM (AN HOL SYSTEM)
[3]    ⍝THE CALL IS LINE 0 WHERE PLΔPES IS THE CATENATION OF
[4]    ⍝PL, THE VECTOR OF MEAN ARRIVAL RATES OF PRIORITY CLASSES,
[5]    ⍝WITH ES, THE VECTOR OF MEAN SERVICE TIME OF CLASSES.
[6]    ⍝PES2 IS THE VECTOR OF SECOND MOMENTS OF SERVICE TIME.
[7]    N←⍴PES2
[8]    PLAMBDA←N↑PLΔPES
[9]    PES←(-N)↑PLΔPES
[10]   LAMBDA←+/PLAMBDA
[11]   PROB←PLAMBDA÷LAMBDA
[12]   ES←+/PROB×PES
[13]   ES2←+/PROB×PES2
[14]   PU←0,+\PLAMBDA×PES
[15]   DIV1←N↑(1-PU)
[16]   DIV2←(-N)↑(1-PU)
[17]   PWQ←0.5×LAMBDA×ES2÷DIV1×DIV2
[18]   PW←PWQ+PES
[19]   PL←PLAMBDA×PW
[20]   PLQ←PLAMBDA×PWQ
[21]   WQ←+/PROB×PWQ
[22]   LQ←LAMBDA×WQ
[23]   W←WQ+ES
[24]   L←LAMBDA×W
[25]   'ES= ',⍕ES
[26]   'ES2= ',⍕ES2
[27]   'LAMBDA= ',⍕LAMBDA
[28]   'RHO= ',⍕RHO
[29]   'WQ= ',⍕WQ
[30]   'W= ',⍕W
[31]   'LQ= ',⍕LQ
[32]   'L= ',⍕L
[33]   'THE AVERAGE QUEUEING TIMES FOR THE '
[34]   'THE RESPECTIVE PRIORITY CLASSES ARE:'
[35]   PWQ
```

B.24 APL Programs

```
[36]    'THE CORRESPONDING SYSTEM TIMES ARE:'
[37]    PW
[38]    'THE AVERAGE NUMBER IN THE SYSTEM FROM THE'
[39]    'RESPECTIVE CLASSES IS:'
[40]    PL
[41]    'THE CORRESPONDING MEAN NUMBER QUEUEING IS:'
[42]    PLQ
      ∇

      ∇ Z←C RBCA A;B
[1]     Z←A÷1+A
[2]     →0×ιC=1
[3]     Z←B÷C+B←A×((C-1)RBCA A)
      ∇

      ∇ R←U ROUNDS A
[1]    ⍝ROUNDS FUNCTION FROM THE APL HANDBOOK OF TECHNIQUES
[2]    ⍝U IS A SCALAR OR CONFORMABLE STRUCTURE OF SPECIFIED UNITS
[3]     R←(×A)×U×⌊0.5+|A÷U
      ∇

      ∇ Z←SORT X
[1]     Z←X[⍋X]
      ∇

      ∇ STAT X
[1]     XBAR←MEAN X
[2]     S2←(+/(X-XBAR)*2)÷(N←ρX)-1
[3]     S←S2*0.5
[4]     'THE SAMPLE MEAN IS ',⍕XBAR
[5]     'THE SAMPLE VARIANCE IS ',⍕S2
[6]     'THE SAMPLE STANDARD DEVIATION IS ',⍕S
      ∇

      ∇ Z←C WBCA A
[1]     Z←÷+/×\1,⌽(ιC)÷A
      ∇

      ∇ Z←WH K1ΔK2ΔK3;K1;K2;K3;C2;X;Y;V;MU1;MU2;Q1;Q2
[1]    ⍝WH CALCULATES THE PARAMETERS FOR A TWO STAGE HYPEREXPONENTIAL
[2]    ⍝RANDOM VARIABLE WHICH HAS THE FIRST THREE MOMENTS EQUAL TO
[3]    ⍝K1, K2, AND K3, RESPECTIVELY.   THE CALL IS 'WH K1, K2, K3'.
[4]    ⍝THE OUTPUT IS THE VECTOR Q1ΔQ2ΔMU1ΔMU2  NEEDED AS THE LEFT
[5]    ⍝PARAMETER BY THE TWO STAGE HYPEREXPONENTIAL
[6]    ⍝DISTRIBUTION FUNCTION H2ΔDIST AND BY THE FUNCTION MOMENTS AS
[7]    ⍝THE RIGHT PARAMETER.  MOMENTS COMPUTES THE FIRST, SECOND AND
[8]    ⍝THIRD MOMENTS OF A TWO STAGE HYPEREXPONENTIAL DISTRIBUTION.
[9]     K1←K1ΔK2ΔK3[1]
[10]    K2←K1ΔK2ΔK3[2]
[11]    K3←K1ΔK2ΔK3[3]
[12]    C2←(K2÷K1*2)-1
[13]    →ERROR IF C2<1
[14]    X←(K1×K3)-1.5×K2*2
[15]    →ERROR IF X<0
[16]    Y←K2-2×K1*2
[17]    V←(((X+1.5×Y*2)-3×(K1*2)×Y)*2)+18×(K1*2)×Y*3
[18]    MU1←÷((X+(1.5×(Y*2))+(3×(K1*2)×Y))+V*0.5)÷6×K1×Y
[19]    MU2←÷((X+(1.5×(Y*2))+(3×(K1*2)×Y))-V*0.5)÷6×K1×Y
[20]    Q1←(K1-(÷MU2))÷((÷MU1)-(÷MU2))
[21]    Q2←1-Q1
[22]    Z←Q1,Q2,MU1,MU2
[23]    →0
[24] ERROR:'WHITT''S CRITERIA IS NOT SATISFIED'
[25]    →0
      ∇

      ∇ X←WΔT Z
[1]    ⍝WΔT CALCULATES THE NORMALIZED RESPONE TIME, THAT IS, THE
[2]    ⍝MEAN RESPONSE TIME DIVIDED BY THE MEAN THINK TIME, FOR THE
[3]    ⍝FINITE PROCESSOR SHARING MODEL WHEN THE NUMBER OF TERMINALS
[4]    ⍝IS WHAT KLEINROCK CALLS THE SATURATION NUMBER.   THE Z
[5]    ⍝IN THE CALL IS THE RATIO OF THE MEAN THINK TIME TO MEAN
[6]    ⍝SERVICE TIME.
[7]     X←((1+÷Z)÷(1-(1+Z)WBCU Z))-1
      ∇
```

Appendix C

Queueing Theory Formulas

In a lobby in South Tennessee
Teenage Pollaczek gained his "esprit"
 He watched as some guests
 Made the lineups congest,
Then he left, humming Fi Fo, Fum Fee!

Ben W. Lutek

Table 1, an extension of Table 5.1.2, provides the basic queueing theory notation and definitions for the queueing theory of Chapter 5. Table 2 explains the relationship between the random variables of queueing theory models. Tables 3 through 27 provide queueing theory formulas for the models of Chapter 5. Table 28 provides definitions for queueing theory networks. The remaining tables provide the queueing theory formulas for the models of Chapter 6. APL programs are displayed in Appendix B to implement the formulas for most of the models. All equations refer to steady state values.

C.1   Notation and Definitions

Table 1. Basic Queueing Theory Notation and Definitions

a	Traffic intensity $a = \lambda W_s$ or offered load. The international unit of traffic intensity is the erlang, named for A. K. Erlang, a queueing theory pioneer.
$A[t]$	Distribution function of interarrival time, $A[t] = P[\tau \leq t]$.
b	Random variable describing the busy period for a server.
$B[c, a]$	Erlang's B formula. Probability all servers busy in M/M/c/c system. Also called Erlang's loss formula.
c	Number of servers in service facility.
$C[c, a]$	Erlang's C formula. Probability all servers busy in M/M/c system. Also known as Erlang's delay formula.
C_X^2	Squared coefficient of variation of a positive random variable. $C_X^2 = \dfrac{\text{Var}[X]}{E[X]^2}$.
D	Symbol for constant (deterministic) interarrival or service time. Also used to represent a constant in several formulas.
E_k	Symbol for Erlang-k distribution of interarrival or service time.
$E[N_q \| N_q > 0]$	Expected (mean or average) queue length of nonempty queues.
$E[q \| q > 0]$	Expected (mean or average) queueing time for customers delayed.
FCFS	Symbol for "first-come, first-served" queue discipline.
FIFO	Symbol for "first-in, first-out" queue discipline. Identical with "first-come, first-served."
G	Symbol for general probability distribution of service time. Independence usually assumed.
GI	Symbol for general independent interarrival time distribution.
H_2	Symbol for two-stage hyperexponential distribution. Can be generalized to k stages.
K	Maximum number of customers allowed in queueing system. Also size of population in finite population models.
L	Expected steady state number of customers in the queueing system, $E[N]$.
$\ln(\cdot)$	Natural logarithm function (log to base e.)
L_q	Expected steady state number of customers in the queue, $E[N_q]$.

Table 1. Basic Queueing Theory Notation and Definitions (continued)

L_s	Expected steady state number of customers receiving service, $E[N_s]$.
LCFS	Symbol for "last-come, first-served" queue discipline.
LIFO	Symbol for "last-in, first-out" queue discipline. Same as LCFS.
λ	Mean arrival rate of customers into the system.
λ_a	Actual mean arrival rate *into* a queueing system for which some arrivals are turned away, e.g., the M/M/c/c system.
λ_T	Mean throughput of a computer system measured in transactions or interactions per unit time.
M	Symbol for exponential interarrival or service time.
μ	Mean service rate per server, that is, the mean rate of service completions while the server is busy.
μ_a, μ_b	Parameters of the two-stage hyperexponential distribution of w for the $M/H_2/1$ queueing system.
$N[t]$	Random variable describing the number of customers in the system at time t.
N	Random variable describing the steady state number of customers in the system.
$N_q[t]$	Random variable describing the number of customers in the queue at time t.
N_q	Random variable describing the steady state number of customers in the queue.
N_b	Random variable describing the number of customers served by a server in one busy period.
$N_s[t]$	Random variable describing the number of customers receiving service at time t.
N_s	Random variable describing the steady state number of customers in the service facility.
O	Operating time of a machine in a machine repair queueing model. The time a machine remains in operation after repair before repair is again necessary.
π_a, π_b	Parameters of the distribution function of w for the $M/H_2/1$ queueing system.
$\pi_X[r]$	The rth percentile for random variable X where X can be s, w, q, etc.
$p_n[t]$	Probability there are n customers in the system at time t.
p_n	Steady state probability that there are n customers in the system.

Table 1. Basic Queueing Theory Notation and Definitions (continued)

PRI	Symbol for priority queueing discipline.
PS	Symbol for "processor sharing" queue discipline.
q	Random variable describing the time a customer spends in the queue before service begins.
q_i	A parameter of a hyperexponential random variable.
q'	Random variable describing time a customer who must queue spends in the queue before receiving service. Also called *conditional queueing time*.
RSS	Symbol for queue discipline "random selection for service."
ρ	Server utilization $= \frac{\lambda}{c\mu} = \frac{E[N_s]}{c}$.
s	Random variable describing the service time. $E[s] = \frac{1}{\mu}$.
ρ_i	Utilization of component i in a queueing network. Also used to represent $\frac{\lambda}{\mu_i}$ for some M/G/1 systems.
SIRO	Symbol for "service in random order," which is identical to RSS. It means each customer in queue has the same probability of being served next.
τ	Random variable describing interarrival time. $E[\tau] = \frac{1}{\lambda}$.
U	Symbol for uniform interarrival or service time.
w	Random variable describing the total time a customer spends in the queueing system. $w = q + s$.
$W[t]$	Distribution function of w, $W[t] = P[w \leq t]$.
W	Expected steady state time a customer spends in the system. $W = E[w] = W_q + W_s$.
$W_q[t]$	Distribution function of q, $W_q[t] = P[q \leq t]$.
W_q	Expected steady state time a customer spends in the queue. $W_q = E[q] = W - W_s$.
$W_s[t]$	Distribution function of s, $W_s[t] = P[s \leq t]$.
W_s	Expected customer service time, $E[s] = \frac{1}{\mu}$.

C.2 Relationships between Random Variables

Table 2. Relationships between Random Variables

$a = \dfrac{E[s]}{E[\tau]} = \lambda W_s.$	Traffic intensity in erlangs.
$\rho = \dfrac{a}{c} = \dfrac{\lambda}{c\mu}.$	Server utilization. The probability any particular server is busy.
$w = q + s.$	Total waiting time in the system.
$W = W_q + W_s.$	Mean total waiting time in steady state system.
$N = N_q + N_s.$	Number of customers in steady state system.
$L = \lambda W$	Mean number of customers in steady state system. This formula often called *Little's law.*
$L_q = \lambda W_q.$	Mean number in steady state queue. This formula also called *Little's law.*
$L_s = \lambda W_s$	Mean number of customers receiving service in steady state system. This formula sometimes called *Little's law.*

C.3   M/M/1 Queueing Formulas

Table 3. Formulas for M/M/1 Queueing System

$\rho = \lambda W_s, \quad p_n = P[N = n] = (1 - \rho)\rho^n, \quad n = 0, 1, \cdots.$

$P[N \geq n] = \rho^n \quad n = 0, 1, \cdots. \quad$ See Exercise 2.

$L = E[N] = \lambda W = \dfrac{\rho}{1 - \rho}, \quad \sigma_N^2 = \dfrac{\rho}{(1 - \rho)^2}.$

$L_q = \lambda W_q = \dfrac{\rho^2}{1 - \rho}, \quad \sigma_{N_q}^2 = \dfrac{\rho^2(1 + \rho - \rho^2)}{(1 - \rho)^2}. \quad$ See Exercise 4.

$E[N_q | N_q > 0] = \dfrac{1}{1 - \rho}, \quad \operatorname{Var}[N_q | N_q > 0] = \dfrac{\rho}{(1 - \rho)^2}. \quad$ See Exercise 3.

$W[t] = P[w \leq t] = 1 - \exp\left(\dfrac{-t}{W}\right), \quad P[w > t] = \exp\left(\dfrac{-t}{W}\right).$

$W = E[w] = \dfrac{W_s}{1 - \rho}, \quad \sigma_w^2 = W^2.$

$\pi_w[r] = W \ln\left(\dfrac{100}{100 - r}\right), \quad \pi_w[90] = W \ln 10, \quad \pi_w[95] = W \ln 20$

$W_q[t] = P[q \leq t] = 1 - \rho \exp\left(\dfrac{-t}{W}\right), \quad P[q > t] = \rho \exp\left(\dfrac{-t}{W}\right).$

$W_q = \dfrac{\rho W_s}{1 - \rho}, \quad \sigma_q^2 = \dfrac{(2 - \rho)\rho W_s^2}{(1 - \rho)^2}. \quad$ See Exercise 8.

$\pi_q[r] = \max\left\{ W \ln\left(\dfrac{100\rho}{100 - r}\right), 0 \right\}.$

$\pi_q[90] = \max\{W \ln(10\rho), 0\}, \quad \pi_q[95] = \max\{W \ln(20\rho), 0\}.$

C.4   M/M/1/K Queueing Formulas

Table 4. Formulas for M/M/1/K Queueing System

$$
p_n = \begin{cases} \dfrac{(1-a)a^n}{(1-a^{K+1})} & \text{if } \lambda \neq \mu, \\[4mm] \dfrac{1}{K+1} & \text{if } \lambda = \mu, \end{cases}
$$

for $n = 0, 1, \cdots, K$, where $a = \lambda W_s$.

$\lambda_a = (1 - p_K)\lambda$.   Mean arrival rate into system.

$$
L = \begin{cases} \dfrac{a[1 - (K+1)a^K + Ka^{K+1}]}{(1-a)(1-a^{K+1})} & \text{if } \lambda \neq \mu, \\[4mm] \dfrac{K}{2} & \text{if } \lambda = \mu. \end{cases}
$$

$$
L_q = L - (1 - p_0), \quad q_n = \frac{p_n}{1 - p_K}, n = 0, 1, \cdots, K - 1.
$$

$$
W[t] = 1 - \sum_{n=0}^{K-1} q_n Q[n; \mu t],
$$

where

$$
Q[n; \mu t] = e^{-\mu t} \sum_{k=0}^{n} \frac{\mu t}{k!}.
$$

$$
W = \frac{L}{\lambda_a}, \quad W_q = \frac{L_q}{\lambda_a}.
$$

$$
W_q[t] = 1 - \sum_{n=0}^{K-2} q_{n+1} Q[n; \mu t].
$$

$$
E[q|q > 0] = \frac{W_q}{1 - p_0}, \quad \rho = (1 - p_K)a.
$$

C.5   M/M/c Queueing Formulas

Table 5. Formulas for M/M/c Queueing System

$a = \lambda W_s, \quad \rho = \frac{a}{c}.$

$$p_0 = \left[\sum_{n=0}^{c-1} \frac{a^n}{n!} + \frac{a^c}{c!(1-\rho)} \right]^{-1} = \frac{c!(1-\rho)P[N \geq c]}{a^c}.$$

$$p_n = \begin{cases} \dfrac{a^n}{n!} p_0 & \text{if } n \leq c, \\[3mm] \dfrac{a^n}{c!c^{n-c}} p_0 & \text{if } n \geq c. \end{cases}$$

$$P[N \geq n] = \begin{cases} p_0 \left[\displaystyle\sum_{k=n}^{c-1} \frac{a^k}{k!} + \frac{a^c}{c!(1-\rho)} \right] & \text{if } n < c \\[5mm] p_0 \left[\dfrac{a^c \rho^{n-c}}{c!(1-\rho)} \right] = P[N \geq c]\rho^{n-c} & \text{if } n \geq c \end{cases}$$

$$L_q = \lambda W_q = \frac{aP[N \geq c]}{c(1-\rho)},$$

where

$$P[N \geq c] = C[c,a] = \frac{\dfrac{a^c}{c!}}{(1-\rho) \displaystyle\sum_{n=0}^{c-1} \frac{a^n}{n!} + \frac{a^c}{c!}}.$$

Table 5. Formulas for M/M/c Queueing System (continued)

$$\sigma^2_{N_q} = \frac{\rho C[c,a][1 + \rho - \rho C[c,a]]}{(1 - \rho)^2}.$$

$$L = \lambda W = L_q + a.$$

$$\sigma^2_N = \sigma^2_{N_q} + a(1 + P[N \geq c]). \quad W_q[0] = 1 - P[N \geq c].$$

$$W_q[t] = 1 - P[N \geq c] \exp[-c\mu t(1 - \rho)]. \quad W_q = \frac{P[N \geq c]W_s}{c(1 - \rho)}.$$

$$\sigma^2_q = \frac{[2 - C[c,a]]C[c,a]W_s^2}{c^2(1 - \rho)^2}.$$

$$\pi_q[r] = \max\{0, \frac{W_s}{c(1 - \rho)} \ln\left(\frac{100C[c,a]}{100 - r}\right)\}.$$

$$\pi_q[90] = \max\{0, \frac{W_s}{c(1 - \rho)} \ln(10C[c,a])\}.$$

$$\pi_q[95] = \max\{0, \frac{W_s}{c(1 - \rho)} \ln(20C[c,a])\}.$$

$$W_{q'} = P[q \leq t|q > 0] = 1 - \exp\left(\frac{-ct(1 - \rho)}{W_s}\right), \quad t > 0.$$

$$E[q|q > 0] = E[q'] = \frac{W_s}{c(1 - \rho)}.$$

$$\text{Var}[q|q > 0] = \left(\frac{W_s}{c(1 - \rho)}\right)^2.$$

$$W[t] = \begin{cases} 1 + C_1 e^{-\mu t} + C_2 e^{-c\mu t(1-\rho)} & \text{if } a \neq c - 1 \\ 1 - \{1 + C[c,a]\mu t\}e^{-\mu t} & \text{if } a = c - 1 \end{cases}$$

Table 5. Formulas for M/M/c Queueing System (continued)

where

$$C_1 = \frac{P[N \geq c]}{1 - c(1 - \rho)} - 1,$$

and

$$C_2 = \frac{P[N \geq c]}{c(1 - \rho) - 1}.$$

$$W = W_q + Ws.$$

$$E[w^2] = \begin{cases} \dfrac{2P[N \geq c][1 - c^2(1 - \rho)^2]W_s^2}{(a + 1 - c)c^2(1 - \rho)^2} + 2W_s^2 & \text{if } a \neq c - 1 \\[2ex] 2\{2P[N \geq c] + 1\}W_s^2 & \text{if } a = c - 1 \end{cases}$$

$$\sigma_w^2 = E[w^2] - W^2.$$

$\pi_w[90] \approx W + 1.3\sigma_w, \quad \pi_w[95] \approx W + 2\sigma_w$ (estimates due to James Martin).

C.6 M/M/2 Queueing Formulas

Table 6. Formulas for M/M/2 Queueing System

$a = \lambda W_s, \quad \rho = \frac{a}{2}.$

$p_0 = \dfrac{1 - \rho}{1 + \rho}.$

$p_n = 2p_0\rho^n, \quad n = 1, 2, 3, \cdots$

$P[N \geq n] = \dfrac{2\rho^n}{1 + \rho}, \quad n = 1, 2, \cdots$

$L_q = \lambda W_q = \dfrac{2\rho^3}{1 - \rho^2},$

$P[N \geq 2] = C[2, a]$ is the probability that an arriving customer must queue for service. $P[N \geq 2]$ is given by

$P[N \geq 2] = C[2, a] = \dfrac{2\rho^2}{1 + \rho}.$

$\sigma^2_{N_q} = \dfrac{2\rho^3[(1 + \rho)^2 - 2\rho^3]}{(1 - \rho^2)^2}.$

$L = \lambda W = L_q + a = \dfrac{2\rho}{1 - \rho^2}.$

$\sigma^2_N = \sigma^2_{N_q} + \dfrac{2\rho(1 + \rho + 2\rho^2)}{1 + \rho}.$

$W_q[0] = \dfrac{1 + \rho - 2\rho^2}{1 + \rho}.$

Table 6. Formulas for M/M/2 Queueing System (continued)

$$W_q[t] = 1 - \frac{2\rho^2}{1+\rho} \exp[-2\mu t(1-\rho)]$$

$$W_q = \frac{\rho^2 W_s}{1 - \rho^2}.$$

$$\sigma_q^2 = \frac{\rho^2(1+\rho-\rho^2)W_s^2}{(1-\rho^2)^2}.$$

$$\pi_q[r] = \max\{0, \frac{W_s}{2(1-\rho)} \ln\left(\frac{200\rho^2}{(100-r)(1+\rho)}\right)\}.$$

$$\pi_q[90] = \max\{0, \frac{W_s}{2(1-\rho)} \ln\left(\frac{20\rho^2}{1+\rho}\right)\}.$$

$$\pi_q[95] = \max\{0, \frac{W_s}{2(1-\rho)} \ln\left(\frac{40\rho^2}{1+\rho}\right)\}.$$

$$W_{q'} = P[q \le t | q > 0] = 1 - \exp\left(\frac{-2t(1-\rho)}{W_s}\right), \quad t > 0.$$

$$E[q|q > 0] = E[q'] = \frac{W_s}{2(1-\rho)}.$$

$$\text{Var}[q|q > 0] = \left(\frac{W_s}{2(1-\rho)}\right)^2.$$

$$W[t] = \begin{cases} 1 + \dfrac{1-\rho}{1-\rho^2-2\rho^2}e^{-\mu t} + \dfrac{2\rho^2}{1-\rho-2\rho^2}e^{-2\mu t(1-\rho)} & \text{if } a \ne 1 \\[3ex] 1 - \{1 + \dfrac{\mu t}{3}\}e^{-\mu t} & \text{if } a = 1. \end{cases}$$

Table 6. Formulas for M/M/2 Queueing System (continued)

$$W = W_q + W_s = \frac{W_s}{1 - \rho^2}.$$

$$E[w^2] = \begin{cases} \dfrac{\rho^2[1 - 4(1 - \rho)^2]W_s^2}{(2\rho - 1)(1 - \rho)(1 - \rho^2)} + 2W_s^2 & \text{if } a \neq 1 \\[3mm] \dfrac{10}{3}W_s^2 & \text{if } a = 1 \end{cases}$$

$$\sigma_w^2 = E[w^2] - W^2.$$

$$\pi_w[90] \approx W + 1.3\sigma_w, \quad \pi_w[95] \approx W + 2\sigma_w$$

C.7   M/M/c/c Queueing Formulas

Table 7. Formulas for M/M/c/c Queueing System (M/M/c loss)

$a = \lambda W_s$

$$p_n = \frac{\dfrac{a^n}{n!}}{1 + a + \dfrac{a^2}{2!} + \cdots + \dfrac{a^c}{c!}} \qquad n = 0, 1, \ldots, c.$$

The probability that all servers are busy, p_c, is called Erlang's B formula, $B[c, a]$, and thus,

$$B[c, a] = \frac{\dfrac{a^c}{c!}}{1 + a + \dfrac{a^2}{2!} + \cdots + \dfrac{a^c}{c!}}.$$

$\lambda_a = \lambda(1 - B[c, a])$ is the average arrival rate of customers who actually enter the system. Thus, the true server utilization, ρ, is given by

$$\rho = \frac{\lambda_a W_s}{c}.$$

$$L = \lambda_a W_s.$$

$$W = \frac{L}{\lambda_a} = W_s.$$

$$W[t] = 1 - \exp\left(\frac{-t}{W_s}\right).$$

Table 7. Formulas for M/M/c/c Queueing System (M/M/c loss) (continued)

All of the formulas except the last one are true for the M/G/c/c queueing system. For this system we have

$$W[t] = W_s[t],$$

where $W_s[\cdot]$ is the distribution function for service time.

C.8 M/M/c/K/K Queueing Formulas

Table 8. Formulas for M/M/c/K Queueing System

$a = \lambda W_s.$

$$p_0 = \left[\sum_{n=0}^{c} \frac{a^n}{n!} + \frac{a^c}{c!} \sum_{n=1}^{K-c} \left(\frac{a}{c} \right)^n \right]^{-1}.$$

$$p_n = \begin{cases} \dfrac{a^n}{n!} p_0 & \text{if } n = 1, 2, \ldots, c, \\[2ex] \dfrac{a^n}{c!} \left(\dfrac{a}{c} \right)^{n-c} p_0 & \text{if } n = c+1, \ldots, K. \end{cases}$$

The average arrival rate of customers who actually enter the system is $\lambda_a = \lambda(1 - p_K$

The actual mean server utilization, ρ, is given by

$$\rho = \frac{\lambda_a W_s}{c}.$$

Table 8. Formulas for M/M/c/K Queueing System (continued)

$$L_q = \frac{a^c r p_0}{c!(1-r)^2} \left[1 + (K-c)r^{K-c+1} - (K-c+1)r^{K-c} \right],$$

where

$$r = \frac{a}{c}.$$

$$L = L_q + E[N_s] = L_q + \sum_{n=0}^{c-1} n\, p_n + c \left(1 - \sum_{n=0}^{c-1} p_n \right).$$

By Little's law,

$$W_q = \frac{L_q}{\lambda_a},$$

and

$$W = \frac{L}{\lambda_a}.$$

$$q_n = \frac{p_n}{1 - p_K}, \quad n = 0, 1, 2, \ldots, K-1,$$

where q_n is the probability that an arriving customer who enters the system finds n customers already there.

$$E[q|q > 0] = \frac{W_q}{1 - \sum_{n=0}^{c-1} q_n}.$$

C.9 M/M/∞ Queueing Formulas

Table 9. Formulas for M/M/∞ Queueing System

$a = \lambda W_s.$

$$p_n = \frac{a^n}{n!} e^{-a}, \quad n = 0, 1, \cdots.$$

Since N has a Poisson distribution,

$$L = a \quad \text{and} \quad \sigma_N^2 = a.$$

By Little's law,

$$W = \frac{L}{\lambda} = W_s.$$

Since there is no queueing for service,
$W_q = L_q = 0,$
and
$W[t] = P[w \le t] = W_s[t];$
that is, w has the same distribution as s. All the above formulas are true
for the M/G/∞ queueing system, also.

C.10   M/M/1/K/K Queueing Formulas

Table 10. Formulas for M/M/1/K/K Queueing System

The mean operating time per machine (sometimes called the *mean time to failure*, MTTF) is

$$E[O] = \frac{1}{\alpha}.$$

The mean repair time per machine (by one repairman) is

$$W_s = \frac{1}{\mu}.$$

The probability, p_0, that no machines are out of service is given by

$$p_0 = \left[\sum_{k=0}^{K} \frac{K!}{(K-k)!} \left(\frac{W_s}{E[O]} \right)^k \right]^{-1} = B[K, z],$$

where $B[\cdot, \cdot]$ is Erlang's B formula and

$$z = \frac{E[O]}{W_s}.$$

Then, p_n, the probability that n machines are out of service, is given by

$$p_n = \frac{K!}{(K-n)!} z^{-n} p_0, \quad n = 0, 1, \ldots, K.$$

The formula for p_n can also be written in the form

$$p_n = \frac{\dfrac{z^{K-n}}{(K-n)!}}{\displaystyle\sum_{k=0}^{K} \frac{z^k}{k!}}, \quad n = 0, 1, \ldots, K.$$

$$\rho = 1 - p_0.$$

$$\lambda = \frac{\rho}{W_s}.$$

$$W = \frac{K}{\lambda} - E[O].$$

$$L = \lambda W.$$

$$W_q = W - W_s.$$

Table 10. Formulas for M/M/1/K/K Queueing System (continued)

$$q_n = \frac{(K-n)p_n}{K-L} = \frac{\dfrac{z^{K-n-1}}{(K-n-1)!}}{\displaystyle\sum_{k=0}^{K-1} \dfrac{z^k}{k!}}, \quad n = 0, 1, 2, \ldots, K-1,$$

where q_n is the probability that a machine that breaks down finds n machines in the repair facility.

$$W[t] = P[w \le t] = 1 - \frac{Q(K-1; z+t\mu)}{Q(K-1; \mu)}, \quad t \ge 0,$$

where

$$Q(n; x) = e^{-x} \sum_{k=0}^{n} \frac{x^k}{k!}.$$

$$W_q[t] = P[q \le t] = 1 - \frac{Q(K-2; z+t\mu)}{Q(K-1; z)}, \quad t \ge 0.$$

$$E[q|q > 0] = \frac{W_q}{1 - q_0}.$$

C.11   M/G/1/K/K Queueing Formulas

Table 11. Formulas for M/G/1/K/K Queueing System

___

The mean operating time per machine (sometimes called the *mean time to failure*, MTTF) is

$E[O] = \frac{1}{\alpha}.$

The mean repair time per machine (by one repairman) is

$W_s = \frac{1}{\mu}.$

The probability, p_0, that no machines are out of service is given by

$$p_0 = \left[1 + \frac{KW_s}{E[O]} \sum_{n=0}^{K-1} \binom{K-1}{n} B_n \right]^{-1},$$

where

$$B_n = \begin{cases} 1 & \text{for } n = 0 \\[2mm] \prod_{i=1}^{n} \left(\dfrac{1 - W_s^*[i\alpha]}{W_s^*[i\alpha]} \right) & \text{for } n = 1, 2, \ldots, K-1, \end{cases}$$

and $W_s^*[\theta]$ is the Laplace–Stieltjes transform of s.

$\rho = 1 - p_0.$

$\lambda = \frac{\rho}{W_s}.$

$W = \frac{K}{\lambda} - E[O].$

$L = \lambda W.$

$W_q = W - W_s.$

$L_q = \lambda W_q.$

The derivation of these equations can be found in *Priority Queues* by N. K. Jaiswal, Academic Press, 1968.

___

C.12 M/M/c/K/K Queueing Formulas

Table 12. Formulas for M/M/c/K/K Queueing System

The mean operating time per machine (sometimes called the *mean time to failure*, MTTF) is

$E[O] = \frac{1}{\alpha}$.

The mean repair time per machine (by one repairman) is

$W_s = \frac{1}{\mu}$.

The probability, p_0, that no machines are out of service, is given by

$$p_0 = \left[\sum_{k=0}^{c} \binom{K}{k} z^{-k} + \sum_{k=c+1}^{K} \frac{k!}{c! c^{k-c}} \binom{K}{k} z^{-k} \right]^{-1},$$

where

$$z = \frac{E[O]}{W_s}.$$

Then, p_n, the probability that n machines are out of service is given by

$$p_n = \begin{cases} \binom{K}{n} z^{-n} p_0 & n = 0, 1, \ldots, c \\[2ex] \dfrac{n!}{c! c^{n-c}} \binom{K}{n} z^{-n} p_0 & n = c+1, \ldots, K. \end{cases}$$

$$L_q = \sum_{n=c+1}^{K} (n - c) p_n.$$

$$W_q = \frac{L_q(E[O] + W_s)}{K - L_q}.$$

$$\lambda = \frac{K}{E[O] + W_q + W_s}.$$

$$W = \frac{K}{\lambda} - E[O].$$

$$L = \lambda W.$$

Table 12. Formulas for M/M/c/K/K Queueing System (continued)

$$q_n = \frac{(K-n)p_n}{K-L},$$

where q_n is the probability that a machine which breaks down finds n inoperable machines already in the repair facility. We denote q_n by $q_n[K]$ to emphasize the fact that there are K machines. It can be shown that

$$q_n[K] = p_n[K-1], \quad n = 0, 1, \ldots, K-1.$$

$$p_n[K-1] = \frac{c^c}{c!} \frac{p(K-n-1; cz)}{p(K-1; cz)} p_0[K-1],$$

where, of course,

$$p(k; \alpha) = \frac{\alpha^k}{k!} e^{-\alpha}.$$

$$W_q[t] = P[q \le t] = 1 - \frac{c^c Q(K-c-1; cz) p_0[K-1]}{c! p(K-1; cz)}, \quad t \ge 0,$$

where

$$Q(k; \alpha) = e^{-\alpha} \sum_{n=0}^{k} \frac{\alpha^n}{n!}.$$

$$W[t] = P[w \le t] = 1 - C_1 \exp(-t/W_s) + C_2 \frac{Q(K-c-1; c(z+t\mu))}{Q(K-c-1; cz)}, \quad t \ge 0,$$

where $C_1 = 1 + C_2$ and

$$C_2 = \frac{c^c Q(K-c-1; cz)}{c!(c-1)(K-c-1)! p(K-1; cz)} p_0[K-1].$$

The probability that a machine that breaks down must wait for repair is given by

$$D = \sum_{n=c}^{K-1} q_n = 1 - \sum_{n=0}^{c-1} q_n.$$

$$E[q|q > 0] = \frac{W_q}{D}.$$

C.13 D/D/c/K/K Queueing Formulas

Table 13. Formulas for D/D/c/K/K Queueing System

The mean operating time per machine (sometimes called the *mean time to failure*, MTTF) is

$$E[O] = \frac{1}{\alpha}.$$

The mean repair time per machine (by one repairman) is

$$W_s = \frac{1}{\mu}.$$
$$\rho = \min\{1, \frac{K}{c(1+z)}\},$$
where

$$z = \frac{E[O]}{W_s}.$$

$$\lambda = c\rho\mu = \frac{c\rho}{W_s}.$$

$$W = \frac{K}{\lambda} - E[O].$$

$$L = \lambda W.$$

$$W_q = W - W_s.$$

$$L_q = \lambda W_q.$$

The equations for this model are derived in "A straightforward model of computer performance prediction" by John W. Boyse and David R. Warn in *ACM Comput. Surveys*, **7(2)**, (June 1972).

C.14   M/G/1 Queueing Formulas

Table 14. Formulas for M/G/1 Queueing System

The z-transform of N, the steady-state number of customers in the system, is given by

$$g_N(z) = \sum_{n=0}^{\infty} p_n z^n = \frac{(1-\rho)(1-z)W_s^*[\lambda(1-z)]}{W_s^*[\lambda(1-z)] - z},$$

where $W_s^*[\theta]$ is the Laplace–Stieltjes transform of the service time s. The Laplace–Stieltjes transforms of w and q are given by

$$W^*[\theta] = \frac{(1-\rho)\theta W_s^*[\theta]}{\theta - \lambda + \lambda W_s^*[\theta]},$$

and

$$W_q^*[\theta] = \frac{(1-\rho)\theta}{\theta - \lambda + \lambda W_s^*[\theta]},$$

respectively. Each of the three transforms above is called the *Pollaczek–Khintchine transform equation* by various authors. The probability, p_0, of no customers in the system has the simple and intuitive equation $p_0 = 1-\rho$, where the server utilization $\rho = \lambda W_s$. The probability that the server is busy is $P[N \geq 1] = \rho$.

$$W_q = \frac{\lambda E[s^2]}{2(1-\rho)} = \frac{\rho W_s}{1-\rho}\left(\frac{1+C_s^2}{2}\right) \quad \text{(Pollaczek's formula)}.$$

$$L_q = \lambda W_q.$$

$$\sigma_{N_q}^2 = \frac{\lambda^3 E[s^3]}{3(1-\rho)} + \left(\frac{\lambda^2 E[s^2]}{2(1-\rho)}\right)^2 + \frac{\lambda^2 E[s^2]}{2(1-\rho)}.$$

$$E[q|q>0] = \frac{W_s}{1-\rho}\left(\frac{1+C_s^2}{2}\right).$$

$$E[q^2] = 2W_q^2 + \frac{\lambda E[s^3]}{3(1-\rho)}.$$

$$\sigma_q^2 = E[q^2] - W_q^2.$$

$$W = W_q + W_s.$$

Table 14. Formulas for M/G/1 Queueing System (continued)

$L = \lambda W = L_q + \rho.$

$$\sigma_N^2 = \frac{\lambda^3 E[s^3]}{3(1 - \rho)} + \left(\frac{\lambda^2 E[s^2]}{2(1 - \rho)}\right)^2 + \frac{\lambda^2(3 - 2\rho)E[s^2]}{2(1 - \rho)} + \rho(1 - \rho).$$

$$E[w^2] = E[q^2] + \frac{E[s^2]}{1 - \rho}.$$

$\sigma_w^2 = E[w^2] - W^2.$

$\pi_w[90] \approx W + 1.3\sigma_w, \quad \pi_w[95] \approx W + 2\sigma_w.$

Table 15. Formulas for $M/H_2/1$ Queueing System

The z-transform of the steady-state number in the system, N, is given by

$$g_N(z) = \sum_{n=0}^{\infty} p_n z^n = C_1 \frac{z_1}{z_1 - z} + C_2 \frac{z_2}{z_2 - z},$$

where z_1 and z_2 are the roots of the equation

$\rho_1\rho_2 z^2 - (\rho_1 + \rho_2 + \rho_1\rho_2)z + 1 + \rho_1 + \rho_2 - \rho = 0,$

where

$\rho = \lambda W_s,$

$\rho_i = \dfrac{\lambda}{\mu_i}, \quad i = 1, 2,$

$C_1 = \dfrac{(z_1 - 1)(1 - \rho z_2)}{z_1 - z_2},$

and

$C_2 = \dfrac{(z_2 - 1)(1 - \rho z_1)}{z_2 - z_1}.$

It follows from the formula for $g_N(z)$ above that

$p_n = C_1 z_1^{-n} + C_2 z_2^{-n}, \quad n = 0, 1, \cdots.$

Table 15. Formulas for $M/H_2/1$ Queueing System (continued)

In particular, $p_0 = 1 - \rho$.
$$P[N \geq n] = C_1 \frac{z_1^{-n+1}}{z_1 - 1} + C_2 \frac{z_2^{-n+1}}{z_2 - 1}.$$
It follows from this formula that
$$P[N \geq 1] = \rho.$$
$$W_q[t] = P[q \leq t] = 1 - C_5 e^{-at} - C_6 e^{-bt}, \ t \geq 0,$$
where $a = -\zeta_1$, $b = -\zeta_2$, for ζ_1, ζ_2 the roots of
$$\theta^2 + (\mu_1 + \mu_2 - \lambda)\theta + \mu_1\mu_2(1 - \rho) = 0,$$
and where
$$C_5 = \frac{\lambda(1 - \rho)\zeta_1 + \rho(1 - \rho)\mu_1\mu_2}{a(\zeta_1 - \zeta_2)}$$
and
$$C_6 = \frac{\lambda(1 - \rho)\zeta_2 + \rho(1 - \rho)\mu_1\mu_2}{b(\zeta_2 - \zeta_1)}.$$
$$W_q = \frac{\lambda E[s^2]}{2(1 - \rho)} = \frac{\rho W_s}{1 - \rho}\left(\frac{1 + C_s^2}{2}\right). \text{ (Pollaczek's formula)}$$
$$E[q|q > 0] = \frac{W_s}{1 - \rho}\left(\frac{1 + C_s^2}{2}\right).$$
$$E[q^2] = 2W_q^2 + \frac{\lambda E[s^3]}{3(1 - \rho)}.$$
In this formula we substitute
$$E[s^3] = \frac{6q_1}{\mu_1^3} + \frac{6q_2}{\mu_2^3}.$$
Then
$$\sigma_q^2 = E[q^2] - W_q^2.$$
$$W[t] = P[w \leq t] = 1 - \pi_a e^{-\mu_a t} - \pi_b e^{-\mu_b t}, \quad t \geq 0,$$
where
$$\pi_a = C_1\frac{z_1}{z_1 - 1},$$
$$\pi_b = C_2\frac{z_2}{z_2 - 1},$$

Table 15. Formulas for $M/H_2/1$ Queueing System (continued)

$\mu_a = \lambda(z_1 - 1)$,

and

$\mu_b = \lambda(z_2 - 1)$.

$W = W_q + W_s$.

$E[w^2] = E[q^2] + \dfrac{E[s^2]}{1 - \rho}$,

where, of course,

$E[s^2] = \dfrac{2q_1}{\mu_1^2} + \dfrac{2q_2}{\mu_2^2}$.

$\sigma_w^2 = E[w^2] - W^2$.

$C_w^2 = \dfrac{E[w^2]}{W^2} - 1$.

$L_q = \lambda W = \dfrac{\rho^2}{1 - \rho}\left(\dfrac{1 + C_s^2}{2}\right)$.

$\sigma_{N_q}^2 = \dfrac{\lambda^3 E[s^3]}{3(1 - \rho)} + \left(\dfrac{\lambda^2 E[s^2]}{2(1 - \rho)}\right)^2 + \dfrac{\lambda^2 E[s^2]}{2(1 - \rho)}$.

$L = \lambda W = L_q + \rho$.

$\sigma_N^2 = \dfrac{\lambda^3 E[s^3]}{3(1 - \rho)} + \left(\dfrac{\lambda^2 E[s^2]}{2(1 - \rho)}\right)^2 + \dfrac{\lambda^2(3 - 2\rho)E[s^2]}{2(1 - \rho)} + \rho(1 - \rho)$.

Table 16. Formulas for M/Gamma/1 Queueing System

Since s has a gamma distribution,

$$E[s^n] = \frac{\beta(\beta+1)\cdots(\beta+n-1)}{\alpha^n}, \quad n = 1, 2, \cdots.$$

Since

$$C_s^2 = \frac{1}{\beta},$$

this means that

$$E[s^2] = W_s^2(1 + C_s^2),$$
$$E[s^3] = W_s^3(1 + C_s^2)(1 + 2C_s^2),$$

and

$$E[s^n] = W_s^n \prod_{k=1}^{n-1}(1 + kC_s^2), \quad n = 1, 2, \cdots.$$

The probability, p_0, of no customers in the system is given by
$p_0 = 1 - \rho$,
where $\rho = \lambda W_s$. The probability the server is busy is
$P[N \geq 1] = \rho$.

$$W_q = \frac{\lambda E[s^2]}{2(1-\rho)} = \frac{\rho W_s}{1-\rho}\left(\frac{1+C_s^2}{2}\right) \quad \text{(Pollaczek's formula)}.$$

$$L_q = \lambda W_q.$$

$$\sigma_{N_q}^2 = \frac{\rho^2(1+C_s^2)}{2(1-\rho)}\left[1 + \frac{\rho^2(1+C_s^2)}{2(1-\rho)} + \frac{2\rho(1+2C_s^2)}{3}\right].$$

$$E[q|q > 0] = \frac{W_s}{1-\rho}\left(\frac{1+C_s^2}{2}\right).$$

$$E[q^2] = 2W_q^2 + \frac{\rho W_s^2(1+C_s^2)(1+2C_s^2)}{3(1-\rho)}.$$

$$\sigma_q^2 = E[q^2] - W_q^2.$$

$$W = W_q + W_s.$$

Table 16. Formulas for M/Gamma/1 Queueing System (continued)

$L = \lambda W = L_q + \rho.$

$$\sigma_N^2 = \frac{\rho^3(1 + C_s^2)(1 + 2C_s^2)}{3(1 - \rho)} + \left(\frac{\rho^2(1 + C_s^2)}{2(1 - \rho)}\right)^2 + \frac{\rho^2(3 - 2\rho)(1 + C_s^2)}{2(1 - \rho)} + \rho(1 - \rho).$$

$$E[w^2] = E[q^2] + \frac{W_s^2(1 + C_s^2)}{1 - \rho}.$$

$$\sigma_w^2 = E[w^2] - W^2.$$

$$\pi_w[90] \approx W + 1.3\sigma_w, \quad \pi_w[95] \approx W + 2\sigma_w.$$

Table 17. Formulas for $M/E_k/1$ Queueing System

Since s has an Erlang-k distribution,

$$E[s^n] = \left(1 + \frac{1}{k}\right)\left(1 + \frac{2}{k}\right)\cdots\left(1 + \frac{n-1}{k}\right) W_s^n, \quad n = 1, 2, \cdots.$$

Thus,

$$E[s^2] = W_s^2\left(1 + \frac{1}{k}\right),$$

and

$$E[s^3] = W_s^3\left(1 + \frac{1}{k}\right)\left(1 + \frac{2}{k}\right).$$

The probability, p_0, of no customers in the system is given by

$p_0 = 1 - \rho,$

where $\rho = \lambda W_s$. The probability the server is busy is given by

$P[N \geq 1] = \rho.$

$$W_q = \frac{\lambda E[s^2]}{2(1 - \rho)} = \frac{\rho W_s}{1 - \rho}\left(\frac{1 + \dfrac{1}{k}}{2}\right) \quad \text{(Pollaczek's formula)}.$$

$L_q = \lambda W_q.$

$$\sigma_{N_q}^2 = \frac{\rho^2(1 + k)}{2k(1 - \rho)}\left[1 + \frac{\rho^2(1 + k)}{2k(1 - \rho)} + \frac{2\rho(k + 2)}{3k}\right].$$

Table 17. Formulas for $M/E_k/1$ Queueing System (continued)

$$E[q|q > 0] = \frac{W_s}{1 - \rho} \left(\frac{1 + \dfrac{1}{k}}{2} \right).$$

$$E[q^2] = 2W_q^2 + \frac{\rho W_s^2(k + 1)(k + 2)}{3k^2(1 - \rho)}.$$

$$\sigma_q^2 = E[q^2] - W_q^2.$$

$$W = W_q + W_s.$$

$$L = \lambda W = L_q + \rho.$$

$$\sigma_N^2 = \frac{\rho^3(k + 1)(k + 2)}{3k^2(1 - \rho)} + \left(\frac{\rho^2 \left(1 + \dfrac{1}{k} \right)}{2(1 - \rho)} \right)^2 + \frac{\rho^2(3 - 2\rho) \left(1 + \dfrac{1}{k} \right)}{2(1 - \rho)} + \rho(1 - \rho).$$

$$E[w^2] = E[q^2] + \frac{W_s^2 \left(1 + \dfrac{1}{k} \right)}{1 - \rho}.$$

$$\sigma_w^2 = E[w^2] - W^2.$$

$$\pi_w[90] \approx W + 1.3\sigma_w, \quad \pi_w[95] \approx W + 2\sigma_w.$$

Table 18. Formulas for M/D/1 Queueing System

Since s has a constant distribution,

$E[s^n] = W_s^n, \quad n = 1, 2, \cdots.$

The generating function, $g_N(z)$, for the steady-state number of customers in the system is given by

$$g_N(z) = \frac{(1 - \rho)(1 - z)}{1 - ze^{\rho(1-z)}}.$$

If we assume

$|ze^{\rho(1-z)}| < 1,$

we can expand $g_N(z)$ in the geometric series

$$g_N(z) = (1 - \rho)(1 - z) \sum_{j=0}^{\infty} \left[ze^{\rho(1-z)} \right]^j.$$

Hisashi Kobayashi, in his book, *Modeling and Analysis*, Addison-Wesley, Reading MA, 1978, proved that, by comparing the coefficients of z^n in the above series and in the definition

$$g_N(z) = \sum_{n=0}^{\infty} p_n z^n,$$

it can be shown that

$p_0 = 1 - \rho,$

$p_1 = (1 - \rho)(e^\rho - 1),$

and

$$p_n = (1 - \rho) \sum_{j=1}^{n} \frac{(-1)^{n-j} (j\rho)^{n-j-1} (j\rho + n - j) e^{j\rho}}{(n - j)!} \quad n = 2, 3, \cdots.$$

The distribution function of q is given by

$$W_q[t] = \sum_{n=0}^{k-1} p_n + p_k \left(\frac{t - (k - 1)W_s}{W_s} \right),$$

where

$(k - 1)W_s \leq t < kW_s, \quad k = 1, 2, \ldots.$

Table 18. Formulas for M/D/1 Queueing System (continued)

Thus,

$$W_q[0] = p_0.$$

$$W_q = \frac{\rho W_s}{2(1 - \rho)}.$$

$$W[q|q > 0] = \frac{W_s}{2(1 - \rho)}.$$

$$E[q^2] = 2W_q^2 + \frac{\rho W_s^2}{3(1 - \rho)}.$$

$$\sigma_q^2 = E[q^2] - W_q^2.$$

$$L_q = \lambda W_q = \frac{\rho^2}{2(1 - \rho)}.$$

$$\sigma_{N_q}^2 = \frac{\rho^3}{3(1 - \rho)} + \left[\frac{\rho^2}{2(1 - \rho)}\right]^2 + \frac{\rho^2}{2(1 - \rho)}.$$

$$W[t] = \begin{cases} 0 & \text{for } t < W_s \\ \displaystyle\sum_{n=0}^{k-1} p_n + p_k \left(\frac{t - kW_s}{W_s}\right) & \text{for } t \geq W_s, \end{cases}$$

where

$$kW_s \leq t < (k + 1)W_s, \quad k = 1, 2, \ldots.$$

$$W = W_q + W_s.$$

$$E[w^2] = E[q^2] + \frac{W_s^2}{1 - \rho}.$$

$$\sigma_w^2 = E[w^2] - W^2$$

$$L = \lambda W = L_q + \rho.$$

$$\sigma_N^2 = \frac{\rho^3}{3(1 - \rho)} + \left(\frac{\rho^2}{2(1 - \rho)}\right)^2 + \frac{\rho^2(3 - 2\rho)}{2(1 - \rho)} + \rho(1 - \rho).$$

C.15 GI/M/1 Queueing Formulas

Table 19. Formulas for GI/M/1 Queueing System

The steady state probability that an arriving customer will find the system empty, π_0, is the unique solution of the equation $1 - \pi_0 = A^*[\mu\pi_0]$ such that $0 < \pi_0 < 1$, where $A^*[\theta]$ is the Laplace–Stieltjes transform of τ. The steady state number of customers in the system, N has the distribution $\{p_n\}$, where $p_0 = P[N = 0] = 1 - \rho$, and $p_n = \rho\pi_0(1 - \pi_0)^{n-1}$, $n = 1, 2, \cdots$, with

$$L = \frac{\rho}{\pi_0}, \quad \text{and} \quad \sigma_N^2 = \frac{\rho(2 - \pi_0 - \rho)}{\pi_0^2}.$$

$$L_q = \frac{(1 - \pi_0)\rho}{\pi_0}.$$

$$\sigma_{N_q}^2 = \frac{\rho(1 - \pi_0)(2 - \pi_0 - \rho(1 - \pi_0))}{\pi_0^2}.$$

$$E[N_q | N_q > 0] = \frac{1}{\pi_0}.$$

$$W = \frac{W_s}{\pi_0}.$$

$$W[t] = P[w \le t] = 1 - \exp(-t/W).$$

$$\pi_w[r] = W \ln\left[\frac{100}{100 - r}\right].$$

$$\pi_w[90] = W \ln 10, \quad \pi_w[95] = W \ln 20.$$

$$W_q = (1 - \pi_0)\frac{W_s}{\pi_0}.$$

$$\sigma_q^2 = (1 - \pi_0^2)\left(\frac{W_s}{\pi_0}\right)^2.$$

$$W_q[t] = P[q \le t] = 1 - (1 - \pi_0)\exp(-t/W).$$

$$\pi_q[r] = \max\left\{0, W \ln\left(\frac{100(1 - \pi_0)}{100 - r}\right)\right\}.$$

q', the queueing time for those who must, has the same distribution as w.

Table 20. π_0 versus ρ for GI/M/1 Queueing Systems[1]

ρ	E_2	E_3	U	D	H_2	H_2
0.100	0.970820	0.987344	0.947214	0.999955	0.815535	0.810575
0.200	0.906226	0.940970	0.887316	0.993023	0.662348	0.624404
0.300	0.821954	0.868115	0.817247	0.959118	0.536805	0.444949
0.400	0.724695	0.776051	0.734687	0.892645	0.432456	0.281265
0.500	0.618034	0.669467	0.639232	0.796812	0.343070	0.154303
0.600	0.504159	0.551451	0.531597	0.675757	0.263941	0.081265
0.700	0.384523	0.424137	0.412839	0.533004	0.191856	0.044949
0.800	0.260147	0.289066	0.284028	0.371370	0.124695	0.024404
0.900	0.131782	0.147390	0.146133	0.193100	0.061057	0.010575
0.950	0.066288	0.074362	0.074048	0.098305	0.030252	0.004999
0.980	0.026607	0.029899	0.029849	0.039732	0.012039	0.001941
0.999	0.001333	0.001500	0.001500	0.001999	0.000600	0.000095

C.16   GI/M/c Queueing Formulas

Table 21. Formulas for GI/M/c Queueing System

Let $\pi_n, n = 0, 1, 2, \cdots$ be the steady state number of customers that an arriving customer finds in the system. Then

$$\pi_n = \begin{cases} \sum_{i=n}^{c-1} (-1)^{i-n} \binom{i}{n} U_i, & n = 0, 1, \ldots, c - 2 \\ \\ D\omega^{n-c}, & n = c - 1, c, \ldots, \end{cases}$$

where ω is the unique root of the equation $\omega = A^*[c\mu(1 - \omega)]$ such that $0 < \omega < 1$, where $A^*[\theta]$ is the Laplace–Stieltjes transform of τ,
$g_j = A^*[j\mu], \quad j = 1, 2, \ldots, c,$

---

[1]For the H_2 distribution described in the next-to-last column, $q_1 = 0.4$, $\mu_1 = 0.5\lambda$, $\mu_2 = 3\lambda$. The H_2 distribution described in the last column was generated by Algorithm 3.2.2 of Chapter 3 with $C_\tau^2 = 20$. Thus, $q_1 = 0.024405$, $\mu_1 = 2q_1\lambda$, and $\mu_2 = 2q_2\lambda$.

Table 21. Formulas for GI/M/c Queueing System (continued)

$$
C_j = \begin{cases}
1, & j = 0, \\
\displaystyle\prod_{i=1}^{j} \left(\frac{g_i}{1 - g_i} \right), & j = 1, 2, \dots, c,
\end{cases}
$$

$$
D = \left[\frac{1}{1 - \omega} + \sum_{j=1}^{c} \frac{\dbinom{c}{j}}{C_j(1 - g_j)} \left(\frac{c(1 - g_j) - j}{c(1 - \omega) - j} \right) \right]^{-1},
$$

and

$$
U_n = D C_n \sum_{j=n+1}^{c} \frac{\dbinom{c}{j}}{C_j(1 - g_j)} \left(\frac{c(1 - g_j) - j}{c(1 - \omega) - j} \right), \quad n = 0, 1 \dots, c - 1.
$$

$$
W_q[t] = P[q \le t] = 1 - P[q > 0] e^{-c\mu(1-\omega)t}, \ t \ge 0,
$$

where

$$
P[q > 0] = \frac{D}{1 - \omega}. \quad W_q = \frac{D W_s}{c(1 - \omega)^2}. \quad E[q | q > 0] = \frac{W_s}{c(1 - \omega)}.
$$

If $c(1 - \omega) \ne 1$, then

$$
W[t] = P[w \le t] = 1 + (G - 1)e^{-\mu t} - G e^{-c\mu(1-\omega)t}, \ t \ge 0,
$$

where

$$
G = \frac{D}{(1 - \omega)[1 - c(1 - \omega)]}.
$$

When $c(1 - \omega) = 1$, then

$$
W[t] = P[w \le t] = 1 - \left[1 + \frac{D\mu t}{1 - \omega} \right] e^{-\mu t}, \ t \ge 0.
$$

We also have

$$
W = W_q + W_s.
$$

$$
p_0 = 1 - \frac{\lambda W_s}{c} - \lambda W_s \sum_{j=1}^{c-1} \pi_{j-1} \left(\frac{1}{j} - \frac{1}{c} \right).
$$

$$
p_n = \begin{cases}
\dfrac{\lambda W_s \pi_{n-1}}{n}, & n = 1, 2, \dots, c - 1 \\[2mm]
\dfrac{\lambda W_s \pi_{n-1}}{c}, & n = c, c + 1, \cdots.
\end{cases}
$$

C.17   M/G/1 Priority Queueing

Table 22. Formulas for M/G/1 Queueing System (classes, no priorities)

There are n customer classes. Customers from class i arrive in a Poisson pattern with mean arrival rate $\lambda_i, i = 1, 2, \ldots, n$. Each class has its own general service time with $E[s_i] = 1/\mu_i$, and finite second and third moments $E[s_i^2]$, $E[s_i^3]$. All customers are served on a FCFS basis with no consideration for class. The total arrival stream to the system has a Poisson arrival pattern with

$$\lambda = \lambda_1 + \lambda_2 + \cdots + \lambda_n.$$

The first three moments of service time are given by

$$W_s = \frac{\lambda_1}{\lambda} E[s_1] + \frac{\lambda_2}{\lambda} E[s_2] + \cdots + \frac{\lambda_n}{\lambda} E[s_n],$$

$$E[s^2] = \frac{\lambda_1}{\lambda} E[s_1^2] + \frac{\lambda_2}{\lambda} E[s_2^2] + \cdots + \frac{\lambda_n}{\lambda} E[s_n^2],$$

and

$$E[s^3] = \frac{\lambda_1}{\lambda} E[s_1^3] + \frac{\lambda_2}{\lambda} E[s_2^3] + \cdots + \frac{\lambda_n}{\lambda} E[s_n^3].$$

By Pollaczek's formula,

$$W_q = \frac{\lambda E[s^2]}{2(1 - \rho)}.$$

The mean time in the system for each class is given by
$$W_i = W_q + E[s_i], \quad i = 1, 2, \ldots, n.$$
The overall mean customer time in the system, W, is given by

$$W = \frac{\lambda_1}{\lambda} W_1 + \frac{\lambda_2}{\lambda} W_2 + \cdots + \frac{\lambda_n}{\lambda} W_n.$$

The variance of queueing time is given by

$$\sigma_q^2 = \frac{\lambda E[s^3]}{3(1 - \rho)} + \frac{\lambda^2 \left(E[s^2]\right)^2}{4(1 - \rho)^2}.$$

The variance of system time by class is given by
$$\sigma_{w_i}^2 = \sigma_q^2 + \sigma_{s_i}^2, \quad i = 1, 2, \ldots, n.$$
The second moment of w by class is
$$E[w_i^2] = \sigma_{w_i}^2 + W_i^2, \quad i = 1, 2, \ldots, n.$$

Table 22. Formulas for M/G/1 Queueing System (classes, no priorities) (continued)

Thus, the overall second moment of w is given by

$$E[w^2] = \frac{\lambda_1}{\lambda} E[w_1^2] + \frac{\lambda_2}{\lambda} E[w_2^2] + \cdots + \frac{\lambda_n}{\lambda} E[w_n^2],$$

and

$$\sigma_w^2 = E[w^2] - W^2.$$

The standard M/G/1 formulas now hold for L_q, L, $\sigma_{N_q}^2$, and σ_N^2. Martin's estimates for the 90th and 95th percentiles of w can also be used.

Table 23. Formulas for M/G/1 Nonpreemptive Priority Queueing System (also known as a head-of-the-line queueing system (HOL))

There are n customer classes. Class 1 customers receive the most favorable treatment; class n customers receive the least favorable treatment. Customers from class i arrive in a Poisson pattern with mean arrival rate $\lambda_i, i = 1, 2, \ldots, n$. Each class has its own general service time with $E[s_i] = 1/\mu_i$, and finite second and third moments $E[s_i^2]$, $E[s_i^3]$. Customers are served on a nonpreemptive priority basis. The total arrival stream to the system has a Poisson arrival pattern with

$$\lambda = \lambda_1 + \lambda_2 + \cdots + \lambda_n.$$

The first three moments of service time are given by

$$W_s = \frac{\lambda_1}{\lambda} E[s_1] + \frac{\lambda_2}{\lambda} E[s_2] + \cdots + \frac{\lambda_n}{\lambda} E[s_n],$$

$$E[s^2] = \frac{\lambda_1}{\lambda} E[s_1^2] + \frac{\lambda_2}{\lambda} E[s_2^2] + \cdots + \frac{\lambda_n}{\lambda} E[s_n^2],$$

and

$$E[s^3] = \frac{\lambda_1}{\lambda} E[s_1^3] + \frac{\lambda_2}{\lambda} E[s_2^3] + \cdots + \frac{\lambda_n}{\lambda} E[s_n^3].$$

Table 23. Formulas for M/G/1 Nonpreemptive Priority Queueing System (continued)

Let
$$a_j = \lambda_1 E[s_1] + \lambda_2 E[s_2] + \cdots + \lambda_j E[s_j], \quad j = 1, 2, \ldots, n$$
and note that
$$a_n = a = \lambda W_s.$$
The mean queueing time for each class is given by
$$W_{q_j} = E[q_j] = \frac{\lambda E[s^2]}{2(1 - a_{j-1})(1 - a_j)},$$
$$j = 1, 2, \ldots, n, \quad a_0 = 0.$$
The mean number of customers in queue j is
$$L_{q_j} = \lambda_j W_{q_j}, \quad j = 1, 2, \ldots, n.$$
The overall mean queueing time, W_q, is given by
$$W_q = \frac{\lambda_1}{\lambda} E[q_1] + \frac{\lambda_2}{\lambda} E[q_2] + \cdots + \frac{\lambda_n}{\lambda} E[q_n].$$
The mean time spent in the system for each class is
$$W_j = E[w_j] = E[q_j] + E[s_j], \quad j = 1, 2, \ldots, n,$$
and the mean number in each customer class in the system is
$$L_j = \lambda_j W_j, \quad j = 1, 2, \ldots, n.$$
The mean overall customer time in the system is
$$W = W_q + W_s.$$
The mean number of customers of all classes queueing for service is
$$L_q = \lambda W_q,$$
and in the system is
$$L = \lambda W.$$
The variance of time in the system for each customer class is given by
$$\sigma_{w_j}^2 = \sigma_{s_j}^2 + \frac{\lambda E[s^3]}{3(1 - a_{j-1})^2(1 - a_j)}$$
$$+ \frac{\lambda E[s^2]\left(2\sum_{i=1}^{j} \lambda_i E[s_i^2] - \lambda E[s^2]\right)}{4(1 - a_{j-1})^2(1 - a_j)^2}$$
$$+ \frac{\lambda E[s^2]\sum_{i=1}^{j-1} \lambda_i E[s_i^2]}{2(1 - a_{j-1})^3(1 - a_j)}, \quad j = 1, 2, \ldots, n.$$

Table 23. Formulas for M/G/1 Nonpreemptive Priority Queueing System (continued)

The overall variance of w is given by

$$\sigma_w^2 = \frac{\lambda_1}{\lambda}\left[\sigma_{w_1}^2 + W_1^2\right] + \frac{\lambda_2}{3}\lambda\left[\sigma_{w_2}^2 + W_2^2\right]$$

$$+\cdots+\frac{\lambda_n}{\lambda}\left[\sigma_{w_n}^2 + W_n^2\right] - W^2.$$

The variance of queueing time by customer class is given by

$$\sigma_{q_j}^2 = \sigma_{w_j}^2 - \sigma_{s_j}^2, \quad j = 1, 2, \ldots, n.$$

Then, by Theorem 2.7.2(d),

$$E[q_j^2] = \sigma_{q_j}^2 + W_{q_j}^2, \quad j = 1, 2, \ldots, n.$$

Therefore,

$$E[q^2] = \frac{\lambda_1}{\lambda} E[q_1^2] + \frac{\lambda_2}{\lambda} E[q_2^2] + \cdots + \frac{\lambda_n}{\lambda} E[q_n^2].$$

Finally,

$$\sigma_q^2 = E[q^2] - W_q^2.$$

Table 24. Formulas for M/G/1 Preemptive Priority Queueing System

There are n customer classes. Class 1 customers receive the most favorable treatment; class n customers receive the least favorable treatment. Customers from class i arrive in a Poisson pattern with mean arrival rate $\lambda_i, i = 1, 2, \ldots, n$. Each class has its own general service time with $E[s_i] = 1/\mu_i$, and finite second and third moments $E[s_i^2]$, $E[s_i^3]$. The priority system is preemptive resume, which means that if a customer of class j is receiving service when a customer of class $i < j$ arrives, the arriving customer preempts the server and the customer who was preempted returns to the head of the line for class j customers. The preempted customer resumes service at the point of interruption upon reentering the service facility. The total arrival stream to the system has a Poisson arrival pattern with

$$\lambda = \lambda_1 + \lambda_2 + \cdots + \lambda_n.$$

The first three moments of service time are given by

$$W_s = \frac{\lambda_1}{\lambda}E[s_1] + \frac{\lambda_2}{\lambda}E[s_2] + \cdots + \frac{\lambda_n}{\lambda}E[s_n],$$

$$E[s^2] = \frac{\lambda_1}{\lambda}E[s_1^2] + \frac{\lambda_2}{\lambda}E[s_2^2] + \cdots + \frac{\lambda_n}{\lambda}E[s_n^2],$$

Table 24. Formulas for M/G/1 Preemptive Priority Queueing System
(continued)

$$E[s^3] = \frac{\lambda_1}{\lambda} E[s_1^3] + \frac{\lambda_2}{\lambda} E[s_2^3] + \cdots + \frac{\lambda_n}{\lambda} E[s_n^3].$$

Let $a_j = \lambda_1 E[s_1] + \lambda_2 E[s_2] + \cdots + \lambda_j E[s_j], \quad j = 1, 2, \ldots, n$
and note that
$$a_n = a = \lambda W_s.$$
The mean time in the system for each class is

$$W_j = E[w_j] = \frac{1}{1 - a_{j-1}} \left[E[s_j] + \frac{\sum\limits_{i=1}^{j} \lambda_i E[s_i^2]}{2(1 - a_j)} \right],$$

$$a_0 = 0, \quad j = 1, 2, \ldots, n.$$
The corresponding mean time in the queue is
$$W_{q_j} = E[w_j] - E[s_j], \quad j = 1, 2, \ldots, n.$$
The mean number of customers in queue j is
$$L_{q_j} = \lambda_j W_{q_j}, \quad j = 1, 2, \ldots, n.$$
The overall mean queueing time, W_q, is given by
$$W_q = \frac{\lambda_1}{\lambda} E[q_1] + \frac{\lambda_2}{\lambda} E[q_2] + \cdots + \frac{\lambda_n}{\lambda} E[q_n].$$
The mean number in each customer class in the system is
$$L_j = \lambda_j W_j, \quad j = 1, 2, \ldots, n.$$
The mean overall customer time in the system is
$$W = \frac{\lambda_1}{\lambda} W_1 + \frac{\lambda_2}{\lambda} W_2 + \cdots + \frac{\lambda_n}{\lambda} W_n = W_q + W_s.$$
The mean number of customers of all classes queueing for service is
$$L_q = \lambda W_q,$$
and in the system is
$$L = \lambda W.$$

Table 24. Formulas for M/G/1 Preemptive Priority Queueing System
(continued)

The variance of time in the system for each customer class is given by

$$
\sigma_{w_j}^2 = \frac{\sigma_{s_j}^2}{(1 - a_{j-1})^2} + \frac{E[s_j] \sum_{i=1}^{j-1} \lambda_i E[s_i^2]}{(1 - a_{j-1})^3}
$$

$$
+ \frac{\sum_{i=1}^{j} \lambda_i E[s_i^3]}{3(1 - a_{j-1})^2(1 - a_j)} + \frac{\left(\sum_{i=1}^{j} \lambda_i E[s_i^2]\right)^2}{4(1 - a_{j-1})^2(1 - a_j)^2}
$$

$$
+ \frac{\left(\sum_{i=1}^{j} \lambda_i E[s_i^2]\right)\left(\sum_{i=1}^{j-1} \lambda_i E[s_i^2]\right)}{2(1 - a_{j-1})^3(1 - a_j)} \qquad a_0 = 0,\ j = 1, 2, \ldots, n.
$$

The overall variance of w is given by

$$
\sigma_w^2 = \frac{\lambda_1}{\lambda}\left[\sigma_{w_1}^2 + W_1^2\right] + \frac{\lambda_2}{\lambda}\left[\sigma_{w_2}^2 + W_2^2\right]
$$

$$
+ \cdots + \frac{\lambda_n}{\lambda}\left[\sigma_{w_n}^2 + W_n^2\right] - W^2.
$$

The variance of queueing time by customer class is given by
$$
\sigma_{q_j}^2 = \sigma_{w_j}^2 - \sigma_{s_j}^2, \quad j = 1, 2, \ldots, n.
$$

Then, by Theorem 2.7.2(d),
$$
E[q_j^2] = \sigma_{q_j}^2 + W_{q_j}^2, \quad j = 1, 2, \ldots, n.
$$

Therefore,
$$
E[q^2] = \frac{\lambda_1}{\lambda}E[q_1^2] + \frac{\lambda_2}{\lambda}E[q_2^2] + \cdots + \frac{\lambda_n}{\lambda}E[q_n^2].
$$

Finally,
$$
\sigma_q^2 = E[q^2] - W_q^2.
$$

Table 25. M/G/1 Processor-Sharing Queueing System

The Poisson arrival stream has an average arrival rate of λ and the average service rate is μ. The service time distribution is general with the restriction that its Laplace transform is rational, with the denominator having degree at least one higher than the numerator. Equivalently, the service time, s, is Coxian. The priority system is processor-sharing, which means that if a customer arrives when there are already $n-1$ customers in the system, the arriving customer (and all the others) receive service at the average rate μ/n. Then $p_n = \rho^n(1-\rho)$, $n = 0, 1, \ldots$, where $\rho = \lambda/\mu$. We also have

$$L = \frac{\rho}{1-\rho}, \quad E[w|s=t] = \frac{t}{1-\rho}, \quad \text{and} \quad W = \frac{W_s}{1-\rho}.$$

Finally,

$$E[q|s=t] = \frac{\rho t}{1-\rho}, \quad \text{and} \quad W_q = \frac{\rho W_s}{1-\rho}.$$

Table 26. M/G/c Processor-Sharing Queueing System

The Poisson arrival stream has an average arrival rate of λ. The service time distribution is general with the restriction that its Laplace transform is rational, with the denominator having degree at least one higher than the numerator. Equivalently, the service time, s, is Coxian. The priority system is processor-sharing, which works as follows. When the number of customers in the service center, N, is less than c, then each customers is served simultaneously by one server; that is, each receives service at the rate μ. When $N > c$, each customer simultaneously receives service at the rate $c\mu/N$. We find that just as for the M/G/1 processor-sharing system, W and the probability distribution of the steady state number of customers in the system are independent of the form of the service time distribution and depend only upon the mean, W_s. However, the distribution function of w cannot in general be obtained. It is shown by Lavenberg and Sauer in *Computer Performance Modeling Handbook*, (edited by Lavenberg), Academic Press, 1983, that the formulas for p_n, L, σ_N^2 W, and W_q for the M/G/c processor-sharing queueing system are exactly the same as those for the M/M/c queueing system.

C.18 M/M/c Priority Queueing

Table 27. M/M/c Nonpreemptive (HOL) Queueing System

There are n priority classes with each class having a Poisson arrival pattern with mean arrival rate λ_i. Each customer has the same exponential service time requirement. Then the overall arrival pattern is Poisson with mean

$$\lambda = \lambda_1 + \lambda_2 + \cdots + \lambda_n.$$

The server utilization

$$\rho = \frac{\lambda W_s}{c} = \frac{\lambda}{c\mu},$$

$$W_{q_1} = \frac{C[c, a]W_s}{c(1 - \lambda_1 W_s/c)},$$

and

$$W_{q_j} = \frac{C[c, a]W_s}{c\left[1 - \left(W_s \sum_{i=1}^{j-1} \lambda_i\right)/c\right]\left[1 - \left(W_s \sum_{i=1}^{j} \lambda_i\right)/c\right]}, \quad j = 2, \ldots, n.$$

$W_j = W_{q_j} + W_s, \quad j = 1, 2, \ldots, n.$
$L_{q_j} = \lambda_j W_{q_j}, \quad j = 1, 2, \ldots, n.$
$L_j = \lambda_j W_j, \quad j = 1, 2, \ldots, n.$
$W_q = \dfrac{\lambda_1}{\lambda} + \dfrac{\lambda_2}{\lambda} + \cdots + \dfrac{\lambda_n}{\lambda}.$
$L_q = \lambda W_q.$
$W = W_q + W_s.$
$L = \lambda W.$

C.19   Queueing Networks

Table 28. Queueing Network Notation and Definitions

C	Number of customer classes.
D_{ck}	Average service time provided by service center k for each class c customer while it is in the network. $D_{ck} = V_{ck} \times S_{ck}$.
D_k	Average service time for all customers at service center k. $D_k = \sum_c D_{ck}$.
$E[t]$	Expected (mean or average) think time of a user at an interactive terminal or workstation. $E[t] = T$.
K	Number of service (resource) centers in the network.
L_{ck}	Average number of class c customers at service center k.
L_k	Average number of customers of all classes at service center k.
$L_{q_{ck}}$	Average number of class c customers queueing for service at service center k.
L_{q_k}	Average number of customers of all classes queueing for service at service center k.
λ_{ck}	Average throughput of class c customers at service center k.
λ_k	Average throughput of customers of all classes at service center k. Some authors use this symbol to denote the throughput of class k customers.
λ	Average system throughput. The average rate that customers pass through the system. Usually refers to the rate that interactive users are serviced or the rate that batch jobs are processed.
μ_{ck}	Average service rate (per visit) of class c customers at service center k. $\mu_{kc} = 1/S_{ck}$.
N_c	Number of class c customers in a closed queueing network.
N	Number of customers of all classes in a closed queueing network. Average number of interactive users. Average number of batch jobs in process.

Table 28. Queueing Network Notation and Definitions (continued)

n_{ck}	Number of class c customers at service center k. Often used to describe the state of the network. Not the average value.
n_k	Number of customers of all classes at service center k. Often used to describe the state of the network. Not the average value.
$p_k(n_k)$	Marginal probability there are n_k customers of all classes at service center k.
p_{kj}	Probability a customer who completes service at service center k will branch to service center j.
ρ_{ck}	Utilization of service center k by class c customers.
ρ_k	Utilization of service center k by customers of all classes. Thus, the total utilization of the center.
S_{ck}	Average service time per visit of class c customers at service center k.
T	Average think time. $T = E[t]$.
V_{ck}	Visit ratio of class c customers at service center k. Thus, the average number of visits a class c customer makes to service center k.
W_{ck}	Average time a class c customer spends at service center k (queueing for and receiving service). This is V_{kc} times the average time per visit to service center k.
W_c	Average time a class c customer spends in the network. $W_c = \sum_k W_{ck}$.

Table 29. Finite Population Queueing Model of Interactive System

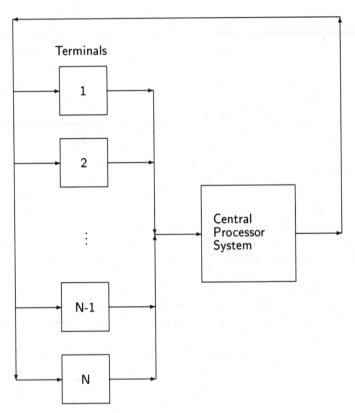

Figure 6.1.1. Finite Population System.

The CPU may have an exponential distribution, in which case the system is a machine repair model with one service center. The other alternative is that the CPU service time is general with the restriction that its Laplace transform is rational, with the degree of the denominator at least one higher than the numerator. Equivalently, the service time is Coxian. When the service time is general, the queue discipline must be processor-sharing. For this model, since the operating time for machines corresponds to think time, with average think time $E[t] = T$, we have, by Little's formula, the mean

Table 29. Finite Population Queueing Model of Interactive System (continued)

response time

$$W = \frac{N}{\lambda} - T.$$

But $\lambda = \rho/W_s$ so the mean response time can be written as

$$W = \frac{NW_s}{\rho} - T,$$

where

$$\rho = 1 - p_0,$$

and

$$p_0 = \left\{ \sum_{k=0}^{N} \frac{N!}{(N-k)!} \times \left(\frac{W_s}{T} \right)^k \right\}^{-1} = B[N, z],$$

where $B[\cdot, \cdot]$ is Erlang's B formula and

$$z = \frac{T}{W_s}.$$

C.20 Graph of $C[c, a]$

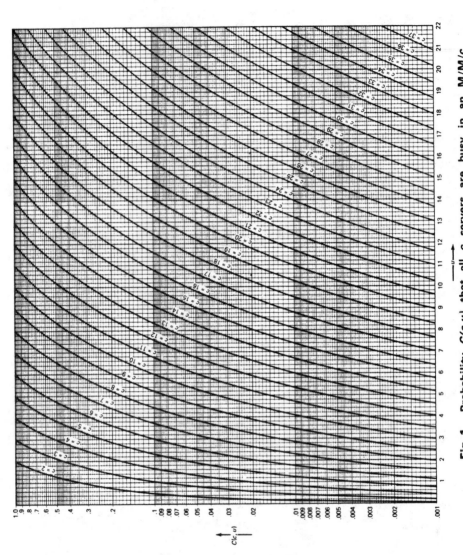

Fig. 1 Probability $C(c, u)$ that all c servers are busy in an M/M/c queueing system versus traffic intensity $u = \lambda E[s]$.

C.21 Graph of $B[c, a]$

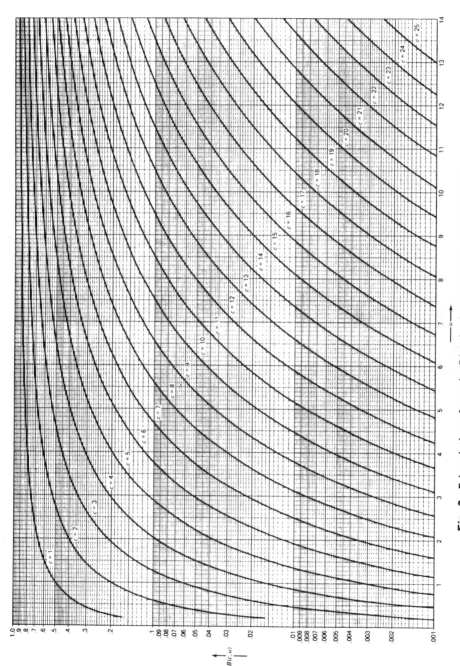

Fig. 2 Erlang's loss formula $B(c,u)$, the probability that all c servers are busy in an M/M/c loss system.

C.22   Graph of ρ for Machine Repair

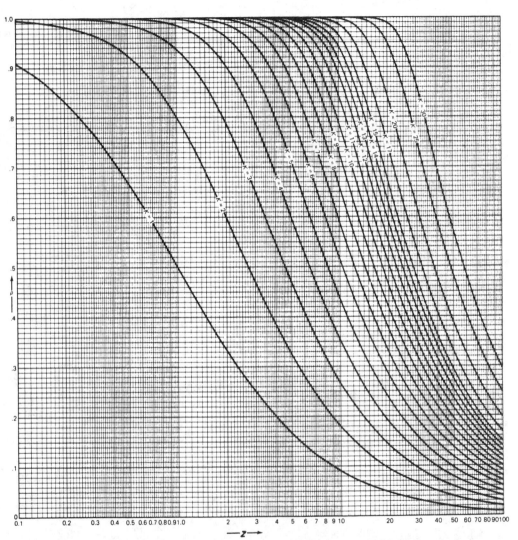

Fig. 3 Server utilization ρ as a function of K and $z = E[\bigcirc]/E[s] = \mu/\alpha$ for the M/M/1/K/K queueing system (machine repair).

Appendix D

Mathematica Programs

All the *Mathematica* programs mentioned in the book are listed here. The programs were written using Version 1.2 of *Mathematica*. The queue4 package contains the **Exact** program for the solution of closed, multiclass, BCMP queueing models of computer systems, plus the subfunctions that are used by that program. In addition, queue4 contains the **Approx** program, which gives the approximate solution of closed, multiclass, BCMP queueing models of computer systems. All the other *Mathematica* programs mentioned in the book are in the queue3 package.

The Stephen Wolfram book [2] which comes with *Mathematica*, is of course, must reading for using this powerful system. Roman Maeder's book [1] was also very useful in writing the packages.

Cited References

[1] Roman Maeder, *Programming in Mathematica*, Addison-Wesley, Redwood City, CA, 1990.

[2] Stephen Wolfram, *Mathematica: A System for Doing Mathematics by Computer*, Addison-Wesley, Redwood City, CA, 1988.

D.1 *Mathematica* **Programs**

```
BeginPackage["queue3'"]
queue3::usage="This is a collection of functions used in this book."
gammadist::usage="Computes the value of the gamma distribution using
the notation in this book."
hyperg::usage="The pmf of the hypergeometric distribution."
```

```
hypergdist::usage="The distribution function of the hypergeometric
distribution."
mmc::usage="Computes the performance statistics for the M/M/c queueing system."
mmck::usage="Computes the performance statistics for the M/M/c/K queueing
system."
bca::usage="Erlang's B (loss) formula."
cca::usage="Erlang's C (delay) formula."
cent::usage="Computes the performance statistics for the central server
model with fixed MPL N."
online::usage="Computes the performance statistics for terminal system
with FESC to replace the central server model of the computer system.
subcent can be used to calculate the rates needed as input."
subcent::usage="Computes the throughput for a central server model with
fixed MPL."
Begin["queue3'private'"]
gammadist/: gammadist[alpha_, beta_, t_] :=
    Gamma[beta, 0, alpha*t]/Gamma[beta]

hyperg[n_, nsucc_, ntot_, k_] :=
        Binomial[nsucc, k]*Binomial[ntot - nsucc, n - k]/Binomial[ntot, n]

hypergdist[n_, nsucc_, ntot_, x_] :=
        Sum[hyperg[n, nsucc, ntot, k], {k, 0, Min[x, nsucc, ntot]}]

bca[c_, a_]:=
Block[{t=1,n},
    Do[t= a t /(n+ a t), {n, 1, c}];
        Return[t]
]
cca[c_,a_]:=
Block[ {n, b, cca},
    rho =a/c ;
    b=a/(1+a);
    For[n=2, n<=c, n++,    b= a*b/(n+a*b)] ;
    cca=b/(rho*b+1-rho);
    Return[{cca}] ;
]

mmc[ c_,   x_,   s_ ] :=
(* x is lambda, s is mean service time *)
Block[ {n, a, w, rho, b, cca,  QueueTime },
    a = x s ;
    rho =a/c ;
    b=a/(1+a);
    For[n=2, n<=c, n++,    b= a*b/(n+a*b)] ;
    cca=b/(rho*b+1-rho);
    QueueTime = cca*s/(c*(1-rho)) ;
    w=QueueTime+s;
(* QueueTime is mean time in the queue *)
(* w is mean time in the system *)
    Return[{QueueTime, w, rho,  cca}] ;
]
```

```
mmck[ c_, k_, x_, s_ ] :=
Block[ {cf, kf, n, a, p, p0, pk, q, Xa, QueueTime },
    cf = Floor[c] ;
    kf = Floor[k] ;
    a = x s ;
    p = 1 ;
    p0 = 1 ;
    q = 0 ;
    For[n=1, n<=cf, n++,  p = p a/n ; p0 = p0 + p  ] ;
    For[n=c+1, n<=kf, n++,  p = p a/cf ; p0 = p0 + p ; q = q + (n-cf) * p  ] ;
    pk = p / p0 ;
    Xa = x * (1-pk) ;
    QueueTime = (q / p0 ) / Xa ;
    w=QueueTime + s;
    (* pk is the probability a customer is lost *)
    (* Xa is true throughput *)
    (* QueueTime is mean time in queue *)
    (* w is mean time in system *)
    Return[{pk, Xa, QueueTime, w}] ;
]


cent[k_,N_, D_]:=
(* central server model *)
(* k is number of service centers *)
(* N is MPL, D is service demand vector *)
Block[{L, w, wn, n, lambdan, rho},
L=Table[0, {k}];
For[n=1, n<=N, n++, w=D*(L+1); wn=Apply[Plus,w]; lambdan=n/wn;
L=lambdan w; rho=lambdan D];
(* lambdan is mean throughput *)
(* wn is mean time in system *)
(* L is vector of number at servers *)
(* rho is vector of utilizations *)
Return[{lambdan, wn, L, rho}];
]


online/: online[srate_, N_, T_]:=
Block[{n, w,s},
m=Length[srate];
x=Table[Last[srate], {N-m}];
nsrate=Join[srate, x];
q=Join[{1}, Table[0, {N-1}]];
s=0;
q0=1;
For[n=1, n<=N, n++,
w=0;
For[j=1, j<=n, j++,
w=w+(j/nsrate[[j]])*If[j>1, q[[j-1]], q0];
lambda=n/(T+w)];
s=0;
For[j=n, j>=1, j--,
```

```
q[[j]]=(lambda/nsrate[[j]])*If[j>1, q[[j-1]],q0];
s=s+q[[j]]];
q0=1-s
];
qplus=Join[{q0},q];
(* lambda is mean throughput *)
(* w is mean response time *)
(* qplus is vector of conditional probabilities *)
Return[{lambda,w, qplus}];
]


subcent/: subcent[k_,N_, D_]:=
Block[{L, w, wn, n, lambdan, rho},
L=Table[0, {k}];
For[n=1, n<=N, n++, w=D*(L+1); wn=Apply[Plus,w]; lambdan=n/wn;
L=lambdan w; rho=lambdan D];
Return[{lambdan}];
]


End[] (* end 'private' context *)
EndPackage[]



BeginPackage["queue4'"]

queue4::usage="This is a collection of queueing theory programs used
to model computer systems."
Exact::usage="Exact[ Nc_?VectorQ, Tc_?VectorQ, Dck_?MatrixQ ]
computes the performance statistics for a closed, multiclass BCMP
queueing system where Nc is the population vector, Tc is the
vector of think times and Dck is the matrix of service demands
by workload class.  The output consists of four performance vectors:
Wc the vector of response times by class, Lambdac the vector of
throughputs by class, Lk the vector of number of customers at each
service center, and Rhok the vector of the total utilization of each
service center."
Approx::usage=" Approx[ Nc_?VectorQ, Tc_?VectorQ, Dck_?MatrixQ,
epsilon_Real ] computes the approximate performance statistics for
a closed, multiclass BCMP queueing system where Nc is the population
vector, Tc is the vector of think times and Dck is the matrix of
service demands by workload class.  The output consists of four
performance vectors: Wc the vector of response times by class, Lambdac
the vector of throughputs by class, Lk the vector of number of
customers at each service center, and Rhok the vector of the total
utilization of each service center."

Begin["queue4'private'"]

FixPerm[ numC_, PVc_, Nc_ ] :=
Block[ {i, m = PVc },
       For[i=numC, i>1, i--,
         If[m[[i]] > Nc[[i]],
```

```
                m[[i-1]]=m[[i-1]]+m[[i]]-Nc[[i]] ;
                m[[i]]=Nc[[i]] ]];
        If[ m[[1]] > Nc[[1]], {}, m]
]


FirstPerm[ numC_, Nc_, n_ ] :=
Block[ {m},
    m = Table[ 0, { numC } ] ;
    m[[numC]] = n ;
    FixPerm[numC, m, Nc ]
]


NextPerm[ numC_, Nc_, PVc_ ] :=
Block[ {m=PVc, i=numC, j},

        While[ m[[i]] == 0, i-- ] ;
        If[i==1, Return[{}] ] ;

        m[[i]]-- ;
        i-- ;
        While[ (i >= 1) && (m[[i]] == Nc[[i]]), i-- ] ;
        If[i < 1, Return[{}] ] ;
        m[[i]]++ ;

        For[j=i+1, j<numC, j++,
            m[[numC]] = m[[numC]] + m[[j]] ;
            m[[j]] = 0 ] ;
        FixPerm[numC, m, Nc ]
]


Exact[ Nc_?VectorQ, Tc_?VectorQ, Dck_?MatrixQ ] :=

Block[ { n, PVc, PVcm1, Wck, Wc, Lambdac, Lck1, Lck2, Ltmp, Rhok, Lk,
         numC = Length[Nc], numK = Dimensions[Dck][[2]],
         Ntot, Wtot, zVectorK },

    zVectorK = N[ Table[0, {numK}] ] ;
    Ntot = Sum[ Nc[[c]], {c, 1, numC} ] ;
    Lck1[ Table[0, {numC}] ] = zVectorK ;

    For[n=1, n <= Ntot, n++,
        PVc = FirstPerm[numC, Nc, n ] ;
        While[PVc!= {},
                Wck = Table[(PVcm1 = PVc ;
                            If[ PVcm1[[c]] > 0,
                                PVcm1[[c]]-- ;
                                Dck[[c]] * ( 1 +
                                If[OddQ[n], Lck1[PVcm1], Lck2[PVcm1]]),
                                zVectorK]),
                        { c, 1, numC} ] ;

                Lambdac = Table[ ( Wtot = Tc[[c]] + Apply[Plus, Wck[[c]] ] ;
```

```
                        If[ Wtot > 0, PVc[[c]] / Wtot, 0 ] ),
                {c, 1, numC} ] ;

            Ltmp = Lambdac . Wck ;
            If[OddQ[n], Lck2[PVc]=Ltmp, Lck1[PVc]=Ltmp ] ;

            PVc = NextPerm[numC, Nc, PVc ] ];
        If[OddQ[n], Clear[Lck1], Clear[Lck2]]
    ] ;

  Wc = Apply[Plus, Wck, 1 ] ;
  Rhok = Lambdac . Dck ;
  Lk = Lambdac . Wck ;

  Return[{Wc,Lambdac,Lk,Rhok }] ;

] /; Length[Nc] == Length[Tc] == Length[Dck]



Approx[ Nc_?VectorQ, Tc_?VectorQ, Dck_?MatrixQ, epsilon_Real ] :=

Block[ { Flag, Wck, Lambdac, newL, Lk, Lck, Wc, Lk, Rhok,
         numC = Length[Nc], numK = Dimensions[Dck][[2]] },

  Lck = N[Table[ Nc[[c]]/numK, {c, 1, numC}, {k, 1, numK} ] ] ;
  Flag = True ;
  While[Flag==True,

      Lk = Apply[Plus, Lck ] ;
      Wck = Table[ Dck[[c,k]] *
            (1 + Lk[[k]] - Lck[[c,k]] + ((Nc[[c]]-1)/Nc[[c]]) Lck[[c,k]] ),
            {c, 1, numC}, {k, 1, numK} ] ;
      Wc = Apply[Plus, Wck, 1 ] ;
      Lambdac = Nc / ( Tc + Wc) ;

      Flag = False ;
      Lck = Table[(newL = Lambdac[[c]] Wck[[c,k]] ;
                   If[ Abs[ Lck[[c,k]] - newL ] >= epsilon, Flag=True] ;
                   newL), {c, 1, numC}, {k, 1, numK} ] ;
  ] ;

(* Compute final results *)

  Rhok = Lambdac . Dck ;
  Lk = Lambdac . Wck ;
  Return[{Wc,Lambdac,Lk,Rhok }] ;

] /; Length[Nc] == Length[Tc] == Length[Dck]

End[] (* end 'private' context *)
EndPackage[]
```

Appendix E

Answers to Exercises

E.1 Answers to Exercises

Chapter 2

1. (a) 0.3645. (b) 0.4275. (c) 0.8585. (d) 0.0665. **5.** $1 - (13/6^4) =$ 1283/1296. **8.** 0.99978. **9.** (a) Three. (b) Four. **10.** Four.
12. (a) $1 - \left(\frac{5}{6}\right)^4 = 671/1296 = 0.5177469.$ (b) $1 - \left(\frac{35}{36}\right)^{24} = 0.491403876.$
13. $\binom{47}{2} \div \binom{50}{2} = 0.88244898.$ **14.** (a) $36 \div \binom{52}{5} = 36/2,598,960 =$ 0.000013852. (b) $(13 \times 48) \div \binom{52}{5} = 0.000240096.$ (c) $\left[\binom{4}{2} \times 13 \times \binom{4}{3} \times 12\right] \div$ $\binom{52}{5} = 0.001440576.$ (d) $\left[4 \times \left(\binom{13}{5} - 10\right)\right] \div \binom{52}{5} = 5108 \div \binom{52}{5} = 0.00196540.$ (e) $(10 \times 4^5 - 40) \div \binom{52}{5} = 0.003924647.$ **15.** $\left[\binom{13}{5} \times 4^5\right] \div \binom{52}{5} = 0.50708283.$
16. 0.501177394. **17.** (a) $\left[13 \times \binom{4}{2} \times \binom{12}{3} \times 4^3\right] \div \binom{52}{5} = 0.422569028.$
(b) $\left[\binom{13}{2} \times \binom{4}{2}^2 \times 11 \times 4\right] \div \binom{52}{5} = 0.047539016.$ (c) $\left[13 \times 4 \times \binom{12}{2} \times 4^2\right] \div$ $\binom{52}{5} = 0.021128451.$ **18.** (a) $1 \div \binom{52}{5} = 1.574769 \times 10^{-12}.$ (b) $\binom{39}{13} \div \binom{52}{13} =$ 0.012790948. (c) $4 \div \binom{52}{13} = 6.299078 \times 10^{-12}.$ **22.** (a) 3/11. (b) 5/13.
(c) 3/7. **23.** (a) 0.83. (b) 0.6944. (c) 0.60. **24.** (a) 1/3. (b) 1/2.
(c) No. **25.** (a) 0.39375. (b) $0.125 \div 0.3975 = 0.31746.$ **26.** (a) 0.315.
(b) $0.12 \div 0.315 = 0.38095.$ **27.** 4/9. **29.** (a) 1/6. (b) 1/30. (c) 3/10.
31. (a) Let p_n be the probability of no match. Then we have the following table.

n	p_n	$|p_n - (1 - e^{-1})|$
2	0.500	0.132
3	0.667	0.035
4	0.625	0.007

32. (a) $5/8 = 0.625$. (b) $1/4$. (c) $1/24$. **33.** (a) 18 (b) 10. **35.** (a) 1,2,3,4 (b) $p(1) = 20/35$, $p(2) = 10/35$, $p(3) = 4/35$, $p(4) = 1/35$. **37.** mean $= 3.951425$, standard deviation $= 2.29543$. **38.** (a) No. (b) $75/216, 15/216, 1/216$. (c) -$0.0787. **40.** $P[A_2|A] = 0.2950$, $P[A_3|A] = 0.1475$, $P[A_4|A] = 0.2212$. **41.** (a) $p_X(0) = \frac{1}{2} = p_X(1)$, $p_Y(0) = \frac{1}{8}, p_Y(1) = \frac{3}{8}, p_Y(2) = \frac{1}{2}$. (b) $p_{X|Y=2}^{(0)} = \frac{1}{4}$, $p_{X|Y=2}^{(1)} = \frac{3}{4}$. (c) No. $p(x,y) \neq p_X(x)p_Y(y)$ for at least one x and one y. (d) $E[X] = 1/2, E[Y] = 11/8$, $\text{Var}[X] = 1/4$, $\text{Var}[Y] = 31/64$. (e) $p_Z(0) = \frac{1}{8}, p_Z(1) = \frac{1}{4}, p_Z(2) = \frac{1}{4}, p_Z(3) = \frac{3}{8}$. **44.** $x_p = E[X] + \sigma\sqrt{\frac{1}{p} - 1}$. **46.** $C = \left[e^{-\alpha}\sum_{k=0}^{N}\alpha^k/k!\right]^{-1}$. **47.** 94 bytes. **48.** (a) Method B. (b) Yes. **49.** (a) 96 (b) 3,734. **51.** (a) $\frac{4!48!/(12!)^4}{52!/(13!)^4} = 0.1054981993$. (b) $\left[\binom{48}{13}/\binom{52}{13}\right]^3 = 0.028043904$. (c) $1 - q^7$ where $q = 1 - P[A]$. **53.** (a) $\binom{35}{13}/\binom{39}{13} = 0.18176$. (b) $1 - \binom{35}{13}/\binom{39}{13} - 4\binom{35}{12}/\binom{39}{13} = 0.4073$. **54.** (a) $2\binom{23}{10}/\binom{26}{13} = 11/50$. (b) $2\binom{23}{12}/\binom{26}{13} = 13/50$. **55.** $24/\left[\binom{52}{13}\binom{39}{13}\binom{26}{13}\right] = 4.473877 \times 10^{-28}$. **56.** (a) $(1 - p^2)^2$. (b) $(1 - p^2)^2 p + (1 - p)$. (c) 0.8789, 0.96973. **58.** Probability Kollossal overbooked 0.36473; Teeny 0.34868. **61.** (a) $f_X(x) = \frac{3}{2} - x$. $f_Y(y) = \frac{3}{2} - y$. No. (b) $f_{X|Y}(x|y) = (2 - x - y)/(\frac{3}{2} - y)$. $f_{Y|X}(y,x) = (2 - x - y)/(\frac{3}{2} - x)$. (c) $E[X|Y = y] = (3y - 4)/(6y - 9)$. $E[Y|X = x] = (3x - 4)(6x - 9)$. **62.** (a) $f_X(x) = e^{-x}$. $f_Y(y) = e^{-y}$. Yes. (b) $f_{X|Y}(x|y) = e^{-x}.f_{Y|X}(y|x) = e^{-y}$. (c) 1. 1. **63.** $585/1326 = 90/204 = 45/102 = 0.44118$. **64.** $4/9$. **65.** 0.5. **66.** $2/3$. **67.** $1 - 1/2! + 1/3! - \cdots - 1/52! \approx 1 - e^{-1}$. The two numbers are equal to at least 10 decimal places. **68.** $11 to John, $5 to Mark. **69.** $1/24, 3/24 = $0.125. **70.** $1/3$. **71.** (a) $1/3$ (b) $3/1$ **72.** $3/8$ **74.** (a) 0.0345. (b) 0.57534. **75.** (a) $2/5$ (b) $15.00.

Chapter 3

1. $405/1024$. 0.08789. **2.** $3/16$. 0.10057. $1 - q^4 = 0.56419$. **3.** 0.498184. **4.** $p(k) = q^{k-1}p$. $E[X] = 1/p$. $\text{Var}[X] = q/p^2$. **5.** 2.5. $q^4 = 0.1296$. **6.** (a) 0.62419. (b) 0.69927. **7.** 300. **9.** 0.01134. **10.** 0.1008. **12.** 0.5665. **13.** $371/455$. **14.** 0.74190. **15.** (a) $p_k = \binom{4}{k}\binom{2}{3-k}/\binom{6}{3}$, $k = 1, 2, 3$. (b) $4/5$. **16.** 0.013635. 0.01416. **17.** (a) 0.04979. Normal approx. 0.05391 Binomial approx. 0.05322. (b) 0.05008. Normal approx. 03615. Binomial approx. 0.05072. **18.** 0.0053878. **19.** 0.669524. **20.** $E[X] = rq/p$. $\text{Var}[X] = rq/p^2$. **21.** 0.994555. **22.** 1.0. **23** 0.323324. **26.** (a) $k = 9, 10$. Max. prob. 0.12511. (b) 0.71429. (c) 0.75 by Chebyshev. Normal approx. 0.96012. **27.** $\exp(-2.4) = 0.09072$. **29.** 0.0902. **30.** 0.136691. Normal approx. 0.13567.

32. 0.14197. Poisson approx. 0.14288. **33.** 0.0026646. Normal approx. 0.00104. **34.** (a) Binomial with $n = 500$, $p = 1/365$. For (b), (c), see table.

Ex. 34. Ch. 3

k	p_k	Poisson
0	0.25366	0.25414
1	0.34844	0.34814
2	0.23883	0.23845
3	0.10892	0.10888
4	0.03718	0.03729
5	0.01013	0.01022

35. (a) 0.77687. (b) 0.13534. **37.** (a) 8 sec. (b) 0.265. (c) 0.242424. (d) No. **38.** $\ln(100) = 4.61$ or about 5 raisins. **43.** Both have mean 20. (a) Variance 40. (b) Variance 34. **44.** 0.8926. **45.** Erlang-5 with mean $1/90$. **47.** (a) Gamma with parameters 5 and α. (b) 0.4405. **48.** 108,241. **51.** (a) 38,416. (b) 36,880. **57.** (a) 0.8487961172. (b) 0.8571234605. **58.** (a) (i) 0.91044. (ii) 0.74107. (iii) 0.89973. (iv) 0.94950. (b) (i) 0.915. (ii) 0.73936. (iii) 0.89936. (iv) 0.95264. **59.** (a) $\beta = 0.25$. $\alpha = 1/40 = 0.025$. $E[X^3] = 45,000$. $P[X \leq 15] = 0.804799$. (b) $q_1 = 0.1127016654$. $q_2 = 1-q_1$. $\mu_1 = 0.02254033308$. $\mu_2 = 0.1774596669$. $E[X^3] = 60,000$. $P[X \leq 15] = 0.8576827036$. (c) $q_1 = 0.7390457219$. $q_2 = 1 - q_1$. $\mu_1 = 0.3673320053$. $\mu_2 = 0.03266799469$. $E[X^3] = 45,000$. $P[X \leq 15] = 0.8371453882$. **60.** $q_1 = 0.3486818046$. $q_2 = 1 - q_1$. $\mu_1 = 0.373986598$. $\mu_2 = 10 - \mu_1$. **61.** Same solution as Example 3.2.9. **65.** Poisson with parameter λp. **66.** (a) Poisson with parameter λp. (b) $E[Y] = 1 = \text{Var}[Y]$. **67.** (a) Poisson with parameter $\lambda = 30$ people. (b) $E[Y] = \text{Var}[Y] = 30$ people. **68.** $K = 2, f(t) = 2e^{-2t}, E[X^3] = 3/4$. **69.** $E[X] = 1, P[X = E[X]] = 0, \text{Var}[X] = 1$.

Chapter 4

1. No. **2.** (a)

$$\begin{bmatrix} 0.5 & 0.3 & 0.2 \\ 0.2 & 0.5 & 0.3 \\ 0.2 & 0.2 & 0.6 \end{bmatrix}$$

(b) $(0.26, 0.37, 0.37)$. $(0.285506, 0.326605, 0.387889)$. **3.** All are recurrent. **4.** Classes $\{0, 1\}$, $\{2, 3\}$, $\{4\}$. First two classes are recurrent. Last is transient. **5.** 0.8677. **6.** 0.1988. **8.** $P[N(t) = k] = \exp(-75)\frac{(75)^k}{k!}$.

Chapter 5

1. $16 + 60 = 76$ people. **2.** (b) $\mu = \lambda \alpha^{-1/n}$. (c) $\mu = 27.1442$. **5.** Cost per eight-hour day with I. M. Slow is \$1,320; with I. M. Fast \$600. Hire Fast fast. **6.** (a) 1/3 (b) 1.5. (c) 0.062959. (d) 7.2 minutes. **7.** 2.486028 minutes. 10. (a) 1.5625. (b) 6.25 minutes. (c) 8.75 minutes. (d) 2.1875. (e) 0.375. (f) 0.390625. (g) 2.6667. **11.** (1) 1 min. (2) 4 min. (3) $\rho = 0.25$. (4) 1/3. (5) $0.25^3 = 0.015625$. (6) $e^{-3.75} = 0.023517746$. (7) 11.983. **15.** (a) 1.4274 jobs per sec. (b) 2.457 jobs per sec., and 4.07 sec. (c) 7.006 sec. (d) 0.999197. **19.** Log is right because $W_{q_{M/M/2}} < W_{q_{M/M/1}}$. Bot is right because $W_{M/M/1} < W_{M/M/2}$. Both are right! If users (people) experience the waiting and the service separately, Log's system would probably be preferable because people seem to dislike waiting for service even more than slow service once it begins. If queueing and service is not separately felt by the user, then Bot's system is best. This is usually the case for computer systems. **20.** (a) 5/9. (b) 8.993 minutes. (c) 0.3747. (d) 48.993 minutes. (e) 2.1414. (f) 0.17266. (g) 0.133226752. **21.** (a) 0.7576. (b) 0.2424. (c) 65.45 minutes. (d) 5.45. (e) 0.43841. (f) 0.0909. **22.** (i) $W_q = 45.75$ minutes, $L_q = 6.86$ customers, $L = 11.36$ customers, and probability of not queueing is 0.2375. (ii) 7. (iii) 0.2172, 0.7828, 32.61 minutes, 4.89 customers, 0.4999, 0.0105. **24.** (a) 94.43754953 minutes. (b) 93.89452 minutes. **26.** (a) The APL function HAM implements Algorithm HM by first calling the APL function BBCA to calculate $B[c, a]$. HAM then returns $C[c, a]$ and sets the global variable B to $B[c, a]$. (b) See the APL function RBCA. (c) BBCA and RBCA both yield $B[15, 5] = 0.0001572562863$. The timing routines of APL*PLUS show that each function requires 270 milliseconds on an 8MHZ IBM PC AT. The one-line APL function WBCA requires only 50 milliseconds! The function RBCA is a memory grabber and will give an APL WS FULL error for relatively small values of c. **27.** (a) 4 agents. (b) All times in seconds. $W_q = 5.37$. $W = 185.37$. $E[q|q > 0] = 72$. $\pi_q[90] = 28.79$. The probability that $w \leq 3.5$ minutes is 0.6758. $P[N \geq 5] = 0.028$. **28.** (a) 0.32195. (b) 8.4756. (c) 0.1353. (d) \$14.48. **29** (a) 9. (b) 5.549 lines. (c) 0.5135. **30.** (a) $W = 30.000005$ min. $\pi_w[90] = 69.0776$ min. $L = 24.500004$ engineers. $W_q = 5 \times 10^{-6}$ min. $\pi_q[90] = 0$. min. $L_q = 4 \times 10^{-6}$ engineers. (b) $W = 30$ min. $\pi_w[90] = 69.0776$ min. $L = 24.5$ engineers. $W_q = \pi_q[90] = L_q = 0$. **31.** (e) $W = 30$ seconds. K^{**} is 12.5 terminals.

Chapter 6

1. 38.5 seconds. **2.** 5 seconds. **3.** 0.48 trans. per sec. **4.** 2.5 sec. **5.** 5.56. **6.** $p_0 = 0.521307$, $\rho = 0.478693$, $\lambda = 0.957386$, $W = 0.890216$. **7.**